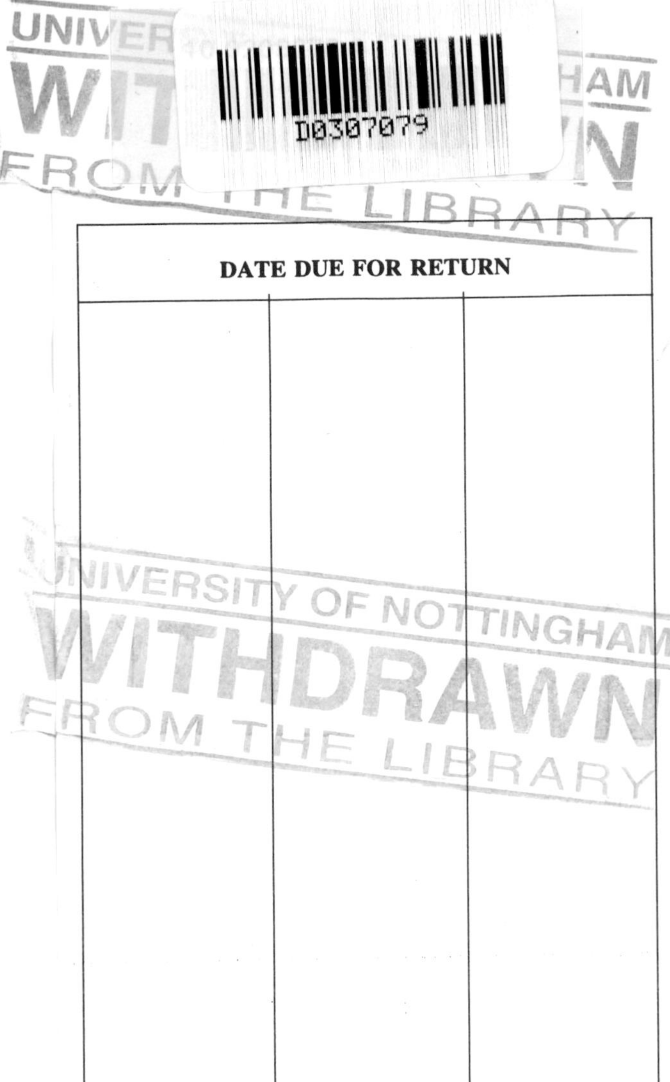

PROGRESS IN BRAIN RESEARCH

VOLUME 125

VOLUME TRANSMISSION REVISITED

Other volumes in PROGRESS IN BRAIN RESEARCH

PROGRESS IN BRAIN RESEARCH

VOLUME 125

VOLUME TRANSMISSION REVISITED

EDITED BY

L.F. AGNATI

Department of Human Physiology, University of Modena, Via Campi 287, I-41100 Modena, Italy

K. FUXE

Department of Neuroscience, Karolinska Institute, 171 77 Stockholm, Sweden

C. NICHOLSON

Department of Physiology and Neuroscience, New York University, School of Medicine, 550 First Avenue, New York, NY 10016, USA

E. SYKOVÁ

Department of Neuroscience, 2nd Medical Faculty, Charles University and Institute of Experimental Medicine, Academy of Sciences of the Czech Republic Vídeňská 1083, 142 20 Prague 4, Czech Republic

ELSEVIER

AMSTERDAM – LAUSANNE – NEW YORK – OXFORD – SHANNON – SINGAPORE – TOKYO

2000

ELSEVIER SCIENCE B.V.
Sara Burgerhartstraat 25
P.O. Box 211, 1000 AE Amsterdam, The Netherlands

First edition 2000

Library of Congress Cataloging in Publication Data
A catalog record from the Library of Congress has been applied for.

ISBN: 0-444-50314-5 (volume)
ISBN: 0-444-80104-9 (series)

∞ The paper used in this publication meets the requirements of ANSI/NISO Z39.48-1992 (Permanence of Paper).
Printed in The Netherlands.

List of Contributors

L.F. Agnati, Department of Human Physiology, University of Modena, Via Campi 287, I-41100 Modena, Italy

T. Bartfai, Hoffmann-La Roche Ltd., Dept: PRPN, Bldg. 69/415, CH-4070 Basel, Switzerland

M. Benoit-Marand, CNRS UMR 5541, Laboratoire d'Histologie Embryologie, UFR 2, BP 28, Université Victor Segalen Bordeaux 2, 33076 Bordeaux, France

B. Bloch, CNRS UMR 5541, Laboratoire d'Histologie Embryologie, UFR 2, BP 28, Université Victor Segalen Bordeaux 2, 33076 Bordeaux, France

F.E. Bloom, Department of Neuropharmacology, The Scripps Research Institute, 10550 North Torrey Pines Road, La Jolla, CA 92037, USA

S. Bröer, Fachbereich Biologie/Chemie, Universität Bremen, 28334 Bremen, Germany

J.B. Burie, CNRS UMR 5466 UFR M.I. 2S, Université Victor Segalen Bordeaux 2, France

M.R. Celio, Institute of Histology and General Embryology, University of Fribourg, CH-1705 Fribourg, Switzerland

C. Chavkin, Department of Pharmacology, University of Washington, Box 357280, Seattle WA 98195-7280, USA

K.C. Chen, Department of Physiology and Neuroscience, New York University School of Medicine, 550 First Avenue, New York, NY 10016, USA

A. Chvátal, Department of Neuroscience, Institute of Experimental Medicine AS CR, Vídeňská 1083, 142 20 Prague 4, Czech Republic

J.A. Coles, INSERM U438, NMR Bioclinique, CHU, Pavillon B, BP 217, 38043 Grenoble cedex 09, France

L. Descarries, Départements de Pathologie et Biologie Cellulaire et de Physiologie, Centre de Recherche en Sciences Neurologiques, Faculté de Médecine, Université de Montréal, C.P. 6128, Succ. Centre-ville, Montréal, PQ H3C 3J7, Canada

L. Di Paola, Department of Biomedical Sciences, Section of Physiology, University of Modena and Reggio Emilia, Via Campi 287, I-41100 Modena, Italy

A.W. Duggan, Department of Anesthesia and Pain Management, University of Sydney, Royal North Shore Hospital, St. Leonards, NSW 2065, Australia

B. Dumartin, CNRS UMR 5541, Laboratoire d'Histologie Embryologie, UFR 2, BP 28, Université Victor Segalen Bordeaux 2, 33076 Bordeaux, France

R. Ferrari, Department of Biomedical Sciences, Section of Physiology, University of Modena and Reggio Emilia, Via Campi 287, I-41100 Modena, Italy

S. Ferré, Department of Neuroscience, Karolinska Institute, S-171 77 Stockholm, Sweden

U. Flögel, Fachbereich Biologie/Chemie, Universität Bremen, 28334 Bremen, Germany

K. Fuxe, Department of Neuroscience, Karolinska Institute, S-171 77 Stockholm, Sweden

F. Gonon, CNRS UMR 5541, Laboratoire d'Histologie Embryologie, UFR 2, BP 28, Université Victor Segalen Bordeaux 2, 33076 Bordeaux, France

S. Grillner, Nobel Institute for Neurophysiology, Department of Neuroscience, Karolinska Institute, SE 171 77 Stockholm, Sweden

M. Höistad, Department of Neuroscience, Karolinska Institute, S-171 77 Stockholm, Sweden

S. Hrabětová, Department of Physiology and Neuroscience, New York University School of Medicine, 550 First Avenue, New York, NY 10016, USA

M. Ibrahim, Institute of Histology and General Embryology, University of Fribourg, CH-1705 Fribourg, Switzerland

M. Jaber, CNRS UMR 5541, Laboratoire d'Histologie Embryologie, UFR 2, BP 28, Université Victor Segalen Bordeaux 2, 33076 Bordeaux, France

A. Jansson, Department of Neuroscience, Karolinska Institutet, S-171 77 Stockholm, Sweden

K. Kaila, Department of Biosciences, Division of Animal Physiology, University of Helsinki, PO Box 17, FIN-00014 Helsinki, Finland

J. Kehr, Department of Neuroscience, Karolinska Institute, S-171 77 Stockholm, Sweden

D.M. Kullmann, Department of Clinical Neurology, Institute of Neurology, UCL, Queen Square, London WC1N 3BG, UK

M. Lehman, Department of Cell Biology, Neurobiology and Anatomy, University of Cincinnati, College of Medicine, 231 Bethesda Avenue, Cincinnati, OH 45267, USA

D. Leibfritz, Fachbereich Biologie/Chemie, Universität Bremen, 28334 Bremen, Germany

A. Lippoldt, Centrum für Molekulare Medizin, 13122 Berlin-Buch, Germany

P.J. Magistretti, Institute of Physiology, Lausanne University School of Medicine, 7 rue du Bugnon, 1005 Lausanne, Switzerland

P. Marcaggi, INSERM U394, Institute François Magendie, 1, rue Camille Saint-Saëns, 33077 Bordeaux cedex, France. Present address: Department of Physiology. University College London, Gower Street, London WC1E 6BT, UK

T. Mazel, Department of Cellular Neurophysiology, Institute of Experimental Medicine ASCR, Vídeňská 1083, 142 20 Prague 4, Czech Republic

N. Mechawar, Départements de Pathologie et Biologie Cellulaire and Centre de Recherche en Sciences Neurologiques, Faculté de Médecine, Université de Montréal, Montréal, PQ H3C 3J7, Canada

C. Meier, Fachbereich Biologie/Chemie, Universität Bremen, 28334 Bremen, Germany

C. Nicholson, Department of Physiology and Neuroscience, New York University School of Medicine, 550 First Avenue, New York, NY 10016, USA

R. Nieuwenhuys, Papehof 25, 1391 BD Abcoude, The Netherlands

S.-O. Ögren, Department of Neuroscience, Karolinska Institute, S-171 77 Stockholm, Sweden

D. Parker, Nobel Institute for Neurophysiology, Department of Neuroscience, Karolinska Institute, SE 171 77 Stockholm, Sweden

L. Pellerin, Institute of Physiology, Lausanne University School of Medicine, 7 rue du Bugnon, 1005 Lausanne, Switzerland

J. Pfeuffer, Fachbereich Biologie/Chemie, Universität Bremen, 28334 Bremen, Germany

V.M. Pickel, Department of Neurology and Neuroscience, Weill Medical College of Cornell University, 411 East 69th St., KB 410, New York, NY 10021, USA

Š. Prokopová-Kubinová, Department of Neuroscience, Institute of Experimental Medicine AS CR, Vídeňská 1083, 142 20 Prague 4, Czech Republic

B.R. Ransom, Department of Neurology, University of Washington School of Medicine, 1959 NE Pacific, Room RR650, Box 356465, Seattle, WA 98195-6465, USA

M.E. Rice, Departments of Neurosurgery, Physiology and Neuroscience, New York University School of Medicine, 550 First Avenue, New York, NY 10016, USA

R. Silver, Department of Psychology, Barnard College and Columbia University, 1190 Amsterdam Avenue, New York, NY 10027, USA

I. Strömberg, Department of Neuroscience, Karolinska Institute, S-171 77 Stockholm, Sweden

E. Syková, Department of Neuroscience, 2nd Medical Faculty, Charles University and Institute of Experimental Medicine, Academy of Sciences of the Czech Republic Vídeňská 1083, 142 20 Prague 4, Czech Republic

L. Tao, Department of Physiology and Neuroscience, New York University School of Medicine, 550 First Avenue, New York, NY 10016, USA

C. Torri, Department of Biomedical Sciences, Section of Physiology, University of Modena and Reggio Emilia, Via Campi 287, I-41100 Modena, Italy

L. Vargová, Department of Neuroscience, Institute of Experimental Medicine AS CR, Vídeňská 1083, 142 20 Prague 4, Czech Republic

C. Véga, INSERM U394, Institute François Magendie, 1, rue Camille Saint-Saëns, 33077 Bordeaux cedex, France. Present address: Institut de Physiologie, Université de Lausanne, 7 rue du Bugnon, 1005 Lausanne, Switzerland.

D. Viggiano, Istituto di Anatomia Umana, Seconda Università, Via L. Armanni 5, 80138 Napoli, Italy

J. Voipio, Department of Biosciences, Division of Animal Physiology, University of Helsinki, PO Box 17, FIN-00014 Helsinki, Finland

I. Voříšek, Department of Neuroscience, Institute of Experimental Medicine AS CR, Vídeňská 1083, 142 20 Prague 4, Czech Republic

M.J. Zigmond, Department of Neurology, S-526 Biomedical Science Tower, University of Pittsburgh, Pittsburgh, PA 15213, USA

I. Zini, Department of Biomedical Sciences, Section of Physiology, University of Modena and Reggio Emilia, Via Campi 287, I-41100 Modena, Italy

M. Zoli, Department of Biomedical Sciences, Section of Physiology, University of Modena and Reggio Emilia, Via Campi 287, I-41100 Modena, Italy

Opening address

The four of us, Agnati, Nicholson, Syková and myself, are delighted to have had the privilege of organising this international Wenner-Gren Center (WGC) symposium on *Volume Transmission Revisited.*

The theme of our meeting started out more than a century ago with the Nobel laureate Golgi, but after then it disappeared from view until the seventies when, especially Descarries, suggested that central monoamine neurons could communicate with other neurons without recourse to classical synaptic specialisations. In the same period Nicholson pointed out, in the framework of Golgi's view, that the microenvironment between neurons could represent a communication channel. These and other observations caused a 'blurring' of the Cajal–Sherrington's paradigm on interneuronal communication in the brain and stimulated several proposals and discussions, as Shepherd has noted in his beautiful book: '*Foundations of the Neuron Doctrine*'.

Against this background in the eighties, Agnati, Fuxe and collaborators suggested a broader view, which can be summarised by stating that the classical Cajal–Sherrington paradigm on intercellular communication in brain had to be revised. Our work aimed not only to give experimental proof of long distance intercellular communication mediated by messages flowing in the extracellular space, but also to try to define objective criteria for classification of the different modes of intercellular communication in the brain.

Thus, the volume transmission concept that emerged from the first WGC symposium organised at the end of September 1989 has subsequently acquired depth and clarity. We meet again after 10 years to critically assess the validity of the volume transmission concept and to evaluate the new findings that will be presented. Beyond that, we seek to offer a starting point to neuroscientists wishing to investigate an entirely new field: the functional meaning of the different modes of intercellular communication in the brain and how they cooperate in the integrative tasks of the brain.

Kjell Fuxe

Acknowledgements

This book is based on a Wenner-Gren International Symposium, which was organized by Luigi F. Agnati, Kjell Fuxe, Charles Nicholson and Eva Syková and was generously supported by the Wenner-Gren Foundation.

Contents

III. Glia-neuronal signaling

IV. Monoamines and VT

V. The wider world of VT – from ions to peptides

VI. Summary

SECTION I

Conceptual basis of VT

L. F. Agnati, K. Fuxe, C. Nicholson and E. Syková (Eds.)
Progress in Brain Research, Vol 125

CHAPTER 1

Volume transmission as a key feature of information handling in the central nervous system possible new interpretative value of the Turing's B-type machine*

L. F. Agnati**[1] and K. Fuxe[2]

[1] *Department of Biomedical Sciences, Section of Physiology, University of Modena and Reggio Emilia, Modena, Italy*
[2] *Department of Neuroscience Karolinska Institutet, Stockholm, Sweden*

Introduction

In the present paper some aspects of the information handling in the Central Nervous System (CNS) will be analysed in the frame of the so called *Volume Transmission* (VT), i.e. of a widespread mode of intercellular communication that occurs in the extracellular fluid (ECF) of the brain and in the cerebrospinal fluid (CSF) (Agnati et al., 1986; Fuxe and Agnati 1991; Nicholson and Sykova, 1998; Temkin, 1999; Zoli et al., 1999). Five aspects will be examined, the first three on the basis of deductions from the available evidence and from the characteristics of the VT, the fourth on the basis of recent experimental evidence gathered in our laboratory, and the last one on the basis of the analogy between VT communication channels and the *connection devices* in the '*B-type unorganised machine*' proposed by Turing in 1948 (Turing, 1950, 1994; Copeland and Proudfoot, 1999).

Thus, the following points to enlighten the importance of VT for the information handling in the CNS will be discussed:

(1) the concept of *complex cellular network* that should supersede the concept of neural network as the set of cellular elements that allows the CNS to carry out its integrative actions.
(2) the concept of *polymorphic networks* (Getting and Denkin, 1985; Dickinson et al., 1990; Harris-Warrick, 1991; Meyrand et al., 1991) in the perspective of VT.
(3) the functional meaning of monoamine systems revised in the frame of the Kravitz proposal of "a biasing action of amine systems on behavioural outputs" (Bloom, 1984; Kravitz, 1988).
(4) kinetic aspects of the communication and elaboration of information in the complex cellular networks. Particular relevance will be attributed to pulses of the VT-signal in the microenvironment of its receptor.
(5) the kinetics of conveying the VT-signal to the target cell. The absence of a clearcut dichotomy between elaboration and communication (as it has been until now suggested in the modelling of the CNS) will be shown. Thus, it will be suggested that, at least in some instances, the two processes can take place simultaneously. This newly uncovered feature of the CNS will be interpreted as a way for

---

* Dedicated to Professor Eugenio Ferrari, chair of Neurology University of Bari Italy, on his 72th birthday. Professor Ferrari was one of the first neurologists to underline the vast clinical implications of the Volume Transmission concept.

** Corresponding author: e-mail: agnati@tin.it

some of its networks to operate according to the computational scheme of a *Turing's B-type unorganised machine* (Turing, 1950; Turing, 1994; Copeland and Proudfoot, 1999).

Premises

When intercellular communication is considered, it is possible to identify two types of cells on the basis of their functional role in the communication process. The *source cell* capable of emitting a signal and the *target cell* capable of recognising and decoding the signal. Hence, a pure functional condition is both necessary and sufficient to define intercellular communication in the living being. This functional condition will be referred to as the *basic assumption for intercellular communication* (BAIC; see Fig. 1).

As discussed by Shepherd and Jacobson in their beautiful books (Shepherd, 1991; Jacobson, 1993) and by our group in a previous paper (Agnati and Fuxe, 1997), neuroscience is still based on the Cajal and Sherrington's paradigma (*CS paradigm*: for a definition of the term 'paradigm' see Note 1). The CS paradigm maintains that the intercellular communication relevant for the integrative tasks of the CNS is the *interneuronal communication* which takes place *if and only if* the source cell and the target cell are connected by means of a synaptic contact.

During the last 15 years, our group has developed a theory based on the two classical opposite paradigms of interneuronal communication, the Cajal and Sherrington's paradigm and the Golgi's paradigm (see Figs 2 and 3). According to our theory, any cell type presents in the CNS can contribute to the integrative brain actions. Hence, not only interneuronal communication should be considered, but rather intercellular communication in the brain. Furthermore, it has been proposed that it is possible to distinguish two main modes for intercellular communication in the CNS: *Wiring Transmission* and *Volume Transmission* (see Figs 2 and 3; Agnati et al., 1986; Fuxe and Agnati, 1991; Agnati et al., 1992; Zoli et al., 1999;).

The **Wiring Transmission (WT)** is a point to point intercellular communication characterized by the transmission of the signal along a 'non leaking

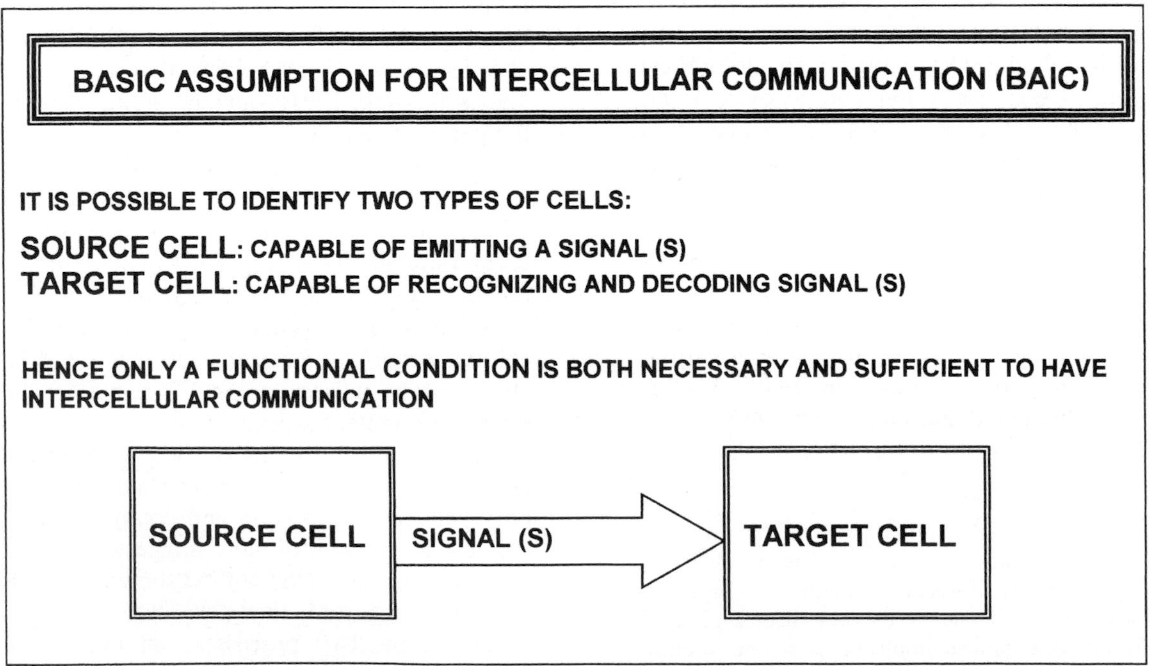

Fig. 1. Schematic representation of the basic assumption for intercellular communication.

channel', which connects the source cell with the target cell. The signal diffuses in the extracellular space (ECS) for distances no larger than the width of the synaptic cleft (width of about 30–50 nm). This physical arrangement allows a comparatively stable connection, and favours both the fast propagation of the messages and the safety of the transmission. According to this definition synaptic transmission (Note 2), gap-junctions and membrane juxtapositions represent different types of WT (see Fig. 2).

The **Volume Transmission (VT)** is a one to many intercellular communication characterized by the diffusion of the signal within the ECS for distances longer than the synaptic cleft (Fig. 3). The communication channel is a 'leaking channel' and hence the signal can diffuse out to invade large volumes of the ECS (see also the pioneering book of Nieuwenhuys (Nieuwenhuys, 1985)). From a functional point of view, VT leads to a highly divergent and often slow and unsafe transmission. However, there are some specializations in the VT that in part abolish these disadvantages. Preferential pathways for the migration of signals in the ECS have been demonstrated. In addition, the CSF (see Lehman and Silver, this volume) can work as a 'vector' to convey signals from the source cells to remote target cells. This type of VT has been called 'asynaptic endocrine-like' (Agnati et al., 1995a, b), when the release of the signal is from any cell structure except the synaptic terminal. Otherwise it has been called exosynaptic VT (Descarries, personal communication). Short range intercellular communication has been called perisynaptic VT (see Gonon, this volume) when the release site is the synaptic terminal and corresponds to the terms

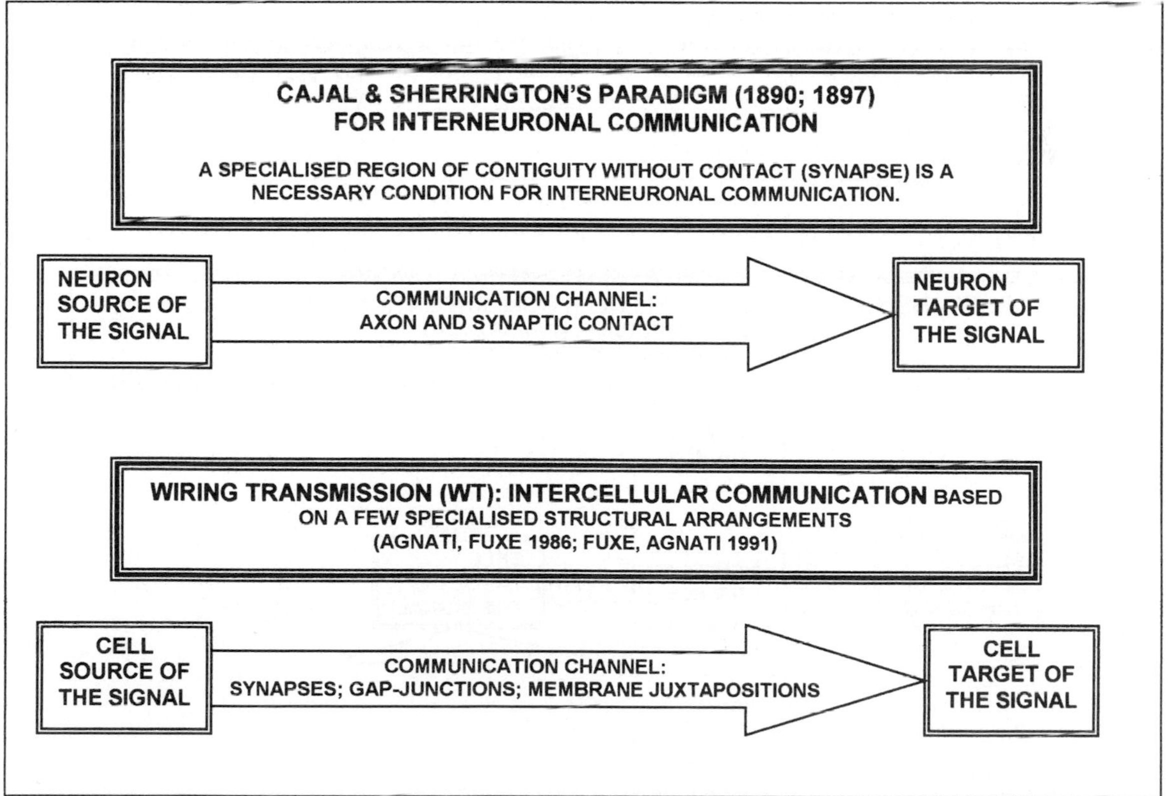

Fig. 2. Upper panel: schematic representation of the Cajal and Sherrington's paradigm for interneuronal communication. Lower panel: schematic representation of the Wiring Transmission (WT) concept for intercellular communication. For further details see text.

spillover and synaptic cross talk (see Kullmann, this volume). Otherwise it has been called asynaptic autocrine and paracrine VT which involve shorter and/or intermediate distance ranges. Electronic currents can also work as short distance VT signals (Nicholson, 1979; Agnati et al., 1995a, b; Zoli et al., 1999).

The definitions of both WT and VT are consistent with the BAIC. Intercellular communication in the CNS is a phenomenon similar to that present in other body tissues. In particular, the structural constraint (the synapse) imposed by the CS paradigm is only one of the suitable criteria to distinguish WT from VT, but certainly not a necessary condition to have intercellular communication, since neurons can communicate even if they are not connected by synaptic contacts. A similar view was proposed by Golgi more than one century ago (Golgi, 1891) by remarking that the ECS is filled up with an electrolytic solution (II. Class conductor according to Volta (see Volta's collected papers, 1967)) and therefore electrical signals (ion currents) can flow from one neuron to another one even if they are not in close proximity.

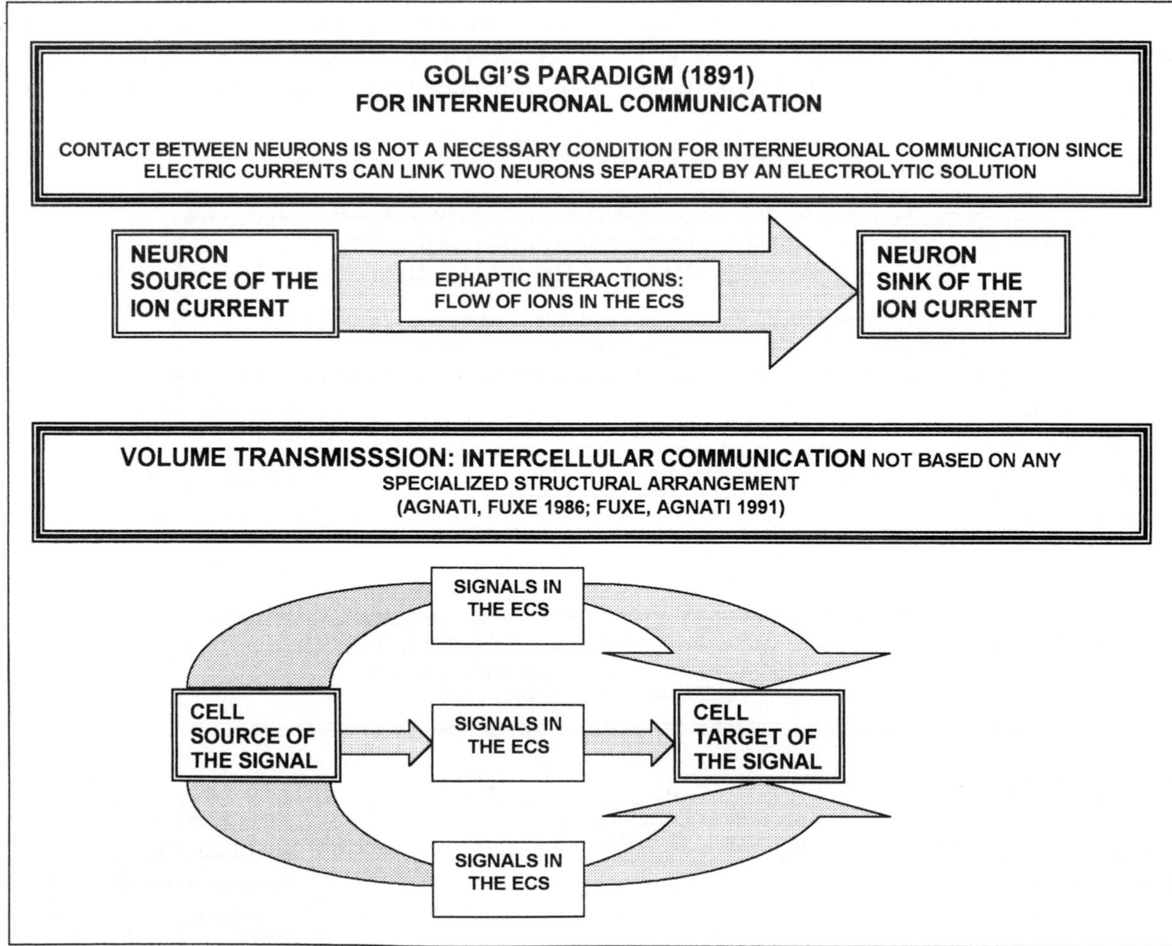

Fig. 3. Upper panel: schematic representation of the Golgi's paradigm for interneuronal communication. Lower panel: schematic representation of the Volume Transmission (VT) concept for intercellular communication. For further details see text.

Subjects specifically examined

(1) The concept of complex cellular networks

The BAIC states that any cell type can participate in an assembly of functionally interconnected cells, as long as it can release signals which are decoded by other cells of the assembly.

Experimental data demonstrate that, in the CNS, cross talks do occur between various cell types such as neurons, astrocytes, microglia, and ependymal cells. Thus, in the study of the integrative actions of the CNS it may be more relevant to consider not only networks made by neurons (*Neural Networks*), but rather *Complex Cellular Networks*, which can be defined as the set of cells of various types that exchange signals in a certain volume of brain tissue and, thanks to this cross talk, are capable of integrating inputs to give out appropriate outputs.

Let us examine the intercellular communication in the complex cellular networks in particular the one taking place by means of VT. To this aim the following characteristics of the VT will be examined:

- the chemico-physical nature of the VT-signals.
- communication channels between the source and the target cells and physical processes involved in the migration of the VT-signals through the channels (Nicholson, 1999).
- functional features of the VT-signal decoding processes.

These points underline three elements that are of paramount importance for intercellular communication taking place via VT: the signal, the channel and the system of signal recognition and decoding.

(1a) Features of the VT-signals

As discussed above (see also Fig. 4), ions (protons, sodium, potassium, calcium, etc.) transmitters and growth factors are important VT-signals (see Golgi, 1891; Nicholson, 1979; Olson et al., 1991; Agnati et al., 1994; Zoli et al., 1999). Our group has mainly studied monoamines (which belong to classical transmitters), neuropeptides, carbon dioxide and nitric oxide (which belong to gas signals) as VT-signals (see also Bjelke et al., 1994; Schneider et al., 1994; Dawson and Snyder 1994; Duggan, 1995; MacMillan et al., 1998; Snyder et al., 1998; Zoli et al., 1999).

Some chemico-physical characteristics of the VT-signals markedly affect intercellular communication. It is important to characterise a VT-signal according to its solubility, steric hindrance, charge, acid-base reaction, and possibility of participating to the formation of hydrogen bonds. As far as solubility is concerned, lipophilic signals (such as nitric oxide) can freely diffuse in the CNS by crossing the cell membranes, while hydrophilic signals (such as peptides) remain confined in the ECS. The other characteristics of the signal (molecular weight, shape, charge, acid-base reaction and capability of giving hydrogen bonds) influence the diffusion of the hydrophilic signals through the tiny intercellular spaces where the carbohydrate residues of the glycocalyx bulge out from the cell membranes (Alberts et al., 1994; Kleinsmith and Kish 1995).

(1b) Communication channels and physical processes involved in the migration of the VT-signal along the channels

As previously discussed (Agnati et al., 1994; Agnati et al., 1995; Agnati and Fuxe, 1997; Zoli et al., 1999) there are preferential pathways for VT-signal migration in the CNS, such as the para-axonal and the para-vascular routes. Para-axonal pathways result from the fact that diffusion and/or flow is favoured along the direction of myelinated axon bundles. A para-vascular route has been demonstrated by observing that there is a para-vascular fluid circulation that allows a rapid solute transport throughout the brain (Rennels et al., 1985).

In the present paper we would like to stress a general feature of the communication channels, namely the fact that the ECS is filled up not simply by extracellular fluid, but also by extracellular matrix (Note 3).

Furthermore, all cell membranes are covered by the glycocalyx and some neural cells by the so called perineural net (Note 4). This material can increase the tortuosity of the communication channel (Note 5). Thus, VT-signals are 'filtered' by the extracellular matrix during their migration in the ECS and by the glycocalyx in close proximity of the cell membrane (i.e. of the receiving-decoding system) and, in some cases, by the perineural net (Note 6).

Three main processes are involved in the migration of the VT-signals along the communication channels:

- *diffusion*: passive transport due to migration of molecules from a region of higher concentration to a region of lower concentration.
- *solvent drag*: mass movement of solvent (bulk flow (Note 7)) which carries solutes dissolved in it (see Nicholson, this volume).
- *convection* (Note 8): mass movement of solvent due to the presence of a temperature or pressure gradient. The movement of the solvent goes along with the migration of the solutes dissolved in it.

The diffusion process is important mainly for short distances as it allows only a slow rate of migrations (Note 9). On the contrary, solvent drag in the brain may represent a very important process both for short and long distance migrations of signals. A special case is the transport of signals by the CSF which can represent an effective vector to carry signals to far located brain areas (Greitz, 1993; Agnati et al., 1994; Agnati et al., 1995; Nicholson and Sykova, 1998; Nicholson, 1999; Lehman and Silver, this volume). Finally, data have been reported on the existence not only of hydrostatic and osmotic pressure gradients but also of temperature gradients between different brain areas (Delgado and Hanai, 1966). Regional cerebral temperatures are determined by local neural heat

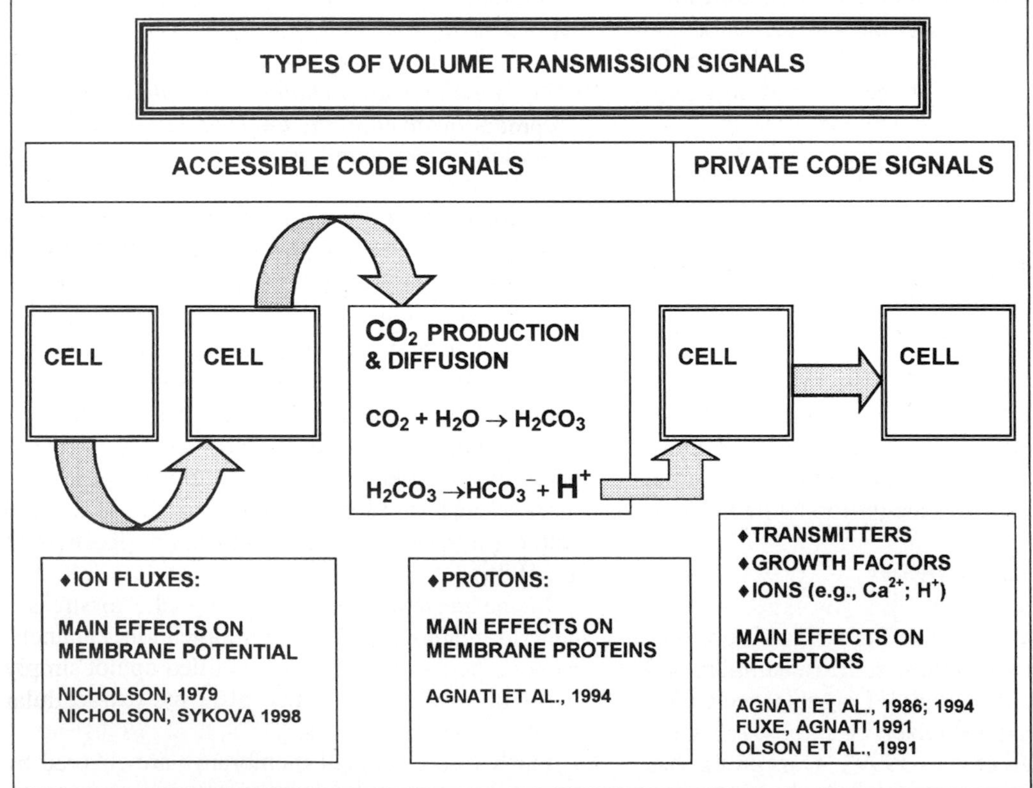

Fig. 4. Schematic representation of different signals used for Volume Transmission intercellular communication. For further details see text.

production, local blood flow and temperature of the arterial blood perfusing the brain (Hayward and Baker, 1968; Hayward and Baker, 1969). The existence of temperature gradients implies also that convective movements are likely to be in operation in the CNS favouring the migration of VT-signals (Agnati et al., 1994).

Therefore, temperature gradients are functionally important not only in view of the fact that several thermosensitive synaptic networks have been found in the brain (also outside of the hypothalamus see, e.g. Cevolani and Parmeggiani, 1995) but also since they could affect VT-signal migration. These gradients are influenced by different physiological conditions such as the sleep-wakefulness cycle, environmental temperatures and posture (Azzaroni and Parmeggiani, 1995). Hence, physiological (as well pathological) conditions can induce subtle changes in the migration of VT-signals by causing a new regional distribution of brain temperatures.

(1c) Functional features of the VT-signal decoding

It may be important to classify VT-signals on the basis of the features of their decoding system, i.e. of the recognition processes and of the type of response that they trigger in the target cell.

We suggest that a first basic distinction is between VT-signals that are recognised by molecules (*receptors*) that selectively bind them, and VT-signals that are not selectively recognised by receptors or as stated in a previous paper (Agnati et al., 1994) between *private-code signals* and *accessible-code signals* (see Fig. 4).

Private–code signals (selectively recognised by receptors); three types of VT-signals belong to this class:

- *transmitters and ions*: the formation of the transmitter-receptor complex induces primarily changes either in the electrical state of the plasma membrane of the target cell (signal action mediated by rapid *ionotropic receptors*) or in the metabolic state of the target cell (signal action mediated by slow *metabotropic G-protein coupled receptors*). Not only amino acids, amines, peptides, gases but also ions can work as transmitters. Thus, evidence has been obtained that some ions (such as Ca^{2+} and H^{+}) can be selectively recognised by receptor molecules and induce functional changes in the target cell mediated by a specific decoding mechanism (Note 10).
- *growth factors*: the formation of the growth factor-receptor complex induces primarily, but not exclusively, changes either in cell trophism, or in cell differentiation and replication.

Accessible-code signals (not selectively recognised by receptors); two types of VT-signals belong to this class:

- *ions*: thanks to ions, electrotonic currents occur in the CNS and these currents can affect the membrane potential of the target cell. These represent local effects (Nicholson, 1979; Nicholson and Sykova, 1998).
- *carbon dioxide*: this is a proteiform signal since it shows both the high diffusion capability of this gas and the specific (see Note 10) and unspecific (e.g. on protein conformation) actions of the proton (Alberts et al., 1994) which is formed by hydration of CO_2 and the subsequent dissociation of the carbonic acid (Chesler, 1990; Agnati et al., 1995; Amos and Chesler, 1998). Furthermore, CO_2 can favour the formation of α-carbamates from the unprotonated form of some L-amino acids (e.g. cysteine) which are present in the ECS. These compounds are powerful NMDA agonists (Max, 1991; Agnati et al., 1994).

(2) The concept of polymorphic networks

This is a well established concept (Getting and Denkin, 1985; Harris-Warrick, 1991) that is worthwhile to analyse in the frame of VT. Getting and Denkin first defined the *polymorphic network* as the moment-to-moment reconfiguration of a single network to produce several motor patterns (Fig. 5). This concept can be enlarged to the reconfiguration of other circuits beside those involved in the motor control. A consequence of the polymorphic network concept is the fact that neurons cannot be assigned to a certain network. On the contrary, they may take part to one functional circuit under a set of circumstances and be part in another when the conditions change. The switching process can be triggered by widespread or focal actions of modulators acting on the network. The widespread action can result from VT-signals, which can diffuse to flood large CNS volumes and, according to the mixture of VT-signals present in the ECS of a certain network, a different functional circuit can be created. Thus, from a single anatomically defined network, a large number of functional circuits can be created by VT-signals acting e.g. on the efficacy of the various synapses and/or on the release and decoding mechanisms of other VT-signals (Fig. 6).

Also in the case of focal actions VT-signals may be important. In fact, it is possible to surmise that VT-signals by acting on a 'switch' neuron can change the circulation of the information in a network either by excluding part of it or connecting two otherwise separated networks or simply by altering the order of activation of the computational elements of the network.

Therefore, VT can deeply change the computational characteristics of entire networks causing a 'functional rewiring' and hence it can represent a very important mechanism for brain plasticity.

(3) On the functional meaning of monoamine systems

Monoamine pathways are highly divergent neuronal systems that innervate large portions of the brain through collaterals (Fuxe, 1965). This diffuse branching pattern appears to preclude functions

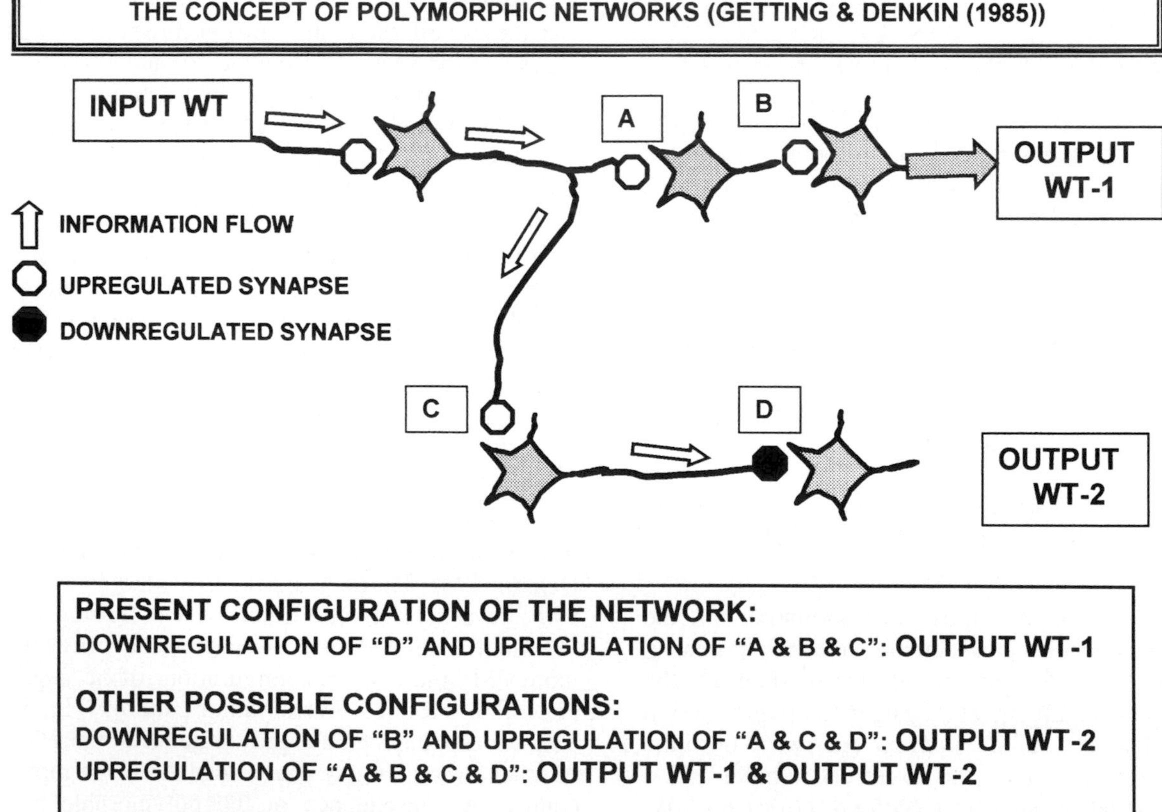

Fig. 5. Schematic representation of the concept of polymorphic network. For further details see text.

involved in transmitting specific information about space or time, and favours the view that these neuronal systems are especially involved in slower and global adjustments of the excitability of entire brain areas throughout the CNS (Bloom, 1984; Shepherd, 1988; Brodal, 1998). This action is likely of basic importance to cause:

(1) sensitisation, i.e. a gain-setting process of different types of receptors in several brain areas enriched with receptors for monoamines (Kravitz, 1988). This modulatory influence makes it possible for far located brain areas to function in an integrated way (Nieuwenhuys, 1985; Agnati, 1998), to produce a new pattern of behaviour or to enhance or diminish the effectiveness of an existing pattern (Yoshioka et al., 1995; Tanaka, 1999).

(2) appropriate matching between the performance of a particular behaviour and the motivational state of the subject (Kuypers, 1982; Agnati, 1998).

Experimental evidence supports the view that this widespread modulatory action of monoamines may occur to a large extent via VT (Fuxe and Agnati, 1991; Zoli and Agnati, 1996; Agnati, 1998). Monoamine release and diffusion into the ECS lead to an appropriate functional rewiring (see above the concept of *polymorphic networks*) in several far away located networks which are involved in carrying out a particular behaviour in order to associate it with the appropriate motivational state (Fig. 7; Agnati, 1998).

A demonstrative case may be that of a director of an orchestra. In his performance, sensory process-

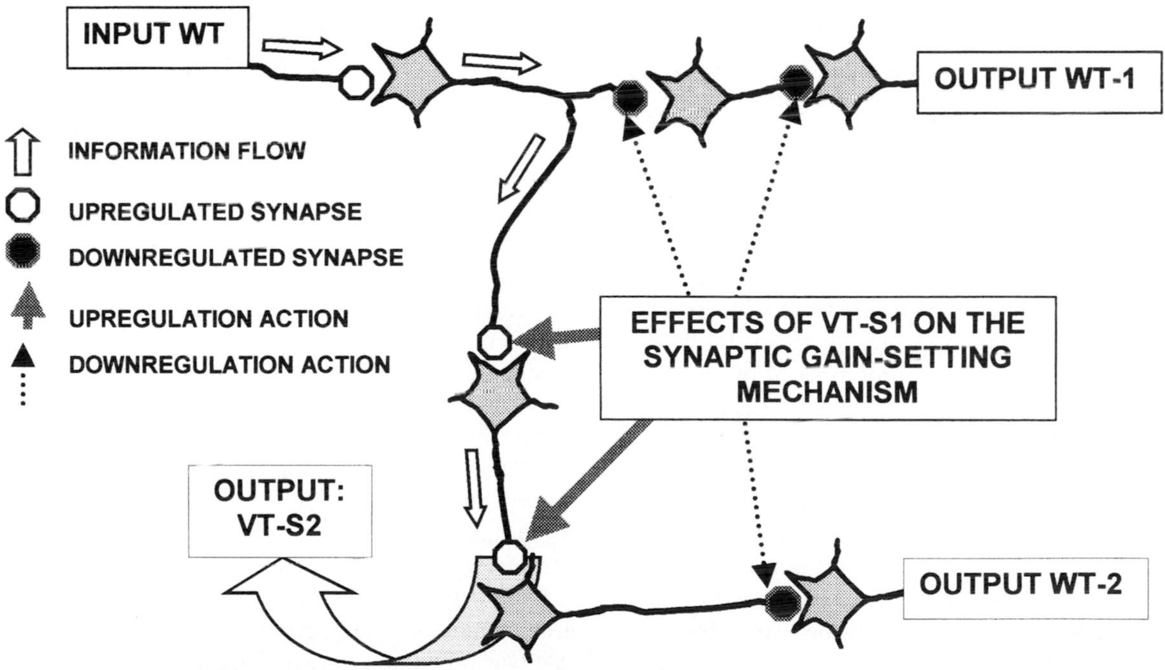

Fig. 6. Schematic representation of the possible rewirings of a polymorphic network caused by a Volume Transmission signal. For further details see text.

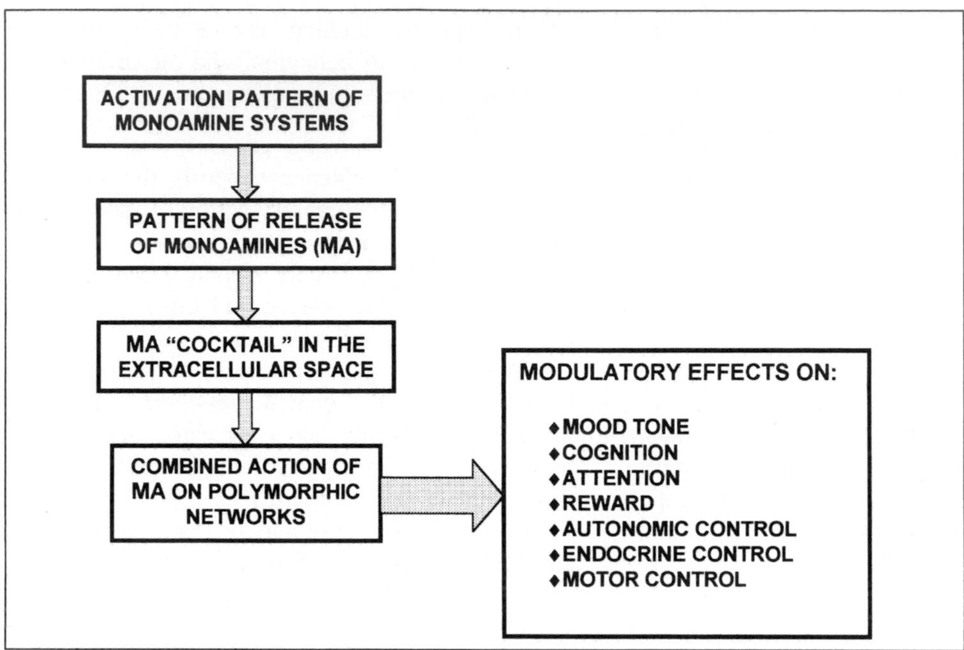

Fig. 7. Schematic representation of the possible modulatory effects of monoamine systems on different integrated responses of the central nervous system. For further details see text.

ing, motor control and emotional state have to be reset and properly harmonised. It has been shown for other demanding performances that a preparatory release of monoamines occurs before the beginning of the performance. Monoamine sensitive polymorphic networks present in several brain regions (such as neocortex, limbic areas, brain stem and spinal cord) respond to the 'cocktail' of monoamines diffusing in the ECS with a functional rewiring. This phenomenon may explain how the conductor reaches a special cognitive and emotional state, which is associated with a new tuning of his/her autonomic and neuroendocrine system. This widespread but well finalised reset of networks allows the optimisation of his/her performance (Fig. 8).

Thus, according to our hypothesis when the performance starts the networks that mediate sensory processing, those that store motor programs (Georgopulos, 1994) as well as the limbic areas are flooded by VT-signals (mainly monoamines) which favour a unique assembly and reading of the appropriate motor programs and a new tuning of the autonomic and neuroendocrine controls. This complex, widespread and not stereotyped reset may explain why the conductor can every time give a performance potentially different from the previous ones both for the emphasis given to some passages and the inspiration with which he directs the entire opera.

Since emotions are deeply rooted in the central autonomic areas (Coon, 1995), it is possible to gather hints on his/her peculiar (inspired) state by monitoring the changes in the autonomic control that occur immediately before and during his performance, mainly by recording cardiovascular parameters (Collet et al., 1997).

One of these parameters has been recorded in a great conductor, Herbert von Karajan. His heart rate was registered in two different conditions once when he was piloting his private aeroplane and the other one when he was directing the *Overture No. 3 Leonora* of Ludwig van Beethoven. Heart rate showed more ample fluctuations in the crucial passages of the overture than during even the most difficult phases of piloting (such as landing) and was unrelated to the physical effort of directing the orchestra (Critchley and Henson, 1987).

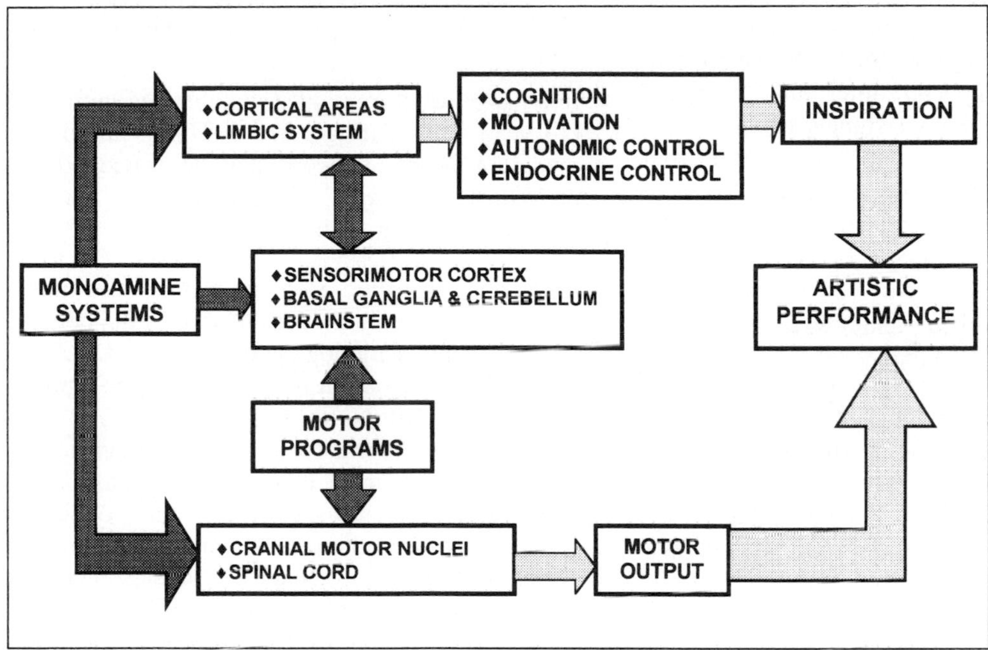

Fig. 8. Schematic representation of the possible intervention of monoamine systems on the artistic performance of, e.g. a director of an orchestra. For further details see text.

(4) Kinetic aspects characterising the VT-signalling

Some features of the VT can be more easily grasped by comparison with a similar phenomenon, the endocrine communication. Endocrinology has clearly established that at least three features should be considered in hormonal communication (Goodman, 1988; Wilson et al., 1998):

- *hormone levels* in the microenvironment of the receptor which must be above a threshold to stimulate receptors.
- *rhythms of hormone secretion* which should follow a certain pattern (frequency and amplitude) to provide appropriate phasic stimulations of the receptors.
- *sensitivity of the hormone receptors* which should be in the right range to respond to the action of the hormone. For this purpose it has to be remarked that a hormone, reaching several target cells with the same average concentration, can activate only some of these cells or even trigger different responses in the same cell according to the pattern of oscillations of its concentration in the receptor microenvironment. It should also be noted that, in the study of the pituitary-ovarian axis, it has been demonstrated that both decreases and increases in the rate of pulse administration of GnRH result in lower LH secretion. This result indicates the dependence of the pituitary cells not simply on a rhythmic input of GnRH, but rather from a precise pattern of signalling (Van Cauter and Aschoff, 1989).

According to our view, the same features are true also for VT. In fact, it may be stated that the following kinetics are of importance for VT:

(1) rate of synthesis and release of VT-signals by the source cell. In the case of neurons the rate of release is mainly a function of the firing rate.

(2) features of the process that conveys the VT-signal to the target cell. By modifying the signal migration the extracellular matrix and the glycocalyx/perineural net can alter the previous kinetics.

(3) modulatory changes in the sensitivity of the molecular decoding system of the target cell. These changes depend, inter alia, from the concentration of the VT-signals in the micro-environment of the receptors, and hence from the two previously mentioned factors.
(4) characteristics of the effector system of the target cell. The response of the target cell mainly depends on the sensitivity of the decoding mechanism, and hence on all the kinetics mentioned above.

The kinetic aspects of the VT have until now been overlooked. However, as shown in a paper published in the present book by our group, microdialysis data strongly suggest that DA release in striatum displays oscillations (see Fig. 9) that may have physiological and pathological relevance, just as in the case of hormone secretion (Van Cauter and Aschoff, 1989).

(5) The Turing B-type unorganised machine

In a paper which remained unpublished until 1968 (14 years after Turing's death) Turing proposed a model of neural network that he called '*B-type unorganised machine*'. This network is made of two elements (Fig. 10):

- *artificial neurons*: each neuron has two inputs and one output and every neuron in the network executes the same logical operation 'NOT AND' (NAND). The output is 1 if either of the inputs is 0 (see Fig. 10) (Note 11).
- *connecting elements* bearing a 'modifier device' which can be under two states: the 'pass mode' in which the message along the connecting element passes through it unmodified, and the 'interrupt mode' in which the message along the connecting element is destroyed. The modifier device has two inputs through which its state can be changed into the other one (see Fig. 10).

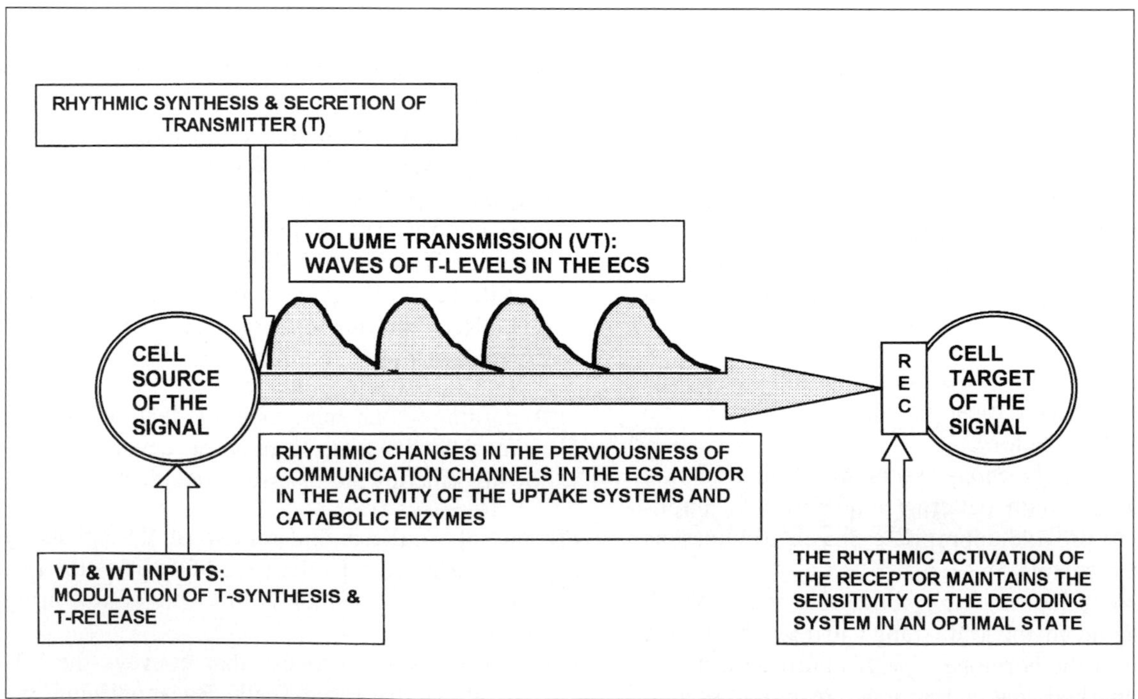

Fig. 9. Schematic representation of a deduction that can be drawn draw from the analogy between Volume Transmission and endocrine transmission. In particular, the emphasis is placed on the pulsatory release of the transmitter (VT-signal) in the extracellular space (ECS). For further details see text.

Neurons can be connected in any pattern, but each neuron–neuron connection must pass through a modifier device. Switching the modifiers from one mode to the other enables the network to be trained. On this basis Turing (1948, quoted by Copeland and Proudfoot, 1999) suggested that "the cortex of an infant is an unorganised machine, which can be organised by suitable interfering training".

Reconsidering the VT intercellular communication in the CNS networks, it becomes clear that the filtering action of the extracellular matrix (as well as of the glycocalyx/perineural net) can operate as a connection modifier by allowing the passage of the message or by destroying it (or simply preventing its migration). Hence, the communication channels in the ECS, at least in some instances (Note 12), may behave as the connection devices of the '*B-type unorganised machine*' of Turing. Therefore, it is possible that a clear-cut dichotomy between elaboration and communication in the CNS does not exist. At least in some instances the two processes can take place simultaneously.

We would like to further develop our thesis based on the following two concepts:

(1) *the VT concept*. The intercellular communication can occur via channels in the ECS which can be in the 'pass mode' or in the 'interrupt mode' according, to e.g. changes in the extracellular matrix.
(2) *the B—type unorganised machine*. If the communication channels between the cells in a network can be set either in the 'pass mode' or in the 'interrupt mode' the network can be trained.

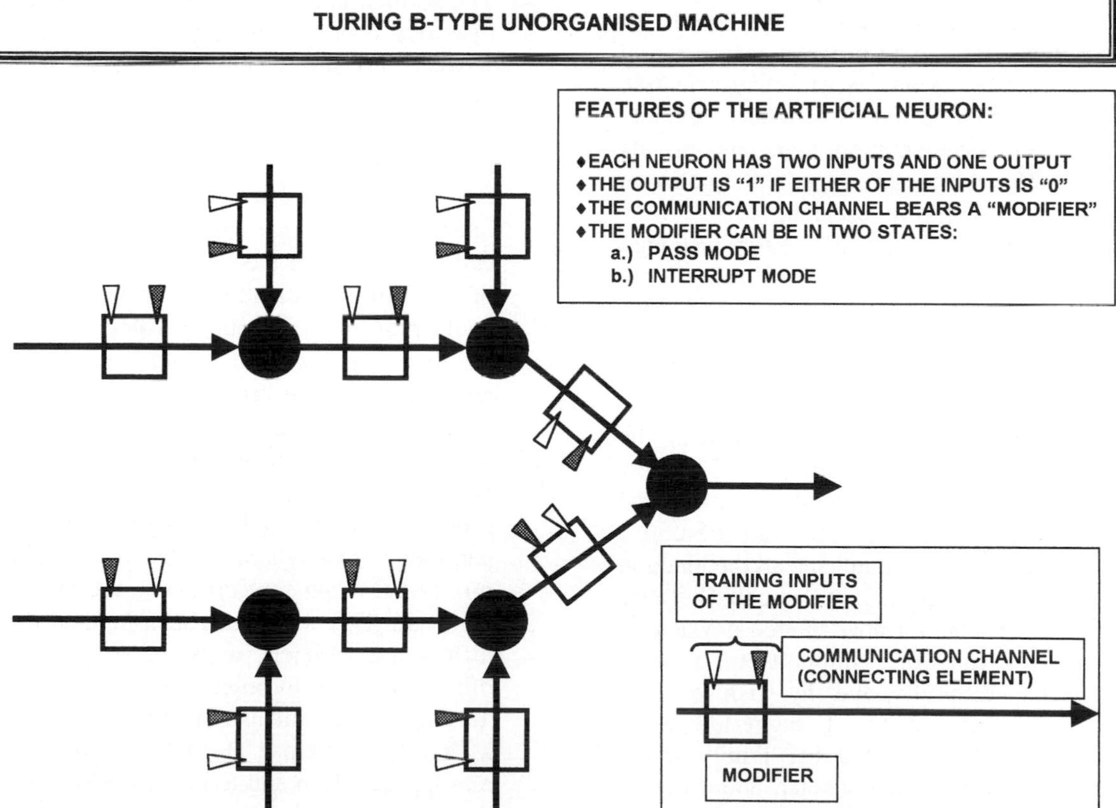

Fig. 10. Schematic representation of an artificial neuronal network based on the Turing's B-type unorganised machine. For further details see text.

It is possible to deduce that CNS networks can be trained by suitable changes in the perviousness of the channels for intercellular communication. Hence, we can reach the surprising conclusion that learning in the CNS networks can take place not only by instructions acting on the cells (classically by changing the 'weights' (i.e. efficacies) of the synapses in the circuit) but also by changing the characteristics of the extracellular matrix for example by focal changes in its capability of transmitting or destroying a signal flowing through it (Note 12 and Note 13).

Notes

(1) According to Kuhn (1962) 'normal science' is firmly based upon one or more past scientific achievements that the scientific community acknowledges for a time as supplying the foundations for its further practice. Some of these achievements acquire the status of 'paradigm'. By this term Kuhn indicates that some accepted examples of actual scientific practice (which include law, theory, application, and instrumentation together) provide models from which spring particular coherent traditions of scientific research.

(2) In his classical book of neurophysiology Fulton (1943) reports the following note about the origin of the term 'synapse': In a recent personal communication Sherrington writes: "You enquire about the introduction of the term 'synapse'"; it happened thus. M. Forster had asked me to get on with the Nervous System part (Part iii) of a new edition of his 'Text of Physiol.' for him. I had begun it, and had not got far with it before I felt the need of some name to call the junction between nerve-cell and nerve-cell (because the place of junction now entered physiology as carrying functional importance). I wrote him of my difficulty, and my wish to introduce a specific name. I suggested using 'syndesm'. He consulted his Trinity friend Verrall, the Euripidean scholar, about it, and Verrall suggested 'synapse' and as that yields a better adjectival form, it was adopted for the book.

(3) Extracellular matrix is composed by glycosaminoglycan chains of proteoglycans which form gels of varying pore size and charge density. This organized meshwork is in close association with the surface of the cells that produce it and it can be remodeled by matrix metalloproteases under physiological and pathological conditions. It has also been shown that extracellular matrix can bind diffusible proteins (e.g. growth factors) and small ions (e.g. calcium ion). It may be surmised that the nucleus of the cell has an organizing influence not only on the cell cytoplasm, but also on the extracellular space around the cell by regulating in a more or less precise way geometry, composition and plastic changes of this complex environment. Therefore, there are bidirectional influences between the extracellular space and the cell.

(4) The perineural net (discovered by Golgi more than one century ago, e.g. see Celio et al., 1998; Celio, this volume) represents a complex of extracellular matrix molecules (secreted both by neurons and glia) interposed between the meshwork of glial processes covering the surface of the neurons. This interrupted envelope surrounds not only the cell body but also the second and third order of dendritic arborizations. Perineuronal nets could play many functional roles such as to stabilize synapses and create microenvironments in the close proximity of the neuronal membranes.

(5) The process of diffusion of a molecule in the brain is sensitive to ECS structure. This sensitivity can be evaluated by measuring the parameter that characterizes the diffusion process: the apparent diffusion coefficient in the brain (D^*) which can be compared to the diffusion coefficient in water or in a very dilute gel (D) through the tortuosity (λ). Tortuosity is a dimensionless number defined as $\lambda = (D/D^*)$; since $D \geqslant D^*$, it follows that $\lambda \geqslant 1$ (Nicholson and Sykova, 1998). Further developments of the concept of tortuosity have been suggested (Rusakov and Kullmann, 1998).

(6) As pointed out by A. Temkin (personal communication) there is a general problem: how the loss and distortion of the information carried by the signal depend on the distance between the source and target cells and on the structure of the medium between them?

(7) Bulk flow is a mass movement of a fluid which results from either hydrostatic or osmotic pressure gradients.

(8) Convection (word of latin origin: *cum* (with) *vehere* (to transport)) is the mass movement of solvent along with the migration of the solutes in it dissolved. The fluid (vector) moves in response to a gradient of temperature or of pressure. This is a broader use of the term with respect its common use in physics, where it is used only for heat exchange; however, in physiology books it is stated that *heat exchange by convection is analogous to the transfer of solutes by bulk flow* (e.g. see Patton et al., 1989); and also that *'convection' refers to the transport of solute molecules entrained in a stream of* fluid (Greger and Windhorst, 1996); finally, that *the fast transfer of solutes dissolved in the blood is a process of convection* (Levick, 1991).

(9) By simple diffusion, a small transmitter reaches postsynaptic receptors in a few microseconds, but takes more than 16 min to reach a target located 1 mm away (Shepherd, 1979). For more recent data see Nicholson, this volume.

(10) Recently, a G protein-coupled receptor which binds Ca^{2+} has been identified and cloned. This evidence proves that calcium ions can also function as a first messenger in several body tissues including the brain, where at hippocampal level this Ca^{2+} receptor regulates Ca^{2+} and K^+-permeable ion channels (Brown et al., 1998; Brown, 1999). Similarly, H^+-gated cation channels have been demonstrated throughout the brain supporting the view that the proton can function as a first messenger (Waldmann et al., 1999).

(11) Turing selected the logical function NAND because every other Boolean operation can be accomplished by a group of NAND neurons.

(12) It should be noted that this analogy can be useful to describe some features of the information handling only in some networks or in parts of them, since intercellular communication can take place also via WT and, even in the case of the VT, the action of the extracellular matrix can not simply be that of allowing the passage of the message or of destroying it, but is much more complex. In fact, it has been shown that extracellular matrix enzymes are capable of releasing from a long peptide chain a set of different small peptides each of which triggers a response in target networks. For example, at least seven different peptides can be formed from POMC (pro-opiomelanocortin) and from these at least other five active fragments can originate in the ECS. The formation and quantitative ratios among these different peptides depend on the brain areas, the electrical activity of the neurons that synthesize the POMC and from the chemico-physical characteristics of the ECS (e.g. the pH). In this way a set of responses is generated and we have called this set of functionally related responses a *syndromic response* (Agnati et al., 1990; Agnati, 1998). In any case it should be kept in mind that the VT concept has profound implications for the mathematical modelling of the information processing in the CNS (Temkin, 1999).

(13) It should be considered that these changes in the perviousness of the channels may not be kept for a long time, but only for the period necessary to lead to the formation of other types of engrams.

References

Agnati, L.F., Fuxe, K., Zoli, M., Zini, I., Toffano, G. and Ferraguti, F. (1986) A correlation analysis of the regional distribution of central enkephalin and beta-endorphin immunoreactive terminals and of opiate receptors in adult and old male rats. Evidence for the existence of two main types of communication in the central nervous system: the volume transmission and the wiring transmission. *Acta Physiol. Scand.*, 128: 201–207.

Agnati, L.F., Zoli, M., Merlo Pich, E., Benfenati, F. and Fuxe, K. (1990) Aspects on neural plasticity in the central nervous system. VII. Communication and computation in the central nervous system. *Neurochem. Int.*, 16: 479–500.

Agnati, L.F., Bjelke, B. and Fuxe, K. (1992) Volume Transmission in the Brain. Do brain cells communicate solely through synapses? A new theory proposes that information also flows in the extracellular space. *Am. Sci.*, 80: 362–374.

Agnati, L.F., Cortelli, P., Biagini, G. and Fuxe, K. (1994) Different classes of volume transmission signals exist in the central nervous system and are affected by metabolic signals, temperature gradients and pressure waves. *NeuroReport*, 6: 9–12.

Agnati, L.F., Bjelke, B. and Fuxe, K. (1995a) Volume versus Wiring Transmission in the brain: a new theoretical frame for neuropsychopharmacology. *Med. Res. Rev.*, 15: 33–45.

Agnati, L.F., Zoli, M., Strömberg, I. and Fuxe, K. (1995b) Intercellular communication in the brain: Wiring versus Volume Transmission. *Neuroscience*, 69: 711–726.

Agnati, L.F. and Fuxe, K. (1997) The impact of histological techniques in revealing brain function. Volume transmission: from fluorescence histochemistry to confocal laser microscopy. In: K. Fuxe, T. Hökfelt, L. Olson, D. Ottoson, A. Dahlström and A. Björklund (Eds), *Molecular Mechanisms of Neuronal Communication*, pp 251–277.

Agnati, L.F. (1998) *Il cervello dell'uomo tra scienza e cultura*, Casa Editrice Ambrosiana, Milano.

Alberts, B., Bray, D., Lewis, J., Roberts, K. and Watson, J.D. (1994) *Molecular Biology of the Cell*, Garland.

Amos, B.J. and Chesler, M. (1998) Characterization of an intracellular alkaline shift in rat astrocytes triggered by metabotropic receptors. *J. Neurophysiol.*, 79: 695–703.

Azzaroni, A. and Parmeggiani, P.L. (1995) Postural and sympathetic influences on brain cooling during the ultradian wake-sleep cycle. *Brain Res.*, 671: 78–82.

Bjelke, B., Strömberg, I., O'Connor, W.T., Andbjer, B., Agnati, L.F. and Fuxe, K. (1994) Evidence for Volume Transmission in the dopamine denervated striatum of the rat after unilateral 6–0HDA microinjections. Studies with systemic D-amphetamine treatment. *Brain Res.*, 662: 11–24.

Bloom, F. (1984) The functional significance of neurotransmitter diversity. *Am. J. Physiol.*, 246: C184–C194.

Brodal, P. (1998) *The Central Nervous System: Structure and Function*. Oxford University Press, Oxford.

Brown, E.M., Chattopadhyay, N., Vassilev, P.M. and Herbert, S.C. (1998) The calcium-sensing receptor (CaR) permits Ca^+ to function as a versatile extracellular first messenger. *Recent Prog. Horm. Res.*, 53: 257–280.

Brown, E.M. (1999) The physiology and pathophysiology of the extracellular calcium-sensing receptor. *Am. J. Med.*, 106: 238–253.

Celio, M.R., Spreafico, R., De Biasi, S. and Vitellano-Zuccarello, L. (1998) Perineuronal nets: past and present. *TINS*, 21: 510–515.

Cevolani, D. and Parmeggiani, P.L. (1995) Responses of extrahypothalamic neurons to short temperature transients during the ultradian wake-sleep cycle. *Brain Res. Bull.*, 37: 227–232.

Chesler, M. (1990) The regulation and modulation of pH in the nervous system. *Prog. Neurobiol.*, 34: 401–427.

Collet, C., Vernet-Maury, E., Delhomme, G. and Dittmar, A. (1997) Autonomic nervous system response patterns specificity to basic emotions. *J. Auton. Nerv. Syst.*, 62: 45–57.

Coon, D. (1995) *Introduction to Psychology*, West Publishing Company, Minneapolis.

Copeland, B.J. and Proudfoot, D. (1999) Alan Turing's forgotten ideas in computer Science. *Scientific American*, pp. 77–81.

Critchley, M. and Henson, R.A. (Eds) (1987) *La Musica e il Cervello*. Piccin, Padova.

Dawson, T.M. and Snyder, S.H. (1994) Gases as biological messengers: nitric oxide and carbon monoxide in the brain. *J. Neurosci.*, 14: 5147–5159.

Delgado, J.M.R. and Hanai, T. (1966) Intracerebral temperatures in free-moving cats. *Am. J. Physiol.*, 211: 755–769.

Dickinson, P.S., Mecsas, C. and Marder, E. (1990) Neuropeptide fusion of two motor-pattern generator circuits. *Nature*, 344: 155–158.

Duggan, A.W. (1995) Release of neuropeptides in the spinal cord. *Prog. Brain Res.*, 104: 197–223.

Fulton, J.F. (1943) Physiology of the nervous system. Oxford University Press, Oxford.

Fuxe, K. (1965) Evidence for the existence of monoamine containing neurons in the central nervous system. The distribution of monoamine terminals in the central nervous system. *Acta Physiol. Scand.*, 64 (Suppl. 247): 39–85.

Fuxe, K., Wikström, A.C., Okret, S., Agnati, L.F., Härfstrand, A., Yu, Z.Y., Granholm, L., Zoli, M., Vale, W. and Gustafsson, J.Å. (1985) Mapping of glucocorticoid receptor immunoreactive neurons in the rat tel- and diencephalon using a monoclonal antibody against rat liver glucocorticoid receptor. *Endocrinology*, 117: 1803–1812

Fuxe, K., Cintra, A., Härfstrand, A., Agnati, L.F., Kalia, M., Zoli, M., Wikstrom, A.C., Okret, S., Aronsson, M. and Gustafsson, J.A. (1988) Central glucocorticoid receptor immunoreactive neurons: new insights into the endocrine regulation of the brain. In: W.F. Ganon, M.F. Dallman and J.L. Roberts (Eds), The hypothalamic-pituitary-adrenal axis revisited, *Ann. New York Acad. Sci.*, Vol. 512, pp. 362–393.

Fuxe, K. and Agnati, L.F. (Eds) (1991) *Volume Transmission in the Brain, Novel Mechanisms for Neural Transmission, Advances in Neuroscience*, Vol. 1, Raven Press, New York.

Georgopoulos, A.P. (1994) Behavioural neurophysiology of the motor cortex. *J. Lab. Clin. Med.*, 124: 766–774.

Getting, P.A. and Denkin, M.S. (1985) Tritonia swimming: a model system for integration within rhythmic motor systems In: *Model neural networks and Behavior*. Selverston Edition, Plenum Press, New York.

Golgi, C. (1891) *La rete nervosa diffusa degli organi centrali del sistema nervoso*. Suo significato fisiologico. Rendiconti Regio Istituto Lombardo, Serie 11, Vol. 24, Fascicolo Vlll, IX Milano.

Goodman, H.M. (1988) *Basic Medical Endocrinology*. Raven Press, New York.

Greger, R. and Windhorst, U. (Eds) (1996) *Comprehensive Human Physiology*. Springer, Berlin.

Greitz, D. (1993) *Cerebrovascular fluid circulation and associated intracranial dynamics, a radiographic investigation using MR imaging and radionucleide cistrnography.* PhD Thesis, Stockholm.

Harris-Warrick, R.M. (1991) Modulation of neural networks for behavior. *Ann. Rev. Neurosci.*, 14: 39–57.

Hayward J.N. and Baker M.A. (1968) Role of cerebral arterial blood in the regulation of brain temperature in the monkey. *Am. J. Physiol.*, 215:389–403.

Hayward J.N. and Baker M.A. (1969) A comparative study of the role of the cerebral arterial blood in the regulation of brain temperature in five mammals. *Brain Res.*, 16: 417–440.

Jacobson, M. (1993) *Foundations of Neuroscience.* Plenum Press, New York.

Kandel, E., Schwartz, J.H. and Jessel, T.M. (1991) *Principles of Neural Science.* Elsevier, New York.

Kleinsmith, L.J. and Kish, V.M. (1995) *Principles of Cell and Molecular Biology.* Harper Collins.

Kravitz, E.A. (1988) Hormonal control of behavior: amines and the biasing of behavioral output in lobsters. *Science*, 241: 1775–1781.

Kuhn, T.S. (1962) *The Structure of Scientific Revolutions.* Chicago University Press, Chicago.

Kuypers, H.G.J.M. (1982) A new look at the organization of the motor system. *Prog. Brain. Res.*, 57: 381–403.

Levick, J.R. (1991) *Cardiovascular Physiology.* Butterworth, London.

MacMillan, S.J.A., Mark, M.A. and Duggan, A.W. (1998) The release of beta-endorphin and the neuropeptide-receptor mismatch in the brain. *Brain Res.*, 794: 127–136.

Max, B. (1991) This and That: the neurotoxicity of carbon dioxide. *TIPS*, 12: 408–411.

Meyrand, P., Simmers, J. and Moulins, M. (1991) Construction of a pattern-generating circuit with neurons of different networks. *Nature*, 351: 60–63.

Nicholson, C. (1979) Brain cell microenvironment as a communication channel. In: F.G. Worden and F.O. Schmitt (Eds), *The Neurosciences: Fourth Study Program*, MIT Press Cambridge, MA, pp. 457–476.

Nicholson, C. and Sykova, E. (1998) Extracellular space structure revealed by diffusion analysis. *TINS*, 21: 207–215.

Nicholson, C. (1999) Signals that go with the flow. *TINS*, 22: 143–145.

Nieuwenhuys, R. (1985) *Chemoarchitecture of the brain.* Springer-Verlag, Berlin.

Olson, L., Wetmore, C., Strömberg, I. and Ebendal, T. (1991) Endogenous and exogenous Nerve Growth Factor in the Central Nervous System. In: K. Fuxe and L.F. Agnati (Eds), *Volume Transmission in the Brain, Novel Mechanisms for Neural Transmission, Advances in Neuroscience*, Vol. 1, Raven Press, New York, pp. 455–462.

Patton, H.D., Fuchs, A.F., Hille, B., Scher, A.M. and Steiner, R. (1989) *Textbook of Physiology.* W.B. Saunders Company, Philadelphia.

Rennels, M.L., Gregory, O.R., Blaumanis, O.R., Fujimoto, K. and Grady, P.A. (1985) Evidence for a 'paravascular' fluid circulation in the mammalian central nervous system, provided by the rapid distribution of tracer protein throughout the brain from the subarachnoid space. *Brain Res.*, 326: 47–63.

Rusakov, D.A. and Kullmann, D.M. (1998) Geometric and viscous components of the tortuosity of the extracellular space in the brain. *Proc. Natl. Acad. Sci.*, 95: 8975–8980.

Schneider, J.S., Rothblat, D.S. and Di Stefano, L. (1994) Volume transmission of dopamine over large distances may contribute to recovery from experimental parkinsonism. *Brain Res.*, 643: 86–91.

Shepherd, G.M. (1979) *The Synaptic Organization of the Brain.* Oxford University Press, Oxford.

Shepherd, G.M. (1988) *Neurobiology.* Oxford University Press, Oxford.

Shepherd, G.M. (1991) *Foundations of the Neuron Doctrine.* Oxford University Press, Oxford.

Snyder, H.S., Jaffrey, S.R. and Zakhary, R. (1998) Nitric oxide and carbon monoxide: parallel roles as neural messengers. *Brain Res. Rev.*, 26: 167–175.

Tanaka, M. (1999) Emotional stress and characteristics of brain noradrenaline release in the rat. *Ind. Health*, 37: 143–156.

Temkin, A. Ya. (1999) *Some ideas on information processing, thinking and genetics.* Tel-Aviv University Press, Tel Aviv.

Turing, A.M. (1950) Computing Machinery and Intelligence. Mind LIX: 443–460.

Turing, A.M. (1994) *Collected works of A.M. Turing: Mechanical Intelligence.* Elseviere Science Publishers B.V. (Italian Translation: Lolli G., Dazzi N., Alan M. Turing: Intelligenza meccanica, Bollati Boringhieri, Torino).

Van Cauter, E. and Aschoff, J. (1989) Endocrine and other biological rhythms. In L.J. Degroot (Ed.), *Endocrinology*, W.B. Saunders Company, Philadelphia, pp. 2658–2705.

Volta, A. (1967) Scritti scelti (a cura di Ghozzi M.), UTET, Torino.

Waldmann R., Champigny G., Lingueglia E., De-Weille jr., Heuteaux C. and Lazdunski M. (1999) H^+-gated cation channels. *Ann. NY Acad. Sci.*, 868: 67–76.

Wilson, J.D., Foster, D.W., Larsen, P.R. and Kroneberg, H. (1998) *Williams Textbook of Endocrinology.* W.B. Saunders Company, Philadelphia.

Yoshioka, M., Matsumoto, M., Togashi, H. and Saito, H. (1995) Effects of conditioned fear stress on 5-HT release in the rat prefrontal cortex. *Pharmacol. Biochem. Behav.*, 51: 515–519.

Zoli, M. and Agnati, L.F. (1996) Wiring and Volume Transmission in the central nervous system: the concept of closed and open synapses. *Prog. Neurobiol.*, 49: 363–380.

Zoli, M., Jansson, A., Syková, E., Agnati, L.F. and Fuxe, K. (1999) Intercellular communication in the central nervous system. The emergence of the Volume Transmission concept and its relevance for neuropsychopharmacology. *TIPS*, 20: 142–150.

L. F. Agnati, K. Fuxe, C. Nicholson and E. Syková (Eds.)
Progress in Brain Research, Vol 125

CHAPTER 2

Integration of wiring transmission and volume transmission

Floyd E. Bloom

Department of Neuropharmacology, The Scripps Research Institute, 10550 North Torrey Pines Road, La Jolla, CA, 92037, USA

Introduction

A decade has passed since the first Wenner–Gren Symposium on Volume Transmission (VT). Looking back at that volume recently, and in particular at my attempt then to offer some perspectives on the evidence for and against the imaginative concepts put forth by Agnati and Fuxe, was revealing. Many concepts were discussed quite heatedly then, such as whether monoamine or neuropeptide-containing fibers do or do not make synapses or release non-synaptically to act at sites remote from the time and place of release. These issues of experimental and descriptive fine structure and physiology have now been studied sufficiently to provide open-minded observers with evidence that accepts both possibilities within the range of normal interneuronal and glia-neuronal operations.

Thus, by direct serial section analysis of immunocytochemical localizations, monoamine containing nerve terminals do make specialized contacts with their target neurons, including serotonin (de Lima et al., 1988), as well as dopamine terminals (Sesack and Pickel, 1990). These systems, like the prototypical wiring transmission synapses, form selective convergent synapses with other specifically wired afferent systems. For example, the serotonin type 3 receptor, with an ionophoric molecular configuration is expressed almost exclusively in the rat forebrain by GABA-immunoreactive interneurons (Morales et al., 1996). Moreover, direct immunolocalizations of the receptor proteins themselves, have revealed that the receptors for monoamines (see Lewis and Sesack, 1997 for review) as well as many neuropeptides (see Milner and Pickel, 1986; Pickel et al., 1993; Van Bockstaele et al., 1995; Van Bockstaele et al., 1996) are clustered post-synaptically as conventionally portrayed co-transmitters with their amino acid partners.

On the other hand, as has been clear for many years, another then-revolutionary concept, presynaptic terminals can clearly regulate the synthesis and release of their own transmitters through auto-receptors for their own transmitter (see Xu et al., 1998 for recent extensions and molecular mechanisms). Presynaptic regulation by auto and heteroreceptors certainly demands that significant numbers of transmitter molecules escape from actions within the synaptic cleft to act at non-synaptic, but pre-synaptic receptors. Substantial physiological and pharmacological evidence also strongly supports the ability of co-existing neuropeptides to affect the release of their monoamine partners (see Bartfai et al., 1988). Indeed, recent iterations of the Volume Transmission Hypothesis (Zoli et al., 1998, 1999) emphasize the complementarity of wiring and volume transmission operations.

Moreover, one rarely finds now, studies devoted to what a decade ago was a quite fascinating endeavor, and one of the oft-cited justifications for

*Corresponding author. Tel.: 858-784-9730

support of the concept of volume transmission, the so-called 'ligand-receptor mismatch'. The progress of the intervening decade has repeatedly documented that ligand binding can often be misleading. Novel ligands like the amidated lipids anandamide (Crawley et al., 1993; Vogel et al., 1993) and oleamide (Cravatt et al., 1995) have changed expectations of what neurotransmitters may be as much as did the pronounced synaptic regulatory actions of nitric oxide, carbon monoxide and hydrogen sulfide acting as local gases (see references cited by Zoli et al., 1999). Novel receptors with minor variations also continue to be uncovered at a steady pace, such as unexpected ligands (Spina et al., 1996) and receptors (Lovenberg et al., 1995a, b) within the corticotropin releasing factor family and new members of the somatostatin family (de Lecea et al., 1996, 1997). As a result, one would hesitate to conclude that we know either all receptors or their complete inventory of ligands. Lastly, when release and synaptic overflow of transmitters are measured in vivo, even the classical amino acid transmitters GABA and Glutamate can give evidence of synaptic escape (Murphy and Maidment, 1999; Walwyn et al., 1999) especially under pathophysiological perturbations.

This observer continues to conceive of the nervous system as largely operating by 'wiring transmission' (WT) through those well defined but modifiable circuits that neurocytologists have so enjoyed mapping and measuring. Nevertheless, there is substantial evidence that the extracellular spaces can bring to wired synapses factors capable of regulating synaptic function. Such factors are very clear under certain extreme physiological conditions. For example, the reversible states of hepatic encephalopathy and of uremia are well known medically (see Aminoff, 1995), and can only be explained through the ability of urea and NH_3 to penetrate the blood brain barrier and diffuse by volume transmission through the extracellular spaces. In this revisit to the integrative possibilities of 'WT' and 'VT' under conditions closer to the normal physiological range, two aspects will be highlighted based on recent studies of my colleagues: (1) pre-synaptic actions of peptide receptors on release of amino acid or amine transmitters; (2) actions of inflammatory cytokines on synaptic function (both reported in greater detail elsewhere (Raber and Bloom, 1996; Raber et al., 1998; Campbell et al., 1999).

Regulated release of transmitters mediated by pre-synaptic actions

Somatostatin was one of the unexpected neuropeptides encountered by Guillemin and colleagues in their pursuit of the classical hypothalamic hypophysiotrophic hormones (Vale et al., 1975, 1976), and has provided our Department many pleasurable opportunities to map and assess many functional properties relevant to the integration of wiring and volume transmission.

Among the notable early points were the ability of somatostatin to hyperpolarize hippocampal pyramidal neurons, through several novel mechanisms, including an augmentation of the muscarinic response mechanism, the so-called '*m*-current', which is classically closed by acetylcholine (Pittman and Siggins, 1981; Siggins et al., 1982; Moore et al., 1988). Studies of these somatostatin-responding neurons with combined electrophysiology and immunocytochemistry revealed a very consistent finding. Most pyramidal neurons responded to somatostatin with a hyperpolarization associated with marked slowing of spontaneous discharge and reduced input resistance. In addition, somatostatin-containing fibers were always in close proximity to the somatostatin responsive cell identified by intracellular filling with Lucifer yellow. These occurrences are presumed to represent synaptic terminations of somatostatin elements on a pyramidal neuron that are likely to elicit membrane hyperpolarizations. Subsequent identification of a novel cortical and hippocampal enriched peptide with unquestionable sequence similarities to somatostatin revealed that the cortistatin had contrasting and selective effects on synaptic responses as well on sleep regulation (de Lecea et al., 1996).

More recently, Siggins and his co-workers have extended their studies of somatostatin and find that somatostatin's ability to hyperpolarize hippocampal pyramidal neurons are abolished by phospholipase A2 inhibitors and that arachidonic acid mimics all the effects of somatostatin on hippocampal pyramidal neurons. In addition, when the *m*-current effect is blocked by lipoxygenase inhibitors, both somatostatin and arachidonic acid elicit another outward current that is not blocked by either lipoxygenase or cyclooxygenase inhibitors, suggesting a direct role of arachidonic acid itself distinct from the *m*-current effect. (Schweitzer et al., 1993). In fact, subsequent efforts revealed that the necessary enzyme for these actions of arachidonic acid, the 5-Lipoxygenase-activating protein, an 18-kDa integral membrane protein is expressed (by in situ hybridization and immunocytochemistry) in various regions of the rat brain, including the CA1 pyramidal neurons of the hippocampus. Selective inhibition of this activating protein with the compound MK–886 (0.25–1 microM) prevents the somatostatin-induced augmentation of the hippocampal m- current. (Lammers et al., 1996).

Most recently, the effects of somatostatin have been investigated in still greater detail by using the MK866 to prevent the effects on the *m*-current. Under these conditions, Siggins and colleagues were able to determine that an additional action of this peptide is to open a voltage-insensitive leak current, not an inwardly rectifying K^+ current as reported in other neuron types (Schweitzer et al., 1998). Given these separable and potent effects of somatostatin, it is perhaps not surprising that in their latest work, Tallent and Siggins (1999) have reported that it is a potent anti-epileptic substance. In fact, they find that somatostatin can dramatically curtail paroxysmal discharges in hippocampus in two different experimental epilepsy models, (GABA-mediated neurotransmission inhibited by bicuculline and removal of the Mg^{2+} block of the *N*-methyl-D-aspartate (NMDA) glutamate receptors (Tallent and Siggins, 1999). Somatostatin appears to act specifically to reduce recurrent excitation between CA3 neurons because it can selectively depress excitatory influences through pre-synaptic actions on the associational/commissural pathway from the contralateral hippocampus, but not the excitations elicited by activation of the mossy fiber pathways from the dentate granule cells. These results suggest that somatostatin receptors could be a unique, selective clinical target for treatment of limbic seizures. Since somatostatin fibers do not co-exist with glutamate in these excitatory afferents, these effects must require localized volume transmission.

Cytokines regulate wiring transmission by volume transmission

The cytokines comprise a large and diverse family of polypeptide regulators, produced widely throughout the body by cells of assorted embryological origin. In general, these regulatory peptides have multiple functions attributable to specific actions when evaluated under control conditions in vitro. In vivo, the effects of the cytokines are known to be further regulated by the conditions imposed by other simultaneously acting cytokines, interacting as a network with variable effects leading to synergistic, additive, or even opposing actions. Within the immune system, macrophages and activated T lymphocytes are major producers of cytokines, such as the interleukins (IL) IL-1α, IL-1β IL-6, Interferon α (IFN-α) and Transforming Growth Factor β (TGF-β).

These peptide signals exhibit often-unpredictable interactions when evaluated for their pathophysiological involvement in specific inflammatory conditions in vivo. In our joint efforts to understand the basis for early pathophysiological changes in the brains of HIV-infected subjects, we have developed animal models for lentivirus infections, and assessed the actions of various cytokines acutely on transmitter release properties in vitro, and in an in vivo transgenic mouse model. IL-1β, IL2, IL6, and IFN-α will each enhance the release of the neuropeptide arginine vasopressin (AVP) in slices of rat hypothalamus and amygdala. TGF-β selectively blocks the ability of acetylcholine to release AVP from hypothalamus or amygdala, but has no effects on the release stimulated by other cytokines. IFN-α, but not TGF-β will also activate corticotropin releasing hormone release. As with AVP release, TGF-β selectively blocks the acetylcholine-stimulated corticotropin releasing hormone release in both amygdala and hypothalamus (for further details, see Raber and Bloom, 1994; 1996; Raber et al., 1994, 1997). These combined in vitro actions would suggest that CNS cytokine actions should upregulate the hypothalamic pituitary adrenal axis. In a transgenic mouse model with increased astrocytic expression and release of the cytokine IL6, the HPA axis is upregulated, and the effect seems attributable to adrenocortical hypersensitization to ACTH (Raber et al., 1998).

In addition to these influences on transmitter release, our colleagues have also examined the electrophysiological consequences of one of the cytokines, namely IFN-α (see Campbell et al., 1999) known to have effects on electroencephalographic activity in humans and experimental animals. To do so, they studied the CA1 population excitatory postsynaptic potentials in hippocampal slices from two lines of transgenic mice overexpressing IFN-α in astrocytes, and thus affecting neurons and their properties by volume transmission. Compared to control, non-transgenic littermates, these mice showed a pronounced hyperexcitability using a double blind protocol suggesting that these mice exhibit a bias toward seizure vulnerability. Such an outcome could be mediated by reduced numbers of inhibitory interneurons (anatomical and biochemical data indicate mainly that the cholinergic projections to and within the hippocampus are significantly reduced in these mice) or other covert distressed compensatory mechanisms. Paired-pulse facilitation (a well known test of presynaptic facilitation to the second of a pair of stimulation pulses delivered at short intervals) was also enhanced in the interferon overexpressing mice. Increases in this presynaptic facilitation are generally attributed to decreased excitatory transmitter release as a consequence of reduced Ca^{2+} accumulation within presynaptic terminals. Thus, the hyperexcitability may reflect the severely disrupted Ca^{2+} homeostasis in these mice directly interfering with transmitter release. These conclusions are supported by finding that the transgenic mice also show significant enhancements of excitability in two other electrophysiological tests, namely post-tetanic potentiation and long term potentiation, perhaps also attributable to abnormalities in the Ca^{2+} homeostasis necessary for the expression and maintenance of LTP.

Due to this dramatic seizure phenotype in the high expresser GIFN-39 mice, further behavioral studies involved the lower expresser GIFN-12 line. Mice were examined in a non-conditional spatial discrimination procedure. This task measures the ability of the animal to discriminate between right and left and to initiate a response to avoid a negative outcome. When examined at 7, 27 and 42 weeks of age, the interferon overexpressing transgenic mice exhibited significantly worse performance in the non-conditional spatial discrimination task overall when compared with wild type controls. Thus in this pathophysiological simulation of a volume transmitted effect, inflammatory cytokines can profoundly interfere with normal wiring type synaptic events.

Conclusion

As stated above, in this author's view, wiring transmission is the main form of chemical communication in the brain under most physiological conditions. However, substantial opportunity would seem to exist for the localized diffusion of transmitter, and transmitted-elicited signals such as gases, to the releasing pre-synaptic terminal or to adjacent terminals within the immediate locale. Evidence for functionally meaningful diffusion of peptides to more remote targets (see Duggan, this volume) is much more difficult to accept without some new principles by which such trans-synaptic cross-talk can be viewed as other than background noise. Under physiological conditions, lipid mediators, such as arachidonic acid, anadamide, and oleamide may well diffuse long distances, as certainly do the neurosteroids, adrenal steroids and gonadal steroids. However, such distant actions for these mediators are still circumscribed to selected responsive neurons by the restricted expression of their receptors. Lastly, under pathological conditions, such as brain infections, hypoxia or injury, cytokines can interfere with wiring transmission, and these effects are most easily explained by attribution to volume transmission channels. As we look forward to the next revisit to the stimulating hypothesis of volume transmission, the questions for which answers will be sought are: how pervasive, how distant, and how functional?

References

Aminoff, M. (1995) *Neurology and General Medicine*, Churchill Livingstone, New York.

Bartfai, T., Iverfeldt, K., Fisone, G. and Serfozo, P. (1988) Regulation of the release of coexisting neurotransmitters. *Annu. Rev. Pharmacol. Toxicol.*, 28: 285–310.

Campbell, I.L., Krucker, T., Steffensen, S., Akwa, Y., Powell, H.C., Lane, T., Carr, D.J., Gold, L.H., Henriksen, S.J. and Siggins, G.R. (1999) Structural and functional neuropathology in transgenic mice with CNS expression of IFN-alpha [In Process Citation]. *Brain Res.*, 835: 46–61.

Cravatt, B.F., Prospero-Garcia, O., Siuzdak, G., Gilula, N.B., Henriksen, S.J., Boger, D.L. and Lerner, R.A. (1995) Chemical characterization of a family of brain lipids that induce sleep. *Science*, 268: 1506–1509.

Crawley, J.N., Corwin, R.L., Robinson, J.K., Felder, C.C., Devane, W.A. and Axelrod, J. (1993) Anandamide, an endogenous ligand of the cannabinoid receptor, induces hypomotility and hypothermia in vivo in rodents. *Pharmacol. Biochem. Behav.*, 46: 967–972.

de Lecea, L., Criado, J.R., Prospero-Garcia, O., Gautvik, K.M., Schweitzer, P., Danielson, P. E., Dunlop, C.L., Siggins, G.R., Henriksen, S.J. and Sutcliffe, J.G. (1996) A cortical neuropeptide with neuronal depressant and sleep-modulating properties. *Nature*, 381: 242–245.

de Lecea, L., del Rio, J.A., Criado, J.R., Alcantara, S., Morales, M., Danielson, P.E., Henriksen, S.J., Soriano, E. and Sutcliffe, J.G. (1997) Cortistatin is expressed in a distinct subset of cortical interneurons. *J. Neurosci.*, 17: 5868–5880.

de Lima, A.D., Bloom, F.E. and Morrison, J.H. (1988) Synaptic organization of serotonin-immunoreactive fibers in primary visual cortex of the macaque monkey. *J. Comp. Neurol.*, 274: 280–294.

Lammers, C.H., Schweitzer, P., Facchinetti, P., Arrang, J.M., Madamba, S.G., Siggins, G.R. and Piomelli, D. (1996) Arachidonate 5-lipoxygenase and its activating protein: prominent hippocampal expression and role in somatostatin signaling. *J. Neurochem.*, 66: 147–152.

Lewis, D.A. and Sesack, S.R. (1997) Dopamine systems in the primate brain. In F.E. Bloom, A. Bjorklund and T. Hokfelt (Eds), *The Primate Nervous System*, Part I, Elsevier, Amsterdam, pp. 263–375.

Lovenberg, T.W., Chalmers, D.T., Liu, C. and De Souza, E.B. (1995a) CRF2 alpha and CRF2 beta receptor mRNAs are differentially distributed between the rat central nervous system and peripheral tissues. *Endocrinology*, 136: 4139–4142.

Lovenberg, T.W., Liaw, C.W., Grigoriadis, D.E., Clevenger, W., Chalmers, D.T., De Souza, E.B. and Oltersdorf, T. (1995b) Cloning and characterization of a functionally distinct corticotropin- releasing factor receptor subtype from rat brain. [published erratum appears in *Proc. Natl. Acad. Sci. USA* 1995 Jun 6;92(12):5759]. *Proc. Natl. Acad. Sci. USA*, 92: 836–840.

Milner, T.A. and Pickel, V.M. (1986) Neurotensin in the rat parabrachial region: ultrastructural localization and extrinsic sources of immunoreactivity. *J. Comp. Neurol.*, 247: 326–343.

Moore, S.D., Madamba, S.G., Joels, M. and Siggins, G.R. (1988) Somatostatin augments the M-current in hippocampal neurons. *Science*, 239: 278–280.

Morales, M., Battenberg, E., de Lecea, L. and Bloom, F.E. (1996) The type 3 serotonin receptor is expressed in a subpopulation of GABAergic neurons in the rat neocortex and hippocampus. *Brain Res.*, 731: 199–202.

Murphy, N.P. and Maidment, N.T. (1999) Orphanin FQ/nociceptin modulation of mesolimbic dopamine transmission determined by microdialysis. *J. Neurochem.*, 73: 179–186.

Pickel, V.M., Chan, J. and Sesack, S.R. (1993) Cellular substrates for interactions between dynorphin terminals and dopamine dendrites in rat ventral tegmental area and substantia nigra. *Brain Res.*, 602: 275–289.

Pittman, Q.J. and Siggins, G.R. (1981) Somatostatin hyperpolarizes hippocampal pyramidal cells in vitro. *Brain Res.*, 221: 402–408.

Raber, J. and Bloom, F.E. (1994) IL-2 induces vasopressin release from the hypothalamus and the amygdala: role of nitric oxide-mediated signaling. *J. Neurosci.*, 14: 6187–6195.

Raber, J. and Bloom, F.E. (1996) Arginine vasopressin release by acetylcholine or norepinephrine: region- specific and cytokine-specific regulation [published erratum appears in *Neuroscience* 1996 Jul;73(2):623]. *Neuroscience*, 71: 747–759.

Raber, J., Koob, G.F. and Bloom, F.E. (1997) Interferon-alpha and transforming growth factor-beta 1 regulate corticotropin-releasing factor release from the amygdala: comparison with the hypothalamic response. *Neurochem. Int.*, 30: 455–463.

Raber, J., Pich, E.M., Koob, G.F. and Bloom, F.E. (1994) IL-1 beta potentiates the acetylcholine-induced release of vasopressin from the hypothalamus in vitro, but not from the amygdala. *Neuroendocrinology*, 59: 208–217.

Raber, J., Sorg, O., Horn, T.F., Yu, N., Koob, G.F., Campbell, I.L. and Bloom, F.E. (1998) Inflammatory cytokines: putative regulators of neuronal and neuro- endocrine function. *Brain Res. Brain Res. Rev.*, 26: 320–326.

Schweitzer, P., Madamba, S., Champagnat, J. and Siggins, G.R. (1993) Somatostatin inhibition of hippocampal CA1 pyramidal neurons: mediation by arachidonic acid and its metabolites. *J. Neurosci.*, 13: 2033–2049.

Schweitzer, P., Madamba, S.G. and Siggins, G.R. (1998) Somatostatin increases a voltage-insensitive K^+ conductance in rat CA1 hippocampal neurons. *J. Neurophysiol.*, 79: 1230–1238.

Sesack, S.R. and Pickel, V.M. (1990) In the rat medial nucleus accumbens, hippocampal and catecholaminergic terminals converge on spiny neurons and are in apposition to each other. *Brain Res.*, 527: 266–279.

Siggins, G.R., McGinty, J.F., Morrison, J.H., Pittman, Q.J., Zieglgansberger, W., Magistretti, P.J. and Gruol, D.L. (1982) The role of neuropeptides in the hippocampal formation. *Adv. Biochem. Psychopharmacol.*, 33: 413–422.

Spina, M., Merlo-Pich, E., Chan, R.K., Basso, A.M., Rivier, J., Vale, W. and Koob, G.F. (1996) Appetite-suppressing effects of urocortin, a CRF-related neuropeptide. *Science*, 273: 1561–1564.

Tallent, M.K. and Siggins, G.R. (1999) Somatostatin acts in CA1 and CA3 to reduce hippocampal epileptiform activity. *J. Neurophysiol.*, 81: 1626–1635.

Vale, W., Brazeau, P., Rivier, C., Brown, M., Boss, B., Rivier, J., Burgus, R., Ling, N. and Guillemin, R. (1975) Somatostatin. *Rec. Prog. Horm. Res.*, 31: 365–397.

Vale, W., Rivier, C., Brown, M., Leppaluoto, J., Ling, N., Monahan, M. and Rivier, J. (1976) Pharmacology of hypothalamic regulatory peptides. *Clin. Endocrinol.* (Oxf), 5(Suppl): 261S–273S.

Van Bockstaele, E.J., Branchereau, P. and Pickel, V.M. (1995) Morphologically heterogeneous met-enkephalin terminals form synapses with tyrosine hydroxylase-containing dendrites in the rat nucleus locus coeruleus. *J. Comp. Neurol.*, 363: 423–438.

Van Bockstaele, E.J., Colago, E.E., Cheng, P., Moriwaki, A., Uhl, G.R. and Pickel, V.M. (1996) Ultrastructural evidence for prominent distribution of the mu-opioid receptor at extrasynaptic sites on noradrenergic dendrites in the rat nucleus locus coeruleus. *J. Neurosci.*, 16: 5037–5048.

Vogel, Z., Barg, J., Levy, R., Saya, D., Heldman, E. and Mechoulam, R. (1993) Anandamide, a brain endogenous compound, interacts specifically with cannabinoid receptors and inhibits adenylate cyclase. *J. Neurochem.*, 61: 352–355.

Walwyn, W.M., Ta-Haung, J., Ackerson, L., Maidment, N.T. and Edgerton, V.R. (1999) Extracellular glutamate in the dorsal horn of the lumbar spinal cord in the freely moving rat during hindlimb stepping [In Process Citation]. *Pharmacol. Biochem. Behav.*, 63: 581–588.

Xu, Z.Q., Lew, J.Y., Harada, K., Aman, K., Goldstein, M., Deutch, A., Haycock, J.W. and Hokfelt, T. (1998) Immunohistochemical studies on phosphorylation of tyrosine hydroxylase in central catecholamine neurons using site- and phosphorylation state- specific antibodies. *Neuroscience*, 82: 727–738.

Zoli, M., Jansson, A., Sykova, E., Agnati, L.F. and Fuxe, K. (1999) Volume transmission in the CNS and its relevance for neuropsychopharmacology. *Trends Pharmacol. Sci.*, 20: 142–150.

Zoli, M., Torri, C., Ferrari, R., Jansson, A., Zini, I., Fuxe, K. and Agnati, L.F. (1998) The emergence of the volume transmission concept. *Brain Res. Rev.*, 26: 136–147.

L. F. Agnati, K. Fuxe, C. Nicholson and E. Syková (Eds.)
Progress in Brain Research, Vol 125

CHAPTER 3

Ultrastructural evidence for diffuse transmission by monoamine and acetylcholine neurons of the central nervous system

Laurent Descarries*[,1] and Naguib Mechawar[2]

Départements de pathologie et biologie cellulaire[1,2] et de physiologie[1], and Centre de recherche en sciences neurologiques[1,2], Faculté de médecine, Université de Montréal, Montréal, H3C 3J7, Canada

Introduction

The present book highlights the view that, in addition to synaptic transmission, i.e. chemical transmission taking place at morphologically differentiated sites of membrane specialization, many central nervous system (CNS) neurons are capable of spreading transmitter more broadly, to various cellular targets, through what has come to be designated as 'volume transmission'. In mammalian CNS, this mode of neuronal communication was first envisaged on the basis of electron microscopic observations on autoradiographically identified serotonin (5-HT) and noradrenaline (NA) axon terminals (varicosities**) in adult rat cerebral cortex, which often lacked the membrane junctional complexes that are the hallmark of synapses (Descarries et al., 1975; Beaudet et al., 1976; Descarries et al., 1977). The proposal was then made that the monoamine (MA) transmitter released from such varicosities might not exert its effects solely on restricted areas of postsynaptic membrane specialization, but also diffuse in tissue and thus reach relatively distant targets endowed with receptive elements. It was also pointed out at the time that the largely asynaptic character of these innervations was hardly compatible with the preconceived notion of a fixed pattern of neuronal circuitry. Because of the known intra-axonal mobility of inner constituents (e.g., vesicular organelles and mitochondria), it was actually postulated that the varicosities themselves, lying free in the neuropil, were subjected to incessant movements of translocation and/or reshaping along their parent fibers, constantly modifying their position as release sites in relation to their immediate microenvironment. Thus, at the ultrastructural as well as cytological and anatomical levels, such neuronal systems appeared ideally built for achieving sustained and adapted modulation and/or coordination of vast neuronal ensembles, in addition to cell-to-cell, point-to-point, direct transfer of information.

By the time of the first symposium on Volume Transmission, in September 1989, much of the early autoradiographic evidence in favor of the largely asynaptic character of the MA innervations

*Corresponding author. Tel.: (514) 343-7070; Fax: (514) 343-5755; e-mail: descarrl@ere.umontreal.ca

** For lack of better words, the eponyms 'terminal' and 'varicosity' are used interchangeably throughout this chapter, to designate axonal enlargements containing aggregated synaptic vesicles and endowed or not with morphologically defined membrane specializations of synaptic contact (junctional complex).

in adult rat cerebral cortex had already been confirmed by detailed and systematic quantitative electron microscope immunocytochemical analyses (review in Descarries et al., 1991). Immunocytochemical data had also been produced in different laboratories regarding the ultrastructural relationships of dopamine (DA), NA and 5-HT axon varicosities in other regions of the rat brain (for reviews, see Beaudet and Descarries, 1987; Descarries et al., 1988; Soghomonian et al., 1988; Maley et al., 1990), but examination of the ultrastructure of these terminals in primate cerebral cortex had just begun (DeFelipe and Jones, 1988; De Lima et al., 1988). Moreover, despite the availability of specific antibodies against the biosynthetic enzyme of acetylcholine (ACh), choline acetyltransfererase (ChAT; Crawford et al., 1982; Eckenstein and Thoenen, 1982; Levey et al., 1983), only qualitative information had then been acquired about the relational features of cholinergic terminals in either the cerebral cortex or selected brain regions of rat, cat, dog, ferret or primate (for reviews, see Houser, 1990; Umbriaco, 1995). The immunocytochemical visualization of some of the receptors for the monoamines or ACh at the subcellular level was also in its early beginnings.

In the past ten years, considerable progress has been made on each of these fronts. As reviewed in the present chapter (see also Smiley, 1996), the relational features of MA axon terminals have been examined in the cerebral cortex of different mammalian species, including monkey and man. Other CNS regions in rat and other mammals have also been explored, notably the hippocampus and neostriatum. In some of these studies, attempts were made to identify the synaptic targets of the identified terminals and/or to visualize the subcellular distribution of the receptors for their transmitter. Detailed ultrastructural investigations have been carried out on the ACh innervation of the cerebral cortex, hippocampus, neostriatum and olfactory bulb in rat, as well as the cerebral cortex in monkey and man. Lastly, some of these results have been brought together to propose the hypothesis of an ambient extracellular level of transmitter(s) in the CNS, at least in those anatomical regions richly innervated by the monoamines or ACh.

Monoamine innervations

Most of the information gathered in recent years on the fine structural features and relationships of central MA axon terminals has been obtained by means of electron microscopic immunocytochemistry with specific antibodies against the biosynthetic enzymes, tyrosine hydroxylase (TH) and dopamine-β-hydroxylase (DBH), or against DA, NA or 5-HT themselves, coupled to proteins with aldehydes to form immunogenic conjugates. In some of these studies, primary fixation of the CNS could thus be carried out by perfusion with relatively high concentrations of aldehydes, which improves the ultrastructural preservation of the tissue. Moreover, perhaps even more importantly, efforts were made to examine the immunostained axon terminals in serial as well as single thin sections for electron microscopy, allowing for detailed scrutiny of most when not the entire volume of varicosities and direct estimates of the frequency with which these terminals are engaged in synaptic junction (synaptic incidence; Fig. 1). In other instances, such estimates were inferred by linear transformation of the relationship between the frequency of observed synaptic junctions and the number of thin sections available for examination. A third approach was to determine synaptic incidence from the frequency in single thin sections by means of the stereological formula of Beaudet and Sotelo (1981). This formula takes into account the average size of varicosity profiles, the length of their junctional complexes and the thickness of the sections, to predict the probability of seeing a synapse if there is one on every varicosity. The synaptic incidence is then inferred by comparison with this predicted value. There is good evidence that such extrapolated incidences are reliable, providing that the single section sampling of varicosities is of sufficient size. On numerous occasions, both a serial and a single thin section sampling of the same species of varicosities were indeed shown to yield similar values (e.g. Soghomonian et al., 1989; Oleskevich et al., 1991; Descarries et al., 1992; Ridet et al., 1992, 1993; Descarries et al., 1996). Furthermore, almost identical values were found by Umbriaco et al. (1994) in a large population of varicosities examined across their entire volume and which were also treated as a randomized single section sample.

Cerebral cortex and hippocampus

Except for discordant reports from one laboratory (Papadopoulos et al., 1987a, b, 1989a, b; Parnavelas and Papadopoulos, 1989 ; Paspalas and Papadopoulos, 1996), all other studies in which the synaptic incidence of MA terminals was determined in cerebral cortex of rat, monkey and man have confirmed the partly or largely asynaptic character of these innervations (see Table 1).

The DA innervation is undoubtedly the most synaptic, as now documented in the medial prefrontal, suprarhinal and occipital cortex of rat. Except in the study by Papadopoulos et al. (1989a), the junctions made by cortical DA varicosities have always been described as relatively small and mostly symmetrical. All available observations suggest important regional and perhaps laminar differences in the frequency with which these varicosities make synaptic junction (see Van Eden et al., 1987; Séguéla et al., 1988; Papadopoulos et al., 1989b; Verney et al., 1990). Attempts at identifying the partners of synaptic DA varicosities in rat anteromedial, prefrontal and motor cortex have indicated that both pyramidal cell dendritic trees and a subpopulation of intrinsic GABA neurons were contacted (Verney et al., 1990). A

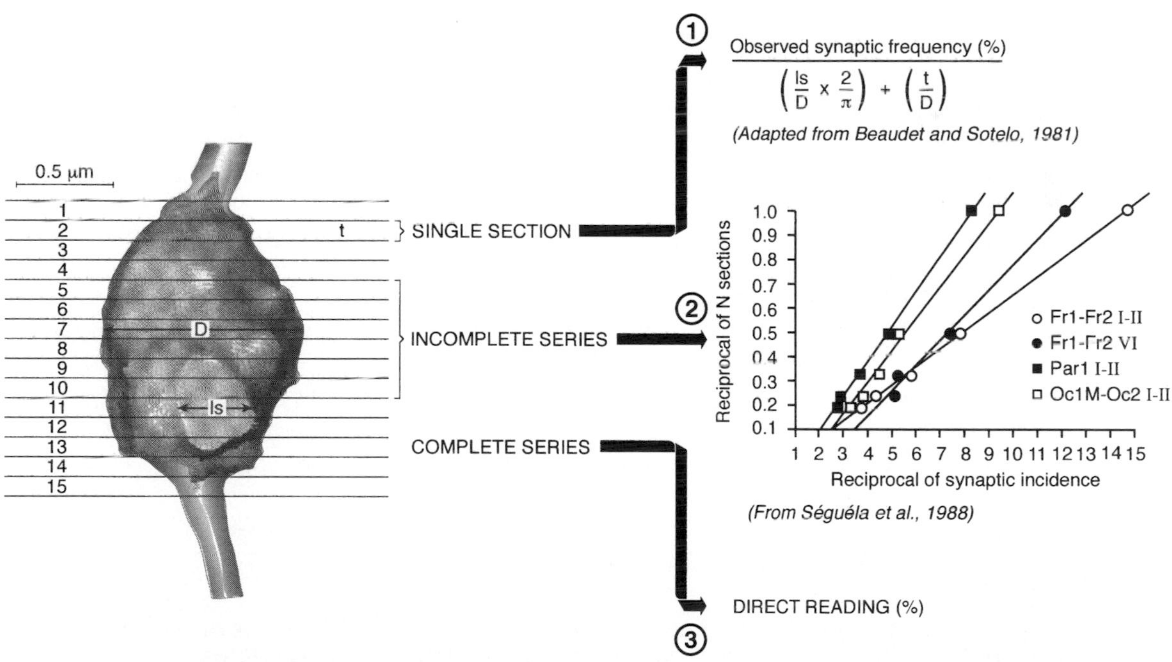

Fig. 1. Schematic representation of current methods for determining the frequency with which a given population of axon varicosities makes synaptic specializations (synaptic incidence). As illustrated by the three-dimensional reconstruction on the left (modified from Umbriaco et al., 1994), the immunostained terminals are visualized as sectional profiles, 0.5–1 μm in diameter (D), in ultrathin sections, approximately 0.1 μm in thickness (t). When present, the junctional complex occupies a small portion of the surface of varicosities (width: ls). From single sections (1), synaptic incidence may be extrapolated by means of the stereological formula of Beaudet and Sotelo (1981), as explained in the text. When partial or incomplete series of sections are available (2), synaptic incidence may be inferred by linear transformation of the relationship between the frequency of observed synaptic junctions and the number of sections available for examination. When complete series are available (3), synaptic incidence may be determined by direct observation.

TABLE 1

Synaptic incidence of monoamine (DA, NA, 5-HT) and acetylcholine (ACh) axon terminals in mammalian CNS

		DA	NA	5-HT	ACh
Cerebral cortex					
Rat	medial prefrontal	93%[1]	—	—	—
	suprarhinal	56%[1]	—	—	—
	frontal	—	7%[2]	32%[3], 22%[4]	14%[5]
	parietal	—	17% or 26%[6]	46%[3]	14%[7]
	occipital	90%[8]		37%[3]	—
	entorhinal	—	—	18%[4]	9%[9]
Cat	auditory	—	—	3%[10]	—
Monkey	sensory-motor	—	—	2–3%[11]	—
	prefrontal	39%[12]	18%[13]	23%[14]	44%[15]
	entorhinal	20%[16]	—	—	—
Human	anterolateral temporal	5–10%[17]	—	—	67%[18]
Hippocampus *(Rat)*					
	CA1		16%[19]	23%[19], 12%[4]	7%[19]
	CA3		—	19%[20]	—
	DG		—	24%[20]	—
Neostriatum *(Rat)*		30–40%[21]		10%–13%[22]	9%[23]
Other brain regions *(Rat)*					
	suprachiasmatic nucleus	—	—	48%[24]	—
	supraoptic nucleus	—	—	38%[24]	—
	substantia nigra, compacta		—	50%[25]	—
	substantia nigra, reticulata		—	~ 100%[25]	—
	dorsal periaqueductal grey	—	—	23%[26]	—
	lateral geniculate nucleus	—	—	42%[27]	—
	basal forebrain (NHL)	—	—	46%[28]	—
	superior colliculus				
	superficial layers	—	—	55%[29]	—
	deep layers	—	—	38%[29]	—
	ventrolateral nucleus of thalamus	—	—	62%[29]	—
Spinal cord *(Rat)*					
	dorsal horn, cervical	34%[30]	25%–29%[31,32]	37%[32]	—
	dorsal horn, thoracic	76%[30]	—	—	—
	central canal	67%[30]		—	—
	intermediolateral cell column	> 100%[30]	87%[31]	> 100%[33]	—
	ventral horn	~ 100%[30]	85%[31]	95%[34]	—

The synaptic incidences (%) were obtained by extrapolation from single thin sections or incomplete series of serial sections, or by observation of varicosities entirely examined in serial sections. The latter values were given priority whenever available. The superscripts refer to the following studies: [1] Séguéla et al., 1988; [2] Cohen et al., 1997; [3] Séguéla et al., 1989; [4] Cohen et al., 1995; [5] Chédotal et al., 1994; [6] Séguéla et al., 1990; [7] Umbriaco et al., 1994; [8] Papadopoulos et al., 1989b; [9] Vaucher and Hamel, 1995; [10] DeFelipe et al., 1991; [11] DeFelipe and Jones, 1988; [12] Smiley and Goldman-Rakic, 1993; [13] Aoki et al., 1998; [14] Smiley and Goldman-Rakic, 1996; [15] Mrzljak et al., 1995; [16] Erickson et al., 2000; [17] Smiley et al., 1992; [18] Smiley et al., 1997; [19] Umbriaco et al., 1995; [20] Oleskevich et al., 1991; [21] Descarries et al., 1996; [22] Soghomonian et al., 1989; [23] Contant et al., 1996; [24] Boulaich et al., 1994; [25] Moukhles et al., 1997; [26] Lovick et al., 2000; [27] Dinopoulos et al., 1995; [28] Dinopoulos et al, 1997; [29] Dori et al., 1998; [30] Ridet et al., 1992; [31] Rajaofetra et al., 1992; [32] Ridet et al., 1993; [33] Poulat et al., 1992; [34] Ridet, 1994.

similar conclusion has been reached with respect to the targets of synaptic DA terminals in the prefrontal, motor and entorhinal cortex of rhesus and cynomolgus monkeys (Goldman-Rakic et al., 1989; Sesack et al., 1995a, b, 1998; Smiley et al., 1996; Erickson et al., 2000). In their study of rhesus monkey prefrontal cortex, Smiley et al. (1996) observed a synaptic incidence of 39% for 153 DA-immunostained varicosities completely examined in serial sections across their entire volume. They therefore concluded that, at least in this part of monkey cortex, DA presumably exerted effects via release from non-synapsing varicosities and extracellular diffusion, in addition to synaptically targeted actions on specific neuronal populations. In addition, a 9.8% proportion of synaptic DA varicosities has been recently reported for single sections of TH-immunostained terminals in the entorhinal cortex of cynomolgous monkey (Erickson et al., 1999). Assuming that these DA varicosities are of the same size as in rat, this should correspond to a synaptic incidence of about 20%.

Interestingly, there has also been an investigation of the distribution and synaptic connectivity of DA immunoreactive axons in human cortical tissue removed from the temporal pole and anterior lateral temporal surface (Brodmann's area 38 and anterior part of area 21) for tumor excision or the treatment of epilepsy (Smiley et al., 1997). Identification of synapses made by these varicosities was then reported to be "a tedious process," in which "approximately 100–200 immunoreactive profiles (e.g. 10–20 processes followed through 10 serial sections) had to be surveyed in order to find a convincing example." Although these terminals appeared to be slightly larger than their counterparts in the monkey (Smiley and Goldman-Rakic, 1996), such a frequency in serial sections was highly suggestive of a low synaptic incidence (5–10%). Also noteworthy in this unique ultrastructural study was the fact that no axo-axonic specialization was found, even though an earlier report had mentioned (but not illustrated) the possible existence of such contacts in the medial and orbital prefrontal cortex of rat (Van Eden et al., 1987). It would be interesting to know if the synaptic incidence of presumed DA varicosities is similarly low in human cortical tissue resected for the treatment of partial seizures due to dysplasia. In two such cases, an increased density of TH-immunostained fibers has indeed been reported by Trottier et al. (1994) in the area of seizure propagation around dysplastic foci (area of seizure onset).

Presumably for methodological reasons, much less additional information has been gained on the relational features of the NA innervation in cerebral cortex since the extensive study by Séguéla et al. (1990) in the rat. Depending on the stringency of the criteria used in identifying the junctional complex, the synaptic incidence of cortical NA terminals was then evaluated at 17% and 26%, in a pooled sample of serially-sectioned NA-immunostained varicosities from the upper layers of primary motor (frontal), somatosensory (parietal) and visual (occipital) cortex. This was quite different from the earlier reports by Papadopoulos et al. (1987a, 1989a), who had concluded that 88% and 89% of 125 and 800 serially sectioned NA-immunostained varicosities from all layers of the fronto-parietal and visual cortex made a synapse. It has never been determined if this discrepancy was due to a sampling bias or different criteria for defining synapses. In similarly prepared material from unspecified regions of rat cerebral cortex, Paspalas and Papadopoulos (1996) have later described direct contacts of NA varicosities with oligodendrocytes that frequently exhibited "a distinct accumulation of electron-dense material resembling membrane differentiations found in symmetrical synapses." In a concurrent immunocytochemical study on the astroglial and vascular relationships of NA terminals in rat fronto-parietal cortex, Cohen et al. (1997) did not observe such contacts and evaluated the synaptic incidence at 7.1% from a population of 130 NA-immunostained profiles examined in single thin sections. These authors also noted the frequent juxtaposition of NA varicosities to astrocytic processes, a finding of particular interest in view of the earlier double immunolabeling study of Aoki (1992), in the rat visual cortex, which had shown occasional contacts (appositions) between presumed NA terminals (TH-immunostained) and astrocytic leaflets displaying β-adrenoreceptor immunoreactivity.

Using a similar approach, Aoki et al. (1998) have recently examined the topological relationship between DBH-immunostained terminals and α2A receptor immunoreactivity in the dorsolateral prefrontal cortex of cynomolgous monkey. Only 1 in 12 single sections of 564 NA varicosities from two monkeys were found to display morphologically identifiable synapses, and none of 79 profiles from layer VI included in that sampling. This should correspond to an average synaptic incidence of about 18% if these varicosities are of the same size as in rat. Not only were α2A receptors visualized on dendritic spines bearing synaptic specializations, but also at sites along axons, dendritic shafts and astrocytic processes lacking identifiable junctions. The suggestion was therefore made that these receptors might be activated by diffuse transmission, and that noradrenergic modulation in prefrontal cortex should involve effects on glia, dendritic shafts and axons as well as synaptic interactions with the spines of pyramidal neurons.

As was the case for the NA innervation, the detailed study by Séguéla et al. (1989) of the 5-HT innervation in rat frontal, parietal and occipital cortex did not support the earlier contention by Papadopoulos et al. (1987a, b) that this cortical innervation is entirely synaptic (see Parnavelas and Papadopoulos, 1989). Values of 36% and 28% were obtained for the superficial and deep layers of the frontal cortex, respectively, and 46% and 37% for the parietal and the occipital cortex; a subsequent study by Cohen et al. (1995) yielded similar results. From a large population of single profiles of tryptophane hydroxylase (TpOH)-immunostained varicosities remote from the microvessels, synaptic incidences of 22% and 18% were then extrapolated for the fronto-parietal and the entorhinal cortex, respectively (Table 1).

Following an earlier study of visual cortex in which only profiles of 5-HT-immunostained varicosities endowed with a synaptic junction had been characterized (De Lima et al., 1988), the synaptic incidence of cortical 5-HT terminals was determined in both the sensorimotor (fronto-parietal) and prefrontal cortex of the monkey. In the sensorimotor cortex of cynomolgous monkey, DeFelipe and Jones (1988) found only 5 of 191 serially sectioned 5-HT-immunostained terminals making a synapse. Furthermore, around four pyramidal cells which showed basket formations in this material, they did not detect a single unequivocal specialization at the point of contact (apposition) by the 5-HT terminals. Similarly, in cat auditory cortex, DeFelipe et al. (1991) observed only four synapses made by 135 serially sectioned 5-HT-immunostained varicosities, 110 of which belonged to basket formations around GABA neurons. In layers I, III and V of the prefrontal cortex of rhesus monkey, Smiley and Goldman-Rakic (1996) reported that only 23% of 213 5-HT-immunostained varicosities completely viewed in serial sections formed identifiable synapses. As previously shown in the rat (Séguéla et al., 1990), these 5-HT synapses were consistently asymmetric. Their preferred targets in the monkey were dendritic shafts which usually belonged to somata with the morphological features of interneurons when followed in serial sections. In fact, only 8% of such postsynaptic shafts were classified as pyramidal dendrites, further indicating that interneurons are the major recipients of 5-HT synapses. Ultrastructural data is still lacking about the 5-HT innervation of human cortex, even though Trottier et al. (1996) have already described a 5-HT hyperinnervation in addition to the altered pattern of catecholaminergic innervation previously reported in their two cases of focal cortical dysplasia.

In rat hippocampus, a systematic comparison of the ultrastructural features of NA, 5-HT, ACh and GABA terminals was carried out in the stratum radiatum of CA1 (Umbriaco et al., 1995). Relational features comparable to those in cerebral cortex were observed for all four species of terminals, with synaptic incidences of 16% and 23% extrapolated from single thin sections for the NA and the 5-HT varicosities, respectively (Table 1). This 23% value for the 5-HT endings was remarkably similar to previous estimates of 19.5% from single sections in the stratum radiatum and 18% from serial sections in the oriens layer of CA3, and 24% from single sections of the dentate gyrus (Oleskevich et al., 1991). It was also in keeping with the 12% incidence reported by Cohen et al. (1995) for TpOH-immunostained varicosities from the stratum lacunosum moleculare of the CA1 sector of dorsal hippocampus. As in the cerebral cortex, subpopulations of GABA neurons appear to be the preferred target of the synaptic 5-HT varicosities in rat hippocampus (Freund et al., 1990; Halasy et al., 1992). Here again, however, the low incidence of synapses formed by NA and 5-HT varicosities leads one to believe that these afferents exert their effects largely through diffuse transmission (for a recent review, see Vizi and Kiss, 1998).

Also worthy of note are the observations by Daszuta et al. (1991) indicating that after reinnervation and hyperinnervation of the previously 5-HT-denervated adult hippocampus by grafted midbrain 5-HT neurons, the proportion of 5-HT-immunostained terminals engaged in synaptic contact in both the CA3 or the dentate sector of the outgrowth zone remains essentially the same as in intact tissue. Since the neurons which then reinnervated the hippocampus were not necessarily destined to this region, these results suggest that the territory of innervation is the principle determinant of the frequency with which ingrowing 5-HT fibers make synaptic junction.

Neostriatum

Two studies carried out in the last ten years have provided definitive evidence for the partially synaptic character of both the DA and 5-HT innervation in adult rat neostriatum (Table 1); these original papers should be consulted for exhaustive reference to earlier literature. In the study of the DA innervation, results gathered over a 15 year period were finally published regarding the ultrastructural features of neostriatal axon terminals identified either by uptake autoradiography after intraventricular administration of [^{3}H]DA (single thin sections) (Fig. 2), or by DA immunocytochemistry in serial as well as single thin sections (Fig. 3) (Descarries et al., 1996); all three approaches yielded comparable values of synaptic incidence. Whether from the paraventricular zone or from a mediodorsal portion of the rostral half of neostriatum and examined in single or serial thin sections, 60–70% of these DA varicosities were found to be asynaptic, and only 30–40% endowed with a synaptic membrane differentiation; the junctional complexes formed by striatal DA varicosities were invariably symmetrical. Approximately two thirds of these DA synapses were made on dendritic branches, one third on dendritic spines, and only 2–3% on cell bodies, as also found previously in a study of synaptic TH-immunolabeled boutons in the sensorimotor territory of squirrel monkey striatum (Smith et al., 1994).

These data in the rat were interpreted as indicative of a dual mode of transmission, i.e. diffuse as well as synaptic, for DA in neostriatum. Moreover, considering the extreme density of this innervation (see Doucet et al., 1986), the possibility was envisaged that a basal extracellular level of DA be permanently maintained around all cellular elements, owing to the spontaneous and evoked release from this multitude of asynaptic as well as synaptic varicosities. A subsequent report by Hanley and Bolam al. (1997) has added to the significance of this study. These investigators demonstrated that the synaptology of the nigrostriatal projection identified by TH immunostaining was essentially the same in both the patches of hyperdense and the matrix of dense striatal DA innervation. Thus, both diffuse and synaptic DA transmission are likely to be at play throughout the striatum, to participate in neostriatal and basal ganglia function(s) (and dysfunctions).

In the second study (Descarries et al., 1992), further characterization of the neostriatal 5-HT innervation was achieved during an ultrastructural investigation of the 5-HT hyperinnervation which takes place in adult rat neostriatum following its neonatal DA denervation by intraventricular administration of 6-hydroxydopamine. In normal controls, a synaptic frequency of 8% was extrapolated from single thin sections of 5-HT-immunostained varicosities in the dorsal third of rostral neostriatum, confirming earlier values of 13% and 10%, respectively obtained from single section autoradiographs of the paraventricular neostriatum after [^{3}H]5-HT uptake in vivo and serial sections from its mid-dorsal third after 5-HT-immunostaining (Soghomonian et al., 1988). Again, this proportion did not change (6% vs. 8%) in spite of a twofold increase in the number of 5-HT axon terminals pervading the rostral portion of the neostriatum endogenously 5-HT hyperinnervated by rapheo-striatal 5-HT neurons.

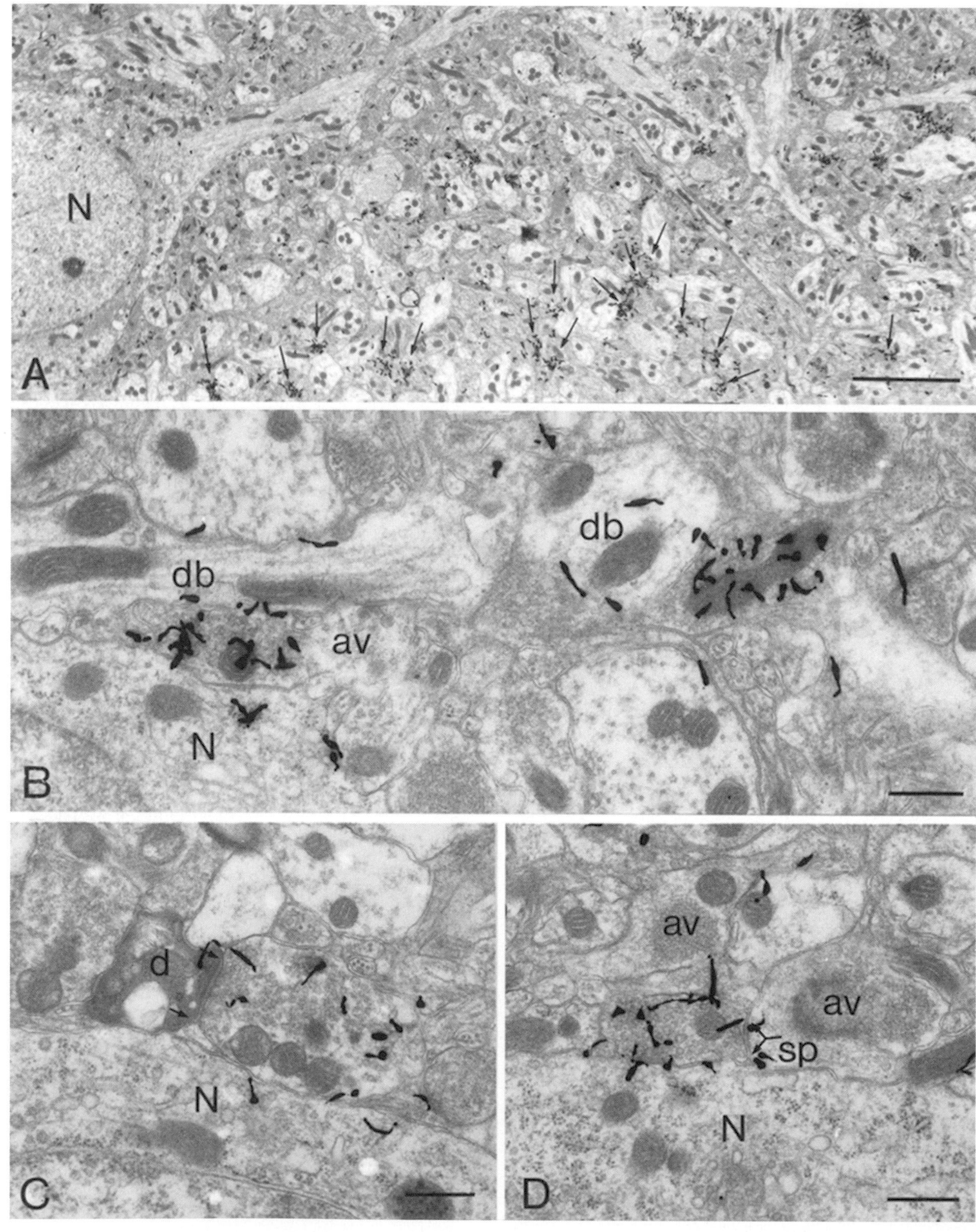
N
A
db
db
av
N
B
d
N
C
av
av
sp
N
D

Also of pertinence in the present context is the fact that knowing the synaptic incidence of DA and 5-HT terminals in adult rat neostriatum allows for estimating the actual number of these terminals engaged in synaptic contact. The density of these two innervations has indeed been measured in number of axon varicosities per cubic mm of tissue (Doucet et al., 1986; Mrini et al., 1995). Thus, it may be calculated that the 30–40% proportion of DA varicosities making a synaptic junction corresponds to as many as $3–4 \times 10^7$ and $5–7 \times 10^7$ DA synapses per mm^3 in the DA islands and the matrix of the dorsolateral neostriatum, respectively, and $1.8–2.4 \times 10^7$ per mm^3 in the ventromedial neostriatum. The DA synapses are therefore likely to represent 1.8% to 7% of all striatal synapses. The synaptic 5-HT varicosities are 40 to 140 times less numerous. In the rostral neostriatum, for example, their 10% proportion corresponds to some 4.8×10^5 synapses per mm^3, i.e. approximately 1 in every 2000 striatal synapses.

Other brain regions

Data on the synaptic incidence of 5-HT terminals have also been obtained from many other regions of adult rat brain (Table 1). A study by Boulaich et al. (1994) has extrapolated values of 48% and 38% from single section of 5-HT-immunostained varicosities in the *suprachiasmatic and supraoptic nuclei of hypothalamus*. Furthermore, these proportions were then shown to be maintained following 5-HT denervation and either partial 5-HT reinnervation (suprachiasmatic) or hyperinnervation (supraoptic) by grafted embryonic 5-HT neurons. Thus, the synaptic incidence appears to remain constant in conditions of only limited as well as of excessive 5-HT reinnervation by 5-HT grafts.

In the *substantia nigra*, which receives one of the densest 5-HT innervation in CNS, a study by Moukhles et al. (1997) combined a quantitative evaluation of the density of this innervation by in vitro uptake autoradiography with [^{3}H]5-HT and its ultrastructural characterization by electron microscopic 5-HT immunocytochemistry. Further indications of regional variability were thus obtained. In the pars reticulata, where their number amounts to 9×10^6 per mm^3, virtually all 5-HT varicosities were found to form synapses, whereas in the pars compacta, where the 5-HT innervation is slightly less dense ($6 \times 10^6/mm^3$), this proportion was only 50%. This work also provided the basis for a comparative evaluation by fast-scan voltammetry of 5-HT release and uptake in a region of CNS where all 5-HT terminals are endowed with synaptic contact, as opposed to the dorsal raphe nucleus, a somatodendritic region with rare synaptic incidence (Bunin et al., 1998). The results supported the existence of diffuse 5-HT transmission in both regions, indicating that this mode of transmission might well apply to synaptic as well as asynaptic MA terminals.

Another region of interest is the *dorsal periaqueductal grey matter*, in which 5-HT has been shown to have anxiolytic (antiaversive) effects (Beckett et al., 1992; Nogueira and Graeff, 1995). In this

Fig. 2A–D. Electron microscope autoradiographs from rat paraventricular neostriatum after intraventricular administration of tritiated dopamine ([^{3}H]DA). Following prolonged autoradiographic exposure (42 weeks) and revelation with the physical developer paraphenelenediamine, accumulations of small and thin silver grains are detected over DA axon terminals (varicosities) laden with [^{3}H]DA. (A) At very low magnification ($\times$3 500), these labeled sites are visible in large number (e.g. arrows) throughout the neuropil, between cell bodies which, in vast majority, exhibit the morphological features of spiny type I projection neurons (e.g. N). Scale bar: 5 μm. (B–D) illustrate ultrastructural characteristics of the labeled DA terminals at a higher magnification. Scale bars: 0.5 μm. In (B), two elongated [^{3}H]DA-labeled varicosities display a rather uniform population of mostly round, small and clear vesicles, associated with a mitochondrion. Neither of these profiles shows any membrane differentiation even suggestive of a synaptic contact. As most of their labeled counterparts, these two varicosities are directly apposed to dendritic branches (db); the left-sided one is also apposed to a neuronal cell body (N) and to another axonal varicosity (av), unlabeled, containing dense core vesicles. In (C), a larger [^{3}H]DA-labeled varicosity contains several mitochondria in addition to aggregated small vesicles. It forms a distinct symmetrical synaptic junction (between arrows) on a degenerating profile (D), presumably dendritic, while being also juxtaposed to a neuronal cell body (N). In (D), another synaptic [^{3}H]DA-labeled varicosity is also directly apposed to a neuronal cell body (N). It makes a small symmetrical junction (between arrows) with the dendritic spine on its right (sp), also contacted synaptically by an unlabeled axonal varicosity (av). Above the labeled varicosity, another unlabeled axonal varicosity (av) shows a cristalline array of small clear vesicles. (Reproduced with permission from Descarries et al., 1996).

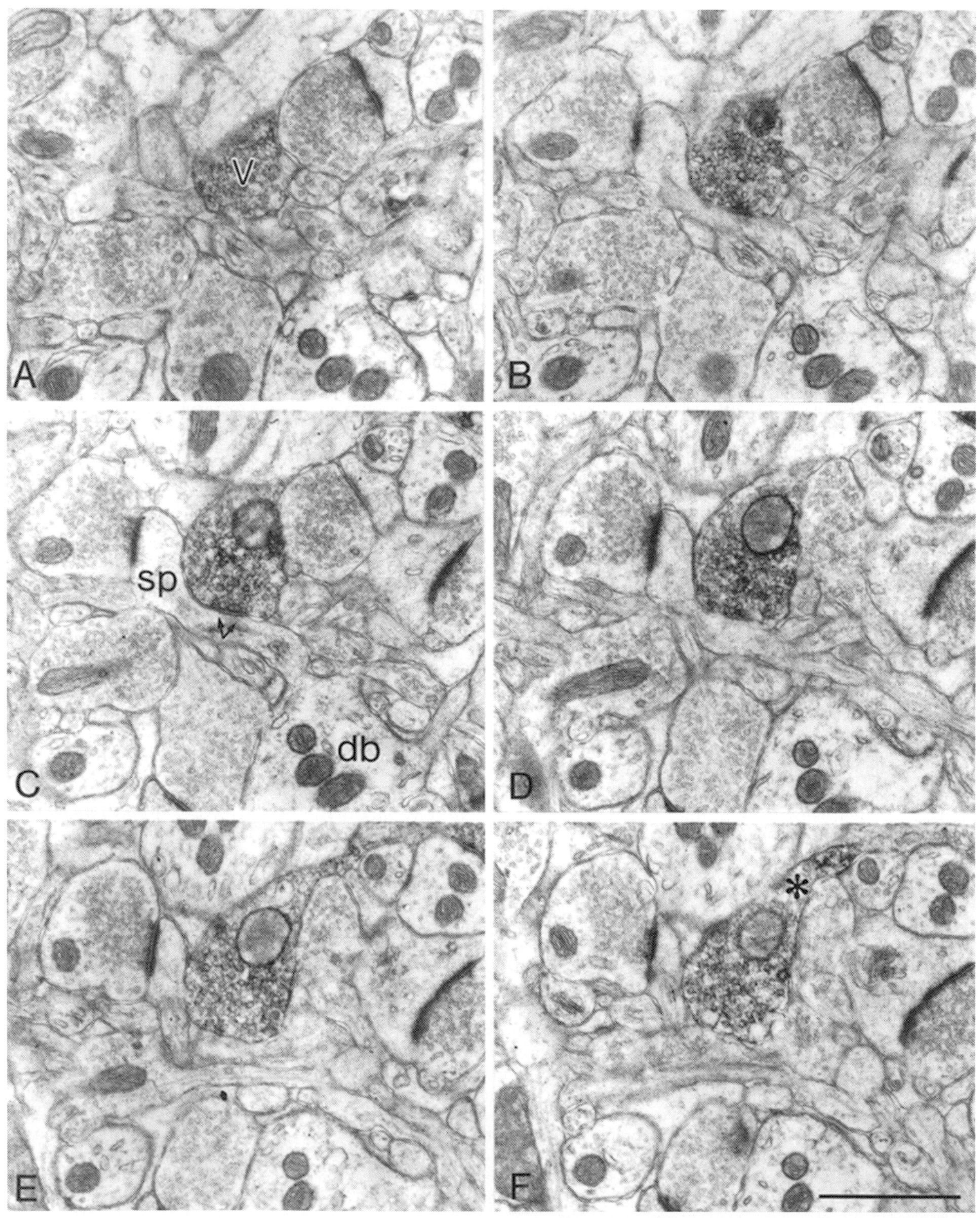
A
V
B
C
sp
db
D
E
F
*

region, the proportion of 5-HT-immunostained varicosities making a synaptic junction was recently evaluated at 23% in serial thin sections (Lovick et al., 2000). To our knowledge, it has not been determined if the 5-HT innervation displays similar features in the ventral periaqueductal grey, where it may be more directly involved in anti-nociception.

Data on serially-sectioned 5-HT-immunostained varicosities were also obtained from several regions of adult rat CNS in the course of developmental studies. The results at early postnatal ages must be regarded with caution, in view of the strong indications that numerous CNS neurons which do not synthetize 5-HT, including the thalamocortical projection neurons, take up and store 5-HT during the postnatal period in rodents (Lebrand et al., 1996, 1998; Cases et al., 1998; Hansson et al., 1998). These neurons can then be mistaken for 5-HT cells when labeled by ligand binding autoradiography or 5-HT immunocytochemistry. However, results obtained after the age of 1 month are reliable because of the complete disappearance of this transient and partial 5-HT phenotype. Thus, a synaptic incidence of 42% may be inferred from the proportion reported by Dinopoulos et al. (1995) for single sections of 5-HT-immunostained terminals from the *lateral geniculate nucleus* in 35 day-old rat. Similarly, in adult rat, these authors have estimated the synaptic incidence of 5-HT terminals at 46%, 55%, 38% and 62% in the *basal forebrain* (nucleus of the horizontal limb of the diagonal band of Broca), superficial and deep layers of *superior colliculus* and *ventrolateral nucleus of thalamus*, respectively (Dinopoulos et al., 1997; Dori et al., 1998). In the superficial gray layer of the superior colliculus of the hamster, a value of 73% has been reported for serially-sectioned TH-immunostained profiles which were presumably noradrenergic, as most such profiles also stained for DBH in this region (Arce et al., 1994).

Spinal cord

Data have also become available regarding all three MA innervations in adult rat spinal cord (Table 1). The spinal DA innervation has been examined in serial thin sections after DA-immunostaining (Ridet et al., 1992). Despite differences between anatomical levels and regions of the cord (Ridet et al., 1992), it was described as the most junctional. At cervical level, 34% of the DA varicosities were synaptic in the dorsal horn and 46% in the ventral horn. At thoracic level, the proportion reached 76% in the dorsal horn and >100% in ventral horn, presumably because of the frequent occurrence of more than one synapse per DA terminal in this latter location. Around the central canal and in the intermediolateral cell column at thoracic level, 67% and >100% of the DA varicosities were synaptic, respectively.

The NA and 5-HT innervations also differed markedly between dorsal and ventral horn. In dorsal horn, whether estimated from single thin sections at lumbar and sacral levels or serial thin sections at cervical level, the synaptic incidence of NA-immunostained varicosities was in the order of 25–29% (Rajaofetra et al., 1992; Ridet et al., 1993). Around the central canal, the intermediolateral cell column and the ventral horn, it was much higher, at about 85%. Similarly, the 5-HT innervation in dorsal horn displayed a synaptic incidence of 37% in serial sections (Ridet et al., 1993), which apparently showed little variations in the different laminae of the dorsal horn or at different spinal cord levels (Marlier et al., 1991). In the intermediolateral cell column and anterior horn, however, the synaptic incidence of the 5-HT-immunostained varicosities was estimated at

Fig. 3A–F. Six from a series of 10 thin sections across a DA immunoreactive axonal varicosity from the mediodorsal neostriatum (diaminobenzidine labeling). This DA varicosity (V in (A)) makes synaptic contact on the neck of a dendritic spine (sp in (C)) which receives an asymmetrical synapse on its head. The junctional complex formed by the DA varicosity on the spine neck is visible in (B) to (D) (between arrows in (C)). In (C), the synaptically contacted spine neck may be observed in continuity with its parent dendrite (db). The large asymmetrical contact on the head of the spine (visible from (B) to (F)) is formed by an unlabeled varicosity containing small round and clear vesicles. In (E) and (F) (asterisk in (F)), the thin axonal segment leading into the DA varicosity may be observed. Scale bar: 1 μm. (Reproduced with permission from Descarries et al., 1996).

>100% (Poulat et al., 1992) and 95% (Ridet, 1994), respectively. Interestingly, as early as 1983, Maxwell et al. had reported a remarkably low synaptic incidence of 4% for 5-HT endings serially sectioned in the marginal zone (layer I of dorsal horn) of the rat spinal cord. These authors had concluded from this observation that non-synaptic and diffuse transmission might be implicated in the antinociceptive effects of the descending 5-HT system. It should be pointed out that such data are quite distinct from earlier ultrastructural results in the cat, in which all the emphasis had been placed on the synaptic relationships of the 5-HT system in dorsal horn (Ruda and Gobel, 1980; Ruda et al., 1982; Light et al., 1983; Glazer and Basbaum, 1984).

Acetylcholine innervations

It is only in the past ten years that an antibody against rat brain choline acetyltransferase (ChAT) has been available, which is sensitive enough to detect ACh axon terminals in vast number throughout the CNS of different mammals (Cozzari et al., 1990). Many studies with other ChAT antibodies had previously allowed description of some of the intrinsic and relational features of these terminals in rat, cat, or monkey brain (for reviews, see Houser et al., 1990; Umbriaco, 1995), but none had been aimed at a quantitative evaluation of their synaptic and/or appositional relationships (see in particular Armstrong, 1986; De Lima and Singer, 1986; Aoki and Kabak, 1992). As reviewed below, this has since been achieved for three major regions of the rat brain: cerebral cortex, hippocampus and neostriatum. An extensive report on rat olfactory bulb has also been published, as well as two studies with the previously available antibodies that managed to assess the synaptic incidence of the ACh terminals in monkey and human cerebral cortex.

Cerebral cortex and hippocampus

The ACh innervation of cerebral cortex is known to arise mainly from the nucleus basalis magnocellularis of Meynert (substantia innominata, Ch4), but also in rodents from intracortical neurons (Eckenstein and Thoenen, 1983; Levey et al., 1984; Lauterborn et al., 1993; Schafer et al., 1998) that may give origin to as many as 20% of these terminals (Eckenstein and Baughman, 1987). The intrinsic and relational features of ACh axon varicosities have been examined in detail in the primary somatosensory area (S1 or Par 1) of adult rat parietal cortex (Umbriaco et al., 1994). Eight hundred and twelve ChAT-immunostained terminals from the different layers of Par1 were scrutinized in long, uninterrupted series of thin sections across their whole volume. Several were actually reconstructed in three dimensions (see Fig. 8 in Umbriaco et al., 1994); about 200 were visualized as two or three varicosities along the same axon.

A totally unexpected finding was that, in every layer of Par1 cortex, only a low proportion of these ACh varicosities displayed a synaptic membrane differentiation (junctional complex). As indicated in Table 1, their mean synaptic incidence across all layers was 14%. Only layer V showed a slightly higher proportion (21%). In general, cortical ACh varicosities were relatively small, averaging 0.57 μm in diameter. Those bearing a synaptic junction were slightly larger than their non synaptic counterparts (0.67 μm in diameter). Both junctional and non-junctional ACh varicosities could be observed on the same axons. The junctional complexes formed by these terminals were single, almost always symmetrical, and occupied a small fraction of the total surface of varicosities (<3%). Synaptic ACh varicosities usually contacted dendritic branches (76%), less often spines (24%), and none were seen on cell bodies.

Two subsequent investigations confirmed the low synaptic incidence of cortical ACh varicosities in the rat (Chédotal et al., 1994; Vaucher and Hamel, 1995). From single thin sections, values of 14% and 9% were then extrapolated for the frontoparietal cortex and the perirhinal cortex, respectively.

An extensive ChAT immunocytochemical study has also been performed by Mrzljak et al. (1995) in the prefrontal cortex of rhesus monkey. These authors then reported that among 100 serially sectioned ChAT immunoreactive boutons at the border of layers II and III, only 44% made synaptic contact. Fifty-six percent were without any visible junctional specialization, even if frequently juxtaposed to dendrites or spines receiving asymmetrical synapses. More recently, Smiley et al. (1997) carried out a similar study on two samples of human anterior temporal lobe removed at surgery. These investigators found 67% of 42 varicosities from layers I and II endowed with small but identifiable synaptic specializations. It remains to be determined whether such variations of synaptic incidence in cortex reflect sampling biases, regional differences or species differences. In any event, these results allow the inference that, in primates as well as rat cortex, the various modulatory actions of ACh are likely to depend on both synaptic and non-synaptic terminals delivering ACh to proximal and remote targets.

The data on the regional and laminar synaptic incidence of the cortical ACh innervation in rat parietal cortex assume further significance when viewed in relation with measurements of the density of this innervation, as expressed not only in length of ChAT-immunostained axons, but also in number of ChAT-immunostained varicosities per cubic mm of tissue (see Mechawar and Descarries, 1999). Thus, in the parietal cortex, the length of ChAT-immunostained axons per mm^3 of tissue is found to range from 13.33 m (layer V) to 9.47 m (layer IV) and to average of 11.94 m for the whole cortical thickness. The corresponding numerical densities for varicosities ($\times 10^6$ per mm^3) range from 4.80 (layer V) to 3.49 (layer IV), with an average of 4.37 for the whole cortex. It may then be calculated that the parietal cortex is innervated by a total length of 23.89 m of ACh axons bearing 8.73×10^6 varicosities underneath a cortical surface of 1 mm^2. Based on the known proportions of synaptic ACh varicosities in each cortical layer (Umbriaco et al., 1994), only 1.38×10^6 of these varicosities would be synaptic (15.86%), for average numerical densities of 0.69×10^6 synaptic and 3.67×10^6 asynaptic ACh varicosities per mm^3 of cortex. Thus, synaptic ACh varicosities are likely to represent less than 1 in 1 450 of all cortical synapses. If release of ACh were to take place from synaptic varicosities only, such a small fraction would certainly make it difficult to account for the amounts of ACh collected from the cortical surface or concentrations measured intracortically with microdialysis.

In the previously mentioned study in which NA-, 5-HT- and ChAT- were compared to GAD-immunostained terminals in the stratum radiatum of CA1 (Umbriaco et al., 1995), the synaptic incidence of hippocampal ACh axon varicosities was found to be 7% and thus even lower than that of their NA and 5-HT counterparts. Again, it would seem difficult to account for the current ACh measurements in hippocampus, if this transmitter were to be released only by synaptic varicosities. As in neocortex, the strong prevalence of asynaptic ACh varicosities in hippocampus could also account for the remarkable capacity of this innervation to regrow from grafts of cholinergic neurons after previous lesions, and then restore behavioral deficits in spatial learning and memory (e.g. Low et al., 1982; Nilsson et al., 1990).

Neostriatum

The ACh innervation of neostriatum is essentially intrinsic, being issued from a fraction of the large interneurons which represent less than 2% of the total neuronal population of this region (McGeer et al., 1971; Woolf, 1991). Yet, the ChAT immunoreactivity in neostriatum is one of the strongest in the brain. Prior to our 1996 study (Contant et al., 1996), one could only imagine how profuse an axonal network might account for such labeling. This became immediately apparent when observing the multitude of ChAT-immunostained varicosities which pervaded the neostriatal neuropil in low power electron micrographs (Fig. 4A). As illustrated in Figs. 4B and 4C, these varicosities are particularly small, as their mean diameter (0.43 μm) was significantly less than that of their cortical congeneers and of striatal DA or 5-HT varicosities (all averaging about 0.6 μm in diameter). Once again, a vast majority were found to be asynaptic. In single thin section, only 2.7% exhibited a junctional complex compared with 57% for unlabeled varicosity profiles selected at random from the same electron micrographs. Stereological extrapolation to the whole volume of these ACh varicosities indicated a real synaptic incidence of less than 10% (Table 1), whereas that for the surrounding unlabeled varicosities was greater than 100% because of the frequent occurrence of

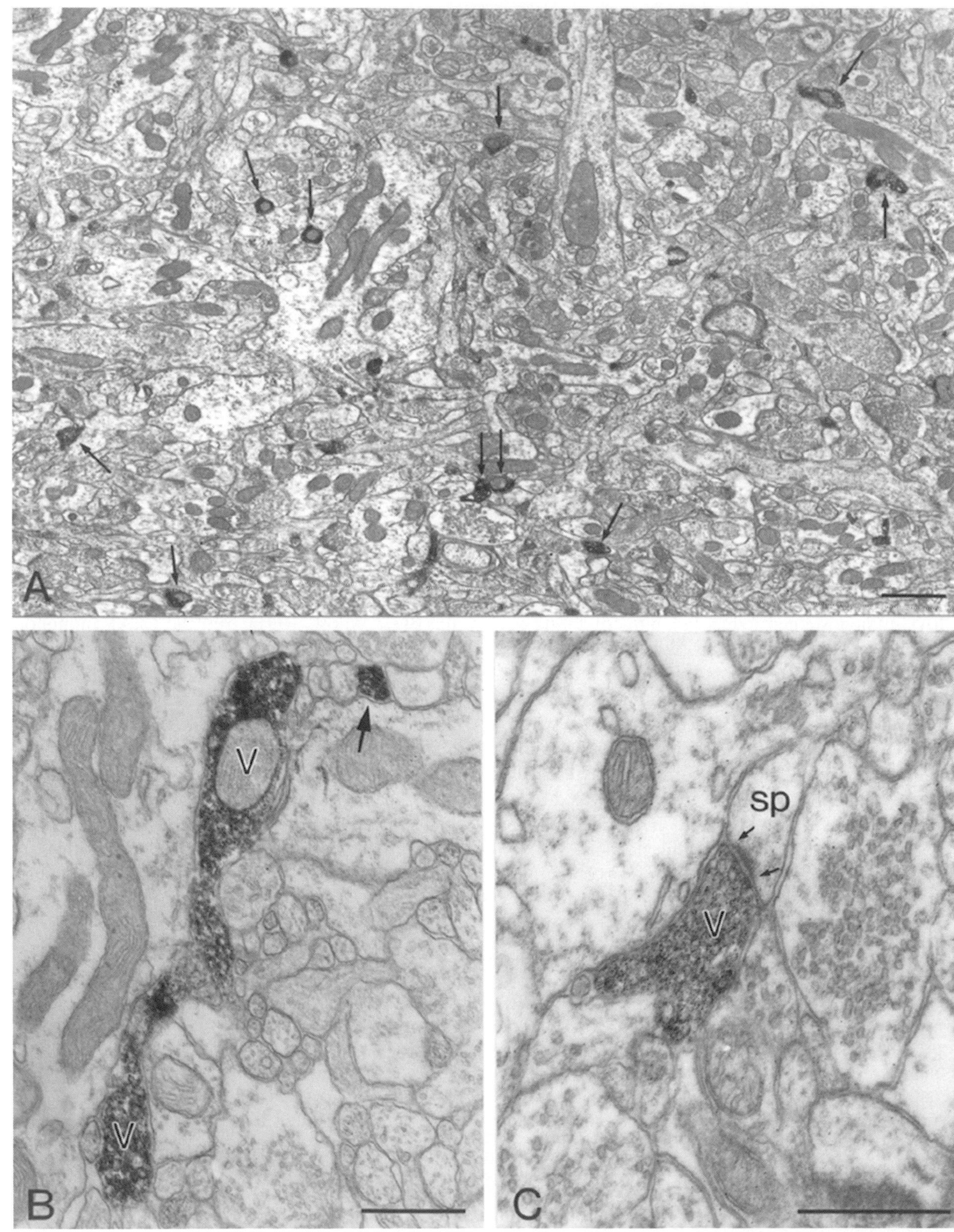
A
V
V
B
sp
V
C

multiple junctions. Direct apposition of ChAT-immunostained varicosities (Fig. 4A) was not uncommon, reflecting the extreme density of this ACh innervation. Striatal ACh varicosities were also frequently found juxtaposed to unlabeled varicosities, many of which were synaptic, accounting for preterminal effects of ACh on the release of other tranmitters or vice versa (for references, see Contant et al., 1996).

Olfactory bulb

Another ultrastructural study has emphasized the asynaptic as well as synaptic character of an ACh innervation in adult rat CNS. In the main olfactory bulb, Kasa et al. (1995) have described a dual mode of innervation by ACh fibers, whereby ChAT- immunopositive terminals made occasional synapses with interneurons but never on relay cells. In the glomerular layer, notably, periglomerular cell dendrites were observed to receive asymmetrical synapses from ACh boutons, but the dendrites of mitral and tufted cells, although closely approached, were never contacted synaptically. In all other layers, occasional synaptic contacts were made on gemmules and/or dendritic spines of granule cells, but the vast majority of ChAT-immunopositive fibers were reported as 'thin and with most of their varicosities (50 to 1) apparently lacking any synaptic specialization'.

Ambient level of monoamines and acetylcholine in the CNS

An interesting complement to the diffuse transmission paradigm has been recently proposed on the basis of some of the above observations, at least for brain regions rich in one or other of the largely asynaptic MA or ACh innervations (Descarries et al., 1995, 1996, 1997; Contant et al., 1996; Descarries, 1998). According to this hypothesis, a low level of ambient transmitter would be maintained throughout the extracellular space by the spontaneous and evoked release from predominantly asynaptic axon terminals, and presumably also by spillover from the minority that are junctional. According to this hypothesis, reuptake and enzymatic degradation of the monoamines would primarily serve to keep their ambient level within physiological limits rather than totally eliminate them from synaptic clefts and the extracellular space. Brain acetylcholinesterase (AChE), which predominantly consists of the tetrameric G4 isoform, would play a similar role toward ACh, as suggested by the fact that, in skeletal muscle, this G4 form is concentrated outside and around endplates (Gisiger and Stephens, 1988) and, in contrast to the A12 form, does not significantly contribute to the rapid removal of ACh from synaptic clefts (for detailed discussion, see Descarries et al., 1997). In this context, individual release events would in fact correspond to local and transient fluctuations over this background level, i.e. signals superimposed on a preexisting state of information already determined, at least in part, by the same transmitter.

In many parts of CNS, microdialysis data seems consistent with the existence of low concentration, resting, ambient levels of monoamines and/or ACh in the extracellular space. Spontaneous dialysis outputs in the nanomolar range have been repeatedly measured for DA, NA, 5-HT and ACh in

Fig. 4A–C. Electron micrograph from the center of neostriatum after ChAT immunostaining (diaminobenzidine labeling). (A) At low magnification ($\times$ 10 000), the density of this ACh innervation may be appreciated. Within this single field, numerous sectional profiles of ChAT-immunostained axonal varicosities are seen (e.g. arrows). These immunostained profiles are generally smaller than their unlabeled counterparts. They are occasionally juxtaposed to one another (pair of vertical arrows). Scale bar: 1 μm. (B) and (C) illustrate some of the intrinsic and relational features of neostriatal ChAT-immunostained terminals. Scale bars: 0.5 μm. In (B), two immunostained varicosities (V) filled with small vesicles are seen in continuity with their intertwining parent axons. In this section, the upper varicosity exhibits a mitochondrion, as usual devoid of immunostaining. Also note the presence of a few small vesicles in the parent axon. Even though both of these varicosities are juxtaposed to a dendritic branch, neither shows any membrane differentiation suggestive of synaptic junction. A small immunoreactive axon segment is visible at arrow. In (C), the ChAT-immunoreactive varicosity (V) displays a synaptic membrane specialization (between thin arrows) on a dendritic spine (sp). This asymmetrical junction is characterized by linearity of the apposed membranes, slight enlargement of the intercellular space and the presence of a postsynaptic density. (Reproduced with permission from Contant et al., 1996).

cerebral cortex, hippocampus or neostriatum of freely moving rat, even in the absence of MA reuptake blockers or AChE inhibitors (e.g. Smith et al., 1992; Testylier et al., 1996; Portas et al., 1998; Rowley et al., 1998). Experimental and theoretical models have been proposed to evaluate the extracellular diffusion of transmitters and answer the often-asked question of the distance that can be reached by these molecules in living brain (e.g. Nicholson and Syková, 1998). In the case of DA, it is usually agreed that a 10 μm distance might be attained within 50 milliseconds (Wightman and Zimmerman, 1990). This is no small domain of influence at the subcellular level. Based on currently available estimates of the density of DA innervation in striatum (Doucet et al., 1986), and the observation of a comparable density of ACh innervation, a sphere of striatal neuropil with a radius of 10 μm should contain about 400 DA and 400 ACh axon terminals, five to ten times more unidentified terminals, and at least several thousand dendritic spines. If DA or ACh diffuse that far, one can easily understand why there may be such a variety of receptor subtypes for these transmitters, and presumably different subtypes on any single neuron.

A low ambient level of transmitter could regulate the expression and/or functional state of high-affinity receptor subtypes located on neurons releasing the corresponding transmitter (autoreceptors) or other neurons (heteroreceptors), glial cells and microvessels. Changes of this ambient level, affecting these widely distributed receptors, could be the ones to mediate many of the behavioral effects produced by drugs acting on the MA transporters, the choline transporter or AChE. The existence of the ambient level could also explain why some of the motor and cognitive deficits attributed to losses of MA or of ACh in CNS become manifest only when a major proportion of release sites have disappeared and even low background levels are no longer maintained. Similarly, it could account for the beneficial effects of substitution and pharmacological therapies, or of grafts of non-neuronal cells engineered to release transmitter (e.g. Winkler et al., 1995; Dickinson-Anson et al., 1998), which obviously take place in the absence of restored synaptic connectivity.

Concluding remarks

As illustrated by the present symposium, a wealth of evidence has accumulated to confirm the reality and heuristic value of the 'volume transmission' paradigm. In its largest acceptance, this diffuse mode of transmission may now be considered to apply not only to neuronal populations characterized by a paucity of synaptic junctions, but also to wholly junctional ones, such as the glycine-, GABA- and glutamate-containing neurons, which display spillover of their transmitter (see Chapter 21). Using immunoelectron microscopy, many of the receptors for these aminoacidergic transmitters, as well as for monoamines, ACh and neuropeptides, have been visualized in extrasynaptic membrane locations, on neuronal somata, dendrites, axons and axon terminals, as well as glia and endothelial cells, often remote from the corresponding release sites (see Chapter 14). The spread of various molecules, including DA, 5-HT, neuropeptides, cytokines and growth factors, has been documented in intact brain tissue following in vivo administration (see Chapter 7). The ambient level hypothesis has provided a framework for explaining some of the sustained and regulatory as well as trophic-like effects of transmitters on a prolonged time scale. Experimental techniques are being developed to investigate such effects and broad transmission by diffusion may now be envisaged not only in terms of cell-to-cell transfer of information, but also of coordination and synchronization of the activity of vast cellular ensembles, (e.g. Riehle et al., 1997; Fisahn et al., 1998; Testylier et al., 1999). A striking example of this progress is to be found in the current thinking on the implication of ACh in cortical functions and dysfunctions (Perry et al., 1999). Initially regarded as the canonical transmitter, exciting cortical neurons much as it did muscle fibers at the neuromuscular junction, ACh is now viewed as a major integrator of neuronal activity, whose widespread actions in the human cortex might in fact be essential for the maintenance of conscious experience. The 'volume transmission' paradigm has been central to this evolution.

Acknowledgements

The authors thank Drs. Michel Geffard (Bordeaux, France), Boyd K. Hartman (Minneapolis, MN) and Costantino Cozzari (Rome, Italy) for generous gifts of antibodies. They also acknowledge the technical assistance of Sylvia Garcia and K.C. Watkins with the electron microscopy, and are grateful to Gaston Lambert for photographic work. Their research work is currently supported by grant MT–3544 to L.D. and a studentship to N.M. from the Medical Research Council of Canada.

List of abbreviations

ACh	acetylcholine
AChE	acetylcholinesterase
DBH	dopamine-β-hydroxylase
ChAT	choline acetyltransferase
CNS	central nervous system
GABA	gamma-aminobutyric acid
MA	monoamine
NA	noradrenaline
TH	tyrosine hydroxylase
5-HT	5-hydroxytryptamine (serotonin)
TpOH	tryptophane hydroxylase

References

Aoki, C. (1992) β-adrenergic receptors: astrocytic localization in the adult visual cortex and their relation to catecholamine axon terminals as revealed by electron microscopic immunocytochemistry. *J. Neurosci.*, 12: 781–792.

Aoki, C. and Kabak, S. (1992) Cholinergic terminals in the cat visual cortex: ultrastructural basis for interaction with glutamate-immunoreactive neurons and other cells. *Vis. Neurosci.*, 8: 177–191.

Aoki, C., Venkatesan, C., Go, C.G., Forman, R. and Kurose, H. (1998) Cellular and subcellular sites for noradrenergic action in the monkey dorsolateral prefrontal cortex as revealed by the immunocytochemical localization of noradrenergic receptors and axons. *Cereb. Cortex*, 8: 269–277.

Arce, E.A., Bennett-Clarke, C.A. and Rhoades, R.W. (1994) Ultrastructural organization of the noradrenergic innervation of the superficial gray layer of the hamster's superior colliculus. *Synapse*, 18: 46–54.

Armstrong, D.M. (1986) Ultrastructural characterization of choline acetyltransferase-containing neurons in the basal forebrain of rat: evidence for a cholinergic innervation of intracerebral blood vessels. *J. Comp. Neurol.*, 250: 81–92.

Beaudet, A. and Descarries, L. (1976) Quantitative data on serotonin nerve terminals in adult rat neocortex. *Brain Res.*, 111: 301–309.

Beaudet, A. and Descarries, L. (1987) Ultrastructural identification of serotonin neurons. In: H.W.M. Steinbusch (Ed.), *Monoaminergic Neurons: Light Microscopy and Ultrastructure*, John Wiley & Sons, Chichester, pp. 265–313.

Beaudet, A. and Sotelo, C. (1981) Synaptic remodeling of serotonin axon terminals in rat agranular cerebellum. *Brain Res.*, 206: 305–329.

Beckett, S.R.G., Lawrence, C.A., Marsden, C.A. and Marshall, P.W. (1992) Attenuation of chemically-induced defence responses by 5-HT_1 receptor agonists administered into the periaqueductal grey. *Psychopharmacology*, 108: 110–114.

Boulaich, S., Daszuta, A., Geffard, M. and Bosler, O. (1994) Synaptic connectivity of serotonin graft efferents in the suprachiasmatic and supraoptic nuclei of the hypothalamus. *Exp. Brain Res.*, 101: 353–364.

Bunin, M.A. and Wightman, R.M. (1998) Quantitative evaluation of 5-hydroxytryptamine (Serotonin) release and uptake: an investigation of extrasynaptic transmission. *J. Neurosci.*, 18: 4854–4860.

Cases, O., Lebrand, C., Giros, B., Vitalis, T., De Maeyer, E., Caron, M.G., Price, D.L., Gaspar, P. and Seif, I. (1998) Plasma membrane transporters of serotonin, dopamine, and norepinephrine mediate serotonin accumulation in atypical locations in the developing brain of monoamine oxidase A knock-outs. *J. Neurosci.*, 18: 6914–6927.

Chédotal, A., Umbriaco, D., Descarries, L., Hartman, B.K. and Hamel, E. (1994) Light and electron microscopic immunocytochemical analysis of the neurovascular relationships of choline acetyltransferase (ChAT) and vasoactive intestinal polypeptide (VIP) nerve terminals in the rat cerebral cortex. *J. Comp. Neurol.*, 343: 57–71.

Cohen, Z., Ehret, M., Maître, M. and Hamel, E. (1995) Ultrastructural analysis of tryptophan hydoxylase immunoreactive nerve terminals in the rat cerebral cortex and hippocampus: their associations with local blood vessels. *Neuroscience*, 66: 555–569.

Cohen, Z., Molinatti, G. and Hamel, E. (1997) Astroglial and vascular interactions of noradrenaline terminals in the rat cerebral cortex. *J. Cereb. Blood Flow Metab.*, 17: 894–904.

Contant, C., Umbriaco, D., Garcia, S., Watkins, K.C. and Descarries, L. (1996) Ultrastructural characterization of the acetylcholine innervation in adult rat neostriatum. *Neuroscience*, 71: 37–947.

Cozzari, C., Howard, J. and Hartman, B.K. (1990) Analysis of epitopes on choline acetyltransferase (ChAT) using monoclonal antibodies (Mabs). *Soc. Neurosci. Abstr.*, 16: 200.

Crawford, G.D., Correa, L. and Salvaterra, P.M. (1982) Interaction of monoclonal antibodies with mammalian choline acetyltransferase. *Proc. Natl Acad. Sci. USA*, 79: 7031–7035.

Daszuta, A., Chazal, G., Garcia, S., Oleskevich, S. and Descarries, L. (1991) Ultrastructural features of serotonin neurons grafted to adult rat hippocampus: an immunocytochemical analysis of their cell bodies and axon terminals. *Neuroscience*, 42: 793–811.

DeFelipe, J. and Jones, E.G. (1988) A light and electron microscopic study of serotonin-immunoreactive fibers and terminals in the monkey sensory-motor cortex. *Exp. Brain Res.*, 71: 171–182.

DeFelipe, J., Hendry, S.H.C., Hashikawa, T. and Jones, E.G. (1991) Synaptic relationships of serotonin-immunoreactive terminal baskets of GABA neurons in the cat auditory cortex. *Cereb. Cortex*, 1: 117–133.

De Lima, A.D. and Singer, W. (1986) Cholinergic innervation of the cat striate cortex: A choline acetyltransferase immunocytochemical analysis. *J. Comp. Neurol.*, 250: 324–338.

De Lima, A.D., Bloom, F.E. and Morrison, J.H. (1988) Synaptic organization of serotonin-immunoreactive fibers in primary visual cortex of the macaque monkey. *J. Comp. Neurol.*, 274: 280–294.

Descarries, L. (1998) The hypothesis of an ambient level of acetylcholine in the central nervous system. *J. Physiol. (Paris)*, 92: 215–220.

Descarries, L. and Umbriaco, D. (1995) Ultrastructural basis of monoamine and acetylcholine function in CNS. *Semin. Neurosci.*, 7: 309–318.

Descarries, L., Doucet, G., Lemay, B., Séguéla, P. and Watkins, K.C. (1988) Structural basis of cortical monoamine function. In: M. Avoli, T.A. Reader, R.W. Dykes and P. Gloor (Eds), *Neurotransmitters and Cortical Function. From Molecules to Mind*, Plenum Press, New York, pp. 321–332.

Descarries, L., Gisiger, V. and Steriade, M. (1997) Diffuse transmission by acetylcholine in the CNS. *Progr. Neurobiol.*, 53: 603–625.

Descarries, L., Séguéla, P. and Watkins, K.C. (1991) Nonjunctional relationships of monoamine axon terminals in the cerebral cortex of adult rat. In: K. Fuxe and L.F. Agnati (Eds), *Volume Transmission in the Brain: Novel Mechanisms for Neural Transmission*, Raven Press, New York, pp. 53–62.

Descarries, L., Soghomonian, J.-J., Garcia, S., Doucet, G. and Bruno, J.P. (1992) Ultrastructural analysis of the serotonin hyperinnervation in adult rat neostriatum following neonatal dopamine denervation with 6-hydroxydopamine. *Brain Res.*, 569: 1–13.

Descarries, L., Umbriaco, D., Contant, C. and Watkins, K.C. (1995) A new hypothesis of acetylcholine (ACh) function in densely ACh-innervated regions of the brain. *IVth IBRO World Congr.Neurosci. Abstr.*, A3.15, p. 101.

Descarries, L., Watkins, K.C., Garcia, S., Bosler, O. and Doucet, G. (1996) Dual character, asynaptic and synaptic, of the dopamine innervation in adult rat neostriatum: a quantitative autoradiographic and immunocytochemical analysis. *J. Comp. Neurol.*, 375: 167–186.

Dickinson-Anson, H., Aubert, I., Gage, F.H. and Fisher, L.J. (1998) Hippocampal grafts of acetylcholine-producing cells are sufficient to improve behavioural performance following a unilateral fimbria-fornix lesion. *Neuroscience*, 84: 771–781.

Dinopoulos, A., Dori, I.E. and Parnavelas, J.G. (1995) Serotonergic innervation of the lateral geniculate nucleus of the rat during postnatal development: a light and electron microscopic immunocytochemical analysis. *J. Comp. Neurol.*, 363: 532–544.

Dinopoulos, A., Dori, I.E. and Parnavelas, J.G. (1997) The serotonin innervation of the basal forebrain shows a transient phase during development. *Dev. Brain Res.*, 99: 38–52.

Dori, I.E., Dinopoulos, A. and Parnavelas, J.G. (1998) The development of the synaptic organization of the serotonergic system differs in brain areas with different functions. *Exp. Neurol.*, 154: 113–125.

Doucet, G., Descarries, L. and Garcia, S. (1986) Quantification of the dopamine innervation in adult rat neostriatum. *Neuroscience*, 19: 427–445.

Eckenstein, F. and Thoenen, H. (1982) Production of specific antisera and monoclonal antibodies to choline acetyltransferase: characterization and use for identification of cholinergic neurons. *EMBO J.*, 1: 363–368.

Eckenstein, F. and Thoenen, H. (1983) Cholinergic neurons in the rat cerebral cortex demonstrated by immunohistochemical localization of choline acetyltransferase. *Neurosci. Lett.*, 36: 211–215.

Eckenstein, F. and Baughman R.W. (1987) Cholinergic innervation in cerebral cortex. In: E.G. Jones and A. Peters (Eds), *Cerebral Cortex. Further Aspects of Cortical Function, Including Hippocampus*, Vol. 6, Plenum Press, New York, pp. 129–160.

Erickson, S.L., Sesack, S.R. and Lewis, D.A. (2000) The dopamine innervation of monkey entorhinal cortex: postsynaptic targets of tyrosine hydroxylase terminals. *Synapse*, 36: 47–56.

Fisahn, A., Pike, F.G., Buhl, E.H. and Paulsen, O. (1998) Cholinergic induction of network oscillations at 40 Hz in the hippocampus in vitro. *Nature*, 394:186–189.

Freund, T.F., Gulyas, A.I., Acsady, L., Gorcs, T. and Toth, K. (1990) Serotonergic control of the hippocampus via local inhibitory interneurons. *Proc. Natl Acad. Sci. USA*, 87: 8501–8505.

Gisiger, V. and Stephens, H.R. (1988) Localization of the pool of G_4 acetylcholinesterase characterizing fast muscles and its alteration in murine muscular dystrophy. *J. Neurosci. Res.*, 19: 62–78.

Glazer, E.J. and Basbaum, A.I. (1984) Axons which take up [^{3}H]serotonin are presynaptic to enkephalin immunoreactive neurons in cat dorsal horn. *Brain Res.*, 298: 386–391.

Goldman-Rakic, P.S., Leranth, C., Williams, S.M., Mons, N. and Geffard, M. (1989) Dopamine synaptic complex with pyramidal neurons in primate cerebral cortex. *Proc. Natl Acad. Sci. USA*, 86: 9015–9019.

Halasy, K., Miettinen, R., Szabat, E. and Freund, T.F. (1992) GABAergic interneurons are the major postsynaptic targets of median raphe afferents in the rat dentate gyrus. *Eur. J. Neurosci.*, 4: 144–153.

Hanley, J.J. and Bolam, J.P. (1997) Synaptology of the nigrostriatal projection in relation to the compartmental organization of the neostriatum in the rat. *Neuroscience*, 81: 353–370.

Hansson, S.R., Mezey, É. and Hoffman, B.J. (1998) Serotonin transporter messenger RNA in the developing rat brain: early expression in serotonergic neurons and transient expression in non-serotonergic neurons. *Neuroscience*, 83: 1185–1201.

Houser, C.R. (1990) Cholinergic synapses in the central nervous system: studies of the immunocytochemical localization of choline acetyltransferase. *J. Electron Microsc. Tech.*, 15: 2–19.

Kasa, P., Hlavati, I., Dobo, E., Wolff, A., Joo, F. and Wolff, J.R. (1995) Synaptic and non-synaptic cholinergic innervation of the various types of neurons in the main olfactory bulb of adult rat: immunocytochemistry of choline acetyltransferase. *Neuroscience*, 67: 667–677.

Lauterborn, J.C., Isackson, P.J., Montalvo, R. and Gall, C.M. (1993) In situ hybridization localization of choline acetyltransferase mRNA in adult rat brain and spinal cord. *Mol. Brain Res.*, 17: 59–69.

Lebrand, C., Cases, O., Adelbrecht, C., Doye, A., Alvarez, C., El Mestikawy, S., Seif, I. and Gaspar, P. (1996) Transient uptake and storage of serotonin in developing thalamic neurons. *Neuron*, 17: 823–835.

Lebrand, C., Cases, O., Wehrlé, R., Blakely, R.D., Edwards, R. H. and Gaspar, P. (1998) Transient developmental expression of monoamine transporters in the rodent forebrain. *J. Comp. Neurol.*, 401: 506–524.

Levey, A.I., Armstrong, D.M., Atweh, S.F., Terry, R.D. and Wainer, B.H. (1983) Monoclonal antibodies to choline acetyltransferase: production, specificity, and immunohistochemistry. *J. Neurosci.*, 3: 1–9.

Levey, A.I., Wainer, B.H., Rye, D.B., Mufson, E.J. and Mesulam, M.-M. (1984) Choline acetyltransferase-immunoreactive neurons intrinsic to rodent cortex and distinction from acetylcholinesterase-positive neurons. *Neuroscience*, 13: 341–353.

Light, A.R., Kavookjian, A.M. and Petrusz, P. (1983) The ultrastructure and synaptic connections of serotonin-immunoreactive terminals in spinal laminae I and II. *Somatosens. Res.*, 1: 33–50.

Lovick, T.A., Parry, D.M., Stezhka, V.V. and Lumb, B.M. (2000) Serotonergic transmission in the periqueductal grey matter in relation to aversive behaviour: morphological evidence for direct modulatory effects on identified output neurones. *Neuroscience*, 95: 763–772.

Low, W.C., Lewis, P.R., Bunch, S.T., Dunnett, S.B., Thomas, S.R., Iversen, S.D., Björklund, A. and Stenevi, U. (1982) Functional recovery following neural transplantation of embryonic septal nuclei in adult rats with septohippocampal lesions. *Nature*, 300: 260–262.

Maley, B.E., Engle, M.G., Humphreys, S., Vascik, D.A., Howes, K.A., Newton, B.W. and Elde, R.P. (1990) Monoamine synaptic structure and localization in the central nervous system. *J. Electr. Microsc. Tech.*, 15: 20–33.

Marlier, L., Sandillon, F., Poulat, P., Rajaofetra, N., Geffard, M. and Privat, A. (1991) Serotonergic innervation of the dorsal horn of rat spinal cord: light and electron microscopic immunocytochemical study. *J. Neurocytol.*, 20: 310–322.

Maxwell, D.J., Leranth, C. and Verhofstad, A.A.J. (1983) Fine structure of serotonin-containing axons in the marginal zone of the rat spinal cord. *Brain Res.*, 266: 253–259.

McGeer, P.L., McGeer, E.G., Fibiger, H.C. and Wickson V. (1971) Neostriatal choline acetylase and cholinesterase following selective brain lesions. *Brain Res.*, 35: 308–314.

Mechawar, N. and Descarries, L. (1999) Quantitative data on the cholinergic innervation in adult rat parietal cortex. *Soc. Neurosci. Abstr.*, 25: 190.

Moukhles, H., Bosler, O., Bolam, J.P., Vallée, A., Umbriaco, D., Geffard, M. and Doucet, G. (1997) Quantitative and morphometric data indicate precise cellular interactions between serotonin terminals and postsynaptic targets in rat substantia nigra. *Neuroscience*, 76: 1159–1171.

Mrini, A., Soucy, J.-P., Lafaille, F., Lemoine, P. and Descarries, L. (1995) Quantification of the serotonin hyperinnervation in adult rat neostriatum after neonatal 6-hydroxydopamine lesion of nigral dopamine neurons. *Brain Res.*, 669: 303–308.

Mrzljak, L., Levey, A.I. and Goldman-Rakic, P.S. (1993) Association of m1 and m2 muscarinic receptor proteins with asymmetric synapses in the primate cerebral cortex: morphological evidence for cholinergic modulation of excitatory neurotransmission. *Proc. Natl Acad. Sci. USA*, 90: 5194–5198.

Mrzljak, L., Pappy, M., Leranth, C. and Goldman-Rakic, P.S. (1995) Cholinergic synaptic circuitry in the macaque prefrontal cortex. *J. Comp. Neurol.*, 357: 603–617.

Nicholson, C. and Syková, E. (1998) Extracellular space structure revealed by diffusion analysis. *Trends Neurosci.*, 21: 207–215.

Nilsson, O.G., Brundin, P. and Björklund, A. (1990) Amelioration of spatial memory impairment by intrahippocampal grafts of mixed septal and raphe tissue in rats with combined cholinergic and serotonergic denervation of the forebrain. *Brain Res.*, 515: 193–206.

Nogueira, R.L. and Graeff, F.G. (1995) Role of 5-HT receptors in the modulation of dorsal periaqueductal gray matter of the rat. *Pharmac. Biochem. Behav.*, 52: 1–6.

Oleskevich, S., Descarries, L, Watkins, K.C., Séguéla, P. and Daszuta, A. (1991) Ultrastructural features of the serotonin innervation in adult rat hippocampus: an immunocytochemical description in single and serial thin sections. *Neuroscience*, 42: 777–791.

Papadopoulos, G.C., Parnavelas, J.G. and Buijs, R.M. (1987a) Monoaminergic fibers form conventional synapses in the cerebral cortex. *Neurosci. Lett.*, 76: 275–279.

Papadopoulos, G.C., Parnavelas, J.G. and Buijs, R.M. (1987b) Light and electron microscopic immunocytochemical analysis of the serotonin innervation of the rat visual cortex. *J. Neurocytol.*, 16: 883–892.

Papadopoulos, G.C., Parnavelas, J.G. and Buijs, R.M. (1989a) Light and electron microscopic immunocytochemical analysis of the noradrenaline innervation of the rat visual cortex. *J. Neurocytol.*, 18: 1–10.

Papadopoulos, G.C., Parnavelas, J.G. and Buijs, R.M. (1989b) Light and electron microscopic immunocytochemical analysis of the dopamine innervation of the rat visual cortex. *J. Neurocytol.*, 18: 303–310.

Parnavelas, J.G. and Papadopoulos, G.C. (1989) The monoaminergic innervation of the cerebral cortex is not diffuse and non-specific. *Trends Neurosci.*, 12: 315–319.

Paspalas, C.D. and Papadopoulos, G.C. (1996) Ultrastructural relationships between noradrenergic nerve fibers and non-neuronal elements in the rat cerebral cortex. *Glia*, 17: 133–146.

Perry, E., Walker, M., Grace, J. and Perry, R. (1999) Acetylcholine in mind: a neurotransmitter correlate of consciousness. *Trends Neurosci.*, 22: 273–280.

Portas, C.M., Bjorvatn, B., Fagerland, S., Gronli, J. Mundal, V., Sorensen, E. and Ursin, R. (1998) On-line detection of extracellular levels of serotonin in dorsal raphe nucleus and frontal cortex over the sleep/wake cycle in freely moving rat. *Neuroscience*, 83: 807–814.

Poulat, P., Marlier, L., Rajaofetra, N. and Privat, A. (1992) 5-Hydroxytryptamine, substance P and thyrotropin-releasing hormone synapses in the intermediolateral cell column of the rat thoracic spinal cord. *Neurosci. Lett.*, 136: 19–22.

Rajaofetra, N., Ridet, J.-L., Poulat, P., Marlier, L., Sandillon, F., Geffard, M. and Privat, A. (1992) Immunocytochemical mapping of noradrenergic projections to the rat spinal cord with an antiserum against noradrenaline. *J. Neurocytol.*, 21: 481–494.

Ridet, J.-L. (1994) *Organisation ultrastructurale et plasticité des systèmes monoaminergiques dans la moelle épinière du rat.* Thèse de doctorat. Université de Montpellier II – Sciences et Techniques du Languedoc.

Ridet, J.-L., Sandillon, F., Rajaofetra, N., Geffard, M. and Privat, A. (1992) Spinal dopaminergic system of the rat: light and electron microscopic study using an antiserum against dopamine, with particular emphasis on synaptic incidence. *Brain Res.*, 598: 233–241.

Ridet, J.-L., Rajaofetra, N., Teilhac, J.R., Geffard, M. and Privat, A. (1993) Evidence for non-synaptic serotonergic and noradrenergic innervation of the rat dorsal horn and possible involvement of neuron-glia interactions. *Neuroscience*, 52: 143–157.

Riehle, A., Grun, S., Diesmann, M. and Aertsen, A. (1997) Spike synchronization and rate modulation differentially involved in motor cortical function. *Science*, 278: 1950–1953.

Rowley, H.L., Kilpatrick, I.C., Needham, P.L. and Heal, D.J. (1998) Elevation of extracellular cortical noradrenaline may contribute to the antidepressant activity of zotepine: an in vivo microdialysis study in freely moving rats. *Neuropharmacology*, 37: 937–944.

Ruda, M.A. and Gobel, S. (1980) Ultrastructural characterization of axonal endings in the substantia gelatinosa which take up [^{3}H]serotonin. *Brain Res.*, 184: 57–83.

Ruda, M.A., Coffield, J. and Steinbusch, H.W.M. (1982) Immunocytochemical analysis of serotonergic axons in laminae I and II of the lumbar spinal cord of the cat. *J. Neurosci.*, 2: 1660–1671.

Schafer, M.K., Eiden, L.E. and Weihe, E. (1998) Cholinergic neurons and terminal fields revealed by immunohistochemistry for the vesicular acetylcholine transporter. I. Central nervous system. *Neuroscience*, 84: 331–359.

Séguéla, P., Watkins K.C. and Descarries, L. (1988) Ultrastructural features of dopamine axon terminals in the anteromedial and suprarhinal cortex of adult rat. *Brain Res.*, 442: 11–22.

Séguéla, P., Watkins, K.C. and Descarries, L. (1989) Ultrastructural relationships of serotonin axon terminals in the cerebral cortex of the adult rat. *J. Comp. Neurol.*, 289: 129–142.

Séguéla, P., Watkins, K.C., Geffard, M. and Descarries, L. (1990) Noradrenaline axon terminals in adult rat neocortex: an immunocytochemical analysis in serial thin sections. *Neuroscience*, 35: 249–264.

Sesack, S.R., Bressler, C.N. and Lewis, D.A. (1995a) Ultrastructural associations between dopamine terminals and local circuit neurons in the monkey prefrontal cortex: a study of calretinin-immunoreactive cells. *Neurosci. Lett.*, 200: 9–12.

Sesack, S.R., Snyder, C.L. and Lewis, D.A. (1995b) Axon terminals immunolabeled for dopamine or tyrosine hydroxylase synapse on GABA-immunoreactive dendrites in rat and monkey cortex. *J. Comp. Neurol.*, 363: 264–280.

Sesack, S.R., Hawrylak, V.A., Melchitzky, D.S. and Lewis, D. A. (1998) Dopamine innervation of a subclass of local circuit neurons in monkey prefrontal cortex: ultrastructural analysis of tyrosine hydroxylase and parvalbumin immunoreactive structures. *Cereb. Cortex*, 8: 614–622.

Smiley, J.F. (1996) Monoamines and acetylcholine in primate cerebral cortex: what anatomy tells us about function. *Rev. Brasil. Biol.*, 56 (Suppl. 1) 153–164.

Smiley, J.F. and Goldman-Rakic, P.S. (1993) Heterogeneous targets of dopamine synapses in monkey prefrontal cortex demonstrated by serial section electron microscopy: a laminar analysis using the silver-enhanced diaminobenzidine sulfide (SEDS) immunolabeling technique. *Cereb. Cortex*, 3: 223–238.

Smiley, J.F. and Goldman-Rakic, P.S. (1996) Serotonergic axons in monkey prefrontal cerebral cortex synapse predominantly on interneurons as demonstrated by serial section electron microscopy. *J. Comp. Neurol.*, 367: 431–443.

Smiley, J.F., Williams, S.M., Szigeti, K. and Goldman-Rakic, P.S. (1992) Light and electron microscopic characterization of dopamine-immunoreactive axons in human cerebral cortex. *J. Comp. Neurol.*, 32: 325–335.

Smiley, J.F., Morrell, F., Mesulam, M.-M. (1997) Cholinergic synapses in human cerebral cortex: an ultrastructural study in serial sections. *Exp. Neurol.*, 144: 361–368.

Smith, A.D., Olson, R.J. and Justice, J.B. Jr. (1992) Quantitative microdialysis of dopamine in the striatum: effect of circadian variation. *J. Neurosci. Meth.*, 44: 33–41.

Smith, Y., Bennett, B.D., Bolam, J.P., Parent, A. and Sadikot, A.F. (1994) Synaptic relationships between dopaminergic afferents and cortical thalamic input in the sensorimotor territory of the striatum in monkey. *J. Comp. Neurol.*, 344: 1–19.

Soghomonian, J.-J., Beaudet, A. and Descarries L. (1988) Ultrastructural relationships of central serotonin neurons. In:

N.N. Osborne and M. Hamon (Eds), *Neuronal Serotonin*, John Wiley & Sons, London, pp. 57–92.

Soghomonian, J.-J., Descarries, L. and Watkins, K.C. (1989) Serotonin innervation in adult rat neostriatum. II. Ultrastructural features: a radioautographic and immunocytochemical study. *Brain Res.*, 481: 67–86.

Testylier, G. and Dykes, R.W. (1996) Acetylcholine release from frontal cortex in the waking rat measured by microdialysis without acetylcholinesterase inhibitors: effects of diisopropylfluorophosphate. *Brain Res.*, 740: 307–315.

Testylier, G., Maalouf, M., Butt, A.E., Miasnikov, A.A. and Dykes, R.W. (1999) Evidence for homeostatic adjustments of rat somatosensory cortical neurons to changes in extracellular acetylcholine concentrations produced by iontophoretic administration of acetylcholine and systemic diisopropylfluorophosphate treatment. *Neuroscience*, 91: 843–870.

Trottier, S., Evrard, B., Biraben, A. and Chauvel, P. (1994) Altered patterns of catecholaminergic fibers in focal cortical dysplasia in two patients with partial seizures. *Epilepsy Res.*, 19: 161–179.

Trottier, S., Evrard, B., Vignal, J.P. and Chauvel, P. (1996) The serotonergic innervation of the cerebral cortex in man and its changes in focal cortical dysplasia. *Epilepsy Res.*, 25: 79–106.

Umbriaco, D. (1995) *Caractérisation ultrastructurale de l'innervation cholinergique du cortex cérébral et de l'hippocampe. Études immunocytochimiques en microscopie électronique chez le rat adulte*. Thèse de doctorat, Université de Montréal.

Umbriaco, D., Watkins, K.C., Descarries, L., Cozzari, C. and Hartman, B.K. (1994) Ultrastructural and morphometric features of the acetylcholine innervation in adult rat parietal cortex: an electron microscopic study in serial sections. *J. Comp. Neurol.*, 348: 351–373.

Umbriaco, D., Garcia, S., Beaulieu, C. and Descarries, L. (1995) Relational features of acetylcholine, noradrenaline, serotonin and GABA axon terminals in the stratum radiatum of adult rat hippocampus (CA1). *Hippocampus*, 5: 605–620.

Van Eden, C.G., Hoorneman, E.M.D., Buijs, R.M., Matthijssen, M.A.H., Geffard, M. and Uylings, H.B.M. (1987) Immunocytochemical localization of dopamine in the prefrontal cortex of the rat at the light and electron microscopical level. *Neuroscience*, 22: 849–862.

Vaucher, E. and Hamel, E. (1995) Cholinergic basal forebrain neurons project to cortical microvessels in the rat: electron microscopic study with anterogradely transported *Phaseolus vulgaris* leucoagglutinin and choline acetyltransferase immunocytochemistry. *J. Neurosci.*, 15: 7427–7441.

Verney, C., Alvarez, C., Geffard, M. and Berger, B. (1990) Ultrastructural double-labelling study of dopamine terminals and GABA-containing neurons in rat anteromedial cerebral cortex. *Eur. J. Neurosci.*, 2: 960–972.

Vizi, E.S. and Kiss, J.P. (1998) Neurochemistry and pharmacology of the major hippocampal transmitter systems: synaptic and non-synaptic interactions. *Hippocampus*, 8: 566–607.

Wightman, R.M. and Zimmerman, J.B. (1990) Control of dopamine extracellular concentration in rat striatum by impulse flow and uptake. *Brain Res. Rev.*, 15: 135–144.

Winkler, J., Suhr, S.T., Gage, F.H., Thal, L.J. and Fisher, L. J. (1995) Essential role of neocortical acetylcholine in spatial memory. *Nature*, 375:484–487.

Woolf, N.J. (1991) Cholinergic systems in mammalian brain and spinal cord. *Progr. Neurobiol.*, 37: 475–524.

L. F. Agnati, K. Fuxe, C. Nicholson and E. Syková (Eds.)
Progress in Brain Research, Vol 125

CHAPTER 4

Comparative aspects of volume transmission, with sidelight on other forms of intercellular communication[1]

Rudolf Nieuwenhuys*

Papehof 25, 1391 BD Abcoude, The Netherlands

I. Introduction

Our ideas about intercellular communication in the central nervous system have changed drastically during the last several decades; around 1965 everything seemed to be simple and clear. There were two main types of neuronal elements: ordinary neurons and neurosecretory neurons, and the elements of the first type could be subdivided into excitatory and inhibitory neurons. The phenomena of excitation and inhibition were physiologically well understood and appeared to have clear ultrastructural substrates. The number of known central neurotransmitters was extremely small, including hardly more than acetylcholine and noradrenalin, and it was thought that a brain could manage with only two such substances, an excitatory and an inhibitory one.

After the 1960s a number of notable new insights emerged.

(1) It appeared that the number of central neurotransmitters is much larger than two, and a new category of neuroactive compounds – the neuropeptides – was discovered.
(2) Certain neurochemicals, not directly involved in the process of synaptic transmission in the strict sense, are able to influence this process in various ways. These so-called neuromodulators may be active at the presynaptic side, e.g. by affecting the amount of transmitter released and the time course of transmitter release, as well as on the postsynaptic side, e.g. by regulating the sensitivity of the receptors.
(3) Not only the synaptic clefts, but also the entire extracellular space, which encompasses about 20% of the total volume of the brain, forms a continuous, though tortuous, communication channel.
(4) The ventricular system also provides a conduit for informational substances.
(5) Synapses now appear to be open devices; their functioning may be influenced by substances originating from beyond the synaptic cleft. Conversely the messenger molecules discharged within the clefts may diffuse to adjacent synapses and to non-synaptic domains of the extracellular space.
(6) The (generally 'peptidergic') contents of large dense-cored vesicles (LDCVs), as well as those of small clear vesicles (SCVs) can be released extrasynaptically.
(7) Receptor complexes may be localized in regions beyond synaptic specializations.

The insights just enumerated have led to the concept that intercellular communication in the

[1] This chapter is dedicated to Mrs Jooske de Sonnaville, Mrs. Elsbeth Zwart, Mrs Hansje Hartman and Mr. George Meerburg, 'treasurers' of the unique 'Artis'-Library, Faculty of Science, University of Amsterdam. Without their help, this work could not have been realized.

*Corresponding author: Tel.: +31-294-287204; Fax: +31-294-287206; e-mail: sbakker@nki.nl

brain can be grouped into two broad classes: wiring transmission and volume transmission (Fuxe and Agnati, 1991; Zoli and Agnati, 1996; Zoli et al., 1998). Wiring or 'classical' transmission is characterized by a single 'transmission' channel made by cellular structures and with a region of discontinuity no larger than a synaptic cleft. Volume transmission (also designated as non-synaptic, parasynaptic or paracrine transmission) is characterized by the diffusion of a messenger substance from a cell source into the extracellular fluid for a distance larger than the synaptic cleft (Fig. 1).

In the present chapter the morphological evidence for the occurrence of volume transmission in vertebrates (section II) and invertebrates (section III) will be reviewed. In the final section (IV), the assembled data will be surveyed and placed in a comparative and a functional perspective.

The following morphological features will be used as criteria for the occurrence of volume transmission (Fig. 2):

(1) Presence of multiple extrasynaptic exocytoses, as revealed by the TARI (standing for Tannic Acid Ringer Incubation) technique. This technique is well suited to the ultrastructural demonstration of exocytosis (Buma et al., 1984). Tannic acid strongly binds to exocytotically released proteinaceous substances, as well as to heavy metal (lead and uranyl) ions during postfixation of tissues. As tannic acid does not penetrate plasma membranes, only extracellularly located substances are stained. Moreover, during TARI treatment, exocytosis proceeds, but the exteriorized contents of the secretory granules are immediately captured by tannic acid and do not diffuse away into the extracellular space (Fig. 2A, B).
(2) Presence of evident exocytoses in electron micrographs (Fig. 2C).
(3) The consistent finding that in synaptic terminals LDCVs are only present at a distance from the membrane specializations (i.e. the so-called 'stand-back' phenomenon: Fig. 2D).
(4) Presence of accumulations of SCVs and/or LDCVs in axonal varicosities which, when followed through serial sections, appear to make no synaptic contact with other neuronal elements (Fig. 2E).
(5) Location of receptor complexes at neuronal membrane surfaces not associated with synaptic junctions (Fig. 2F, G).
(6) Location of transmitter release sites far from their putative target receptors, a phenomenon designated as transmitter-receptor mismatch.

II. Vertebrates

1. Mammals

The occurrence of volume transmission in the central nervous system of mammals is well established. All the illustrations showing various phenomena pertaining to volume transmission, assembled in Fig. 2, have been taken from publications on mammals. Moreover, many examples of pronounced peptide-peptide receptor mismatch have been observed in mammalian brains (for review, see Herkenham, 1987).

The regions in which phenomena have been observed, indicating volume transmission, range

Fig. 1. Synaptic and non-synaptic release of neuromediators in the mammalian central nervous system. N2 represents an 'ordinary' neuron; its axonal system may have terminals of types a, b or c. Terminals of types a and b contain small clear vesicles; they make typical synaptic contacts with other neurons and hence are involved in wiring transmission (w). Terminals of type c contain, in addition to small clear vesicles, numerous large, dense-cored vesicles; the small vesicles are closely apposed to the presynaptic membrane and form part of a classical synaptic contact. The large dense-cored vesicles, on the other hand, consistently stand back from the presynaptic membrane and release their products at non-synaptic sites of the terminal membrane. N1 represents a monoaminergic neuron with thin, highly branched varicose axons; the varicosities are filled with small, clear vesicles. Some of these varicosities make typical synaptic contacts with adjacent neuronal elements (w), but most of them are involved in volume transmission (v). Some branches of N2 pierce the ependymal lining of the ventricular cavity (vc). Their secretion products, which are released in the liquor cerebrospinalis (a.k.a. cerebrospinal fluid, CSF), may influence synaptic processes at considerable distances. N3 represents a neurosecretory neuron. The width of the extracellular space (ecs) has been exaggerated; the arrows, which indicate the flow of neuromediator molecules, point to synaptic and non-synaptic membrane receptors (modified from Nieuwenhuys, 1996).

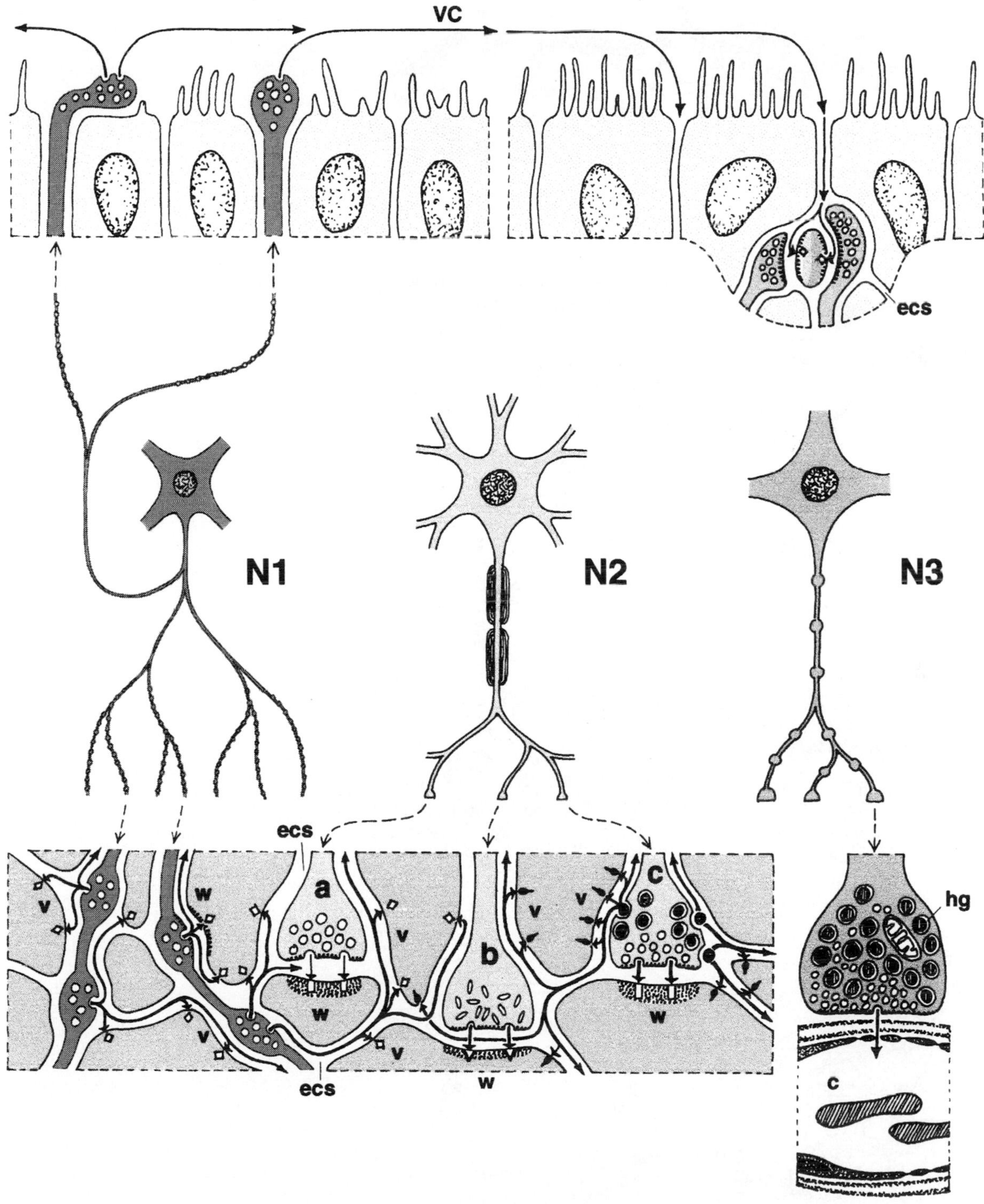
vc
ecs
N1
N2
N3
ecs
a
b
c
v
w
ecs
w
hg
c

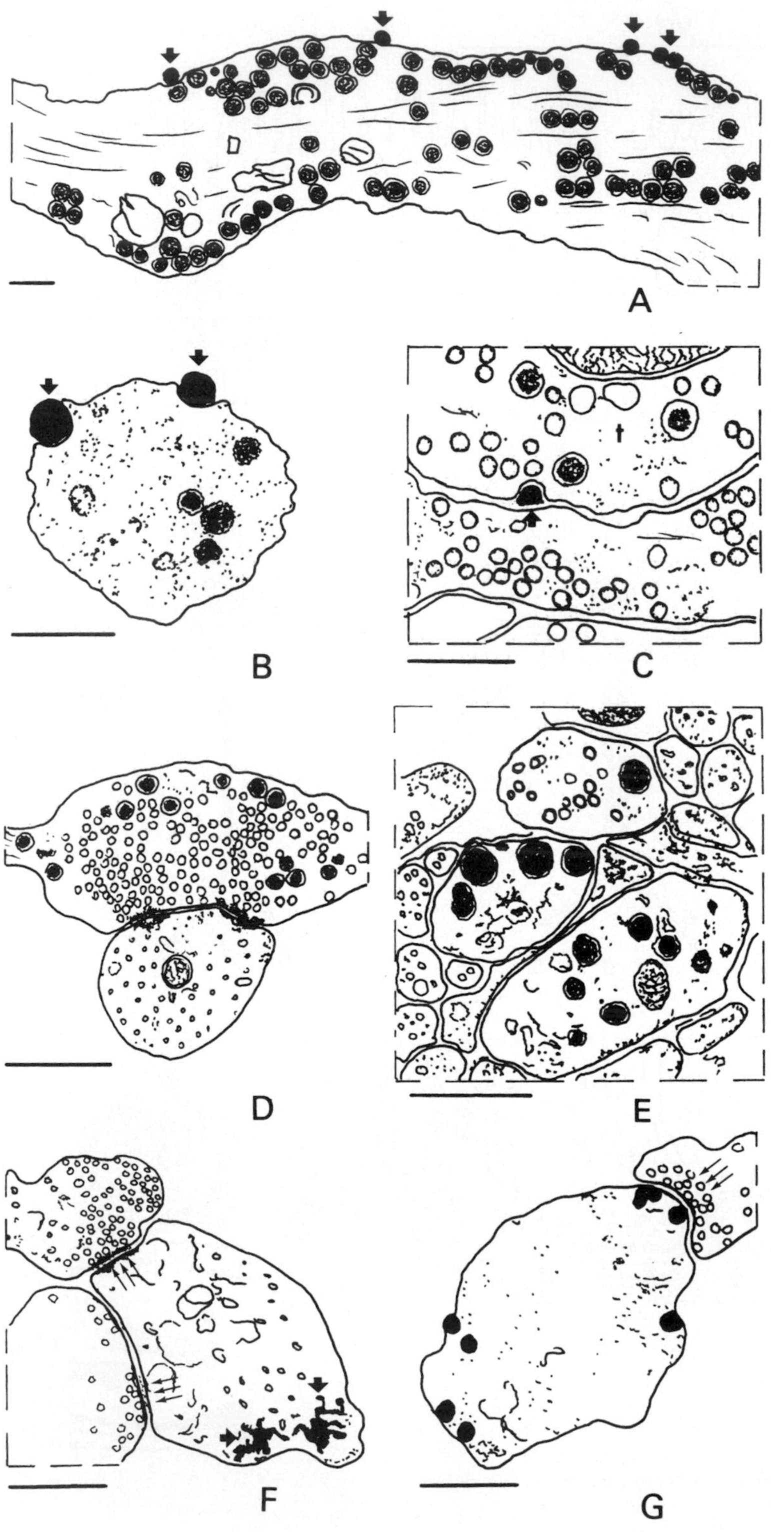
A
B
C
D
E
F
G

from the spinal cord to the telencephalon, and include the following structures:

- the *superficial zone of the spinal dorsal horn* (e.g. Carlton et al., 1987: CGRP-containing LDCVs, present in large terminals, typically 'stand back' from synaptic specializations; Marlier et al., 1991: presence of numerous 'asynaptic' serotoninergic varicosities);
- the *trigeminal subnucleus caudalis* (Zhu et al., 1986: numerous exocytoses from LDCVs outside the active synaptic zones of terminals; Fig. 2C);
- the *dorsal vagal complex* (Baude and Shigemoto, 1988: substance P receptors of the NK1 type preferentially located at non-synaptic sites of neuronal membranes);
- the *substantia nigra* (Jaffe et al., 1998: dopamine is released extrasynaptically from somata and dendrites);
- the *ventral tegmental area* (Dana et al., 1989: neurotensin-binding sites are mostly not associated with synaptic functions; Fig. 2F);
- the *periaqueductal gray* (Clements et al., 1985: serotoninergic fibres rarely make specialized contacts);
- the *supraoptic nucleus* (Pow and Morris, 1989; Morris and Pow, 1991: using the TARI technique it was shown that the large cells in this nucleus exocytose peptidergic vesicles from unspecialized parts of their axons, as well as from their dendrites; Fig. 2A);
- the *suprachiasmatic nucleus* (Castel et al., 1990: typical 'stand-back' position of vasopressinergic LDCVs in many synapses);
- the *neostriatum*, the *hippocampus* and the neocortex (Descarries et al., 1997: low frequency of synaptic membrane differentiations on cholinergic axon terminals; similar findings have also been reported for noradrenergic and serotoninergic axons in these structures: see Descarries and Mechawar, this volume; moreover, Yung et al., 1995 found that in the neostriatum a high proportion of dopamine receptors are on neuronal membranes at non-synaptic sites: Fig. 2G).

In 1985, the present author introduced a new overall functional subdivision of the mammalian brain (Nieuwenhuys, 1985; cf. also: Nieuwenhuys, 1996; Nieuwenhuys et al., 1989, 1998). According to this subdivision, the brain comprises a sensorimotor and cognitomotor domain, which grossly occupies the lateral part of the brain, and a medial domain that may be designated as the greater limbic system. It was hypothesized that within the latter system, non-synaptic interneuronal communication plays a prominent role. The lateral domain encompasses the great sensory systems, the association system, the pyramidal tract and the cerebellum (Fig. 3). Its fibre systems are compact and well myelinated. The medial domain structurally, chemically and functionally differs from the lateral domain. It comprises units designated as the core of the neuraxis and the median and lateral paracores. The core of the neuraxis extends throughout the central nervous system and is situated mostly close to the ventricular cavity (Fig. 4). It is characterized by the presence of: (1)

Fig. 2. Ultrastructural evidence for volume transmission. A. Longitudinal section of a magnocellular supraoptic nucleus dendrite of the rat, containing numerous large dense-cored vesicles (LDCVs), many of which lie close to the plasmolemma. Several exocytotic figures (arrows) are visible (from Pow and Morris, 1989; TARI technique). B. Exocytoses (arrows) in a vasopressinergic varicosity in the median eminence of the rat (from Buma and Nieuwenhuys, 1987; TARI technique). C. Terminal (t) in superficial zone of the trigeminal subnucleus caudalis of the rat, containing small clear vesicles (SCVs), large dense cored vesicles (LDCVs) and an exocytotic profile (arrow) (from Zhu et al., 1986). D. Axonal varicosity in the periaqueductal gray of the rat, making synaptic contact with a dendrite. The varicosity contains both SCVs and LDCVs; the latter are typically situated at a distance from the synaptic membrane specializations (from Buma et al., 1992). E. Axonal profiles with LDCVs in the periaqueductal grey of the rat. On serial sectioning profiles of this type showed no traces of synaptic specializations (from Buma et al., 1992). F. Cross-sectioned dendritic shaft and two abutting axon terminals observed in an autoradiogram from a section through the area tegmentalis ventralis of the rat, incubated with 125I-NT, a ligand that specifically labels neurotensin binding sites. Neither of the two silver grains (thick arrows) includes one of the abutting synapses (thin arrows) (from Dana et al., 1989). G. Cross-sectioned dendrite receiving synaptic input from a bouton, observed in a section through the neostriatum of the rat, in which D1 dopamine receptors are immunocytochemically labeled. Most of the immunoparticles are not associated with the synaptic site (from Yung et al., 1995). All of the figures were drawn from electron micrographs published by the authors cited. Scale bars: 0.5 μm.

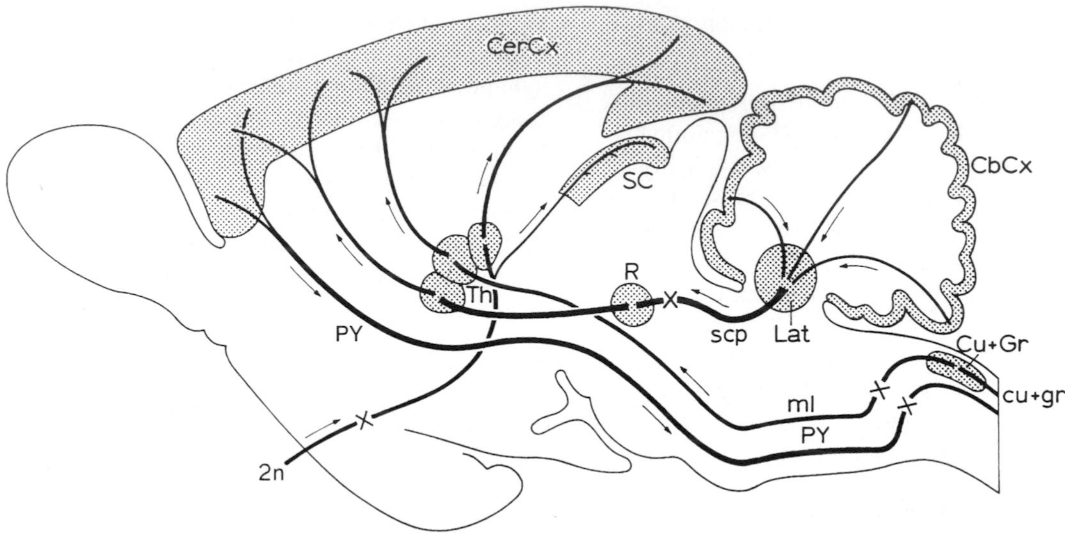

Fig. 3. 'Lateral domain systems' in the brain of the rat. Cbcx, cerebellar cortex; CerCx, cerebral cortex; Cu, nucleus cuneatus; cu, fasciculus cuneatus; Gr, nucleus gracilis; gr, fasciculus gracilis; Lat, lateral cerebellar nucleus; ml, medial lemniscus; Py, pyramidal tract; R, nucleus ruber; SC, superior colliculus; scp, superior cerebellar peduncle; Th, thalamus; 2n, optic nerve (reproduced from Nieuwenhuys et al., 1989).

numerous diffuse grisea, (2) enormous amounts of thin, unmyelinated varicose axons, many of which are arranged in diffuse fibre systems, (3) large numbers of different neuromediators, particularly neuropeptides, (4) large numbers of neurons which concentrate estrogen and androgen hormones, and (5) circumventricular organs, which monitor the concentrations of hormones and other informational substances in the liquor cerebrospinalis. The core region is involved in the organization of specific, motivated or goal-oriented behaviours, directly aimed at the maintenance of homeostasis and the survival of the individual (organism) and of the species. The median and lateral paracores represent extensions of the core at the level of the brain stem. The median paracore includes the raphe nuclei, whereas the (bilateral) lateral paracore is formed by a ventrolaterally extending lamella of tissue (Fig. 5). Both paracores contain sets of monoaminergic cells giving rise to networks of fibres that pervade virtually all grisea of the neuraxis, i.e. the serotoninergic neurons in the median paracore and the catecholaminergic cells in the lateral paracore.

The hypothesis that non-synaptic interneuronal communication plays an important role in the organization and regulation of behaviour by the core and paracore regions, is based on the supposition that the myriad thin, monoaminergic and peptidergic fibres which characterize these regions preferentially use this mode of release (for details cf. Nieuwenhuys, 1985, 1996). It is substantiated by ultrastructural findings in two typical core regions, viz. the median eminence (Buma and Nieuwenhuys, 1987, 1988: presence of numerous non-synaptic release sites; Fig. 2B) and the periaqueductal gray (Buma et al., 1992: presence of numerous asynaptic varicosities; Fig. 2E; typical 'stand-back' position of LDCVs in numerous synapses; Fig. 2D). The hypothesis is further supported by the findings of Baude and Shigemoto (1998), Castel et al. (1990), Clements et al. (1985), Dana et al. (1989), Marlier et al. (1991), Pow and Morris (1989), Morris and Pow (1991) and Zhu et al. (1986), mentioned above.

From a survey of the ultrastructural data discussed in this section, it may be concluded that volume transmission is most probably very common in the greater limbic system of mammals, and that this mode of interneuronal communications also occurs in the neostriatum, the hippocampus and the neocortex of this group.

2. Non-mammalian vertebrates

The fact that volume transmission is common in mammals and very prominent in invertebrates (see section III) makes it likely that this mode of neurotransmission also occurs in the central nervous system of non-mammalian vertebrates. However, morphological evidence supporting this idea is, as far as I am aware, confined to the hypothalamo-hypophysial system of the clawed toad, *Xenopus laevis*, the prepacemaker nucleus and the cerebellum of electric teleosts, and the spinal cord of the lamprey.

The pars intermedia of the pituitary of the clawed toad *Xenopus laevis* contains melanotrope cells, which are responsible for the process of background adaptation. When *Xenopus* is placed on a black background, the melanotrope cells release α-MSH (α-melanophore-stimulating hormone), which causes pigment dispersion in dermal melanophores, leading to darkening of the skin. On a white background, α-MSH release is inhibited, and the animal turns pale. The inhibition of α-MSH release is effected by synapses in the pars intermedia, which contain small, clear vesicles as well as large dense-cored vesicles. These synapses are nerve terminals of a group of neurons in the suprachiasmatic nucleus. The terminals contain three different neuromediators, viz. GABA, dopamine and neuropeptide Y, which all inhibit the release of α-MSH. It has been shown that the small, clear vesicles contain GABA, whereas the large dense-cored vesicles store dopamine and neuropeptide Y. Because the small vesicles cluster in the vicinity of the synaptic membrane specializations of the terminals, whereas the dense-cored vesicles typically stand back from these specializations, it seems likely that GABA is released synaptically,

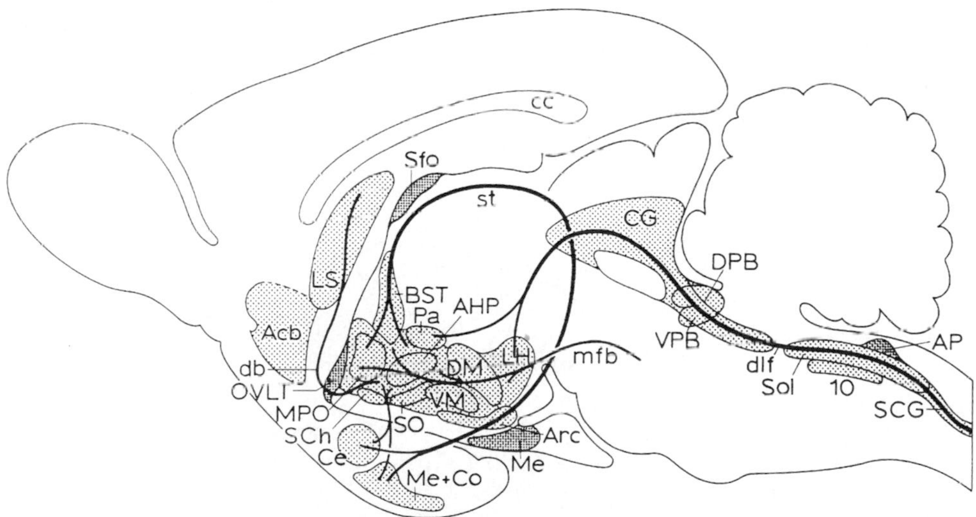

Fig. 4. Core of the central nervous system of the rat. Acb, accumbens nucleus; AHP, anterior hypothalamic area; AM, amygdaloid complex; AP, area postrema; Arc, arcuate hypothalamic nucleus; A1, A2 etc., noradrenergic cell groups; BST, bed nucleus of the stria terminalis; cc, corpus callosum; Ce, central amygdaloid nucleus; CG, central (periaqueductal) gray; Co, cortical amygdaloid nucleus; CSu, central superior nucleus; C1, C2 etc., adrenergic cell groups; DB, nucleus of the diagonal band of Broca; db, diagonal band of Broca; dlf, dorsal longitudinal fasciculus; DM, dorsomedial hypothalamic nucleus; DPB, dorsal parabrachial nucleus; DR, dorsal raphe nucleus; f, fornix; fr, fasciculus retroflexus; Hi, hippocampus; IP, interpeduncular nucleus; LH, lateral hypothalamic area; lpcb, lateral paracore bundle; LS, lateral septal nucleus; ME, median eminence; Me, medial amygdaloid nucleus; mfb, medial forebrain bundle; MHb, medial habenular nucleus; MnR, median raphe nucleus; MPO,medial preoptic nucleus; MS, medial septal nucleus; OVLT, organum vasculosum laminae terminalis; Pa, paraventricular nucleus; PHC, preoptic-hypothalamic continuum; RLi, rostral linear nucleus of the raphe; RMg, raphe magnus nucleus; ROb, raphe obscurus nucleus; RPa, raphe pallidus nucleus; SCG, spinal central gray; SCh, suprachiasmatic nucleus; SfO, subfornical organ; sm, stria medullaris of the thalamus; SO, supraoptic nucleus; Sol, nucleus of the solitary tract; st, straia terminalis; VM, ventromedial hypothalamic nucleus; VPB, ventral parabrachial nucleus; 10, dorsal motor vagal nucleus (reproduced from Nieuwenhuys et al., 1989).

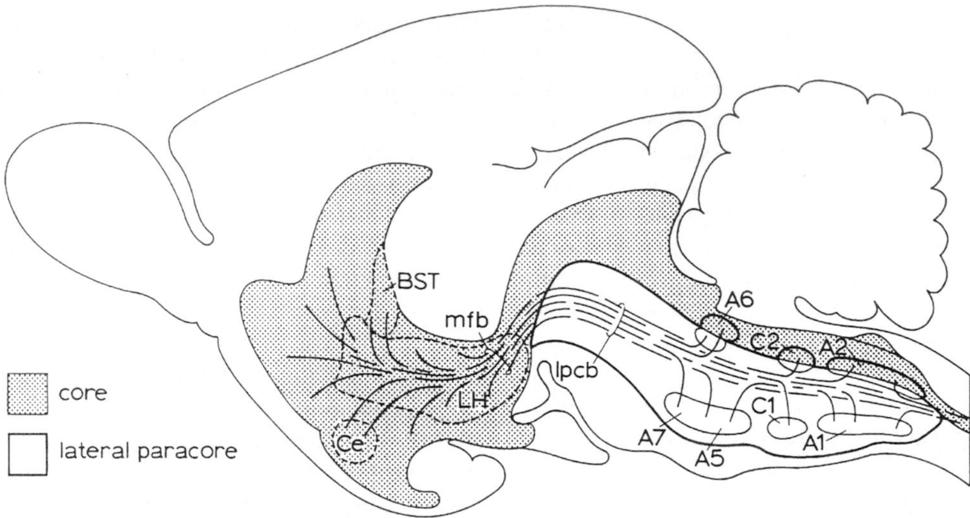

Fig. 5. Core and lateral paracore of the brain of the rat; for abbreviations, see legend to Fig. 4 (reproduced from Nieuwenhuys et al., 1989).

but that dopamine and neuropeptide Y are secreted extrasynaptically (Berghs and Roubos, 1996; Roubos, 1997; Fig. 6A).

The activity of the electric organ in gymnotiform electric fish is under the control of a cascade of centres, among which are a rhombencephalic pacemaker nucleus and a diencephalic prepacemaker nucleus. Zupanc (1991) studied the vesicular content of the synaptic boutons making contact with prepacemaker neurons in the knifefish, *Eigen-*

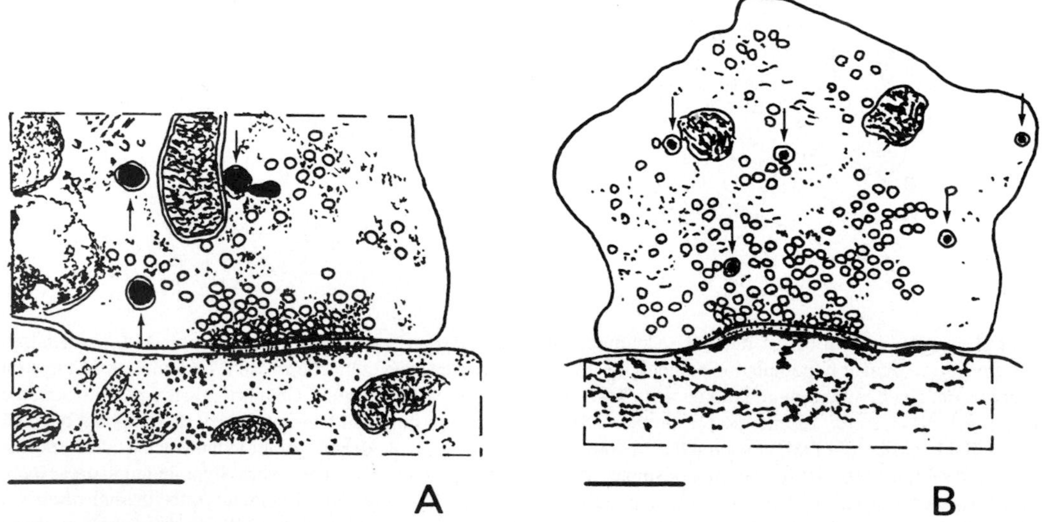

Fig. 6. Nerve endings containing SCVs as well as LDCVs; whereas many of the SCVs are clustered at the presynaptic membrane, the LDCVs (arrows) are dispersed through the terminal and avoid the synaptic area. A. Synaptic contact (top) on a melanotrope cell (bottom) in the pituitary pars intermedia of the clawed toad, *Xenopus laevis* (Roubos, 1997). B. Bouton on a neuron in the diencephalic prepacemaker nucleus of the weakly electric knifefish, *Eigenmannia* (Zupanc, 1991). The figures were drawn from electron micrographs published by the authors cited. Scale bars: 0.5 μm.

mannia. It appeared that these terminals contain large numbers of small, clear vesicles, among which large dense-cored vesicles are sparsely scattered (Fig. 6B). Quantitative analysis revealed that the dense-cored vesicles are not randomly distributed. In these boutons, the number of dense-cored vesicles found in a 200 nm-wide band stretching along the active zone was much lower than the number expected if these vesicles were randomly distributed within the whole profile.

As mentioned in the introduction, discrepancies between the location of the release sites of neuromediators, and the location of their receptors are important signs of volume transmission. The investigations by Stroh and Zupanc (1993) and Zupanc et al. (1994) have shown that in the cerebellum of gymnotiform fish such a discrepancy or 'mismatch' exists between the localization of somatostatin-immunoreactive cells and the distribution of somatostatin-binding sites. In the caudal part of the cerebellum, immunoreactive cell bodies and fibres, which could be detected only by highly sensitive techniques and immunostaining, occurred only in restricted areas. In contrast, the density of somatostatin-binding sites in the caudal cerebellum was shown to be the highest in the whole brain and binding sites appeared to be abundant even in regions where no somatostatin immunoreactivity had been detected at all.

The fourth and final example of morphological evidence for volume transmission in non-mammalian vertebrates occurs in the spinal cord of the lamprey. The lamprey spinal cord is a flattened, ribbon-like structure, in which a thin lamella of gray matter is surrounded by a fibre zone (Nieuwenhuys and Nicholson, 1998; Fig. 7A). Apart from large motor neurons, the gray matter contains several types of morphologically and functionally well-defined interneurons. The fibre zone, which is composed of axons of highly varying diameters, can be bilaterally subdivided into a dorsal, a lateral and a ventral funiculus. The ventral funiculus contains a bundle of coarse, so-called Müller fibres. These fibres, which originate from a group of large neurons in the brain stem, form the principal encephalospinal system.

The lamprey central nervous system has proved a useful model for studying the cellular basis of central nervous mechanisms. During the last few decades, Grillner and his associates have carried out a comprehensive programme of combined morphological and physiological investigations. From this endeavour, considerable insights have been gained into the mechanism of locomotion and its control (for reviews, see: Grillner, this volume; Grillner et al., 1991, 1995; Nieuwenhuys and Nicholson, 1998; Orlovsky et al., 1999). Within the context of this research programme, Christenson et al. (1990, 1991) studied the structure and function of serotoninergic systems in the cord of the lamprey. These studies have shown that varicose serotoninergic fibres are present in all three spinal funiculi (Fig. 7B). The fibres in the dorsal funiculi

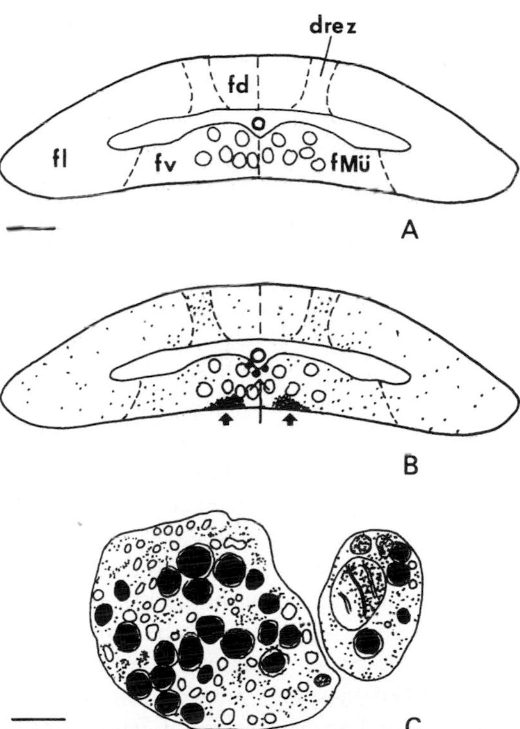

Fig. 7. The spinal cord of the lamprey. A. Diagrammatic transverse section, showing the subdivision of the fibre zone. Scale bar: 100 μm. B. The distribution of serotoninergic cells and fibres. Thin arrow indicates group of serotoninergic perikarya; thick arrows point to bilateral dense ventromedial serotoninergic fibre plexuses. (Drawn from a photomicrograph in Van Dongen et al., 1985). C. Two varicosities in the ventromedial plexus, containing LDCVs as well as smaller agranular vesicles. Scale bar: 0.2 μm. (Drawn from an electron micrograph in Christenson et al., 1990.)

stem from neurons in the spinal ganglia, whereas those in the lateral funiculi originate mainly from the rhombencephalic cell groups. The fibres in the ventral funiculus, which form a dense plexus along the ventromedial part of the cord, have a local origin. They arise from a strip of small neurons situated ventral to the central canal (Fig. 7B). Serotoninergic axonal varicosities were studied in all three funiculi of the cord using immunohistochemistry followed by an analysis of serotonin-immunoreactive varicosities and they appeared to be identical in all regions. The reaction product was found scattered throughout the cytoplasm, with a clear predilection for the numerous large dense-core vesicles present in these varicosities. Since no unlabeled varicosities with large granular vesicles were found in the ventromedial serotoninergic plexus after immunostaining, it was concluded that all varicosities with large granular vesicles belong to serotoninergic neurons in this part of the cord. Comparison of the immunostained material with normal material from this region revealed the presence of small pleomorphic agranular vesicles together with the large granular vesicles in the unstained material (Fig. 7C). Such small vesicles could, however, not be demonstrated in labeled varicosities. It was assumed that the strong immunostaining had obscured the presence of the small pleomorphic vesicles.

Light microscopical studies have shown that motoneurons and several types of interneurons extend dendrites into the ventromedial serotoninergic plexus. In electrophysiological experiments these neurons were shown to respond to extracellular application of serotonin with a marked reduction of the amplitude of the after-hyperpolarization. Moreover, excitatory post-synaptic potentials evoked in spinal motoneurons by stimulation of descending Müller axons were reduced in amplitude by serotonin.

Because (1) none of the numerous serotoninergic varicosities studied in the lamprey cord were seen to make synaptic contact with surrounding neuronal elements, not even when followed through serial sections, and (2) the release of serotonin has been demonstrated in this cord by using serotonin-uptake blockers (Christenson et al., 1989), it may be concluded that the serotoninergic fibres in the spinal cord of the lamprey exert their influence on surrounding neuronal structures by volume transmission.

Recently, Parker and Grillner (1999) and Ullström et al. (1999) studied the influence of some neuropeptides present in the dorsal root entrance zone (substance P, neuropeptide Y, bombesin) on the spinal cord of the lamprey. It appeared that all of these neuropeptides exert a modulatory influence on one or several components of the locomotor network. The morphological substrates of these actions have, however, not yet been studied at the ultrastructural level.

III. Invertebrates

1. Introductory note

The invertebrates are often considered as an appendix of the vertebrates. This is quite unjustified. In a recent survey of life on earth by Margulis and Schwartz (1998) the Kingdom Animalia is divided into 37 phyla, no less than 36 of which are occupied by invertebrates (Table 1). In what follows, the morphological evidence for the occurrence of volume transmission in a number of invertebrate phyla (marked with an asterisk in Table 1) will be discussed. Since many readers of this volume may be unfamiliar with invertebrate neuroanatomy, these discussions are prefaced by brief general descriptions of the nervous system of the various phyla. In the final section of this chapter I shall attempt to place volume transmission in a comparative and a functional perspective. Therefore, in this section: (1) relevant data on the occurrence of other forms of interneuronal communication will be included, and (2) special attention will be paid to the structural relations in the first three phyla of the animal kingdom, i.e. the Placozoa, the Porifera and the Cnidaria.

2. Placozoa

The phylum Placozoa is represented today by a single marine species, *Trichoplax adhaerens*, a flattened, disc-like animal with a diameter varying from 0.2–1.0 mm (Barnes and Harrison, 1991; Grell and Ruthmann, 1991; Fig. 8A). The entire

TABLE 1

Subdivision of Animal Kingdom in 37 Phyla (Margulis and Schwartz, 1998)

*A–1	Placozoa – trichoplaxes	*A–20	Mandibulata – insects, centipedes, millipedes
*A–2	Porifera – sponges	*A–21	Crustacea – lobsters, crabs
*A–3	Cnidaria – hydroids and jellyfish	*A–22	Annelida – annelid worms
A–4	Ctenophora – comb jellies	A–23	Sipuncula – peanut worms
*A–5	Platyhelminthes – flatworms	A–24	Echiura
A–6	Gnathostomulida – jaw worms	A–25	Pogonophora – beard worms, tube worms
A–7	Rhombozoa	*A–26	Mollusca
A–8	Orthonectida	A–27	Tardigrada – water bears
A–9	Nemertina – ribbon worms	*A–28	Onychophora – velvet worms
*A–10	Nematoda – thread worms, round worms	A–29	Bryozoa – moss animals
A–11	Nematomorpha – Gordian worms, horsehair worms	A–30	Brachiopoda – lampshells
A–12	Acanthocephala – thorny-headed worms	A–31	Phoronida
A–13	Rotifera	*A–32	Chaetognatha – arrow worms
A–14	Kinorhyncha	A–33	Hemichordata – enteropneusts, pterobranchs
A–15	Priapulida	*A–34	Echinodermata – sea urchins, sea stars, sea cucumbers
A–16	Gastrotricha	*A–35	Urochordata – ascidians, larvaceans, thaliaceans
A–17	Loricefera	*A–36	Cephalochordata – lancelets
A–18	Entoprocta	*A–37	Craniata – craniate chordates
*A–19	Chelicerata – horseshoe crabs, spiders		

surface of *Trichoplax* is ciliated, but the density of the cilia is much higher on the ventral side than on the dorsal side. With the ventral cilia the animal moves across the seabed; histologically, a dorsal epithelium, a ventral epithelium and an interspace can be distinguished (Fig. 8B). The interspace consists of a fluid-filled compartment in which star-shaped, ramifying fibre cells are embedded; the dorsal epithelium consists of thin, flattened cells. The ventral epithelium is composed of tall, ciliated cylinder cells among which non-ciliated gland cells are interspersed. Specialized muscle and nerve cells are lacking; their functions apparently are fulfilled by the fibre cells (Grell and Ruthmann, 1991). These elements are connected to each other and to both epithelia by branched extensions, forming a three-dimensional meshwork (Grell, 1971). They are contractile and enable the animal to form a pocket on the ventral side over food particles. The contacts between the fibre cells are marked by disc-shaped osmiophilic structures, which are flanked on both sides by large cistern-like vesicles. Some further vesicles may be found nearby (Figs. 8B–D). In some cases the cellular membranes of the participating fibre cells were seen to pass uninterruptedly over into each other along the sides of the discs. Grell and Benwitz (1974), who first described the specializations just discussed, considered these as precursors of synapses ('prosynapses'). They pointed out that the changes of shape observed in *Trichoplax* require the coordinated action of the fibre cells and that the prosynapses may well play a role in this coordination. The so-called junctional plugs, observed by Mackie and Singla (1983) in the remarkable, syncytial hexactinellid sponges closely resemble the placozoan discs (see Fig. 11). Mackie and Singla (1983) did not consider these structures as forerunners of synapses, because synapses are specialized regions of contact between independent cells, not bridges between cells.

The fluid-filled interspace of *Trichoplax* is directly comparable to the primary body cavity of the early embryonic blastula stage of many animal groups (Bütschli, 1884; Grell, 1971). It is not known whether this intercellular space or internal environment is used by messenger materials as a communication channel.

3. *Porifera (sponges)*

Porifera are sessile, aquatic, mostly marine animals, ranging in size from a few millimeters to more than two meters. They are named after the numerous small pores that pierce their body wall; in simple forms the body is flask-shaped (Fig. 9A). The external openings of the pores are called ostia. Internally, the pores end in the central body cavity,

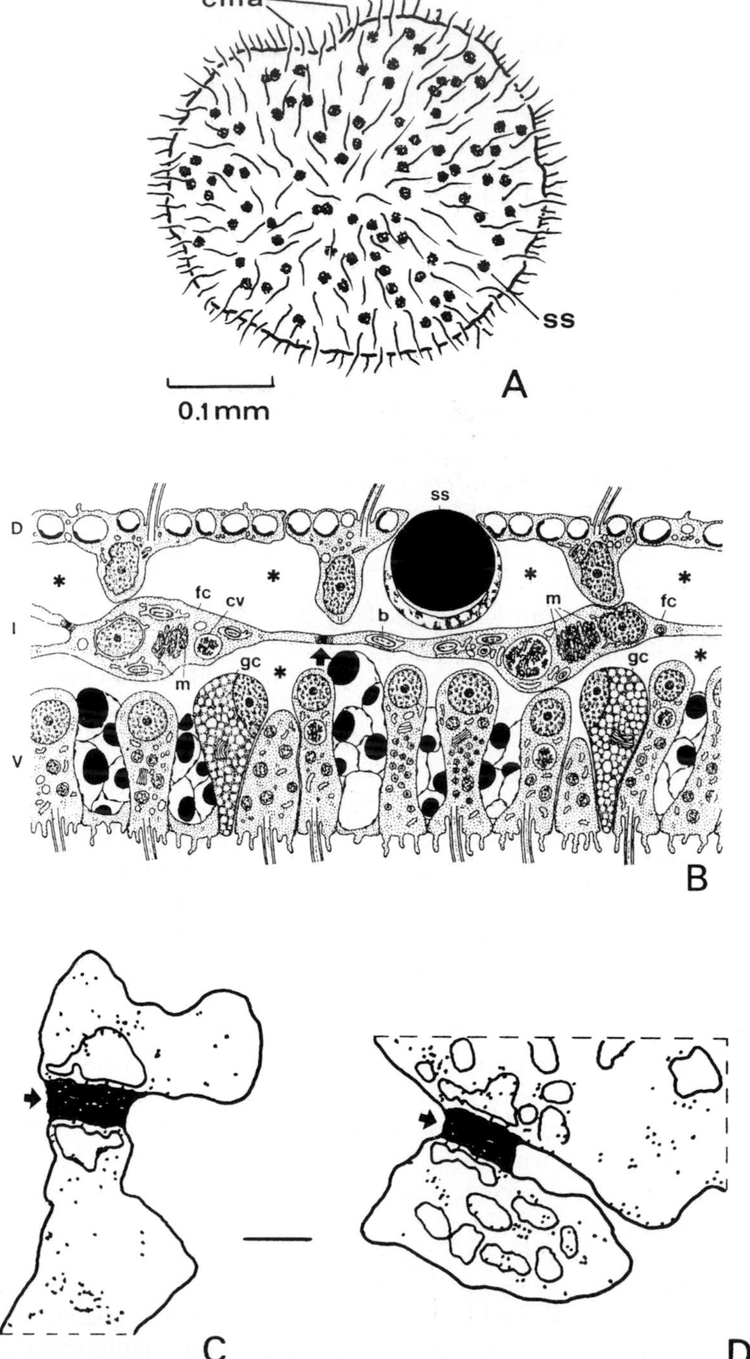

Fig. 8. The placozoan *Trichoplax adhaerens*. A. Dorsal view. B. Semidiagrammatic cross section through part of the animal (reproduced from Grell, 1981). C, D. Disc-like structures(dls) connecting fibre cells (after Grell and Benwitz, 1974); bar: 0,2μm. D, dorsal epithelium; I, interspace; V, ventral epithelium; b, bacterium in endoplasmic cisterna; cv, concrement vacuole; fc, fiber cell; gc, gland cell; m, mitochondrial complex; ss, shiny spheres (lipid droplets); arrows, disc-like connections between fibre cells; asterisks: extracellular space.

the spongocoel or atrium. The latter has a large opening or osculum at the top. In more complex sponges many oscula may be present.

Sponges are typical filter feeders; small organic particles are filtered from the water that flows through the sponge body. As indicated by the arrows in Fig. 9A, water enters the sponge through the ostia and leaves by the osculum.

Sponges are composed of an outer and an inner epithelium and an intervening mesenchyme (Fig. 9B; for details see Simpson, 1984; Harrison and De Vos, 1991). The outer epithelium or pinacoderm

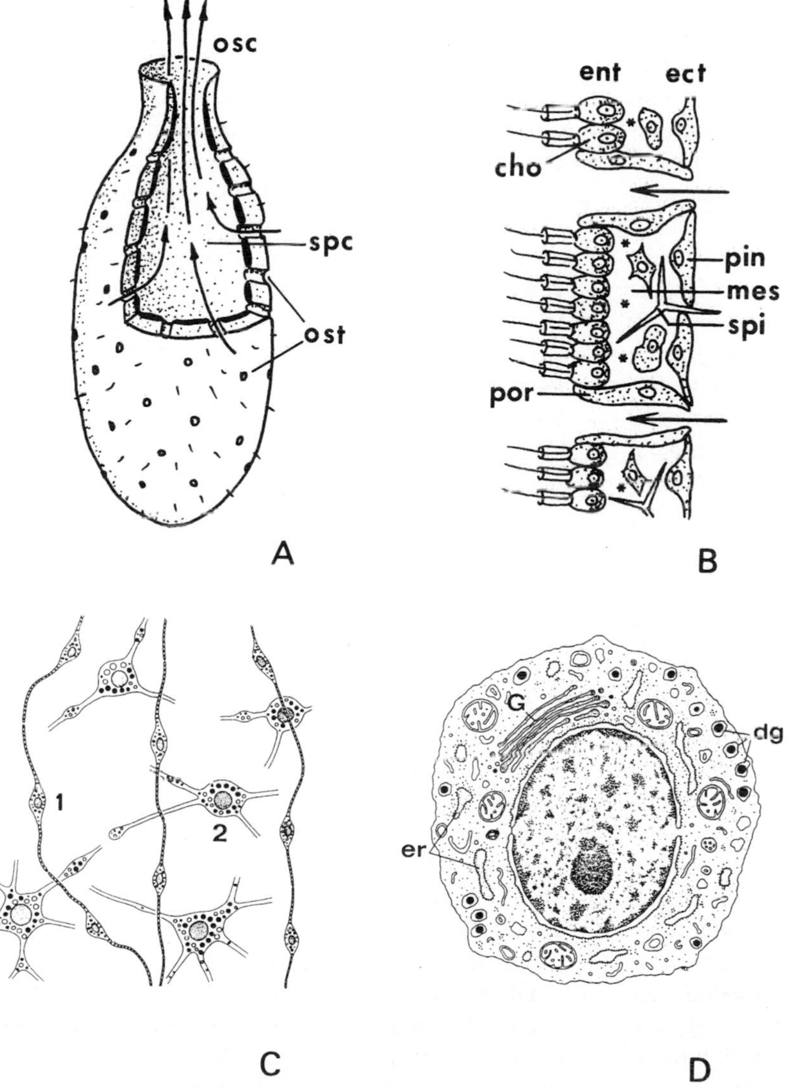

Fig. 9. Aspects of the morphology of sponges. A. The gross structure of a simple sponge; the arrows indicate the direction of the water currents through the animal. B. A portion of the wall of a simple sponge. C. Catecholamine-containing cells in the mesenchyme of the sponge *Sycon*. The small, bipolar cells (1) predominantly contain noradrenaline, the large, multipolar elements (2) adrenaline (Lentz, 1968). D. Granule-containing cell in the mesenchyme of *Sycon*. The granules are most abundant at the periphery of the cell (Lentz, 1968). cho, choanocyte; dg, dense granula; ect, ectoderm; ent, entoderm, er, rough endoplasmatic reticulum; G, Golgi apparatus; mes, mesenchyme or mesohyl; osc, osculum; ost, ostia; pin, pinacocyte; por, porocyte; spi, spicule; asterisks: extracellular space.

consists of flat pinacocytes, which do not exhibit tight junctions or basal laminae. In some groups the pinacocytes form superficial flat expansions, while the somata are located deeper and protrude into the mesenchyme. The inner epithelium or choanoderm is formed by choanocytes. These elements possess a flagellum whose base is surrounded by a microvillar collar. The choanocytes cause the propulsion of water through the sponge body and are also involved in the capture and digestion of food particles. In simple, so-called asconoid sponges, the choanocytes line the central cavity of the vase-like body (indicated by the black strips in Fig. 9A), but in the more complex leucenoid sponges these elements are confined to small chambers. A system of water canals connects these flagellated chambers with the pores and with the central atrium; in many sponges, the pores are surrounded by one or more contractile porocytes. In leucenoid sponges, pinacocytes are not confined to the external surface, but also line the inhalant and exhalant water canals of the flagellated chambers; these elements are called endopinacocytes. Remarkably, in glass sponges (Hexactinellida) the pinacoderm as well as the choanoderm are largely syncytial.

The space enclosed by the pinacocytes and the choanocytes is occupied by mesenchymatous tissue or mesohyl, composed of a glycoprotein-rich ground substance in which spongin fibres, skeletal spicules and a population of largely amoeboid cells are embedded. The needle-like spicules, which are either calcareous or silicious, form a skeletal framework which supports the soft tissue mass and enables the sponge to grow to a considerably size.

The mesenchyme cells come from epithelial cells that detach themselves from the epithelium, become amoeboid and wander into the intervening space as primitive amoebocytes. The amoebocytes are also called archaeocytes, because they can differentiate into various cell types, including food-storing thesocytes, spongin-secreting cells, spicule-producing sclerocytes, gonocytes and myocytes. The latter are contractile, fusiform elements, which are ultrastructurally characterized by the presence of cytoplasmic myofilaments; they are concentrated around the oscula and other openings.

According to some authors (for references see Simpson, 1984), the mesohyl also harbours collencytes, i.e. presumptive pinacocytes that are also capable of secreting collagen fibrils. The question as to whether neurons or neuroid cells are present in the mesohyl will be considered below. It should be emphasized that the mesenchyme cells form a dynamic population in a temporal as well as a spatial sense; temporal because many of them are involved in dynamic transitions, and spatial because they have the capacity to move freely through the mesohyl and to migrate into new growth areas.

As mentioned, sponges are sessile organisms and hence their behavioural repertoire is limited. Activities and responses of sponges consist of contractions leading to localized or general changes in shape, closing and opening of the ostia and oscula and the propulsion of water by the flagellated choanocytes. The stimuli leading to contractions of the body and of the openings are the same and include mechanical and electrical stimulation, changes in temperature and harmful chemicals. The alterations in shape occur as a result of contractions of pincocytes and myocytes, whereas in the closure of the openings contractions of porocytes and of rings of myocytes are involved. All the movements performed by sponges are very slow, often taking minutes to complete (Bullock and Horridge, 1965; Lentz, 1968).

It has long been thought that the flow of the water through sponges is regulated by the contractile states of ostia and oscules rather than by alterations of flagellar activity. However, the experiments of Lawn et al. (1981) and Mackie et al. (1983) have shown that in glass sponges, flagellar activity can be synchronously stopped by local mechanical or electrical stimulation.

Global contraction of pinacocytes and myocytes, leading to changes in shape, coordinated closure of oscules and the arrest of flagellar beating, all suggest the presence of an underlying system for the generation and conduction of signals through large areas of the body of sponges (Simpson, 1984). Possible morphological substrates of such a system have been repeatedly reported in the literature. For a good review of the older literature the reader is referred to Jones (1962); here I will confine myself to the studies by Pavans de Ceccatty (1955, 1959, 1960, 1966a, b, 1974a, b) and Lentz (1966, 1968).

Using silver impregnation techniques, Pavans de Ceccatty (1955, 1960) identified 'nervous type cells' in a large number of species; fusiform, triangular and multipolar elements were observed. Two kinds of processes were distinguished: fine, short ones, corresponding to dendrites, and a single coarse, axon-like process. The processes were seen to terminate on pinacocytes, choanocytes, muscle type cells, as well as on other neuroid cells. Many processes appeared to end freely in the large intercellular spaces. Neuromuscular cells, thought to have conducting and contractile functions, were also identified. Radially oriented, spindle-shaped sensory cells were observed in the walls of canals of some species (Pavans de Ceccatty, 1959).

Lentz (1966) established the presence of acetyl cholinesterase, monoamine oxidase, adrenaline, noradrenaline, serotonin and neurosecretory substance in some cells of Sycon with histochemical techniques. Neurosecretory substance was demonstrated with the chrome hematoxylin and the Alcian blue stain were employed. All of the techniques employed revealed the most intense staining or localization of reaction product in cells of two distinct types: small spindle-shaped bipolar cells, and large multipolar elements (Fig. 9C). Both cell types were situated in the mesenchyme.

The small bipolar cells are very numerous and concentrated around the osculum. In the collar region between the osculum and the first ostia, the cells are mainly oriented in the long axis of the sponge and interconnect to form a network. There are few spindle-shaped cells between the canals, but their number increases again near the base of the sponge.

The large multipolar cells were found in all regions, but appeared to be most numerous in the collar below the osculum. Most of the processes of these large elements terminated freely in the intercellular spaces. Reaction products were usually localized to small granules, vesicles or large droplets within the somata and their processes. The latter often formed small varicosities. The bipolar cells stained strongly for noradrenaline while the large multipolar cells showed a more intense staining reaction for adrenaline. The functional considerations which Lentz (1968, pp. 26–27) attached to the description of these chemically defined cells are worth quoting in full:

> "The simple responses of which sponges are capable are usually thought to be due to the activity of independent effectors such as the myocytes of the osculum. However, if these effectors are sensitive to biologically active neurohumors, the cells in sponges containing these substances might be able to affect or modify the activity of the effectors. The chemically specialized cells are loosely arranged and often appear to terminate in intercellular spaces. Whatever control these cells might have on effectors is probably diffuse, slow acting, and not effective over great distances. It is not known, however, whether neurohumors can alter the activity of sponge effectors or are released in response to external stimuli. In the absence of this evidence, these cells cannot be regarded as nerve cells. It is more reasonable, at the present time, to consider them a recognized sponge cell type (e.g. amoebocyte, collencyte) that has some neural specializations."

Lentz (1968) also studied the fine structure of *Sycon* in the electron microscope. In the mesohyl of this species he observed two types of cells, which both were characterized by the presence of a well-developed Golgi apparatus and of numerous membrane-bounded granules: large cells, irregular in shape with several cytoplasmatic processes, and small ellipsoid elements. The membrane-bounded granules in the large cells were 110–170 nm in diameter, whereas those in the small cells were about 80 nm in diameter. In the small elements the granules appeared to be most abundant at the periphery of the cell (Fig. 9D). Granule-containing cells or processes may be situated adjacent to other elements, but synapses or synapse-like structures were not observed. Lentz (1968) considered it likely that the cells just described correspond to those demonstrated with histochemical techniques and are specialized for the production and exocytosis of neurohumors and neurosecretory substance.

Pavans de Ceccatty (1966a, b) studied the sponges *Tethya lyncurium* and *Hippospongia com-*

munis at the ultrastructural level. In the mesohyl of *Tethya* he found numerous irregular, ramifying cells, containing accumulations of microfilaments, together forming a mesynchymal network. According to him, these network cells correspond with the argyrophilic elements previously observed in silver preparations. Several types of connections were observed between the cells of the mesenchymal network, including appositions of adjacent plasma membranes. These connections were sometimes associated with accumulations of dense core vesicles in the cytoplasm of one cell adjacent to the junction (Fig. 10A), accumulations of dense material on the inner sides of opposing membranes, coinciding with small irregular densities bridging the extracellular space (Fig. 10B) and punctual differentiations on one side with local reduction of the width of the extracellular space (Fig. 10C). Endopinacocytes were found to articulate with bouton-like endings of the processes of mesenchyme cells (Fig. 10D).

In *Hippospongia*, Pavans de Ceccatty (1966b) observed that mesenchyme cells often protrude with a cytoplasmatic extension into an indentation on the surface of another mesenchyme cell (Fig. 10E, F). These press-button-like articulations contain dense irregular granules 20–50 nm in diameter and vesicles 60–150 nm in diameter. Desmosomes associated with intracytoplasmatic filaments were observed between endopinacocytes (Fig. 10G).

Pavans de Ceccatty (1974a, b) postulated the presence of the following three types of conduction and coordination pathways in sponges:

(1) Humoral pathways via the extracellular space.
(2) Mobile cellular coordination pathways following the transitory contacts and consecutive exchanges performed by amoeboid mesenchymal cells.
(3) Fixed pathways composed of pinacocytes and mesenchymal network cells. These pathways are particularly strongly developed in thick-walled sponges, like *Tethya* and *Hippospongia* (Fig. 11). The intercellular connections (Fig. 11) are considered to be the functional links in their coordinating system. The cells of the mesenchymal networks are fusiform and show muscle-like differentiation in accord with the high level of contractile activities of these sponges. However, these elements also contain secretory granula, thus combining muscular and nervous specializations. Because of this hybrid character of the mesenchyme cells, the participation of epithelial pinacocytes and its slowness of conduction, the coordinating network (and its constituent elements) of sponges are designated as neuroid rather than neuronal.

In hexactinellid sponges, most cells form a syncytium (Mackie and Singla, 1983). Choanocytes are lacking in these animals; instead, so-called ‘collar bodies’, encompassing flagellum, collar and basal cytoplasm, are produced as outgrowths of choanoblasts. The collar bodies remain connected to their mother cells; however, in the junctional zones, highly remarkable perforate septa or junctional plugs are formed (Mackie and Singla, 1983; Fig. 12A). These complex disc-like structures, which are probably formed by the Golgi apparatus, show 6.0–8.0 nm pores. Many show a trilaminar structure (Fig. 12B) and membrane-bounded saccules have been seen embedded in some of them (Fig. 12C). Similar plugs have been observed by Grell and Benwitz (1974) between the contractile ‘fibre cells’ of the placozoan *Trichoplax adhaerens* (Fig. 8C, D).

Also in hexactinellid sponges, the bulk of the cellular material is formed by a syncytial trabecular tissue in which the somata of cells of several different cell types, including archaeocytes and choanoblasts, are embedded. Plugged junctions are not confined to the bridges between choanoblasts and collar bodies, but also occur throughout this trabecular syncytium. The plugs are thought to form a filter or partial barrier limiting the translocation of materials between differentially specialized portions of the syncytial cytoplasm (Mackie and Singla, 1983; Fig. 12A).

It has already been mentioned that in hexactinellial sponges, flagellar activity, and consequently the feeding water-current can be arrested by local mechanical or electrical stimulation. Mackie et al. (1983) considered it likely that the trabecular syncytium forms the conducting system by which this response is coordinated. The nature of the propagation event in hexactinellids is

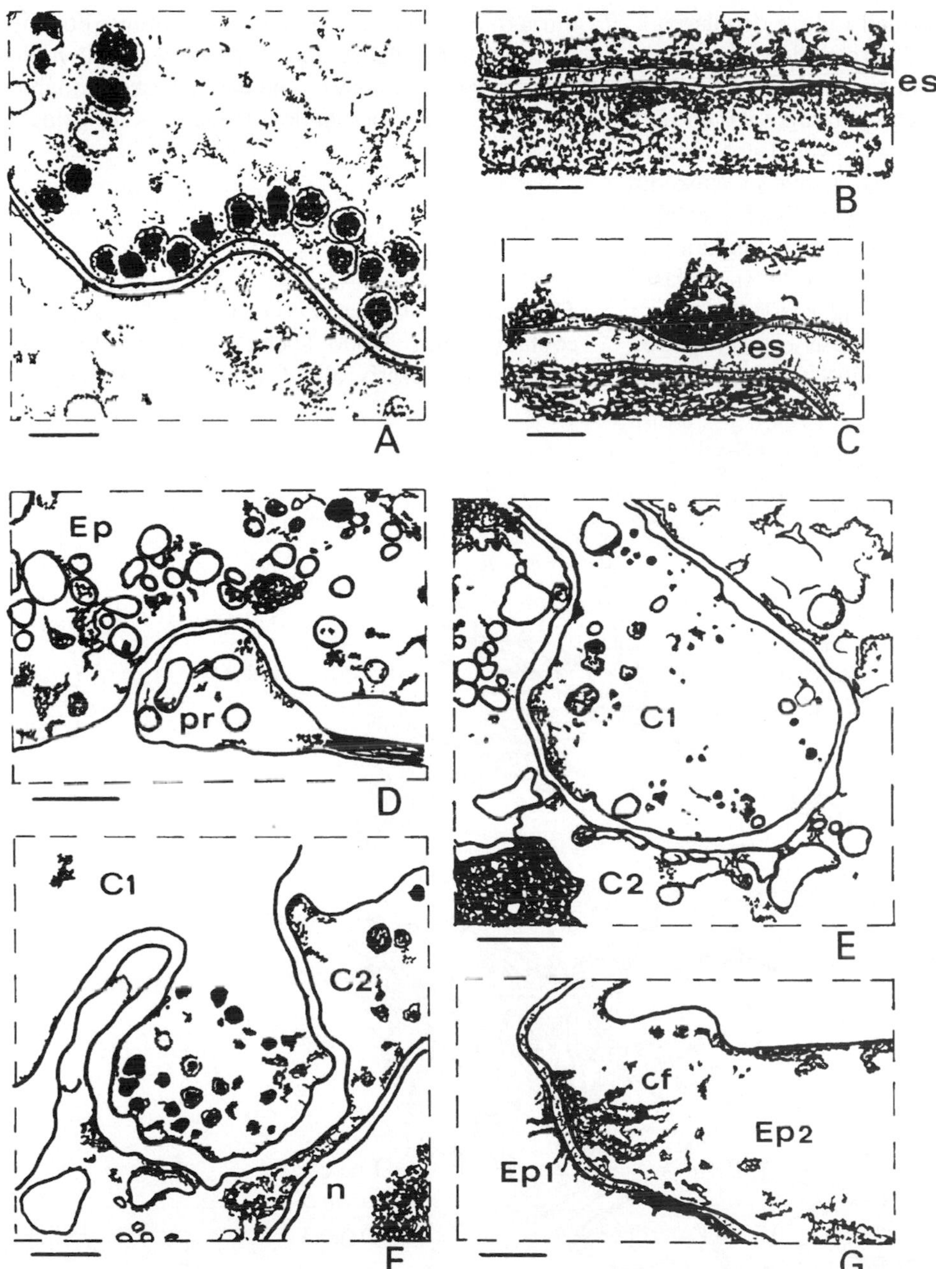

Fig. 10. Possible functional contacts between cellular elements in the mesenchyme of the sponges *Tethya lyncurium* (A, B, C, D) and *Hippospongia communis* (E, F, G), drawn from electron micrographs published by Pavans de Ceccatty (1966a: A-D; 1966b: D-G). A. Accumulation of dense-core vesicles in a mesenchyme cell close to its contact with another mesenchyme cell; bar: 0.2 μm. B. Intercellular contact zone with densities adjacent to both membranes. There are irregular bridges in the extracellular space: es; bar: 0.1 μm. Punctate differentiation on one side with local reduction of the extracellular space: es; bar: 0.1 μm. D. Bouton-like ending of a process of a mesenchyme cell (pr), making contact with an endopinacocyte (Ep); bar: 0.5 μm. E. Polarized junction between two mesenchyme cells, C1 and C2; bar: 0.5 μm. F. Polarized junction between two mesenchyme cells, C1 and C2, close to a nucleus, *n*. The process formed by C1 contains numerous granules bar: 0.2 μm. G. Desmosome between two endopinacocytes, Ep1 and Ep2, with perpendicular cytoplasmatic filaments, cf; bar: 0.2 μm.

not known, but if it is assumed to be a depolarization of the cell membrane, current flow along the trabecular syncytium would be unimpeded by any membrane barriers. The plugged bridges would not act as such, because their pores would not hinder the movement of small ions acting as current carriers (Mackie and Singla, 1983; Mackie et al., 1983). As far as known, in the animal kingdom a syncytial organization is confined to the glass sponges (and, possibly, the placozoans). This unique organization enables these organisms to conduct behaviourally meaningful signals rapidly through their tissue. The junctional plugs probably represent filtering devices that allow small molecules to pass, but restrict the passage of most larger molecules and organelles (Mackie and Singla, 1983).

4. Cnidaria

The phylum Cnidaria (Formerly: Coelenterata) contains some 9,000 species, all of which are aquatic, and nearly all of which are marine. The

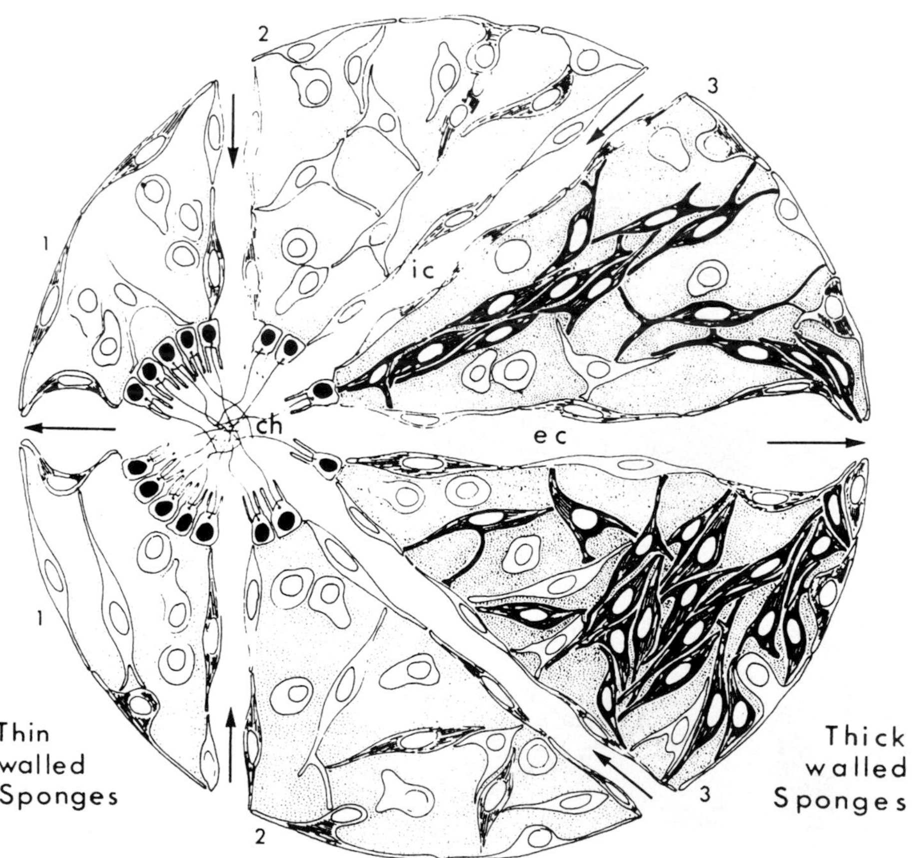

Fig. 11. Diagrammatic representation of hypothesized conduction pathways in sponges; the organism is in direct contact with water by its exopinacocytes on the external surface, its endopinacocytes covering the incurrent (ic) and excurrent (ec) canals, and by the central choanocytes (ch). These cells enclose the mesohyl, consisting of amoeboid and fixed cells embedded in an extracellular matrix. Three types of sponges are depicted: (1) thin-walled with slightly developed mesohyl, (2) those with moderately developed mesohyl, and (3) those with thick walls and strongly developed mesohyl cells. Three types of conduction and coordination pathways are supposed to be present: (1) humoral extracellular pathways, (2) mobile cellular pathways following transitory contacts of wandering cells, and (3) permanent fixed pathways via networks and bundles of contractile cells (black). The pinacoderms are supposed to participate in pathways of all three types. (Reproduced from Pavans de Ceccatty, 1974a).

four classes of Cnidaria are the Hydrozoa, including the fresh-water hydras (Fig. 13A), the Anthozoa, containing the sea anemones (Fig. 13B), the Scyphozoa or true jelly fishes and the Cubozoa or box jelly fishes; all cnidarians are radially symmetrical. Their body walls surround a sac-like digestive cavity, the coelenteron or gastrocoel, opening only through the mouth; tentacles typically surround the latter structure (Fig. 13). In all cnidarians the tentacles and the perioral region are replete with stinging cells, called cnidoblasts, each containing a nematocyst, i.e. a capsule including a spirally coiled hollow thread. Upon stimulation the nematocyst explodes and the thread is forcibly turned inside out, thus becoming transformed into a sting that pierces the body of prey or aggressor and injects a poisonous substance contained in the capsule.

Cnidarians exhibit two body forms, polypoid and medusoid. The polypoid form is sessile and tends to be cylindrical, with mouth and tentacles facing upward. The medusoid form or jelly fish, is disk- or bell-shaped and swims free, mouth downward with tentacles trailing.

Cnidarians are composed of an outer epidermis and an inner gastrodermis; these two cell layers are separated from each other by a fibrous mesogloea (Fig. 14A). The ectoderm contains cells of various types, including the cnidoblasts already discussed, interstitial cells, epitheliomuscular cells and sensory cells. Interstitial cells are undifferentiated and capable of differentiating into other cell types, including neurons (Lentz, 1965a).

The epitheliomuscular cells are provided with contractile processes, arising perpendicularly from the bases of their somata. These processes are longitudinally arranged and form a continuous muscle sheet, which is attached to the surface of the mesogloea. Epitheliomuscular cells prevail in hydroid polyps; in other cnidarians separate muscle cells occur at the bases of the epithelial cells (Lentz, 1968).

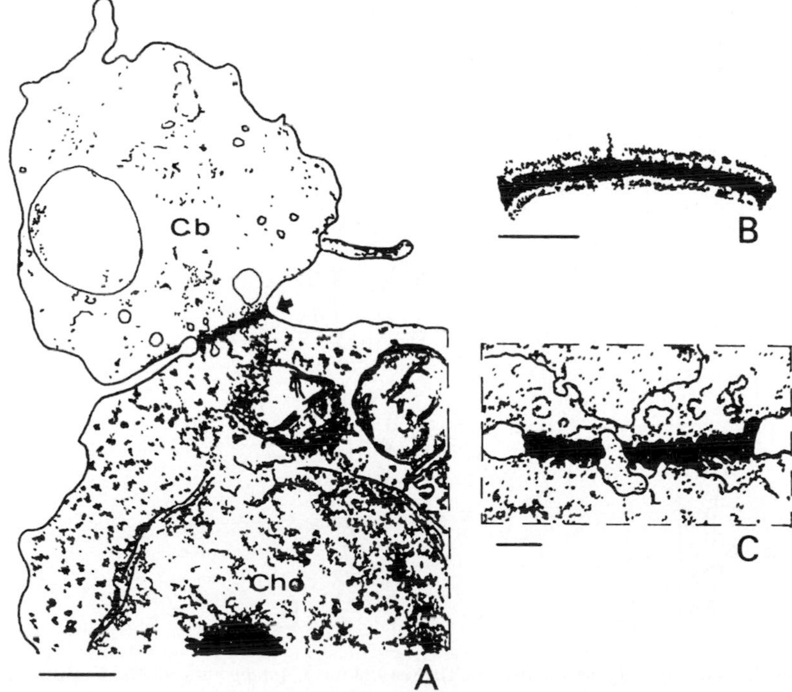

Fig. 12. Perforate septal partitions or plugs in cytoplasmatic bridges in the hexactinellid sponge *Rhabdocalyptus dawsoni*, drawn from electron micrographs, published by Mackie and Singla (1983). A. Plug (arrow) separating a collar body (Cb) from its choanoblast (Cho). Note that the cytoplasm on the two sides of the plug differs markedly; bar: 0.5 μm. B. Plug, showing trilaminar structure; bar: 0.1 μm. C. Plug traversed by a membraneous saccule; bar: 0.1 μm.

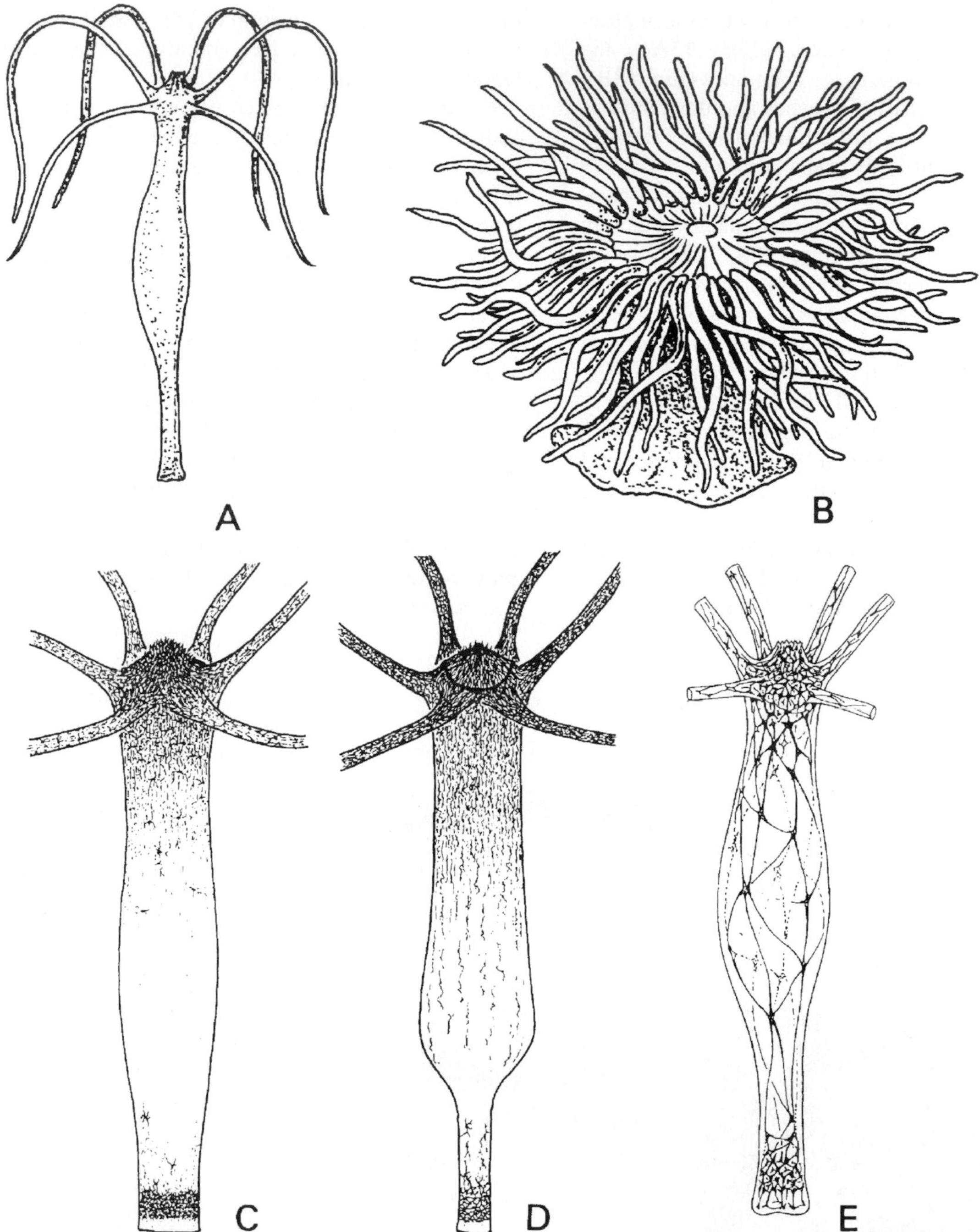

Fig. 13. Aspects of the morphology of cnidarians. A. The polyp, *Hydra vulgaris*. B. The sea anemone, *Annemonia sulcata*. C. Drawing of the nervous system of *Hydra vulgaris* stained with an antiserum against the sequence RFamide. There are two central condensations of neurons, one around the mouth opening and one around the peduncle (from Grimmelikhuijzen, 1985). D. A similar drawing showing the nervous system of *Hydra oligactis*. This species has a dense plexus of immunoreactive processes in the region between the mouth opening and the base of the tentacles (hypostome). The plexus of neurons in the peduncle is less well developed (from Grimmelikhuijzen, 1985). E. Semidiagrammatic representation of the organization of the nervous system in a hydroid polyp.

The slender, elongated sensory cells are particularly numerous on the tentacles and in the foot; they have an apical specialization consisting of one to five sensory hairs. One or two axonal processes arise from the basal side of these cells; these fibres join the tangentially oriented plexus at the bases of the epithelial cells (Fig. 14A). Westfall and Kinnamon (1978) studied the ultrastructure and the synaptic contacts of the sensory cells in the tentacles of *Hydra* (Fig. 14B). They found that the axons of these elements show occasional varicosities containing accumulations of dense-cored vesicles (Fig. 14B: dcv). The sensory cells form synaptic contacts with ganglion cells, cnidoblasts and epitheliomuscular cells; a single sensory cell may innervate one or more epitheliomuscular cells. Remarkably, the synapses with the cnidocytes appeared to differ from those with ganglion cells and epitheliomuscular cells, by always having clear vesicles rather than dense-cored vesicles at the junction.

Because the sensory cells just discussed are provided with axonal processes, according to the present author they should be designated as neurosensory cells. Lentz (1968; Fig. 14A: NsC), however, used this term exclusively for deeply situated elements, having a cilium extending to the surface of the epidermis.

The entoderm or gastroderm is composed of columnar digestive and gland cells, and occasional neurosensory cells (Fig. 14A); the contractile processes of the epitheliomuscular cells are oriented circularly rather than longitudinally in this inner layer.

The mesogloea shows considerable variations in thickness; in *Hydra* it is a thin lamella (Fig. 14A), but in medusae it is coarse and comprises the bulk of the animal.

The cnidarians are commonly designated as the most simply organized or the most primitive animals possessing a true nervous system (Hanström, 1928; Ariëns Kappers, 1929; Bullock and Horridge, 1965). This nervous system consists of ganglion cells that, with their processes, form a continuous network or plexus at the base of the ectodermal and gastrodermal epithelia (Fig. 14A). In the older literature it is often stated that the cnidarian nervous system is diffusely organized, but the observation of Lentz (1968), that in Hydra, ganglion cells are concentrated in the proximal fourth of the tentacles, hypostome and base, has been fully confirmed by the immunohistochemical studies of Grimmelikhuijzen (1985). Using antisera to the neuropeptide sequence Arg-Phe-amide (RFamide), this author visualized the ectodermal nervous system in whole mounts of several Hydra species. *Hydra vulgaris* appeared to have a strong agglomeration of sensory neurons and processes around the mouth opening and a densely packed collar of neurons in the peduncle (Fig. 13C). In *Hydra oligactis*, moreover, a conspicuous nerve ring lying at the border of the hypostome and tentacles was found (Fig. 13D). Comparable concentrations of neuronal tissue have been found in representatives of the other cnidarian groups (Grimmelikhuijzen and Westfall, 1995). An adult hydra has 120,000–130,000 cells among which are some 6,000 neurons and neurosensory cells (Bode et al., 1973).

The ganglion cells of the cnidarian nervous system are spindle-shaped, tri- or multipolar, and those situated beyond the neural concentrations are provided with long, straight processes (Fig. 13E) called neurites; many of these neurites contain bulbous enlargements along their length. Westfall (1973) studied the ganglion cells in the nervous system of *Hydra littorialis*, using electron microscopy; she found that, except for their basal location, these elements closely resemble the sensory cells (Fig. 14B, D). Both types of cells have a cilium, store dense-cored vesicles at non-synaptic regions (and, hence, are 'neurosecretory' and/or involved in volume transmission) and form chemical synapses with both effector cells (cnidoblasts and epitheliomuscular cells) and other ganglion cells. Westfall proposed that these primitive, apparently multifunctional nerve cells may resemble the primitive stem cells that evolutionarily gave rise to the sensory cells, motor neurons, interneurons, and neurosecretory cells of higher animals. However, it seems unlikely that all cnidarian neurons have these multifunctional characteristics. Thus, it has been established that in the sea anemone *Calliactis parasitica* many neurons lack the cilium observed in the ganglion cells of Hydra (Grimmelikhuijzen and Westfall, 1995). Saripalli and Westfall (1996) isolated nerve cells from the tentacles of the same sea anemone species; they were able to distinguish 33 different forms that could be allocated to several types of neurosensory cells and ganglion cells.

At the beginning of the 20th century, it was generally believed that the cellular elements forming the nerve net in cnidarians show protoplasmatic continuity, i.e. form a syncytium. Towards the end of the exclusively light microscopic era, however, the concept prevailed that this net is composed of separate individual cells whose processes do not fuse (Pantin, 1952; Bullock and Horridge, 1965). Electron microscopical investigations have confirmed that cnidarian neurons are generally not syncytial, and have shown the presence of neuro-neuronal as well as neuro-effector synapses (Horridge and Mackay, 1962; Westfall et al., 1971, 1997, 1998, 1999; Peteya, 1973; Westfall, 1973, 1987; Westfall and Kinnamon, 1978; Quagglia and Grasso, 1986; Westfall and Savyar, 1997). These synapses are characterized by the presence of a pair of dense parallel membranes, with a uniformly wide intercellular cleft containing granular or filamentous intracleft material, and a presynaptic aggregation of synaptic vesicles (Westfall, 1987; Fig. 15). Most of these synapses are of the en-passant type, and in many, though not all, the vesicles have dense cores. The synapses are generally small in area and the number of vesicles per synapse is less than in other phyla. Two-way, or symmetrical synapses, showing accumulations of vesicles on both sides of the synaptic cleft, are common, particularly in jelly fish, and it has been shown that these junctions transmit excitation in both directions, with the same synaptic delay (Anderson, 1985). Peteya (1973) observed remarkable multipolarized neuro-neuronal synapse complexes, composed of three or four morphologically polarized zones, in the sea anemone *Cereantheopsis americanus* (Fig. 15B). In many synapses some vesicles are connected with the presynaptic membrane by dense material, but to my knowledge exocytoses have not been observed in cnidarians so far. Most neuromuscular synapses contain dense-cored vesicles, just like the neuro-neuronal contacts (Westfall et al., 1973; Westfall and Savyar, 1997; Fig. 15F), and it has been demonstrated that in some the dense cores contain FRamide-like peptides (Koizuma et al., 1989; Westfall and Grimmelikhuijzen, 1993). Most neuro-nematocyte synapses have clear vesicles (Fig. 14B, D: NC, N^D, N^S). Using electron microscopy of serially sectioned tentacles from the sea anemone *Aiptasia pallida*, Westfall et al. (1999) located and characterized two types of neuro-spirocyte synapses. Clear vesicles were observed at 10 synapses and dense-cored vesicles at five synapses. One sequential pair of synapses appeared to include a neuro-neuronal synapse with dense-cored vesicles and a neuro-spirocyte synapse with clear vesicles; a second synapse on the same cell had dense-cored vesicles (Fig. 14C). The occurrence of either clear or dense-cored vesicles at neuro-spirocyte synapses suggest that at least two types of neurotransmitter substances control the discharge of spirocysts in sea anemones (Westfall et al., 1999).

Peteya (1973) observed that in the sea anemone, *Ceriantheopsis americanus*, the extracellular space within the nerve plexus is large and irregular

Fig. 14. Aspects of the histology and cytology of cnidarians. A. Semidiagrammatic section through the wall of *Hydra*; sensory cells, neurosensory cells and ganglion cells in black (from Lentz, 1968). B. Schematic drawing showing ultrastructural features and types of synaptic contacts of a sensory cell from the tentacle of *Hydra* (from Westfall and Kinnamon, 1978). C. Similar drawing of a spirocyte of the sea anemone, *Aiptasia pallida*, showing sequential neuro-neuro-spirocyte synapses with neuro-neuronal dense-cored and neuro-spirocyte clear vesicles plus a second neuro-spirocyte synapse with dense-cored vesicles (from Westfall et al., 1999). D. Similar drawing of a ganglion cell from the tentacle of Hydra, indicating various cytoplasmic features and types of synaptic contacts with other cells (from Westfall, 1973). A. Polarized synapse with axon; Cd, cnidoblast; c, sensory cilium; ce, centrioles; cv, coated vesicles; DC, digestive cell; dcv, dense-cored vesicles; EMC, EMc, epitheliomuscular cell; es, epitheliomuscular cell sheath; GC, polarized synapse with ganglion cell; Gc, ganglion cell; GlC, gland cell; g, Golgi complex; gl, glycogen; IC, interstitial cell; l, lipid droplet; Me, mesogloea; MP, muscular processes; m, mitochondrium, mf, myelin figure; mt, microtubules; mv, microvilli; mfb, multivesicular body; NC, N^D, N^S, polarized synapses with different types of nematocytes; NE, two-way somatic neuro-neuronal synapse; NsC, neurosensory cell; n, nucleus; pg, pigment granules; r, rootlets; rer, rough endoplasmatic reticulum; ri, ribosomes; SC, sensory cell; sj, septate junction; st, stereocilium; v, vacuole.

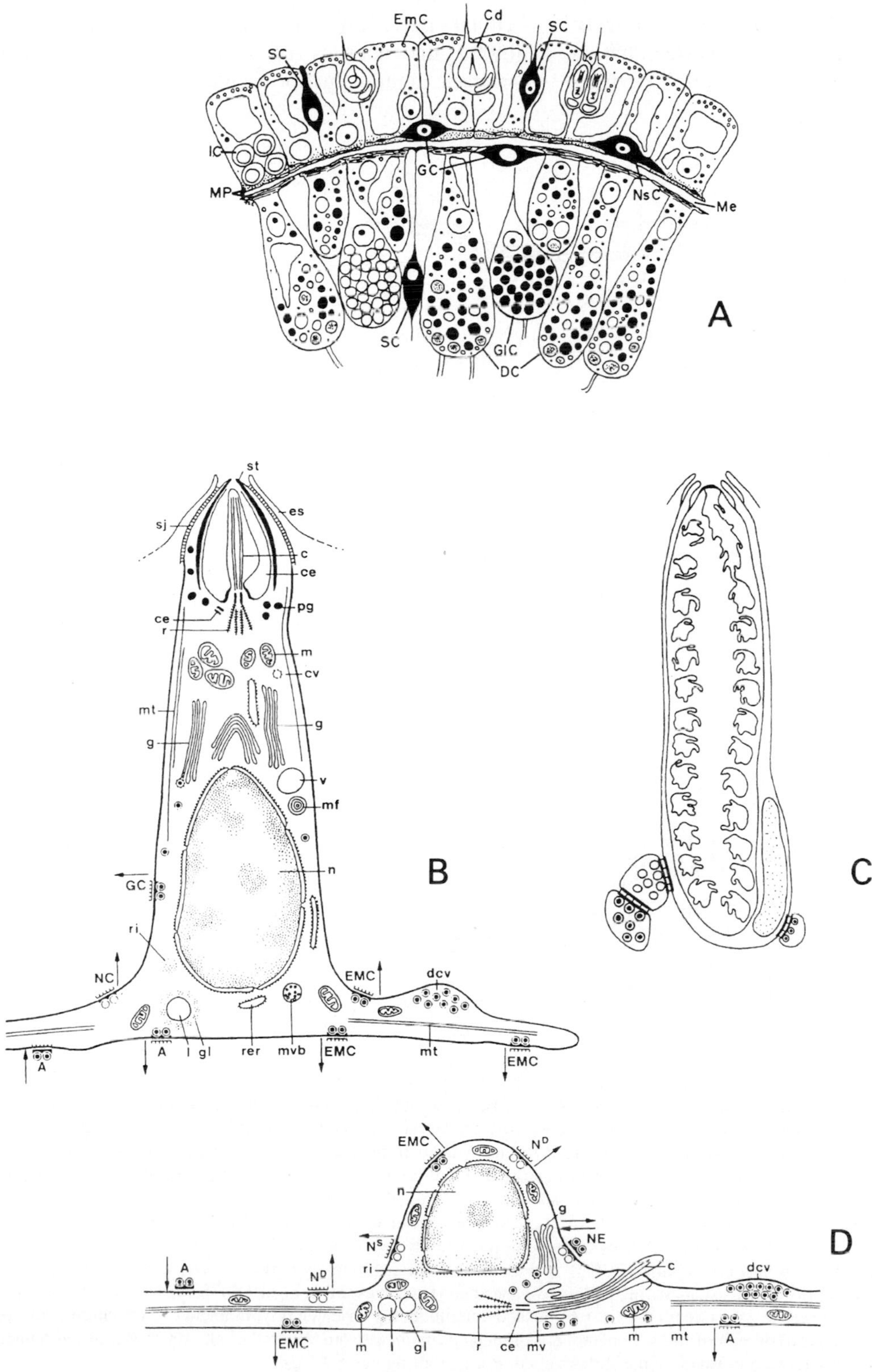
EmC
Cd
SC
SC
IC
GC
MP
NsC
Me
A
SC
GlC
DC
st
es
sj
c
ce
ce
pg
r
m
cv
mt
g
g
v
mf
n
B
GC
ri
NC
EMC
dcv
A
A
l
gl
rer
mvb
EMC
mt
EMC
C
EMC
N^D
n
g
NE
N^S
D
A
N^D
ri
c
dcv
EMC
m
l
gl
r
ce
mv
m
mt
A

ranging in width from 20 nm to a few microns at non-synaptic sites. Lentz (1968) reported that in hydra many neurites terminate freely in the extracellular space. It has already been mentioned that in hydra the neurites of neurosensory cells and neurons contain fusiform enlargements filled with

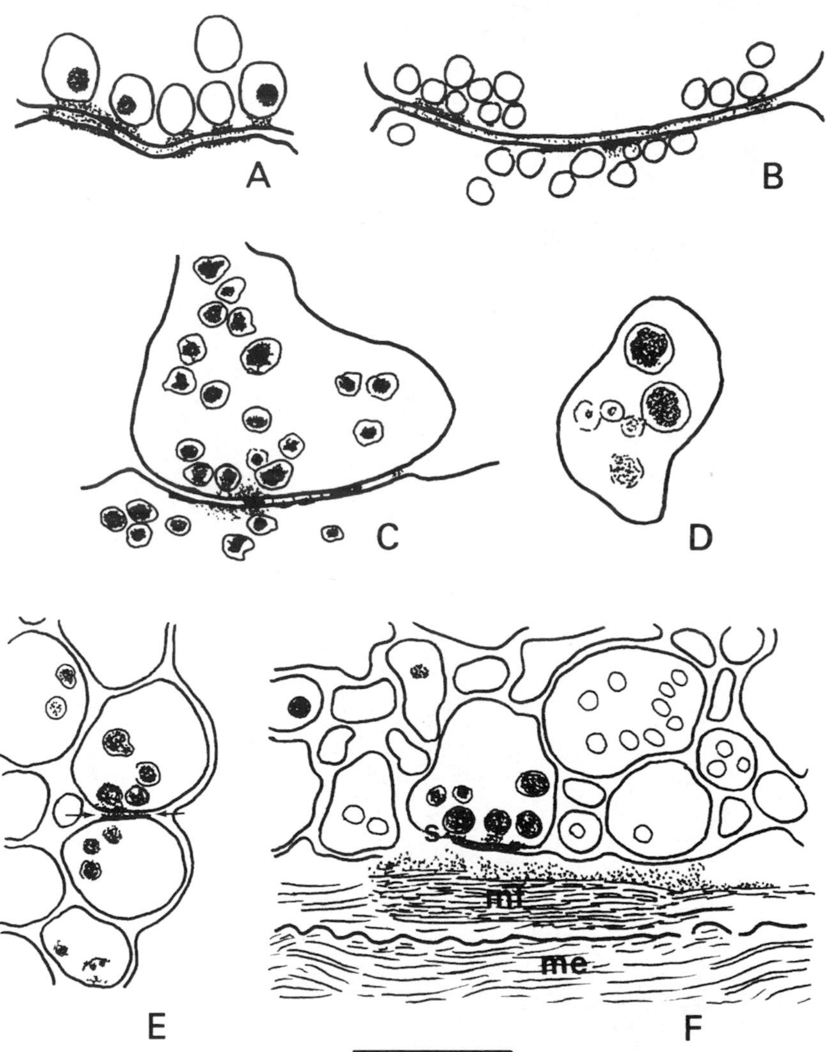

Fig. 15. Vesicle-containing profiles and synapses in cnidarians. A. Synapse between two neurites in a tentacle of Hydra. Serial sections revealed that vesicles that appear clear also contain a dense core; five vesicles are attached to the fibrous coat on the presynaptic membrane (from Westfall et al., 1971). B. Multipolarized neuro-neuronal synapse in the column of the sea anemone, *Cereantheopsis americanus* (from Peteya, 1973). C. Bidirectional synapse between the terminal bouton of a nerve fibre and the soma of a neuron in the sea anemone, *Actinia equina mediterranea* (from Quagglia and Grasso, 1986). D. Profile in the tentacle nerve plexus of the sea anemone, *Antopleura elegantissima*, containing two large dense-cored vesicles. Incubation with an antiserum against Antho-RFamide followed by immunogold labeling revealed that the Antho-RFamide immunoreactivity is specifically located in the granular cores of the large vesicles (from Westfall and Grimmelikhuijzen,1993). E. Polarized neuro-neuronal synapse (arrows) with two dense-cored vesicles in the tentacle gastrodermis of the sea anemone, *Calliactis parasitica* (from Westfall et al., 1997). F. Neuromuscular synapse (s), with granular vesicles, between a cross section of a neurite and longitudinally oriented myofilaments (mf) adjacent to the mesogloea (mg) in the tentacle gastrodermis of the sea anemone, *Calliactis parasitica* (from Westfall et al., 1997). All of the figures were drawn from electron micrographs published by the authors cited. Bar (for all figures): 0.5 μm.

dense-cored vesicles (Fig. 14B, D: dcv). Similar aggregations of dense-cored vesicles have been observed in serially traced nerve processes of sea anemones (Westfall and Grimmelikhuijzen, 1993; Fig. 15D). These accumulations of dense-cored vesicles are assumed to be associated with non-synaptic or paracrine release sites (Grimmelikhuijzen and Westfall, 1995).

Gap junctions – i.e. pairs of parallel closely apposed membranes separated by a 2–3-nm-wide cleft, representing sites of electric communication – have been observed between neurons in the hypostome, peduncle and basal disc of Hydra (Westfall et al., 1980; Kinnamon and Westfall, 1982). The marginal nerve rings of hydrozoan medusae consist of electrically coupled neurons, which function as single large elements (Spencer and Arkett, 1984).

The data just reviewed suggest that all forms of inter-neuronal communication observed in the animal kingdom are represented in cnidarians.

As regards neurochemistry, Lentz (1968) mentioned that the following enzymes and neurohumors have been identified using histochemical techniques in the nervous system of hydra: acetyl cholinesterase, monoamine oxydase, noradrenaline, adrenaline and serotonin. Neurosecretory cells were identified in cnidarians by staining with various techniques, including Gabe's paraldehyde fuchsin method and the periodic acid Schiff reaction. The electron microscopical studies in hydra by Lentz (1965b) have shown that, during regeneration, neurosecretory granules are released from nerve terminal. The same author (Lentz, 1965c) also demonstrated that the neurosecretory substances thus released contain form-regulating substances. Schaller and Bodenmüller (1987) isolated four growth- and differentiation-controlling substances from hydra tissue: an activator and an inhibitor of head formation and an activator and an inhibitor of foot formation. All four of these substances appeared to be produced by nerve cells, and the two activators were shown to be peptides. Because a vascular system is lacking in cnidarians, neurohormones can attain their targets in these animals only via the extracellular space.

Due to the immunohistochemical and biochemical studies by Grimmelikhuijzen and collaborators we know that the nervous systems of cnidarians produce large amounts of several neuropeptides (for reviews see Grimmelikhuijzen and Westfall, 1995; Grimmelikhuijzen et al., 1996).

In the nervous system of hydra, neurons which react with antisera to cholecystokinin, substance P, neurotensin, bombesin, oxytocin/vasopressin and FMRFamide-like peptides have been found (Grimmelikhuijzen et al., 1982; Grimmelikhuijzen, 1985). The neurons displaying these immunoreactivities appeared to have their own characteristic distribution patterns (Figs. 13C, D, 16). However, in many regions of hydra more than one type of peptidergic neuron occurs, and it has been shown that oxytocin-like immunoreactivity coexists with bombesin-like immunoreactivity in neurons in several regions, including the tentacles and the

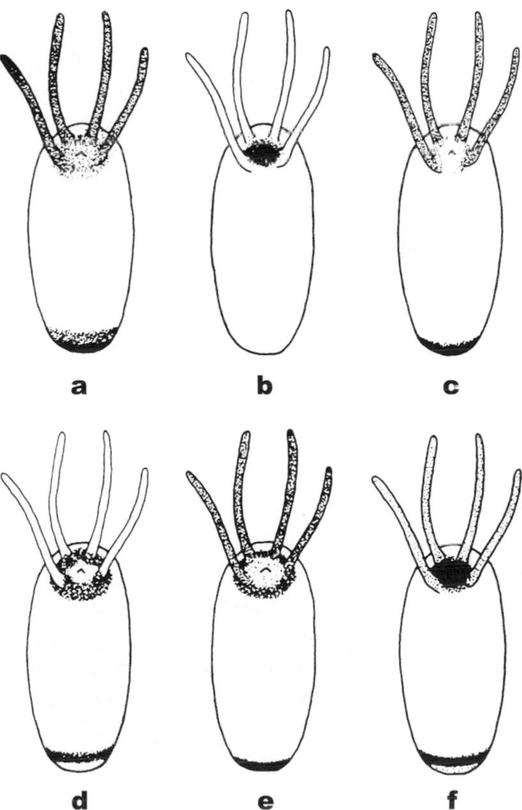

Fig. 16. Schematic drawings showing the locations of ganglion cells in *Hydra*, containing (a) oxytocin/vasopressin-, (b) cholecystokinin-, (c) substance P-, (d) neurotensin-, (e) bombesin-and (f) FMRFamide-like immunoreactivity (from Grimmelikhuijzen et al., 1982).

basal disk. On the other hand, oxytocin-like immunoreactivity never coexists with FMRFamide-immunoreactivity, although the neurons showing these immunoreactivities are often found to be closely intermingled (Grimmelikhuijzen, 1983).

From the sea anemone *Anthopleura elegantissima*, 17 different neuropeptides were isolated, and these peptides appeared to be located in at least six distinct sets of neurons (Grimmelikhuijzen and Westfall, 1995; Grimmelikhuijzen et al., 1996). So, with respect to their peptide content, the neurons in this sea anemone show a remarkable differentiation; the biosynthesis of the cnidarian neuropeptides has also been studied. Most recently, Mitgutsch et al. (1999) found that *Hydra magnipapillata* has three distinct genes coding for preprohormones A, B and C, each yielding a characteristic set of Hydra-RFamide (Arg-Phe-NH_2) neuropeptides, and a fourth gene coding for a preprohormone that yields various Hydra-LWamide (Leu-Trp-NH_2) neuropeptides. They demonstrated that each of these four genes is specifically expressed in a different subset of neurons in the ectoderm of adult Hydra. Reinscheid and Grimmelikhuijzen (1994) found in the sea anemone *Renilla köllikeri* an RFamide neuropeptide precursor protein containing no less than 36 copies of Antho-RFamide.

With the aid of immunogold electron microscopy together with antisera to RFamide carboxyterminal peptide fragments, it has been demonstrated that in cnidarians, peptides of this type are specifically located in the granular cores of neuronal dense-cored vesicles. Thus, Koizuma et al. (1989), studying ganglion cells in the foot of *Hydra littoralis* found that immunoreactive dense-cored vesicles were present in the soma, long neurites and terminals of these neurons. An axon terminal ending on the muscular base of an epitheliomuscular cell contained an accumulation of labeled dense-cored vesicles, suggesting a possible function of RFamide-like peptides in neuromuscular transmission. Singla and Mackie (1991) demonstrated FMRFamide-like immunoreactivity in neuronal dense-cored vesicles in the hydromedusa *Aglantha digitale*, but they emphasized the absence of this immunoreactivity in small, clear vesicles at synapses. Antho-RFamide is a neuropeptide isolated from the sea anemone *Anthopleura elegantissima*; using antisera against this neuropeptide, Westfall and Grimmelikhuijzen (1993) studied the nerve plexus in the tentacles of *Anthopleura*. They found immunoreactivity in neuronal dense-cored vesicles that were associated with two-way neuro-neuronal synapses and also with non-synaptic release sites (Fig. 15D).

The various neuropeptides isolated from *Anthopleura* have been shown to be biologically active and to exert characteristic inhibitory or excitatory actions on ectodermal or entodermal muscle preparations from sea anemones. Some of these peptides appeared to act directly on isolated muscle cells and are, hence, to be considered as neuromuscular transmitters (cf. Grimmelikhuijzen and Westfall, 1995 and Grimmelikhuijzen et al., 1996 for references and details).

From the data just summarized, Grimmelikhuijzen and his associates draw the conclusion that neuropeptides play a crucial role in signal transmission in the cnidarian nervous system. Because they remained unable to demonstrate acetylcholine, catecholamines or serotonin in neurons of Hydra (Grimmelikhuijzen, 1986; Grimmelikhuijzen and Westfall, 1995), they claimed that these substances do not generally occur in cnidarians and, hence, that ‘classical’ transmitters are not the evolutionary ‘oldest’ neurotransmitters (Grimmelikhuijzen and Westfall, 1995, p. 12). Rather, they believed that “evolutionary ‘old’ nervous systems use peptides as transmitters” (Grimmelikhuijzen and Westfall, 1995, p. 7). Since peptides with the C-terminal sequence Gly-Arg-Phe-NH_2 have been shown to be ubiquitous in the Cnidaria, Grimmelikhuijzen and Westfall (1995) considered these types of peptides as good candidates for being the first transmitters.

Some data cast doubt on the peptide saga just summarised. Mackie (1990) stated that there is abundant evidence for the presence of catecholamines and serotonin in cnidarian tissues and for their roles in modulating behaviour, but that these substances have not yet been shown to function at the level of the individual synapse. Although the transmitter picture is still quite incomplete, Mackie (1990) considered it highly probable that both aminergic and peptidergic transmitter systems occur in cnidarians. Anderson (1985), who studied the physiology of bidirectional chemical synapses in the hydrozoan *Cyanea* and the scyphozoan *Aglantha*, pointed out that the extremely short synaptic delays found in these synapses strongly suggest the action of 'fast' classical transmitters. It is also noteworthy that most of the – supposedly fast-acting – synapses made by neurosensory cells and neurons with nematocysts and related cells, contain small, clear vesicles (Westfall, 1973; Westfall and Kinnamon, 1978; Westfall et al., 1998, 1999; Fig. 14B–D). Most recently, immunohistochemical evidence for the presence of noradrenaline, adrenalin and serotonin in ganglion cells of sea anemones has been presented (Elliott et al., 1998; Westfall et al., 1999). It may be concluded that in cnidarians both peptidergic and aminergic transmitter systems occur, and that there is a vacancy for at least one fast, excitatory ionotropic neurotransmitter.

5. Platyhelminthes

The Platyhelminthes or flatworms are bilaterally symmetrical, dorsoventrally flattened leaf- or ribbon-shaped animals (Fig. 17). Contrary to the cnidarians, the flatworms have a distinct middle layer of mesodermal tissue, which fills the region between body wall and digestive tract; the digestive tract has only one opening; a vascular system and endocrine glands are lacking.

The Platyhelminthes are usually subdivided into three classes: Turbellaria, Trematoda and Cestoda. The turbellarians are free living, ciliated animals; many are freshwater, some are marine and a few dwell in damp places on land. This large and form-

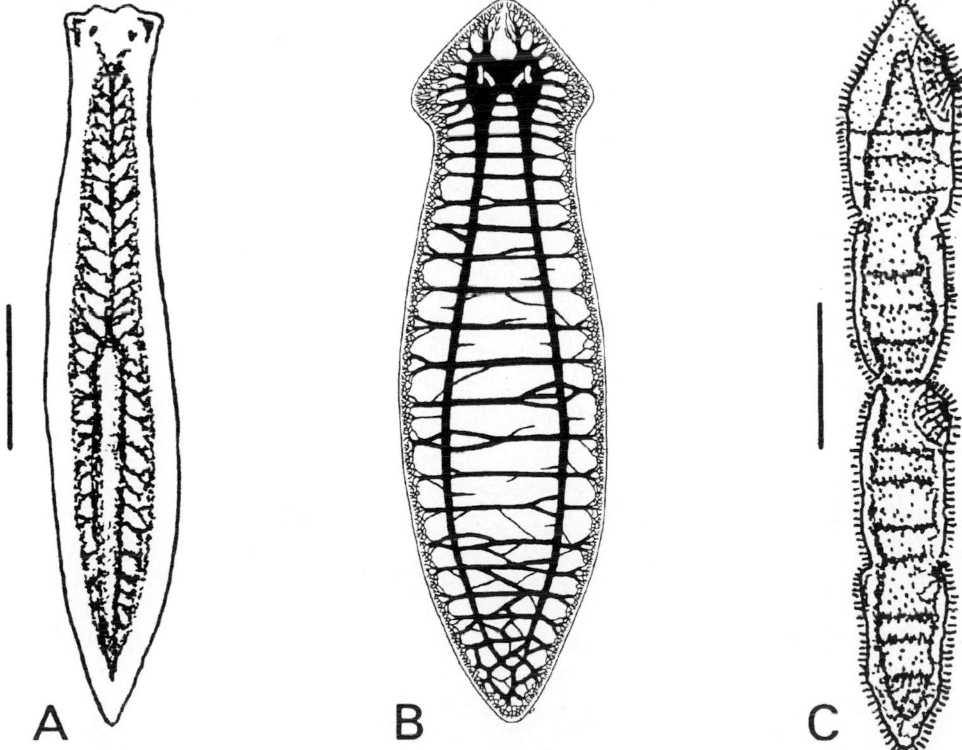

Fig. 17. Platyhelminthes (flatworms). A. The turbellarian, *Dendrocoelum lacteum*. B. The nervous system of a turbellarian (reproduced from Lentz, 1968). C. Another turbellarian, *microstomum spec.*, in division (redrawn from Storch and Welsch, 1997).

rich class encompasses eight orders, among which the much discussed Acoela, which lack a digestive tract (see below), the Seriata, to which the genera *Dendrocoelum* (Fig. 17A), *Planaria* and *Bdelloura* (see Fig. 18) belong, and the Macrostomida (e.g. *Microstomum*: Fig. 17C).

The Trematoda or flukes and the Cestoda or tapeworms are all parasitic.

The turbellarian flatworms have generally been considered to stand at the base of all Bilateria (Bogitsch and Harrison, 1991); recently, the monophyletic or polyphyletic status of the Phylum Platyhelminthes has been under discussion. According to Carranza et al. (1997) and Ruiz-Trillo et al. (1999) sequence data of 18S ribosomal DNA genes of acoel flatworms and other metazoans indicate that the acoels do not belong to the Platyhelminthes, but rather represent the extant members of the earliest divergent Bilateria. The following features of their nervous system also suggest a deep gap between the Acoela and the other Platyhelminthes. (1) Acoels have only a single, intraepithelial nerve plexus, whereas in other flatworms at least two plexuses are present; this intraepithelial plexus is considered to be directly comparable to the nervous system of cnidarians (Rieger et al., 1991). (2) Their very small brain ganglia lack a central neuropil and are largely occupied by a group of FMRF-immunoreactive neurons, which form a remarkable gland-like structure (Reuter et al., 1998a; Raikova et al., 1998). Such a gland-like structure is not observed in other flatworms. (3) Acoels have three to five pairs of longitudinal nerve cords, connected by an irregular network of transverse fibres; a regular orthogon pattern, so characteristic for other flatworms (Figs. 17B, 18) could not be detected (Raikova et al., 1998).

The nervous system of flatworms comprises: (1) peripheral plexuses, i.e. meshworks of nerve fibres and neurons interconnected with the central nervous system, (2) one or several pairs of longitudinal nerve cords, generally interconnected by transverse commissures, in a ladder-like configuration, and (3)

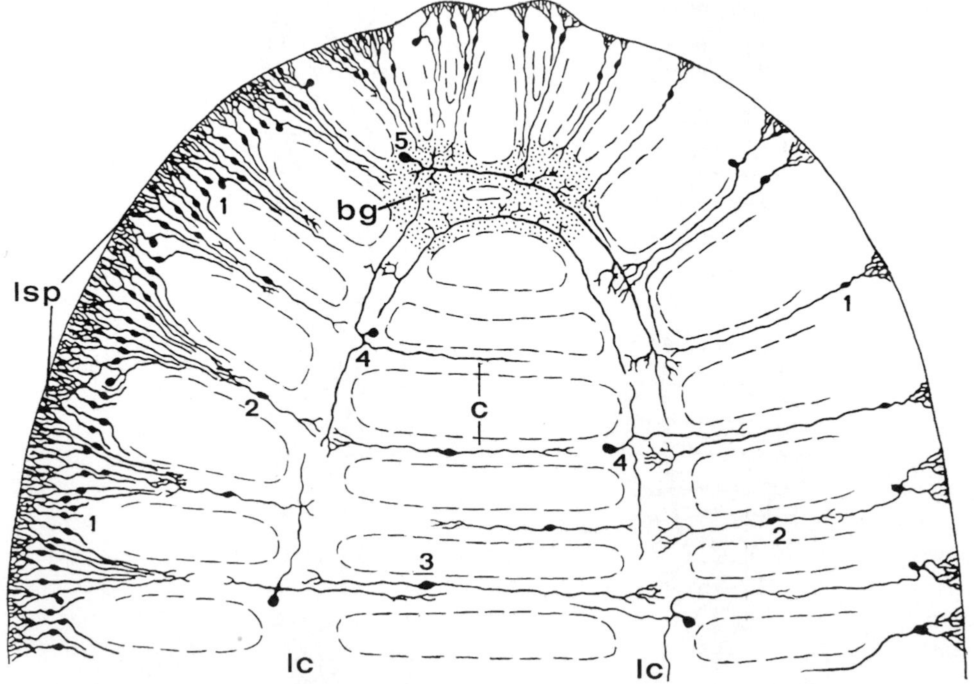

Fig. 18. Composite drawing of neurons observed in Golgi-preparations of the rostral part of the turbellarian, *Bdelloura candida*. The outline of the nervous system is indicated with thin dashed lines. Modified from Hanström (1928). bg, brain ganglion; c, commissures; lc, longitudinal nerve cords; lsp, lateral sensory plexus; 1, 2 etc. cell types discussed in the text.

a brain, consisting of a pair of cephalic ganglia (Figs. 17B, 18). The presence of a true brain is one of the five important innovative features that make their first appearance in the Phylum Platyhelminthes (the other four are: bilateral symmetry, tripoblastic structure, organ-grade construction and presence of a specially developed excretory system). It is assumed that the nerve cords and brain ganglia have differentiated out of the diffuse nerve plexuses, and that the accumulation of sense organs, among which ocelli (i.e. light sensitive pigment-cup eyespots) and statocysts, in the head region has led to the formation of the brain.

The number of nerve plexuses vary from intra- and/or subepidermal plexuses in the simpler flatworms to submuscular and specialized pharyngeal, intestinal and genital plexuses in more complex forms. The longitudinal nerve cords and the transverse commissures constitute in most flatworms a characteristic orthogon pattern (Reisinger, 1925, 1972; Hanström, 1928; Figs. 17B, 18). In species like *Bothrioplana semperi*, four longitudinal nerve cords (dorsal, lateral, ventrolateral and ventral) are present on both sides, but in many species the number of longitudinal cords is smaller than eight. Reuter and collaborators (Joffe and Reuter, 1993; Reuter and Gustafsson, 1995; Reuter et al., 1998b) remarked upon the considerable differences in size of the longitudinal cords among the various flatworm species and advocated a subdivision into main and minor nerve cords. The (usually ventrolaterally situated) main cords were defined as the pair of cords that: (1) starts with the strongest roots in the brain, (2) consists of wide fibre bundles and (3) has serotoninergic and catecholaminergic marker neurons.

Brains in turbellarians show a wide range in form, in relative size and in the number of accumulated neurons (Bullock and Horridge, 1965; Rieger et al., 1991). In general (with the exception of the Acoels), the neuronal perikarya form an outer zone, which surrounds an inner core of neuropil; the brain ganglia are interconnected by one or several commissures (Fig. 18).

Reference to Fig. 18 shows that the turbellarian nervous system is composed of neurons of different types, including: (1) primary sensory neurons, the peripheral processes of which form a dense plexus, (2) secondary sensory neurons, (3) commissural cells, (4) central elements, provided with ascending, descending as well as associative axonal branches, and (5) brain cells, with an axon that has many collaterals and descends in the contralateral main nerve cord. According to Hanström (1928) the latter cells are comparable to the giant neurons of nemertine- and annelid worms, and may well be involved in the coordination of movements.

Light microscopical studies with ordinary cytological techniques have shown that the brain ganglia of turbellarians contain, apart from numerous unipolar cells, also bipolar and multipolar neurons. The somata of the unipolar elements lie mostly in the outer zone or rind of the ganglia (Rieger et al., 1991).

Several authors, among them Lentz (1968) and Reuter and Gustafsson (1995) have studied the cellular elements in the nervous system of flatworms with the electron microscope. In the nervous system of planaria, Lentz (1968) distinguished the following cell types.

(1) Ganglion cells. These elements comprise the major portion of the brain ganglia of *Procotyla*, and are tightly packed around the central neuropil; they are irregular in shape, with three to eight processes extending from the perikaryon. Small clear vesicles, about 50 nm in diameter abut from the ends of the lamellae of the Golgi apparatus.
(2) Nerve cells containing a large number of small vesicles, 50 nm in diameter; cells of this type have been observed in the brain ganglia of *Dugesia*. Most of the vesicles have contents of low density, but some are filled with moderately dense material or have a dense core.
(3) Cells, containing dense-cored vesicles, which can be subdivided on the basis of size and density of the contents of their vesicles; these cells are found in the periphery of the brain ganglia and are also distributed throughout the nerve cords. Three cell types could be distinguished: (a) cells containing numerous vesicles of about 80 nm in diameter with a homogenous, very dense core (Fig. 19A); (b) cells in which vesicles of about the same diameter with contents of medium density prevail, and (c)

cells characterized by the presence of numerous large vesicles, of about 120 nm in diameter; the cores of most of these vesicles are of medium density although a few are very dense (Fig. 19B). Lentz (1968) considered the contents of the large vesicles identical to the

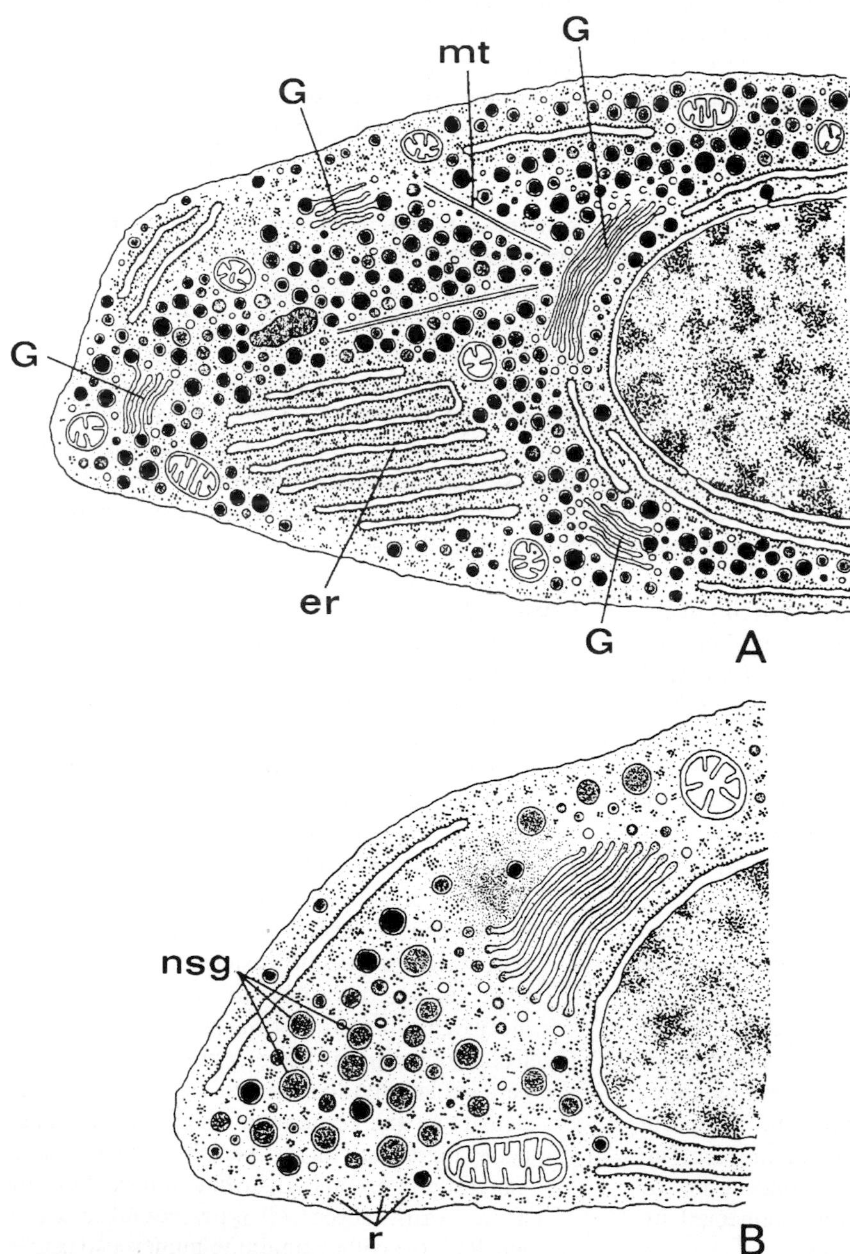

Fig. 19. Neuronal perikarya in the brain ganglion of a turbellarian. A. Cell, containing membrane-bounded granules, most of which are about 80 nm in diameter. B. Neurosecretory cell, characterized by the presence of large dense-cored vesicles, the largest of which are about 120 nm in diameter. Reproduced from Lentz (1968). er, ergastoplasma; G, Golgi complexes; mt, microtubuli; nsg, neurosecrete granules; r, free ribosomes.

neurosecretory material demonstrated in the nervous system of planaria with light microscopical techniques. On this account the designated elements described under (c) and depicted in Fig. 19B as neurosecretory cells. Rieger et al. (1991) pointed out that because turbellarians lack a circulatory system, neurosecretory substances are thought to be released in the extracellular space. However, according to Reuter and Gustafsson (1995) the term neurosecretory cell has to be reserved for elements, that release their material in the blood stream.

The classification of flatworm neurons, presented by Reuter and Gustafsson (1995) is, just like that of Lentz (1968), mainly based on the size and contents of their vesicles. They distinguished: (1) neurons containing dense-cored vesicles, varying in size between 50 and 140 nm; (2) neurons containing large dense vesicles or elementary granules, with a diameter of 50–200 nm and larger; (3) sensory, usually bipolar neurons, and (4) ganglion cells containing solely small clear vesicles, occurring only in advanced large flatworms. Reuter and Gustafsson (1989, 1995) emphasized that neurons containing dense-cored vesicles dominate the ultrastructural picture of the flatworm nervous system. These elements represent, in their opinion, an archaic cell type that may have given rise to the different types of neurons in advanced flatworms, as well as to those in higher metazoans.

The central neuropil of the brain ganglia is composed of tightly packed fibres, most of which contain large numbers of vesicles (Fig. 20); in some fibres, vesicles of one type predominate, but in others there is a considerable intermixing. Small clear vesicles often occur in tight clusters (Lentz, 1968). Figure 18 shows that axons of sensory cells, neural cord cells and collaterals of brain ganglion cells all participate in the formation of the central neuropil.

The neuronal release sites show a considerable morphological variation in flatworms. Following Reuter (1981) and Reuter and Gustafsson (1989, 1995), they may be categorized as conventional synapses, synapse-like structures and non-synaptic release sites.

Conventional synapses are characterized by the presence of accumulations of presynaptic vesicles, parallel synaptic membranes with pre- and postsynaptic densities and a distinct, uniformly wide synaptic cleft containing dense material (Reuter and Gustafsson, 1989). They can be divided into several types, including single, shared, en-passant (Fig. 21A) and multiple synapses (Reuter and Gustafsson, 1989, 1985; Reuter and Palmberg, 1990). In shared synapses a single presynaptic terminal makes contact with two adjacent postsynaptic elements; the synapses of the types just mentioned are all polarized. Unpolarized, bidirectional synapses have also been observed in flatworms (Reuter and Palmberg, 1989; Bedini and Lafranchi, 1991). These contacts are characterized by clusters of vesicles and membranous and paramembranous densities in opposite terminals on both sides of the synaptic cleft. In most conventional flatworm synapses, the synaptic vesicles are of the small clear type, but in some dense-cored vesicles are closely apposed to the presynaptic membrane and, hence, their contents may be assumed to be involved in wiring transmission. Reuter et al. (1990) demonstrated with the aid of the immunogold technique that in *Microstomum lineare* some of these synaptic dense-cored vesicles contain material immunoreactive to RFamide.

Synapse-like structures lacking the membrane specializations of conventional synapses are very common in the nervous system of *Microstomum lineare* and other simple turbellarians (Reuter 1981, 1990; Reuter and Gustafsson, 1995). They are characterized by rows or clusters of dense-cored vesicles close to the axolemma (Fig. 21B, C). This type of contact seems to be intermediate between true synapses and non-synaptic release (Reuter and Gustafsson, 1989).

Non-synaptic or paracrine release sites occur in all parts of the nervous system of flatworms. Reuter (1981) observed in *Microstomum lineare* axonal varicosities devoid of membrane specializations, containing large dense-cored vesicles; such a varicosity is shown in Fig. 21D. It will be seen that some of the vesicles are in close contact with the axolemma; in one the membrane surrounding the core has fused with the axolemma, thus forming a so-called omega profile. Similar non-synaptic release sites have been observed in the cerebral ganglion of the turbellarian, *Dendrocoelum lacteum* (Golding and Bayraktaroglu, 1984) and in the tapeworm, *Diphyllobothrium dendriticum* (Reuter and Gustafsson, 1989). Finally, the expelled contents of dense-cored vesicles have been visualized at non-synaptic release sites with the TARI-technique in the turbellarians *Stenostomum leucops* (Reuter and Palmberg, 1990; Fig. 21E, F) and *Promonotus schultzei* (Reuter, 1991).

Small, clear vesicles have often been regarded as cholinergic, small dense-cored vesicles as aminergic and large dense-cored vesicles as peptidergic, but according to Reuter and Gustafsson (1995) this

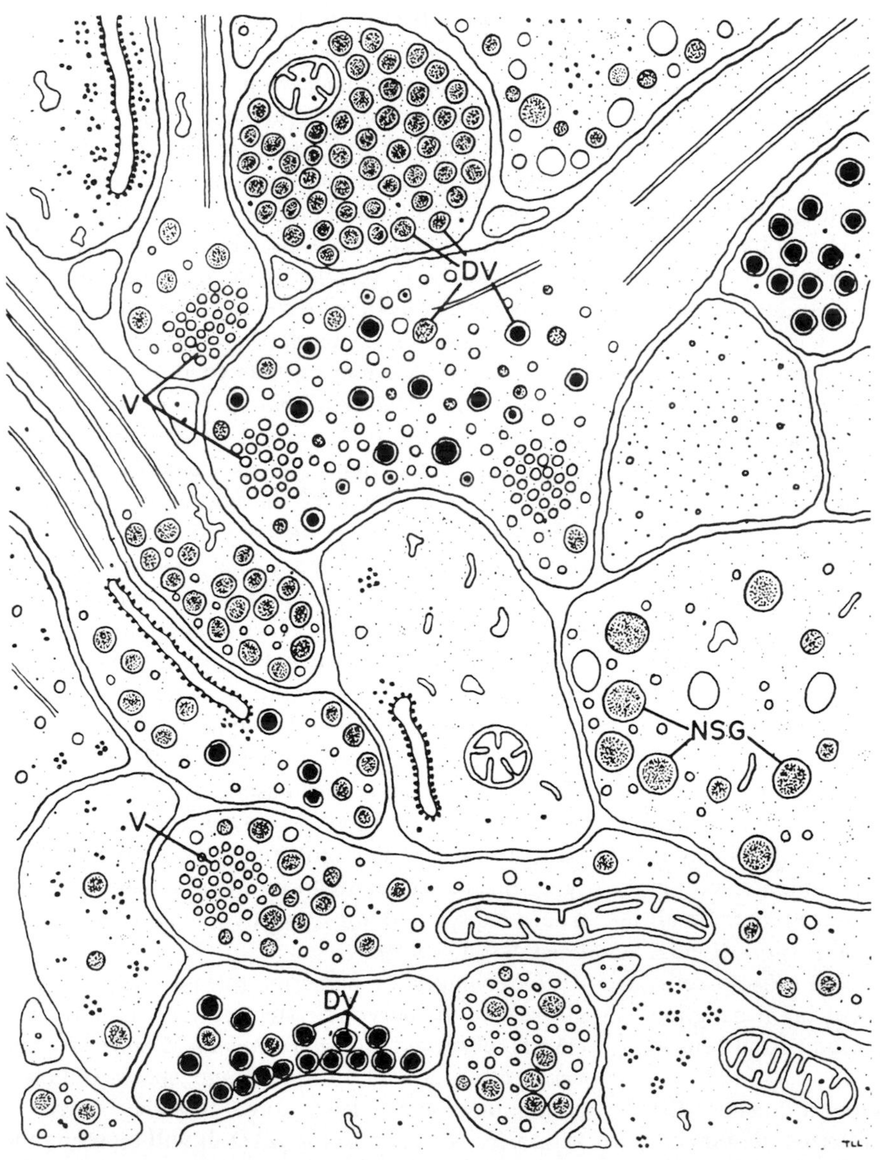

Fig. 20. Neuropil in the brain ganglion of the turbellarian, *Procotyla fluviatilis*; neuronal processes of different kinds are closely packed. Several types of vesicles are present, including small clear vesicles (V), dense-cored vesicles (DV) and large neurosecretorygranules (NSG). Various combinations of vesicles and granules occur in individual processes. Reproduced from Lentz (1968).

categorization does not hold. We simply do not know the nature of the contents of the small vesicles, and the immuno-electron microscopical studies by Reuter et al. (1990) have shown the presence of an RF-amide-like peptide in small dense-cored vesicles. As regards the three modes of neuromediator release – synaptic, synaptoid and non-synaptic – occurring in the flatworm nervous system, two questions may be asked: (1) what is the morphological identity of the neurons, using these different modes of communication, and (2) what is the functional context in which these elements operate? Unfortunately, the answers to both of these questions are unknown, save for the fact that peptidergic neurons may be involved in both synaptic and non-synaptic transmission (Reuter et al., 1990),

The rather voluminous literature on the various neuroactive substances occurring in the nervous system of flatworms will not be discussed here (fore reviews see Reuter and Gustafsson, 1989, 1995). Suffice to mention that acetylcholine, the monoamines dopamine, noradrenaline, serotonin and histamine, and several neuropeptides, including a β-endorphin-like peptide and several growth factors, have all been demonstrated in flatworms; moreover, several native platyhelminth neuropeptides have been isolated. These include neuropeptide F (NPF) and a whole series of FMRF amide-related peptides (FaRPs) (cf. Mäntylä et al., 1998 for references and details). Evidence for the presence of amino acids has not been presented so far; however, it has been demonstrated that GABA and glycine depress the activity in the longitudinal

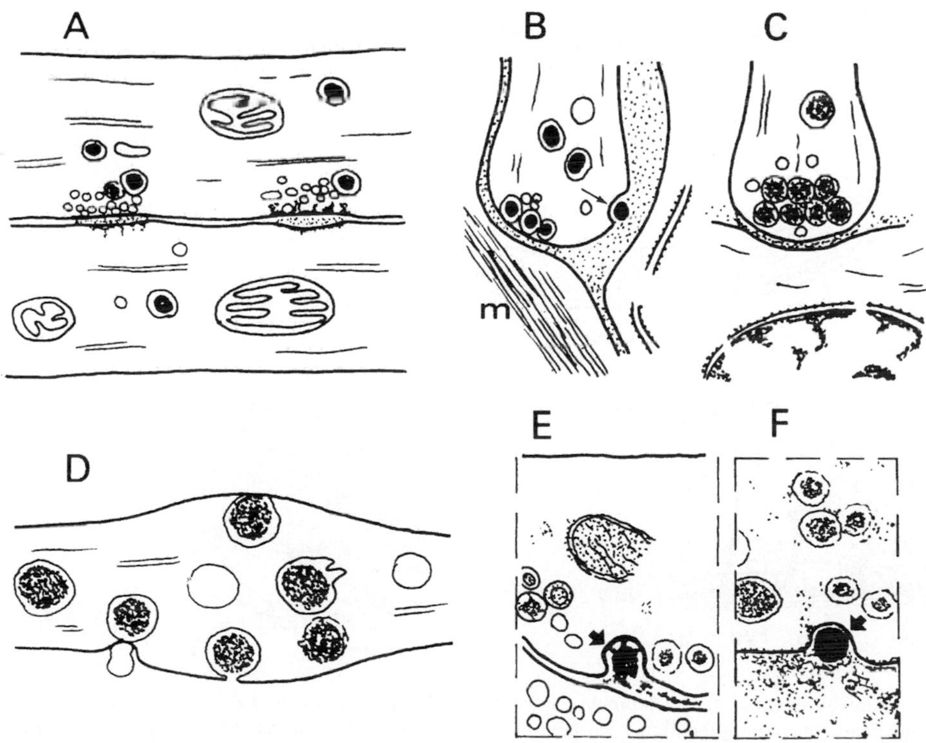

Fig. 21. Diagrams showing different types of synapses and release sites in flatworms. A. Two en-passant synapses. B. Synapse-like structure between nerve terminal and muscle (m); note extruded granular core (arrow). C. Synapse-like structure on neuronal soma. D. Axonal varicosity, containing large dense-cored vesicles, several of which are in close contact with axolemma. E, F. Non-synaptic exocytoses of large dense-cored vesicles, as revealed by the TARI technique (arrows). Figures A–D are redrawn from Reuter (1981); Figs E, F are drawn from electron micrographs in Reuter and Palmberg (1990).

nerve cords in the turbellarian *Notoplana*, while glutamate and aspartate act as excitatory neurotransmitters on the same structures (Koopowitz, 1989).

Recently, it has been demonstrated that plexuses of serotonin- and NPF-immunoreactive cells and fibres extend throughout the nervous system of turbellarians (Reuter et al., 1995a, b). The isolated position of Acoela (see above) was emphasised by the facts that: (1) a cluster of serotonin-immunoreactive neurons typical of other turbellarians is lacking in this group, and (2) the serotoninergic fibres which connect the longitudinal nerve cords form an irregular plexus rather than an orthogon pattern (Raikova et al., 1998). FMRF-immunoreactivity and serotonin-immunoreactivity are restricted to separate neurons (Reuter, 1988), but NPF and FaRPs labels may be co-localized (Mäntylä et al., 1998).

6. Nematoda

The Nematoda are tapering cylindrical, transparent worms, ranging from 0.1 mm to 9 meters in length (Fig. 22A). Although the number of species described in the scientific literature is only about 20,000, estimates of the actual number range from 40,000 to 10 million (Blaxter 1998). The nematodes are, just like the platyhelminths, triplobastic. Their digestive tract forms a tube complete with mouth and anus; cilia are lacking.

The central nervous system consists of a conspicuous circumpharyngeal nerve ring, with a number of associated ganglia (Figs. 22B, C). The

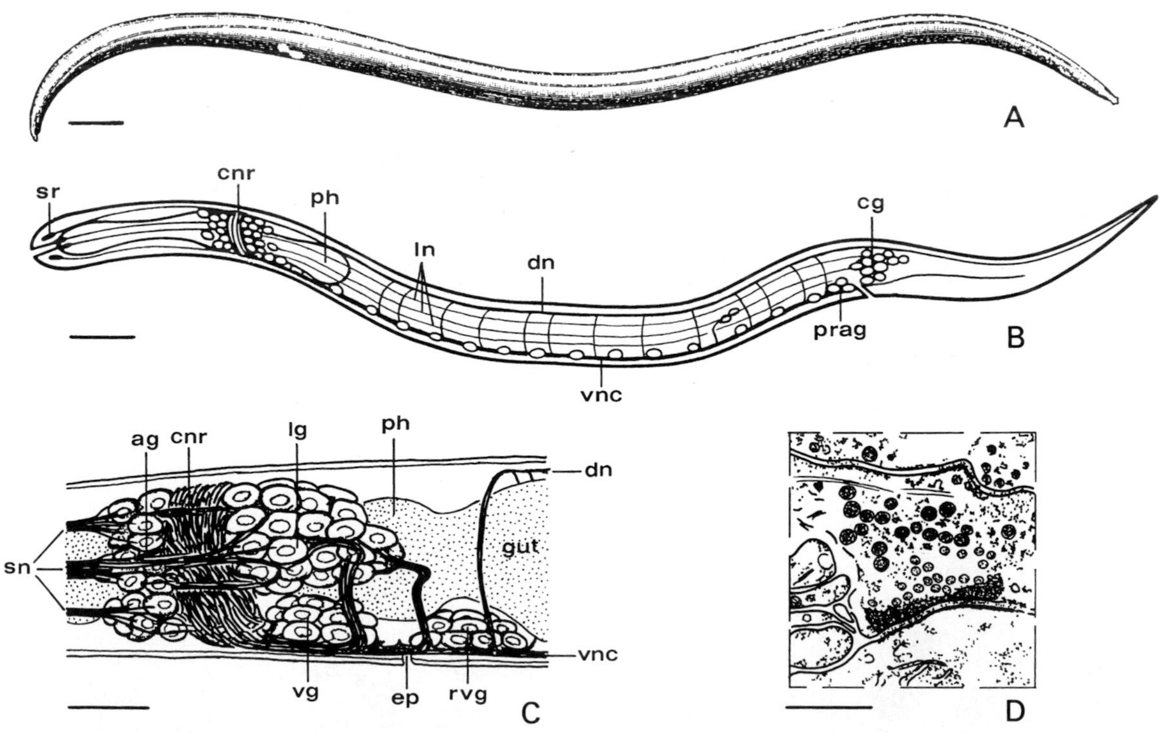

Fig. 22. Nematodes (roundworms or thread worms). A. The common parasitic roundworm, *Ascaris lumbricoides*. The rostral end is to the right. Scale bar, 1 cm. B. Schematic diagram of the nervous system of *Caenorhabditis elegans*. Modified from Durbin (1987). Scale bar, 50 μm. C. The brain ganglia of *C. elegans*. Redrawn from Albertson and Thomson (1976). D. A synapse from *C. elegans*; small and large vesicles can be distinguished. The large vesicles, which are somewhat darker than the small ones, are situated at a distance from the synaptic cleft. Drawn from an electron micrograph in Rand and Nonet (1987). Scale bar, 0.25 μm. ag, anterior ganglion; cg, caudal ganglion; dn, dorsal nerve; ep, excretory pore; lg, lateral ganglion; ln, lateral nerves; cnr, circumpharyngeal nerve ring; ph, pharynx; prag, preanal ganglion; rvg, retrovesicular ganglion; sn, sensory nerves; sr, sensory receptors; vn, ventral nerve cord; vg, ventral ganglion.

nerve ring consists of interacting processes of neurons situated within and beyond the brain ganglia. The number of brain ganglia varies. Goldschmidt (1908, 1909) distinguished in *Ascaris lumbricoides* eleven ganglia, together containing 162 cells. In *Caenorhabditis elegans* the neurons situated in front of the nerve ring are collectively called the anterior ganglion. Directly behind the ring there are a pair of lateral ganglia, a ventral ganglion, and further back, the retrovesicular ganglion (Albertson and Thomson, 1976; White et al., 1986; Fig. 2C). Sensory nerves extend rostrally from the ganglionic complex, to contact mechanosensory and chemosensory structures in the tip of the snout. A coarse ventral nerve, a much thinner dorsal nerve and a variable number of lateral nerves extend caudally from the ganglionic complex. The ventral nerve contains a row of nerve cell bodies and, hence, deserves to be designated as a nerve cord. Circumferential bundles at irregular distances complete a hexagon pattern (Fig. 22B). In the caudal region there is a preanal ganglion and a pair of caudal ganglia (Fig. 22B).

In 1965, Sydney Brenner introduced the hermaphrodite, free-living nematode *Caenorhabditis elegans* as a model organism for studying the genetic basis of development and behaviour (see e.g. Brenner, 1973). Salient advantages of this animal include small size (1.5 mm as an adult), transparency, rapid life cycle (4 days at 20°C), a small genome (only 20 times that of *E. coli* and anatomical simplicity (Riddle et al., 1997). Collectively, these properties have contributed to the following three notable accomplishments: (1) determination of the exact lineage of the 959 cells that make up he adult animal (Sulston et al., 1983), (2) reconstruction of the complete circuitry of the nervous system (White et al., 1986), and (3) charting the complete genome. The latter task was recently completed by a large group of investigators (the *C. elegans* Sequencing Consortium, 1998), a monumental achievement, on account of which 1998 was aptly proclaimed "the year of the worm" (Plasterk, 1999). It was established that the genome of C. elegans encompasses 97 mega basepairs encoding 19,099 genes.

In the above-mentioned paper of White et al (1986), the structure and connectivity of the nervous system of *C. elegans* has been deduced from reconstructions of electron micrographs of serial sections. The running head, reading 'The mind of a worm', is much more sweeping than the neutral title of the paper. The main results of this meticulous study may be summarized as follows.

The nervous system of the adult *C. elegans* has a total complement of 302 neurons, which are arranged in an essentially invariant structure. These neurons are extremely simple: cell bodies are typically 2–5 μm in diameter and have one or at the most a few processes of about 0.1 μm in diameter. The processes run parallel to each other in processing bundles, making and receiving en-passant synapses along their length. The most important processing bundles are the circumpharyngeal ring and the dorsal and ventral nerve cords (Figs. 22B, C).

On the basis of similarities in morphology and connections, 118 classes of neurons could be distinguished, with 1 to 13 members in each class; primary sensory neurons and central neurons are mostly in bilaterally symmetric pairs.

The nervous system contains approximately 5,000 chemical synapses between neurons, 700 gap junctions and 2,000 neuromuscular junctions. The chemical synapses are characterised by a conspicuous presynaptic membrane thickening (Fig. 22D; cf. also Rand and Nonet, 1997). Most vesicles in *C. elegans* synapses are spherical, 35 nm in diameter, and have lightly staining interiors. Some classes of neurons have a second type of vesicles coexisting with vesicles of the first type. These vesicles have dense cores and their size ranges from 37 nm to 53 nm. "The dark-cored vesicles generally seem to be excluded from the region immediately adjacent to the presynaptic specialization, which contains only the smaller type of vesicle" (White et al., 1986, p. 28; Fig. 22D).

The flow of information is fairly direct from primary sensory neurons to motor neurons, with no more than two or three intervening central elements as a rule. Moreover, direct synaptic contacts between primary sensory neurons and motoneurons have frequently been observed.

The study by White et al. (1986), just reviewed, encompasses 340 pages and contains numerous photomicrographs of sections accompanied by tracings and a number of circuit diagrams, one of which is reproduced in Fig. 23B. The bulk of the paper is occupied by a database, in which the topography and the synaptic inputs and outputs of the individual neurons of *C. elegans* are recorded. The descriptions of each of the neurons is accompanied by some low-power electron micrographs ($\times$25,000) of its synaptic connections. Closely related studies on the ultrastructure of the nervous system of *C. elegans* include: Ware et al. (1975: the nerve ring), Ward et al. (1975: the anterior sensory anatomy), Albertson and Thomson (1976: the pharynx) and Hall and Russel (1991: the posterior nervous system).

Up to now, there have only been two attempts to provide a complete description of the neuronal circuitry of an animal and, remarkably enough, both pertain to a nematode. The analysis of White and his colleagues was preceded by Goldschmidt's (1908, 1909) heroic attempt to chart the neuronal circuits of the roundworms *Ascaris lumbricoides* and *Ascaris megalocephala* on the basis of serially sectioned, light microscopical material. (These studies also had a remarkable subtitle, in translation: "An attempt to penetrate into the structure of a simple nervous system.") One of the resultant diagrams is reproduced in Fig. 23A. It was Goldschmidt's aim to provide a basis for neurophysiological investigations. Until his death in 1958, he remained convinced that his results could point to definite ideas concerning the working mechanism of the nervous system of *Ascaris* (cf. Kuhlenbeck, 1967; footnote 9 on p. 53). However, in actuality, Goldschmidt's work never led to any functional studies. His – as we know now – quite erroneous main conclusion, that in the nervous system everything is continuously connected with everything else, was criticized in his own time, e.g. Hanström (1928, p. 121).

So a major flaw of Goldschmidt's model is that it contains connections that are not there. The model of White et al. (1986) is focused on wiring transmission; it provides detailed information on the distribution of chemical synapses and gap junctions over the surface of the various neurons. However, by including data on the occurrence of dense-cored vesicles in the terminals of the various neurons, it hints at least at possible sites of volume transmission. In their database the neurons are characterized with three-letter codes, and for all elements it isspecified whether their terminals contain, apart from small vesicles, many (ADL, PVQ), some (a large number, among which AFD, ASH, ASJ, ASK), a few (AWA) or occasional (IL2) dense-cored vesicles. These data have already been proved to be relevant for the interpretation of the functioning of certain neurons (see e.g. the studies of Maricq et al., 1995 and Hart et al., 1995 on the ASH sensory neurons).

The evidence for the occurrence of volume transmission in nematodes is so far only indirect and confined to the observations already discussed that in *C. elegans* many synapses contain dense-cored vesicles, and that in all of these synapses these dense-cored vesicles typically stand back from the presynaptic membrane (Fig. 22D).

As regards neuromediators, the presence of acetylcholine, the amino acids GABA and glutamate, the monoamines dopamine, octopamine ad serotonin, and of a large (and steadily increasing) number of neuropeptides, has been demonstrated in nematodes (for review, see Brownlee and Fairweather, 1999). Thus, in *C. elegans* more than 50 different FMRFamide-related peptides have been shown to be present. It is remarkable that in *Ascaris* FMRFamide-specific staining predominates with up to 75% of the neurons (Brownlee et al., 1996), whereas in *C. elegans* only 10% of the neurons display FMRF immunoreactivity (Schinkmann and Li, 1992). It would be interesting to know whether these immunoreactive neurons correspond to the elements, the terminal of which contains dense-cored vesicles.

7. Chelicerata, Mandibulata and Crustacea

The Phyla: Chelicerata, Mandibulata and Crustacea are commonly united in a higher category, the Arthropoda. In what follows the mandibulates will be discussed first and next brief consideration will be given to the chelicerates and crustaceans.

The *Mandibulata* include the Insecta or Hexapoda and the Myriapoda; the insects are the most successful and diverse invertebrates. Their numbers, both of species and of individuals, certainly far exceed those of any other group. More than 700,000 species are known but estimates run as high as 10 million. The insects include many groups of familiar creatures, like cockroaches, grasshoppers, dragon flies, bugs, butterflies, flies, beetles, ants and bees. A single species, the common cockroach *Blatta orientalis*, is depicted in Fig. 24A. The Myriapoda (many feet) comprise about 3,000 species; their number of trunk segments varies from 15 to 177.

The central nervous system of insects comprises a dorsal brain, bilateral circumesophageal connectives and a ventral cord of segmented ganglia (Fig. 24B).

The brain displays three subdivisions or lobes, protocerebrum, deutocerebrum and tritocerebrum (Fig. 24C). Each of these lobes is composed of a peripheral zone of densely packed cell bodies, surrounding a central core occupied by fibres and neuropil; the neurons in the brain are mostly unipolar. A single process, the primary neurite, emanates from the soma and gives rise to numerous fine branches. Golgi staining, immunohistochemistry and cobalt electrophoretic injections in individual cells may all reveal the subtle beauty of insect neurons. Many participate with different sets of ramifications in different neuropil zones and it is often hard to make a distinction between dendritic and axonal branches.

The protocerebrum represents the largest part of the brain. It is bilaterally connected to the rostrolaterally-situated optic lobes by large optic peduncles. Visual information entering the facets of the compound eyes is relayed through a series of neuropil regions before reaching the protocrebrum. These neuropil regions, which make up the bulk of the optic lobe, are known as the lamina, medulla and lobule (Fig. 24C). The output neurons of the optic lobe project by several tracts to so-called optic foci in the protocerebrum and, by a commissural bundle, to the contralateral optic lobe (Fig. 24C).

The protocerebrum contains several neuropil masses of which the corpora pedunculata or mushroom bodies and the median central body complex are the most prominent (Fig. 24C). The bilateral mushroom bodies are large and highly differentiated in social insects; they have two major regions, the cap-shaped calyx and the stalk-shaped peduncle; in many species separate medial and lateral calyces can be distinguished. The peduncle bifurcates at its lower end into an anteriorly directed α lobe and a posteriorly directed β lobe (Fig. 24C). Lesion and stimulation experiments have shown that the mushroom bodies are involved in olfactory learning, memory formation, as well as in the control of complex motor repertoires. Erber et al. (1987) advanced the hypothesis that these structures serve the temporal integration, concatenation and comparison of sensory information. Interestingly, the structural organization of mushroom bodies shows striking parallels with the neuronal architecture of the mammalian hippocampus, a telencephalic region known to be implicated in spatial learning (Schürmann, 1987; Strausfeld et al., 1995). The central body complex comprises several interconnected neuropils; it receives mostly indirect inputs from many sensory organs, and is involved in the control of motor activity (Homberg, 1987).

The deutocerebrum is strongly connected with the antennae, hence its bilateral halves are also known as the antennal lobes; both of these lobes comprise a smaller dorsal part and a larger ventral part. The antennae are sensory organs provided with olfactory as well as mechanosensory receptors. The olfactory receptors project to the ventral part of the antennal lobe, whereas the dorsal part receives input from the mechanoreceptors; the dorsal part also contains the motor neurons innervating the muscles that move the antenna. Many insects rely on olfaction as a principal sensory modality; volatile chemical cues are often required to help insects orient towards and locate appropriate food sources and mating partners. High chemical selectivity and sensitivity are especially characteristic of the olfactory receptors responsible for detecting sex pheromones (Christensen and Hildebrand, 1987). The structure of the ventral part of the antennal lobe is characterized by a series of spherical islands of synaptic neuropil, called glomeruli; such glomeruli are also found in the olfactory bulbs, i.e. the primary olfactory centres of vertebrates; efferents of the antennal lobes project mainly to the mushroom bodies in the protocerebrum.

The tritocerebrum consists of two symmetrical lobes bounded dorsally by the deutocerebrum and

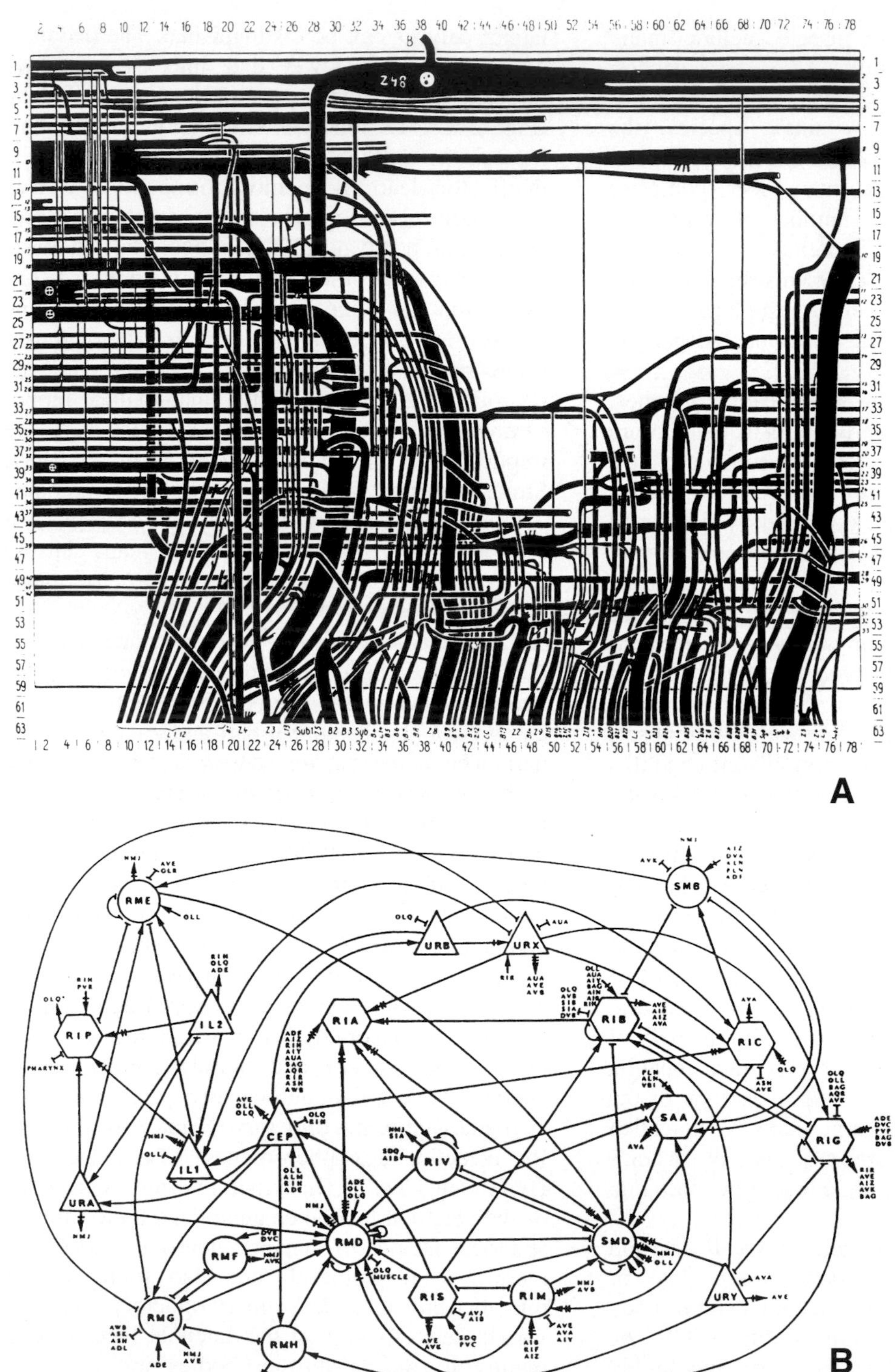

ventrally tapering into the peri-esophageal connectives; it gives rise to nerves projecting to the oral region (the labral nerves) and to the alimentary canal. Commissural connections connect the halves of the three main divisions of the brain with their fellows on the opposite side (Fig. 24C).

The ventral cord consists essentially of segmentally arranged pairs of ganglia, longitudinally joined by connectives and transversely by commissures. However, this 'rope ladder' pattern, in most insects, is blurred by fusions, both between left and right ganglia, as well as by concentrations in the longitudinal direction. The large subesophageal ganglion is the first ganglion of the ventral cord (Figs. 24B, 25); it develops from two, three or more neuromeres and supplies the mouth-parts. Many of the data presented in the previous survey are derived from Bullock and Horridge (1965) and from Burrow's (1996): 'The Neurobiology of an Insect Brain', a comprehensive monograph devoted to the structure and function of the central nervous system of locusts and grasshoppers.

Figure 25 summarizes the most important neuroendocrine structures in insects. It shows that the axons of neurosecretory neurons situated in the protocerebrum, the frontal ganglion and the subesophageal ganglion converge upon the corpora cardiaca. The corpora cardiaca are paired neurohemal organs situated just behind the brain and consist of storage lobes and glandular (secretory) lobes. The storage lobes contain the densely packed terminals of the various groups of neurosecretory cells. The glandular lobes consist of intrinsic neurosecretory cells that synthesize, store and release a number of peptide hormones, among which the adipokinetic hormones (AKH). When released into the hemolymph, most notably after the start of flight, the AKH hormones enhance protein synthesis, heart rate, and the mobilization of lipids from fat bodies and carbohydrates from flight muscles. The complexity of the interaction between the nervous system and the endocrine system may be apparent from the fact that in the somata of the pars intercerebralis of the protocerebrum (L and M in Fig. 25), and in the terminals in the storage lobes of the corpora cardiaca, no less than 25 different peptides have been localized. Complex events such as metamorphosis and reproduction are coordinated by the concerted release of different sets of hormones (Schooneveld, 1998). Some of the neurosecretory cells innervate and form terminal branches in the corpora allata, a pair of non-nervous endocrine glands that are closely associated with the corpora cardiaca; the corpora allata secrete juvenile hormone, which plays an important role in development and in maturation.

As regards neuromediators, most of the so-called classical neurotransmitters, among which acetylcholine, the amino acids glutamic acid, aspartic acid, GABA, glycine and taurine, and the monoamines dopamine, noradrenaline, octopamine, histamine and serotonin, have all been demonstrated to play a role in the insect nervous system. There is a remarkable similarity between the various monoaminergic systems in insects and in vertebrates; in both groups, relatively small ensembles of monoaminergic neurons supply vast areas of the brain with profusely ramifying axonal branches (Nässel, 1991).

In insects, as in other groups, the neuropeptides constitute a fast-growing class of neuroactive compounds. In this group alone more than fifty neuropeptides belonging to a number of peptide families have been sequenced. Seven years ago no less than 38 neuropeptides were known alone for the locust, *Locusta migratoria* (Schoofs et al., 1993).

Histochemical studies using antibodies raised against the various neuropeptides often yield beautiful Golgi-like pictures. With the aid of this technique an enormous number of purely descriptive papers dealing with the various

◄

Fig. 23. Neuronal circuitry in the central nervous system of nematodes. A. Plan of the fibre composition and connections in the circumpharyngeal ring of *Ascaris megalocephala*, prepared by Goldschmidt (1909). The figure is much reduced; the original measures 30 × 35 cm. B. Diagram of the circuitry associated with motoneurons in the nerve ring of *C. elegans*, reproduced from White et al. (1986). Neurons are characterized by three-letter codes. Sensory neuron classes are represented by triangles, interneurons by hexagons and motoneurons by circles. Connections made via gap junctions are indicated by double T-shaped symbols, those via chemical synapses with arrows. The latter are graded according to their prominence on a scale of 1 to 4 by small cross-hatches on the arrows.

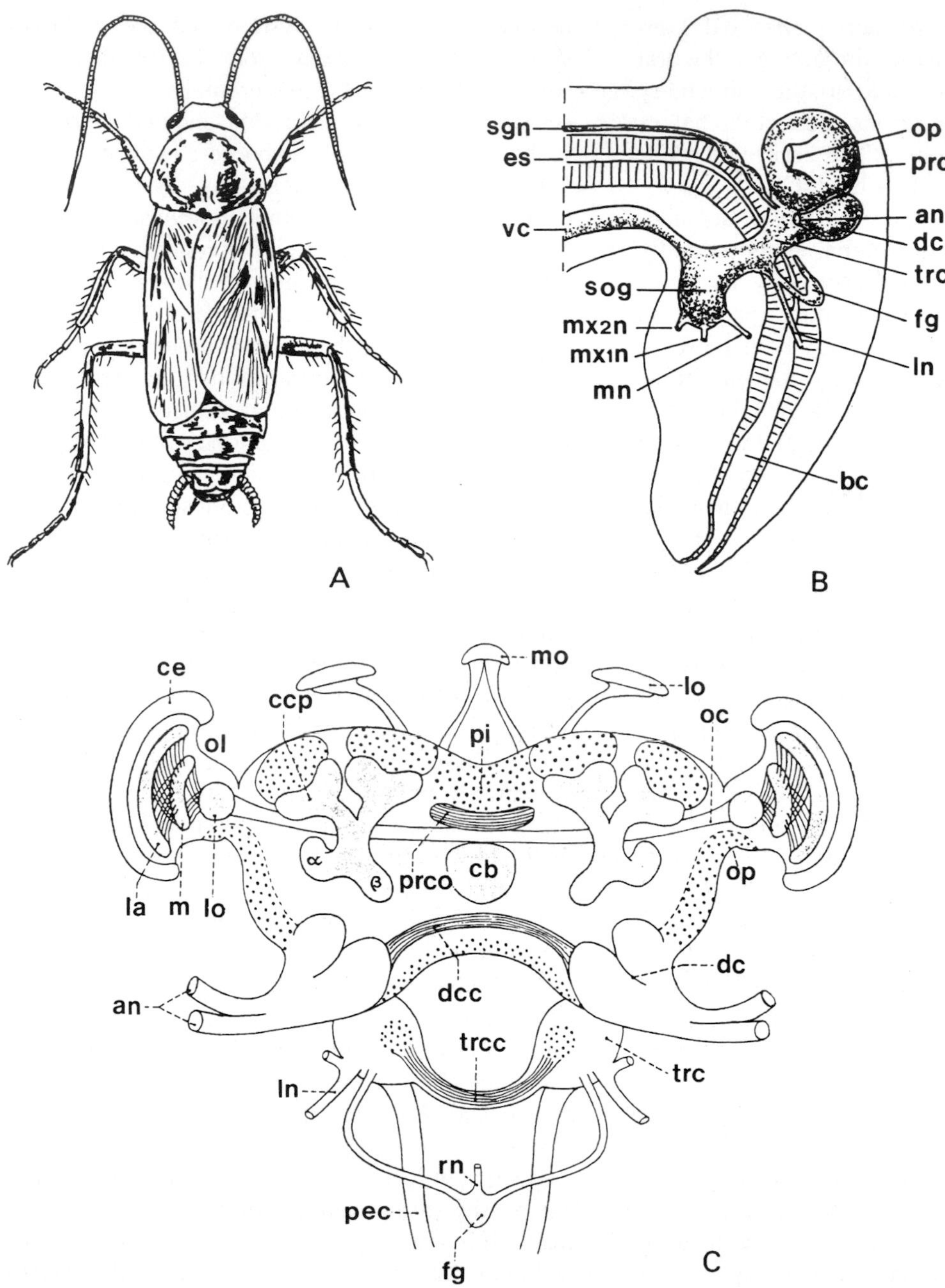

Fig. 24. Insects and their brains. A. The common cockroach, *Blatta orientalis*. Redrawn from Storch and Welsch (1997). B. The head of the same species with the nervous system in position. Redrawn from Bütschli (1910). C. Diagram showing the position of the main regions in the brain of insects. Modified from Grassé (1975). an, antennal nerves; bc, buccal cavity; cb, central body; ccp, calyx of corpus pedunculatum; ce, compound eye; dc, deutencephalon; dcc, dc, deutocephalic commissure; es, esophagus; fg, frontal ganglion; la, lamina; ln, labral nerve; lo, lobule; m, medulla; mn, mandibular nerve; mo, median ocellus; mx1n, mx2n, nerves of the first and second maxilla; oc, optic commissure; ol, optic lobe; op, optic peduncle; pi, pars intercerebralis; pec, peri-esophageal connective; prc, protocerebrum; rn, recurrent nerve; sgn, stomatogastric nerve; sog, subesophageal ganglion; trc, tricocerebrum; trcc, tritocerebral commissure; vc, ventral cord.

peptide-immunoreactive neuronal populations in insects have been produced. After having carefully reviewed the pertinent literature, the plethora of data available elicited from Burrows (1996, p. 228) the following remarks: "And so continues the seemingly ever-expanding descriptions of peptides.

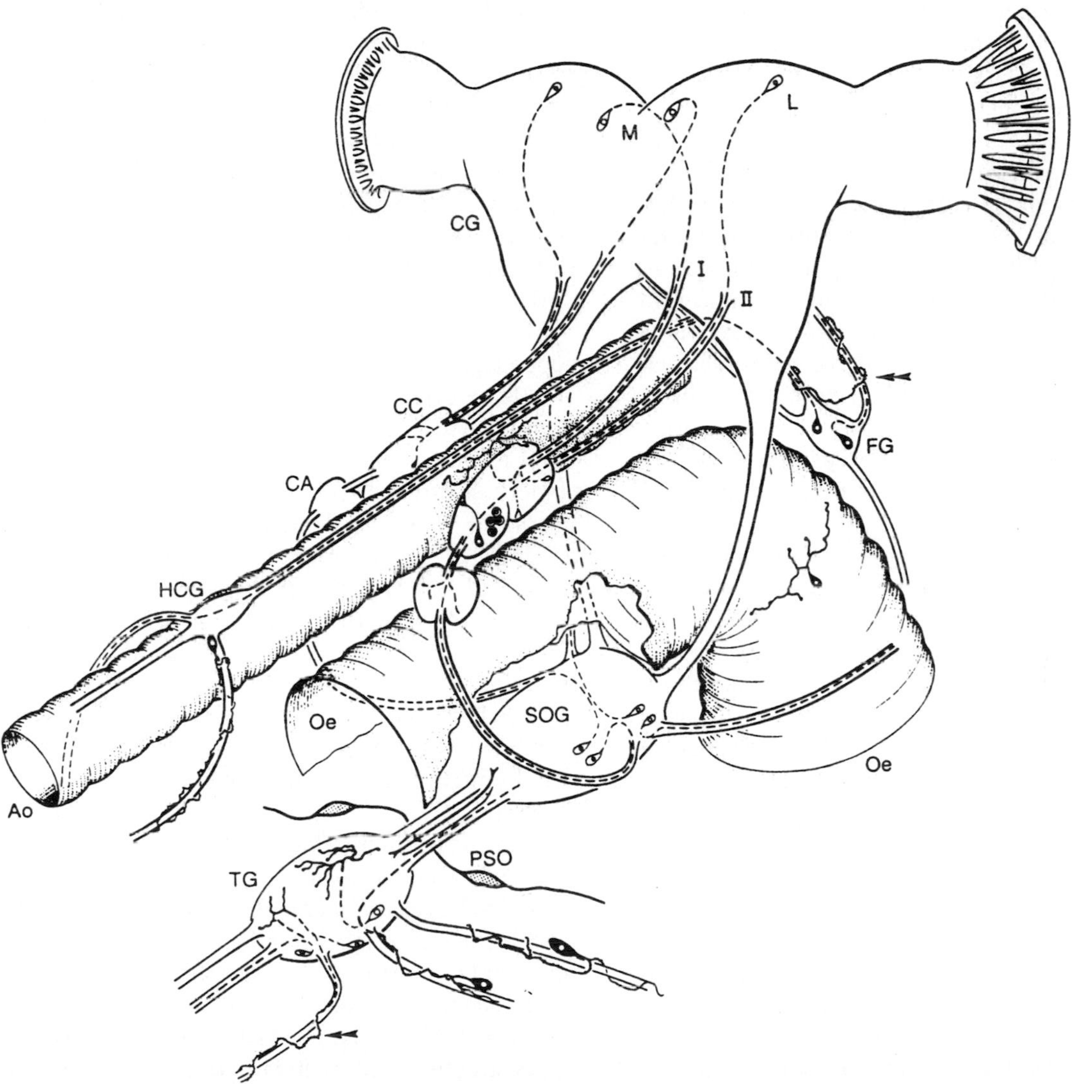

Fig. 25. The neuroendocrine system and the main neurosecretory pathways (dashed lines) in a generalized insect. The cerebral ganglia (CG), subesophageal ganglion (SOG), first thoracic ganglion (TG), frontal ganglion (FG) and hypocerebral ganglion (HCG) are shown. Also shown are two neurohemal organs, the bilateral corpora cardiaca (CC) and corpora allata (CA), and parts of the aorta (Ao) and oesophagus (Oe). Neurosecretory neurons are found in the intercerebral part of the protocerebrum, where they form conspicuous medial (M) and lateral (L) groups, in the ganglia of the ventral nerve cord (SOG, TG), in FG, as well as along some peripheral nerves. The axons of the neurosecretory cells run to the neurohemal organs for release of secretions into the hemolymph. The axons of the medial and lateral intercerebral cell groups pass to the corpora cardiaca via two separate nerves, the nervi corporis cardiaci I and II (I, II). As shown in the diagram, some fibres originating from the intercerebral cell groups pass through the corpora cardiaci and terminate in the corpora allata. The surfaces of nerves originating from FG, SOG, TG and other ventral ganglia also store and release secretions produced in these ganglia (double arrowheads). Ventral perisympathetic organs (PSO) contain secretions from TG and other ventral ganglia.Reproduced from Schooneveld (1998).

Nearly all of them have, as yet, undefined actions, and the identity and actions of the neurons in which the immunoreactivity is located are also largely unknown." And further: "The lesson from the abundance of papers on the fluorescence of amines, and from the descriptions of neurosecretory cells revealed with a large number of stains and the relatively small increment of understanding that accrued, is there for us all to ponder before more descriptive papers are written."

Co-localization of different neuromediators has been found in various combinations in insect neurons. Thus, in the locust, arginin vasopressin and a peptide immunoreactive to an antiserum raised against FLRF-amine are co-localized in a single pair of neurons located in the subesophageal ganglion (Evans and Cournil, 1990). Co-localization of RFamide-like peptides with taurine (Stevenson, 1999) and GABA (Homberg et al., 1999), respectively, have also been reported for locust neurons.

Synaptic junctions have been widely recognized in all insect neuropils (Grassé, 1975; Westfall, 1987; Strausfeld and Meinertshagen, 1998), but there is also ample evidence for the ubiquity of non-synaptic or volume transmission in the central nervous system of this group. Buma and Roubos (1986) studied the cerebral and subesophageal ganglia of the cockroach, Periplaneta americana (Fig. 24A, B) using the TARI technique. In both ganglia exocytosis in synapses was only rarely seen, but non-synaptic release sites, which showed numerous exocytosis phenomena, were frequent (Fig. 26A); up to 30 exocytoses per profile of a nerve fibre were counted. Schooneveld (1998) observed the profiles of axonal branches containing both small clear vesicles and large dense-cored vesicles in the protocerebrum of *L. migratoria*. The small vesicles were concentrated in the vicinity of distinct presynaptic membrane-specializations, but the dense-cored vesicles occupied a typical 'stand-back' position. Serotoninergic neurons situated in the posterior part of the protocerebrum of the blow fly *Calliphora* invade the optic lobe and spread over the neuropil layers of that structure (Fig. 24C). Nässel (1991) found that the varicosities associated with these fibres do not show synaptic specializations in the lamina, and concluded that the action of serotonin must by paracrine action in this region. Similar observations were made on serotoninergic neurons situated in the antennal lobe of the moth, *Manduca sexta* (Sun et al., 1993).

The TARI technique also facilitated the study of release phenomena in the neuro-endocrine organs of insects. Thus, Pow and Golding (1987) found that aminergic terminals in the storage lobe of the corpus cardiacum of the locust *Schistocerca gregaria* show numerous sites of non-synaptic exocytosis, and Khan and Buma (1985) made similar observations in the corpus allatum of the Colorado beetle, *Leptinotarsa decemlineata* (Fig. 26B).

The *Chelicerata* (named according to their first appendages, the chelicerae), which number about 70,000 described species, include three classes, the Merostomata or horseshoe crabs, the Pycnogonidae or sea spiders and the Arachnida, to which the spiders, scorpions, mites and ticks belong. Chelicerates have segmented bodies, joined appendages and a chitinous exoskeleton in common with other arthropods, but they lack the antennae of insects and crustaceans. The head and thorax of chelicerates are fused to a single unit, the cephalothorax.

The horseshoe crab *Limulus polyphemus*, to which I will confine myself here, reaches a length of 60 cm (Fig. 27A). This remarkable animal has been virtually unchanged since it dwelt in shallow Silurian seas, more than 400 million years ago (Margulis and Schwartz, 1998). It was on the compound eye of *Limulus* that Hartline and Ratliff (1957) carried out their fundamental investigations on lateral inhibition.

The nervous system of *Limulus* is extremely condensed (Fig. 27A, B). The ensuing notes on the macrostructure and microstructure of this nervous system are based on the studies of Fahrenbach and Chamberlain (1987), Wegerhof and Breidbach (1995) and Fahrenbach (1999).

The brain is a large, nearly spherical body, up to 7 mm in diameter in an adult, with a slightly concave dorsal surface. On the anterior surface, the optic nerves from the compound lateral eyes, the two median ocellar eyes and the rudimentary ventral eyes, enter dorsally. The massive circumesophageal connectives exit caudally from the brain; five coarse neural nerves, innervating the extremities, emanate from these connectives (Fig. 27A, B).

The protocerebrum is large and occupies the bulk of the brain; because antennae are lacking the deutocerebrum is feebly developed. The tritocerebrum, which gives rise to the cheliceral nerves, is situated ventrally adjacent to the rostral end of the circumesophageal connectives. The protocerebrum is largely occupied by masses of neuropil, of which the central body and the huge bilateral corpora pedunculata are the most prominent; the neuropil masses just mentioned were long considered homologous with their namesakes in insects. However, Breidbach (1995) recently arrived at the conclusion that these structures are non-homologous in Chelicerata and Mandibulata; he considers their striking structural similarity as the outcome of convergent development.

Fahrenbach (1999) studied the ultrastructure of several neuropil masses in the brain of *Limulus*, including the superficial or laminar zone in the optic lobe, the neuropils in the peduncular and central bodies, and the so-called undifferentiated

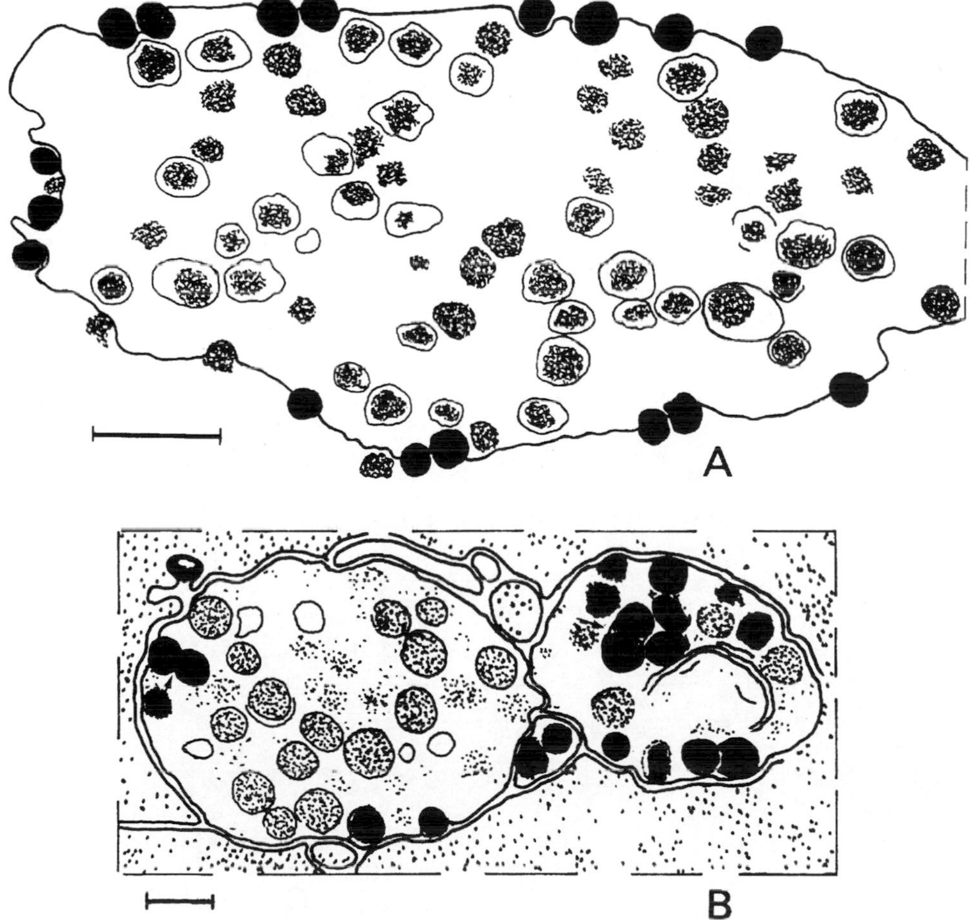

Fig. 26. Non-synaptic exocytoses in insects visualized with the TARI technique. A, Axonal profile in the subesophageal ganglion of the cockroach, *Periplaneta americana*. The profile is filled with dense-cord vesicles and shows numerous exocytoses Drawn from an electron micrograph in Buma and Roubos (1986). Scale: 0.5 μm. B. Profiles of some neurosecretory axons in a corpus allatum of the Colorado beetle, *Leptinotarsa decemlineata*. The profiles contain numerous secretion granula; the dark granula are in exocytosis, causing some of them to fuse together (arrows). The fact that the exocytotic figures do not always appear in contact with the axolemma is due to the tangential plane of sectioning. Scale: 0.2 μm. Drawn from an electron micrograph from Khan and Buma (1985).

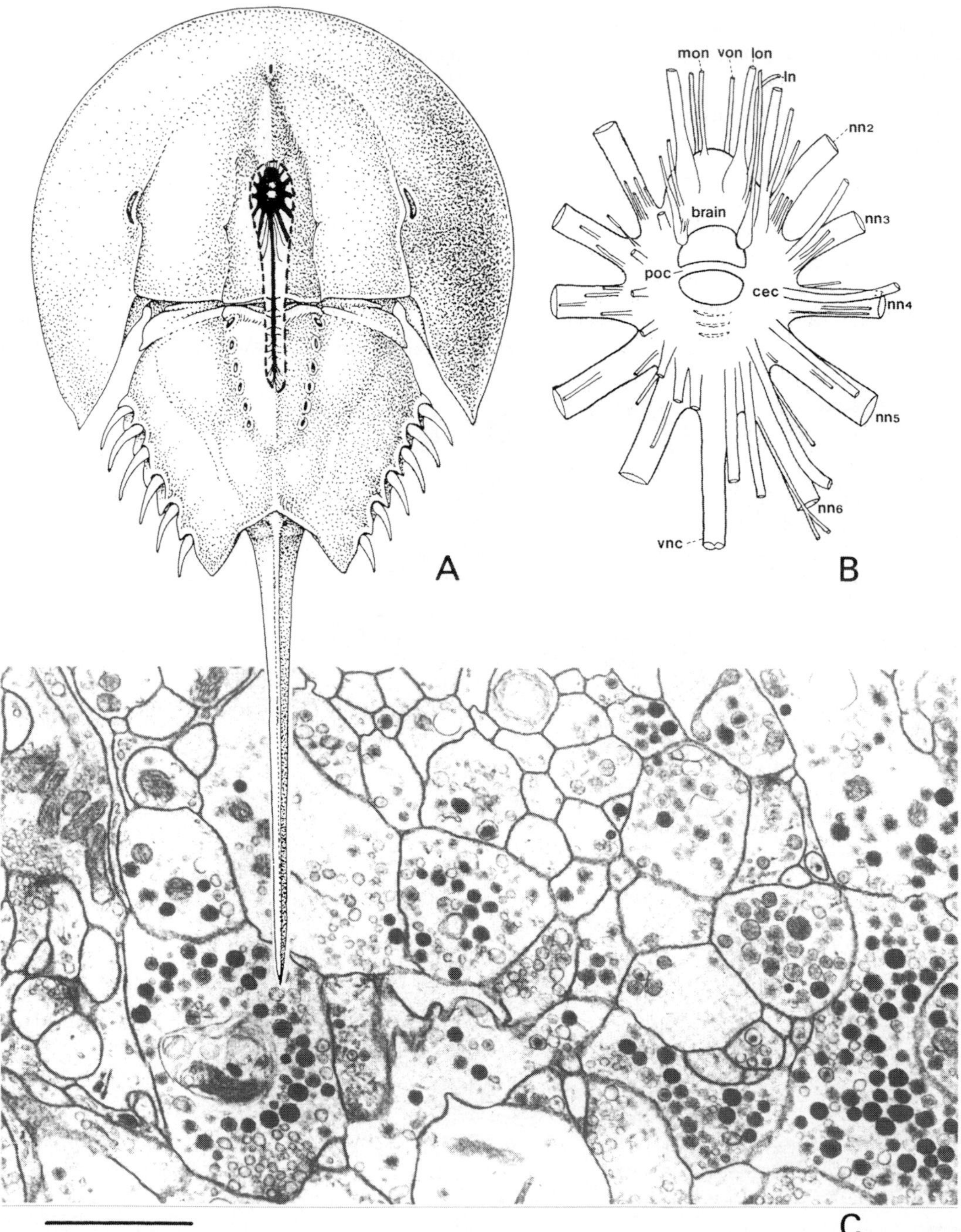

Fig. 27. The horseshoe crab, *Limulus polyphemus*. A. Dorsal view showing the position of the central nervous system. Partly based on Bullock and Horridge (1965: Fig. 21.51). B. The rostral part of the central nervous system. Modified from Hanström (1928). C. Neuropil in the central body of the brain. Scale: 1 μm. Reproduced from Fahrenbach and Chamberlain (1987). cec, circumesophageal connective; lon, lateral optic nerve; ln, lateral nerve; mon, median optic nerve; nn 2–6, neural nerves (to the extremities); poc, pre-oral commissure; vnc, central nerve cord; von, ventral optic nerve.

neuropil of the protocerebrum. He noticed that the axonal profiles in the superficial optic lobe contain almost exclusively small clear vesicles, but that in the other neuropils the constituent fibres contain mixed populations of clear and dense-cored vesicles in seemingly unending variations (Fig. 27C). In most synapses small clear vesicles are clustered against the presynaptic membrane, with dense-cored vesicles at a distance. However, in some synapses dense granula or dense-cored vesicles prevail and also occupy juxtamembraneous positions.

The defining feature of the *Phylum Crustacea* is the possession of two pairs of antennae on the head; mandibulates have one pair and chelicerates none (Margulis and Schwartz, 1998). Many familiar animals belong to the Crustacea phylum, among them crabs, crayfish, lobsters and brine shrimps. In the Crustacea, two basic body forms prevail, depending on whether the abdomen is long and extended or short and flexed under the animal. Crayfish and lobsters represent the long-bodied form, while crabs represent the short-bodied form (Govind, 1992). These general body forms are clearly reflected in the gross structure of the ventral nerve cords of these animals. In long-bodied forms these cords show basically the well-known rope ladder-like configuration, although the left and right segmental hemiganglia are often fused. However, in the short-bodied crabs the thoracal ganglia have condensed to a single complex; the abdomen and the abdominal ganglia are much reduced in these animals.

The dorsally situated crustacean brain is connected with the ventral cord by two long circumesophageal connectives. It can be subdivided into protocerebrum, deutocerebrum, and tritocerebrum, all of which contain one or more central neuropil masses; just as in insects, neurosecretory cells are found in all parts of the crustacean central nervous system. The sinus gland is the principal neurohemal organ in crustaceans; this organ, which is situated on the dorsal surface of the protocerebrum, is composed of numerous neurosecretory axon terminals, clustered about an irregularly shaped blood sinus.

Ultrastructural studies on the central nervous system of crustaceans have yielded several data that are of direct relevance for the central theme of the present review. Bunt (1969) studied the neurosecretory terminals in the sinus gland of the crayfish, *Procambarus clarkii*; all of these terminals were found to contain large electron dense granules and small clear vesicles. Characteristic granule morphologies allowed a classification into five different types of terminals. Bunt visualized numerous exocytoses of large dense granula; moreover, she demonstrated in one type of terminal (her type C) a remarkably sharp segregation of the small clear vesicles and the large dense granula (Fig. 28A). Cuadras (1989) analyzed the ultrastructure of the abdominal ganglia of the same species. He found that: (1) the axon terminals which make up the central neuropil of these ganglia all contain mixtures of large dense-cored vesicles and small clear vesicles in different proportions; (2) the small clear vesicles are, contrary to the large dense-cored ones, usually associated to synaptic membrane specializations, and (3) the dense-cored vesicles release their content at non-synaptic loci, lacking any structural specialization (Fig. 28B). Similar observations were made by Schürmann et al. (1991) in the neuropil of the protocerebral central body of crayfish, Cherax destructur. With the aid of the TARI technique these authors visualized numerous non-synaptic exocytoses of dense-cored vesicles in this neuropil.

8. *Mollusca*

With more than 50,000 described species, the remarkably diverse Phylum Mollusca is the third largest phylum after Mandibulata and Chelicerata. Most molluscs have an internal or an external shell, a muscular foot, and an unsegmented soft body (Margulis and Schwartz, 1998). Molluscs are grouped into seven classes, of which the Bivalvia (mussels and their relatives), Gastropoda (snails and their relatives) and Cephalopoda (squids and octopuses and their relatives) are the most important. The Gastropods form by far the largest class and comprise abundant sea snails and all terrestrial molluscs.

The basic morphological plan of the molluscan central nervous system is a series of about six pairs of well-defined ganglia, one pair being supraeso-

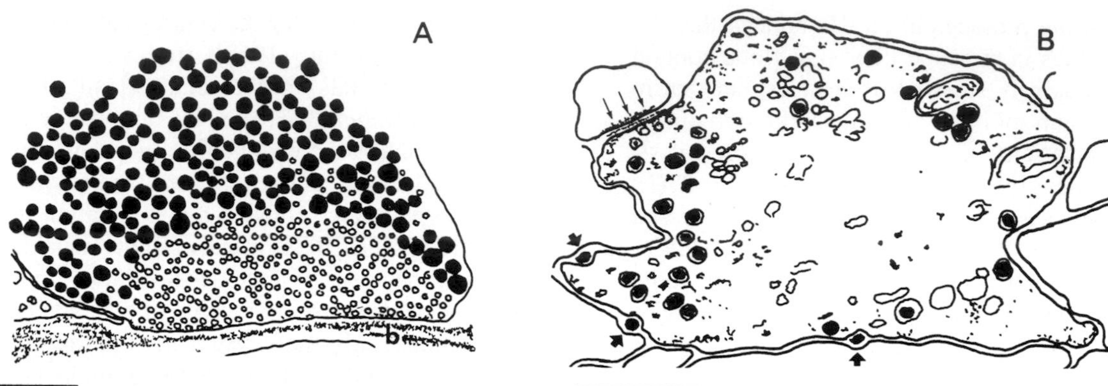

Fig. 28. Axon terminals in crustaceans. A. Neurosecretory terminal in sinus gland of the crayfish *Procambarus clarkii*, containing well-segregated populations of small clear vesicles and large secretion granula, abutting on the basal lamina (b) lining the blood sinus. Bar scale: 0.5 μm. Drawn from an electron micrograph published by Bunt (1969). B. Terminal in the neuropil of the abdominal ganglion of the same species, containing both small clear vesicles and large dense-cored vesicles. Some dense-cored vesicles have released their contents at non-synaptic loci (thick arrows); tThe thin arrows indicate a synapse. Scale bar: 0.5 μm. Drawn from an electron micrograph in Cuadras (1989).

phageal (i.e. the cerebral ganglia), one pedal and primitively elongated, a pleural, a buccal, a branchial or intestinal, and a visceral pair. All of these ganglia communicate by a system of connectives and commissures, which complete a nervous ring around the esophagus (Bullock and Horridge, 1965). In all of the ganglia, a peripheral zone occupied by neuronal perikarya and a central core, consisting of fibres and neuropil, can be distinguished; the nervous system of cephalopods is the largest and most complicated of any of the invertebrate nervous systems. The various ganglia, though still discernable, have fused into a single complex; this complex also encompasses a number of centres, which cannot be distinguished in other molluscs (Budelmann, 1995).

In what follows I will confine myself almost entirely to a single species, belonging to a group of snails, known as the pulmonates, the pond snail, *Lymnaea stagnalis* (Fig. 29A). There are two reasons for this focus: (1) due to a comprehensive 30-year research programme, of the Department of Zoology of the Free University in Amsterdam, coordinated by Jan Lever, Joos Joosse, Theo de Vlieger and the late Harry Boer, the structure and function of the nervous system of this species are exceptionally well known and, as will appear below, (2) *Lymnaea* is the absolute champion of volume transmission.

In pulmonates the head is strongly differentiated and bears one or two pairs of tentacles; there are paired eyes at the bases of the single pair of tentacles or at the termini of the posterior pair of tentacles. Behind the exploring head, the body is largely comprised of digestive and reproductive organs encased in a muscular body wall and shell and a foot that can protrude from the shell (Luchtel et al., 1997). As shown in Fig. 29B, the central

►

Fig. 29. The pond snail, *Lymnaea stagnalis*. A. Holotype. B. The central nervous system in position. C. Diagram showing the location of the neurosecretory cell groups; the pedal ganglia and the ventral parts of the cerebral ganglia are turned to the sides. Modified from Wendelaar Bonga (1970). *Brain parts*: BUC, buccal ganglion; CC, 18, cerebral commissure; CER, cerebral ganglion; COM, commissural lobe; LAT, 17, lateral lobe; PAR, parietal ganglion; PED, pedal ganglion; PLE, pleural ganglion; SUP, superior lobe; VISC, visceral ganglion; *nerves, commissures and connectives*: 1, nuchal nerve; 2, optic nerve; 3, tentacular nerve; 4, superior frontal lip nerve; 5, median lip nerve; 6, cerebrobuccal connective; 7, penial nerve; 8, subcerebral commissure; 9, statocyst nerve; 10, left pallial nerve; 11, cutaneous pallial nerve; 12, anal nerve; 13, intestinal nerve; 14, genital nerve; 15, right internal pallial nerve; 16, right external pallial nerve; *neurosecretory cell groups*: LGC, light green cells; MDC, mediodorsal cells (a); LDC, laterodorsal cells (b); BGC, bright green cells; DGC, dark green cells; YGC, yellow green cells; dro, droplet cells; can, canopy cell; B, B cells; CDC, caudodorsal cells; YC, yellow cells; LYC, light yellow cells.

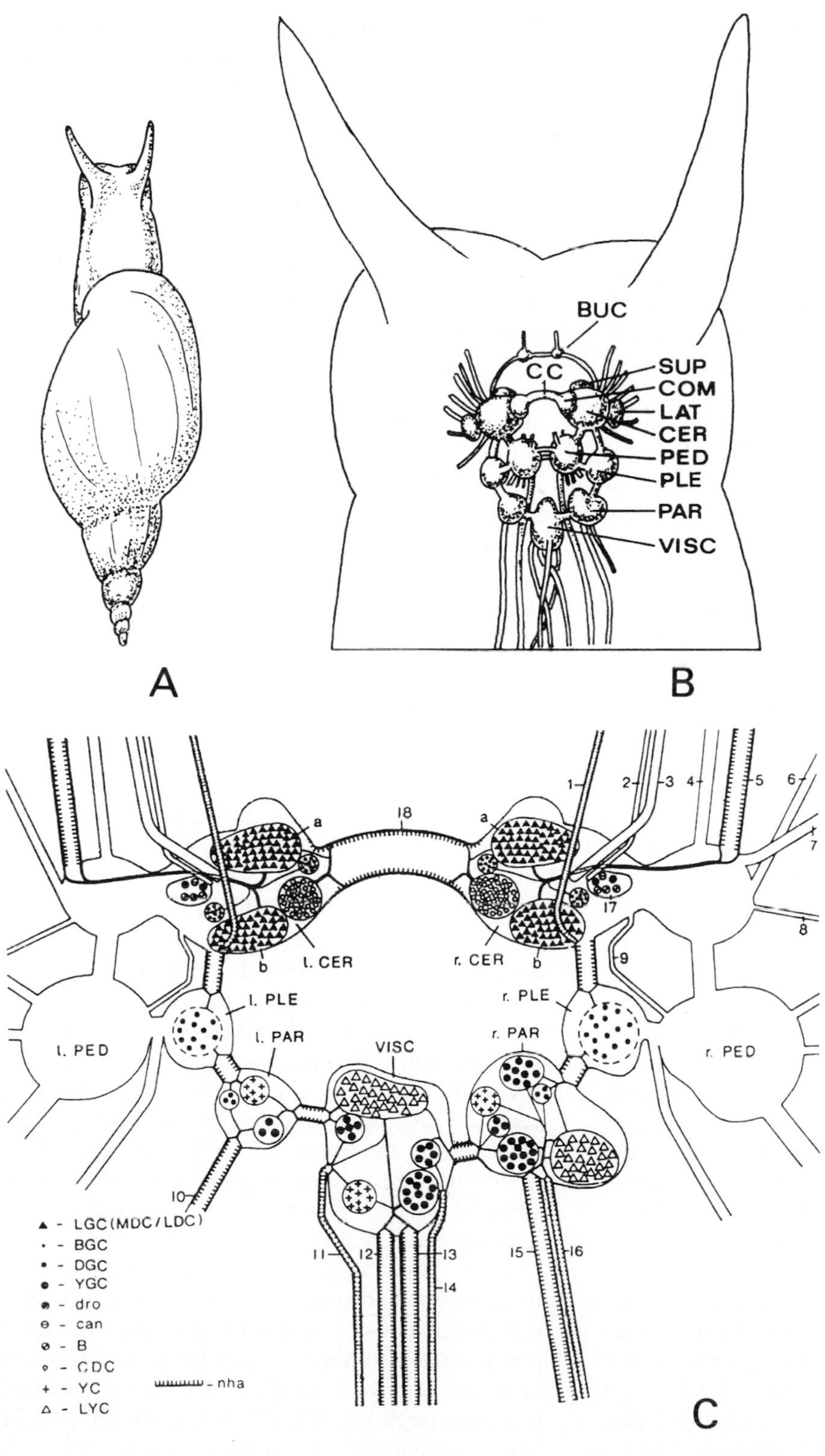
BUC
CC
SUP
COM
LAT
CER
PED
PLE
PAR
VISC
A
B
a
18
a
1
2
3
4
5
6
7
8
17
9
b
l. CER
r. CER
b
l. PLE
r. PLE
l. PED
l. PAR
VISC
r. PAR
r. PED
10
▲ - LGC(MDC/LDC)
· - BGC
• - DGC
● - YGC
dro
can
B
CDC
+ - YC
Δ - LYC
nha
11
12
13
14
15
16
C

nervous system closely adheres to the general molluscan pattern described above, with the reservations that the visceral ganglion is unpaired and that certain regions of the cerebral ganglia are differentiated into separate lobes (Wendelaar Bonga, 1970).

The neuro-endocrine system is well-developed in pulmonates; in *Lymnaea* about 20 different types of neurosecretory cells, distributed over almost all of the circumesophageal ganglia, have been described. Cells of the same type are generally concentrated in distinct groups; the distribution of the most important groups of neurosecretory cells is indicated in Fig. 29C. The neurohemal release sites are formed by axon terminals located in the periphery of commissures, connectives and nerves (marked in Fig. 29C by small dashes oriented perpendicularly to the pertinent structures). They are separated from the large cephalopedal lymph sinus only by a thin, permeable perineurium; the latter sinus, which surrounds almost the entire brain, forms part of a semi-open vascular system, containing the hemolymph.

Three prominent types of neuroendocrine elements in Lymnaea, the caudodorsal cells, the light yellow cells and the light green cells, will now be briefly discussed. (The data presented are derived from the excellent review of Luchtel et al., 1997, to which the reader is referred for references and details.)

The *light green cells* (LGC) form a dorsomedial and lateral cluster in the two cerebral ganglia. As indicated in Fig. 29C, the axons of these elements join in the central neuropil and project to the neurohemal zone in the periphery of the median lip nerves, where they release their neuropeptide content by exocytosis. It has been experimentally established that LGC produce a growth-stimulating hormone; recent molecular biological studies have shown that they actually produce four different insulin-like neuropeptides, the molluscan insulin-related peptides (MIP-I, -II, -III and -V).

The *light yellow cells* (LYC) form large clusters in the visceral ganglion and in the right parietal ganglion (Fig. 29C). Like other neurosecretory cell groups in the visceral complex, they have a vast central neurohemal area, their axonal endings being located in the periphery of the connective, linking the two ganglia involved, and of the nerves numbered 11–16 in Fig. 29C. Axon ramifications of LYC were also observed in the neuropils of all central ganglia; peripheral branches of their axonal systems appeared to innervate muscle cells of the aorta and the ureter papilla. Electrophysiological experiments using fine electrodes on the LYC of intact freely moving animals have shown that these cells fire spontaneous bursts of spikes. Increased burst was correlated with a number of behaviours, as for instance egg laying, copulation and feeding. These observations indicate that the LYC subserve functions common to all of these behaviours, as e.g. the regulation of blood pressure.

LGC and LYC are named so on the basis of their staining with Alcian blue/Alcian yellow (Wendelaar Bonga, 1970).

The *caudodorsal cells* (CDC) are located in the caudodorsal portion of the two cerebral ganglia (Fig. 29C); they form clusters of about 70 cells in the right and 30 in the left ganglion. All CDC are electrotonically coupled by gap junctions; within each CDC cluster, two types of CDC can be distinguished, dorsal and ventral CDC. The dorsal CDC possess only one axon, which passes via the ipsilateral 'loop area' to the neurohemal contact area in the periphery of the large cerebral commissure (Fig. 30A). Ventral CDC have two axons, one of which decussates to the contralateral 'loop area', via the cerebral commissure, and then passes to the neurohemal area; the second axon passes directly to the neurohemal are (Fig. 30A). The neurohemal contact region of the CDC, i.e. the outer compartment of the commissure, is separated from the inner compartment of the commissure by a layer of glia cells. The decussating axons of the ventral CDC form a dense plexus of collateral branches in the inner compartment (Fig. 30A), where they release their peptide contents in a non-synaptic fashion (Buma and Roubos, 1986; Schmidt and Roubos, 1987; Fig. 30B). On the basis of cDNA and in situ hybridization studies, it has been established that the CDC express two neuropeptide genes, called CDCH-I and -II. These genes encode different yet related prohormones, each containing a number (10/11) of predicted bioactive peptides. (CDCH is the 36 amino acid-long ovulation hormone.) Each of the

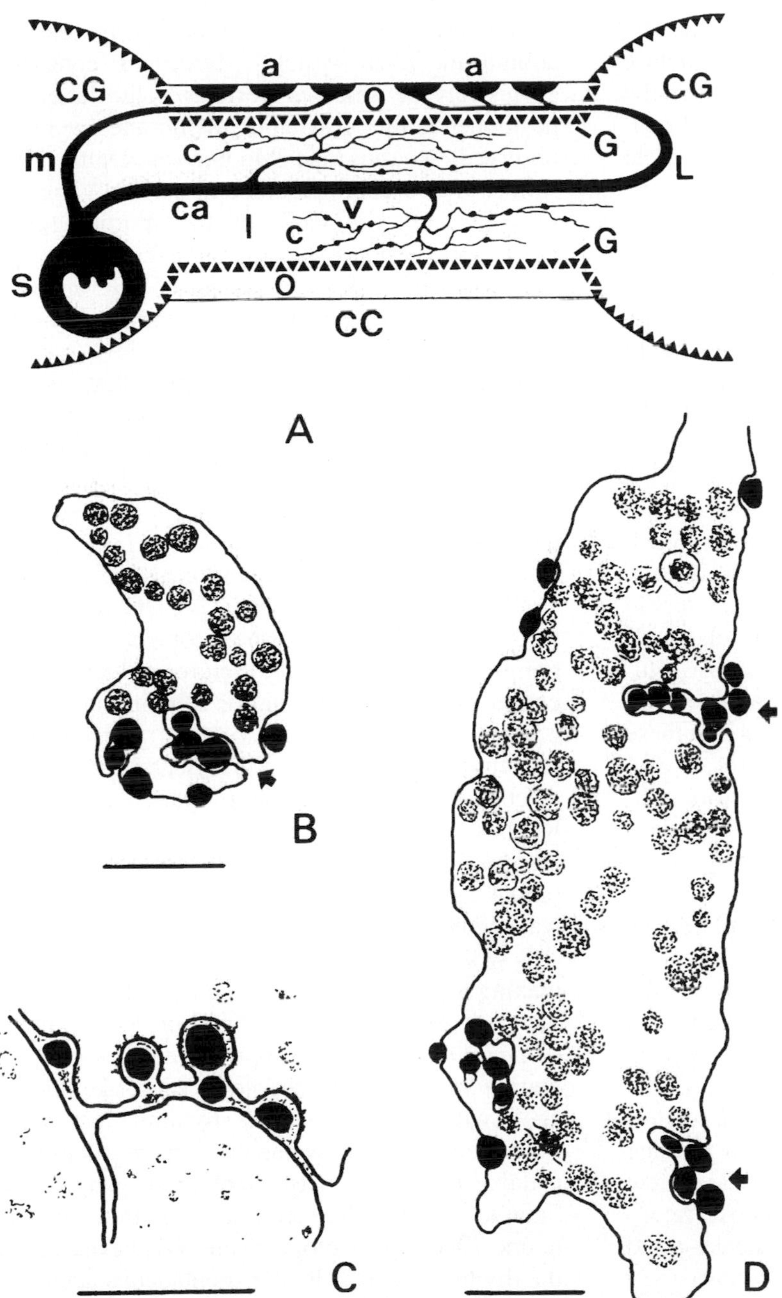

Fig. 30. Non-synaptic release of neuropeptides in the brain of the pond snail, *Lymnaea stagnalis*. A. Soma and axonal ramifications of a ventral caudodorsal cell. Reproduced from Schmidt and Roubos (1987). B.–D. Non-synaptic release visualized with the TARI technique. For location of cells and structures mentioned, see Fig. 29C. Arrows indicate multiple exocytoses. B. Terminal of a caudodorsal cell in the centre of the cerebral commissure. Scale bar: 0.5 μm. C. Exocytoses in the neuropil of the visceral ganglion. Scale bar: 0.25 μm. D. Axonal profile in the neuropil of the right cerebral ganglion. Scale bar: 0.5 μm. B.-D. are drawn from electron micrographs in Buma and Roubos (1986). a, axon terminals; c, collaterals; ca, crossing axon; CC, cerebral commissure; CG, cerebral ganglion; G, glial sheath; l, inner compartment; L, loop area; m, main axon; O, outer compartment; S, soma; v, varicosities.

neuropeptides produced by CDC has a specific function in one particular phase of the reproduction process, as e.g. the transport of the eggs through the genital tract, the supply of protein for the fabrication of the protective egg capsule, the mechanical cleaning of the substrate for oviposition and the deposition of the eggs. CDC exhibit different states of electrical activity, including a resting state, lasting one to several days, and an active state of about sixty minutes; the latter is characterized by sustained electrical activity. Experiments with freely moving animals have shown that egg laying is always preceded by an active state of the CDC; during the active state, the release of neuropeptides by exocytosis from the neurohemal axonal endings of the CDC is high. The CDC of *Lymnaea* are structurally and functionally equivalent to the much studied bag cells of the giant marine snail, *Aplysia*.

I will not deal with the various neuromediators in molluscs; suffice to mention that several 'classical' transmitters, including acetylcholine, GABA, glutamate, dopamine, octopamine, histamine and serotonin, play a role in signal transduction in these animals, and that a variety of neuropeptides are also active in their nervous system. In the brain of *Lymnaea*, more than 60 different neuropeptides have been identified and the great majority of the neurons are peptidergic (Schot, 1984).

Several authors, among them Coggeshall (1967), Tremblay et al. (1979) and Roubos and Moorer-van Delft (1979) have studied the ultrastructure of the central nervous system of snails. Coggeshall, who studied the abdominal ganglion in *Aplysia*, noticed a paucity of typical synapses. In the neuropil of the same structure, Trembley et al. (1979) distinguished four zones: (1) zones of fibres of passage coursing together in the neuropil and making few synaptic contacts; (2) zones of neurosecretory fibres containing large granules and dense-cored vesicles, again making few synaptic contacts; (3) zones with a great variety of synaptic contacts between medium-sized and small profiles, and (4) glomerular zones. Many fibres appeared to contain mixtures of clear and dense-cored vesicles. They observed that in profiles with mixtures of clear and large dense-cored vesicles, it is often only the clear vesicles which agglomerate towards the differentiated membranes, and that in such cases the large dense-cored vesicles lie as a peripheral halo around the clear vesicles. Terminals containing exclusively dense-cored vesicles – the latter also positioned at the presynaptic membrane –were also observed. Roubos and Moorer-van Delft (1979) studied the cerebral, pleural and parietal ganglia of *Lymnaea*. On the basis of the morphology of presynaptic vesicles, they distinguished seven types of true synapses; the neurosecretory LGC and CDC appeared to receive a particularly complex synaptic input.

It has already been indicated that volume transmission plays a very prominent role in the brain of snails. With the aid of the TARI technique, Buma and Roubos (1986; cf. also Buma 1989) demonstrated abundant non-synaptic exocytoses in the following neuronal structures in *Lymnaea* (for location of the structures mentioned, see Figs. 29C and 30A): (a) terminals of LYC, CDC and LGC in their respective neurohemal contact zones; (b) terminals of CDC in the centre of the cerebral commissure (Fig. 30B) and of LYC in the neuropil of the pleural ganglion, and (c) terminals of unidentified neurons in the visceral (Fig. 30C), cerebral (Fig. 30D) and pleural ganglia. Buma and Roubos noticed that in the various ganglia non-synaptic release sites are much more numerous (about 5–10 times) than synapses; multiple release was regularly observed (Fig. 30B, D).

I will conclude this section on molluscs by making a brief reference to the remarkable experiments of Arshavski et al. (1988) on the gastropod, *Clione limacina*. This approximately 4-cm long marine snail has small bilateral wing-like appendages, with which it makes rhythmic locomotory movements; these movements are controlled by the pedal ganglia. When isolated, these ganglia continue to generate the locomotory rhythm for many hours. Electrophysiological analysis revealed that the rhythm is caused by the spontaneous activity of command neurons located in the ganglia. Arshavsky et al. inserted a recording microelectrode into a particular visually-identifiable motor neuron of the isolated ganglia; then, the impaled neuron was extracted from the ganglion by moving the microelectrode. When the isolated motoneuron was replaced between the other ganglion cells, it immediately resumed its rhythmic activity. Arshavski et al. (1988) concluded from these experiments that the command neurons in the pedal ganglia of *Clione* can exert action on motoneurons non-synaptically, i.e. by volume transmission.

9. Remaining non-chordate invertebrate Phyla

Apart from the Phyla discussed so far, studies on members of some other non-chordate invertebrate Phyla, viz. Annelida, Onychophora, Chaetognatha and Echinodermata, have also yielded data relevant to the theme of this review. Hence, these Phyla will now be briefly discussed.

The *Annelida* differ from Platyhelminthes (flatworms) and Nematoda (roundworms) in that they are segmented; their segments are visible externally as ring-like bulges (Fig. 31A). The grooves between the segments coincide with internal compartments, often separated by transverse septa; the digestive tube, principal blood vessels and nerve cords pass lengthwise through many or all segments. Other organs, such as ganglia, nerves, circular blood vessels and excretory organs are serially repeated in most segments.

About 15,000 species of annelids are grouped into three classes, the mostly marine Polychaeta, the terrestrial and freshwater Oligochaeta (to which the earthworm, *Lumbricus terrestris* belongs) and the Hirudinea, or leeches. All of the annelids, except for the leeches, have chitinous lateral bristles, called setae, on each segment; these appendages are used for locomotion or to anchor the annelid to its substrate or burrow. Sense organs are concentrated in the prostomium, a specialized rostral segment (Fig. 31A).

The annelid central nervous system comprises a bi-lobed, dorsally situated supra-esophageal or cerebral ganglion, and a ventral rope ladder-like system of paired longitudinal cords and metameric pairs of ganglia, each with a commissure (Bullock and Horridge, 1965). In many annelids the paired cords and ganglia have fused in the median plane. The cerebral ganglion and the first ganglia of the ventral cords, i.e. the subpharyngeal ganglia, are connected by circumpharyngeal connectives (Fig. 31A). A conspicuous constituent of the nerve cord in many oligochaetes and polycheates, but not in leeches, is the system of giant fibres; these giant fibres, each of which consists of a number of closely packed axons, runs along the cord from the pharyngeal region to the posterior segments. In *Lumbricus*, three giant fibres are situated dorsally in the single ventral cord. As in most other

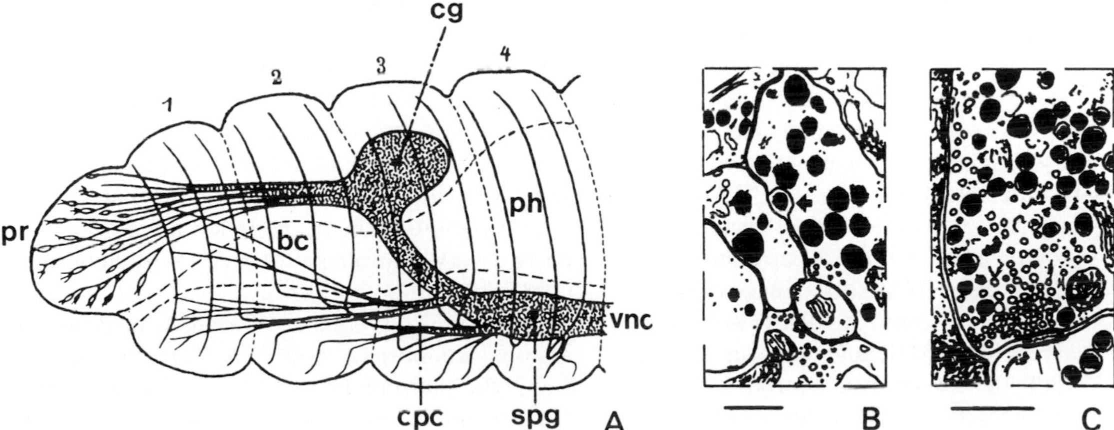

Fig. 31. The earthworm, *Lumbricus terrestris*. A. The rostral part of the nervous system in lateral view; modified from Hanström (1928). B. Fibres in the neuropil of the cerebral ganglion. Arrow, site of exocytosis. Scale bar: 0.5 μm. Drawn from an electron micrograph in: Golding and Bayaraktaroglu (1984). C. Synaptic terminal containing small clear vesiclesand large dense-cored vesicle. The small vesicles are focused on the membrane thickenings (arrows), while the large vesicles 'stand back' from the synaptic junction. Scale bar: 0.5 μm. Drawn from an electron micrograph in: Golding and Pow (1988). bc, buccal cavity; cg, cerebral ganglion; cpc, circumpharyngeal connective; ph, pharynx; pr, prostomium; spg, subpharyngeal ganglion; vnc, ventral nerve cord; 1–4, segments.

invertebrates, the peripheral zone of the ganglia is occupied by neuronal perikarya, whereas their central cores are mainly composed of fibres and masses of neuropil. Remarkably, in the cerebral ganglia of several families of annelids, corpora pedunculata, i.e. stalked bodies resembling those of arthropods (cf. Fig. 24C), have been observed. The annelid cerebral ganglia contain several groups of neurosecretory cells. The axons of these elements pass in bundles to the ventral surface of the brain ganglia, where their terminals abut the neural lamella, i.e. the non-cellular stroma that lines the brain. Together with an adjacent coelomic blood sinus and local blood vessels, the neurosecretory terminals form the cerebrovascular or infracerebral complex. This complex also includes intrinsic neurosecretory neurons, which are located external to the neural lamella; because these neurons were initially interpreted as glandular, the complex is also known as the 'infracerebral gland'.

Golding and collaborators (Golding and May, 1982; Golding and Bayrakteroglu, 1984; Golding and Pow, 1988; Golding, 1992, 1994) studied the ultrastructure of the cerebral and ventral ganglia of the polychaete *Nereis diversicolor* and the oligochaete *Lumbricus terrestris*. Their results, which were partly obtained using the TARI technique, may be summarized as follows. The processes, constituting the neuropil in these ganglia, are richly endowed with dense-cored vesicles, which show great variety in size and density of the cores (Fig. 31B). The clear vesicles show less variation and, unlike the dense-cored vesicles, they are usually thickly clustered adjacent to sites of specialized contacts with other elements (Fig. 31C). Most of the dense-cored vesicles in terminals 'stand back' from the synaptic junction and occupy the more peripheral regions (Fig. 31C). Sites of exocytosis of the contents of dense-cored vesicles were frequently observed (Fig. 31B) and were invariably associated with the expansive, morphologically undifferentiated regions of nerve terminals. "Not one of the hundreds of exocytotic figures we have encountered has involved discharge of the contents of a secretory granule into a synaptic cleft" (Golding and Bayrakteroglu, 1984, p. 1278). On the basis of these results and of similar observations in other invertebrates (including flatworm and mollusc species) Golding et al. concluded that neurons typically possess two dichotomous, secretory mechanisms for the discharge of neurochemical mediators: spatially specific synaptic and diffuse non-synaptic exocytosis. Golding (1994) later added parasynaptic or juxtasynaptic exocytosis as a third mode.

Golding (1992) also studied the neurohemal contact zone of *N. diversicolor*, using the TARI technique; exocytosis from typical neurosecretory terminals appeared to be particularly common adjacent to the neural lamella.

The small Phylum *Onychophora* (velvet worms) comprise about 100 species, all terrestrial; onychophorans are just 15 to 200 mm long and have 13–43 pairs of unjointed legs (Fig. 32A). Their heads are well developed and bear one pair of antennae; small vesicular eyes are situated at the base of each antenna. Onychophorans have features in common with the arthropods as well as with the annelids, and were long considered as 'missing links' between these two groups. Ballard et al. (1992) concluded from a cladistic analysis of 12S ribosomal DNA that the Onychophora should be included in the arthropods.

The onychophoran brain is well developed and caudally connected to the ventral nerve cords by massive circumesophageal connectives (Fig. 32B). There is embryological evidence that the brain is formed from three neuromeres (proto-, deuto- and tritocerebrum), but the boundaries between these segments are not externally visible in the adult. The two ventral cords are interconnected by numerous transverse commissures, but are spatially widely separated (Fig. 32B). The following notes on the microstructure of the onychophoran brain are based on two very informative overviews by Schürmann (1987, 1995).

In the brain, a number of well-defined neuropil regions are present, which are clearly separated from the surrounding tissue by glial sheaths. Schürmann identified no less than nine of such neuropil compartments in the onyhophoran brain, compared to about seven in insects. Most of these appeared to correspond, in both structure and connectivity, with insect neuropils, and were named accordingly. Thus, optic lobe laminae, mushroom bodies, a central body and bilateral

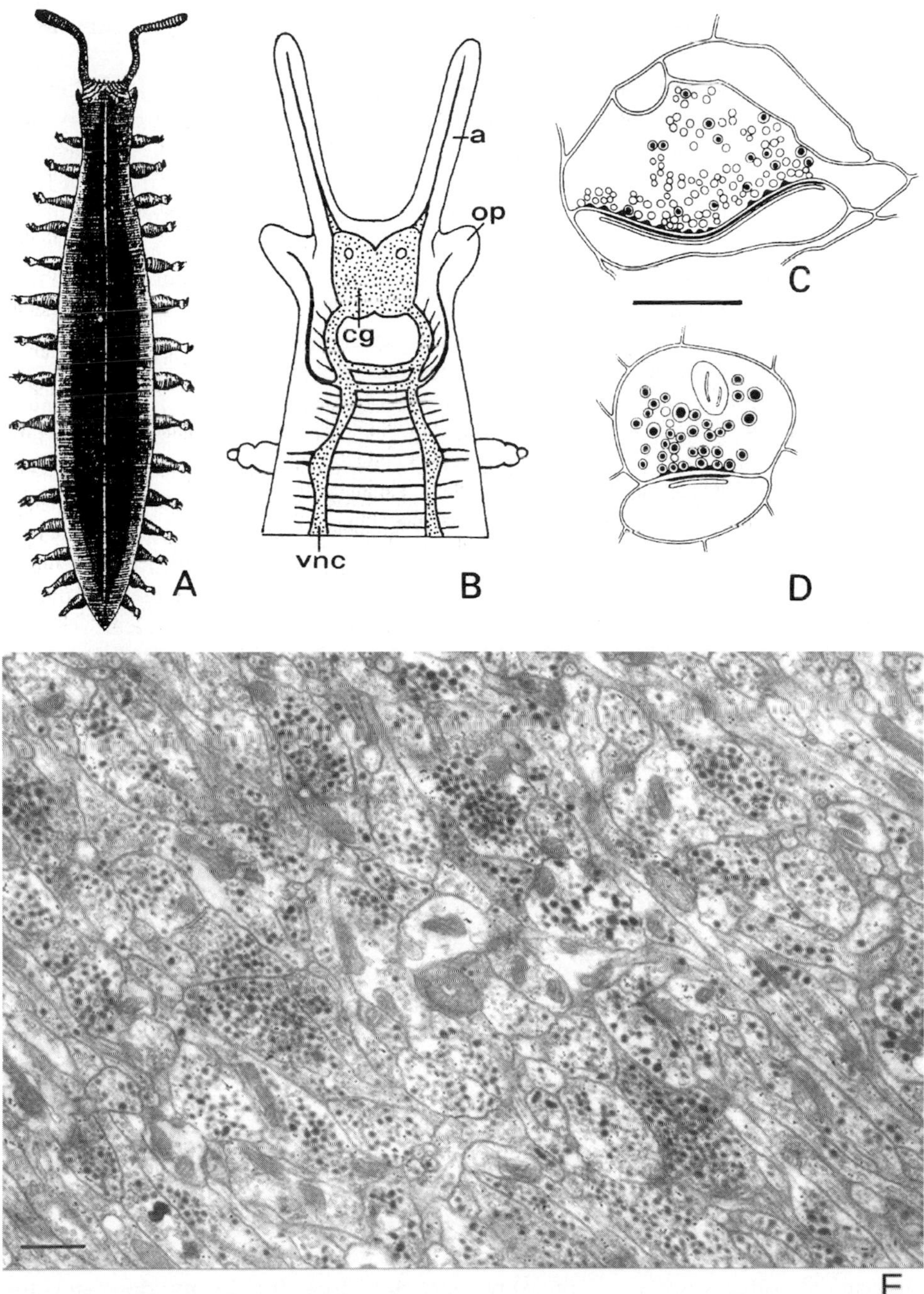

Fig. 32. Onychophora. A. The velvet worm, *Peripatus Novae-Zealandiae*. B. The rostral part of the nervous system of *Peripatus capensis* in position. Modified from Hanström (1928). C,D. Synapses in the central nervous system of *Peripatoides leuckartii*. Scale bar: 0.5 μm. Reproduced from Schürmann (1995). C. Synapse with accumulation of clear vesicles and some dense-cored vesicles. D. Synapse with dense-cored vesicles at the presynaptic membrane. a, antenna with antennal nerve; cg, cerebral ganglion; op, oral papilla with oral papillary nerve; vnc, ventral nerve cord. E. Neuropil in the central body of *Peripatoides leuckarti*. Many axons are packed with dense-cored vesicles, scale: 0.5 μm. Reproduced from Schürmann (1995).

antennal glomeruli neuropils could be distinguished; within these neuropil masses many different types of synapses and synaptic arrangements were observed. Most synapses contained both clear and dense-cored vesicles, the latter generally in 'stand-back' position (Fig. 32C), but occasionally also apposed to the presynaptic membrane (Fig. 32D). Schürmann (1995) also depicted a neuromuscular synapse complex in which six nerve terminals impinge on a single end-plate structure. Remarkably, all of these terminals contain almost exclusively small clear vesicles. In the neuropil of the central body, Schürmann noticed numerous fibres packed with dense-cored vesicles and a paucity of synaptic contacts (Fig. 32E); he characterized the onychophoran central body as a key structure, appropriately positioned to modulate vast parts of the brain.

Chaetognatha are marine worms distinguished by the presence of horizontally projecting fins and, at the anterior end, two groups of moveable grasping spines (Fig. 33A). Chaetognathes, also known as arrow worms, are extremely rapid, active predators who use their spines for capturing prey; they detect prey mainly with vibration sensors. Paralytic neurotoxins, released from pores adjacent to the mouth, prevent prey from escaping. Chaetognathe species range in length from 0.5 to 15 cm; about 70 species have been described; their affinities are obscure.

The central nervous system of chaetognathes consists of six ganglia in the head and an exceptionally large ventral ganglion in the trunk (Fig. 33A). The dorsally situated cerebral ganglion is linked by the main connectives to the ventral ganglion, and to paired vestibular ganglia on either side of the esophagus. A short connective links the two vestibular ganglia with small esophageal ganglia on either side of the esophagus; a caudal commissure interconnects the two vestibular ganglia. From the middle of this caudal commissure a small connective passes rostrally to the minuscule subesophageal ganglion, which is visible but not labeled in Fig. 33A (Goto and Yoshida, 1987; Bone and Goto, 1991; Shinn, 1997). The cerebral ganglion coordinates the activities of the other ganglia; the vestibular ganglion is thought to control operation of the grasping spines, whereas the ventral ganglion is involved in the control of swimming.

The cerebral ganglion contains a large central mass of neuropil, which is bordered dorsally and laterally by a layer of densely packed perikarya; a glial sheath separates the neuropil from the cell layer. Ultrastructural studies have shown that in the neuropil most of the fibre profiles are closely apposed without any intervening glial material, and that many of these fibres are packed with dense-cored vesicles. Numerous synapses with thickened post-synaptic membranes and clusters of clear vesicles apposed to the presynaptic membranes exist within the central neuropil (Rehkämper and Welsch, 1985; Goto and Yoshida, 1987; Fig. 33B). In the large ventral ganglion, the neuropil is bordered laterally by a zone of perikarya; this neuropil is mainly composed of thin fibres with a diameter of 1 μm or less. There are many synaptic contacts, all of which contain small clear vesicles some 40 nm in diameter. A few fibre profiles in this ganglion contain larger (80–90 nm) dense-cored vesicles, but synapses with vesicles of this type have not been seen (Bone and Pulsford, 1984; Goto and Yoshida, 1987).

The *Echinodermata* are radially symmetrical, pentameric marine animals, the body of most of which is composed of a central disc or body and arms of varying length extending from the disc. The central disc harbours the intestinal tract, an elaborate system of coelomic spaces, and the main part of the nervous system; frequently the arms branch, the branching points being characteristically in a proximal position. A unique feature of echinoderm organization is a well developed water circulatory system; this system, which is a derivative of the coelom, contains seawater filtered through a special sieve plate. It is connected to numerous tube feet that project through the body wall; these tube feet are operated by hydraulic pressure and are used for locomotion and for grasping. An internal calcium carbonate skeleton protects the arms and central body. Although the adults display radial symmetry, they develop from tiny, bilaterally symmetrical larvae, such as the wonderful, transparent pluteus larvae of echinoids and ophiurids, or similar larvae of the other echinoderm groups. These larvae have bands of ectodermal cells with long cilia that run bilaterally along their body edges and are used for swimming.

The echinoderms, which include about 7,000 extant species, are subdivided into the following five classes: Crinoidea or sea lilies (Fig. 34A), Asteroidea or starfishes (Fig. 3B), Ophiuroidea or serpent stars, Echinoidea or sea urchins and Holothuroidea or sea cucumbers.

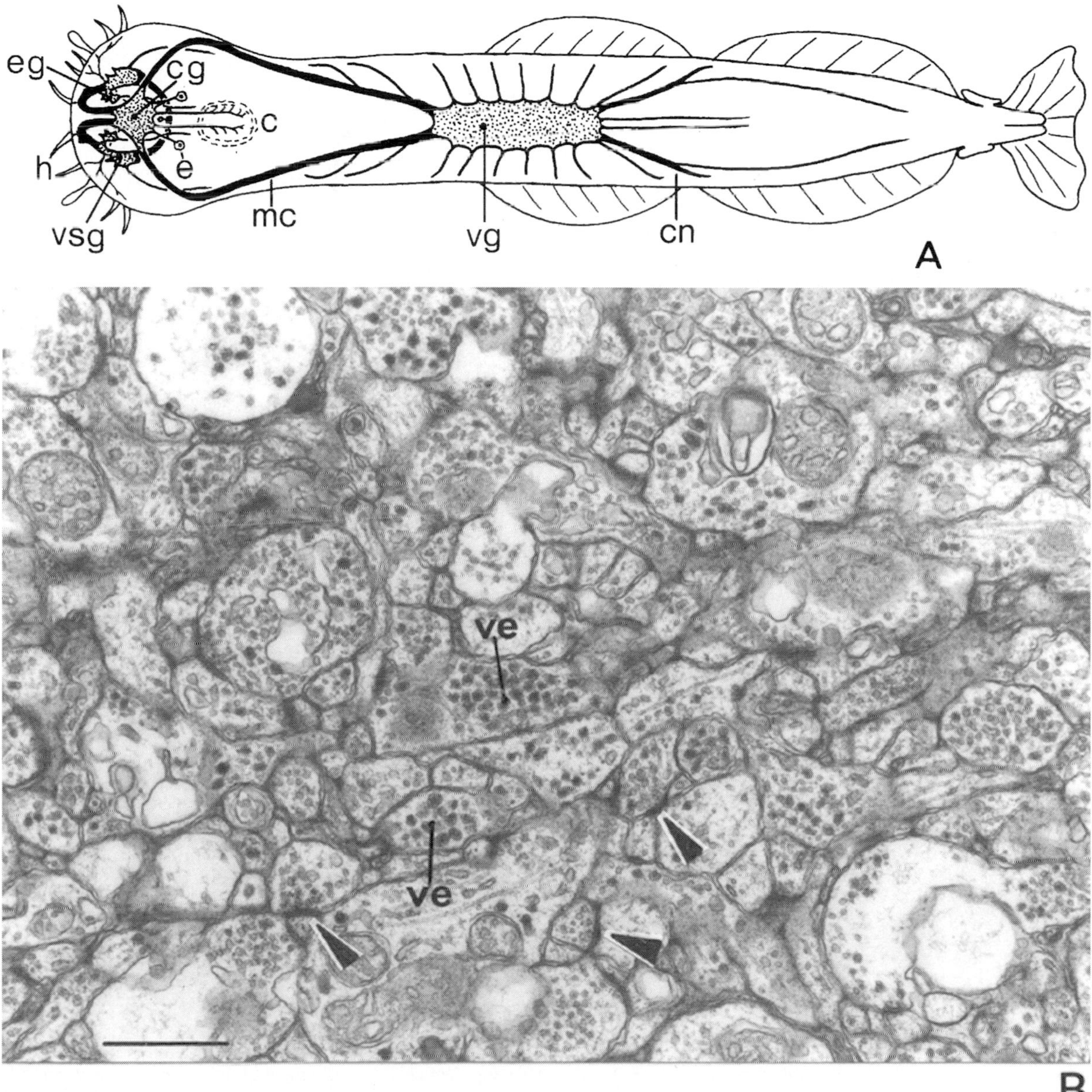

Fig. 33. Chaetognatha (arrow worms). A. Diagram showing the general plan of the chaetognath nervous system. Dorsal view. Combined from figures by Bütschli (1910) and Goto and Yoshida (1987). B. Neuropil in the cerebral ganglion of *Aidanosagitta crassa* showing dense-cored vesicles (ve) in neurites and synaptic contacts between neurites (arrowheads). Bar: 0.5 μm. Reproduced from Goto and Yoshida (1987). c, corona; cg, cerebral ganglion; cn, caudal nerve; e, eye; eg, esophageal ganglion; h, hooks; mc, main connective; vg, ventral ganglion; vsg, vestibular ganglion.

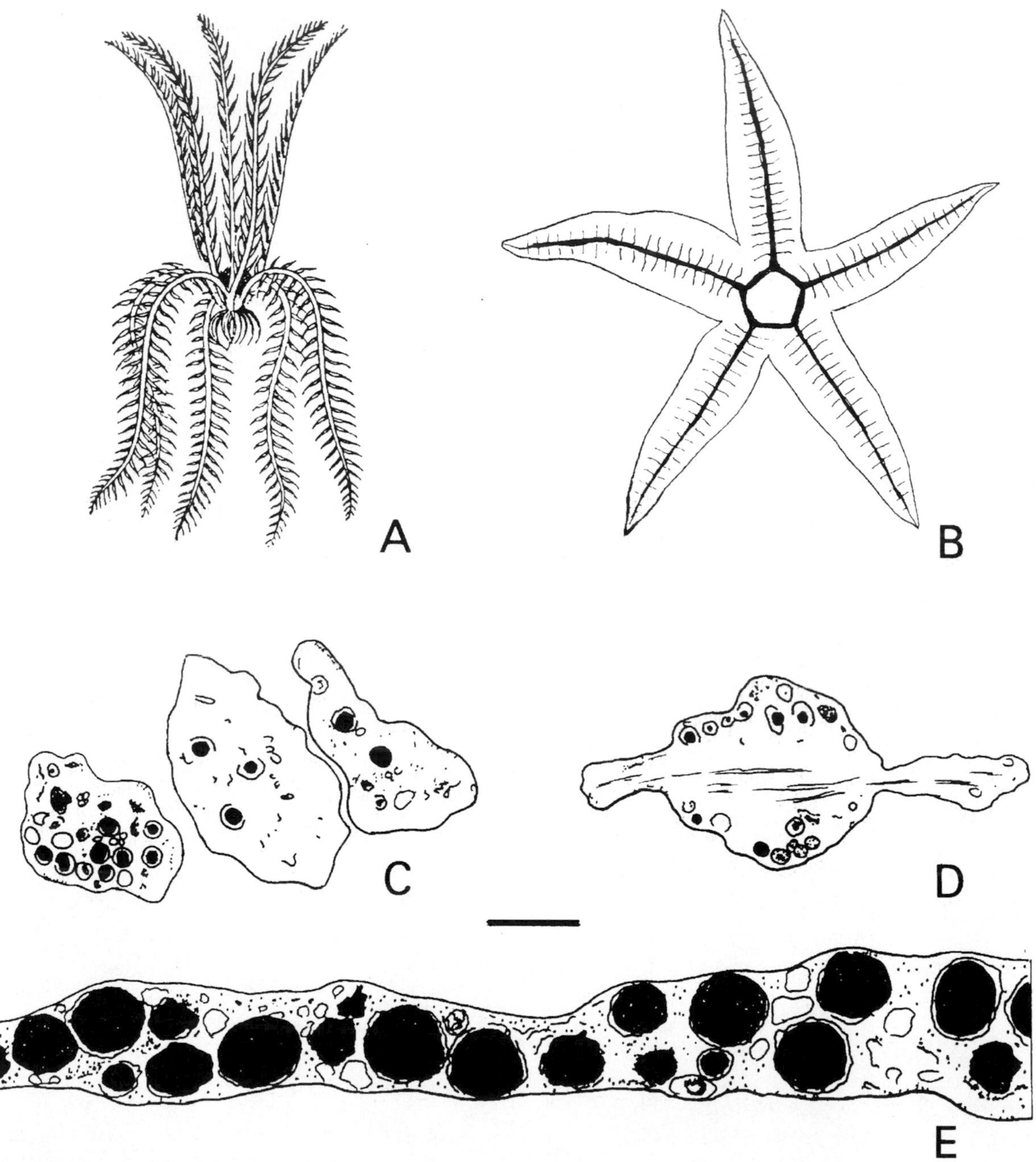

Fig. 34. Echinodermata. A. The crinoid, *Antedon bifida*. B. Diagram showing the general arrangement of the ectoneural nerve strands in a starfish. Modified from Kuhlenbeck (1967). C-E. Fibres in the nervous system of Antedon bifida. Bar: 0.5 μm. Drawn from electron micrographs in: Heinzeller and Welsch (1994). C. Fibres in the neuropil of the ectoneural plexus. D. Varicosity in the ectoneural plexus traversed by neurotubules. E. Fibre in the hyponeural ring commissure containing extremely large, dark granules.

The echinoderm nervous system consists of two major parts, the ectoneural nervous system and the hyponeural nervous system.

The ectoneural system comprises a circumoral nerve ring and radial cords that branch off from the circumoral ring and extend throughout the length of each arm (Fig. 34B). The circumoral nerve ring consists of circumferentially running axons connecting the radial nerve cords and local areas of neuropil associated with the visceral innervation, motor output and hyponeural ganglia (Cobb, 1987). The older literature often states that the circumoral ring represents a coordinating centre, but according to Cobb (1995, p. 411), there is no evidence whatsoever for such a central function: "The idea that it acts as any sort of central control can be totally ruled out since it simply does not have the structure for this." The ectoneural tissue of the radial nerve cords consists of complex ganglia associated with each segment and interganglionic connectives; there are segmental side branches, passing perpendicularly out from the cords to the periphery (Cobb, 1987; Fig. 34B). The ectoneural system is peripherally connected with an extensive, ganglionated basi-epithelial plexus, into which sensory axons feed from widespread receptor cells.

The hyponeural nervous system, which is only clearly differentiated in asteroids, ophiurids and holothuroids, is purely motor; its constituent neurons are situated on the internal surfaces of the circumoral ring and radial cords.

Cobb (1987, 1989, 1995) has pointed out that the ectoneural and hyponeural systems have a different origin, and that this developmental difference is clearly reflected in the mutual relations of the two systems, as well as in the relations of both with the musculature. His observations may be summarized as follows. (1) The ectoneural system is of ectodermal origin, but the hyponeural system is of mesodermal origin. (2) All ectodermally derived tissue is separated from the remaining tissues by a basement membrane. (3) At no point does the ectoneural system penetrate this basement membrane, consequently ectoneural nerve endings do not directly connect with the hyponeural system and cannot directly innervate muscle. (4) In both systems, varicose vesicle-filled axons of the ectoneural nervous system, which abut the basement membrane, are influencing their neural or muscular targets by volume transmission. The contents of their synaptic vesicles, once released, are assumed to diffuse across the basement membrane. Cobb and collaborators have published several ultrastructural studies, substantiating the remarkable relations just described, e.g. Cobb 1985a, b, 1987; Cobb and Stubbs, 1981, 1982; Stubbs and Cobb, 1981, to which the reader is referred for details. Here, I confine myself to mentioning that small, clear vesicles have been found to be present mainly at the direct neuromuscular junctions of the hyponeural system, and that many nerve fibres are filled with vesicles of different types and appear to be varicose with no synaptic specializations.

The ultrastructure of the nervous system of the crinoid *Antedon bifida* (Fig. 34A) has been thoroughly studied by Heinzeller and Welsch (1994); they found that the nervous tissue of this species, first and foremost, consists of neuropil. Within this neuropil they noticed fibres containing dense-cored vesicles in various concentrations (Fig. 34C), among them numerous varicose fibres (Fig. 34D); thin fibres, densely packed with exceptionally large membrane-bound granula were occasionally observed (Fig. 34E).

10. Chordate invertebrates

The chordate invertebrates encompass the Phyla Tunicata and Cephalochordata, both of which are highly relevant for the central theme of this review.

Tunicates are soft-bodied marine animals, encased in a gelatinous or leathery tunic; this tunic, which gives the Phylum its name, is a secretion product of the epidermis. Within the tunic is the body that contains a large perforated pharynx or branchial sac; this organ filters phytoplankton from incoming sea water and exchanges oxygen and carbon dioxide from the seawater.

The phylum *Tunicata* is divided into three classes, the Ascidiacea, the Appendicularia and the Thaliacea. In what follows I will confine myself to the ascidians or sea squirts, which are by far the most familiar and best-studied tunicates; they are represented in the recent fauna by about 2,000 widely distributed species. Adult ascidians, like *Ascidia canina*, are bag-like, sessile animals with two funnel-like processes, the incurrent siphon and the excurrent siphon (Fig. 35E). Via the incurrent siphon, feeding and respiratory current of water enters the large pharynx, which is perforated by numerous gill slits; the pharynx is surrounded by the atrium, a cavity that is directly connected with

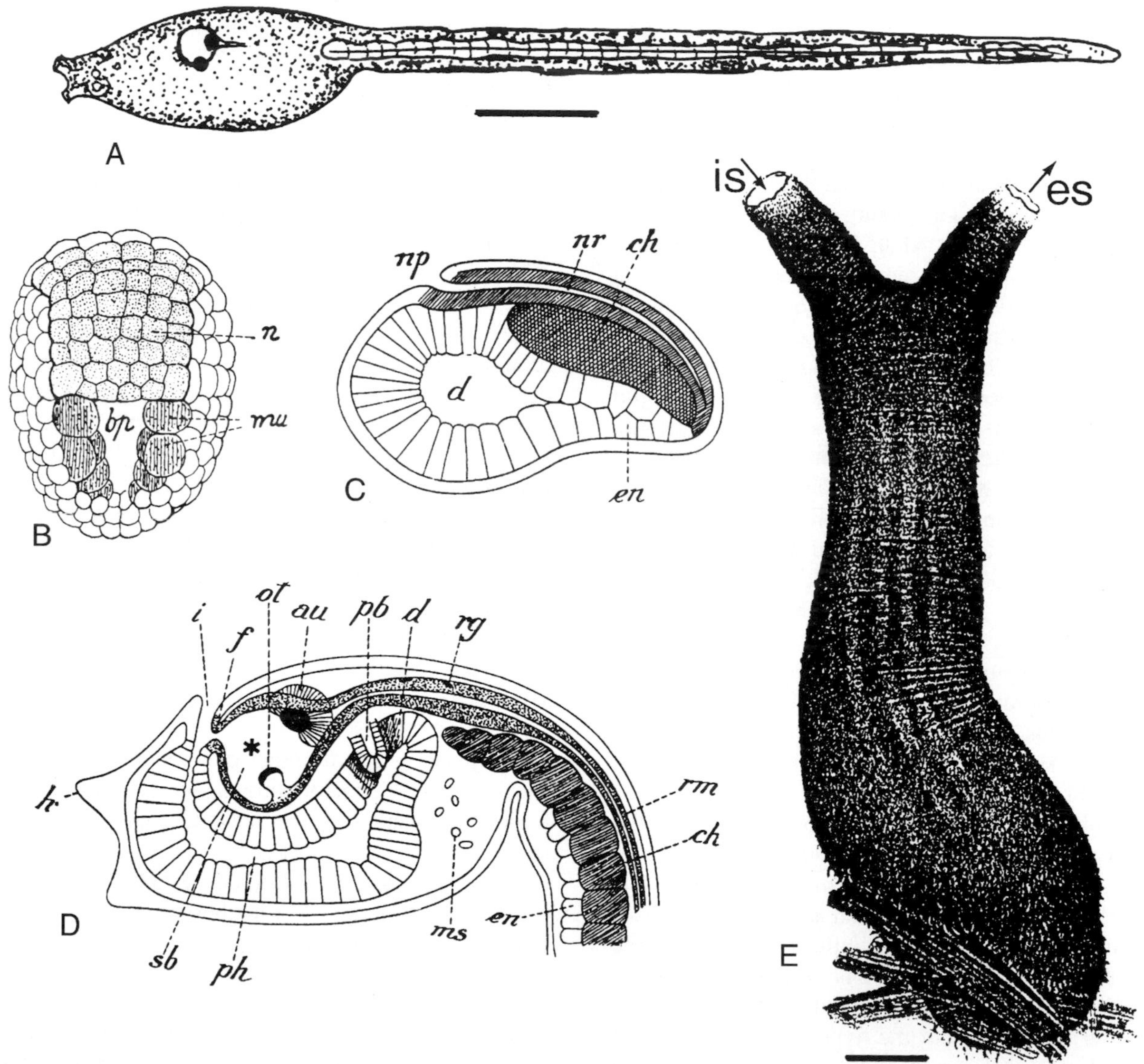

Fig. 35. Urochordata. A. Lateral view of a 19–20 h. tadpole of the ascidian *Ciona intestinalis*; the animal is transparent. In the head region the prosencephalic ventricle with two pigmented cells, the dorsal and ventral melanocytes can be seen. The notochord runs down the centre of the entire tail. Bar: 0.1 mm. Largely based on a photograph in Katz (1983). B. A late gastrula stage of *Styela partita* in dorsal view. Reproduced from Korschelt (1936). Diagrammatic median sections through ascidian embryos. Reproduced from Heider (1928). C. Older embryo. D. Embryo just before hatching. E. *Ascidia canina*, adult. Bar: 1 cm. Reproduced from Kupffer (1870). au, eye; bp, blastopore; ch, notochord; d, gut; en, subchordal entoderm; es, excurrent siphon; f, ciliated groove; h, adhesive papillae; i, mouth; is, incurrent siphon; ms, mesenchyme cells; mu, myoblasts; n, neural plate; np, anterior neuropore; nr, neural tube; ot, otolith; pb, peribranchial cavity; ph, pharynx; rg, trunk ganglion; rm, spinal cord; sb, brain vesicle.

the excurrent siphon. The feeding current is filtered through the pharynx wall and the food passes on into the alimentary system, while the filtered water passes into the atrium and leaves by the exhalent siphon. The atrium also serves as an exit for faeces, because the alimentary canal discharges in it. The feeding current is generated by cilia on the pharynx; so, the ascidians are typical filter-feeders, just like the sponges (Section III. 2). However, unlike the latter, they have a true central nervous system, consisting of a small cerebral ganglion, strategically located midway between the two siphons. Nerves, all mixed sensory and motor, radiate from the ganglion. Histologically, the ascidian cerebral ganglion is a typical invertebrate ganglion with a central core of fibres and neuropil and an outer zone of nerve cell bodies.

The organization of the cerebral ganglion of the ascidian *Polyandrocarpa misakiensis* has been studied by Koyama and Kusunoki (1993); they noticed the presence of numerous synapses, the size of which appeared to be variable, and some had dense contents. Two observations of Koyama and Kusunoki deserve special mention: (1) Contrary to what is seen in the ganglia of other invertebrates, the peripheral, perikaryal zone of the ganglion of P. misakiensis appeared to contain numerous axosomatic, somatoaxonic and somatosomatic synapses. (2) The ganglion is largely surrounded by blood sinuses; some neurons were observed to send their processes directly into the lumen of these sinuses.

All of this, however interesting, was not the specific reason for including the ascidians in this review. The reason for their inclusion is that ascidians have an indirect development; they pass through a larval stage that differs in almost every respect from the adult. These larvae are free-swimming, they look like tiny amphibian tadpoles (Fig. 35A) and, most remarkably, they have a notochord and a dorsally situated, hollow central nervous system (Fig. 35C, D) that develops from a neural plate (Kowalevski, 1866; Kupffer, 1870; Fig. 35B). This mode of development of the central nervous system differs radically from that in all non-chordate invertebrates and closely resembles that in vertebrates; in all non-chordates the central nervous system develops from multiple zones or patches of proliferative ectoderm. The presumptive neural cells detach from these anlagen and migrate inwardly to participate in the formation of the ganglia; hence, all invertebrate ganglia and ganglionic complexes are solid structures (Korschelt, 1936). In vertebrates, on the contrary, the central nervous system develops from a continuous dorsal neural plate, which rolls in and becomes transformed into a hollow, tube-shaped neuraxis. The cavity enclosed by the neural walls becomes transformed into a system of interconnected fluid spaces providing an important, conduit for informational substances and, hence for volume transmission (Fig. 1).

The central nervous system of larval ascidians consists of a widened, thin-walled vesicle, which caudally tapers into the anlage of the spinal cord; the ventricular cavity of the brain vesicle (marked with an asterisk in Fig. 35C) is externally visible in the transparenttadpole (Fig. 35A). The brain vesicle is often designated as 'sensory vesicle', because parts of its otherwise mainly ependymal wall are differentiated into an otolith which serves as a balance organ, a single ocellus and a presumed hydrostatic pressure receptor. Nicol and Meinertshagen (1991) established that the average number of cells in the larval central nervous system of *Ciona intestinalis* is 335; the correspondence of the central nervous system of larval ascidians with that of vertebrates is substantiated by striking similarities in the expression patterns of a number of regulatory genes (Di Gregorio and Levine, 1998). Thus, the rostral expression limits of Hox3 and Hox5 genes, which in vertebrates are situated in the rhombencephalon, fall in ascidian tadpoles, in the transitional area of brain vesicle and spinal cord (Locassio et al., 1999).

If the ventricular system of ascidian larvae plays a role in volume transmission it is doing so only briefly. After one or a few days of swimming the larvae adhere with their perioral adhesive papillae (Fig. 35A) to a rock and undergo extensive metamorphosis into the sessile adult (Fig. 35E). Tail and notochord are resorbed and the tubular dorsal nervous system is replaced by a small ganglion.

The small and very homogeneous Phylum *Cephalochordata* is made up of 23 species, all of which are marine; although cephalochordates resemble fishes, they have a much simpler organization than these craniates (Fig. 36A). There is no true head, and paired sense organs are entirely lacking; their body axis is formed by a well-developed notochord, extending from the very tip of the rostrum to the end of the tail (Fig. 36B). As their body shape suggests, cephalochordates are able to swim effectively; the propulsive force is provided by the patterned contractions of the body musculature, which is segmented into a series of some 60 myotomes on either side of the body (Fig. 36A). Cephalochordates live in shallow waters in various regions of the world; they spend most of their time buried in the sand, with only their rostral end protruding above the surface. In this position they feed by extracting small particles from a stream of water drawn in by ciliary action; so, just like adult ascidians, cephalochordates are typical filter feeders and they share a very large, perforated pharyngeal basket with that group. Pharyngeal gill slits, a notochord and a hollow, dorsally situated central nervous system are the three defining features of chordates (tunicates,cephalochordates and craniates). We have seen that in tunicates, the notochord and the tubular neuraxis disappear during metamorphosis, but all three persist in adult cephalochordates. As shown by Fig. 36B, in adult cephalochordates the central nervous system is represented by a cord, situated directly dorsally to the notochord and extending throughout almost the entire extent of the body. A striking difference with all craniates is the complete absence of a frontal accumulation of nervous tissue; as can be seen in Fig. 36C, the rostral end of the neural cord tapers and does not show any enlargements.

All living cephalochordates resemble each other closely, and with reference to an old generic name, all are designated as amphioxus.

In what follows, I will not dwell on the structure of the adult cephalochordate nervous system (for review, see: Nieuwenhuys, 1998); rather, I will draw attention to the recent studies by Lacalli and collaborators (Lacalli et a., 1994; Lacalli, 1994, 1996a, b; Lacalli and Hou, 1999; Lacalli and Kelly, 2000), which are highly relevant for this review.

Lacalli and collaborators studied the rostral part of the central nervous system of 12.5-day-old larvae of the lancelet, *Branchiostoma floridae*, by making graphical and three-dimensional reconstructions from serial electron micrographs. So far as methodology and approach are concerned, these meticulous studies are comparable with the work of White et al. (1986) on the tiny nematode *Caenorhabditis elegans* (see Section III. 6).

A midsagittal view of the rostral part of the neuraxis of a 12.5-day larva of B. floridae is shown in Fig. 36D. Figure 36E is a circuit diagram of the same region in larvae of corresponding age, but also yields some information on the position and extent of certain structures. With the aid of these two figures, Lacalli's results on the larval amphioxus brain may be summarized as follows.

The larval cerebral vesicle can be subdivided into a wider rostral part and a narrower caudal part; the boundary between these two parts is marked by a conspicuous cluster of secretory infundibular cells. These cells, which are provided with backward projecting cilia, are responsible for producing Reissner's fibre, a fibrous strand also found in the vertebrate neuraxis. The front part of the cerebral vesicle is tubular with a cylindrical central canal, opening anteriorly through a neuropore; just beneath the neuropore is a cup-shaped pigment spot that forms part of the frontal eye complex. Apart from the pigment spot, this complex includes four rows of cells (R1–R4 in Fig. 36D, E); the first two rows are visual receptor cells and the remaining rows are occupied by neurons and include internuncial as well as long-axoned projection elements. The cells forming the frontal eye complex are tall and flask-shaped and the same holds true for the elements lining the remainder of the ventral and ventrolateral walls of the anterior cerebral vesicle; the more dorsal cells are flatter and less numerous. Just in front of the infundibular region, there is a cluster of distinctive cells with highly swollen, matrix-filled cilia; this cluster has been tentatively interpreted by Lacalli (1996a) as a gravity receptor or balance organ.

Beginning near the front of the posterior cerebral vesicle, the ventricular cavity expands dorsally to accommodate the lamellar body, a sausage-shaped mass of membranous lamellae that extend the whole length of the posterior cerebral vesicle. The lamellae arise in parallel stacks from the sides of

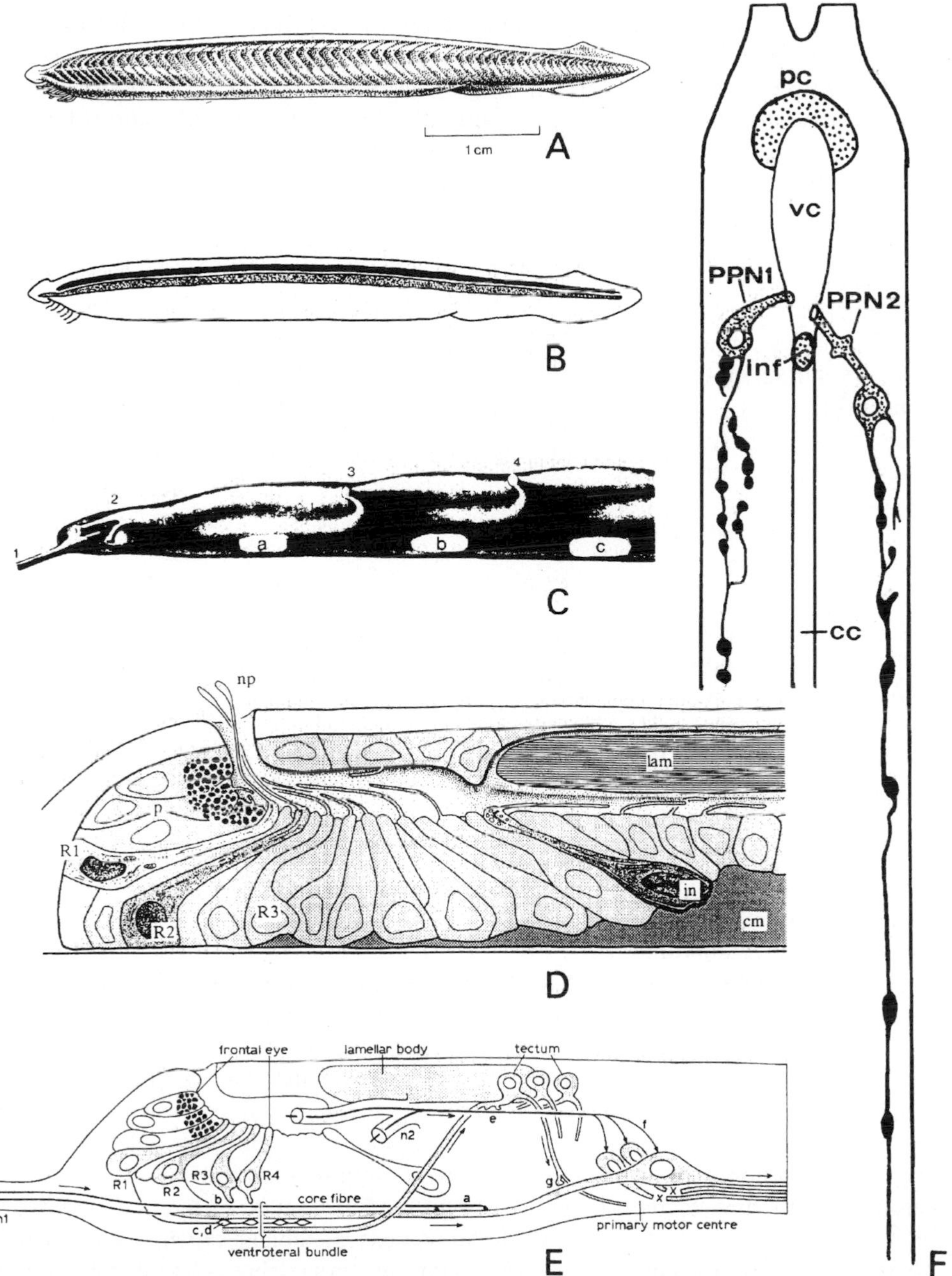

Fig. 36. Cephalochordata. A. The lancelet, *Branchiostoma lanceolatum*. B. Diagram showing the central nervous system (black) and the notochord (dotted) of the lancelet in position. C. The rostral part of the central nervous system of *B. lanceolatum* in lateral view. D. Median section through the rostral part of the brain of a 12.5-day-old larva of *B. floridae*. E. Diagram showing the pattern of connectivity in the brain of a 12.5-day larva of *B. floridae*. Synaptic contacts between the various elements are indicated with the letters a-g. F. Graphical reconstruction of a PPN1 and a PPN2 cell in dorsal view. Figs. A, C, D, E are reproduced from Nieuwenhuys (1998). cc, central canal; cm, principal commissure; in, inf, infundibular organ; lam, lamellar body; n1,2, first and second nerves; np, neuropore; p, pc, pigment cell cluster; PPN1,2, preinfundibular projection neurons types 1 and 2; R1-R4, transversely oriented cell rows; vc, ventricular cavity.

cilia belonging to cells arranged in a row along each side of the lamellar body. The lamellar body terminates just in front of a cluster of motor neurons and large projection neurons, together forming the primary motor centre (Fig. 36E). In the dorsal region of the posterior cerebral vesicle there is a bilateral group of small elements, which Lacalli (1996b) referred to as tectal cells. The long dendrites of these cells project forward to the region of the frontal eye and are apparent targets for terminals from receptor cells in that organ; the axons of the tectal cells pass ventrocaudally and enter the primary motor centre (Fig. 36E).

Lacalli compared the structural relations observed in the larval amphioxus with those in adult craniotes, as well as with those in certain larval invertebrates and he suggested the following homologies with vertebrates:

(1) The frontal eye represents, on account of its position, structure and connections, the counterpart of the paired eyes, the optic stalks and the chiasmatic ridge of craniotes.
(2) The group of infundibular cells represents the infundibular region of the craniate brain. The fact that the infundibular cells are responsible for producing Reissner's fibre in amphioxus, whereas this filament in vertebrates is formed by the dorsal subcommissural organ, does not nullify this homology, because in embryonic teleosts the substance of the fibre comes partly from the infundibular region (Olsson, 1993).
(3) The area of high columnar cells situated between the frontal eye and the infundibular cells corresponds to the craniate hypothalamus.
(4) The lamellar body may be homologized, on grounds of both position and structure, to the pineal organ of anamniotes.
(5) The file of small neurons situated dorsally, directly caudal to the lamellar body, is designated as tectum because, with regard to its position, afferent and efferent connections, it corresponds to the centre of the same name in craniates.
(6) The motoneurons forming part of the primary motor centre, which innervate the first two somites, may be homologous to the cells innervating the extrinsic eye muscles in vertebrates. In contrast, the somewhat more caudally situated large projection neurons within that centre presumably correspond to the most rostral giant reticulospinal elements in the lamprey.

So, according to Lacalli, the brains of lancelets and vertebrates are sufficiently similar in structure that detailed comparisons can be made between them. Identifying antecedents for basic chordate features among invertebrates is, in his opinion, more speculative, and requires that comparisons be made between organisms that are very dissimilar. However, embryological investigations have shown that the larvae of several invertebrate groups, including the echinoderms, hemichordates and tunicates have many more features in common with craniates than their respective adult stages. The ascidians, briefly discussed in section III.g and illustrated in Fig. 35, are a striking case in point. This remarkable feature led Garstang (1894) and Garstang and Garstang (1926) to the theory that it is larvae of (one of) the invertebrate groups mentioned, which during evolution, might have become transformed into the chordates. Within the context of this theory, Lacalli noted the following similarities between the larvae of amphioxus on the one hand, and those of other invertebrate groups on the other hand.

(1) A considerable proportion of the cerebral vesicle in young amphioxus larvae is occupied by sensory organs, viz. the frontal eye complex, the lamellar body and the preinfundibular balance organ. This compares with the brain vesicle of ascidian tadpoles, which also has three sense organs (section III.g).
(2) Embryos and larvae of different invertebrate Phyla possess an apical organ, which marks their anterior pole; the cells of this organ, which are often ciliated, form a distinct, superficially situated, cluster or plate. The function of these organs is not fully understood, but they are assumed to be sensory structures of some type. The chordate counterpart of the apical organs was entirely unknown, but recently Lacalli et al. (1994) have suggested that the preinfundibular portion of the

cerebral vesicle in amphioxus is derived from the apical plate, and, consequently, the craniote hypothalamus represents an internalized derivative of a once superficially situated sensory structure. The question of what all this has to do with volume transmission will be addressed in the final section of the present review.

The ventrolateral parts of the rostral neuraxis is bilaterally occupied by a fibre compartment, known as the ventrolateral bundle or ventrolateral nerve tract (Figs. 36E, 37); however the structure and composition of this bundle will not be discussed here. Suffice to mention that within its confines, incoming sensory fibres from peripheral sensory cells and descending axons from a variety of neurons situated in the preinfundibular zone enter into functional contact with remarkably long and coarse dendrites ascending from tectal and tegmental neurons (Fig. 36E). Lacalli and Kelly (2000) observed that, in addition to the clements already mentioned, two groups of highly characteristic

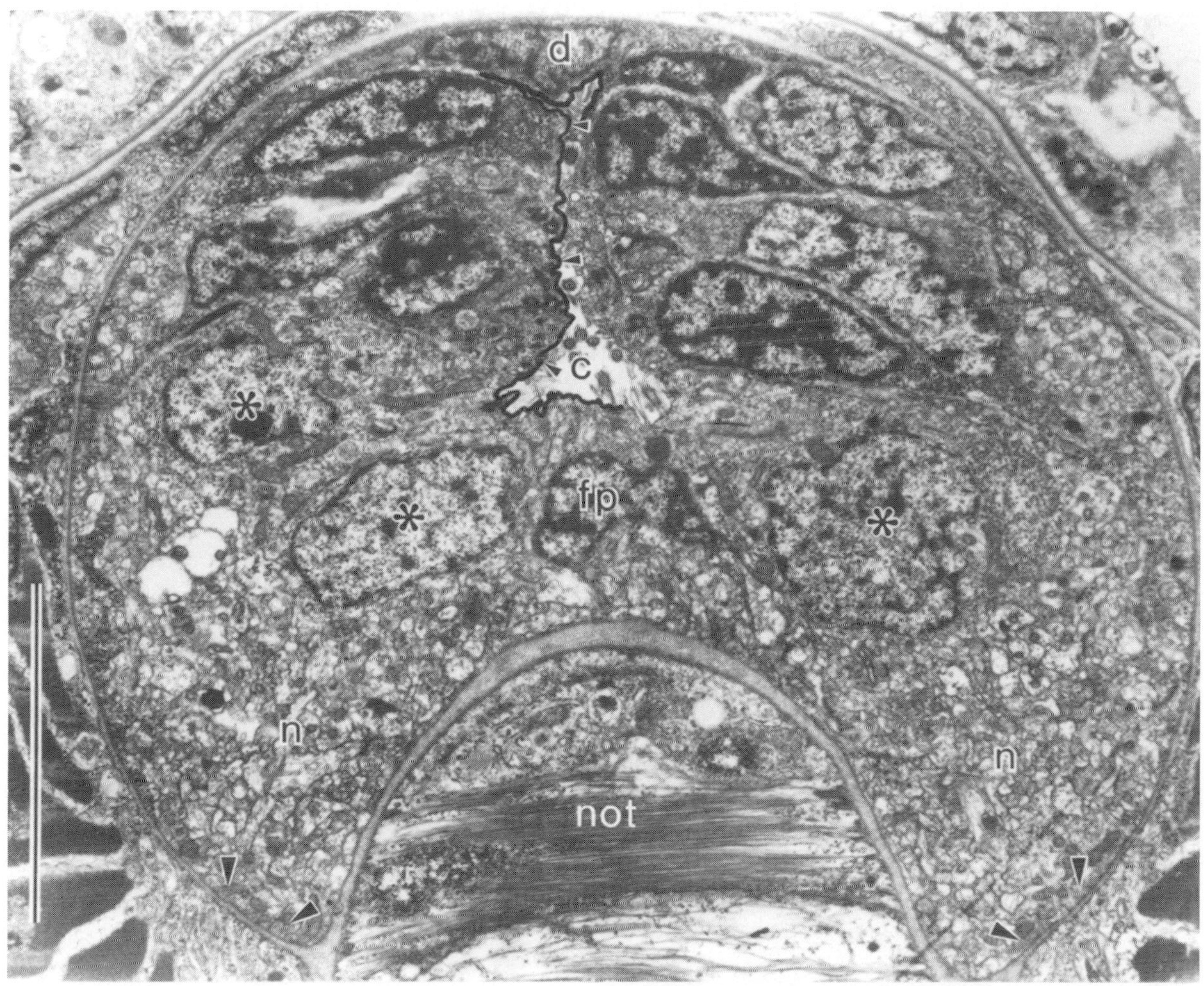

Fig. 37. Transverse section through the caudal brain region of a 14-day larva of *Branchiostoma floridae*. The surface of the central canal is traced on one side for emphasis (solid line, small arrows). It is capped at the top by a single dorsal cell (d). The ventral midline is occupied by a single floorplate cell (fp). The cells marked by an asterisk are neurons. The ventrolateral nerve tracts (n) are shown, each with a superficial zone in whichneuromuscular junctions are made (between arrowheads). Scale bar: 5 μm. Reproduced from Lacalli et al. (1994).

large neurons, designated by them as preinfundibular projection neurons type I and type II (PPNI, PPNII), also contribute to the ventrolateral bundle. These neurons are situated in the caudal part of the cerebral vesicle; their apical processes contact the ventricular cavity just in front of those of the infundibular cells, hence their name; both cell types send long varicose, caudally directed axons into the ventrolateral bundle (Fig. 36F). The varicosities of both are filled with vesicles, and make small junctions with adjacent elements. Lacalli considers it likely that these junctions are mainly for adhesion, because (1) they form similar junctions with non-neuronal elements, and (2) there are no concentrations of vesicles near the contact points.

Eight to ten PPNI cells were counted on each side; many have multiple axons. Their varicosities, which are large and irregularly shaped, contain small clear vesicles and both small and large dense-cored vesicles, all of them intermingled (Figs. 38A, C).

There are only four PPNII cells, two on each side. The axons of these elements extendfurther caudally than those of the PPNI cells, and their varicosities, which are much larger than those of PPNI elements (Fig. 36F), are jammed with tiny

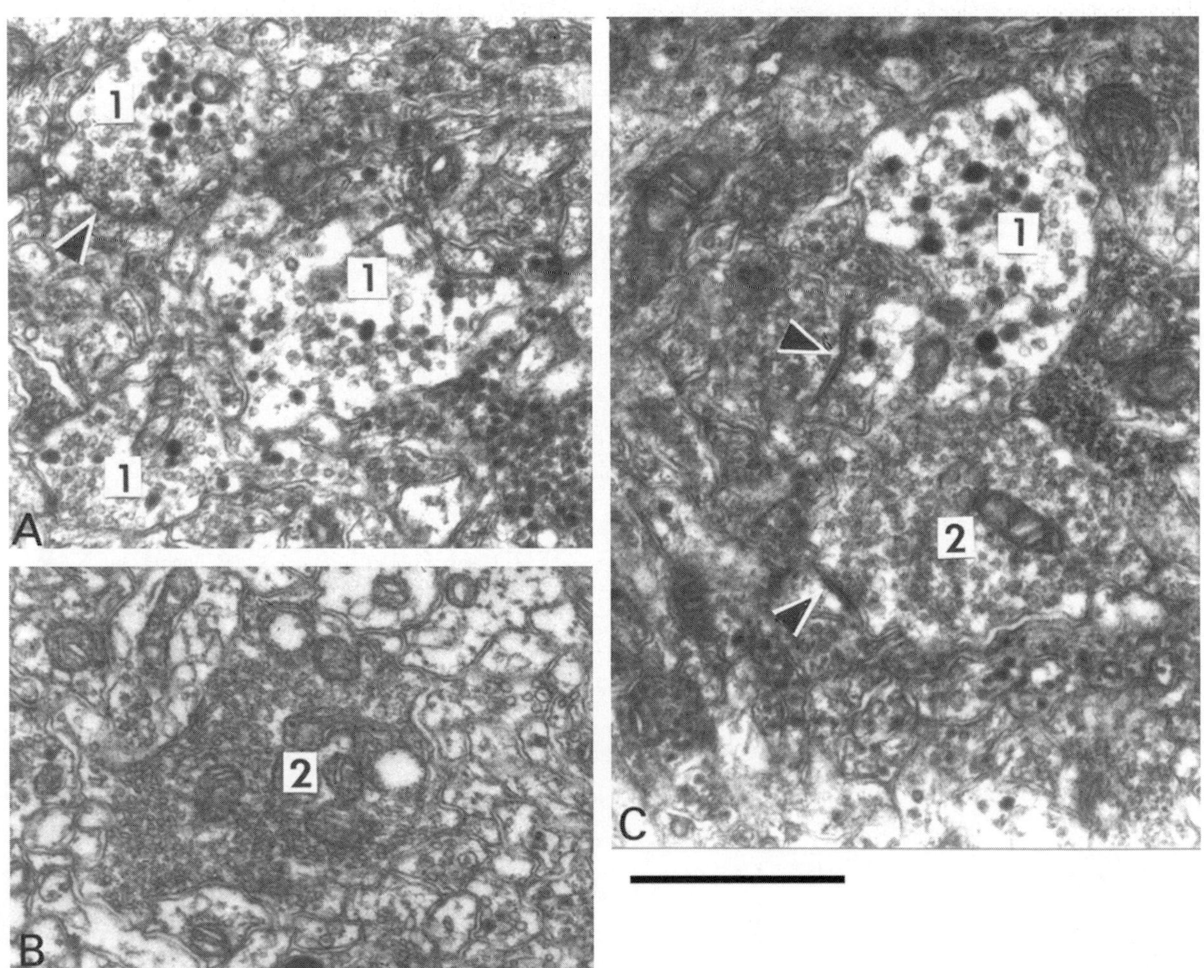

Fig. 38. Axonal varicosities of PPN1 (1) and PPN2 (2) cells in the central nervous system of a 12.5-day larva of *B. floridae*. Synaptic contacts are indicated by arrowheads. Courtesy of Dr.T. Lacalli. Scale bar: 1 μm.

clear vesicles and almost nothing else (Fig. 38A, C).

Studying the connectivity of the neuronal elements in the rostral cord of amphioxus larvae, Lacalli and Kelly (2000) noticed that the major target for most of the axons, whether they originate inside or outside the cord, are the dendrites of six large projection neurons situated in the primary motor centre (PMC; Fig. 36E). Examining nerve terminals associated with the PMC dendritic fields, they observed that most of those with synapses belong to fibres originating outside the nerve cord. The majority of those belonging to cells inside the cord appeared to be non-synaptic, a fact which – according to Lacalli and Kelly (2000) – implies that there is a considerable input via paracrine transmission to the PMC neurons.

Having completed the discussion of the central nervous system of the lancelet, I have come full circle, because the cephalochordates, to which the lancelet belongs, are generally considered to be the closest relatives of the vertebrates, with which this review opened.

IV. Conclusions and comments

In the preceding sections, morphological correlates of volume transmission observed in the various vertebrate and invertebrate groups have been discussed. This final section will draw some conclusions from the assembled data, and provide some general comments.

(1) *There can be no doubt that volume transmission occurs widely throughout the animal kingdom, but for most groups our knowledge concerning place, prevalence and function of this mode of intercellular communication is at best fragmentary.*

Table 2 provides an overview of the occurrence of volume transmission in the animal kingdom, as evidenced by the six criteria specified in section I; these data fully warrant the first half of the conclusion presented above. The cautionary second half has to do with the nature of the data available; most of the data on invertebrates are based on preliminary reconnaissances, rather than on systematic in-depth analyses. Reliable data on quantitative aspects can only be derived either from painstaking work in serial thin sections for electron microscopy or stereological extrapolation from data obtained in single or short series of thin sections. As far as vertebrates are concerned, only Descarries and collaborators (e.g. Descarries et al., 1991, 1997; Descarries and Mechawar, present volume) have made extensive application of these techniques in the rat. In invertebrates, to my knowledge, the serial section approach has only been extensively utilized by Westfall and Savyar

TABLE 2

Occurrence of morphological correlates of volume transmission in animal Phyla

		1	2	3	4	5	6
A–1	Placozoa – trichoplaxes						
A–2	Porifera – sponges						
A–3	Cnidaria – hydroids and jellyfish		+	+	+		
A–5	Platyhelminthes – flatworms	+	+	+	+		
A–10	Nematoda – thread worms, round worms			+			
A–11	Chelicerata – horseshoe crabs, spiders			+	+		
A–20	Mandibulata – insects, centipedes, millipedes	+	+	+	+		
A–21	Crustacea – lobsters, crabs	+	+	+	+		
A–22	Annelida – annelid worms	+	+	+			
A–26	Mollusca	+	+	+	+		
A–28	Onychophora – velvet worms			+	+		
A–32	Chaetognatha – arrow worms			+	+		
A–34	Echinodermata – sea urchins, sea stars, sea cucumbers				+		
A–36	Cephalochordata – lancelets			+			
A–37	Craniata – craniate chordates	+	+	+	+	+	I

Criteria: 1, extrasynaptic exocytoses TARI; 2, extrasynaptic exocytoses; 3, dense-cored vesicles 'stand back'; 4, asynaptic varicosities; 5, extrasynaptic receptors; 6, transmitter-receptor mismatch.

(1997: tentacles of the sea anemone, *Calliactis parasitica*), White et al. (1986: *Caenorhabditis elegans*) and Lacalli and Kelly (2000: larval *Branchiostoma floridae*). Moreover, in many invertebrate Phyla, the very useful TARI technique has never been applied. Finally, it should be emphasized that in most studies on invertebrates, the parent neurons of the fibres or profiles showing signs of volume transmission, remained unidentified. Needless to say, the lack of such basic information makes it almost impossible to begin to place the fragmentary anatomical data in any functional context. In light of all this, most of the conclusions drawn below must be considered as tentative.

(2) *Neurons typically possess two secretory mechanisms for the discharge of neurochemical mediators: diffuse, peptidergic, non-synaptic exocytosis and spatially specific 'classical' transmitter-mediated, synaptic exocytosis, but these two mechanisms should be considered only as prominent elements in a spectrum of functional configurations.*

The distinction emphasized in the first part of this statement has been made by numerous authors, among them Nieuwenhuys (1985), Thureson-Klein et al. (1988), Fuxe and Agnati (1991), Golding (1994) and Zupanc (1996). Columns 1 and 2 in Table II show that non-synaptic release of the contents of dense-cored vesicles has been observed in various groups. Indeed, configurations such as the one shown diagrammatically in Fig. 39A are very common in the central nervous system of both vertebrates and invertebrates; some caution should be exercised with the generalization that dense-cored vesicles contain neuropeptides. However, the facts that the TARI reaction specifically blackens peptide-like substances and that in many cases the peptide nature of the contents of dense-cored vesicles has been demonstrated by immuno-electron microscopy, justifies the conclusion that dense-cored vesicles are the main storage organelle for neuropeptides.

Typical chemical synapses, characterized ultrastructurally by a pair of parallel membranes with a uniformly wide intercellular cleft and a presynaptic aggregation of small clear vesicles, are very common in the central nervous system of both invertebrates and vertebrates (Fig. 39B; for review of invertebrates synapses see Westfall, 1987), and the same holds true for configurations like the one depicted in Fig. 39C. As indicated by column 3 of Table II, and illustrated by Figs. 2D, 6A, B, 21A, 22D, 28A, 31C, 32C and 33, such synaptic

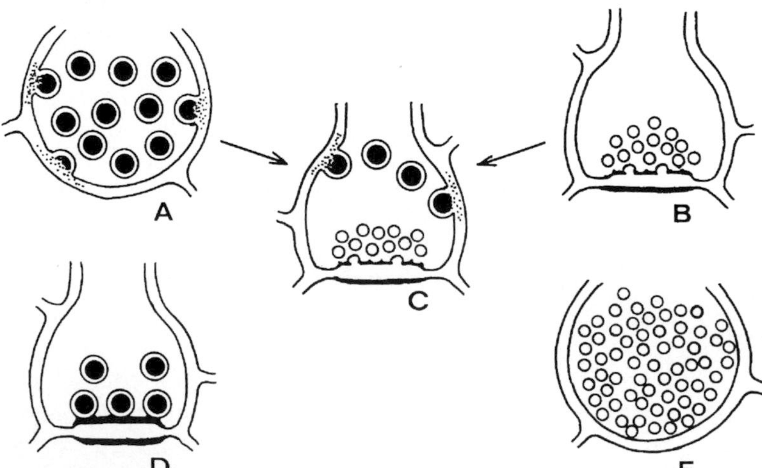

Fig. 39. Neuronal secretory mechanisms. A. Profile containing dense-cored vesicles, showing non-synaptic release sites. B. A typical fast, ionotropic synapse, showing an accumulation of small clear vesicles at the presynaptic membrane. C. Terminal, containing small synaptic vesicles and large dense-cored vesicles in typical 'stand-back' position. D. A typical peptidergic synapse. E. Profile densely filled with small clear vesicles, showing no synaptic membrane specializations.

terminals with dense-cored vesicles in 'stand-back' position, have been seen in many groups. It is well known that many neurons produce, store and release more than one type of messenger molecules, and that the configuration diagrammatically shown in Fig. 39C represents the morphological substrate of this diversity. Many peptidergic neurons contain multiple rather than a single neuropeptide, and in most of these neurons the neuropeptide(s) coexist(s) with one or two classical neurotransmitters, such as dopamine, noradrenaline, serotonin, acetylcholine or GABA. A single example may illustrate this feature. In the clawed toad *Xenopus laevis*, cells in the suprachiasmatic nucleus innervate the α-MSH cells in the pars intermedia of the pituitary; the terminals of the suprachiasmatic cells contain small, clear synaptic vesicles and large dense-cored vesicles in 'stand-back' position (Fig. 6A). Three different inhibitory neuromediators have been shown to be present in these terminals, viz. GABA, stored in the small-clear vesicles and dopamine and neuropeptide Y (co-stored in the dense-cored vesicles). It has been demonstrated that all three of these inhibitory substances play their own specific role in the regulation of the activity of the α-MSH cells. For a thorough review on the contents and functions of multiple-messenger neurons the reader is referred to Hökfelt et al. (1986).

In addition to the three functional configurations discussed above, the morphological substrates of two other modes of inter-neuronal communication, viz. peptidergic synapses (Fig. 39D) and non-synaptic accumulations of small clear vesicles (Fig. 39E), occur frequently.

Several authors (e.g. Thureson-Klein et al., 1988; Golding, 1994; Zupanc, 1996) have pointed out that within the confines of classical synapses (Fig. 39B), exocytoses of dense-cored vesicles are extremely rarely observed. However, in several invertebrate groups, including platyhelminths (Fig. 21B, C), chelicerates, molluscs and onychophorans (Fig. 32D), synapses have been observed in which dense-cored vesicles prevail and also occupy a juxtamembranous position. In cnidarians, the great majority of the neuro-neuronal (Figs. 14B, C, D, 15A, C, E) and neuromuscular (Fig. 15F) synapses show this particular disposition, and an RFamide-like peptide is contained in the dense-cored vesicles of several of these contacts. These observations, and the fact that the cnidarian nervous system contains a variety of different neuropeptides, indicate that in this group signal transmission is mainly, if not exclusively, peptidergic (cf. section III.4).

In mammals, noradrenergic, serotoninergic and cholinergic neurons, situated in the brain stem and basal telencephalon, form richly branching networks of thin, varicose fibres in the striatum, hippocampus and neocortex. The thorough and systematic immuno-electron microscopic studies in the rat by Descarries and collaborators (summarized in Descarries et al., 1991, 1997; cf. also Ridet and Privat, 2000) have shown that the varicosities in these fibres, which are densely packed with small clear vesicles, rarely show synaptic membrane specializations. It was concluded that in the areas studied (and probably in other areas as well), non-synaptic transmission by acetylcholine, noradrenaline and serotonin prevails. Descarries et al. (1991, p. 61) added the intriguing statement: "In the CNS, it remains to be demonstrated that the membrane differentiations often considered as the hallmark of synaptic transmission are more than 'attachment' sites, ensuring some permanence in the structural relation between neuronal partners."

It is remarkable that recently, Lacalli and Kelly (2000) observed somewhat similar varicose fibres in the central nervous system of larval amphioxus (Figs. 36F, 38) and that they also arrived at the conclusion that these fibres are involved in volume transmission.

Varicosities containing dense-cored vesicles and showing no synaptic membrane specializations have been observed in many groups (Table II, column 4). However, since most of these observations were not based on serial section analysis, their strict asynaptic nature is not completely established.

The secretory mechanisms, diagrammatically shown in Figs. 39B, , may be complicated by the morphologically detectable occurrence of exocytoses of dense-cored vesicles in the immediate vicinity of the active zone ('parasynaptic release'; Thureson-Klein et al., 1988; Golding, 1994) and by the morphologically undetectable 'escape' of neurotransmitter molecules out of the synaptic cleft.

Finally, it may be mentioned that during my long and adventurous journey through the animal kingdom, I was confronted with several other substrates of (putative) intercellular communication. For example, the many 'almost' synapses in 'ordinary' sponges (Fig. 10) and the remarkable 'plugged' junctions in the predominantly syncytial hexactinellid sponges (Fig. 12) and placozoans (Fig. 8C, D).

(3) *Volume transmission most probably is dominant in the neuropils of many invertebrates.*

In most invertebrates, the central core of the cerebral ganglia is occupied by one or several neuropil zones; these neuropil zones, which are sealed off from their environment by glial configurations, are filled with tightly packed and closely apposed fibres. Electron microscopical studies have shown that most of the fibres in these neuropils contain dense-cored vesicles of varying size and density, and a general paucity of synaptic differentiations (platyhelminths: Fig. 20; chelicerates: Fig. 27C; crustaceans: Fig. 28B; annelids: Fig. 31B; onychophorans: Fig. 32E; chaetognathes: Fig. 33). Moreover, experiments with the TARI technique have revealed that in mandibulates (Fig. 26A), crustaceans and molluscs (Fig. 30D) the neuropil fibres containing dense-cored vesicles show abundant non-synaptic exocytoses.

(4) *The density of the (putative) morphological substrates of volume transmission in the central nervous system may show considerable differences, both between different species and between the various parts of the central nervous system of the same species.*

As regards differences between species, it may be mentioned that there is an overall abundance of signs of volume transmission in platyhelminths and molluscs, but that such signs are much less abundant in vertebrates. In echinoderms a special type of volume transmission prevails, because in this group the ectoneural nervous system is separated from the hyponeural nervous system and from all other non-ectodermal tissues by a continuous basement membrane (Section III.9).

With regard to differences between regions in the same species it may be remembered that Nieuwenhuys and collaborators subdivided the mammalian brain into two large functional domains. These were the sensorimotor/cognitomotor (Fig. 3) domains and the greater limbic domains (Figs. 4, 5), and evidence was presented suggesting that volume transmission may well play a prominent role in the latter. It should be added that the focus in this concept is on peptidergic non-synaptic neurotransmission, but that the findings of Descarries and collaborators on cholinergic and monoaminergic non-synaptic transmission supplement and extend our results. Without going into details, it may be mentioned that the cells of origin of the monoaminergic and cholinergic projections studied by Descarries and collaborators are all embedded in the greater limbic domain. The evidence favouring our concept and its functional implications have been briefly discussed in Section II.1, and will not be repeated here. Suffice to say that this concept has received little attention so far; It was not mentioned in a historical survey of the development of the concept of volume transmission (Fuxe and Agnati, 1991). In 1993, Herbert advanced ideas concerning the nature of the limbic system, which closely resemble ours, but again our prior work was not referred to; so, putting it positively, our concept has received independent support from Herbert (1993). The applicability of our concept in non-mammalian chordates has never been tested, but recently, Lacalli and Kelly (2000) postulated that certain well-delineable parts of the rostral neuraxis of amphioxus correspond to the greater limbic domain (cf. Section III.10). There are a few data that hint at the possibility that comparable functional subdivisions can also be made in non-chordate invertebrates. Thus, it was found that in insects, the neuropils involved in the processing of visual information (i.e. the lamina, medulla and lobule in Fig. 24C) contain few dense-cored vesicles, whereas such vesicles are abundant in the remaining protocerebral neuropil fields. Similarly, in the neuropil of the large ventral ganglion of chaetognathes (Fig. 33A), which is involved in fast swimming, only a few fibres contain dense-cored vesicles, but in the neuropil of the cerebral ganglion many of the fibres are packed with dense-cored vesicles (section III.g; Fig. 33B).

(5) *The widespread notion that volume transmission is primitive, generalized, sluggish and lacks precision, whereas wiring transmission is advanced, specialized, fast and accurate, is erroneous.*

Notions like the one just stated are akin to the erroneous but ineradicable idea that living organisms can be arranged in a phylogenetic series or scale, ranging from generalized to specialized, from primitive to advanced, from imperfect to perfect, or from low to high. Rather than presenting a long argument, I will quote here three zoologists who have adequately and eloquently stood in opposition to those ideas.

(1) Alfred Sherwood Romer, noted zoologist and paleontologist, once remarked in a lecture that organisms "cannot continue to sit around being generalized ancestors; they must go on with the business of living" (quoted from Gans, 1969).
(2) C.F.A. Pantin (1952, p. 147) opened his Croonian Lecture on 'The elementary nervous system', which was mainly devoted to the nervous organization in sea anemones, with the following general statements. "At the outset let us remember to discard one notion that is apt to creep into discussion of the lower animals. They are not considered as poor and inefficient relics of our own evolutionary past." ". . . we are dealing with living animals and there is nothing inefficient about them. One has only to walk to our Marine Laboratory over the Devonian limestone of Plymouth Hoe to realize the enduring success of the coelenterate type. Natural selection has said so clearly, 'that will do well enough'; it has not yet said that about ourselves."
(3) And finally G.O. Mackie (1990, p 907), in the same vein: "No existing animals have 'elementary' nervous systems if that term implies the existence of crude or inefficient functional adaptations."

What has been said in these quotations about organism and nervous system also holds for structures and intercellular functional contacts. Ideas like that of Anctil (1990), that cnidarians stand as it were at the transition point between the primitive paracrine mode of neurohumoral action and the more advanced neurotransmitter mode, are untenable. In cnidarians, neuropeptides doubtlessly play a prominent role in neuro-neuronal and neuromuscular communication, and it is also known that gap junctional communication is very common in the nervous system of these animals (Green, 1989). However, fast synaptic mechanisms are not lacking in cnidarians; these animals possess a formidable weapon in the form of their nematocysts; when these stinging and poisoning devices are fired, they do so quickly. Remarkably, most of the synapses made by neurons with nematocysts contain exclusively small, clear vesicles, suggesting the presence of a fast, excitatory ionotropic neurotransmitter (Section III.4).

The conspicuous caudodorsal cells in the cerebral ganglia of the pond snail, *Lymnaea stagnalis* play a crucial role in the reproduction process of these animals (Section III.8, Figs. 29C, 30A). These cells receive a complex synaptic input; they are all electronically coupled, and their efferents influence their targets by volume transmission in two adjacent, though strictly separated, compartments.

Finally, the mammalian neocortex is first and foremost an ultracomplex synaptic network (Nieuwenhuys, 1994). However, the investigations of Descarries and collaborators have convincingly shown that subcortical cholinergic and monoaminergic neuronal populations influence the activity of the neocortex mainly by volume transmission, and quite recently physiological evidence has been presented showing that certain types of neocortical inhibitory neurons are electrically coupled. It appeared that this communication through electrical synapses allows excitatory signaling among inhibitory cells and promotes their synchronous spiking (Galarreta and Hesbrin, 1999; Gibson et al., 1999). These three examples show that in the nervous system of the 'very low' cnidarians, in the caudodorsal cells of 'low' snails as well as in the 'highest' centre of the 'highest' animals, electrotonic coupling, volume transmission and wiring transmission are all integrated in the information processing of these various entities.

(6) *The two conduits of volume transmission, i.e. the extracellular space and the ventricular system, both have 'external environment' antecedents.*

All metazoans have an extracellular space; in cross sections through the 'lowest' metazoa, i.e. the placozoans (Fig. 8B) and the sponges (Fig. 9B), this space can be seen clearly. Although for the placozoans, direct evidence is lacking, there can be no doubt that in all metazoans the fluid contained within the extracellular space is used as a carrier for informational substances.

Haldane (1954) advanced the hypothesis that signalling between cells within metazoans evolved from signalling through the external milieu between unicellular organisms. He noted that in certain protozoans, conjugation occurs only between dissimilar mating types, whose identity is signalled to neighboring cells by specific diffusible chemicals. He conjectured that chemical communication by unicellular organisms might well involve mechanisms similar to those whereby neuromediators and hormones are employed for chemical communication between neurons and other cells in metazoans. More recently, Mackie (1990) presented an impressive list of striking similarities between chemical communication systems of unicellular organisms and those of neurons and other internal celltypes. The following examples are derived from Mackie's study, to which the reader is referred for references and details.

(1) Various ciliates have receptors for acetylcholine, catecholamines and other neuroactive substances.
(2) The ciliate *Tetrahymena* produces β-endorphin and shows positive chemotaxis in a gradient of this opiate; this protist possesses a membrane protein closely resembling the opiate receptor found in annelids and vertebrates.
(3) Ion channels are found in bacteria, yeasts and protists; ciliates, like *Paramecium* and *Stylonychia* have many (up to nine) distinct, ion-selective membrane channels, including voltage-, calcium- and mechanically-gated types.
(4) Machemer and Deitmer (1987) have pointed out that ciliates resemble neurons not only in their possession of multiple ion channels, but also in the fact that these channels are not distributed uniformly over the surface. They respond to a variety of environmental stimuli (mechanical, thermal, chemical, photic, ionic) by changing their membrane potential, which then leads to appropriate behavioural responses, by activation of their ciliary organelles.

There are numerous other data indicating that protists are capable of many of the functions that are normally attributed to neurons. Thus, Van Houten (1989) has shown that the basic mechanisms used by protozoans to transduce chemical stimuli are essentially the same as those used by 'higher' invertebrates and vertebrates. In addition, Carr (1989) has pointed out that the signalling molecules, receptors and transducing mechanisms used by fungi and slime molds are remarkably similar to those used by neurons for similar purposes. It may be concluded that cellular elements, once aggregated into a multicellular organisation, did not 'invent' a new mode of intercellular communication, offered by a new compartment, i.e. the extracellular space; rather, these elements continued to fully exploit the faculties inherited from their unicellular past.

There is mounting evidence that the ventricular system provides an important conduit for informational substances (Nicholson, 1999; see also the contribution of Lehman and Silver to this volume). It is well known that the terminals of monoaminergic fibres penetrate into the ventricular cavities in many places and release their contents into the cerebrospinal fluid (Fig. 1). It is also known that peptidergic neurons are frequently located in periventricular areas, and that the peptides released by these neurons readily reach the ventricular cavities. It has been experimentally shown that peptides, migrating via the ventricular system, may reach the extracellular spaces of the periventricular parenchyme throughout the entire brain within a few minutes (Renneis et al., 1985). Finally, it is known that in the diencephalon, a system of CSF-contacting neurons sample the ventricular environment, and that specialized glial elements may well play a similar role in the core and paracore regions of the brainstem (Felten and Sladek, 1983; cf. Nieuwenhuys, 1985).

We have seen that a ventricular system is lacking in non-chordate invertebrates; in these animals the central nervous system is formed by neuroblasts which, detaching themselves from placode-like ectodermal anlagen, aggregate into solid ganglia. Only in chordate invertebrates, like the tunicates (Fig. 35) and the cephalochordates (Fig. 36), and in vertebrates, a neural plate forms a cylindrical roll resulting in a tubular neuraxis, enclosing a continuous ventricular system. Recently, Vigh (1999) as well as Lacalli and collaborators (Lacalli et al., 1994; Lacalli and Kelly, 2000) have expressed the opinion that the CSF-contacting diencephalic cells, which in vertebrates monitor the ventricular environment, originally subserved sensory functions, sampling the outside environment. Vigh remarked that in larval amphioxus the ventricular canal is open to the sea water (Fig. 36D, E) and he might have added that the same holds true for larval tunicates (Fig. 35C, D). Lacalli and collaborators went one step further by claiming that the rostral part of the neural tube of amphioxus derived from the apical plate, a superficially situated sensory structure that performs various functions in invertebrate larvae. Once internalized, the apical plate cells would then have been co-opted for internal functions.

Acknowledgements

I would like to thank Mrs. Inge Eijkhout for her excellent secretarial assistance, Mr. Chris van Huijzen for the preparation of Fig. 1, and Mr Gerben van der Meulen for photographing it. I would also like to thank Drs. Thurston Lacalli and Eric Roubos for stimulating discussions, Dr. Charles Nicholson for critically reading the manuscript and my partner, Mrs. Suzanne Bakker, M.Sc. for moral support, patience and understanding.

References

Albertson, D.G. and Thomson, J.N. (1976) The pharynx of Caenorhabditis elegans. *Phil. Trans. Roy. Soc. (Lond. B)*, 275: 299–325.

Anctil, M. (1990) The antiquity of monoaminergic neurotransmitters: Evidence from Cnidaria. In: P.A.V. Anderson (Ed.), *Evolution of the First Nervous Systems*. Plenum, New York, pp. 141–155.

Anderson, P.A.V. (1985) Physiology of a bidirectional, excitatory, chemical synapse. *J. Physiol.*, 53: 821–835.

Ariëns Kappers, C.U. (1929) *The evolution of the central nervous system*, Bohn, Haarlem, 335 pp.

Arshavsky, Y.I., Deliagina, T.G., Gelfand, I.M., Orlovsky, G.N., Panchin, Y.V., Pavlova, G.A. and Popova, L.B. (1988) Non-synaptic interaction between neurons in molluscs. *Comp. Biochem. Physiol.*, 91C: 199–203.

Ballard, J.W.O., Olsen, G.J., Faith, D.P., Odgers, W.A., Rowell, D.M. and Atkinson, P.W. (1992) Evidence from 12S ribosomal RNA sequences that onychophorans are modified arthropods. *Science*, 258: 1345–1348.

Barnes, R.D. and Harrison, F.W. (1991) Introduction. In: F.W. Harrison (Ed.), *Microscopic Anatomy of Invertebrates*, Vol. 2, New York, Wiley-Liss, pp. 1–12.

Baude, A. and Shigemoto, R. (1998) Cellular and subcellular distribution of substance P receptor immunoreactivity in the dorsal vagal complex of the rat and cat: a light and electon microscope study. *J. Comp. Neurol.*, 402: 181–196.

Bedini, C. and Lanfranchi, A. (1991) The central and peripheral nervous system of Acoela (Plathelminthes): an electron microscopical study. *Acta Zool. (Stockholm)*, 72: 101–106.

Berghs, C.A.F.M. and Roubos, E.W. (1996) Synaptic plasticity in the pars intermedia of Xenopus laevis. *Neuroscience*, 70: 833–841.

Blaxter, M. (1998) Caenorhabditis elegans is a nematode. *Science*, 282: 2041–2046.

Bode, H., Berking, S., David, C.N., Gierer, A., Schaller, H. and Trenkner, E. (1973) Quantitative analysis of cell types during growth and morphogenesis in Hydra. *Arch. Entwickl. Mech. Org.*, 171: 269–285.

Bogitsh, B.J. and Harrison, F.W. (1991) Introduction. In: F.W. Harrison (Ed.), *Microscopic Anatomy of Invertebrates*, Vol. 3: *Plathylminthes and Nemertinea*, Wiley-Liss, New York, pp. 1–5.

Bone, Q. and Goto, T. (1991) The nervous system. In: Q. Bone, H. Kapp, H. and A.C. Pierrot-Bults (Eds), *The Biology of Chaetognaths*, Oxford University Press, Oxford, pp. 18–31.

Bone, Q. and Pulsford, A. (1984) The sense organs and ventral ganglion of Sagitta (Chaetognatha). *Acta Zool. (Stockholm)*, 65: 209–220.

Breidbach, O. (1995) Is the evolution of arthropod brains convergent? In: O. Breidbach and W. Kutsch (Eds), *The Nervous System of Invertebrates: an Evolutionary and ComparAtive Approach*, Birkhäuser, Basel, pp. 383–406.

Brenner, S. (1973) The genetics of behavior. *Brit. Med. Bull.*, 29: 269–271.

Brownlee, D.J.A. and Fairweather, I. (1999) Exploring the neurotransmitter labyrinth in nematodes. *Trends Neurosci.*, 22: 16–24.

Brownlee, D.J.A., Fairweather, I., Holden-Dye, L. and Walker, R.J. (1996) Nematode neuropeptides: localization, isolation and functions. *Parasitol. Today*, 12: 343–351.

Budelman, B.U. (1995) The cephalopod nervous system: what evolution has made of the molluscan design. In: O. Breidbach and W. Kutsch (Eds), *The Nervous System of Invertebrates: an Evolutionary and Comparative Approach*, Birkhäuser, Basel, pp. 115–138.

Bullock, T.H. and Horridge, G.H. (1965) *Structure and function in the nervous system of invertebrates*, Freeman, San Francisco, CA, 2 vols., 1719p.

Buma, P. (1989) Synaptic and non-synaptic release of neuromediators in the central nervous system. *Acta Morphol. Neerl.-Scand.*, 26: 81–113.

Buma, P. and Nieuwenhuys, R. (1987) Ultrastructural demonstration of oxytocin and vasopressin release sites in the neural lobe and median eminence of the rat by tannic acid and immunogold methods. *Neurosci. Lett.*, 74: 151–157.

Buma, P. and Nieuwenhuys, R. (1988) Ultrastructural characterization of exocytotic release sites in different layers of the median eminence of the rat. *Cell Tissue Res.*, 252: 107–114.

Buma, P. And Roubos, E.W. (1986) Ultrastructural demonstration of nonsynaptic release sites in the central nervous system of the snail Lymnaea stagnalis, the insect Periplaneta americana, and the rat. *Neuroscience*, 17: 867–879.

Buma, P., Roubos, E.W. and Buijs, R.M. (1984) Ultrastructural demonstration of exocytosis of neural, neuroendocrine and endocrine secretions with an in vitro tannic acid (TARI-) method. *Histochemistry*, 80: 247–256.

Buma, P., Veening, J., Hafmans, T., Joosten, H. and Nieuwenhuys, R. (1992) Ultrastructure of the periaqueductal grey matter of the rat: an electon microscopical and horseradishperoxidase study. *J. Comp. Neurol.*, 39: 519–535.

Bunt, A.H. (1969) Formation of coated and 'synaptic' vesicles within neurosecretory axon terminals of the crustacean sinus gland. *J. Ultrastruct. Res.*, 28: 411–421.

Burrows, M. (1996) *The neurobiology of an insect brain*, Oxford University Press, Oxford, 682 pp.

Bütschli, O. (1884) Bemerkungen zur Gastraeatheorie. *Morph. Jahrb.*, 9: 415–425.

Bütschli, O. (1910) *Vorlesungen über vergleichende Anatomie*, Engelmann, Leipzig, 644 pp.

Carlton, S.M., McNeill, D.L., Chung, K. and Coggeshall, R.E. (1987) A light and electron microscopic level analysis of calcitonin gene-related peptide (CGRP) in the spinal cord of the primate: an immunohistochemical study. *Neurosci. Lett.*, 82: 145–150.

Carr, W.E.S. (1989) Chemical signaling systems in lower organisms: a prelude to the evolution of chemical communication in the nervous system,. In: P.A.V. Anderson (Ed.), *Evolution of the First Nervous Systems*, Plenum Press, New York, London, pp. 81–94.

Carranza, S., Baguñà, J. and Riutort, M. (1997) Are the Platyhelminthes a monophyletic primitive group? An assessment using 18s rDNA sequences. *Mol. Biol. Evol.*, 14: 485–497.

Castel, M., Feinstein, N., Cohen, S. and Harari, N. (1990) Vasopressinergic innervation of the mouse suprachiasmatic nucleus: an immuno-electron microscopic analysis. *J. Comp. Neurol.*, 298: 172–187.

Christensen, T.A. and Hildebrand, J.G. (1987) Functions, organization, and physiology of the olfactory pathways in the lepidopteran brain. In: A.P. Gupta (Ed.), *Arthropod Brain: Its Evolution, Development, Structure, and Functions*, Wiley, New York, pp. 457–484.

Christenson, J., Cullheim, S., Grillner, S. and Hökfelt, T. (1990) 5-Hydroxytryptamine immunoreactive varicosities in the lamprey spinal cord have no synaptic specializations: an ultrastructural study. *Brain Res.*, 512: 201–209.

Christenson, J., Franck, J. and Grillner, S. (1989) Release of endogenous 5-hydroxytryptamine modulates the central pattern generator in the lamprey spinal cord. *Neurosci. Lett.*, 100: 188–192.

Christenson, J., Wallén, P., Brodin, L. and Grillner, S. (1991) 5-HT systems in lower vertebrate model: ultrastructure, distribution, and synaptic and cellular mechanisms. In: K. Fuxe and L.F. Agnati (Eds), *Volume Transmission in the Brain: Novel Mechanisms for Neural Transmission*, Raven Press, New York, pp. 159–170.

Clements, J.R., Beitz, A.J., Fletcher, T.F. and Mullett, M.A. (1985) Immunocytochemical localization of serotonin in the rat periaqueductal gray: a quantitative light and electron microscopic study. *J. Comp. Neurol.*, 236: 60–70.

Cobb, J.L.S. (1985a) The motor innervation of the oral plate ligament in the brittlestar Ophiura. *Cell Tissue Res.*, 242: 685–688.

Cobb, J.L.S. (1985b) The neurobiology of the ectoneural/hyponeural synapse. *Biol. Bull.*, 168: 432–446.

Cobb, J.L.S. (1987) Neurobiology of the Echinodermata. In: M.A. Ali (Ed.), *Invertebrate Nervous Systems*, Plenum Press, New York, pp. 483–526.

Cobb, J.L.S. (1989) Enigmas of echinoderm nervous systems. In: P.A.V. Anderson (Ed.), *Evolution of the First Nervous System*, Plenum Press, New York, pp. 329–337.

Cobb, J.L.S. (1995) The nervous systems of echinodermata: recent results and newapproaches. In: O. Breidbach and W. Kutsch (Eds), *The Nervous System of Invertebrates: an Evolutionary and Comparative Approach*, Birkhäuser, Basel, pp. 407–424.

Cobb, J.L.S. and Stubbs, T. (1981) The giant neurone system in ophiuroids: I. The general morphology of the radial nerve cords and circumoral ring. *Cell Tissue Res.*, 219: 197–207.

Cobb, J.L.S. and Stubbs, T. (1982) The giant neurone system in ophiuroids: III. The detailed connections of the circumoral ring. *Cell Tissue Res.*, 226: 675–687.

Coggeshall, R.E. (1967) A light and electron microscope study of the abdominal ganglion of Aplysia californica. *J. Neurophysiol.*, 30: 1263–1287.

Cuadras, J. (1989) Non-synaptic release from dense-cored vesicles occurs at all terminal types in crayfish neuropile. *Brain Res.*, 477: 332–335.

Dana, C., Vial, M., Leonard, K., Beauregard, A., Kitabgi, P., Vincent, J.P., Rostène, W. and Beaudet, A. (1989) Electron microscopic localization of neurotensin binding sites in the midbrain tegmentum of the rat: I. Ventral tegmental area and interfascicular nucleus. *J. Neurosci.*, 9: 2247–2257.

Descarries, L., Gisiger, V. and Steriade, M. (1997) Diffuse transmission by acetylcholine in the CNS. *Progr. Neurobiol.*, 53: 603–625.

Descarries, L., Séguéla, P. and Watkins, K.C. (1991) Non-junctional relationships of monoamine axon terminals in the cerebral cortex of adult rat. In: K. Fuxe and L.F. Agnati (Eds), *Volume Transmission in the Brain: Novel Mechanisms for Neural Transmission*, Raven Press, New York, pp. 53–62.

Di Gregorio, A. and Levine, M. (1998) Ascidian embryogenesis and the origins of the chordate body plan. *Curr. Opin. Genet. Dev.*, 8: 457–463.

Durbin, R. (1987) Nematode C. elegans. In: G. Adelman (Ed.), *Encyclopedia of Neuroscience*, Birkhäuser, Boston, pp. 738–740.

Elliott, S.R., Mohan Kumar, P.S. and Westfall, J.A. (1998) Immunoelectron microscopical and neurochemical evidence for some classical neurotransmitters in sea anemones. *Mol. Biol. Cell*, 9: 351a.

Erber, J., Homberg, U. and Gronenberg, W. (1987) Functional roles of the mushroom bodies in insects. In: A.P. Gupta (Ed.), *Arthropod Brain: Its Evolution, Development, Structure, and Functions*, Wiley, New York, pp. 485–511.

Evans, P.D. and Cournil, I. (1990) Co-localization of FLRF-like and vasopressin-like immunoreactivity in a single pair of sexually dimorphic neurones in the nervous system of the locust. *J. Comp. Neurol.*, 292: 331–348.

Fahrenbach, W.H. (1999) Merosomata. In: F.W. Harrison (Ed.), *Microscopic Anatomy of Invertebrates*, Vol. 8A: *Chelicerate Arthropoda*, Wiley-Liss, New York, pp. 21–115.

Fahrenbach, W.H. and Chamberlain, S.C. (1987) The brain of the horseshoe crab, Limulus polyphemus. In: A.P. Gupta (Ed.), *Arthropod Brain: Its Evolution, Development, Structure, and Functions*, Wiley, New York, pp. 63–94.

Felten, D.L. and Sladek Jr, J.R. (1983) Monoamine distribution in primate brain.V. Monoaminergic nuclei: anatomy, pathways and local organization, *Brain Res. Bull.*, 10: 171–284.

Fuxe, K. and Agnati, L.F. (1991) Two principal modes of electrochemical communication in the brain: volume versus wiring transmission. In: K. Fuxe and L.F. Agnati (Eds), *Volume Transmission in the Brain: Novel Mechanisms for Neural Transmission*, Raven Press, New York, pp. 1–9.

Galarreta, M. and Hestrin, S. (1999) A network of fast-spiking cells in the neocortex connected by electrical synapses. *Nature*, 402: 72–75.

Gans, C. (1969) Discussion: some questions and problems in morphological comparison. *Ann. New York Acad. Sci.*, 167: 506–513.

Garstang, S.L. and Garstang, W. (1926) On the development of Botrylloides and the ancestry of the vertebrates. *Proc. Leeds Phil. Lit. Soc. Sci. Sect.*, 1: 81–86.

Garstang, W. (1894) Preliminary note on a new theory of the phylogeny of the Chordata. *Zool. Anz.*, 17: 122–125.

Gibson, J.R., Beierlein, M. and Connors, B.W. (1999) Two networks of electrically coupled inhibitory neurons in neocortex. *Nature*, 402: 75–79.

Golding, D.W. (1992) Polychaeta: nervous system. In: F.W. Harrison (Ed.), *Microscopic Anatomy of Invertebrates*, Vol. 7: Annelida, Wiley-Liss, New York, pp. 153–179.

Golding, D.W. (1994) A pattern confirmed and refined: synaptic, non-synaptic and parasynaptic exocytosis (Review). *BioEssays*, 16: 503–508.

Golding, D.W. and Bayrakteroglu, E. (1984) Exocytosis of secretory granules: a probable mechanism for the release of neuromodulators in invertebrate neuropiles. *Experientia*, 40: 1277–1280.

Golding, D.W. and May, B.A. (1982) Duality of secretory inclusions in neurons: ultrastructure of the corresponding sites of release in invertebrate nervous systems. *Acta Zool. (Stockholm)*, 63: 229–238.

Golding, D.W. and Pow, D.V. (1988) The new neurobiology: ultrastructural aspects of peptide release as revealed by studies of invertebrate nervous systems. In: M.C. Thorndyke and G.J. Goldsworthy (Eds), *Neurohormones in Invertebrates, Society for Experimental Biology*, Seminar Series 33, Cambridge University Press, New York, pp. 7–18.

Goldschmidt, R. (1908) Das Nervensystem von Ascaris lumbricoides und megalocephela, *I.Z. Wiss. Zool.*, 90: 73–136.

Goldschmidt, R. (1909) Das Nervensystem von Ascaris lumbricoides und megalocephela, *II.Z. Wiss. Zool.*, 92: 306–357.

Goto, T. and Yoshida, M. (1987) Nervous system in Chaetognathax. In: M.A. Ali (Ed.), *Nervous System in invertebrates*, NATO ASI series A, Plenum Press, New York, pp. 461–481.

Govind, C.K. (1992) Nervous system. In: F.W. Harrison (Ed.), *Microscopic Anatomy of Invertebrates*, Vol. 10: *Decapod Crustacea*, Wiley-Liss, New York, pp. 395–438.

Grassé, P.-P. (1975) Le système nerveux des insectes. In: P.-P. Grassé (Ed.), *Traité de Zoologie*, Vol. VIII, Masson, Paris, pp. 321–510.

Green, C.R. (1989) Cnidarian gap junctions: structure, function and evolution. In: P.A.V. Anderson (Ed.), *Evolution of the First Nervous Systems*, Plenum Press, New York, London, pp. 3–20.

Grell, K.G. (1971) Trichoplax adhaerens F.E. Schulze und die Entstehung der Metazoen. *Naturw. Rdsch.*, 24: 160–161.

Grell, K.G. (1981) Trichoplax adhaerens and the origin of metazoa. In: *Origine dei Grandi Phyla dei Metazoi*, Acc. Naz. Lincei, Convegno Intern., pp. 107–121.

Grell, K.G. and Benwitz, G. (1974) Spezifische Verbindungsstrukturen der Faserzellen von Trichoplax adhaerens F.E. Schulze. *Z. Naturforsch.*, 29C: 790.

Grell, K.G. and Ruthmann, A. (1991) Placozoa. In: F.W. Harrison (Ed.), *Microscopic Anatomy of Invertebrates*, New York, Wiley-Liss, pp. 13–27.

Grillner, S., Deliagina, T., Ekeberg, Ö., El Manira, A., Hill, R.H., Lansner, A., Orlovsky, G.N. and Wallén, P. (1995) Neural networks that co-ordinate locomotion and body orientation in lamprey. *Trends Neurosci.*, 18: 270–279.

Grillner, S., Wallén, P., Brodin, L. and Lansner, A. (1991) Neuronal network generating locomoter behavior in lamprey: circuitry, transmitters, membrane properties, and simulation. *Annu. Rev. Neurosci.*, 14: 169–199.

Grimmelikhuijzen, C.J.P. (1983) Coexistence of neuropeptides in hydra. *Neuroscience*, 9: 837–845.

Grimmelikhuijzen, C.J.P. (1985) Antisera to the sequence Arg-Phe-amide visualize neuronal centralization in hydroid polyps. *Cell Tissue Res.*, 241: 171–182.

Grimmelikhuijzen, C.J.P. (1986) FMRFamide-like peptides in the primitive nervous systems of coelenterates and complex nervous systems of higher animals. In: G. Stephani (Ed.), *Handbook of Comparative Opioid and Related Neuropeptide Mechanisms*, CRC Press, Boca Raton, pp. 103–115.

Grimmelikhuijzen, C.J.P. and Westfall, J.A. (1995) The nervous system of Cnidarians. In: O. Breidbach and W. Kutsch (Eds), *The Nervous System of Invertebrates: An Evolutionary and Comparative Approach*, Birkhäuser, Basel, pp. 7–24.

Grimmelikhuijzen, C.J.P., Dierickx, K. and Boer, G.J. (1982) Oxytocin/Vasopressin-likeimmunoreactivity is present in the nervous system of hydra. *Neuroscience*, 7: 3191–3199.

Grimmelikhuijzen, C.J.P., Leviev, I. and Carstensen, K. (1996) Peptides in the nervous system of cnidarians: structure, function and biosynthesis. *Int. Rev. Cytol.*, 167: 37–89.

Haldane, J.B.S. (1954) La signalisation animale, *Anneé Biol.*, 58: 89–98.

Hall, D.H. and Russell, R.L. (1991) The posterior nervous system of the nematode Caenorhabditis elegans: serial reconstruction of identified neurons and complete pattern of synaptic interactions. *J. Neurosci.*, 11: 1–2.

Hanström, B. (1928) *Vergleichende Anatomie des Nervensystems der wirbellosen Tiere*, Springer, Berlin, 628 pp.

Harrison, F.W. and De Vos, L. (1991) Porifera. In: F.W. Harrison (Ed.), *Microscopic Anatomy of Invertebrates*, New York, Wiley-Liss, pp. 28–89.

Hart, A.C., Sims, S. and Kaplan, J.M. (1995) Synaptic code for sensory modalities revealed by C. elegans GLR-1 glutamate receptor. *Nature*, 378: 82–85.

Hartline, H.K. and Ratliff, F. (1957) Inhibitory interaction of receptor units in the eye of Limulus. *J. Gen. Physiol.*, 40: 357–376.

Heider, K. (1928) *Entwichlungsgeschichte und Morphologie der Wirbellosen*, Teubner, Berlin, 332p.

Heinzeller, T. and Welsch, U. (1994) Crinoidea. In: F.W. Harrison (Ed.), *Microscopic Anatomy of Invertebrates*, Vol. 14: Echinodermata, Wiley-Liss, New York, pp. 9–148.

Herbert, J. (1993) Peptides in the limbic system: neurochemical codes for co-ordinatedadaptive responses to behaviouraland physiological demand. *Progr. Neurobiol.*, 41: 723–791.

Herkenham, M. (1987) Mismatches between neurotransmitter and receptor localization in brain: observations and implications. *Neuroscience*, 23: 1–38.

Hökfelt, T., Everitt, B., Meister, B. et al. (1986) Neurons with multiple messengers with special reference to neuroendocrine systems, *Rec. Progr. Horm. Res.*, 42: 1–70.

Homberg, U. (1987) Structure and functions of the central complex in insects. In: A.P. Gupta (Ed.), *Arthropod Brain: Its Evolution, Development, Structure and Functions*, Wiley, New York, pp. 347–367.

Homberg, U., Vitzthum, H., Müller, M. and Binkle, U. (1999) Immunocytochemistry of GABA in the central complex of the locust Schistocerca gregaria: identification of immunoreactive neurons and colocalization with neuropeptides. *J. Comp. Neurol.*, 409: 495–507.

Horridge, G.A. and Mackay, B. (1962) Naked axons and symmetrical synapses in coelenterates. *Quart. J. Microsc. Sci.*, 103: 531–541.

Jaffe, E.H., Marty, A., Schulte, A. and Chow, R.H. (1998) Extrasynaptic vesicular transmitter release from the somata of substantia nigra neurons in rat midbrain slices. *J. Neurosci.*, 18: 3548–3553.

Joffe, B.I. and Reuter, M. (1993) The nervous system of Brothriomolus balticus (Proseriata): a contribution to the knowledge of the orthogon in the Plathelminthes. *Zoomorphology*, 113–127.

Jones, W.C. (1962) Is there a nervous system in sponges? *Biol. Rev.*, 37: 1–50.

Katz, M.J. (1983) Comparative anatomy of the tunicate tadpole Ciona intestinalis. *Biol. Bull.*, 164: 1–27.

Khan, M.A. and Buma, P. (1985) Neural control of the corpus allatum in the Colorado potato beetle, Leptinotarsa decemlineata: an electron microscope study utilizing the in vitro tannic acid ringer incubation method. *J. Insect Physiol.*, 31: 639–645.

Kinnamon, J.C. and Westfall, J.A. (1982) Type of neurons and synaptic connections at hypostome-tentacles junctions in Hydra. *J. Morphol.*, 173: 119–128.

Koizuma, O., Wilson, J.D., Grimmelikhuijzen, C.J.P. and Westfall, J.A. (1989) Ultrastructural localization of RF amide-like peptides in neuronal dense-cored vesicles in the peduncle of Hydra. *J. Exp. Zool.*, 249: 17–22.

Koopowitz, H. (1989) Polyclad neurobiology and the evolution of central nervous systems. In: P.A.V. Anderson (Ed.), *Evolution of the First Nervous Systems*, Plenum Press, New York, pp. 315–328.

Korschelt, E. (1936) *Vergleichende Entwicklungsgeschichte der Tiere*. Fischer, Jena, 2 vols, 1314 pp.

Kowalevsky, A.O. (1866) Entwicklungsgeschichte der einfachen Ascidien. *Mém. Acad. Sci. St. Petersburg*, 10: 1–19.

Koyama, H. and Kusunoki, T. (1993) Organization of the cerebral ganglion of the colonial ascidian Polyandrocarpa misakiensis. *J. Comp. Neurol.*, 338: 549–559.

Kuhlenbeck, H. (1967) *The central nervous system of vertebrates*, Vol. 2: *Invertebrates and origin of vertebrates*, Karger, Basel, 364 pp.

Kupffer, C. (1870) Die Stammverwandtschaft zwischen Ascidien und Wirbelthieren. *Arch. mikrosk. Anat.*, 6: 1–58.

Lacalli, T.C. (1994) Apical organs, epithelial domains, and the origin of the chordate central nervous system. *Amer. Zool.*, 34: 544–541.

Lacalli, T.C. (1996a) Frontal eye circuitry, rostral sensory pathways and brain organization in amphioxus larvae: evidence from 3D reconstructions. *Phil. Trans. Roy. Soc. (Lond. B)*, 351: 243–263.

Lacalli, T.C. (1996b) Landmarks and subdomains in the larval brain of Branchiostoma: vertebrate homologs and invertebrate antecedents. *Israel J. Zool.*, 42: S131-S146.

Lacalli, T.C. and Hou, S. (1999) A re-examination of the epithelial sensory cells of amphioxus (Branchiostoma). *Acta Zool. (Stockholm)*, 80: 125–134.

Lacalli, T.C. and Kelly, S.J. (2000) Cerebral vesicle structures in amphioxus larvae: the infundibular balance organ and related aspects of c.v. organization. *Acta Zool. (Stockholm)*, in press.

Lacalli, T.C., Holland, N.D. and West, J.E. (1994) Landmarks in the anterior central nervous system of amphioxus larvae. *Phil. Trans. Roy. Soc. (Lond. B)*, 344: 165–185.

Lawn, I.D., Mackie, G.O. and Silver, G. (1981) Conduction system in a sponge. *Science*, 211: 1169–1171.

Lentz, T.L. (1965a) The fine structure of differentiating interstitial cells in Hydra. *Z. Zellforsch.*, 67: 547–560.

Lentz, T.L. (1965b) Fine structural changes in the nervous system of the regenerating hydra. *J. Exp. Zool.*, 150: 181–194.

Lentz, T.L. (1965c) Hydra: induction of supernumerary heads by isolated neurosecretory granules. *Science*, 150: 633–635.

Lentz, T.L. (1966) Histochemical localization of neurohumors in a sponge. *J. Exp. Zool.*, 162: 171–180.

Lentz, T.L. (1968) *Primitive nervous systems*, Yale University Press, New Haven, 148 pp.

Locascio, A., Aniello, F., Amoroso, A., Manzanares, M., Krumlauf, R. and Branno, M. (1999) Patterning the ascidian nervous system: structure, expression and transgenic analysis of the CiHox3 gene. *Development*, 126: 4737–4748.

Luchtel, D.L., Martin, A.W., Deyrup-Olsen, I. and Boer, H.H. (1997) Gastropoda: Pulmonata. In: F.W. Harrison (Ed.), *Microscopic Anatomy of Invertebrates*, Vol. 6B: *Mollusca II*, Wiley-Liss, New York, pp. 459–718.

Machemer, H. and Deitmer, J.W. (1987) From structure to behaviour: Stylonichia as a model system for cellular physiology. *Progr. Protistol.*, 2: 213–330.

Mackie, G.O. (1990) The elementary nervous system revisited. *Amer. Zool.*, 30: 907–920.

Mackie, G.O. and Singla, C.L. (1983) Studies on hexactinellid sponges: I. Histology of Rhabdocalyptus dawsoni. *Phil. Trans. Roy. Soc. B*, 301: 365–400.

Mackie, G.O., Lawn, I.D. and Pavans de Ceccaty, M. (1983) Studies on hexactinellid sponges: II. Excitability, conduction and coordination of responses in Rhabdocalyptus dawsoni. *Phil. Trans. Roy. Soc. B*, 301: 401–418.

Mäntylä, K., Reuter, M., Halton, D.W., Maule, A.G., Brennan, G.P., Shaw, C. and Gustafsson, M.K.S. (1998) The nervous system of Procerodes littoralis (Maricola,Tricladida): an ultrastructural and immunoelectron microscopical study. *Acta Zool. (Stockholm)*, 79: 1–8.

Margulis, L. and Schwartz, K.V. (1998) *Five kingdoms: an illustrated guide to the phyla of life on earth*, (3rd ed.), W.H. Freeman, New York, 520 pp.

Marlcq, A.V., Peckol, E., Driscoll, M. and Bargmann, C.I. (1995) Mechanosensory signalling in C. elegans mediated by the GLR-1 glutamate receptor. *Nature*, 378: 78–81.

Marlier, L., Sandillon, F., Poulat, P., Rajaofetra, N., Geffard, M. and Privat, A. (1991) Serotonergic innervation of the dorsal horn of rat spinal cord: light and electron microscopic immunocytochemical study. *J. Neurocytol.*, 20: 310–322.

Mitgutsch, C., Hauser, F. and Grimmelikhuijzen, C.J.P. (1999) Expression and developmental regulation of the Hydra-RFamide and Hydra-LWamide preprohormone genes in Hydra: evidence for transient phase of head formation. *Dev. Biol.*, 207: 189–203.

Morris, J.F. and Pow, D.V. (1991) Widespread release of peptides in the central nervous system: quantitation of tannic acid-captured exocytoses. *Anat. Rec.*, 231: 437–445.

Nässel, D.R. (1991) Neurotransmitters and neuromodulators in the insect visual system. *Progr. Neurobiol.*, 37: 179–254.

Nicholson, C. (1999) Signals that go with the flow. *Trends Neurosci.*, 22: 143–145.

Nicol, D. and Meinertzhagen, I.A. (1991) Cell counts and maps in the larval central nervous system of the ascidian Ciona intestinalis (L.). *J. Comp. Neurol.*, 309: 415–429.

Nieuwenhuys, R. (1985) *Chemoarchitecture of the brain.* Springer, Berlin Heidelberg New York, 246 pp.

Nieuwenhuys, R. (1994) The neocortex: an overview of its evolutionary development, structural organization and synaptology. *Anat. Embryol.*, 190: 307–337.

Nieuwenhuys, R. (1996) The greater limbic system, the emotional motor system and the brain. *Progr. Brain Res.*, 107: 551–580.

Nieuwenhuys, R. (1998) Amphioxus. In: R. Nieuwenhuys, H.J. ten Donkelaar and C. Nicholson (Eds), *The Central Nervous System of Vertebrates*, Vol. 1, Springer, Berlin, pp. 365–396.

Nieuwenhuys, R. and Nicholson, C. (1998) Lampreys, petromyzontoidea. In: R. Nieuwenhuys, H.J. ten Donkelaar and C. Nicholson (Eds), *The Central Nervous System of Vertebrates*, Vol. 1, Springer, Berlin, pp. 397–495.

Nieuwenhuys, R., Ten Donkelaar, H.J. and Nicholson, C. (1998) *The Central Nervous System of Vertebrates*. Springer, Heidelberg, 2219 pp.

Nieuwenhuys, R., Veening, J. and Van Domburg, P. (1989) Core and paracores: some new chemoarchitectural entities in the mammalian neuraxis. *Acta Morphol. Neerl.-Scand.*, 26: 131–163.

Nieuwenhuys, R., Voogd, J. and Van Huijzen, C. (1988) *The human central nervous system: a synopsis and atlas*, (3rd rev. ed.), Springer, Berlin Heidelberg New York, 437 pp.

Olsson, R. (1993) Reissner's fiber mechanisms: some common denominators. In: A. Oksche, E.M. Rodríguez and P. Fernández-Llebrez (Eds), *The Subcommissural Organ*. Springer, Berlin, Heidelberg, New York, pp 33–39.

Orlovsky, G.N., Deliagina, T. and Grillner, S. (1999) *Neuronal control of locomotion: from mollusc to man*, Oxford, Oxford University Press, 322 pp.

Pantin, C.F.A. (1952) The elementary nervous system. *Phil. Trans. Roy. Soc. (Lond. B)*, 140: 147–168.

Parker, D. and Grillner, S. (1999) Long-lasting substance-P-mediated modulation of NMDA-induced rhythmic activity in the lamprey locomoter network involves separate RNA- and protein-synthesis-dependent stages. *Eur. J. Neurosci.*, 11: 1515–1522.

Pavans de Ceccaty, M. (1955) Le système nerveux des éponges calcaires et siliceuses. Ann. Sci. Nat. (Zool.), 17: 203–290.

Pavans de Ceccaty, M. (1959) Les structures cellulaires de type nerveux chez Hippospongia communis Lmk. *Ann. Sci. Nat. (Zool.)*, 12: 105–112.

Pavans de Ceccatty, M. (1960) Les structures cellulaires de type nerveux et de type musculaire de l'Éponge silicieuse Tethya lyncurium Lmk. *C.R. Acad. Sci., Paris*, 251: 1818–1819.

Pavans de Ceccaty, M. (1966a) Ultrastructures et rapports des cellules mésenchymateuses de type nerveux de l'éponge Tethya lyncurium. *Ann. Sci. Nat. (Zool.)*, 8: 577–614.

Pavans de Ceccaty, M. (1966b) Connexions cellulaires et jonctions polarisées du réseau intramesenchymateux, chez l'Éponge Hippospongia communis. *C.R. Acad. Sci. Paris*, 263D: 145–147.

Pavans de Ceccaty, M. (1974a) Coordination in sponges: the foundations of integration. *Amer. Zool.*, 14: 859–903.

Pavans de Ceccaty, M. (1974b) The origin of the integrative systems: a change in view derived from research on coeleterates and sponges. *Persp. Biol. Med.*, 17: 379–390.

Peteya, D.J. (1973) A light and electron microscopic study of the nervous system of Ceriantheopsis americanus. *Z. Zellforsch.*, 141: 301–317.

Plasterk, R.H.A. (1999) The year of the worm. *BioEssays*, 21: 105–109.

Pow, D.V. and Golding, D.W. (1987) 'Neurosecretion' by aminergic synaptic terminals in vivo: a study of secretory granule exocytosis in the corpus cardiacum of the flying locust. *Neuroscience*, 22: 1145–1149.

Pow, D.V. and Morris, J.F. (1989) Dendrites of hypothalamic magnocellular neurons release neurohypophysial peptides by exocytosis. *Neuroscience*, 32: 435–439.

Quaglia, A. and Grasso, M. (1986) Ultrastructural evidence for a peptidergic-like neurosecretory cell in a sea anemone. *Oebalia*, 13: 147–156.

Raikova, O.I., Reuter, M., Kotikova, E. and Gustafsson, M.K.S. (1998) A commissural brain! The pattern of 5-HT immunoreactivity in Acoela (Plathelminthes). *Zoomorphology*, 118: 69–77.

Rand, J.B. and Nonet, M.L. (1997) Synaptic transmission. In: D.L. Riddle, T. Blumenthal, B.J. Meyer and J.R. Priess (Eds), *C. Elegans II*, Cold Spring Harbor Laboratory Press, Cold Spring Harbor, NY, pp. 611–643.

Rehkämper, G. and Welsch, U. (1985) On the fine structure of the cerebral ganglion of Sagitta (Chaetognatha). *Zoomorphology*, 105: 83–89.

Reinscheid, R.K. and Grimmelikhuijzen, C.J.P. (1994) Primary structure of the precursor for the anthozoan neuropeptide antho-RFamide from Renilla köllikeri: evidence for unusual processing enzymes. *J. Neurochem.*, 62: 1214–1222.

Reisinger, E. (1925) Untersuchungen am Nervensystem der Bothrioplana semperi Braun. *Z. Morph. Ökol. Tiere*, 5: 119–149.

Reisinger, E. (1972) Die Evolution des Orthogons der Spiralier und das Archicoelomatenproblem. *Z. Zool. Syst. Evolutionsforsch.*, 13: 184–206.

Renneis, M.L., Gregory, T.F., Blaumanis, O.R., Fujimoto, K. and Grady, P.A. (1985) Evidence for a 'paravascular' fluid circulation in the mammalian central nervous system, provided by the rapid distribution of tracer protein throughout the brain from the subarachnoid space. *Brain Res.*, 326: 47–63.

Reuter, M. (1981) The nervous system of Microstomum lineare (Turbellaria, Macrostomida). *Cell Tissue Res.*, 218: 375–387.

Reuter, M. (1988) Development and organization of nervous systems visualized by immunocytochemistry in three flatworm species. *Forsch. Zool.*, 36: 181–184.

Reuter, M. (1990) From innovation to integration: trends of the integrative systems in microturbellarians. *Acta Acad. Aboensis Ser. B*, 50: 121–136.

Reuter, M. (1991) Are there differences between proseriates and lower flatworms in ultrastructure of the nervous system? *Hydrobiologia*, 227: 221–227.

Reuter, M. and Gustafsson, M. (1989) 'Neuroendocrine cells' in flatworms: progenitors to metazoan neurons? *Acta Histol. Cytol.*, 52: 253–263.

Reuter, M. and Gustafsson, M.K.S. (1995) The flatworm nervous system: pattern and phylogeny. In: O. Breidbach and W. Kutsch, W. (Eds), *The Nervous System of Invertebrates: An Evolutionary and Comparative Approach*, Birkhäuser, Basel, pp. 25–59.

Reuter, M. and Palmberg, I. (1989) development and differentiation of neuronal subsets in asexually reproducing Microstomum lincare. Immunohistochemistry of 5-Ht, RF-amide and SCP_B. *Histochemistry*, 91: 123–131.

Reuter, M. and Palmberg, I. (1990) Synaptic and non-synaptic release in Stenostomum leucops. A study of the nervous system and sensory receptors. *Acta Acad. Abonensis Ser. B*, 50(7): 121–136.

Reuter, M., Gustafsson, M.K.S., Lang, J. and Grimmelikhuijzen, C.J.P. (1990) The release sites and targets of nerve cells immunoreactive to RFamide: an ultrastructural study of Microstomum lineare and Diphyllobothrium dendriticum (Plathelminthes). *Zoomorphology*, 109: 303–308.

Reuter, M., Gustafsson, M.K.S., Sahlgren, C., Halton, D.W., Maule, A.G. and Shaw, C. (1995b) The nervous system of Tricladida: II. Neuroanatomy of Dugesia tigrina (Paludicola, Dugesiidae): an immunocytochemical study. *Invertebrate Neurosci.*, 1: 133–143.

Reuter, M., Gustafsson, M.K.S., Sahlgren, C., Halton, D.W., Maule, A.G. and Shaw, C. (1995a) The nervous system of Tricladida: I. Neuroanatomy of Procerodes littoralis (Maricola, Procerodidae): an immunocytochemical study. *Invertebrate Neurosci.*, 1: 113–122.

Reuter, M., Mäntylä, K. and Gustafsson, M.K.S. (1998b) Organization of the orthogon: main and minor nerve cords. *Hydrobiologia*, 383: 175–182.

Reuter, M., Raikova, O.I. and Gustafsson, M.K.S. (1998a) An endocrine brain? The pattern of FMRF-amide immunoreactivity in Acoela (Plathelminthes). *Tissue Cell*, 30: 57–63.

Riddle, D.L., Blumenthal, T., Meyer, B.J. and Priess, J.R. (1997) Introduction to C. elegans. In: D.L. Riddle, T.

Blumenthal, B.J. Meyer and J.R. Priess (Eds), *C. Elegans II*, Cold Spring Harbor Laboratory Press, Cold Spring Harbor, NY, pp. 1–22.

Ridet, J.-L., Privat, A. (2000) Volume transmission. *Trends Neurosci.*, 23: 58–59.

Rieger, R.M., Tyler, S., Smith, J.P.S.I. and Rieger, G. (1991) Plathylminthes: Turbellaria. In: F.W. Harrison and B.J. Bogitsh (Eds), *Microscopic Anatomy of Invertebrates*, Vol. 3: *Plathylminthes and Nemertinea*, Wiley-Liss, New York, pp. 7–140.

Roubos, E.W. (1997) Background adaptation by Xenopus laevis: a model for studying neuronal information processing in the pituitary pars intermedia. *Comp. Biochem. Physiol.*, 118A: 533–550.

Roubos, E.W. and Moorer-van Delft, C.M. (1979) Synaptology of the central nervous system of the freshwater snail Lymnaea stagnalis (L.), with particular reference to neurosecretion. *Cell Tissue Res.*, 198: 217–235.

Ruiz-Trillo, I., Riutort, M., Timothy, D., Littlewood, J., Herniou, E.A. and Baguña, J. (1999) Acoel flatworms: earliest extant bilaterian Metazoans, not members of Platyhelminthes. *Science*, 283: 1919–1923.

Saripalli, L.D. and Westfall, J.A. (1996) Classification of nerve cells dissociated from tentacles of the sea anemone Calliactis parasitica. *Biol. Bull.*, 190: 111–124.

Schaller, H.C. and Bodenmüller, H. (1987) Hydra, neurohormones. In: G. Adelman (Ed.), *Encyclopedia of Neuroscience*, Birkhäuser, Boston, pp. 505–506.

Schinkmann, K. and Li, C. (1992) Localization of FMRFamide-like peptides in Caenorhabditis elegans. *J. Comp. Neurol.*, 316: 251–260.

Schmidt, E.D. and Roubos, E.W. (1987) Morphological basis for nonsynaptic communication within the central nervous system by exocytotic release of secretory material from the egg-laying stimulating neuroendocrine caudodorsal cells of Lymnaea stagnalis. *Neuroscience*, 20: 247–257.

Schoofs, L., Broeck, J.V. and DeLoof, A. (1993) The myotropic peptides of Locusta migratoria: structures, distribution, functions and receptors. *Insect Biochem. Mol. Biol.*, 23: 859–881.

Schooneveld, H. (1998) Neurosecretion. In: F.W. Harrison (Ed.), *Microscopic Anatomy of Invertebrates*, Wiley-Liss, New York, Vol. 11B: Insecta, pp. 467–486.

Schot, L.P.C. (1984) *Immunocytochemical studies on peptidergic neurons in the pond snail Lymnaea stagnalis*. Ph.D. thesis, Free University, Amsterdam, 85 pp.

Schürmann, F.-W. (1987) Histology and ultrastructure of the Onychophoran brain. In: A.P. Gupta (Ed.), *Arthropod Brain: its Evolution, Development, Structure, and Functions*, Wiley, New York, pp. 159–180.

Schürmann, F.-W. (1987) The architecture of the mushroom bodies and related neuropils in the insect brain. In: A.P. Gupta (Ed.), *Arthropod Brain: Its Evolution, Development, Structure, and Functions*, Wiley, New York, pp. 231–264.

Schürmann, F.-W., Sandeman, R. and Sandeman, D. (1991) Dense-core vesicles and non-synaptic exocytosis in the central body of the crayfish brain. *Cell Tissue Res.*, 265: 493–501.

Schürmann, F.W. (1995) Common and special features of the nervous system of Onychophora: a comparison with anthropoda, annelida and some other invertebrates. In: O. Breidbach and W. Kutsch (Eds), *The nervous system of invertebrates: an evolutionary and comparative approach*, Birkhäuser, Basel, pp. 139–158.

Shinn, G.L. (1997) Chaetognatha. In: F.W. Harrison (Ed.), *Microscopic Anatomy of Invertebrates*, Vol. 15: *Hemichordata, Chaetognatha, and the Invertebrate Chordates*, Wiley-Liss, New York, pp. 103–220.

Simpson, T.L. (1984) *The cell biology of sponges*, Springer, New York, 662 pp.

Singla, C.L. and Mackie, G.O. (1991) Immunogold labeling of FMRF-amide-like neuropeptide in neurons of Aglantha digitale. *Can. J. Zool.*, 69: 800–802.

Spencer, A.N. and Arkett, S.A. (1984) Radial symmetry and the organization of central neurones in a hydrozoan jellyfish. *J. Exp. Biol.*, 110: 69–90.

Stevenson, P.A. (1999) Colocalisation of taurine- with transmitter-immunoreactivities in the nervous system of the migratory locust. *J. Comp. Neurol.*, 404: 86–96.

Storch, V. and Welsch, U. (1997) *Systematische Zoologie*, Fischer, Stuttgart, 804 pp.

Strausfeld, N.J. and Meinertzhagen, I.A. (1998) The insect neuron: types, morphologies, fine structure, and relationship to the architectonics of the insect nervous system. In: F.W. Harrison (Ed.), *Microscopic Anatomy of Invertebrates*, Vol. 11B: Insecta, Wiley-Liss, New York, pp. 487–538.

Strausfeld, N.J., Buschbeck, E.K. and Gomez, R.S. (1995) The arthropod mushroom body: its functional roles, evolutionary enigmas and mistaken identities. In: O. Breidbach and W. Kutsch (Eds), *The Nervous System of Invertebrates: An Evolutionary and Comparative Approach*, Birkhäuser, Basel, pp. 351–381.

Stroh, T. and Zupanc, G.K.H. (1993) Identification and localization of somatostatin-like immunoreactivity in the cerebellum of gymnotiform fish, Apteronotus leptorhynchus. *Neurosci. Lett.*, 160: 145–148.

Stubbs, T.R. and Cobb, J.L.S. (1981) The giant neurone system in Ophiuroids: II. The hyponeural motor tracts. *Cell Tissue Res.*, 220: 373–385.

Sulston, J.E., Schierenberg, E., White, J.G. and Thomson, J.N. (1983) The embryonic cell lineage of the nematode Caenorhabditis elegans. *Dev. Biol.*, 64–119.

Sun, X.J., Tolbert, L.P. and Hildebrand, J.G. (1993) Ramification pattern and ultrastructural characteristics of the serotonin-immunoreactive neuron in the antennal lobe of the moth Manduca sexta: a laser scanning confocal and electron microscopic study. *J. Comp. Neurol.*, 338: 5–16.

The C. Elegans Sequencing Consortium (1998) Genome sequence of the nematode C. elegans: a platform for investigating biology. *Science*, 282: 2012–2018.

Thureson-Klein, Å.K., Klein, R.L. Zhu, P.-C. and Kong, J.-Y. (1988) Differential release of transmitters and neuropeptides co-stored in central and peripheral neurons. In: H. Zimmerman (Ed.), *Cellular and Molecular Basis of Synaptic Transmission*. NATO ASI Series, Vol. H21, Springer, Berlin, Heidelberg, 137–151.

Tremblay, J.P., Colonnier, M. and McLennan, H. (1979) An electron microscope study of synaptic contacts in the abdominal ganglion of Aplysia californica. *J. Comp. Neurol.*, 367–390.

Ullström, M., Parker, D., Svensson, E. and Grillner, S. (1999) Neuropeptide-mediated facilitation and inhibition of sensory inputs and spinal cord reflexes in the lamprey. *J. Neurophysiol.*, 81: 1730–1740.

Van Dongen, P.A.M., Hökfelt, T., Grillner, S., Verhofstad, A.A.J., Steinbusch, H.W.M., Cuello, A.c. and Terenius, L. (1985) Immunohistochemical demonstration of some putative neurotransmitters in the lamprey spinal cord and spinal ganglia: 5-hydroxytryptamine-,tachykinin-, and neuropeptide-Y-immunoreactive neurons and fibers. *J. Comp. Neurol.*, 234: 501–522.

Van Houten, J. (1989) Chemoreception in unicellular eukaryotes. In: P.A.V. Anderson (Ed.), *Evolution of the First Nervous Systems*, Plenum Press, New York, London, pp. 343–356.

Vigh, B. (1999) Quoted from Nicholson (1999).

Ward, S., Thomson, N., White, J.G. and Brenner, S. (1975) Electron microscopical reconstruction of the anterior sensory anatomy of the nematode Caenorhabditis elegans. *J. Comp. Neurol.*, 160: 313–338.

Ware, R.W., Clark, D., Crossland, K. and Russell, R.L. (1975) The nerve ring of the nematode Caenorhabditis elegans: sensory input and motor output. *J. Comp. Neurol.*, 162: 71–110.

Wegerhoff, R. and Breidbach, O. (1995) Comparative aspects of the chelicerate nervous systems. In: O. Breidbach and W. Kutsch, W. (Eds), *The Nervous Systems of InverteBrates: An Evolutionary and Comparative Approach*, Birkhäuser, Basel, pp. 159–179.

Wendelaar Bonga, S.E. (1970) Ultrastructure and histochemistry of neurosecretory cells and neurohaemal areas in the pond snail Lymnaea stagnalis. *Z. Zellforsch.*, 108: 190–224.

Westfall, J.A. (1973) Ultrastructural evidence for a granule-containing sensory-motor-interneuron in Hydra littoralis. *J. Ultrastruct. Res.*, 42: 268–282.

Westfall, J.A. (1987) Ultrastructure of invertebrate synapses. In: M.A. Ali (Ed.), *Nervous Systems in Invertebrates*, Plenum Press, New York, pp. 2–28.

Westfall, J.A. and Grimmelikhuijzen, C.J.P. (1993) Antho-RFamide immunoreactivity in neuronal synaptic and nonsynaptic vesicles of sea anemones. *Biol. Bull.*, 185: 109–114.

Westfall, J.A. and Kinnamon, J.C. (1978) A second sensory-motor-interneuron with neurosecretory granules in Hydra. *J. Neurocytol.*, 7: 365–379.

Westfall, J.A. and Sayyar, K.L. (1997) Ultrastructure of neurons and synapses in the tentacle epidermis of the sea anemone Calliactis parasitica. *J. Morphol.*, 232: 207–216.

Westfall, J.A., Kinnamon, J.C. and Sims, D.E. (1980) Neuro-epitheliomuscular cell and neuro-neuronal gap junctions in Hydra. *J. Neurocytol.*, 9: 725–732.

Westfall, J.A., Landers, D.D. and McCallum, J.D. (1998) Different nematocytes have different synapses in the sea anemone Aiptasia pallida (Cnidaria, Anthozoa). *J. Morphol.*, 238: 53–62.

Westfall, J.A., Landers, D.D. and McCallum, J.D. (1999) Ultrastructure of neuro-spirocyte synapses in the sea anemone Aiptasia pallida (Cnidaria, Anthozoa, Zoantharia). *J. Morphol.*, 241: 165–173.

Westfall, J.A., Sayyar, K.L. and Bone, J.K. (1997) Ultrastructure of neurons and synapses in the tentacle gastrodermis of the sea anemone Calliactis parasitica. *J. Morphol.*, 231: 217–223.

Westfall, J.A., Yamataka, S. and Enos, P.D. (1971) Ultrastructural evidence of polarized synapses in the nerve net of Hydra. *J. Cell Biol.*, 51: 318–323.

White, J.G., Southgate, E., Thomson, J.N. and Brenner, S. (1986) The structure of the nervous system of the nematode Caenorhabditis elegans. *Phil. Trans. Roy. Soc. (Lond. B)*, 314: 1–340.

Yung, K.K.L., Bolam, J.P., Smith, A.D., Hersch, S.M., Ciliax, B.J. and Levey, A.I. (1995)Immunocytochemical localization of D_1 and D_2 dopamine receptors in the basal ganglia of the rat: light and electron microscopy. *Neuroscience*, 65: 709–730.

Zhu, P.C., Thureson-Klein, Å. and Klein, R.L. (1986) Exocytosis from large dense cored vesicles outside the active synaptic zones of terminals within the trigeminal subnucleus caudalis: a possible mechanism for neuropeptide release. *Neuroscience*, 19: 43–54.

Zoli, M. and Agnati, L.F. (1996) Wiring and volume transmission in the central nervous system: the concept of closed and open synapses. *Progr. Neurobiol.*, 49: 363–380.

Zoli, M., Torri, C., Ferrari, R., Jansson, A., Zini, I., Fuxe, K. and Agnati, L.F. (1998) The emergence of the volume transmission concept. *Brain Res. Rev.*, 26: 136–147.

Zupanc, G.K.H. (1991) The synaptic organization of the prepacemaker nucleus in weakly electric knifefish, Eigenmannia: a quantitative ultrastructural study. *J. Neurocytol.*, 20: 818–833.

Zupanc, G.K.H. (1996) Peptidergic transmission: from morphological correlates to functional implications. *Micron*, 27: 35–91.

Zupanc, G.K.H., Cécyre, D., Maler, L., Zupanc, M.M. and Quirion, R. (1994) The distribution of somatostatin binding sites in the brain of gymnotiform fish, Apteronotus leptorhynchus. *J. Chem. Neuroanat.*, 7: 49–63.

SECTION II

Diffusion and extracellular space

L. F. Agnati, K. Fuxe, C. Nicholson and E. Syková (Eds.)
Progress in Brain Research, Vol 125

CHAPTER 5

Diffusion of molecules in brain extracellular space: theory and experiment

Charles Nicholson*, Kevin C. Chen, Sabina Hrabětová and Lian Tao

Department of Physiology and Neuroscience, New York University School of Medicine, 550 First Avenue, New York, NY 10016, USA

Introduction

Volume transmission (VT) is an elusive concept. Classical synaptic transmission is localized at morphologically distinct structures formed by pre- and postsynaptic terminals in apposition and takes place in a highly stereotyped way that involves rapid release of distinctive clusters of vesicles. In contrast, VT, except when it occurs as spillover from a classical synapse, lacks a morphological substrate and the mechanisms can be many and varied. Small wonder then, that it has been hard to identify. Despite these difficulties we may say with certainty that diffusion plays a major, and often definitive, role in VT.

Frank Schmitt, one of the pioneers of the concepts of the brain cell microenvironment (Schmitt and Samson, 1969), introduced the term 'informational substance' (Schmitt, 1984). We can agree that VT involves the introduction of an informational substance into the extracellular space (ECS) at some location, the movement of the substance through the ECS, and the action of the substances on one or more targets. This chapter will focus on the movement of informational substances.

*Corresponding author. Tel.: +1 (212) 263 5421; Fax: +1 (212) 689 9060; e-mail: cn7@is.nyu.edu

Some aspects of diffusion theory in the context of the ECS

Diffusion has played a prominent role in biology, indeed one could say that much of diffusion theory arose from the need to solve biological problems. The microscopic description came about through the observation of what is now called Brownian motion after the discoveries of the botanist Robert Brown in 1828 that pollen grains suspended in water underwent constant random motion (Stachel, 1998). In 1905, Einstein wrote papers on the theory special relativity, the photoelectric effect and Brownian motion (Stachel, 1998). A few years later Jean Perrin carried out experiments to verify Einstein's work on Brownian motion (Perrin, 1990). For this and related work Perrin received the Nobel Prize in 1926.

Adolf Fick formulated the macroscopic laws of diffusion in 1855; Fick sought to combine medicine and physics in his career and made important discoveries in both fields (Cussler, 1997). Among many subsequent applications of Fick's Laws of Diffusion was the work of August Krogh on the diffusion of oxygen from capillaries in the brain (Kreuzer, 1982) that was a major element in Krogh's receipt of the Nobel Prize in 1920.

These are but a few of the highlights of the early history of diffusion. Berg (1993) has provided an excellent introduction to random walks and diffusion in biology. Crank (1975) is an invaluable guide to the practical application of macroscopic theoretical results, many of which were derived from the parallel theory of heat conduction by Carlslaw and Jaeger (1959), a book that remains unsurpassed as a source of analytical solutions.

Microscopic and macroscopic viewpoints on diffusion

As the history shows, diffusion can be described from two viewpoints, the microscopic and the macroscopic; one of the satisfying aspects of this branch of biophysics is that it is possible to move between the two levels of description in a seamless manner. At the level of microscopic description, diffusion consists of the random walks of molecules that collide incessantly with water molecules and occasionally with the obstructions formed by cell membranes. With a suitable statistical treatment, the behavior of a large ensemble of random walks can be described by the equations of classical macroscopic diffusion.

Some simple simulations and calculations can illustrate this. Figure 1 shows three sets of simulated random walks in a medium with obstacles and three sets in a free medium without obstructions. In Fig. 1A, 20 particles have been released from a central point in an idealized brain tissue consisting of a number of spherical obstacles or ‘cells’ represented by the dashed circles. Each particle executes 500 small steps in the spaces between the cells, occasionally colliding with a cell wall and being reflected from it; it is assumed that each step is the same length and takes the same time but the direction is random. In Fig. 1B the number of steps that each of the 20 particles executes is doubled to 1000 and finally in Fig 1C, the steps are doubled again to 2000; Figure 1D–F repeat the same calculations without the cells. It is clear that the ‘territory’ being explored by the particles is increasing as they are allowed to spend more time on their travels. How are exploration time and distance traveled related? Because this is a statistical problem it is necessary to think in terms of statistical measures. Classical theory of random walks uses the mean square distance of a particle from its point of release, denoted by $\langle r^2 \rangle$, to measure how particles spread from a source. It can be shown that this quantity is proportional to the time t since release and that the constant of proportionality is $2nD^*$ where n is the dimension of the space and D^* is the apparent diffusion coefficient (Lipinski, 1990; Berg 1993). So in three-dimensional space, one can define a ‘root mean square diffusion distance’, R, with the dimension of length, as

$$R = \sqrt{\langle r^2 \rangle} = \sqrt{6D^*t}. \tag{1}$$

This well-known and often quoted expression has to be interpreted with caution; it means that in some sense the average distance at which a particle is to be found is proportional to the square root of time.

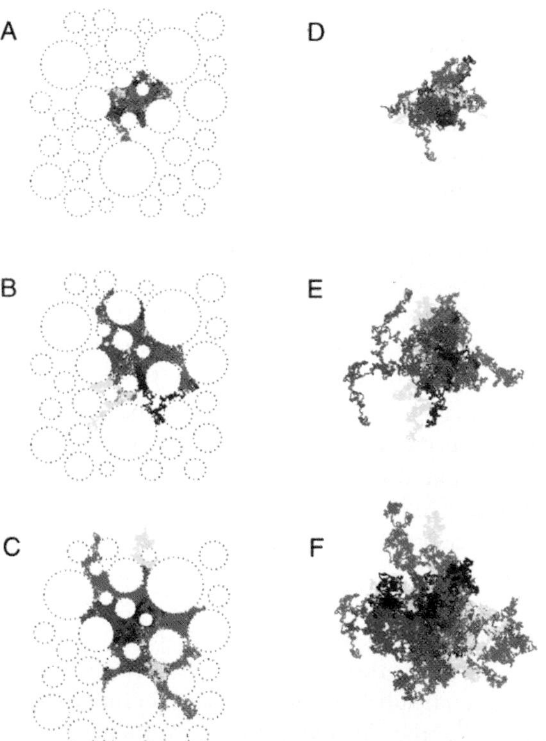

Fig. 1. Microscopic view of diffusion as expressed through random walks. Panels A-C show 20 particles executing a sequence of two-dimensional random walks in the spaces between circular cells; each sequence is assigned a color. Each step is the same length but the direction is chosen at random. In A, the particles make 500 steps, in B they make 1000 steps, in C they make 2000 steps. Panels D–F show a similar sequence of random walks repeated in a ‘free’ medium, that is, one without obstructions. Note how the particles spread further in the absence of hindrance.

Consequently, a particle initially covers a lot of distance from the origin in a short period, but as time goes by, the particle covers less and less distance away from its initial position. But this is 'on the average'; to see what this means one can go to the other perspective on diffusion, the macroscopic viewpoint.

Imagine that a vast number of random walks have been observed over different times. Then a statistical distribution can be constructed for the probability of finding a particle at any distance; for probability we can substitute concentration and it turns out that the distribution is a Gaussian or normal distribution. For particles moving away from a source in a three-dimensional medium, the distribution is (Crank, 1975; Nicholson, 1985; Berg, 1993):

$$C = \frac{P}{\alpha} \frac{1}{(4D^* t\pi)^{3/2}} \exp\left(\frac{-r^2}{4D^* t}\right) \quad (2)$$

where P is the number of moles of substance initially deposited at the origin ($r=0$), α is the fraction of the volume available for the particles to move in, i.e. the space excluding the cells as a fraction of the total space. Figure 2A illustrates the distribution given in Eq. 2 for three different times using $D^* = 7.5 \times 10^{-8}$ cm^2 s^{-1}, which corresponds to the value measured in brain tissue for 70,000 M_r dextran (Nicholson and Tao, 1993). Initially, at time $t=0$, all the substance is concentrated at $r=0$; with time the molecules spread out and the concentration 'collapses'. Note, though, that at all finite times the maximum concentration of molecules remains at $r=0$.

Figure 2B shows an expansion of the falling phase of the curves with the distances corresponding to R (Eq. (1)) for each of the three times indicated by symbols on the curves. It is clear that from this representation that the distance R has no obvious significance; most of the particles are not at R but closer to the origin.

Looking carefully at Fig. 2A, and especially at 2B, it is apparent that another feature is occurring during the collapse of the concentration distribution. Choosing a distance of 40 μm in Fig. 2B it is seen that the lowest concentration occurs on the 15 s curve, the greatest on the 30 s curve and then an intermediate concentration at 60 s. It follows that, if a location is chosen away from the origin and concentration plotted as a function of time, it will have a maximum. This is shown in Fig. 3. It can be established, by differentiating Eq. (2), that the maximum of the concentration curve as a function of time at any given distance occurs at time

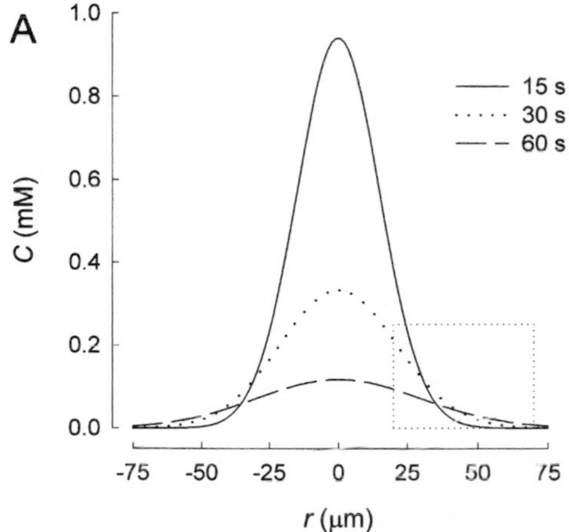

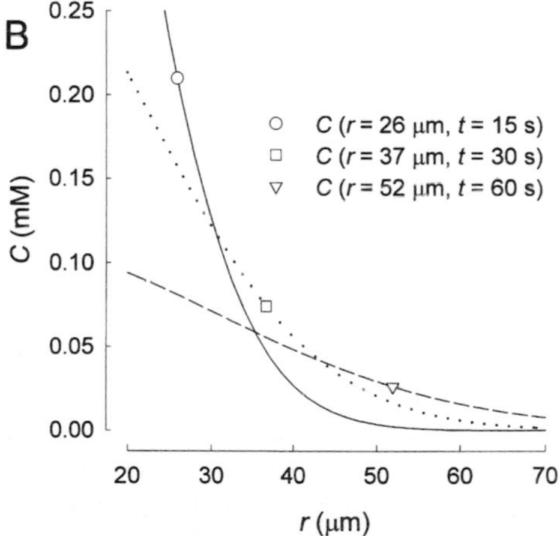

Fig. 2. Macroscopic diffusion. Panel A shows three diffusion profiles calculated for 70,000 M_r dextran at times 15, 30 and 60 s after release from the origin at $r=0$, based on Eq. (2). The apparent diffusion coefficient used was $D^* = 7.5 \times 10^{-8}$ cm^2 s^{-1} and 10^{-14} moles was deposited at the origin at time zero. Panel B shows an expanded plot of the part of A indicated by the dotted box for part of $r>0$ to show how the curves cross over each other as they collapse with time. The three symbols: O, ∇, □, denote the concentrations at the distances $r=(6D^*t)^{1/2}$ for each of the three times; these distances are 26, 37 and 52 μm.

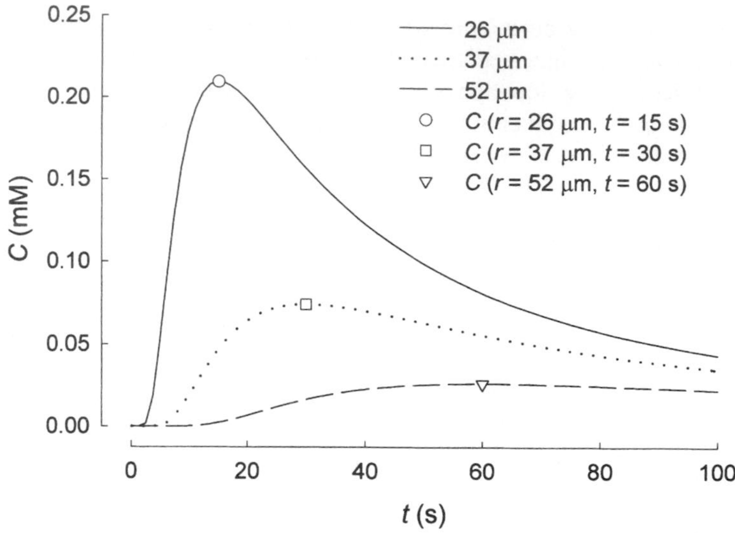

Fig. 3. Concentration as a function of time taken at the three distances (26, 37 and 52 μm) indicated by the symbols Fig. 2B. It is seen that the three maxima occur at 15, 30 and 60 s, i.e. $t=r^2/6D^*$.

$$t=r^2/6D^*. \tag{3}$$

One can say most accurately therefore, that at a given distance r from the instantaneous source, the maximum concentration of molecules will be observed *at that location* at a time t. These locations of maximum concentration, which are the same ones shown in Fig. 2B, are depicted by the same symbols in Fig. 3, where the curves have been drawn for the appropriate distances. Thus a diminishing wave of concentration moves out from the source; but it does not have a constant velocity of propagation, so that talking about the speed of diffusion is not meaningful. The maximum concentration for any distance r can be calculated by substituting Eq. (3) into Eq. (2) to obtain

$$C_{\max}=\frac{P}{\alpha}\frac{0.0736}{r^3}. \tag{4}$$

Surprisingly, $C_{\max}$ is independent of the diffusion coefficient; however it diminishes with the inverse third power of distance.

These elementary considerations show that diffusion is a simple but yet subtle phenomenon. They also show that the often asked question: "how far has a particle gone in time t?" is not a meaningful question unless it is carefully phrased in statistical terms. Finally, these discussions indicate the interplay between the microscopic and the macroscopic perspectives.

Effect of molecular size on diffusion. Later in this paper we shall discuss the behavior of molecules ranging in size from less that 100 M_r up to 100, 000 M_r and beyond. Molecular size affects diffusion profoundly and this can be illustrated by considering a simple extension of the model just discussed. In the previous model we considered all the particles as being deposited at a point at time zero and then released simultaneously, a so-called instantaneous point source. One can imagine the particles being released over a period of time d, by iontophoresis from a micropipette for example, with a constant rate of Q moles per second. The subsequent distribution of molecules in space and time can then be found by an appropriate integration of Eq. (2) (Crank, 1975; Nicholson and Phillips, 1981; Berg, 1993) to obtain the expression:

$$C = \frac{Q}{4\pi D^* r\alpha}\left(\mathrm{erfc}\left(\frac{r}{2\sqrt{D^* t}}\right)\right) \text{ for } t \leq d$$

$$C = \frac{Q}{4\pi D^* r\alpha}\left(\mathrm{erfc}\left(\frac{r}{2\sqrt{D^* t}}\right) - \mathrm{erfc}\left(\frac{r}{2\sqrt{D^*(t-d)}}\right)\right) \text{ for } t > d. \quad (5)$$

In this expression, 'erfc' is the complementary error function, an integral of the Gaussian distribution.

Using Eq. (5), adapting a presentation given by Berg (1993), one can calculate the concentration as a function of time at a distance of 50 μm from the release point varying Q and d so that the total amount released is a constant Qd, chosen here to be 1 pmole. We consider some molecules that will feature later in this paper: TMA+ (tetramethylammonium, $M_r = 74$), dextran ($M_r = 3000$ and 70,000) and BSA (bovine serum albumin, $M_r = 66{,}000$) and use the apparent diffusion coefficients D^*, that we have measured in the brain at 34°C. The results are shown in Fig. 4. In Fig. 4A the duration of the pulse $d = 0.1$ s is very short and the maximum concentration reached is independent of the diffusion coefficient, though the time of the maximum is clearly a function of the apparent diffusion coefficient. These results are explained by realizing that for a very short pulse with measurements on a much longer time scale, we are dealing with what is essentially an instantaneous point source of the type described by Eq. (2). Then $P = Qd$ and the time of the maximum concentration will again be given by Eq. (3), which does depend on D^* while the maximum value will be given by Eq. (4) which is independent of D^*.

As the pulse length increases, the diffusion of small molecules will no longer be accurately approximated by Eq. (2) but will require Eq. (5) while the diffusion of larger molecules will still be accurately modeled by Eq. (2); this situation is shown in Fig. 4B for $d = 10$ s. Finally, if a very long pulse is given ($d = 1000$ s), all molecules will approach a steady state distribution during the pulse, which will be the $t \rightarrow \infty$ limit of Eq. 5,

$$C(r, \infty) = \frac{Q}{4\pi D^* r\alpha}. \quad (6)$$

For the same Q and r, the steady state amplitudes are governed by the reciprocal of D^* while the maximum value of C decreases as the reciprocal of r now.

It is evident that the concentration profiles can be very different for different molecules and this can be attributed both to the different apparent diffusion coefficients and to the duration of the source and its location relative to the observation point. The curves shown in Fig. 4 may well mimic what might be expected under some conditions where VT is taking place.

Diffusion in brain tissue

Two perspectives on diffusion in the ECS. A key concept about diffusion in the ECS is that it provides two different types of information. The first type is the most obvious; it describes how a given molecule moves in the brain. This is largely determined by the apparent diffusion coefficient, D^*, for that molecule, so long as the factors discussed later are not the overriding determinant. The second type of information follows from the idea, outlined above, that the diffusing molecules are executing a vast set of random walks, so that over time they will visit most of the volume in which they are moving. This means that if the sum total of random walks can be visualized, i.e. the concentration profiles can be determined, this can be used to reveal the structure of the underlying volume. Two parameters, the tortuosity and the volume fraction, represent this structural information.

To explore these issues more precisely we turn to the basic equation that describes the process of diffusion in the ECS.

Quantitative description of diffusion in ECS. Experimental work, beginning with Fenstermacher,

Patlak and Rall (for references see Patlak and Fenstermacher 1975; Fenstermacher and Kaye 1988) and continuing with our own studies and those of Syková and colleagues (summarized in Nicholson and Rice, 1991; Syková, 1997; Nicholson and Syková, 1998) has shown that the diffusion properties of the ECS may be encapsulated in a macroscopic partial differential equation (Nicholson and Phillips, 1981; Nicholson and Syková, 1998):

$$\frac{\partial C}{\partial t} = \frac{D}{\lambda^2}\nabla^2 C + \frac{Q}{\alpha} - \mathbf{v}\cdot\nabla C - \frac{f(C)}{\alpha}. \qquad (7)$$

This is the classical diffusion equation arising from Fick's Laws (Crank 1975) modified with some additional parameters. The concentration, C, is a function both of time t and position, the latter represented by x, y, z in a Cartesian coordinate system or r, θ, φ in a spherical coordinate system. The symbols ∇ and ∇^2

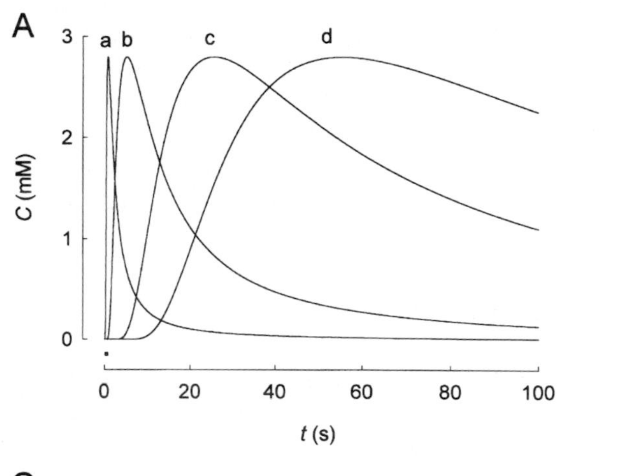

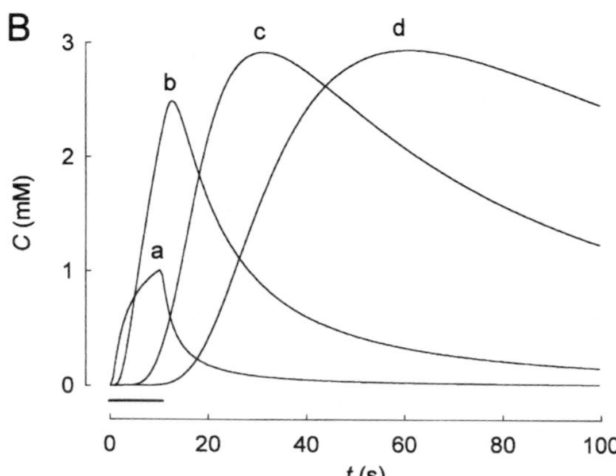

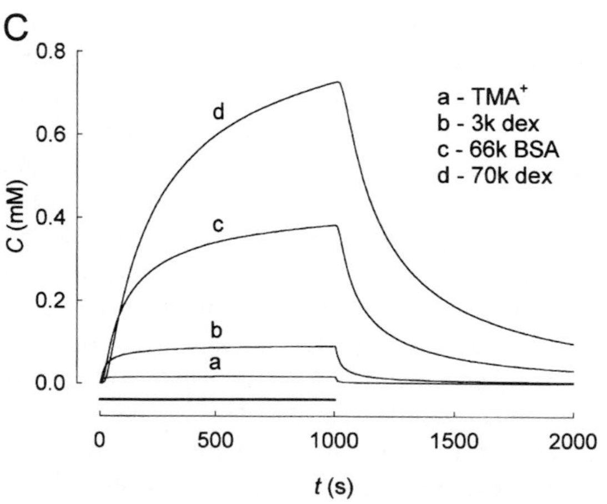

Fig. 4. Effect of molecular size and pulse duration on diffusion profiles. In this simulation, four molecules with different molecular weights, TMA^+ (a, $M_r = 74$), dextran (b, $M_r = 3000$), BSA (c, $M_r = 66{,}000$) and dextran (d, $M_r = 70{,}000$) were released at varying rates Q for different durations d (indicated by bar above abscissa) so that in every case the total amount release was a constant $Qd = 1$ pmole. The concentration of each molecule was calculated, using Eq. (5), at a distance of 50 μm from the source using apparent diffusion coefficients (D^*) that had been previously measured in cortical tissue at 34°C (TMA^+: 4.8×10^{-6}; dextran 3000: 8.1×10^{-7}; BSA: 1.6×10^{-7}; dextran 70,000: 7.5×10^{-8}, all in units of $cm^2\ s^{-1}$). Panel A. Duration of pulse was $d = 0.1$ s. Note that all curves reach a constant amplitude, regardless of molecular weight (i.e. regardless of D^*) because the pulse duration is so short that it can be described by Eq. (2). Panel B. Duration of pulse was $d = 10$ s. Now the smaller molecules behave differently to the larger ones. Panel C. Duration of pulse was $d = 1000$ s (note the different time scale compared to A and B). Here all molecules approach steady-state concentrations during the release period and the amplitude is inversely related to D^*. For further description see text.

symbolize the first and second spatial derivatives in the appropriate coordinate system.

Tissue structure is introduced through two non-dimensional factors, the volume fraction, α, and the tortuosity, λ. Volume fraction is the local ratio of ECS volume to that of total tissue volume. Tortuosity measures how diffusing molecules are hindered by cellular obstructions (Fig. 1). Tortuosity is a composite parameter that may incorporate several geometric effects and possibly interstitial viscosity. It is formally measured by comparing the diffusion coefficients for a molecule in free solution to that in the obstructed medium through the expression $\lambda = (D/D^*)^{1/2}$. The precise relation of this macroscopic quantity to microscopic structure is non-trivial and is the subject of ongoing work (see Chen and Nicholson, 2000b, for further discussion).

The various terms of Eq. (7) have the following interpretations; the term on the left of Eq. (7) represents the way that the concentration of a substance changes with time at any location. The first term on the right of the equal sign brings in the contribution of diffusion itself through the free diffusion coefficient D and tortuosity to form the apparent diffusion coefficient, $D^* = D/\lambda^2$. The next term is the source-term, Q; this describes how a substance is released into the ECS. Experimentally, it may describe local iontophoresis (Nicholson, 1992), or release of molecules by a pressure pulse (Nicholson, 1985); it could also represent spillover from a synapse. The third term represents the contribution of bulk flow, if present. Bulk flow is defined by its velocity vector $\mathbf{v}$ and it forms a scalar product with the concentration gradient ∇C. The final term, $f(C)$, represents loss of material from the extracellular space, typically into cells, across the blood-brain barrier (BBB), or by degradation through enzymatic attack or other processes.

Volume averaging of variables. A somewhat technical issue that has been neglected in this basic discussion of diffusion in the brain is the discontinuous structure of the tissue itself. The brain is essentially a porous medium made up of two intermingled fractions, called here the extracellular and intracellular spaces; in classical porous media theory they would be called the two phases of the medium. In such porous media, any smooth continuous curve drawn through the medium encounters a succession of discontinuous instances of the two phases so that any variable, which is a function of space, is discontinuous. But all the macroscopic theory (such as Eq. (7)) is based on continuous variables and smoothly varying functions. This problem has long been recognized and dealt with to varying degrees of sophistication. One common approach is to average the variables of interest over a so-called Representative Elementary Volume (REV), which has a characteristic length, appropriate to the structure of the medium. This so-called 'volume averaging' can be used to justify the continuous equations and in general one may assume that this has been done successfully. For the present problems the volume averaging justification can be found in Nicholson and Phillips (1981).

Evidence for the importance of different factors in ECS diffusion

Geometry. The geometry of the ECS is a major factor in determining how substances diffuse. As we have noted above, geometry appears in the diffusion equations through the two dimensionless parameters α and λ. Volume fraction is a relatively simple concept and is often called the void fraction in porous media work. It is simply the ratio of extracellular space volume to the entire tissue volume in some chosen REV. Tortuosity, on the other hand, is a more complex idea; it is formally defined as a macroscopic parameter: the square root of the ratio of the free and apparent diffusion coefficients. But we visualize it as the hindrance imposed on molecules executing random walks as they collide with membranes, when we adopt a microscopic perspective. As noted above, the connection between random walks and the partial differential equations of diffusion is not trivial. Another aspect of the inherent complexity of tortuosity is that tissue may have anisotropic properties. That is, there may be different apparent diffusion coefficients associated with different intrinsic axes in the tissue (Rice et al., 1993). This has been found to be the case in several brain regions (Rice et al., 1993; Prokopová et al., 1997; Voříšek and Syková, 1997; Mazel et al., 1998;

Syková et al., this book). In parallel with these results, anisotropy has been discovered in numerous brain regions using diffusion-weighted magnetic resonance imaging (DW-MRI, e.g. Moseley et al., 1990; Neil et al., 1998). When tissue is anisotropic, λ is interpreted as a tensor quantity with three components, each associated with a principal axis of the tissue (e.g. the x, y, z-axes).

Extracellular matrix. It is established that there is an extracellular matrix consisting of proteoglycans and glycosaminoglycans (Margolis and Margolis, 1993; Ruoslahti, 1996). These are all long-chain sugar molecules with many fixed negative charges. Some of these molecules are anchored to membranes and often associated with ion-channels or receptors. Others, such as hyaluronic acid (hyaluronate), are presumed held in the ECS by virtue of their size but otherwise are untethered. Antibodies now permit visualization of components of the matrix (Hobohm et al., 1998; Roitbak and Syková, 1999) and it is especially evident in pericellular nets (Maleski and Hockfield, 1997, Hobohm et al., 1998; Celio et al., 1998; Viggiano and Celio this book). Perhaps the most evident component of the matrix in the brain ECS is chondroitin preoteoglycan (Bertalotto et al., 1996; Bruckner et al., 1998). Despite all these recent data, a crucial issue remains unresolved: how much matrix is actually present in the ECS? Allied to this issue is the one that we are primarily interested in here: does the matrix affect diffusion?

If sufficient matrix is present to affect diffusion, a viscosity function f_η would need to be introduced so that D^* and D would now be related by:

$$D^* = (1/\lambda)^2 (1/f_\eta)^2 D. \qquad (8)$$

At least two theoretical papers have used this type of formulation (El-Kareh et al., 1993; Rusakov and Kullmann 1998a) and discussed the implications. Since the effect of the matrix enters the relationship between D^* and D as a multiplicative viscosity factor, it is impossible to dissect this out directly from a measurement of tortuosity. In principle, all measurements of λ may be contaminated by a viscosity contribution. Therefore the only way to approach the issue of the contribution of the matrix is by some experimental manipulation that either seeks to remove the matrix or, alternatively, accentuates its influence (Syková et al., this book).

Uptake. Uptake is sometimes referred to as clearance and denotes loss of a substance from the ECS, so that a particular molecule no longer participates in the extracellular diffusion process. Note that certain types of fast reversible uptake (see Chapter 14 in Crank, 1975), where the molecule rapidly returns to the diffusion pool, are formally identical to an increase in λ. In most cases, however, uptake requires an additional term in the diffusion equation. In its simplest form, uptake is driven by the difference in concentration between the substance in the ECS and the same substance in another compartment to which the substance has access. The two compartments are separated by some sort of barrier, which has a defined permeability to the substance. Under such conditions one can write an expression for the uptake term

$$f(C) = \alpha k' C \qquad (9)$$

where k' is the uptake constant. With such linear uptake, Eq. (7) can usually be solved analytically, as will be shown; this can apply at cellular membranes, brain interfaces and across the blood vessels that constitute the BBB (Patlak and Fenstermacher, 1975).

In other situations the uptake process is governed by saturable Michaelis–Menten kinetics. Then Eq. (7) becomes non-linear and the analysis much more complex. In this case

$$f(C) = V_{max}\, C/(K_m + C) \qquad (10)$$

where V_{max} is the maximum velocity of uptake and K_m is the Michaelis–Menten constant. A much studied example is the uptake of dopamine; here V_{max} is a measure of the number of amine uptake sites present, while K_m is generally assumed to be a measure of the dissociation constant for binding of dopamine to the membrane uptake sites (Horn 1979). The solution of the diffusion equation with this type of dopamine uptake has been explored in detail by Nicholson (1995).

Bulk Flow. Bulk flow is the concept that interstitial fluid (ISF), similar in composition to CSF, moves through the ECS like a river, carrying informational substances with it. The flow would be vectorial, that is to say, it would move in a particular direction and, like a river, could move substances over long distances with little dispersion. There is, however, some question as to whether the ECS matrix could permit such flow; it is more likely that, if operative, bulk flow is confined to perivascular spaces mentioned below; it is fair to say that the issue of endogenous bulk flow remains unresolved at this time.

The mechanism of bulk flow would be that fluid is secreted across the blood-brain barrier (Bradbury 1979) and makes its way into the CSF or leaves via lymphatics (Cserr and Patlak, 1991). Cserr and collaborators provided the most direct evidence using blue dextran (Cserr and Ostrach, 1974) and HRP (Cserr et al., 1977). This group also measured the efflux of molecules with a variety of molecular weights (Cserr et al., 1981; Szentisvanyi et al., 1984) and showed that they all cleared with a single rate constant. Those efflux experiments argued against diffusion as a clearing mechanism and indicated an average interstitial bulk flow velocity in the range 5.5–14.5 μm min^{-1} (assuming an extracellular volume fraction, $\alpha = 0.2$). Cserr and colleagues (Ichimura et al., 1991), however, thought that much of the bulk flow took place in the restrictive perivascular Virchow–Robin space surrounding the capillaries, rather than through the entire interstitial space. Rennels et al. (1985) used horseradish peroxidase (HRP) to provide further anatomical evidence for the bulk movement of substances in pathways that ran parallel to the vascular system of the brain, possibly aided by vascular pulsation.

Rosenberg et al. (1980), using a different approach to that of Cserr and colleagues, also arrived at a flow velocity of 10.5 μm min^{-1} towards the ventricle in white matter under normal conditions, but gray matter showed flow only under osmotic stress. In contrast, Davson (1970) was able to explain the data existing at that time in terms of the 'CSF sink' hypothesis, i.e. that substances are cleared from brain tissue simply by the diffusion gradient caused by the low concentration of substance in the ventricles. Later, however, Abbott (1992) proposed that some earlier data on clearance of tracers in mammalian brain could best be interpreted by invoking bulk flow.

In our own studies (Bjelke et al., 1995) we showed evidence for the rapid movement of fluorescent dextran over long distances in the periaxonal spaces of the corpus callosum, possibly aided by brain pulsation.

Practical solution of the diffusion equation. If we assume that any matrix contribution will simply alter D^* (or λ, which amounts to the same thing), exclude bulk flow and assume linear uptake, then Eq. (7) can be solved analytically for a source pulse of finite duration d. The solution can be written compactly as (Nicholson and Phillips 1981; Nicholson 1992):

$$C(r, t) = \frac{Q}{8\pi D^* \alpha r}[h(r, t, \theta) - h(r, t, -\theta)] \text{ and}$$

$$h(r, t, \theta) = [g(r, t, \theta) - g(r, t-d, \theta)] \exp(r\theta) \tag{11}$$

where

$$D^* = D/\lambda^2,\ \theta = \sqrt{k'/D^*},\ g(r, t, \theta) = H(t)\mathrm{erfc}(r/2\sqrt{D^*t} + \theta\sqrt{D^*t}).$$

In Eq. (11), the apparent diffusion coefficient is $D^* = D/\lambda^2$. $H(\ )$ is the Heaviside step function. In practical experiments, the source term Q is the rate of release of substance by a point source microelectrode. For iontophoresis $Q = In/F$, where I is the current, n is the transport number for the substance and electrode, and F is Faraday's Electrochemical Equivalent. The duration of the iontophoresis is d.

Uptake of the diffusing substance from the ECS is represented in Eq. (11) by a linear process defined by Eq. (9). When $k' = 0$, Eq. (11) becomes Eq. (5). If in addition $d \rightarrow 0$, while Q remains finite, then Eq. (11) becomes Eq. (2). The additional element in Eq. (11), compared to Eqs. (2) and (5), is the uptake term. Depending on its value, this term somewhat modifies the shape of the diffusion curve, as shown in Fig. 5. Figure 5 also compares diffusion in the brain with that in dilute agarose, an essentially free medium; the striking difference in amplitude reflects the combined effect of volume fraction and tortuosity and is approximately λ^2/α.

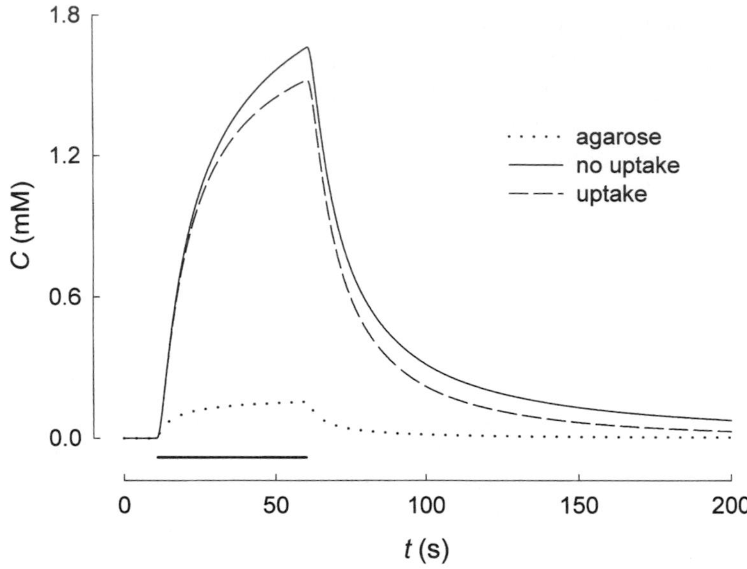

Fig. 5. Effect of uptake on concentration profiles in brain. The solid curve shows the solution to Eq. (11) with $k'=0$ while the dashed line shows the effect of incorporating an uptake of $k'=0.006\ s^{-1}$. These curves were computed for TMA^+ with a volume fraction, $\alpha=0.2$, and a tortuosity of $\lambda=1.6$. The iontophoresis current was 100 nA, applied for 50 s indicated by the horizontal bar beneath the abscissa, electrode transport number $n=0.3$. The dotted curve shows the concentration curve in agarose with same electrode parameters; but $\alpha=1$, $\lambda=1$, $k'=0$. Note the large reduction in amplitude in agarose compared to brain.

When the medium is anisotropic, a modified version of Eq. (11) is required; this is fully described by Rice et al. (1993) and the consequences will be illustrated later in this chapter.

Some recent experimental results

There are several reviews of recent studies of diffusion in brain and spinal cord ECS (Nicholson and Rice, 1986; 1988; 1991; Syková, 1997; Nicholson and Syková, 1998; Nicholson, 1999; Syková et al., this volume) so here we will briefly review some of our recent studies that are relevant to VT.

Diffusion of small ions

Early studies of diffusion of small ions and other molecules in brain tissue used radiotracers (see Fenstermacher and Kaye, 1988; Nicholson and Syková, 1998 for references). The most common method of measuring the diffusion behavior of small ions presently is the TMA^+-method (real-time iontophoretic method). The paradigm consists of the defined release of TMA^+ (Nicholson and Phillips, 1981; Nicholson, 1993) from a source that approximates a point and the measurement of concentration at a known distance as a function of time. The TMA^+ ($M_r=74$) is iontophoresed from a micropipette. An ion-selective microelectrode (ISM) located about 100–150 μm from the release electrode measures the concentration of the ion.

The diffusion parameters are determined by fitting Eq. (11) to the experimental concentration versus time curves. Appropriate calibration and non-linear curve fitting procedures are performed in both brain and dilute gel and the parameters α, λ and k' extracted (several computer programs are available from Nicholson to carry out data acquisition and analysis). TMA^+ is used because it remains predominately extracellular, is small, and does not alter physiological function at the concentrations used (Nicholson and Phillips, 1981). Some other cations and anions can be used also (Nicholson and Phillips, 1981). The ISM can be substituted with a carbon-fiber microelectrode, and by using fast-scan cyclic voltammetry, the diffusion of electroactive substances, such as dopamine can be measured (Rice and Nicholson, 1995; Rice, this volume).

Diffusion properties of in vitro preparations. The TMA^+ method has been used successfully to study diffusion parameters of brain tissue in vitro. The first experiments were performed in guinea pig cerebellar slices (Hounsgaard and Nicholson, 1983), followed by studies in rat hippocampus, neostriatum and neocortex (McBain et al., 1990, Rice and Nicholson, 1991, Pérez-Pinzón et al., 1995, Nicholson et al., 1998).

In the slices prepared from isotropic brain regions, λ and α were reported to be 1.54 and 0.21, respectively, in neostriatum (Rice and Nicholson, 1991) and 1.62 and 0.18, respectively, in neocortex (Pérez-Pinzón et al., 1995). These values are consistent with diffusion parameters obtained in vivo where $\lambda \sim 1.6$ and $\alpha \sim 0.2$ (for review see Nicholson and Syková, 1998). In some parts of brain and spinal cord, however, the tissue structure is anisotropic. Anisotropy has now been shown with diffusion analysis in white matter of the corpus callosum (Voříšek and Syková, 1997a) and of the spinal cord (Prokopová et al., 1997) as well as the hippocampus (Mazel et al. 1998). A very well documented example is the molecular layer of the cerebellum where anisotropy was demonstrated first through impedance measurements (Nicholson and Freeman, 1975; Freeman and Nicholson, 1975) and later using diffusion analysis (Rice et al., 1993). In the latter study, performed on the isolated turtle cerebellum, Rice and co-workers measured tortuosity along three axes in granular and molecular layers (Fig. 6). Anisotropy was detected in the molecular layer only. In the plane parallel to the surface, $\lambda_x = 1.44$ along parallel fibers and $\lambda_y = 1.95$ across parallel fibers. In vertical direction $\lambda_z = 1.58$. The functional consequences of anisotropy were demonstrated by measuring the spatial dynamics of $[Ca^{2+}]_e$ and $[K^+]_e$ changes induced by local application of glutamate, a known cerebellar neurotransmitter (Rice et al., 1993). The changes in

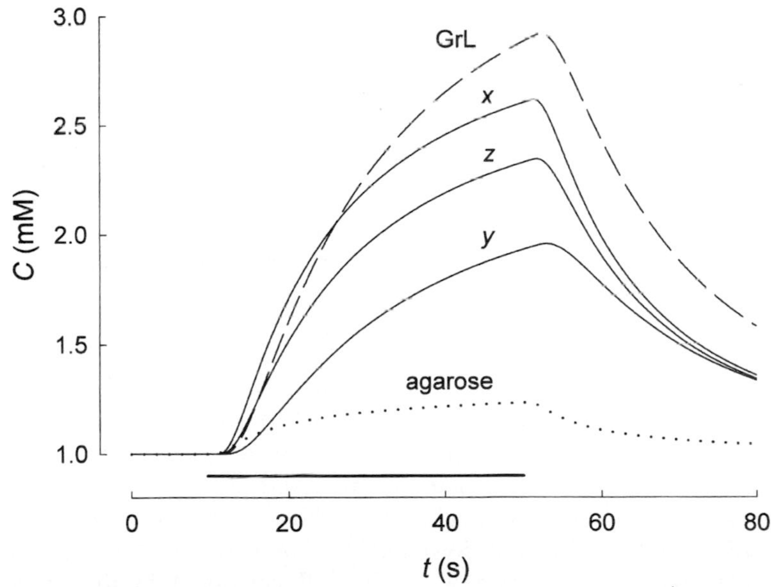

Fig. 6. Anisotropy in isolated turtle cerebellum. Theoretical concentration-time profiles derived from average diffusion parameters in the turtle cerebellum and in agarose. Theoretical records illustrate the relative concentration reached during the iontophoresis of TMA^+ in the *x*-, *y*- and *z*-plane of the molecular layer, the granular layer (GrL) and agarose. The respective tortuosities were: $\lambda_x = 1.44$, $\lambda_y = 1.95$, $\lambda_z = 1.58$, $\lambda_{Gr} = 1.77$, $\lambda_{agarose} = 1$. Volume fraction in the molecular layer was $\alpha = 0.31$ and in the granular layer $\alpha = 0.22$. In agarose $\alpha = 1$. Separation between source and recording locations used to calculate the curves was $r = 120$ μm, transport number $n = 0.5$ and free diffusion coefficient for TMA^+ at 22°C was $D = 9.7 \times 10^{-6}$ cm^2 s^{-1}. Iontophoresis current of 100 nA was applied for 40 s, at the time indicated by the horizontal bar on the graph. The curves are given as linear concentration changes with baselines normalized to 1 mM. The value of linear uptake term, k', used to calculate the brain records was 0.01 s^{-1}. Figure modified from Rice et al. (1993).

the concentration of the ions were most pronounced along the x-axis, e.g. along the path with lowest tortuosity; the differential properties of extracellular space thus can create pathways of preferential informational flow. The cerebellar anisotropy also may explain the original values of λ and α obtained by Hounsgaard and Nicholson (1983) since anisotropy was not taken into account in those initial studies.

Another part of the brain with highly organized morphology is the hippocampus. TMA^+ diffusion analysis uncovered diffusion inhomogeneity in the extracellular space among hippocampal regions (McBain et al., 1990, Pérez-Pinzón et al., 1995). Volume fraction was reported to be as small as 0.12–0.14 in stratum pyramidale of the CA1 region with tortuosity ranging between 1.50–1.67. A more recent in vivo study proposed that the hippocampus is actually anisotropic with a uniform volume fraction of 0.22 (Mazel et al., 1998).

The hippocampus and cerebellum are examples of brain structure where the neuropil forms distinct, structurally well defined, zones. Such arrangements may have several functional implications for volume transmission; the stratum pyramidale in the hippocampus and Purkyně cell layer in the cerebellum form regions with densely packed cell bodies. One can envision that these layers may serve as a kind of diffusion barrier for large molecules. Alternatively, densely packed layers of cell bodies in the hippocampus may form a functionally separate microdomain, capable of modulating the concentration dynamics of ions, neurotransmitters, and neuropeptides within the region. In addition, a recent electron microscopy study demonstrated that astrocytic processes (Ventura and Harris, 1999) shield only 57% of the synapses in the CA1 region. This finding lends morphological support to the phenomena of glutamate spillover and synaptic crosstalk (Barbour and Häusser, 1997; Rusakov and Kullmann, 1998b; Kullmann and Asztely, 1998; Kullmann, this volume). The properties of ECS in the hippocampus thus have the potential to influence the passage of information streams conveyed by major excitatory neurotransmitters between neurons.

In addition to tortuosity and volume fraction, the TMA^+ method also quantifies the loss of this substances from the ECS as a non-specific uptake k' parameter (Nicholson, 1992, see also Eqs. (7), (9) and (11)). In rat neocortex in vivo, a typical value is $k' = 0.006\ s^{-1}$ (Lehmenkühler et al., 1993). This loss is quite small but understanding where the TMA^+ goes has important physiological implications. In the intact animal two possibilities exist: uptake may represent entry of TMA^+ into cells or it may represent loss across the BBB. Diffusion measurements in slices incubated in a submerged-slice-recording chamber detected a non-specific uptake of $0.005\ s^{-1}$ (Pérez-Pinzón et al., 1995). This result seemed to favor cellular uptake and argue against the possibility that k' represents TMA^+ molecules vanishing across the BBB, since brain slice preparations lack a functional blood supply. However, further measurements using an interface chamber slice chamber, which allows better control of the conditions at the slice boundaries and which will be outlined in more detail below, indicated that k' could fall to zero in the center of a slice in such a chamber. These data indicated that the artificial cerebrospinal fluid (ACSF) flowing over the surface, or surfaces, of a slice might mimic the sink effect of vascular perfusion. This lent support to the idea that the uptake observed in the in vivo preparation was most reasonably attributed to clearance across the BBB (Hrabětová and Nicholson, 1999).

In general, an in vitro preparation is preferred for diffusion measurements when substantial manipulation of ACSF composition is required or in studies that depend on prompt drug delivery. The use of brain slices is justified by the fact that the normal parameters of ECS measured in the brain slice are similar to values obtained in vivo. In anoxic conditions, the changes in α and λ are smaller in vitro than those seen during terminal ischemia (Syková, 1997) in vivo, possibly because there is less mechanical constraint in the brain slice.

TMA^+ diffusion measurements in slice preparations are usually performed in the center (i.e. at a depth of 200 μm in a typical 400 μm thick slice) to avoid boundary effects. As noted above, TMA^+ is applied as a short pulse (50 s) and detected with an ISM placed in the same horizontal plane at the distance 100–150 μm. Such an arrangement ensures undistorted diffusion of TMA^+ molecules from the point source (TMA^+ iontophoretic electrode) to the detector (TMA^+–ISM). However, care has to be taken when the measurements are made in close proximity to the slice surfaces or in thin slices. Once the spacing between point source and detector is comparable to or larger than the distance of the electrodes from the slice surface, the conditions at the boundaries have to be taken into account when extracting λ and α from diffusion curves.

In brain slice held in an interface chamber, TMA^+ diffusing away from the point source positioned in the close proximity to the slice surfaces is influenced by the nature of the slice boundaries. In this type of chamber, the upper slice surface is exposed to humidified gas that effectively forms a barrier preventing TMA^+ molecules from leaving the tissue. In contrast, the lower surface neighbors on the flowing ACSF that functions as a sink of molecules reaching this boundary. This arrangement thus creates different conditions at the upper and lower boundaries, which have distinct implications for the interpretation of diffusion measurements made near to slice surfaces and these effects can be modeled.

The results of one such model that attempts to account for the influence of boundary conditions on the diffusion are shown in Fig. 7. These theoretical diffusion curves were obtained by modeling TMA^+ diffusion at the different depths of an isotropic neocortical slice incubated in an interface chamber. It is obvious that the position of the electrode array with respect to the slice surfaces influences the behavior of the TMA^+ molecules. The upper surface of the slice represents an impermeable

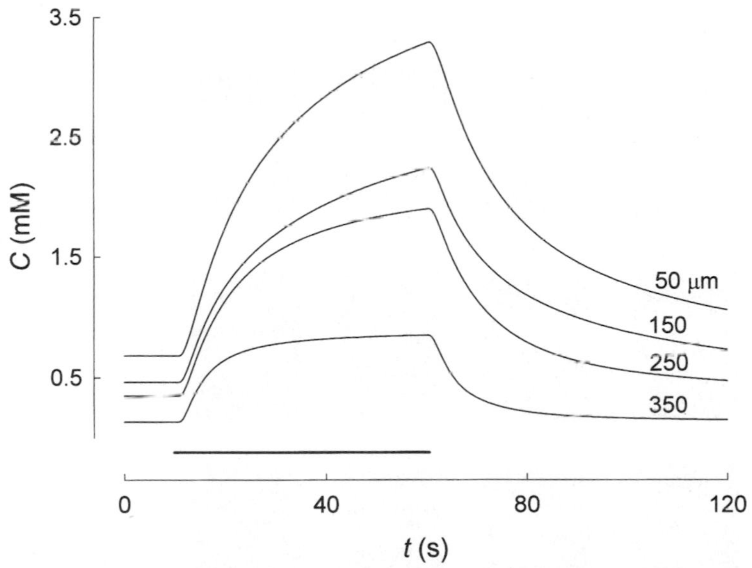

Fig. 7. Model of diffusion curves in slice at different depths; slice assumed 400 μm thick, upper surface at 0 μm. The curves were calculated at the depths indicated below the surface. The model simulated an interface slice chamber with the upper surface exposed to humidified gas and the lower surface bathed in flowing ACSF. Since diffusing molecules are 'reflected' at the upper surface, the curve 50 μm below the surface is increased in amplitude compared to curves further away. In contrast, the curve near the lower surface, at 350 μm, is diminished in size because of the 'sink' effect of the solution at this surface. It was assumed that any molecule crossing the lower boundary was lost from the slice. Curves at 150 and 250 μm are fairly similar as they are less affected by the boundary conditions of the slice. Records are normally made at 200 μm depth. In this model zero uptake was assumed yet the curves exhibit some of the characteristics associated with uptake (c.f. Fig. 5) because of the removal of molecules at the lower surface. Separation between source and recording was 100 μm in a plane parallel to the surfaces of the slice, $\lambda = 1.6$, $\alpha = 0.2$. A constant iontophoretic bias current of 20 nA was assumed at all times, which accounts for the different baselines (see Nicholson and Phillips 1981). For the period between 10 s and 60 s (indicated by horizontal bar) the current was increased by 100 nA. The diffusion coefficient was that for TMA^+ at 34°C. The calculation was based on Eq. (8) in Chapter 10 of Carslaw and Jaeger (1959).

boundary, which reflects diffusing TMA^+ molecules and effectively limits their movement to the half sphere. The consequence is that the diffusion curve near to the upper slice surface has larger amplitude than the ones near the middle of the preparation; in fact right at the upper boundary the amplitude is twice that in the center. In contrast, close to the lower permeable boundary, TMA^+ molecules are lost into the solution flowing underneath the slice; this loss results in the reduction of the amplitude of TMA^+ diffusion curve in the region near to the lower slice surface. The theoretical approach is currently being complemented by experimental observations. Ultimately, we will be able to extract ECS parameters, α and λ, throughout the whole thickness of the preparation; in addition, the model will contribute to the understanding of the conditions influencing the diffusion of many different molecules in the widely used brain slice preparation. Boundary effects should also exist in studies on the intact animal but no systematic experiments or modeling has been done to date.

Diffusion properties of macromolecules

In order to measure the diffusion properties of macromolecules, many of which lack charge, we devised the integrative optical imaging (IOI) method (Nicholson and Tao, 1993). The IOI-method measures D^* (and consequently, λ) for macromolecules tagged with fluorescent probes, in living brain tissue, using epifluorescence microscopy and quantitative image analysis (Fig. 8).

Because the molecules must be imaged, the IOI method is most easily applied to brain slices; fluorescent molecules are ejected into the brain slice with a brief pressure-pulse from a micropipette (see Eq. (2) and Fig. 2), inserted about 200 μm below the surface of a typical 400 μm thick slice, in a submersion chamber on the stage of an upright compound microscope. The diffusing

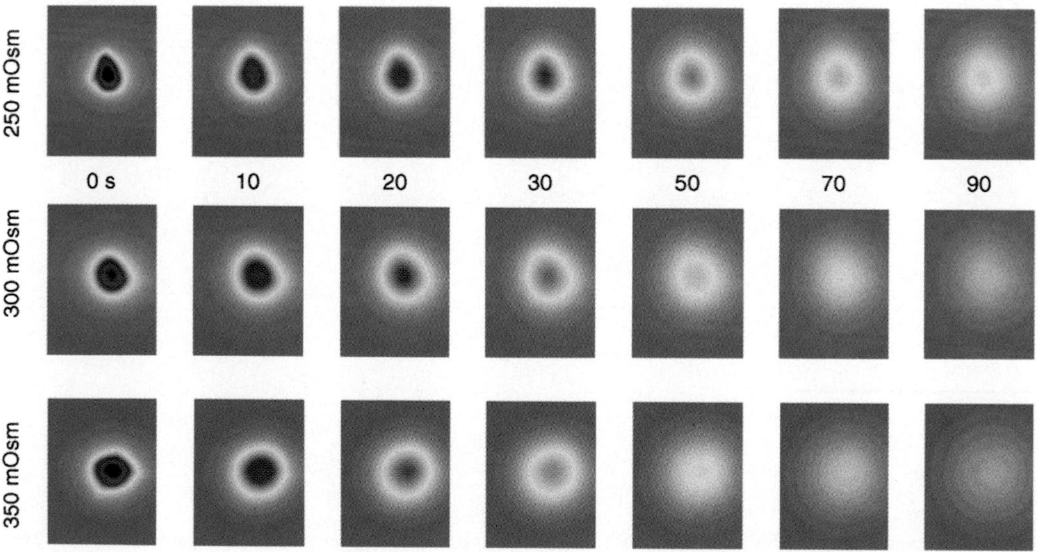

Fig. 8. Images of fluorescent dextran ($M_r = 3{,}000$) diffusing in cortical slice recorded with IOI method. At time zero (0 s) a few nanoliters of the dextran labeled with the fluorescent dye Texas Red® were briefly pressure ejected from a micropipette and a sequence of images were recorded. Each row in the figure shows a sequence of images taken after an ejection with the dye concentration depicted in pseudo color (purple highest concentration, orange lowest). A background image taken before each ejection has been subtracted from the subsequent images. Appropriate analysis of the intensity curves of the images enabled D^* to be obtained; the three sequences were all taken on the same slice but in different solutions; the middle row was the sequence in the normal isotonic solution (300 mOsm); the top row shows the sequence in a mild hypotonic solution (250 mOsm). Note that the intensity changed more slowly than that in isotonic solution; the bottom row shows the sequence in a mild hypertonic solution (350 mOsm) and the intensity changed somewhat more rapidly than that in the isotonic solution. The vertical dimension of each image is 564 μm (modified from Tao 1999).

molecules are imaged with a 10× water-immersion objective, focused on the tip of the ejection micropipette. Before each ejection, an image is taken to record the background; after each ejection, a sequence of 10 images is recorded at an interval of 2–20 seconds using a cooled CCD camera. The images are transferred directly to a PC, where the background is subtracted from each subsequent image. By taking account of how the microscope and objective record in-focus and out-of-focus points, using the point-spread function of the objective, the relative concentration of the diffusing cloud of molecules can be quantified (Tao and Nicholson, 1995). An intensity profile is measured through the ejection point in each image and fitted with the theoretical expression for a diffusion image to extract the apparent diffusion coefficient D^* (Nicholson and Tao, 1993).

Diffusion studies with globular macromolecules. Published studies to date have used dextrans with M_r of 3000; 10,000; 40,000 and 70,000 M_r (Nicholson and Tao, 1993). Further measurements employed lactalbumin with M_r of 14,400, ovalbumin with M_r of 45,000 and bovine serum albumin (BSA) with M_r of 66,000 (Tao and Nicholson, 1996).

All these globular molecules can diffuse through brain tissue but the larger ones produce tortuosities that are significantly greater than those found with TMA^+. For example, 70,000 M_r dextran and 66,000 M_r BSA generate tortuosities of about 2.3 (Table 1); one hypothesis to explain the elevated tortuosity is that the molecules were too large to access some local regions of the ECS. Our recent studies have further explored the issue of macromolecular diffusion by using flexible long-chain synthetic polymers.

Diffusion studies with PHPMA polymers. The synthetic polymers consisted of poly(N-(2-hydroxypropyl) methacrylamide), known as PHPMA. Dr. Karel Ulbrich and co-workers in Prague synthesized these polymers in order to carry drug molecules into tissue (Ulbrich et al., 1997); for use with IOI, Dr. Ulbrich attached FITC or Texas Red® molecules.

We were surprised to discover that PHPMA molecules as large as 1,000,000 M_r were able to diffuse through the ECS with a tortuosity of about 1.5–1.6 (Table 1), the same as that obtained with TMA^+ (Vargová et al., 1998). Moreover, when the PHPMA was conjugated with BSA, so that the configuration became a more bulky globular molecule instead of a long chain, the tortuosity became that of BSA rather than PHPMA (Table 1). We do not yet have a complete explanation for these very interesting findings with such large molecules. An early paper by Preston et al. (1980) showed that long-chain molecules diffusing in polymer networks lost their rotational degree of freedom and apparently moved faster because of this, but later work (Preston et al., 1984) suggested that gravitational forces came into play in the original

TABLE 1

Diffusion parameters of macromolecules in agarose and brain at 34°C

Molecule	M_r (kDa)	Agarose gel (0.3%)	Cortical slice	Tortuosity
		D (10^{-7} cm^2 s^{-1})	D^* (10^{-7} cm^2 s^{-1})	$\lambda = (D/D^*)^{1/2}$
PHPMA-chain§	7.8	9.84	4.34	1.51
	28.0	6.30	2.25	1.67
	47.2	4.65	1.94	1.55
	219.9	2.03	0.75	1.65
	515.0	1.17	0.47	1.58
	1057.0	0.72	0.34	1.46
Dextran*	70.0	3.80	0.75	2.25
Albumin#	66.0	8.29	1.63	2.26
PHPMA-Alb§	175.8	2.34	0.45	2.27

§ Data from Vargová et al. (1998); * Data from Nicholson and Tao (1993); # Data from Tao and Nicholson (1996).

experiments. Long-chain molecules may also be expected to experience reptation (de Gennes, 1975) in constrained environments but this would lead to a rapid reduction in the apparent diffusion coefficient (Doi, 1996; Pluen et al., 1999) which is not seen in our experiments. Neither of these explanations seem able to explain our results.

Diffusion when the ECS is modified

It has long been known that the values of α and λ change under a variety of pathophysiological conditions; these include painful stimuli (Svoboda and Syková, 1991), ischemia and anoxia (Pérez-Pinzón et al., 1995; Voříšek and Syková, 1997b); in these instances it appears that brain cells swell (either glia, neurons or both) and so reduce the size of the ECS while also increasing the tortuosity. In the most extreme cases of terminal ischemia, $\alpha = 0.04$ and $\lambda = 2.3$ (Syková, 1997). From such data, and common intuition, one might assume that there was a functional relationship between α and λ. This is indeed a fundamental question that has arisen many times in the more general context of descriptions of porous media (e.g. Archie, 1942; Sen et al., 1981; Latour et al., 1994). We have tackled this question experimentally by manipulating the size of the ECS through osmotic stress, which can both swell and shrink brain cells.

Studies using osmotic challenge with the TMA^+ method. In our first study (Križaj et al., 1996), diffusion analysis was used to determine α and λ for the ECS and water compartmentalization in the turtle cerebellum exposed to media of different osmolalities. The water compartmentalization was quantified by combining extracellular diffusion analysis with wet-weight and dry-weight measurements.

To vary the ECS, isolated turtle cerebella were immersed in normal, oxygenated, ACSF (302 mOsm), hypotonic saline (238 mOsm) and a series of hypertonic salines (up to 668 mOsm) and the osmolality was varied by altering the NaCl content. Values of α and λ for the ECS of the granule cell layer of the cerebellum were determined using the TMA^+ method (we did not study the molecular layer because of the anisotropy mentioned above, which necessitates extensive additional measurements). The value of α was 0.22 in normal saline, 0.12 in hypotonic medium and 0.60 in the most hypertonic medium while the corresponding values of λ were 1.70, 1.79 and 1.50 respectively.

The water content, defined as ((*wet-weight*)–(*dry-weight*))/*wet-weight*, of a typical isolated cerebellum (including granule cell, Purkyně cell and molecular layers) was 82.9%; it increased to 85.2% in hypotonic saline and decreased to 80.1% in the most hypertonic saline. By combining the measurements of α with water content it was shown that hypotonic solutions caused water to move from the extra- to the intracellular compartment while hypertonic solutions caused water to move from the intra- to extracellular compartment, with only a relatively small changes in total tissue water.

A recent study (Nicholson et al., 1998) again used variations in NaCl content to change osmolarity of the ACSF but in this case neocortical slices from the rat were used. The techniques were similar to those described above, except that TMA^+ diffusion measurements were made in the middle of a submerged slice.

In normal medium (300 mOsm), the average value of λ was 1.69 and of α was 0.24. Reducing osmolarity to 150 mOsm, increased λ to 1.86 and decreased α to 0.12. Increasing osmolarity to 350 mOsm, reduced λ to about 1.67 where it remained unchanged as osmolarity increased further to 500 mOsm. In contrast, α increased steadily to 0.42 as osmolarity increased. These results are shown in Fig. 9, where the 'plateau' behavior of λ is evident.

Recent TMA^+ studies on the isolated spinal cord of rats at different developmental stages with hypotonic solutions (Syková et al., 1999) showed generally similar decreases in α and increases in λ to those described above. Hypertonic solutions were not used in the spinal cord studies, so the plateau behavior in λ could not be investigated.

Studies using osmotic challenge with the IOI method. As described above, different macromolecules, such as dextrans, albumins and PHPMA (Vargová et al., 1998), may have different values of D^*, even when their molecular weights are similar.

This indicates that the apparent diffusion coefficient of a molecule depends not only on its molecular weight but also on other properties, including shape, radius of gyration, electric charge, etc. On the other hand, for a given molecule and tissue, the apparent diffusion coefficient can also be modulated by varying the ECS properties. This has been demonstrated in a recent study (Tao, 1999) on the diffusion properties of fluorescent dextran molecules ($M_r = 3000$) in rat neocortex when the osmolarity in ECS was varied, using changes in NaCl content in the ACSF, as in the TMA^+ studies. Slices from the rat neocortex were again used.

As an example of the effects of osmolarity, Fig. 8 shows the diffusion images of the fluorescent dextran in a cortical slice. In the figure, each of the three rows shows an image sequence taken after an ejection; the three sequences were taken from the same slice but in different solutions. The middle row was the image sequence in the normal isotonic solution (300 mOsm); the top row was the sequence in a mild hypotonic solution (250 mOsm) while the bottom row was the sequence in a mild hypertonic solution (350 mOsm).

The apparent diffusion coefficient D^* of the dextran in the neocortex under normal isotonic condition served as the control value for the experiment and the mean value from all the experiments was $D^* = 6.8 \times 10^{-7}$ cm^2 sec^{-1} at 34°C.

Under a mild hypotonic stress of 250 mOsm the mean value of D^* was 5.3×10^{-7} cm^2 sec^{-1}. When the hypotonic stress was increased, by reducing osmolarity to 200 mOsm, D^* decreased further to $D^* = 4.4 \times 10^{-7}$ cm^2 sec^{-1}; finally with a hypotonic stress of 150 mOsm, D^* decreased further to $D^* = 3.8 \times 10^{-7}$ cm^2 sec^{-1}. In a hypertonic stress of 350 mOsm, the mean value of the apparent diffusion coefficient was $D^* = 7.8 \times 10^{-7}$ cm^2 sec^{-1}, however, as more hypertonic stress was applied, up to 500 mOsm, D^* did not increase further.

The D^* data obtained under hypotonic stresses and under hypertonic stresses were converted to λ-values using a D of 2.12×10^{-6} cm^2 s^{-1} at 34°C, measured previously in 0.3% agarose gel dissolved in 150 mM NaCl (Nicholson and Tao, 1993). These data are plotted along with the TMA^+ results in Fig. 9. It is striking that the same type of plateau behavior is observed in the tortuosity, although with hypotonic solutions the slope is steeper for the dextran compared to the TMA^+.

Taken together, these data show that λ and α behave differently as the geometry of ECS varies; when α decreases, λ increases but when α increases, λ rapidly attains a constant value that is hypothesized to be largely determined by the geometry of the ECS. These data confirm that α and λ behave independently. Further confirmation of this independence comes from the extensive work by Syková and co-workers on X-irradiated brain tissue (Syková et al., 1996), and gliotic tissue (Roitbak and Syková, 1999; Syková et al., 1999; Syková et al., this book).

Many studies have demonstrated that varying the extracellular osmolarity can alter the volume of cells in tissue (Chebabo et al., 1995; Križaj et al., 1996). Tosteson and Hoffman (Tosteson and Hoffman, 1960; Tosteson, 1964) proposed a theoretical model for regulation of cell volume when external NaCl is varied. The model was based on the interaction of electrical and osmotic effects. According to this model, when the rate of the Na-K pump is fixed, the cell volume is inversely proportional to the total extracellular ion concentration. Moreover, when the extracellular ion ratio $[Na^+]_e/[K^+]_e$ is reduced, cell volume also increases (Hoppensteadt and Peskin, 1992). Such a model, however, does not account for the plateau behavior in tortuosity with hypertonic media; it is possible that the extracellular matrix plays some role in this but later in this paper we offer another explanation.

Two examples of modeling: cell shape and spatial buffering of K^+

As more quantitative experimental data have become available about diffusion of substances in the ECS, it has become increasingly feasible to formulate models to help interpret the data. One of the earliest attempts in this direction was the model of Patlak and Fenstermacher (1975) that related the behavior of a radiotracer in the brain (such as radiolabeled sucrose or inulin) to the concentration applied via the ventriculo-cisternal and vascular systems in anesthetized preparations. Later models of diffusion from point sources (Nicholson and Phillips, 1981; Nicholson, 1985; Nicholson, 1992) formed the basis for interpreting the data from the TMA^+ and IOI methods. An extensive model combining both diffusion and membrane cable theory was formulated by Gardner-Medwin to try to account for experiments (Gardner-Medwin, 1983; Gardner-Medwin and Nicholson, 1983) involving the movement of K^+ in brain. A further detailed model demonstrated the effect of Michaelis-Menten kinetics on the diffusion of dopamine (Nicholson, 1995). Recently we have formulated two new models, the first to try to account for the diffusion behavior seen with osmotic challenge and the second to reconsider the issues underlying potassium movement in the brain.

Modeling relation between λ and α during osmotic challenge. From the experiments under osmotic challenge using TMA^+ or dextran, described in the previous section, it was seen that while α monotonically increased with the osmolarity of the bathing solution, λ exhibited an asymmetric behavior. The measured λ increased with decreasing osmolarity in the hypotonic direction but only declined to a constant value with increased osmolarity in the hypertonic direction. Changes in fluid viscosity and hindrance due to molecular size do not provide satisfactory explanations since their effect in obstructing diffusion would be expected to occur in the direction of decreasing *a*. This λ-plateau, however, can be explained on a purely geometric basis providing that the cells are permitted to alter their shape selectively during osmotic challenge (Chen and Nicholson, 2000b).

Our analysis was based on homogenization theory applied to the ECS, regarded as a porous medium composed of repeating structures (Bensoussan et al., 1978; Hornung, 1996). Figure 10 depicts some of the two-dimensional lattice arrangements that we examined. A nominal scale

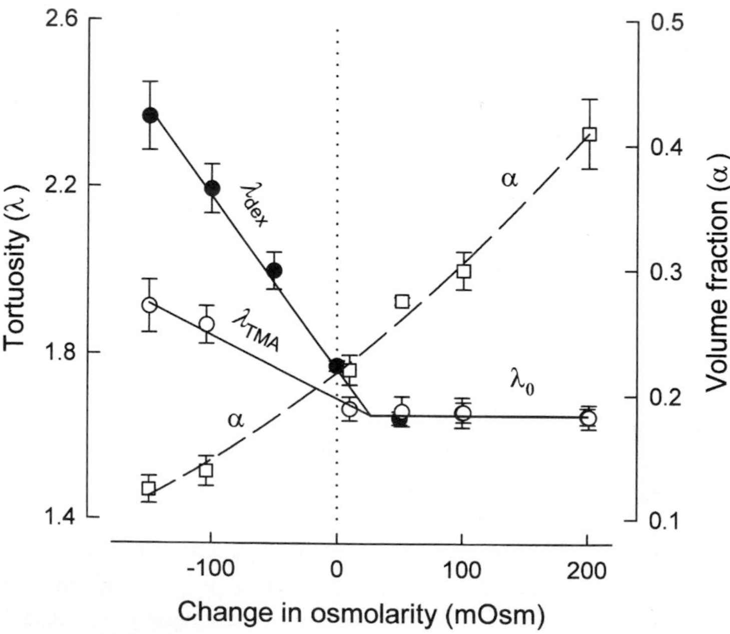

Fig. 9. Effect of osmolarity on volume fraction and tortuosity. Slices were exposed to the different osmolarities (normal value 300 mOsm); volume fraction α measured with TMA^+ varies smoothly with osmolarity. Tortuosity measured with either TMA^+ or dextran (3,000 M_r) shows linear increase with diminishing osmolarity (cells swelling) but quickly reaches a plateau in hyperosmotic solution (cells shrinking) despite the fact that volume fraction continues to increase. Note different slopes of λ-curves in hyposmotic solution for the two different molecules (modified from Nicholson et al., 1998).

was used and the center of each cell was assumed immobile during cellular swelling or shrinkage. To alter the shape of the cell, the corners could be rounded by an arc with a variable radius a. This parameter, together with the intercleft distance $2h$, determined the void area fraction, which we continued to denote as α.

For each arrangement depicted in Fig. 10, a periodic unit, enclosed by dashed lines, can be identified. Assuming that the cell membranes are impermeable to the diffusing substances, the differential equation for a periodic function ω in the void space of the periodic unit, Ω, is given as (Blum et al., 1989; El-Kareh et al., 1993)

$$\nabla^2\omega = 0. \tag{12}$$

The function ω represents the small periodic component of the macroscopic concentration field. Using suitable boundary conditions for ω, and assuming that the direction of the macroscopic diffusion direction is defined by a unit vector, $\mathbf{e}$, the ratio of apparent diffusion coefficient D^* in the direction of $\mathbf{e}$, to the free diffusion coefficient, D, is then estimated (El-Kareh et al., 1993) as

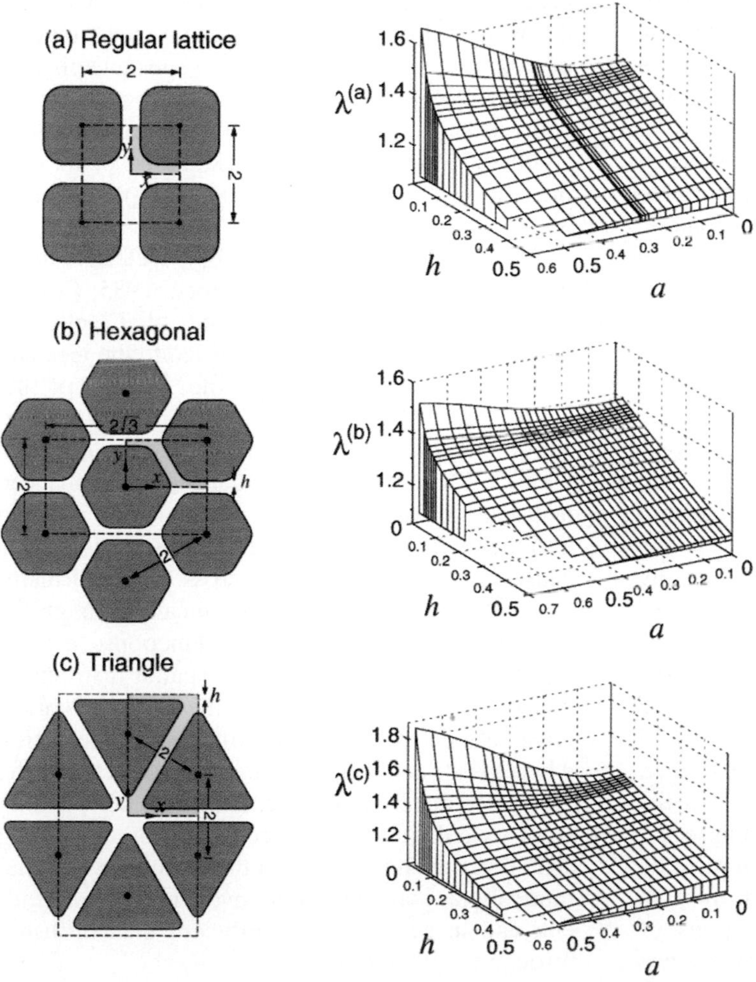

Fig. 10. Lattice models of the ECS and resulting behavior of λ as a function of cell shape and separation. Left panels (a), (b) and (c) show schematic diagrams for three different lattice arrangements. The ECS domains on which ω is solved are colored in yellow. Right hand panels how distributions of λ versus a (radius of curvature of edge of unit cell) and h (separation of cells) for each lattice arrangement depicted on the left (modified from Chen and Nicholson, 2000b).

$$\frac{D^*}{D} = \frac{1}{A} \int_{\Omega} (1 - \mathbf{e} \cdot \nabla\omega)\, d\Omega = \frac{1}{\lambda^2} \qquad (13)$$

where A is the area of the domain Ω.

Each lattice arrangement shown in the left-hand panels of Fig. 10 was solved numerically with a variety of combinations of a and h; the value of λ was then plotted against a and h in right-hand panels of Fig. 10. Note that α is increased as either a or h increased and there exists a common trend for λ to increase with a in regions where h was small. This is because circular cells, in the limit of $h \rightarrow 0$, always maintain residual void space, within which random-walking molecules are likely to be trapped locally. Obviously, this entrapment cannot occur if h is too large, which explains why the rising λ occurs in regions of small h. This tendency for λ to increase in the a–h plane offers a possible explanation of the plateau λ during osmotic challenge, as observed in the results summarized in the previous section (Fig. 9).

In reality, however, the ECS does not conform to a periodic structure with a repeating unit cell shape. To test whether the increase in λ with increasing residual space was a peculiarity of lattice structure, a more complex and realistic ECS domain composed of many cells with random shapes and orientations was constructed (Chen and Nicholson, 2000b). The value of α in this model was incrementally increased from 0.13 to 0.22 by selectively shrinking the cells in their regions of greatest elongation and similar behavior was observed to that seen with the regular lattices. We conclude that, over the α-range tested here, the increasing behavior of λ was independent of cell shape and did not require a lattice structure. Rather, it is a consequence of the enlargement of the residual space trapped between surrounding cells. For this increasing phase of λ to appear, h must be small and the cells must shrink *non-uniformly*, in the sense that cells shrink faster in regions of elongation than in others. This is plausible since brain cells in vivo usually are densely packed, which means that the intercleft distance is small and that the cell shape must be deformed; when brain cells shrink, it is likely that the elongated regions of the cells will shrink first to unload local strain. This is likely because the elongated regions will be where additional non-uniform mechanical strains are loaded due to the small radii of the curvature of the lipid membrane (similar to the radius a in the corners of cells in our study). This selective shrinkage will be expected to continue until an approximately spheroidal shape is regained.

Spatial Buffering of Potassium Ions in Brain Extracellular Space. One reason why TMA^+ has been chosen as our diffusing molecule of choice is because it largely remains in the ECS; when the diffusing substance is endogenous the situation may be more complicated. If the substance is able to move between intracellular and extracellular compartments, a modified diffusion equation with additional terms is needed to take into account such interactions; a particularly good example of this is K^+.

The glial cell membrane is highly selective to K^+ ions (Orkand et al., 1966; Newman, 1985; Chvátal and Syková, this volume) and glial cells are practically unexcitable. This has led to the idea that glial cells are involved in the control of the extracellular K^+ level ($[K^+]_e$) by a passive mechanism called 'spatial buffering' (Orkand et al., 1966). When, as a consequence of enhanced neuronal activity, excessive K^+ ions are released into the interstitial clefts, the local $[K^+]_e$ rises. This causes local depolarization across glial cell membranes that can spread electrotonically through the cell, and possibly also via gap junctions, to more distal regions. The asymmetric spatial distribution of potential difference over the cell membrane elicits a local K^+ influx into the cell in the region where $[K^+]_e$ is raised, and an efflux of K^+ into the ECS from distal glial processes whose surrounding $[K^+]_e$ is still low. This mechanism of dissipating spatial variations of $[K^+]_e$ in the brain ECS via glial intracellular pathways is passive, energy-independent, and in general, more efficient than diffusion through the interstitium.

We re-analyzed (Chen and Nicholson, 2000a) a one-dimensional mathematical model of Gardner–Medwin (1983) that had previously been solved numerically. The schematic diagram illustrating the co-transport of interstitial diffusion and spatial buffering by glial cells is shown in Fig. 11a. For simplicity, a macroscopic approach was adopted by viewing the brain microenvironment as a homogeneous continuum and the glial cells as an equivalent cable, as shown in Fig. 11b.

The one-dimensional governing equations for the glial membrane potential, represented by v, and the dynamics of $[K^+]_e$, represented by c, can be described, in dimensionless form, by

$$\frac{\partial^2 v}{\partial x^2} = g_{Kir}(v - \ln(c)) \tag{14}$$

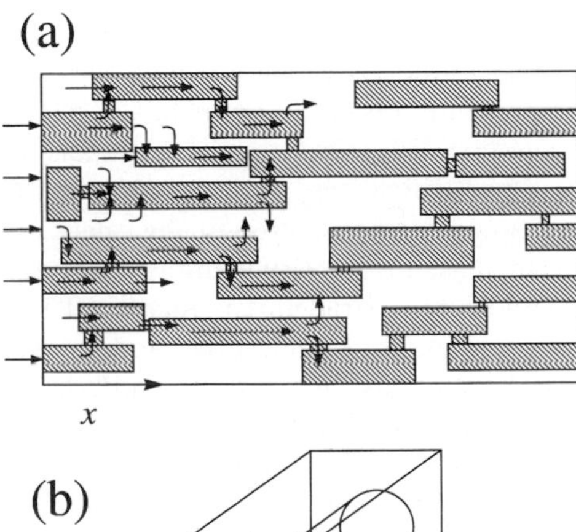

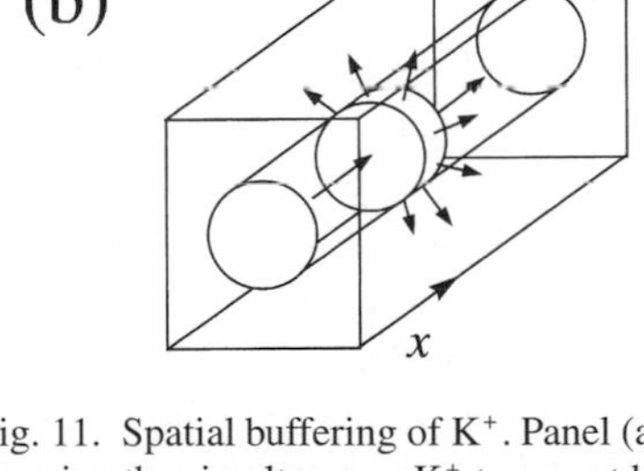

Fig. 11. Spatial buffering of K^+. Panel (a) is schematic drawing showing the simultaneous K^+ transport by diffusion in the ECS (empty area) and spatial buffering via glial cells (shaded area), possibly interconnected by gap junctions. Figure adapted from Gardner–Medwin (1983). For simplicity neurons are not shown. Panel (b) illustrates that when the macroscopic transport is effectively one-dimensional, the complex intracellular and extracellular geometry in the brain tissue is simplified to the form of a unit rectangular slab, in which an equivalent cylindrical cable represents a single glial cell or a collective glial syncytium. The ECS is regarded homogeneous, with an apparent diffusion coefficient D^* that incorporates λ. In both drawings arrows indicate K^+ flows due to either extracellular diffusion or spatial buffering (modified from Chen and Nicholson, 2000a).

and

$$\frac{\partial c}{\partial t} = \frac{\partial^2 c}{\partial x^2} + \rho\frac{\partial^2 v}{\partial x^2} - \varepsilon\left\{\left[\frac{\partial v}{\partial x} + \frac{1-\varepsilon}{\varepsilon} \quad I\right]\frac{\partial c}{\partial x} + \frac{\partial^2 v}{\partial x^2}c\right\} \tag{15}$$

in which g_{Kir} is the glial membrane K^+ conductance, attributed to the inward rectifier, ρ that represents the strength of the spatial buffering relative to the extracellular diffusion, I is an externally imposed electric current, if present, and ε represents the fractional resistance encountered by the electrical current in the ECS. Equation (14) is derived from conventional cable theory after ignoring the transient terms resulting from membrane capacitance (since this contribution takes place too quickly to play any role in spatial buffering). In Eq. (15), the first term on the right-hand side represents the extracellular diffusion, the second term accounts for the contribution of K^+ ions entering or leaving the ECS via glial membranes and the third term indicates the influence of the electrochemical gradient on the extracellular K^+ movement. For more detailed definitions for all the dimensionless parameters see Chen and Nicholson (2000a).

Equations (14) and (15) were solved analytically in a finite domain L. From cable theory, the boundary conditions at $x = 0$ and $x = L$ can be defined in terms of ϕ, a parameter that measures the relative conductance of the glial endfeet at the two boundaries. Endfeet are specialized regions of glial cells, for example in the Müller cells of the retina, that contain a much higher density of potassium channels than does the rest of the cell (Eberhardt and Reichenbach, 1987). We studied what parameter values make the spatial buffering action most efficient and under what conditions the induced K^+ current is optimized. One notable result of the analyses is that it makes explicit how the K^+ current that drives spatial buffering depends on membrane conductance and glial geometric length. We found that if the glial membrane channel density is fixed, to achieve a maximum spatial buffering K^+ current, the glial length should be around one unit of the electrotonic space constant of the glial core conductance. On the other hand, if the glial length is fixed, the optimum spatial buffering K^+ current is also obtained when the membrane conductance is adjusted so that the corresponding electrotonic space constant is equal to the glial dimension. This conclusion is consistent with the results of Eberhardt and Reichenbach (1987).

Implications of diffusion studies for VT

While diffusion is an essential part of classical synaptic transmission, being the mediator of the signal between pre- and postsynaptic sites, it is highly localized (Eccles and Jaeger, 1957) and stereotyped in its action. In contrast, diffusion plays a much more varied and interesting role in VT; our research points to several diffusion-related factors that are likely to be inescapably involved in VT.

Our work and that of others has established that the geometry of the ECS, parameterized in the two dimensionless factors, α and λ, together with the free diffusion coefficient, are major determinants of the way that molecules behave in brain tissue. Geometry may not be the only factor; there is increasing discussion of the role of the extracellular matrix, which would impose a viscous resistance on the movement of molecules and might have many other effects, depending on the specific molecule involved. But we lack experimental evidence for matrix interactions, probably due to inadequate experimental techniques. We can say that for TMA^+ and other small molecules, the measured tortuosity is quite close to that computed from various models based solely on geometry, so that for small molecules, we do not need to introduce the effect of the matrix to account for present results with small molecules. The matrix may still contribute, however, especially in the case of large molecules.

The movement of substances in the ECS is always subject to diffusion at some phase of the transport process. In many instances, diffusion dominates but not always. The migration of substances may be limited primarily by uptake, as in the case of dopamine in the striatum (Wightman and Zimmerman, 1990; Nicholson, 1995; Kehr et. al., this volume). For large molecules, like nerve growth factor (Krewson et al, 1995), that linger for long periods in the ECS due to their small apparent diffusion coefficient, binding to receptors, endocytocis or enzymatic attack may bring migration to an end. Another interesting example is the behavior of K^+, the movement of which is controlled both by diffusion and by the electrical properties of glial cells and glial syncitia. While the main function of this so-called spatial buffering is likely to be protection of neurons from exposure to excess K^+, spatial buffering also provides a new dimension in channeling K^+ within nervous tissue with the potential ability to cause patterns of focal depolarization.

Our work has shown that, for macromolecules, molecular shape is important; large globular molecules, like high-molecular weight dextran, BSA or BSA-PHPMA bulky-polymers give rise to significantly larger tortuosities than do small ions like TMA^+. We presently do not know if this is due to the fact that ECS is composed of channels of different widths, which can exclude large molecules from some pathways through the space, or whether the increased tortuosity signifies increased interaction with the extracellular matrix. These results are thrown into sharp relief by the discovery that linear-PHPMA polymers up to 1 million M_r can move through the ECS with the same tortuosity as TMA^+; at the present time it is not clear how this is accomplished. Whatever the final explanation, these results show that the ECS can discriminate among different molecular species.

We have also spent time exploring the dynamics of the ECS by using osmotic stress to diminish or enlarge the spaces between cells; the most remarkable finding in these studies has been the dissociation between α and λ. Here α behaves as one might have expected, while λ increases rapidly as α decreases but reaches a plateau as the spaces widen. Modeling reveals that a possible explanation of this behavior is that cells not only swell or shrink during osmotic challenge, but also change shape. Consequently, small lakes are created in certain regions of the ECS which have the capacity to temporarily trap diffusing particles; such lakes were long ago illustrated by van Harreveld et al. (1965) and Bondareff and Pysh (1968). This is certainly not the only possible explanation of the data but it is one of the simplest and does not require the introduction of ad hoc properties of the extracellular matrix.

Beyond the specific details, these studies remind us that the ECS is a dynamic structure responding to cellular activity and perhaps under dynamic control; this implies that the ECS is a communication channel whose characteristics can be altered to change the nature of the communication. From this perspective, there are rich and complex properties that await discovery in the near future.

Summary

Volume transmission depends on the migration of informational substances through brain extracellular space (ECS) and almost always involves diffusion; basic concepts of diffusion are outlined from both the microscopic viewpoint based on random walks and the macroscopic viewpoint based on the solution of equations embodying Fick's Laws. In a complex medium like the brain, diffusing molecules are constrained by the local volume fraction of the ECS and tortuosity, a measure of the hindrance imposed by cellular obstacles. Molecules can also experience varying degrees of uptake or clearance. Bulk flow and the extracellular matrix may also play a role.

Examples of recent work on diffusion of tetramethylammonium (molecular weight, 74) in brain slices, using iontophoretic application and ion-selective microelectrodes, are reviewed. In slices, the volume fraction is about 20% and tortuosity about 1.6, both similar to values found in the intact brain. Using integrative optical imaging, results obtained with dextrans and albumins up to a molecular weight of 70,000 are summarized; for such large molecules the tortuosity is about 2.3. Experiments using synthetic long-chain PHPMA polymers up to 1,000,000 molecular weight show that these molecules also diffuse in the ECS but with a tortuosity of about 1.6. Studies with osmotic challenge show that volume fraction and tortuosity do not vary together as expected when the size of the ECS changes; a model is presented that explains the osmotic-challenge on the basis of changes in cell shape. Finally, new analytical insights are provided into the complex movement of potassium in the brain.

Acknowledgements

This work was supported by NIH Grants NS28642 and NS34115.

References

Abbott, N.J. (1992) The apparent 'sink action' of the CSF: are the observations due to secretion and flow of interstitial fluid? *J. Physiol.*, 446: 496P.

Archie, G.E. (1942) The electrical resistivity log as an aid in determining some reservoir characteristics. *Trans. Am. Inst. Min., Metal. Pet. Eng.*, 146: 54–62.

Barbour, B. and Häusser, M. (1997) Intersynaptic diffusion of neurotransmitter. *Trends Neurosci.*, 20: 377–384 (Correction 20: 509)

Bensoussan, A., Lions, J.L. and Papanicolaou, G. (1978) *Asymptotic Analysis for Periodic Structures*. North Holland, New York.

Berg, H.C. (1993) *Random Walks in Biology*. Expanded Edn. Princeton University Press, Princeton, New Jersey.

Bertalotto, A., Manzardo, E. and Guglielmonw, R. (1996) Immunohistochemical mapping of perineuronal nets containing chondriotin unsulfated proteoglycan in the rat central nervous system. *Cell Tissue Res.*, 283: 283–295.

Bjelke, B., England, R., Nicholson, C., Rice, M.E., Lindberg, J., Zoli, M., Agnati, L.F. and Fuxe, K. (1995) Long distance pathways of diffusion for dextran along fibre bundles in brain: relevance for volume transmission. *NeuroReport*, 6: 1005–1009.

Blum, J.J., Lawler, G., Reed, M. and Shin, I. (1989) Effect of cytoskeletal geometry on intracellular diffusion. *Biophys. J.*, 56: 995–1005.

Bondareff, W. and Pysh, J.J. (1968) Distribution of the extracellular space during postnatal maturation of rat cerebral cortex. *Anat. Rec.*, 160: 773–780.

Bradbury, M.W. (1979) *The Concept of a Blood-Brain Barrier*. John Wiley, Chichester.

Bruckner, G., Bringmann, A., Hartwig, W., Koppe, G., Delpech, B. and Brauer, K. (1998) Acute and long-lasting changes in extracellular-matrix chondroitin-sulphate proteoglycans induced by injection of chondroitinase ABC in the adult rat brain. *Exp. Brain Res.*, 121: 300–310.

Carslaw, H.S. and Jaeger, J.C. (1959) *Conduction of Heat in Solids*. (2nd ed.). Clarendon Press, Oxford.

Celio, M.R., Spreafico, R., DeBiasi, S. and Vitellaro-Zuccarello, L. (1998) Perineuronal nets: past and present. *Trends Neurosc.*, 21: 510–514.

Chebabo, S.R., Hester, M.A., Jing, J., Aitken, P.G. and Somjen, G.G. (1995) Interstitial space, electrical resistance and ion concentrations during hypotonia of rat hippocampal slices. *J. Physiol.*, 487: 685–697.

Chen, K.C. and Nicholson, C. (2000a) Spatial buffering of potassium ions in brain extracellular space. *Biophysical J.*, 78: 2776–2797.

Chen, K.C. and Nicholson, C. (2000b) Changes in cell shape create residual extracellular space volume and explain tortuosity behavior during osmotic challenge. *Proc. Natl. Acad. Sci. USA* (in press).

Crank, J. (1975) *The Mathematics of Diffusion*. (2nd ed.). Clarendon Press, Oxford.

Cserr, H.F., Cooper, D.N. and Milhorat, T.H. (1977) Flow of cerebral interstitial fluid as indicated by the removal of extracellular markers from rat caudate nucleus. *Exp. Eye Res.*, 25 (Suppl.): 461–473.

Cserr, H.F., Cooper, D.N., Suri, P.K. and Patlak, C.S. (1981) Efflux of radiolabeled polyethylene glycols and albumin from rat brain. *Am. J. Physiol.*, 240: F319-F328.

Cserr, H.F. and Ostrach, L.H. (1974) Bulk flow of interstitial fluid after intracranial injection of blue dextran 2000. *Exp. Neurol.*, 45: 50–60.

Cserr, H.F. and Patlak, C.S. (1991) Regulation of brain volume under isosmotic and anisosmotic conditions. In: R. Gilleis (Ed.), *Advances in Comparative and Environmental Physiology*, Springer-Verlag, Berlin, pp. 61–80.

Cussler, E.L. (1997) Diffusion: *Mass Transfer in Fluid Systems*. (2nd ed.). Cambridge University Press, Cambridge.

Davson, H. (1970) *A Textbook of General Physiology*. (4th ed.). Vol. 1. Williams and Wilkins, Baltimore.

de Gennes, P.G. (1975) *Scaling Concepts in Polymer Physics*. Cornell University Press, Ithica, NY.

Doi, M. (1996) *Introduction to Polymer Physics*. Clarendon Press, Oxford.

Eberhardt, W. and Reichenbach, A. (1987). Spatial buffering of potassium by retinal Müller (glial) cells of various morphologies calculated by model. *Neuroscience*, 22. p. 687–696.

Eccles, J.C. and Jaeger, J.C. (1957) The relationship between the mode of operation and the dimensions of the junctional regions at synapses and motor end-organs. *Proc. Roy. Soc. B*, 148: 38–56.

El-Kareh, A.W., Braunstein, S.L. and Secomb, T.W. (1993) Effect of cell arrangement and interstitial volume fraction on the diffusivity of monoclonal antibodies in tissue. *Biophys. J.*, 64: 1638–1646.

Fenstermacher, F. and Kaye, T. (1988) Drug 'diffusion' within the brain. *Annals NY Acad. Sci.*, 531: 29–39.

Freeman, J.A. and Nicholson, C. (1975) Experimental optimization of current source-density technique for anuran cerebellum. *J. Neurophysiol.*, 38: 369–382.

Gardner-Medwin, A.R. (1983) Analysis of potassium dynamics in mammalian brain tissue. *J. Physiol.*, 335: 393–426.

Gardner-Medwin, A.R. and Nicholson, C. (1983) Changes of extracellular potassium activity induced by electric current through brain tissue in the rat. *J. Physiol.*, 335: 375–392.

Hobohm, C., Härtig, W., Brauer, K. and Brückner, G. (1998) Low expression of extracellular matrix components in rat brain stem regions containing modulatory aminergic neurons. *J. Chem. Neuroanat.*, 15: 135–142.

Hoppensteadt, F.C. and Peskin, C.S. (1992) Control of cell volume and the electrical properties of cell membranes. In: F.C. Hoppensteadt and C.S. Peskin (Eds), *Mathematics in Medicine and the Life Sciences*, pp. 165–80.

Horn, A.S. (1979) Characteristics of dopamine uptake. In A.S. Horn, J. Korf J. and B.H.C. Westerink (Eds), *The Neurobiology of Dopamine*, Academic, London, pp. 217–235.

Hornung, U. (1996) *Homogenization and Porous Media*. Springer-Verlag, New York.

Hounsgaard, J. and Nicholson, C. (1983) Potassium accumulation around individual Purkinje cells in cerebellar slices from the guinea-pig. *J. Physiol.*, 340: 359–388.

Hrabětová, S. and Nicholson, C. (1999) Origin of non-specfic TMA^+ loss in vivo revealed by content and diffusion analysis in vitro. *Soc. Neurosci. Abstracts*, 25: 1758

Ichimura, T., Fraser, P.A. and Cserr, H.F. (1991) Distribution of extracellular tracers in perivascular spaces of the rat brain. *Brain Res.*, 545: 103–113.

Kreuzer, F. (1982) Oxygen supply to tissues: the Krogh model and its assumptions. *Experentia*, 38: 1415–1426.

Krewson, C.E., Klarman, M.L. and Saltzman, W.M. (1995) Distribution of nerve growth factor following direct delivery to brain interstitium. *Brain Res.*, 680: 196–206.

Križaj, D., Rice, M.E., Wardle, R.A. and Nicholson, C. (1996) Water compartmentalization and extracellular tortuosity after osmotic changes in cerebellum of *Trachemys scripta. J. Physiol.*, 492: 887–896.

Kullmann, D.M. and Asztely, F. (1998) Extrasynaptic glutamate spillover in the hippocampus: evidence and implications. *Trends Neurosci.*, 21: 8–14.

Latour, L.L., Svoboda, K., Mitra, P.P. and Sotak, C.H. (1994) Time-dependent diffusion of water in a biological model system. *Proc. Natl Acad. Sci. USA*, 91: 1229–1233.

Lehmenkühler, A, Syková, E., Svoboda, J., Zilles, K. and Nicholson, C. (1993) Extracellular space parameters in the rat neocortex and subcortical white matter during postnatal development determined by diffusion analysis. *Neuroscience*, 55: 339–351.

Lipinski, H-G. (1990) Monte Carlo simulation of extracellular diffusion in brain tissues. *Phys. Med Biol.*, 35: 441–447.

Maleski, M. and Hockfield, S. (1997) Glial cells assemble hyaluronan-based pericellular matrices in vitro. *Glia*, 20: 193–202.

Margolis, R.K. and Margolis, R.U. (1993) Nervous tissue proteoglycans. *Experientia*, 49: 429–446.

Mazel, T., Šimonova, Z. and Syková, E. (1998) Diffusion heterogeneity and anisotropy in rat hippocampus. *NeuroReport*, 9: 1299–1304.

McBain, C.J., Traynelis, S.F. and Dingledine, R. (1990) Regional variation of extracellular space in the hippocampus. *Science*, 249: 674–677.

Moseley, M.E., Cohen, Y., Kucharczyk, J., Mintorovitch, J., Asgari, H.S., Wendland, M.F., Tsuruda, J. and Norman, D. (1990) Diffusion-weighted MR imaging of anisotropic water diffusion in cat central nervous system. *Radiology*, 176: 439–445.

Neil, J.J., Shiran, S.I., McKinstry, R.C., Schefft, G.L., Snyder, A.Z., Almli, C.R., Akbudak, E, Aronovitz, J.A., Miller, J.P., Lee, B.C.P. and Conturo, T.E. (1998) Normal brain in human newborns: apparent diffusion coefficient and diffusion anisotropy measured by using diffusion tensor MR imaging. *Radiology*, 209: 57–66.

Newman, E.A. (1985) Membrane physiology of retinal glial (Müller) cells. *J. Neurosci.*, 5: 2225–2239.

Nicholson, C. (1985) Diffusion from an injected volume of a substance in brain tissue with arbitrary volume fraction and tortuosity. *Brain Res.*, 333: 325–329.

Nicholson, C. (1992) Quantitative analysis of extracellular space using the method of TMA^+ iontophoresis and the issue of TMA^+ uptake. *Can. J. Physiol. Pharmacol*, 70: S314-S322.

Nicholson, C. (1993) Ion-selective microelectrodes and diffusion measurements as tools to explore the brain cell microenvironment. *J. Neurosci. Meth.*, 48: 199–213.

Nicholson, C. (1995) Interaction between diffusion and Michaelis-Menten uptake of dopamine after iontophoresis in striatum. *Biophy. J.*, 68: 1699–1715.

Nicholson, C. (1999). Structure of extracellular space and physicochemical properties of molecules governing drug movement in brain and spinal cord. In: T.L. Yaksh (Ed.), *Spinal Drug Delivery*, Amsterdam, Elsevier, pp. 253–269.

Nicholson, C. and Freeman, J.A. (1975) Theory of current source-density analysis and determination of conductivity tensor for anuran cerebellum. *J. Neurophysiol.*, 38: 356–368.

Nicholson, C. and Phillips, J.M. (1981) Ion diffusion modified by tortuosity and volume fraction in the extracellular microenvironment of the rat cerebellum. *J. Physiol.*, 321: 225–257.

Nicholson, C. and Rice, M.E. (1986) The migration of substances in the neuronal microenvironment. *Annals NY Acad. Sci.*, 481: 55–71.

Nicholson, C. and Rice, M.E. (1988) Use of ion-selective microelectrodes and voltammetric microsensors to study brain cell microenvironment. In: A.A. Boulton, G.B. Baker and W. Walz (Eds), *Neuromethods: The Neuronal Microenvironment*, Humana, Clifton, NJ, 247–361.

Nicholson, C. and Rice, M.E. (1991) Diffusion of ions and transmitters in the brain cell microenvironment. In: K. Fuxe and L.F. Agnati (Eds), *Volume Transmission in the Brain: Novel Mechanisms for Neural Transmission, Advances in Neuroscience*, 1, Raven Press, New York, pp. 279–294.

Nicholson, C. and Syková, E. (1998) Extracellular space structure revealed by diffusion analysis. *Trends Neurosci.*, 21: 207–215.

Nicholson, C. and Tao, L. (1993) Hindered diffusion of high molecular weight compounds in brain extracellular microenvironment measured with integrative optical imaging. *Biophys. J.*, 65: 2277–2290.

Nicholson, C., Tao, L. and Kume-Kick, J. (1998) Constraints on the extracellular diffusion of molecules. *J. Physiol.*, 511.P: 8S–9S.

Orkand, R.K., Nicholls, J.G. and Kuffler, S.W. (1966) Effect of nerve impulses on the membrane potential of glial cells in the central nervous system of amphibia. *J. Neurophysiol.*, 29: 788–806.

Patlak, C.S. and Fenstermacher, J.D. (1975) Measurements of dog blood-brain transfer constants by ventriculocisternal perfusion. *Am. J. Physiol.*, 229: 877–884.

Pérez-Pinzón, M.A., Tao, L. and Nicholson, C. (1995) Extracellular potassium, volume fraction, and tortuosity in rat hippocampal CA1, CA3, and cortical slices during ischemia. *J. Neurophysiol.*, 74: 565–573.

Perrin, J. (1990) *Atoms*. Ox Bow Press, Woodbridge, Connecticut.

Pluen, A., Metti, P.A., Jain, R.K. and Berk, D.A. (1999) Diffusion of macromolecules in agarose gels: comparison of linear and globular configurations. *Biophys. J.*, 77: 542–552.

Preston, B.N., Laurent, T.C., Comper, W.D. and Checkley, G.J. (1980) Rapid polymer transport in concentrated solutions through the formation of ordered structures. *Nature*, 287: 499–503.

Preston, B.N., Laurent, T.C. and Comper, W.D. (1984) Transport of molecules in connective tissue polysaccharide solutions. In S. Arnott, D.A. Rees and E.R. Morris (Eds), *Molecular Biophysics of the Extracellular Matrix*, Humana, Clifton, New Jersey, pp. 119–170.

Prokopová, Š., Vargová, L. and Syková, E. (1997) Heterogeneous and anisotropic diffusion in the developing rat spinal cord. *Neuroreport*, 8: 3527–3532.

Rennels, M.L., Gregory, T.F., Blaumanis, O.R., Fujimoto, K. and Grady, P.A. (1985) Evidence for a 'paravascular' fluid circulation in the mammalian central nervous system, provided by the rapid distribution of tracer protein throughout the brain from the subarachnoid space. *Brain Res.*, 326: 47–63.

Rice, M.E. and Nicholson, C. (1991) Diffusion characteristics and extracellular volume fraction during normoxia and hypoxia in slices of rat neostriatum. *J. Neurophysiol.*, 65: 264–272.

Rice, M.E. and Nicholson, C. (1995) Diffusion and ion shifts in the brain extracellular microenvironment and their relevance for voltammetric measurements. In: A.A. Boulton, G.B. Baker and R.N. Adams (Eds), *Voltammetric Methods in Brain Systems*. Neuromethods 27, Humana, Totowa, NJ, pp. 27–79.

Rice, M.E., Okada, Y.C. and Nicholson, C. (1993) Anisotropic and heterogeneous diffusion in the turtle cerebellum: implications for volume transmission. *J. Neurophysiol.*, 70: 2035–2044.

Roitbak, T. and Syková, E. (1999) Diffusion barriers evoked in the rat cortex by reactive astrogliosis. *Glia*, 28: 40–48.

Rosenberg, G.A., Kyner, W.T. and Estrada, E. (1980) Bulk flow of brain interstitial fluid under normal and hyperosmolar conditions. *Am. J. Physiol.*, 238: F42-F49.

Ruoslahti, E. (1996) Brain extracellular matrix. *Glycobiology*, 6: 489–492.

Rusakov, D.A. and Kullmann, D.M. (1998a) Geometric and viscous components of the tortuosity of the extracellular space in the brain. *Proc. Natl Acad. Sci. USA*, 95: 8975–8980.

Rusakov, D.A. and Kullmann, D.M. (1998b) Extrasynaptic glutamate diffusion in the hippocampus: ultrastructural constraints, uptake, and receptor activation. *J. Neurosci.*, 18: 3158–3170.

Schmitt, F.O. (1984) Molecular regulators of brain functioning: a new view. *Neuroscience*, 13: 991–1001.

Schmitt, F.O. and Samson, F.E. (1969) The brain cell microenvironment. *Neurosci. Res. Prog. Bull.*, 7: 277–417.

Sen, P.N., Scala, C. and Cohen, M.H. (1981) A self-similar model for sedimentary rocks with application to the dielectric constant of fused glass beads. *Geophysics*, 46: 781–795.

Stachel, J. (1998) *Einstein's Miraculous Year: Five Papers that Changed the Face of Physics*. Princeton University Press, Princeton, New Jersey.

Svoboda, J. and Syková, E. (1991) Extracellular space volume changes in the rat spinal cord produced by nerve stimulation and peripheral injury. *Brain Res.*, 560: 216–224.

Syková, E. (1997) The extracellular space in the CNS: its regulation, volume and geometry in normal and pathological neuronal function. *Neuroscientist*, 3: 28–41.

Syková, E., Svoboda, J., Šimonova, Z., Lehmenkühler, A. and Lassmann, H. (1996) X-irradiation-induced changes in the diffusion parameters of the developing rat brain. *Neuroscience*, 70: 597–612.

Syková, E., Vargová, L., Prokopová, Š. and Šimonova, Z. (1999) Glial swelling and astrogliosis produce diffusion barriers in the rat spinal cord. *Glia*, 25: 56–70.

Szentistvanyi, I., Patlak, C.S., Ellis, R.A. and Cserr, H.F. (1984) Drainage of interstitial fluid from different regions of rat brain. *Am. J. Physiol.*, 246: F835-F844.

Tao, L. (1999) Effects of osmotic stress on dextran diffusion in rat neocortex studied with integrative optical imaging. *J. Neurophysiol.*, 81: 2501–2507.

Tao, L. and Nicholson, C. (1995) The three-dimensional point spread functions of a microscope objective in image and object space. *J. Microscopy*, Oxford 178: 267–271.

Tao, L. and Nicholson, C. (1996) Diffusion of albumins in rat cortical slices and relevance to volume transmission. *Neuroscience*, 75: 839–847.

Tosteson, D.C. (1964) Regulation of cell volume by sodium and potassium transport. In: J.F. Hoffman (Ed.), *The Cellular Functions of Membrane Transport, Society of General Physiologists Symposium*, No. 10, pp. 3–22.

Tosteson, D.C. and Hoffman, J.F. (1960) Regulation of cell volume by active cation transport in high and low potassium sheep red cells. *J. Gen. Physiol.*, 44: 169–194.

Ulbrich, K., Pechar, M., Strohalm, J., Šubr, V. and Rihova, B. (1997) Synthesis of biodegradable polymers for controlled drug release. *Annals NY Acad. Sci.*, 831: 47–56.

Van Harreveld, A., Crowell, J. and Malhotra, S.K. (1965) A study of extracellular space in the central nervous system. *J. Cell Biol.*, 25: 117–137.

Vargová, L., Tao, L., Syková, E., Ulbrich, K., Šubr, V. and Nicholson, C. (1998) Diffusion of large polymers in rat cortical slices measured by integrated optical imaging. *J. Physiol.*, 511.P: 16P

Ventura, R. and Harris, K.M. (1999) Three-dimensional relationships between hippocampal synapses and astrocytes. *J. Neurosci.*, 19: 6897–6906.

Voříšek, I. and Syková, E. (1997a) Evolution of anisotropic diffusion in the developing rat corpus callosum. *J. Neurophysiol.*, 78: 912–919.

Voříšek, I. and Syková, E. (1997b) Ischemia-induced changes in the extracellular space diffusion parameters, K^+, and pH in the developing rat cortex and corpus callosum. *J. Cerebral Blood Flow Metab.*, 17: 191–203.

Wightman, R.M. and Zimmerman, J.B. (1990) Control of dopamine extracellular concentration in rat striatum by impulse flow and uptake. *Brain Res. Rev.*, 15: 135–144.

L. F. Agnati, K. Fuxe, C. Nicholson and E. Syková (Eds.)
Progress in Brain Research, Vol 125

CHAPTER 6

Extracellular space diffusion and pathological states

Eva Syková*, Tomáš Mazel, Lýdia Vargová, Ivan Voříšek and Šárka Prokopová-Kubinová

Department of Neuroscience, 2nd Medical Faculty, Charles University and Institute of Experimental Medicine, Academy of Sciences of the Czech Republic, Vídeňská 1083, 14220 Prague 4, Czech Republic

Introduction

The extracellular space (ECS) is the microenvironment of the nerve cells and an important communication channel (Nicholson, 1979; Syková, 1983; Syková, 1992; Syková, 1997; Nicholson and Syková, 1998). It includes ions, transmitters, metabolites, peptides, neurohormones, other neuroactive substances and molecules of the extracellular matrix, and directly or indirectly affects neuronal and glial cell functions. Neurons and glia release a number of neuroactive substances into the ECS, which diffuse via the ECS to their targets located on nerve as well as glial cells, frequently distant from the release sites. Transmitters which escape from synapses or are released non-synaptically diffuse through the ECS and bind to extrasynaptic, usually high-affinity, binding sites located on neurons, axons and glial cells. This type of extrasynaptic transmission is also called 'diffusion transmission' (neuroactive substances diffuse through the ECS) or 'volume transmission' (neuroactive substances move through the volume of the ECS) (Fuxe and Agnati, 1991; Bach-y-Rita, 1993; Agnati et al., 1995; Syková, 1997; Nicholson and Syková, 1998; Zoli et al., 1999). Populations of neurons can interact both by synapses and by the diffusion of ions and neurotransmitters in the ECS. Since glial cells do not have synapses, their communication with neurons is only mediated by the diffusion of ions and neuroactive substances in the ECS. This mode of communication without synapses provides a mechanism of long-range information processing in functions such as vigilance, sleep, chronic pain, hunger, depression, LTP, LTD, memory formation and other plastic changes in the CNS (Syková, 1997). The size and irregular geometry of diffusion channels in the ECS (tissue tortuosity and anisotropy) substantially affect and/or direct the movement of various neuroactive substances in the CNS (Fig. 1) and thereby

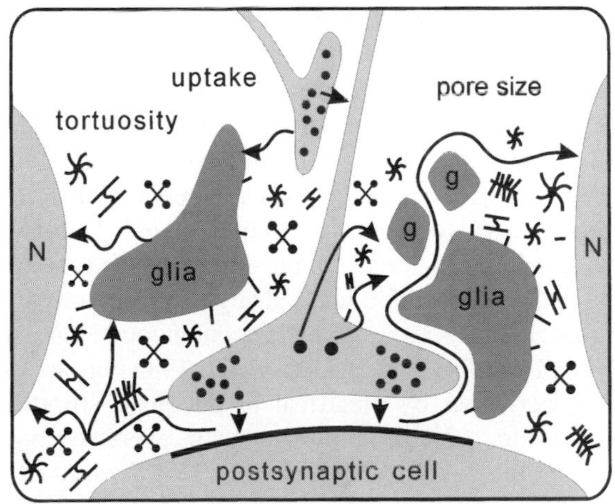

Fig. 1. Schematic of CNS architecture. The CNS architecture is composed of neurons (N), glial cells (glia), neuronal and glial processes (g), molecules of the extracellular matrix and intercellular channels between the cells. This architecture slows down the movement (diffusion) of substances in the brain, which is critically dependent on the ECS diffusion parameters volume fraction (α), tortuosity (λ) and nonspecific uptake (k').

*Corresponding author. Tel.: (+420 2) 475 2204; Fax: (+420 2) 475 2783; e-mail: sykova@biomed.cas.cz modulate neuronal signalling, neuron-glia communication and volume transmission.

Any change in the ECS diffusion parameters could result in the impairment of signal transmission and contribute to functional deficits and to neuronal damage. Dynamic changes in ECS ionic composition, volume and geometry accompany neuronal activity, neuronal loss, glial development and proliferation, aging, CNS injury, anoxia/ischemia, spreading depression, tumors, inflammmation, dyemyelination and many other brain pathological states.

Extracellular space composition

Cellular elements and blood vessels fill about 80% of the total CNS tissue volume and the remaining portion (15–25%) is the extracellular space. ECS ionic changes resulting from transmembrane ionic shifts during neuronal activity depolarize neighboring neurons and glial cells, enhance or depress their excitability, and affect ion channel permeability (Syková, 1983; Walz, 1989; Chesler, 1990; Syková, 1992; Deitmer and Rose, 1996; Syková, 1997). These ionic changes may also lead to the synchronization of neuronal activity and stimulate glial cell function.

In the mammalian CNS, the average ionic constituents of the ECS are basically the same as in the cerebrospinal fluid (mEq/kg H_2O): about 148 Na^+, 130 Cl^-, 3 K^+, 22 HCO_3^-, 2.5 Ca^{2+}, 1.7 Mg^{2+}, 5.3 glucose, 0.9 amino acids, 6.5 urea, osmolality 305, pH 7.27 (rabbit CSF, Dawson and Segal, 1996). However, in vivo measurements with ion-selective microelectrodes have revealed local changes in ECS ionic composition resulting from neuronal activity. The local changes in ion activity are localized to areas of high spontaneous activity (Syková et al., 1974; Syková et al., 1983), in areas being activated by electrical or adequate stimuli, e.g. tactile, visual, auditory, taste aversive and painful stimuli (Singer and Lux, 1975; Svoboda et al., 1988; Johnston et al., 1989; Syková et al., 1990; Syková and Svoboda, 1990), and during pathological states like anoxia, seizures, injury, spreading depression etc. Ionic and volume homeostasis in the CNS is maintained by a variety of mechanisms present in neurons as well as in glial cells. It was shown in a number of studies in vivo as well as *in vitro* that changes in extracellular K^+ concentration ($[K^+]_e$), alkaline and acid shifts in pH_e and a decrease in extracellular Ca^{2+} concentration ($[Ca^{2+}]_e$) accompany neuronal activity in different brain regions (for review see Syková, 1983; Chesler, 1990; Syková, 1992).

Other important chemical components of the ECS are substances involved in metabolism, particularly glucose and dissolved gases (O_2 and CO_2). The presence of HCO_3^- and CO_2 forms a powerful buffering system which controls extracellular and intracellular pH. The ECS also contains free radical scavengers such as ascorbate and glutathione (Rice, 2000), which may counteract some potentially lethal products of oxygen metabolism. In addition, the ECS contains amino acids like glutamate and aspartate, catecholamines, indolamines such as dopamine and serotonin, various opioid peptides, NO and growth hormones. Transmitters in the ECS bind to extrasynaptically located high affinity binding sites on neurons and glia.

The solution in the ECS is, however, not a simple salt solution. It has become apparent that long chain polyelectrolytes, either attached or unattached to cellular membranes, are present in the ECS. The ECS also contains a number of glycosaminoglycans (e.g. hyaluronate), glycoproteins and proteoglycans that constitute the extracellular matrix (ECM). It has been shown that the molecular content of the ECM, e.g. chondroitin sulphate proteoglycan, fibronectin, tenascin, laminin, etc. (Thomas and Steindler, 1995; Celio et al., 1998), dynamically changes during development, aging, wound healing and many pathological processes. ECM molecules are produced by both neurons and glia. These molecules have been suggested to cordon off distinct functional units in the CNS (groups of neurons, axon tracts, and nuclear groups). As shown in Fig. 1, these large molecules can slow down the movement (diffusion) of various neuroactive substances through the ECS. More importantly, these molecules may hinder diffusion of molecules so that they are confined to certain places, while diffusion to other brain regions will be facilitated.

Cells, particularly glia, maintain not only ECS ionic homeostasis but also ECS volume homeostasis (by swelling and shrinking during ionic shifts). They produce various extracellular matrix molecules (Celio et al., 1998) and therefore produce diffusion barriers. Glial cells when hypertrophied or proliferating form diffusion barriers (Syková, 1997; Roitbak and Syková, 1999) and in this way critically affect the permissiveness of the tissue, synaptic as well as volume transmission, activity-dependent synaptic plasticity, neurogenesis and regeneration.

Diffusion parameters of the ECS

The diffusion of substances in a free medium, such as water or diluted agar, is described by Fick's laws. In contrast to a free medium, diffusion in the ECS of the nervous tissue is hindered by the size of the extracellular clefts, the presence of membranes, fine neuronal and glial processes, macromolecules of the extracellular matrix and charged molecules, and also by cellular uptake (Fig. 1). To take these factors into account, it was necessary to modify Fick's original diffusion equations (Nicholson and Phillips, 1981; Nicholson and Syková, 1998; Nicholson et al., this volume). First, diffusion in the CNS is constrained by the restricted volume of the tissue available for the diffusing particles, i.e. by the extracellular space volume fraction (α), which is a dimensionless quantity and is defined as the ratio between the volume of the ECS and the total volume of the tissue ($\alpha = V_{ECS}/V_{TOT}$). It is now evident that the ECS in adult brain amounts to about 20% of the total brain volume, i.e. $\alpha = 0.2$ (Fig. 2, Table 1). Second, the free diffusion

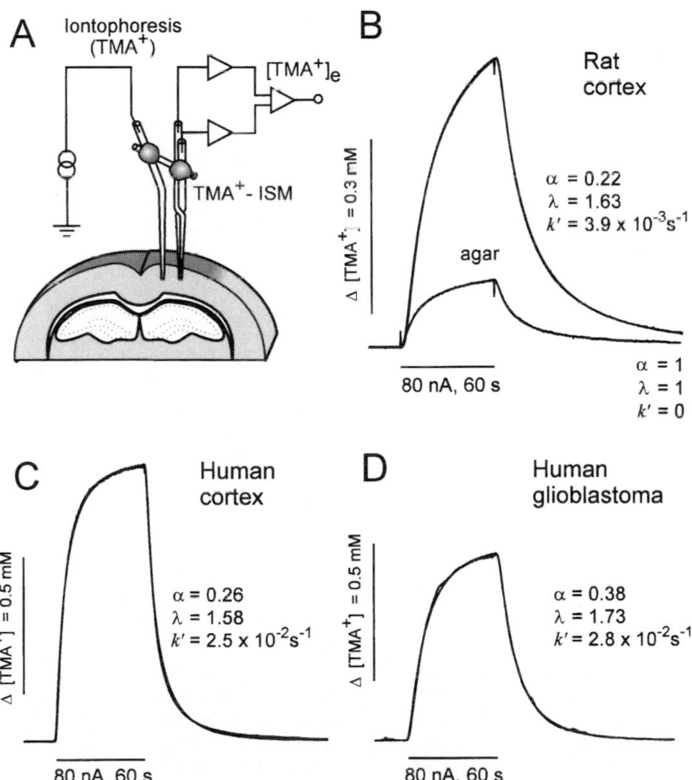

Fig. 2. A: Schema of the experimental arrangement. TMA^+-selective double-barrelled ion-selective microelectrode (ISM) was glued to a bent iontophoresis microelectrode. The separation between electrode tips was 80–200 μm. B: TMA^+ diffusion curves in the rat somatosensory cortex in vivo and free diffusion measured with the same microelectrode array in agar. C: Typical diffusion curve recorded in a slice (submerged slices, 400 μm thick) from the human temporobasal cortex of an 18-year-old woman with pharmacoresistant epilepsy. D: Typical diffusion curve in a slice (400 μm) from a human glioblastoma (grade 4, WHO clasification). For each curve, the ECS diffusion parameters α (volume fraction) and λ (tortuosity) were extracted by appropriate nonlinear curve fitting. Experimental and theoretical curves are superimposed in each case. For each figure the concentration scale is linear.

coefficient (D) in the brain is reduced by the square of the tortuosity (λ). ECS tortuosity is defined as $\lambda = (D/ADC)^{0.5}$, where D is a free diffusion coefficient and ADC is the apparent diffusion coefficient in the brain. As a result of tortuosity (in adult brain λ amounts to 1.55–1.65, Fig. 2), D is reduced to an apparent diffusion coefficient $ADC = D/\lambda^2$. Thus, any substance diffusing in the ECS is hindered by many obstructions. Third, substances released into the ECS are transported across membranes by non-specific concentration-dependent uptake (k'). In many cases, however, these substances are transported by energy-dependent uptake systems that obey nonlinear kinetics (Nicholson, 1995). When these three factors (α, λ and k') are incorporated into Fick's law, diffusion in the CNS is described fairly satisfactorily (Nicholson and Phillips, 1981).

The real-time iontophoretic method is used to determine ECS diffusion parameters and their dynamic changes in nervous tissue in vitro as well as in vivo (Syková, 1997; Nicholson and Syková, 1998). In principle, ion-sensitive microelectrodes (ISM) are used to measure the diffusion of ions to which the cell membranes are relatively impermeable, e.g. TEA^+, TMA^+ or choline. These substances are injected into the nervous tissue by pressure or by iontophoresis from an electrode aligned parallel to a double-barreled ISM at a fixed distance (Fig. 2). Such an electrode array is made by gluing together a pressure or iontophoretic pipette and a TMA^+-sensitive ISM with a tip separation of 60–200 μm. In the case of iontophoretic application, the TMA^+ is released into the ECS by applying a current step of +80–100 nA with a duration of 40–80 sec. The released TMA^+ is recorded with the TMA^+–ISM as a diffusion curve (Fig. 2), which is then transferred to a computer. Values of the ECS volume, *ADC*s, tortuosity and

TABLE 1

Values of volume fraction (α) and tortuosity (λ) obtained with the real-time iontophoretic TMA^+ method in the CNS

Region	Preparation	α	λ (λ_x, λ_y, λ_z)	References
Spinal cord, dorsal horn (rat)	In vivo	0.20–0.21	1.55–1.62	(Syková et al., 1994)
dorsal horn (frog)	In vitro	0.19	1.59	(Prokopová & Syková, in press)
dorsal horn (mouse)	In vitro	0.20	1.82	(Prokopová & Syková, in press)
intermediate region (rat)	In vivo	0.22	1.54	(Šimonová et al., 1996)
ventral horn (rat)	In vivo	0.23	1.46	(Šimonová et al., 1996)
white matter (rat)	In vivo	0.18	1.56[a]	(Šimonová et al., 1996)
filum terminale (frog)	In vitro	0.32	1.42	(Prokopová & Syková, in press)
Neocortex (rat)	In vitro	0.18	1.62	(Pérez-Pinzón et al., 1995)
Neocortex (human)	In vitro	0.21–0.26	1.44–1.58	(Vargová & Syková, unpublished)
Sensorimotor cortex (rat), layer II	In vivo	0.19	1.51	(Lehmenkühler et al., 1993)
layer III	In vivo	0.20	1.63	(Lehmenkühler et al., 1993)
layer IV	In vivo	0.21	1.59	(Lehmenkühler et al., 1993)
layer V	In vivo	0.22	1.62	(Lehmenkühler et al., 1993)
layer VI	In vivo	0.23	1.65	(Lehmenkühler et al., 1993)
Sensorimotor cortex (mice)	In vivo	0.23	1.78	(Mazel & Syková, 1999)
Auditory cortex (rat)	In vivo	0.21	1.52; 1.70; 1.81	(Syková et al., 1999a)
Corpus callosum (rat)	In vivo	0.21	1.47; 1.67; 1.69	(Mazel et al., 1998)
Neostriatum (rat)	In vitro	0.21	1.54	(Rice & Nicholson, 1991)
Hippocampus (rat) CA1 st. pyramidale	In vivo	0.21	1.39; 1,65; 1,80	(Mazel et al., 1998)
CA1 st. radiatum	In vivo	0.22	1.52; 1.61; 1.82	(Mazel et al., 1998)
Dentate gyrus	In vivo	0.22	1.49; 1.59; 1.68	(Mazel et al., 1998)
CA3 st. pyramidale	In vivo	0.22	1.50; 1.62; 1.69	(Mazel et al., 1998)
CA3 st. radiatum	In vivo	0.20	1.53; 1.58; 1.70	(Mazel et al., 1998)
Hippocampus (mice) CA1	In vivo	0.20	1.69[a]	(Mazel & Syková, 1999)
Cerebellum, molecular layer (rat)	In vivo	0.21	1.55[a,b]	(Nicholson & Phillips, 1981)
Cerebellum, molecular layer (turtle)	In vitro	0.31	1.44, 1.95, 1.58	(Rice et al., 1993)
Cerebellum, granular layer (turtle)	In vitro	0.21	1.77	(Rice et al., 1993)

[a] = measurements did not take anisotropy into account

[b] = measurements did not take non-specific TMA_+/TEA^+ uptake into account

non-specific cellular uptake are extracted by a nonlinear curve-fitting simplex algorithm applied to the diffusion curves (see Nicholson et al., this volume).

The other methods that are also used to study ECS volume and geometry, e.g. intrinsic optical signals (IOS), tissue resistance, integrative optical imaging (IOI) and nuclear magnetic resonance (NMR), are less comprehensive because they can measure only relative changes in the ECS diffusion parameters or only some of the three diffusion parameters (Van Harreveld et al., 1971; Matsuoka and Hossmann, 1982; Korf et al., 1988; Nicholson and Tao, 1993; Andrew and MacVicar, 1994). Integrative optical imaging is used to measure the *ADC*s of molecules tagged with fluorescent dye, while recordings of intrinsic optical signals, either light transmittance or light reflectance, are believed to reflect changes in the ECS volume; however, direct evidence is missing. On the other hand, diffusion-weighted NMR methods provide information only about the water diffusion coefficient (Benveniste et al., 1992; Latour et al., 1994; Norris et al., 1994; Van der Toorn et al., 1996). Although the correlation between water diffusion maps and changes in cell volume and ECS diffusion parameters can be very good, it is still not well understood (see below).

Using a light transmittance method (IOS, intrinsic optical signals), we recently found that changes in the ECS volume in brain slices measured by the TMA method have a different time course than those revealed by IOS. Simultaneous measurements using IOS and the TMA method were used to determine the absolute values of the ECS volume fraction α, tortuosity λ and non-specific uptake k' in the dorsal horns of rat spinal cord slices (Vargová et al., 1999). Cell swelling was evoked by a 45 min exposure to hypotonic solutions, elevated potassium or glutamate receptor agonists. Hypotonic solution (160 mOsmol/kg), 50 or 80 mM K^+, or NMDA (10^{-4}) induced a decrease in α of 45–85%, which peaked at 10–15 min, while an increase in IOS of 20–35% peaked in the first five minutes. After the initial peak, IOS quickly decreased to control levels, while the changes in α and λ persisted throughout the application of the test solutions (Fig. 3). There was also no correlation with the changes in ECS tortuosity and uptake. These data show that there is no simple correlation between ECS volume and IOS changes.

Neuroactive substances released constantly into the ECS will accumulate in this limited volume more rapidly than in free solution. Tortuosity (which is absent in a free medium) also causes a greater and more rapid accumulation of released substances. CNS tortuosity reduces the diffusion coefficient for small molecules by a factor of about 2.5 in many CNS regions. Larger molecules (with a relative molecular mass above 10 kDa), have a smaller diffusion coefficient than small molecules and are significantly more hindered in their diffusion, and therefore exhibit larger tortuosity (Nicholson and Tao, 1993; Tao et al., 1995; Tao and Nicholson, 1996). However, even large proteins, e.g. negatively charged globular proteins such as bovine serum albumin (66 kDa), or dextrans of 70 kDa, still migrate through the narrow interstices of brain slices (Nicholson and Tao, 1993; Tao and Nicholson, 1996). Recently, the diffusion properties of two types of rather large copolymers of N-(2-hydroxypropyl)methacrylamide (HPMA), developed as water-soluble anti-cancer drug carriers, were studied in rat cortical slices – HPMA polymeric chains with $M_r = 1{,}000{,}000$ and star-like systems, containing either albumin (179,000 M_r) or immunoglobulin (IgG) (319,000 M_r) in the center with HPMA side branches. Using the integrative optical imaging method and pressure microinjection of fluorescein-tagged polymers, the apparent diffusion coefficients (*ADC*) were obtained in rat cortical slices (Vargová et al., 1998). Surprisingly, long-chain HPMA polymers diffuse through the ECS with the same tortuosity as small molecules such as TMA. However, when the HPMA is configured into a more bulky globular molecule, the tortuosity increases to about 2.3. The tortuosity for long-chain HPMA is always found to be smaller than the tortuosity for globular copolymers (Prokopová et al., 1999). These data show that rather than M_r, the shape of the substance is the limiting factor in its movement through the extracellular space.

Diffusion inhomogeneity and anisotropy

ECS diffusion parameters are different in different parts of the CNS (Table 1). For example, it has been recognized that the TMA^+ diffusion parameters in the sensorimotor cortex of adult rat in vivo are heterogeneous (Lehmenkühler et al., 1993). The mean volume fraction gradually increases from $\alpha = 0.19$ in cortical layer II to $\alpha = 0.23$ in cortical layer VI. These typical differences are apparent in each individual animal. In subcortical white matter (corpus callosum) the volume fraction is always lower than in cortical layer VI, often between 0.19–0.20 (Voříšek and Syková, 1997a; Mazel et al., 1998). There is also a heterogeneity in the spinal cord, the mean values of the volume fraction being highest in the ventral horn and lowest in the white matter (Syková et al., 1994; Šimonová et al., 1996; Prokopová et al., 1997). Similar α values ($\alpha = 0.21$–0.22) have been found throughout the rat brain. In slices from human cortex (temporal and frontal lobe), recently obtained α values were not much different – they ranged between 0.21–0.26 (Fig. 2). Also the α values in other species such as mice, frog and turtle have been in the same range, with the exception of slices from frog filum

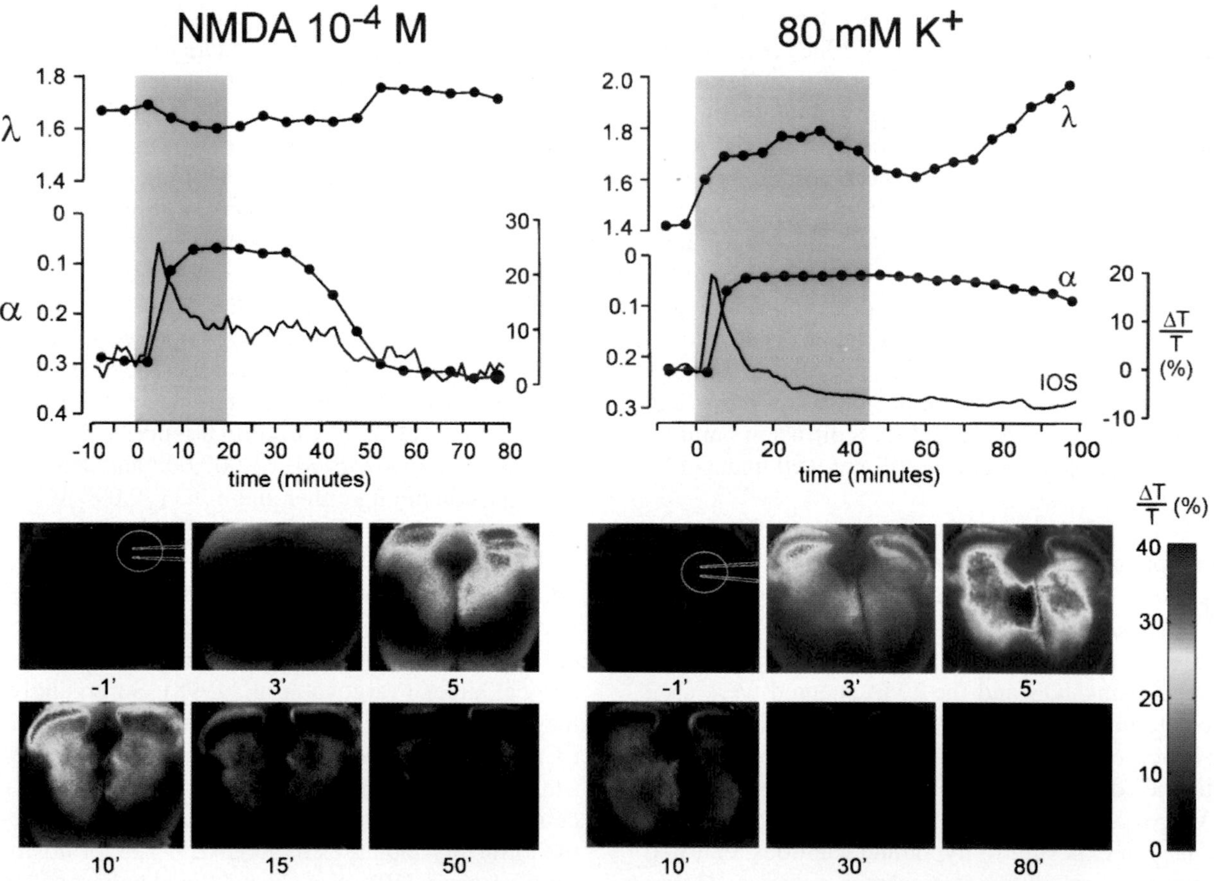

Fig. 3. Simultaneous measurements of the TMA^+ diffusion parameters and intrinsic optical signal (IOS) in spinal cord slices (400 μm thick) of a 14-day-old rat. Changes in TMA diffusion parameters α (ECS volume fraction) and λ (ECS tortuosity) before, during and after perfusion of the slice with NMDA (10^{-4}) or 80 mM K^+ were extracted by appropriate nonlinear diffusion curve fitting and plotted with the same time course as the IOS. Below: IOS images taken 1 minute before application (–1′; control) and 3′, 5′, 10′, 15′, 30′, 50′ and 80′ minutes after application of NMDA or K^+. Cellular, particularly glial, swelling results in an ECS volume decrease and, only in the case of K^+ application, also in a tortuosity increase. While the changes in TMA diffusion parameters persist for the whole application period and long after, the changes in IOS decrease with time. It is also evident that the time course of the IOS signal peaks faster than the changes in ECS volume and tortuosity. There is, therefore, no simple correlation between the IOS signal and ECS volume changes.

terminale ($\alpha = 0.32$) and turtle cerebellar molecular layer ($\alpha = 0.31$) (Table 1).

By introducing the tortuosity factor into diffusion measurements in the CNS, it soon becomes evident that diffusion is not uniform in all directions and that it is affected by diffusion barriers. This so-called anisotropic diffusion preferentially channels the movement of neuroactive substances in the ECS in one direction (e.g. along axons) and may, therefore, be responsible for a certain degree of specificity in volume transmission. Significant differences in tortuosity have been found in various brain regions, showing that the local architecture is different. There is increasing evidence that diffusion in brain tissue is anisotropic; isotropy is defined as the state of constant λ in any direction from a point source, while anisotropy indicates a difference in λ along different axes. To test for anisotropy, the ECS diffusion parameters are measured in three orthogonal axes x, y and z. Indeed, anisotropic diffusion was described using the TMA^+-method in the white matter of the corpus callosum (Fig. 4) and spinal cord (Prokopová et al., 1997; Voříšek and Syková, 1997a) as well as in the gray matter of the molecular layer of the cerebellum (Rice et al., 1993), in the hippocampus (Figs. 4 and 5) (Mazel et al., 1998) and in the auditory (Fig. 4) but not in the somatosensory cortex (Syková et al., 1999a). Using MRI, evidence of anisotropic diffusion in white matter was found in cat brain (Moseley et al., 1990) as well as in human brain (Le Bihan et al., 1993). Therefore, not only the diffusion of molecules such as TMA^+ or dextrans, but even the diffusion of water, is modified by various cellular structures including myelin sheaths. Because of the distinct diffusion characteristics, the extracellular molecular traffic will be different in various brain regions. The anisotropy and inhomogeneity of white and gray matter could enable different modes of diffusion transmission in these regions.

Recently we also obtained evidence that changes in anisotropy during development, aging and pathological states, mediated by the different structure of neurons, dendrites, axons, glial processes, myelin sheaths and extracellular matrix, can affect the migration of substances in the ECS (e.g preferred diffusion in one direction is lost) and may, therefore, account for an impairment of volume transmission (see below).

Diffusion parameters and neuronal activity

Transmembrane ionic fluxes during neuronal activity are accompanied by the movement of water and cellular, presumably particularly glial (see below), swelling. Changes in ECS diffusion parameters (ECS volume decrease, tortuosity increase and *ADC* decrease) are the consequence of activity-related transmembrane ionic shifts and cellular swelling. In the spinal cord of the rat or frog, repetitive electrical stimulation results in an ECS volume decrease from about 0.24 to 0.12–0.17, i.e. the ECS volume decreases by as much as 30–50% (Syková, 1987; Svoboda and Syková, 1991). The ECS volume in the spinal dorsal horn of the rat also decreases by 20–50% after injury of the ipsilateral hind paw evoked by subcutaneous injection of turpentine or after thermal injury. The changes in ECS diffusion parameters persist for many minutes (30 min after electrical stimulation or even 120 min after peripheral injury) after stimulation has ceased, suggesting long-term changes in neuronal excitability, neuron-glia communication and volume transmission.

Role of glia in ECS volume and geometry changes

It was shown in a number of studies that astrocyte swelling is an early event in numerous pathological states, accompanied by an elevation of $[K^+]_e$ (Kimelberg and Ransom, 1986; Kimelberg, 1991; Kimelberg et al., 1992). It was also shown that in the isolated turtle cerebellum exposed to hypotonic medium, volume fraction decreased to 0.12, while in hypertonic medium it increased to as much as 0.60 (Križaj et al., 1996).Cell swelling and astrogliosis (manifested as an increase in GFAP) were also evoked in isolated rat spinal cords of 4–21-day-old rats by incubation in either 50 mM K^+ or hypotonic solution (235 mOsmol/kg). Application of K^+ or hypotonic solution resulted at first in a decrease in the ECS volume fraction and in an increase in tortuosity in spinal gray matter (Fig. 6). These changes resulted from cell swelling, since the total water content (TW) in the spinal cord was unchanged and the changes were blocked in Cl-free solution and slowed down by furosemide and bumetanide (Syková et al., 1999c). During a continuous 45 min application of 50 mM K^+, α and λ often start to return towards control values, apparently due to the shrinkage of previously swollen cells since TW remains unchanged. This return is blocked by the gliotoxin fluoroacetate, suggesting that most of the changes are due to the swelling of glia. A 45 min application of 50 mM K^+ and, to a lesser degree, of hypotonic solution evokes astrogliosis, which persists after washing out these solutions with physiological saline. During astrogliosis λ increases again to values as high as 2.0, while α either returns to or increases above control values (Syková et al., 1999c). This

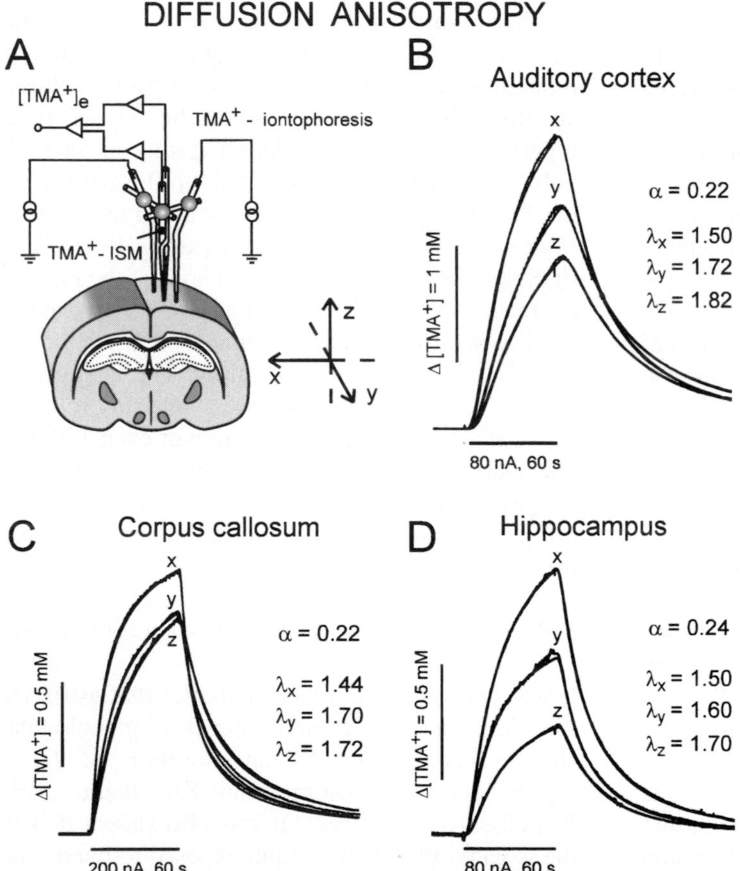

Fig. 4. Anisotropic diffusion in the rat brain. A: Schema of experimental arrangement to study diffusion anisotropy. A TMA^+-selective double-barrelled ion-selective microelectrode (ISM) was glued to two bent iontophoresis microelectrodes. The tips of the 3 pipettes formed a 90° horizontal angle for simultaneous measurements along the x- and y-axes. Similarly, for measurements along the z-axis, one iontophoresis pipette tip was lowered 110–180 μm below the tip of the ISM. B: Anisotropic diffusion in the auditory cortex. C: In the corpus callosum diffusion in the direction perpendicular to the orientation of the axons (y- and z- axes) is compromised by the number of myelin sheaths. D: Anisotropic diffusion in the dentate gyrus of the hippocampus. B, C, D: TMA^+ diffusion curves (concentration-time profiles) were measured along three orthogonal axes (x – mediolateral, y- rostrocaudal, z – dorsoventral). The slower rise in the z than in the y direction and in the y than in the x direction indicates a higher tortuosity and more restricted diffusion. The amplitude of the curves show that TMA^+ concentration, at approximately the same distance from the tip of the iontophoresis electrode, is much higher along the x-axis than along the y-axis and even higher than along the z- axis (λ_x, λ_y, λ_z).This can be explained if we realize that TMA^+ concentration decreases with the 'diffusion distance' from the iontophoretic micropipette and that the real 'diffusion distance' is not r but λr. Note that the actual ECS volume fraction α is 0.22–0.24 and can be calculated only when measurements are done along the x-, y- and z-axes.

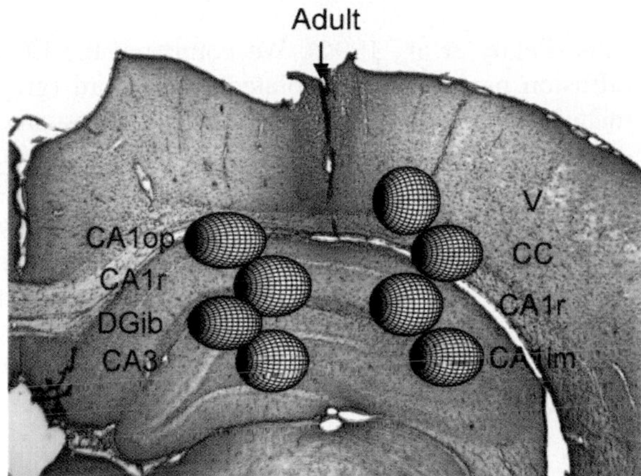

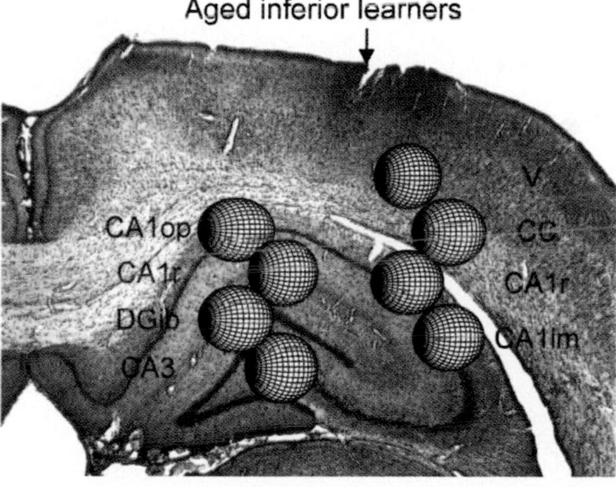

Fig. 5. Diffusion parameters in a young adult (3 months old) and aged (28 months old) rat with a learning deficit in the Morris water maze. Data were recorded in anesthetized animals; microelectrode tracks were verified after the experiments (see arrows). Iso-concentration surfaces for a 1 mM TMA^+ concentration contour 60 seconds after the onset of a 80 nA iontophoretic pulse. The surfaces were generated using the actual values of volume fraction and tortuosity. The ellipsoid represents anisotropic diffusion in a young adult rat. The larger sphere in an aged rat corresponds to isotropic diffusion and to a lower ECS volume fraction. It demonstrates that diffusion from any given source will lead to a higher concentration of substances in the surrounding tissue and a larger action radius in aged rats than in young adults. Anisotropy is almost lost in the aged rat.

persistant increase in λ after washout is also found in white matter. These data show that glial swelling and astrogliosis are associated with a persistant increase in ECS diffusion barriers.

Astrogliosis. Many pathological processes in the central nervous system are accompanied by a loss of cells or neuronal processes, astrogliosis, demyelination, and changes in the extracellular matrix, all of which may affect the apparent diffusion coefficients of neuroactive substances. Several animal models have been developed to study changes in ECS diffusion parameters. Brain injury of any kind elicits reactive gliosis, involving both hyperplasia and hypertrophy of astrocytes, which show intense staining for glial fibrillary acidic protein (GFAP) (Norton et al., 1992). Astrogliosis is also a typical characteristic of cortical stab wounds in rodents (Norton et al., 1992). The lesion is typically accompanied by an ECS volume increase and a substantial tortuosity increase to mean values of α of about 0.26 and λ of about 1.77 (see below) (Roitbak and Syková, 1999).

Similarly, both the size of the ECS (α) and, surprisingly, also λ are significantly higher in cortical grafts than in host cortex, about 0.35 and 1.79, respectively, as is also the case in gliotic cortex after stab wounds. Both α and λ are increased in cortical grafts of fetal tissue transplanted to the midbrain, where severe astrogliosis compared to host cortex is found, but not in fetal grafts placed into a cavity in the cortex, where only mild astrogliosis occurs (Syková et al., 1999b). Another characteristic feature of cortical grafts into midbrain is the variability of α and λ; the different values found at various depths of the grafts correlate with the morphological heterogeneity of the graft neuropil. These measurements show that even when the ECS in gliotic tissue or in cortical grafts is larger than in normal cortex, the tortuosity is still higher, and the diffusion of chemical signals in such tissue may be hindered. Limited diffusion may also have a negative impact on the viability of grafts in host brains. Compared to host cortex, immunohistochemistry shows myelinated patches and a larger number of hypertrophic astrocytes in areas of high λ values, suggesting that more numerous and/or thicker glial cell processes might be a cause of the increased tortuosity.

Diffusion properties in CNS of GFAP+/+ and GFAP–/– mice. To assess the function of glial fibrillary acidic protein (GFAP), the main component of astroglial intermediate filaments, we studied astrocyte membrane properties and the regulation of extracellular space volume in GFAP-negative mice (GFAP–/–). The GFAP gene was disrupted via targeted mutation in embryonic stem cells (Pekny et al., 1995). We compared the ECS diffusion parameters in isolated spinal cord (gray matter) of GFAP–/– and +/+ mice at postnatal days 5, 8 (P5, P8) and in the adult brain (P50–80). We found no difference in ECS diffusion parameters in GFAP–/– vs. +/+ mice at any age or in any brain region. In both groups, λ gradually increases with age to 1.7–1.8 and α decreases to about 0.2 at P50–80 (Table 2). However, during the application of 50 mM K^+ a significantly greater decrease in volume fraction and increase in tortuosity is seen in GFAP +/+ than in –/– mice (Fig. 6).

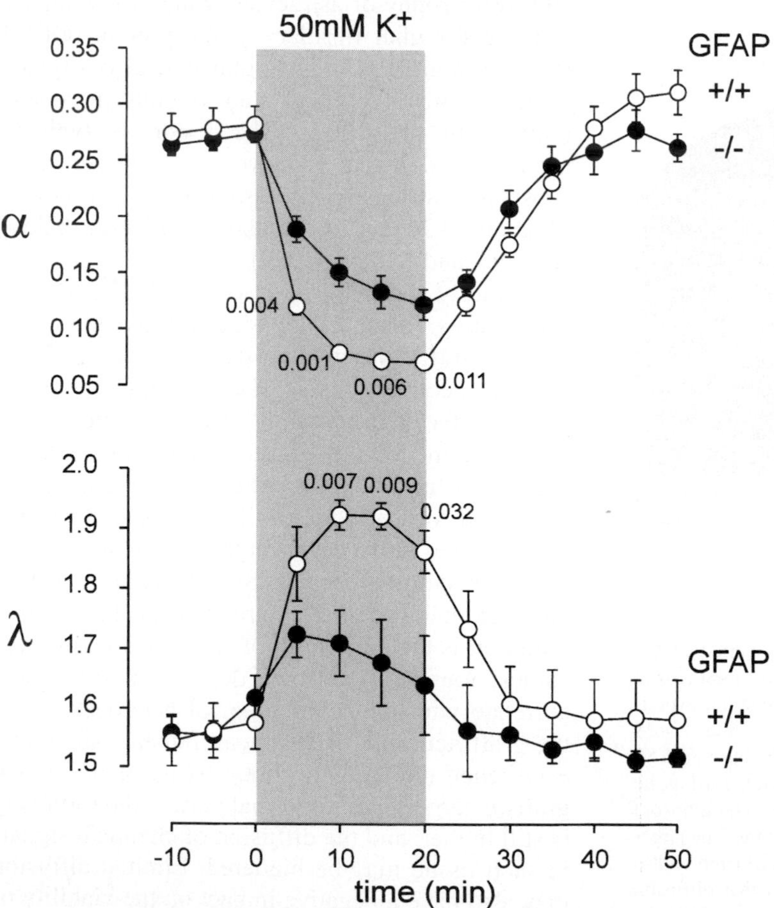

Fig. 6. The effect of 50 mM K^+ on ECS diffusion parameters in the gray matter of mouse spinal cord as measured in GFAP +/+ and GFAP –/– mice (mean ± S.E.M., $n=4$ in each). Each data point represents calculated α and λ values recorded at 5-minute intervals. Note that the time course of ECS volume decrease evoked by cell swelling is slower and smaller in GFAP –/– mice; as well, the tortuosity increase is significantly smaller (p values are shown where the difference is significant).

We have also found that a wave of spreading depression elicited in the cortex of GFAP +/+ mice by a needle prick is faster than that in GFAP –/– mice, although the final maximal values of the ECS volume fraction decrease and tortuosity increase are not significantly different (Mazel and Syková, unpublished results).

Using the whole-cell patch-clamp technique, we compared the reversal potential (V_{rev}) of astrocytes and their responses to elevated K^+ and glutamate. Depolarization of GFAP–/– astrocytes in spinal cord slices by 50 mM K^+ is significantly slower than that of GFAP +/+ ones. Although the maximal amplitude of the response is not significantly different, its peak is reached later in GFAP –/– astrocytes compared to GFAP +/+ astrocytes. The inward currents evoked by 1 mM glutamate in GFAP–/– astrocytes are smaller and slower than in GFAP+/+ ones by about 40%. In other words, the GFAP–/– astrocytes respond more slowly to high K^+ and with smaller inward currents to glutamate (Anděrová et al., 1999).

Our results show that astrocytes in GFAP+/+ and GFAP–/– mice may have different membrane properties, particularly under pathological conditions, e.g. anoxia, ischemia, seizures or repetitive

TABLE 2

Mean values of the ECS diffusion parameters during development, aging and pathological states as measured by the TMA^+-method. All data from rats in vivo

Status	Region	α	λ (λ_x, λ_y, λ_z)	References
Development (rat)	Cortex (P2)	0.36–0.4	1.59–1.68	(Lehmenkühler et al., 1993)
	Corpus callosum (P2)	0.46	1.58	(Lehmenkühler et al., 1993)
	Spinal cord, dorsal horn (P5)*	0.25	1.52	(Prokopová et al., 1997)
	Spinal cord, white matter (P8)*	0.27	1.33; 1.52	(Prokopová et al., 1997)
Aging (rat)	Cortex	0.19	1.57	(Syková et al., 1998b)
	Corpus callosum	0.19	1.48; 1.69; 1.71	(Syková et al., 1998b)
	Hippocampus CA1	0.19	1.49; 1.59; 1.72	(Syková et al., 1998b)
	Dentate gyrus	0.19	1.51; 1.54; 1.69	(Syková et al., 1998b)
	Hippocampus CA3	0.18	1.52; 1.52; 1.66	(Syková et al., 1998b)
Chronic pain (rat)	Spinal cord	0.12	—	(Jendelová & Syková, 1991)
GFAP –/– (mice)	Sensorimotor cortex	0.23	1.69	(Mazel & Syková, 1999)
	Corpus callosum	0.24	1.60 (λ_x)	(Mazel & Syková, 1999)
	Hippocampus CA1	0.25	1.61 (λ_x)	(Mazel & Syková, 1999)
	Spinal cord, dorsal horn*	0.22	1.84	(Prokopová et al., 1998)
SD (rat)	Cortex	0.09	2.01	(Richter et al., 1999)
SD GFAP +/+ (mice)	Cortex	0.05	2.38	(Mazel & Syková, unpublished)
SD GFAP –/– (mice)	Cortex	0.07	1.76	(Mazel & Syková, unpublished)
Hypoxia (rat)	Spinal cord	0.16	1.62	(Syková et al., 1994)
Terminal anoxia (rat)	Spinal cord	0.07	2.2	(Syková et al., 1994)
	Cortex	0.07	1.63[b]	(Lundbaek & Hansen, 1992)
	Cortex	0.06	2.00	(Voříšek & Syková, 1997b)
	Corpus callosum	0.05	2.10[a]	(Voříšek & Syková, 1997b)
Recovery after anoxia (rat)	Spinal cord	0.30	1.63	(Syková et al., 1994)
Hypernatremia (rat)	Cortex	0.10–0.15	1.65	(Cserr et al., 1991)
Astrogliosis-stab wounds (rat)	Cortex	0.30	1.93	(Roitbak & Syková, 1999)
X-irradiation acute state (rat)	Cortex	0.48–0.51	1.42–1.06	(Syková et al., 1996)
X-irradiation chronic state (rat)	Cortex	0.34–0.55	1.56–1.81	(Syková et al., 1996)
EAE (rat)	Spinal cord, dorsal horn	0.28	1.40	(Šimonová et al., 1996)
	Spinal cord, ventral horn	0.47	1.48	(Šimonová et al., 1996)
	Spinal cord, white matter	0.30	1.48	(Šimonová et al., 1996)
Cortical grafts (rat)	Midbrain	0.34	1.78	(Syková et al., 1999b)
Tumor (human)*	Glioblastoma (gr.4–5)	0.34	1.66	(Vargová & Syková, unpublished)

* in vitro preparation
[a] = measurements did not take anisotropy into account
[b] = measurements did not take non-specific TMA^+/TEA^+ uptake into account

stimulation. GFAP as a structural protein apparently may also affect the ability of astrocytes to change their volume during swelling.

Diffusion parameters and extracellular matrix

ECM molecules and other large molecules can also affect the tortuosity of the ECS. Their possible effect on changes in TMA^+ diffusion parameters has been studied in rat cortical slices (Tao et al., 1995; Vargová et al., 1998) and in isolated rat spinal cord (Prokopová et al., 1996). Superfusion of the slice or spinal cord with a solution containing either 40-kDa or 70-kDa dextran or hyaluronic acid (HA) results in a significant increase in λ. In standard physiological solution, λ is about 1.57, while in a 1% or 2% solution of 40-kDA or 70-kDa dextran, λ increases to about 1.72–1.77. Application of a 0.1% solution of HA (1.6×10^6 Da) results in an increase in λ to about 2.0 (Fig. 7A). The α is either unchanged or it decreases by only about 10%, suggesting that these substances have no effect on cell volume and the viability of the preparation.

Modification of the extracellular matrix can also be achieved by enzymatic treatment. Chondroitin sulphate proteoglycans are essential components of the extracellular matrix, forming so-called perineuronal nets surrounding neurons in cortex and hippocampus. There is also increasing evidence that N-CAM, the protein backbone of polysialic acid (PSA), is involved in synaptic plasticity. PSA, which is almost exclusively carried by N-CAM, is a major modulator of cell adhesion and is high in areas of continuous neurogenesis, neuronal migration, neurite extension and synapse formation. It has been found that mice treated with chondroitinase ABC or with antibodies against N-CAM and transgenic mice lacking N-CAM gene have impaired LTP (Becker et al., 1996; Muller et al., 1996). It has also been demonstrated that hydrated PSA influences a sufficiently large volume at the cell surface to exert broad steric effects, and that the removal of PSA causes a detectable change in the intercellular space. By contrast, chondroitin sulphate has been found to have little influence on the intercellular space (Yang et al., 1992).

We therefore used a single intracortical injection

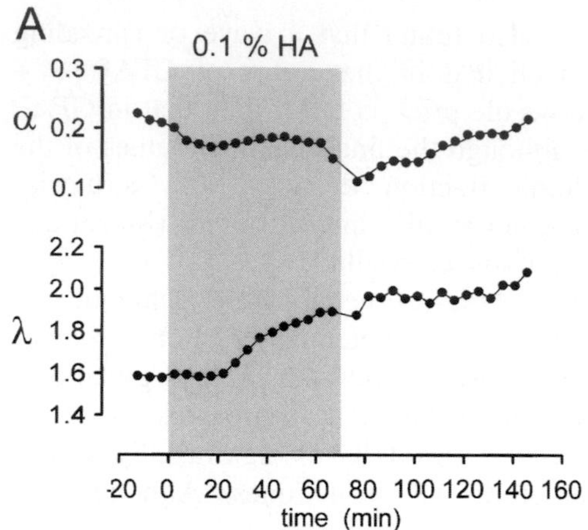

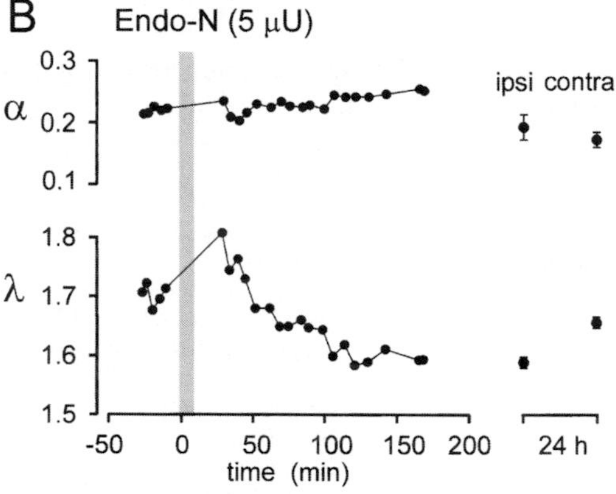

Fig. 7. Typical experiments showing the effect of hyaluronic acid and endo-neuraminidase on ECS volume fraction α and tortuosity λ. A: An isolated spinal cord was incubated in a 0.1% solution of hyaluronic acid. Note the increase in tortuosity with only about a 10% decrease in the ECS volume fraction α. B-Endo-N: Effect of endo-neuraminidase injection in the cortex of adult (3–6-month-old) rats. Two kinds of experiments were done: for acute treatment, the electrode array for TMA^+ diffusion measurement was introduced into the cortex and several control diffusion curves were taken. Clostridium neuraminidase (5 μU) dissolved in 500 nl ACSF was then applied for 10 min by pressure injection using a micropipette glued to the measuring electrode array at an inter-tip distance of 500 μm. ECS diffusion parameters were then followed for approximately 3 hours. In chronic experiments, a similar injection in anesthetized animals was made, the animals were returned to their cages, and diffusion measurements were performed 24 hours after injection. Note that immediately after neuraminidase application there was a transient decrease in ECS volume fraction accompanied by an increase in tortuosity, apparently caused by the injection. After one hour a long-lasting increase in ECS volume fraction and decrease in tortuosity occurred. These changes persisted at 24 hours after Endo-N application. No changes were found in the contralateral hemisphere.

0.5 μl) or chondroitinase ABC (10 μU) and studied the acute (3 hours) and chronic (24 hours) effects of this treatment on ECS diffusion parameters. Fig. 7B shows the effect of endo-N in rat cortex. A significant decrease in tortuosity is already found in the ipsilateral hemisphere 2 hours after injection and persists at 24 hours (Mazel and Syková, unpublished data). There is also a small but significant increase in the ECS volume fraction in the first 3 hours after injection that is not found at 24 hours. Similar changes have also been found in spinal cord slices incubated for 2 hours in endo-N or with chondroitinase (Syková and Prokopová, unpublished data).

A decrease in tortuosity and a loss of anisotropy that might be attributed to changes in the ECM have also been found during aging. This decrease correlates with the disapperance of fibronectin and chondroitin sulphate proteoglycans, forming perineuronal nets around granular and pyramidal cells in the hippocampus of young adult rats (Syková et al.,1998b). These results suggest that bigger molecules such as 40- and 70-kDa dextran, hyaluronic acid and molecules of the extracellular matrix may slow down the diffusion of small molecules such as TMA^+ (74 Da), ions, transmitters, metabolites etc. in the ECS.

ECS in pathological states

Pathological states, e.g. anoxia/ischemia, are accompanied by a lack of energy, seizure activity, the excessive release of transmitters and neuroactive substances, neuronal death, glial cell loss or proliferation, glial swelling, the production of damaging metabolites including free radicals and the loss of ionic homeostasis. Others are characterized by inflammation, edema or demyelination. It is therefore evident that they will be accompanied not only by substantial changes in ECS ionic composition (for review see Syková, 1983; Syková, 1992) but also by various changes in ECS diffusion parameters according to the different functional and anatomical changes.

Anoxia/ischemia

Dramatic K^+ and pH_e changes occur in the brain and spinal cord during anoxia and/or ischemia (Syková, 1992; Syková et al., 1994; Xie et al., 1995). Within 2 minutes after respiratory arrest in adult rats, blood pressure begins to increase and pH_e begins to decrease (by about 0.1 pH unit), while $[K^+]_e$ remains unchanged. With the subsequent blood pressure decrease, the pH_e decreases by 0.6–0.8 pH units to pH 6.4–6.6. This pH_e decrease is accompanied by a steep rise in $[K^+]_e$ to about 50–70 mM; decreases in $[Na^+]_e$ to 48–59 mM, $[Cl^-]_e$ to 70–75 mM, $[Ca^{++}]_e$ to 0.06–0.08 mM, and pH_e to 6.1–6.8; an accumulation of excitatory amino acids; a negative DC slow potential shift; and a decrease in ECS volume fraction to 0.04–0.07. The ECS volume starts to decrease when the blood pressure drops below 80 mm Hg and $[K^+]_e$ rises above 6 mM (Syková et al., 1994).

Figure 8 shows that during hypoxia and terminal anoxia, the ECS volume fraction in rat cortex or spinal cord decreases from 0.30 (note larger ECS volume in P8 rat) to 0.04, while tortuosity increases from 1.5 to about 2.2 (Lundbaek and Hansen, 1992; Syková et al., 1994). The same ultimate changes have been found in all neonatal, adult and aged rats, in grey and white matter, in the cortex, corpus callosum and spinal cord. However, the time course in white matter is significantly slower than in gray matter; and the time course in neonatal rats is about 10 times slower than in adults (Voříšek and Syková, 1997b). Linear regression analysis reveals a positive correlation between the normoxic size of the ECS volume and the time course of the changes. This corresponds to the well-known resistance of the immature CNS and the greater susceptibility of the aged brain to anoxia.

In recent studies using diffusion-weighted ^1H MRS/MRI, the apparent diffusion coefficient of water (ADC_W) was measured during terminal anoxia in rats. Anoxia evokes similar decreases in the apparent diffusion coefficient of ADC_W (measured by the NMR method) and ADC_{TMA} (measured by the TMA method). A comparison of the decreases in ADC_W and ADC_{TMA} in rats 8–9 days of age (P8–9) revealed the same time course, both corresponding to the decrease in ECS volume fraction (Fig. 8). Although water moves freely across the cellular membranes, TMA^+ stays predominantly in ECS. Since the total amount of tissue water is not believed to increase (Latour et al.,

1994; Van der Toorn et al., 1996), our study shows that changes in the *ADC* of brain tissue water measured by diffusion-weighted in vivo MR techniques reflect extra- and intracellular volume changes resulting from a water shift from the extra- to the intracellular compartment.

Full recovery to 'normoxic' diffusion parameters is achieved after successful recovery from severe ischemia (Lehmenkühler et al., 1993). Beginning 5–10 min after this recovery, the ECS volume fraction significantly increases above the 'normoxic' values to an α of about 0.30 (Table 2); λ and k' are not significantly different from the values found under normoxic conditions.

Diffusion parameters of the ECS in gliotic tissue

A stab wound of the rodent brain is a well-characterized and common model of reactive gliosis, which can impose diffusion barriers in the CNS due to the hypertrophy of astrocytic processes and an increased production of extracellular matrix

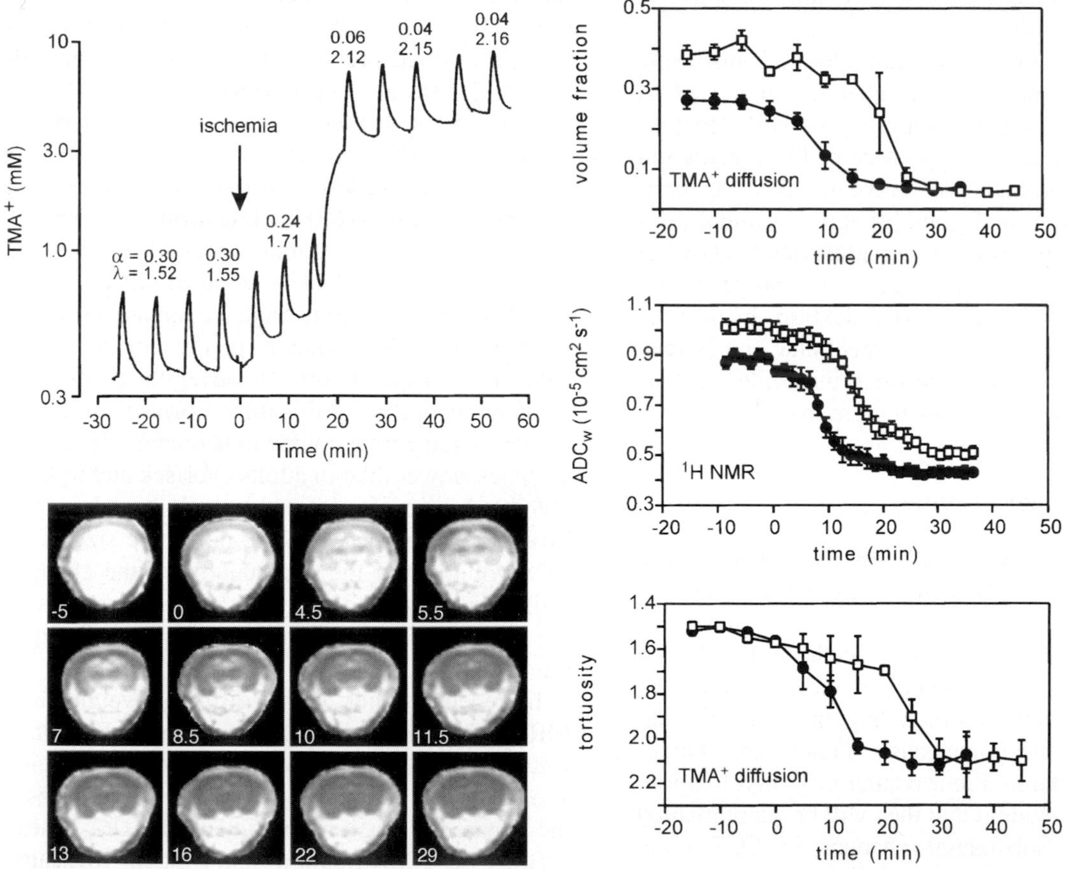

Fig. 8. Upper left: A typical TMA$^+$ diffusion experiment in the cortical gray matter (Layer V) of a 9-day-old rat. The TMA diffusion curves were recorded before and during global ischemia. The TMA$^+$ baseline increases and the superimposed diffusion curves show an increasing amplitude as ischemia progresses, because the scaling is logarithmic. The time after cardiac arrest and the α and λ values are shown with some curves. Below left: Water *ADC* maps of another 9-day-old rat brain before (–5) and at different times after cardiac arrest evoked by an i.p. injection of KCl (time after injection is shown with each record). The maps depict the calculated *ADC* on a pixel-by-pixel basis. The numbers in the lower left corner of the images indicate time. The *ADC* was calculated using diffusion-weighted NMR images obtained with U-FLARE (10 b values ranging up to 1500 s/mm^2) and fitting the intensities of the corresponding pixels in the images to the Stejskal-Tanner equation (see Van der Toon et al., 1996). Right: The time courses of *ADC* decrease, ECS volume decrease and tortuosity increase were found to be very similar (adapted from Van der Toorn et al., 1996).

components (Hatten et al., 1991; Norton et al., 1992; Ridet et al., 1997). We compared two different methods for revealing diffusion changes in rat cortex after injury: the TMA method and diffusion-weighted MR. The TMA and MR measurements were performed in the cortex of adult rats from the 3rd to the 35th day after a unilateral sterile cut through the cortex. Severe astrogliosis is found close to the injury site (up to 1 mm), and mild astrogliosis up to 2 mm from the wound in the ipsilateral cortex, but no astrogliosis is found in the auditory cortex or in the contralateral hemisphere. In contrast to GFAP staining, immunostaining for chondroitin-sulfate preoteoglycan increases in the whole ipsilateral hemisphere (Voříšek et al., 1999; Syková et al., 1999d). The mean values of α, ADC_{TMA} and ADC_W in the contralateral hemisphere were not significantly different from those in non-lesioned, control animals. In the astrogliotic cortex, less than 1 mm distant from the wound, the mean values of α are significantly higher ($\alpha = 0.26$), while the mean values of ADC_{TMA} are lower: (0.42×10^{-5} cm^2 s^{-1}). The more distant from the wound, the less the values of α and ADC_{TMA} differ from control values. On the other hand, ADC_W is significantly lower in the whole ipsilateral hemisphere, particularly in the auditory cortex: $ADC_W = (0.55 \times 10^{-5}$ cm^2 s^{-1}) (Fig. 9). We conclude that an increase in diffusion barriers, manifested by the decrease of both ADC_{TMA} and ADC_W, occurs throughout the entire cortex of the wounded hemisphere without significant changes in ECS volume. The changes are related to astrogliosis, particularly in and closely around the injured area, and to an increase in the extracellular matrix which occurs throughout the entire hemisphere.

These experiments revealed that not only glial swelling and astrogliosis, but also an increase in ECM content, are associated with a long-term increase of diffusion barriers in the ECS, and can, therefore, lead to the impairment of the diffusion of neuroactive substances and of volume transmission (Roitbak and Syková, 1999).

Extracellular space diffusion parameters during development

Compared to healthy adults, ECS diffusion parameters significantly differ during postnatal development (Lehmenkühler et al., 1993; Prokopová et al., 1997; Voříšek and Syková, 1997a; Voříšek and Syková, 1997b). The ECS volume in the cortex is about twice as large ($\alpha = 0.36$–0.46) in the newborn rat as in the adult rat ($\alpha = 0.21$–0.23), while the tortuosity increases with age (Table 2). The reduction in the ECS volume fraction correlates well with the growth of blood vessels. The larger ECS in the first days of postnatal development can be attributed to incomplete neuronal migration, gliogenesis and angiogenesis and to the presence of large extracellular matrix proteoglycans, particularly hyaluronic acid, which due to the mutual repulsion of its highly negatively charged branches occupies a great deal of space and holds cells apart. In rat spinal cord gray matter, α decreases with neuronal development and gliogenesis from postnatal day 4 to 12 by about 15%, while λ significantly increases, showing that the diffusion of molecules becomes more hindered with age. The large ECS channels during development may allow the migration of larger substances (e.g. growth factors) and provide better conditions for cell migration during development. On the other hand, the large ECS in the neonatal brain could significantly dilute ions, metabolites and neuroactive substances released from cells, relative to release in adults, and may be a factor in the prevention of anoxic injury, seizure and spreading depression in young individuals. The diffusion parameters could also play an important role in the developmental process itself; diffusion parameters are substantially different in myelinated and unmyelinated white matter (Prokopová et al., 1997; Voříšek and Syková, 1997a). Isotropic diffusion is found in the corpus callosum and spinal cord white matter of young rats with incomplete myelination. In myelinated spinal cord and corpus callosum, the tortuosity is higher (the apparent diffusion coefficient is lower) when TMA^+ diffuses across the axons than when it diffuses along the fibers (Fig. 4).

Diffusion properties of the nervous tissue after neonatal X-irradiation

The CNS during the early postnatal period is more sensitive to X-irradiation than is the adult nervous

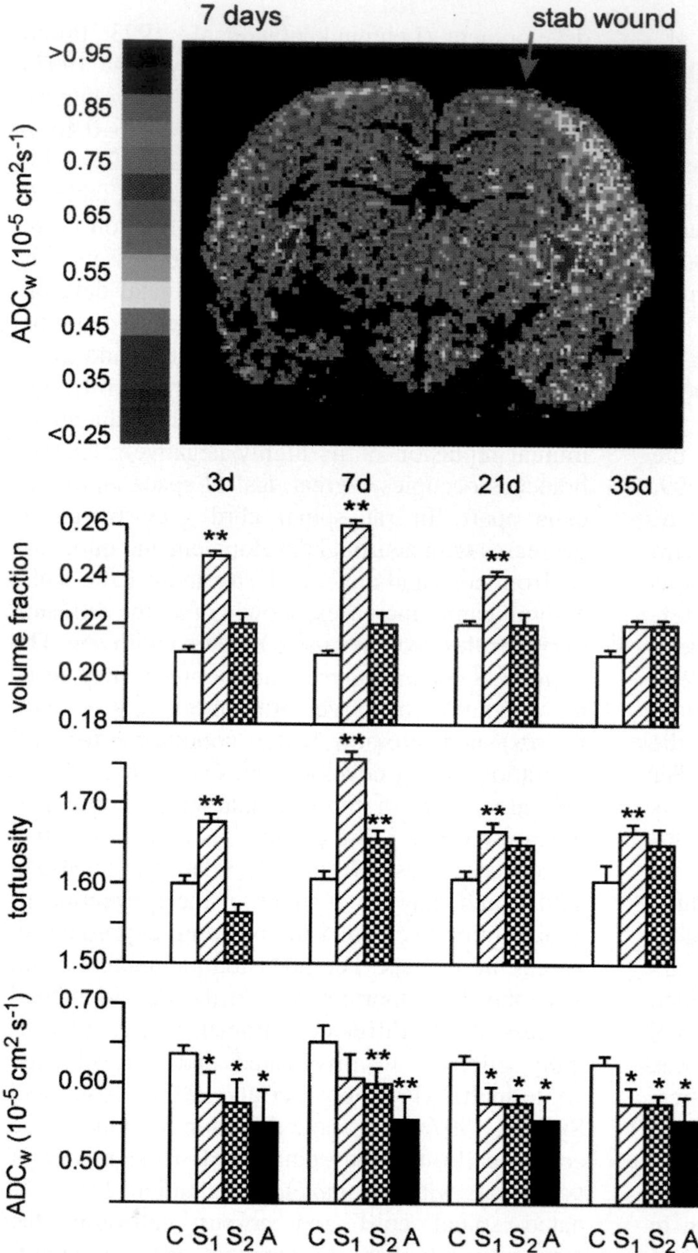

Fig. 9. ECS volume fraction, TMA tortuosity and ADC_W as measured by the TMA method and DW–NMR.The pseudocolour image shows the ADC_W map of a rat brain 7 days after a cortical stab wound. Note that ADC_W is lower in the entire cortex of the wounded hemisphere than in the contralateral hemisphere (C). The bar charts show the ECS volume fraction and tortuosity as measured by TMA method, and ADC_W as measured by DW–NMR. Data were obtained in the cortex of the injured and contralateral hemisphere 3, 7, 21 and 35 days post-lesion (d). All data are expressed as mean ± SEM. The asterisks above columns indicate significant diference from the values in contralateral hemisphere, $* = p < 0.05$, $** = p < 0.01$. The ECS volume fraction, as measured by the TMA method, revealed significant changes only in the vicinity of the wound in the area S1 (up to 1000 μm from the wound); tortuosity is increased in the area S1 as well as in S2 (1000–2000 μm from the wound). Significant changes in ADC_W were found in the entire ipsilateral hemisphere including the auditory cortex (A). This corresponded to an increase in chondroitin sulphate immunostaining (not shown).

system, apparently due to the proliferative potential and increased radiation sensitivity of glial and vascular endothelial cells in the immature nervous system. In experiments on the somatosensory neocortex and subcortical white matter of 1-day-old (P1) rats, X-irradiation at a single dose of 40 Gy results in radiation necrosis with typical early morphological changes in the tissue, namely cell death, DNA fragmentation, extensive neuronal loss, blood-brain-barrier (BBB) damage, activated macrophages, astrogliosis, an increase in extracellular fibronectin, and concomitant changes in all three diffusion parameters. The changes are observed as early as 48 hours post-irradiation and persist at P21 (Syková et al., 1996). Under normal conditions, the volume fraction of the ECS in the cortex is large in newborn rats, $\alpha = 0.35$–0.40, diminishes with age and reaches adult values at P21 (Lehmenkühler et al., 1993). X-irradiation at a single dose of 40 Gy blocks the normal pattern of volume fraction decrease during postnatal development and results in a significant increase in the ECS volume (Syková et al., 1996). The volume fraction in both cortex and corpus callosum increases to about 0.50. This increase persists at 3 weeks after X-irradiation. Tortuosity and non-specific uptake significantly decrease at 48 hours (P2); at P8–P9 they are not significantly different from those of control animals, and both significantly increase with astrogliosis at P10–P21. These data indicate that in chronic lesions, which occur 1–3 weeks after X-irradiation and/or in gliotic tissue, the volume fraction remains elevated, while tortuosity increases due to astrogliosis. Even when X-irradiation at a single dose of 20 Gy is used, resulting in relatively light neuronal damage and loss and BBB damage, it produces similar changes in diffusion parameters as those found with 40 Gy. Less pronounced but significant changes in diffusion parameters are also found in areas adjacent to the directly X-irradiated cortex of the ipsilateral hemisphere and in the contralateral hemisphere.

It is evident that the block of postnatal gliogenesis and the damage to neurons evoked by X-irradiation results in an even greater increase in the ECS volume fraction of nervous tissue than injury evoked by a stab wound. Thus it can substantially contribute to impaired signal transmission, e.g. by diluting ions and neuroactive substances released from cells, and may thus play an important role not only in functional deficits, but also in malfunctions during the developmental processes. Moreover, the increase in ECS volume and tortuosity in the X-irradiated cortex as well as in the contralateral hemisphere suggests that the diffusion of substances can be substantially changed even a long time after mild irradiation.

ECS volume and geometry during inflammation and demyelination

Changes in ECS diffusion parameters can be expected during inflammation, during which brain edema may develop; in an experimental model, inflammation was evoked by an intracerebral inoculation of a weakly pathogenic strain of *Staphylococcus aureus* (Lo et al., 1993). Acute inflammation and an increase in BBB permeability in the abscess region resulted in rather mild changes in the ECS diffusion parameters, i.e. the volume fraction tended to be somewhat larger and the tortuosity somewhat smaller.

Dramatic changes in the ECS diffusion parameters are found in the spinal cord of rats during experimental autoimmune encephalomyelitis (EAE), an experimental model of multiple sclerosis (Šimonová et al., 1996). EAE, which is induced by the injection of guinea-pig myelin basic protein (MBP), results in typical morphological changes in the CNS tissue, namely demyelination, an inflammatory reaction, astrogliosis, BBB damage and paraparesis, at 14–17 days post-injection. Paraparesis is accompanied by increases in α in the dorsal horn, in the intermediate region, in the ventral horn and in white matter from about 0.18 to about 0.30 (Table 2). The λ in the dorsal horn and the intermediate region significantly decreases and k' decreases in the intermediate region and the ventral horn (Šimonová et al., 1996). There is a close correlation between the changes in ECS diffusion parameters and the manifestation of neurological abnormalities.

These results suggest that the expansion of the ECS alters diffusion parameters in inflammatory and demyelinating diseases and may affect the accumulation and movement of ions, neurotransmitters, neuromodulators and metabolites in the CNS in these disorders, possibly by interfering with axonal conduction.

Diffusion properties of malignant tumors

Cancer is the second leading cause of death in many industrialized countries, and malignant brain tumors, particularly the gliomas, are among the deadliest of tumors, since many patients, including children, die within two years. Only recently have basic new findings about tumor cell division, differentiation and migration, the relationship between glial cells and gliomas, the existence of multiple glial precursor cell populations and new insights into the developmental biology of glial cells been made. One of the recently discussed issues is the existence of CNS-specific extracellular matrix proteins, e.g. brain-enriched hyaluronan binding proteins (BEHAB) and brevican, that are expressed at high levels during initial gliogenesis and also in all types and grades of human gliomas (Jaworski et al., 1996). Hyaluronan binding proteins help cells to move through tissue during development and have also been associated with invasive cancers. It has also been suggested that the migration of cells could be critically dependent on their shape and size, their binding to various proteins in the extracellular space (ECS) such as hyaluronan-binding protein that can boost the invasiveness of tumors, and on the size and geometry of the ECS. The delivery of drugs to tumors is affected by the permeability of the blood-brain-barrier, their diffusion through the ECS in normal and malignant tissue, and their side-effects on healthy cells surrounding the tumor. It is therefore crucial to quantify the size, composition and geometry of the extracellular space as these factors critically affect cell migration and the diffusion of substances in the brain (for review see Syková, 1997; Nicholson and Syková, 1998).

Older studies, particularly using sucrose space and electron microscopic methods, showed that the visible extracellular compartment is larger in brain tumors, particularly in gliomas, than in normal brain tissue (Bakay, 1970a, b). However, these methods resulted either in small values for the size of the extracellular space because of tissue shrinkage during fixation and embedding, or did not allow one to measure the absolute values of ECS size and geometry or to follow the diffusion of molecules of different sizes and shapes. Increased extracellular space, along with the above-mentioned extracellular matrix proteins, could allow cells to migrate more easily in tumors and into surrounding tissue. Shrinkage of the extracellular space and extracellular matrix proteins could slow or substantially limit the diffusion and migration of cells.

To quantify extracellular space size, geometry and diffusion properties in tumors (particularly malignant gliomas) TMA measurments were performed in slices from surgically removed pieces of patients' brains. We found that the more malignant the glioma, the more dramatic is the increase in the ECS volume. In many brain slices from glioblastomas (grade 4, WHO classification system), the ECS volume is as large as 37–46% of total tissue volume (Fig. 2). There is also an increase in tortuosity which can be due either to the frequently observed astrogliosis or to changes in the extracellular matrix. It is therefore reasonable to assume that the ECS composition, volume and geometry play an important role in cancer malignancy and invasiveness. The size of the ECS in malignant tumors and their geometry and structure should be considered during therapeutic drug delivery.

ECS diffusion parameters and aging

In the mammalian brain, higher cognitive functions such as learning and memory depend upon the circuits that run through the hippocampus. Until recently, learning deficits during aging have been associated with neuronal degeneration and synaptic inefficiency. However, recent observations of a lack of hippocampal cell loss in aged humans, monkeys and rats (West, 1993; Rapp and Gallagher, 1996; Rasmussen et al., 1996) suggest that age-related functional change in the nervous system may not necessarily be a sign of degenerative pathology. The question thus arises whether learning deficits during aging also involve the impairment of extrasynaptic or 'volume' transmission, i.e. the diffusion of neuroactive substances in the ECS.

The ECS diffusion parameters α, $\lambda_{x,y,z}$ and k' were measured in the cortex, corpus callosum and hippocampus (CA1, CA3 and in dentate gyrus). If diffusion in a particular brain region is anisotropic, then the correct value of the ECS volume fraction cannot be calculated from measurements done only in one direction. For anisotropic diffusion, the diagonal components of the tortuosity tensor are not equal, and generally its non-diagonal components need not be zero. Neverthless, if a suitable frame of reference is chosen (i.e. if we measure in three privileged orthogonal directions), neglecting the non-diagonal components becomes possible, and the correct value of the ECS size can thus be determined (Rice et al., 1993; Mazel et al., 1998). Therefore, TMA^+ diffusion was measured in the ECS independently along three orthogonal axes (x – transversal, y – sagital, z – vertical). In all three regions – cortex, corpus callosum and hippocampus – the mean ECS volume fraction α was significantly lower in aged rats (26–32 months old), ranging from 0.17 to 0.19, than in young adults (3–4 months old) in which α ranged from 0.21 to 0.22 (Table 2). Non-specific uptake k' was also significantly lower in aged rats. Importantly, there is a loss of anisotropy in the aging hippocampus, particularly in the CA3 region and the dentate gyrus (Syková et al., 1998b).

The three-dimensional pattern of diffusion away from a point source can be illustrated by constructing iso-concentration spheres (isotropic diffusion) or ellipsoids (anisotropic diffusion) for extracellular TMA^+ concentration. The surfaces in Fig. 5 represent the locations where TMA^+ concentration first reached 1 mM, 60 sec after its application in the center. The ellipsoid in the hippocampus of the young adult rat reflects the different abilities of substances to diffuse along the x-, y- and z-axes, while the sphere in the hippocampus of the aged rat shows isotropic diffusion. The smaller ECS volume fraction in aged rats is reflected in the spheres being larger than the ellipsoid.

Morphological changes during aging include cell loss, loss of dendritic processes, demyelination, astrogliosis, swollen astrocytic processes and changes in the extracellular matrix. It is reasonable to assume that there is a significant decrease in the *ADC* of many neuroactive substances in the aging brain, which accompanies astrogliosis and changes in the extracellular matrix. One of the explanations of why α in the cortex, corpus callosum and hippocampus of senescent rats is significantly lower than in young adults could be astrogliosis in the aged brain. Increased GFAP staining and an increase in the size and fibrous character of astrocytes have been found in the cortex, corpus callosum and hippocampus of senescent rats, which may account for changes in the ECS volume fraction (Syková et al., 1998b). Other changes could account for the decreases in λ values and for the disruption of tissue anisotropy. In the hippocampus in CA1, CA3, as well as in the dentate gyrus, we have observed changes in the arrangement of fine astrocytic processes. These are normally organized in parallel in the x–y plane (Fig. 10A, B), and this organization totally disappears during aging. Moreover, the decreased staining for chondroitin sulfate proteoglycans and for fibronectin (Fig. 10C, D) suggests a loss of extracellular matrix macromolecules.

Because α is lower in aging rats, we expected some differences in the ECS diffusion parameter changes during ischemia in senescent rats. In fact, the final values of α, λ and k' induced by cardiac arrest are not significantly different between young and aged rats; however, the time course of all the changes is faster in aged animals (Syková et al., 1998b). The accelerated changes in extracellular space volume fraction and tortuosity evoked by ischemia in nervous tissue during aging can contribute to a more rapid impairment of signal transmission, e.g. due to a faster accumulation of ions and neuroactive substances released from cells and their slower diffusion away from the hypoxic/ischemic area in the more compacted ECS brought about by aging.

Our recent study also revealed that the degree of learning deficit during aging correlates with the changes in ECS volume, tortuosity and non-specific uptake (Syková et al., 1998a). The hippocampus is well-known for its role in memory formation, especially declarative memory. It is therefore reasonable to assume that diffusion anisotropy, which leads to a certain degree of specificity in extrasynaptic communication, may play an important role in memory formation. There was a significant difference between mildly and severely behaviorally impaired rats (rats were tested in a

Morris water maze), which was particularly apparent in the hippocampus. The ECS in the dentate gyrus of severely impaired rats was significantly smaller than in mildly impaired rats. Also, anisotropy in the hippocampus of severely impaired rats, particularly in the dentate gyrus, was much reduced, while a substantial degree of anisotropy was still present in aged rats with a better learning performance. Anisotropy might be important for extrasynaptic transmission by channeling the flux of substances in a preferential direction. Its loss may severely disrupt volume transmission in the CNS, which has been suggested to play an important role in memory formation (Syková et al., 1998a). Chodroitine sulphate proteoglycans participate in multiple cellular processes (Hardington and Fosang, 1992; Margolis and Margolis, 1993), including axonal outgrowth, axonal branching and synaptogenesis, which are important for the formation of memory traces.

What is the functional significance of the observed changes in ECS diffusion parameters during aging? We suggest that the alterations in hippocampal diffusion may account for the learning impairment seen in aged animals, either due to their effect on volume transmission (Nicholson and Syková, 1998) or on 'cross-talk' between synapses, which has been suggested to be involved in LTP and LTD (Kullmann et al., 1996; Kullmann, this volume), or on both. Anisotropy, which, particularly in the hippocampus and corpus callosum, may help to facilitate the diffusion of neuro-

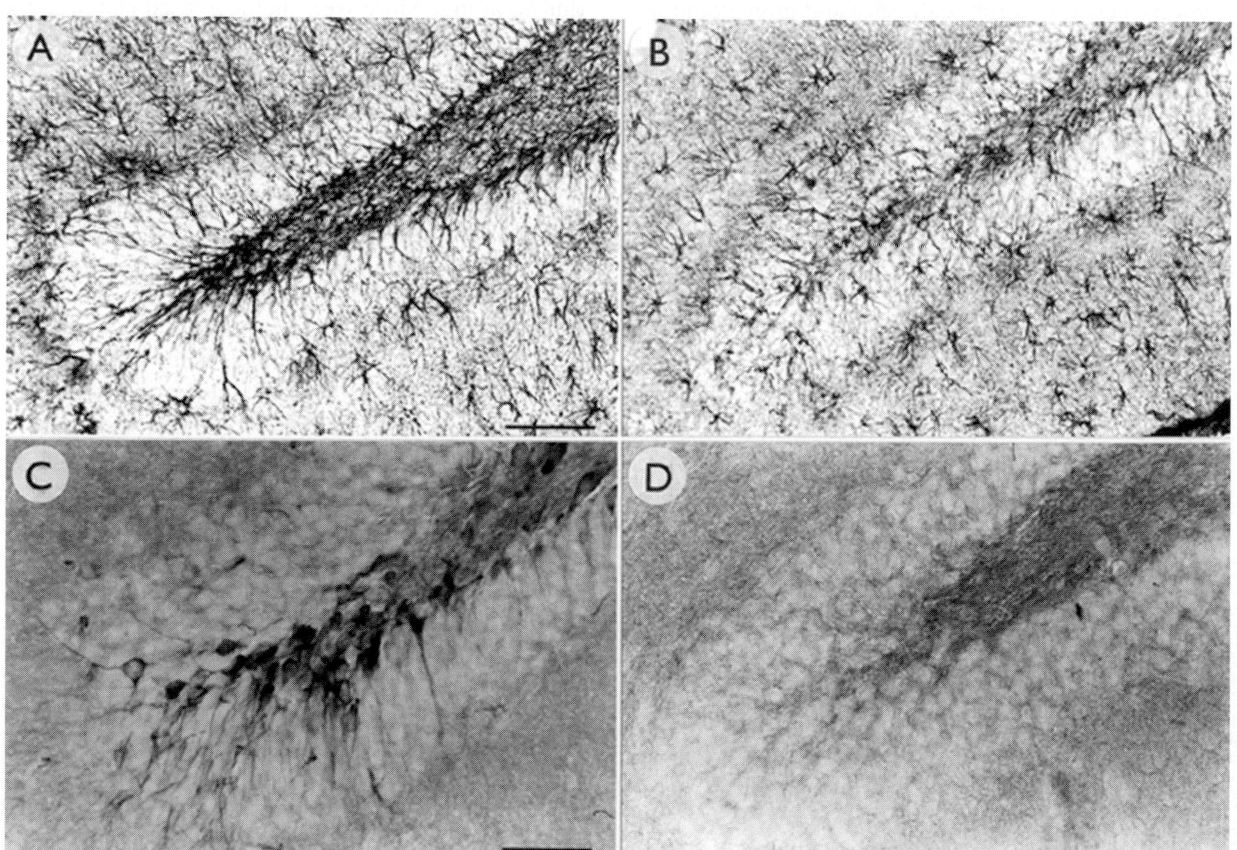

Fig. 10. Structural changes in the hippocampus gyrus dentatus region of aged rats. A: Astrocytes stained for GFAP in a young adult rat; note the radial organization of the astrocytic processes between the neurons (not stained). B: In an aged rat the radial organization of the astrocytic processes is lost. C: Staining for fibronectin in a young adult rat shows densely stained cells, apparently due to perineuronal staining around granular cells. D. In an aged rat the fibronectin staining is lost. Scale bar: in A and B = 100 μm, C and D = 50 μm. (From Syková et al., 1998b, by permission).

transmitters and neuromodulators to regions occupied by their high affinity extrasynaptical receptors, might have crucial importance for the specificity of signal transmission. The importance of anisotropy for the 'spill-over of glutamate', 'cross-talk' between synapses, and for LTP and LTD has been proposed (Kullmann et al., 1996; Asztely et al., 1997). The observed loss of anisotropy in senescent rats could therefore lead to impaired cortical and, particularly, hippocampal function. The decrease in ECS size could be responsible for the greater susceptibility of the aged brain to pathological events such as ischemia, cell death during anesthesia (Syková et al., 1998b), the poorer outcome of clinical therapy and the more limited recovery of affected tissue after insult.

Conclusions

There is increasing evidence that long-term changes in the physical and chemical parameters of the ECS accompany many physiological and pathological states. The 'acute' or relatively fast changes in the size of the intercellular channels are apparently a consequence of cellular (particularly glial) swelling. Abrupt ECS volume decrease may cause cellular or molecular 'crowding' (Zimmerman and Minton, 1993) which can lead to an acute increase in tortuosity. Long-term changes in diffusion would require changes in ECS composition, either permanent changes in the size of the intercellular channels, changes in extracellular matrix molecules, or changes in the number and thickness of cellular processes. Available data suggest that in some pathophysiological states α and λ behave as independent variables. A persistent increase in λ (without a decrease in ECS volume) is always found during astrogliosis and in myelinated tissue, suggesting that glial cells can form diffusion barriers, make the nervous tissue less permissive and play an important role in signal transmission, tissue regeneration and pathological states. This observation has important implications for our understanding of the function of glial cells. The extracellular matrix apparently also contributes to diffusion barriers and to diffusional anisotropy, particularly since its loss, e.g. during aging, correlates with a tortuosity decrease and a loss of anisotropy.

We may conlude that changes in ECS ionic composition, size and geometry may affect: (1) Synaptic transmission (width of synaptic clefts, permeability of ionic channels, concentration of transmitters, dendritic length constant, etc.), (2) Extrasynaptic 'volume' transmission by diffusion (diffusion of factors such as ions, NO, transmitters, neuropeptides, neurohormones, growth factors and metabolites), (3) Neuronal interaction and synchronization, (4) Neuron-glia communication, (5) ECS ionic homeostasis and glial function, (6) Clearance of metabolites and toxic products, (7) Permeability of ionic channels and (8) Regeneration processes. Long-term changes in local CNS architecture including ECS volume and tortuosity can apparently influence plastic changes, LTP or LTD, changes in behaviour and memory formation. Using the methods described in this chapter (not excluding possible new ones) will, in further studies, lead to a fuller understanding of the consequences that altered ECS size, content and geometry may have on nervous system function.

Acknowledgements

Supported by grants VS96–130, GA ČR 307/96/K226, GA ČR 309/97/K048, GA ČR 309/99/0657 and GA ČR 305/99/0655. We would like to thank Dr. M. Pekny and C. Eliasson for providing GFAP − / − mice.

List of abbreviations

ADC	apparent diffusion coefficient
CNS	central nervous system
ECM	extracellular matrix
ECS	extracellular space
GFAP	glial fibrillary acidic protein
HA	hyaluronic acid
HPMA	N-(2-hydroxypropyl)methacrylamide
IOI	integrative optical imaging
IOS	intrinsic optical signals
ISM	ion-selective microelectrode
LTD	long-term depression
LTP	long-term potentiation
NMR	nuclear magnetic resonance
PSA	polysialic acid
TEA	tetraethylammonium

TMA	tetramethylammonium
TW	total water content

References

Agnati, L.F., Zoli, M., Stromberg, I. and Fuxe, K. (1995) Intercellular communication in the brain: wiring versus volume transmission. *Neuroscience*, 69: 711–726.

Anděrová, M., Chvátal, A., Eliasson, C., Pěkný, M. and Syková, E. (1999) Membrane properties and swelling of astrocytes in GFAP–/– mice. *Physiol. Res.*, 48: S52.

Andrew, R.D. and MacVicar, B.A. (1994) Imaging cell volume changes and neuronal excitation in the hippocampal slice. *Neuroscience*, 62: 371–383.

Asztely, F., Erdemli, G. and Kullmann, D.M. (1997) Extrasynaptic glutamate spillover in the hippocampus: dependence on temperature and the role of active glutamate uptake. *Neuron*, 18: 281–293.

Bach-y-Rita, P. (1993) Neurotransmission in the brain by diffusion through the extracellular fluid: a review. *NeuroReport*, 4: 343–350.

Bakay, L. (1970a) The extracellular space in brain tumor. I. Morphological considerations. *Brain*, 93: 693–698.

Bakay, L. (1970b) The extracellular space in brain tumor. II. The sucrose space. *Brain*, 93: 699–707.

Becker, C.G., Artola, A., Gerardy-Schahn, R., Becker, T., Welzl, H. and Schachner, M. (1996) The polysialic acid modification of the neural cell adhesion molecule is involved in spatial learning and hippocampal long-term potentiation. *J. Neurosci. Res.*, 45: 143–152.

Benveniste, H., Hedlund, L.W. and Johnson, G.A. (1992) Mechanism of detection of acute cerebral ischemia in rats by diffusion-weighted magnetic resonance microscopy. *Stroke*, 23: 746–754.

Celio, M.R., Spreafico, R., De Biasi, S. and Vitellaro-Zuccarello, L. (1998) Perineuronal nets: past and present. *Trends Neurosci.*, 21: 510–515.

Chesler, M. (1990) The regulation and modulation of pH in the nervous system. *Prog. Neurobiol.*, 34: 401–427.

Cserr, H.F., De Pasquale, M., Nicholson, C., Patlak, C. Pettigrew, K.D. and Rice, M.E. (1991) Extracellular volume decreases while cell volume is maintained by ion uptake in rat brain during acute hypernatremia. *J. Physiol. (Lond.)*, 442: 277–295.

Dawson, H. and Segal, M.B. (1996) *Physiology of The CFS and Blood–Brain Barriers*. Boca Raton: CRC Press.

Deitmer, J.W. and Rose, C.R. (1996) pH regulation and proton signalling by glial cells. *Prog. Neurobiol.*, 48: 73–103.

Fuxe, K. and Agnati, L.F. (1991) *Volume Transmission in the Brain: Novel Mechanisms for Neural Transmission*. Raven Press, New York.

Hardington, T.E. and Fosang, A.J. (1992) Proteoglycans: many forms and many functions. *FASEB J.*, 6: 861–870.

Hatten, M.E., Liem, R.K. H., Shelanski, M.L. and Mason, C.A. (1991) Astroglia in CNS injury. *Glia*, 4: 233–243.

Jaworski, D.M., Kelly, G.M. and Hockfield, S. (1996) The CNS-specific hyaluronan binding protein, BEHAB, is expressed during periods of glial cell generation and motility. *Sem. Neurosci.*, 8: 391–396.

Jendelová, P. and Syková, E. (1991) Role of glia in K^+ and pH homeostasis in the neonatal rat spinal cord. *Glia*, 4: 56–63.

Johnston, B.M., Patuzzi, R., Syka, J. and Syková, E. (1989) Stimulus-related potassium changes in the organ of Corti of guinea pig. *J. Physiol. (Lond.)*, 408: 77–92.

Kimelberg, H.K. (1991) Swelling and volume control in brain astroglial cells. In: R. Gilles (Ed.), *Advances in Comparative and Environmental Physiology*, Springer-Verlag, Berlin, Heidelberg, pp. 81–117.

Kimelberg, H.K. and Ransom, B.R. (1986) Physiological and pathological aspects of astrocyte swelling. In: S. Federoff and A. Vernadakis (Eds), *Astrocytes: cell biology and pathology of astrocytes*, Academic Press, New York, pp. 129–166.

Kimelberg, H.K., Sankar, P., O'Connor, E.R., Jalonen, T. and Goderie, S.K. (1992) Functional consequences of astrocyte swelling. *Prog. Brain Res.*, 94: 57–68.

Korf, J., Klein, H.C. and Postrema, F. (1988) Increases in striatal and hippocampal impedance and extracellular levels of amino acids by cardiac arrest in freely moving rats. *J. Neurochem.*, 50: 1087–1096.

Križaj, D., Rice, M.E., Wardle, R.A. and Nicholson, C. (1996) Water compartmentalization and extracellular tortuosity after osmotic changes in cerebellum of *Trachemys scripta*. *J. Physiol. (Lond.)*, 492: 887–896.

Kullmann, D.M., Erdemli, G. and Asztely, F. (1996) LTP of AMPA and NMDA receptor-mediated signals: evidence for presynaptic expression and extrasynaptic glutamate spillover. *Neuron*, 17: 461–474.

Latour, L.L., Svoboda, K., Mitra, P.P. and Sotak, C.H. (1994) Time-dependent diffusion of water in a biological model system. *Proc. Natl Acad. Sci. USA*, 91: 1229–1233.

Le Bihan, D., Turner, R. and Douek, P. (1993) Is water diffusion restricted in human brain white matter? An echoplanar NMR imaging study. *NeuroReport*, 4: 887–890.

Lehmenkühler, A., Syková, E., Svoboda, J., Zilles, K. and Nicholson, C. (1993) Extracellular space parameters in the rat neocortex and subcortical white matter during postnatal development determined by diffusion analysis. *Neuroscience*, 55: 339–351.

Lo, W.D., Wolny, A.C., Timan, C., Shin, D. and Hinkle, G.H. (1993) Blood-brain barrier permeability and the brain extracellular space in acute cerebral inflammation. *J. Neurol. Sci.*, 118: 188–193.

Lundbaek, J.A. and Hansen, A.J. (1992) Brain interstitial volume fraction and tortuosity in anoxia. Evaluation of the ion-selective microelectrode method. *Acta Physiol. Scand.*, 146: 473–484.

Margolis, R.K. and Margolis, R.U. (1993) Nervous tissue proteoglycans. *Experientia*, 49: 429–446.

Matsuoka, Y. and Hossmann, K.A. (1982) Cortical impedance and extracellular volume changes following middle cerebral artery occlusion in cats. *J. Cereb. Blood Flow Metab.*, 2: 466–474.

Mazel, T. and Syková, E. (1999) Faster extracellular diffusion in the brain of GFAP–/– mice. *Physiol. Res.*, 48: S95.

Mazel, T., Šimonová, Z. and Syková, E. (1998) Diffusion heterogeneity and anisotropy in rat hippocampus. *NeuroReport*, 9: 1299–1304.

Moseley, M.E., Cohen, Y., Mintorovitch, J., Chileuitt, L., Shimizu, H., Kucharczyk, J., Wendland, M.F. and Weinstein, P.R. (1990) Early detection of regional cerebral ischemia in cats: comparison of diffusion- and T2-weighted MRI and spectroscopy. *Magn. Reson. Med.*, 14: 330–346.

Muller, D., Wang, C., Skibo, G., Toni, N., Cremer, H., Calaora, V., Rougon, G. and Kiss, J.Z. (1996) PSA-NCAM is required for activity-induced synaptic plasticity. *Neuron*, 17: 413–422.

Nicholson, C. (1979) Brain cell microenvironment as a communication channel. In: F.O. Schmitt and F.G. Worden (Eds), *The Neurosciences: Fourth Study Program*, Cambridge, MA: M.I.T. Press. pp. 457–476.

Nicholson, C. (1995) Interaction between diffusion and Michaelis–Menten uptake of dopamine after iontophoresis in striatum. *Biophys. J.*, 68: 1699–1715.

Nicholson, C. and Phillips, J.M. (1981) Ion diffusion modified by tortuosity and volume fraction in the extracellular microenvironment of the rat cerebellum. *J. Physiol. (Lond.)*, 321: 225–257.

Nicholson, C. and Syková, E. (1998) Extracellular space structure revealed by diffusion analysis. *Trends Neurosci.*, 21: 207–215.

Nicholson, C. and Tao, L. (1993) Hindered diffusion of high molecular weight compounds in brain extracellular microenvironment measured with integrative optical imaging. *Biophys. J.*, 65: 2277–2290.

Norris, D.G., Niendorf, T. and Leibfritz, D. (1994) Healthy and infarcted brain tissues studied at short diffusion times: the origins of apparent restriction and the reduction in apparent diffusion coefficient. *NMR in Biomed.*, 7: 304–310.

Norton, W.T., Aquino, D.A., Hosumi, I., Chiu, F.C. and Brosnan, C.F. (1992) Quantitative aspects of reactive gliosis: a review. *Neurochem. Res.*, 17: 877–885.

Pekny, M., Leveen, P., Pekna, M., Eliasson, C., Berthold, C.H., Westermark, B. and Betsholtz, C. (1995) Mice lacking glial fibrillary acidic protein display astrocytes devoid of intermediate filaments but develop and reproduce normally. *EMBO J.*, 14: 1590–1598.

Pérez-Pinzón, M.A., Tao, L. and Nicholson, C. (1995) Extraceullular potassium, volume fraction, and tortuosity in rat hippocampal CA1, CA3 and cortical slices during ischemia. *J. Neurophysiol.*, 74: 565–573.

Prokopová-Kubinová, S. and Syková, E. (in press) Extracellular diffusion parameters in spinal cord and filum terminale of the frog. *J. Neurosci. Res.*

Prokopová, Š., Nicholson, C. and Syková, E. (1996) The effect of 40-kDa or 70-kDa dextran and hyaluronic acid solution on extracelluar space tortuosity in isolated rat spinal cord. *Physiol. Res.*, 45: P28.

Prokopová, Š., Vargová, L. and Syková, E. (1997) Heterogeneous and anisotropic diffusion in the developing rat spinal cord. *NeuroReport*, 8: 3527–3532.

Prokopová, Š., Anděrová, M., Chvátal, A., Elliasson, C., Pekny, M. and Syková, E. (1998) Extracellular space diffusion parameters and membrane properties of astrocytes in GFAP-deficient mice. *Soc. Neurosci. Abstr.*, 24: 56.

Prokopová, Š., Vargová, L., Tao, L., Ulbrich, K., Šubr, V., Nicholson, C. and Syková, E. (1999) Diffusion of polymeric drug carriers in rat cortical slices measured by integrative optical imaging. *Physiol. Res.*, 48: S109.

Rapp, P.R. and Gallagher, M. (1996) Preserved neuron number in the hippocampus of aged rats with spatial learning deficits. *Proc. Natl Acad. Sci. USA*, 93: 9926–9930.

Rasmussen, T., Schliemann, T., Sorensen, J.C., Zimmer, J. and West, M.J. (1996) Memory impaired aged rats: no loss of principal hippocampal and subicular neurons. *Neurobiol. Aging*, 17: 143–147.

Rice, M.E. (2000) Ascorbate regulation and neuroprotective role in the brain. *Trends Neurosci.*, 23: 209–216.

Rice, M.E. and Nicholson, C. (1991) Diffusion characteristics and extracellular volume fraction during normoxia and hypoxia in slices of rat neostriatum. *J. Neurophysiol.*, 65: 264–272.

Rice, M.E., Okada, Y.C. and Nicholson, C. (1993) Anisotropic and heterogeneous diffusion in the turtle cerebellum: Implications for volume transmission. *J. Neurophysiol.*, 70: 2035–2044.

Richter, F., Vargová, L., Mazel, T. and Syková, E. (1999) Extracellular space size and geometry during and after cortical spreading depression in immature and adult rats. *Physiol. Res.*, 48: S112.

Ridet, J.I., Malhotra, S.K., Privat, A. and Gage, F.H. (1997) Reactive astrocytes: cellular and molecular cues to biological function. *Trends Neurosci.*, 20: 570–577.

Roitbak, T. and Syková, E. (1999) Diffusion barriers evoked in the rat cortex by reactive astrogliosis. *Glia*, 28: 40–48.

Šimonová, Z., Svoboda, J., Orkand, R., Bernard, C.C.A., Lassmann, H. and Syková, E. (1996) Changes of extracellular space volume and tortuosity in the spinal cord of Lewis rats with experimental autoimmune encephalomyelitis. *Physiol. Res.*, 45: 11–22.

Singer, W. and Lux, H.D. (1975) Extracellular potassium gradients and visual receptive fields in the cat striate cortex. *Brain Res.*, 96: 378–383.

Svoboda, J. and Syková, E. (1991) Extracellular space volume changes in the rat spinal cord produced by nerve stimulation and peripheral injury. *Brain Res.*, 560: 216–224.

Svoboda, J., Motin, V.G., Hájek, I. and Syková, E. (1988) Increase in extracellular potassium level in rat spinal dorsal horn induced by noxious stimulation and peripheral injury. *Brain Res.*, 458: 97–105.

Syková, E. (1983) Extracellular K^+ accumulation in the central nervous system. *Prog. Biophysiol. Molec. Biol.*, 42: 135–189.

Syková, E. (1987) Modulation of spinal cord transmission by changes in extracellular K^+ activity and extracellular volume. *Can. J. Physiol. Pharmacol.*, 65: 1058–1066.

Syková, E. (1992) *Ionic and volume changes in the microenvironment of nerve and receptor cells*. Springer-Verlag, Heidelberg.

Syková, E. (1997) The extracellular space in the CNS: Its regulation, volume and geometry in normal and pathological neuronal function. *Neuroscientist*, 3: 28–41.

Syková, E. and Svoboda, J. (1990) Extracellular alkaline-acid-alkaline transients in the rat spinal cord evoked by peripheral stimulation. *Brain Res.*, 512: 181–189.

Syková, E., Rothenberg, S. and Krekule, I. (1974) Changes of extracellular potassium concentration during spontaneous activity in the mesencephalic reticular formation of the rat. *Brain Res.*, 79: 333–337.

Syková, E., Kříž, N. and Preis, P. (1983) Elevated extracellular potassium concentration in unstimulated spinal dorsal horns of frogs. *Neurosci. Lett.*, 43: 293–298.

Syková, E., Jendelová, P., Svoboda, J., Sedman,G. and Ng, K.T. (1990) Activity-related rise in extracellular potassium concentration in the brain of 1–3-day-old chicks. *Brain Res. Bull.*, 24: 569–575.

Syková, E., Svoboda, J., Polák, J. and Chvátal, A. (1994) Extracellular volume fraction and diffusion characteristics during progressive ischemia and terminal anoxia in the spinal cord of the rat. *J. Cereb. Blood Flow Metab.*, 14: 301–311.

Syková, E., Svoboda, J., Šimonová, Z., Lehmenkühler, A. and Lassmann, H. (1996) X-irradiation-induced changes in the diffusion parameters of the developing rat brain. *Neuroscience*, 70: 597–612.

Syková, E., Mazel, T., Frisch, C., Šimonová, Z., Hasenöhrl, R.U. and Huston, J.P. (1998a) Spatial memory and diffusion parameters in aged rat cortex, corpus callosum and hippocampus. *Soc. Neurosci. Abstr.*, 24: 1420.

Syková, E., Mazel, T. and Šimonová, Z. (1998b) Diffusion constraints and neuron-glia interaction during aging. *Exp. Gerontol.*, 33: 837–851.

Syková, E., Mazel, T., Roitbak, T. and Šimonová, Z. (1999a) Morphological changes and diffusion barriers in auditory cortex and hippocampus of aged rats. *Assoc. Res. Otolaryngol. Abs.*, 22: 17.

Syková, E., Roitbak, T., Mazel, T., Šimonová, Z. and Harvey, A.R. (1999b) Astrocytes, oligodendroglia, extracellular space volume and geometry in rat fetal brain grafts. *Neuroscience*, 91: 783–798.

Syková, E., Vargová, L., Prokopová, Š. and Šimonová, Z. (1999c) Glial swelling and astrogliosis produce diffusion barriers in the rat spinal cord. *Glia*, 25: 56–70.

Syková, E., Vořišek, I., Tintěra, J., Roitbak, T. and Nicolay, K. (1999d) Water ADC, extracellular space volume and tortuosity in a rat model of injury. *Soc. Neurosci. Abstr.*, 25: 2064.

Tao, L. and Nicholson, C. (1996) Diffusion of albumins in rat cortical slices and relevance to volume transmission. *Neuroscience*, 75: 839–847.

Tao, L., Vořišek, I., Lehmenkühler, A., Syková, E. and Nicholson, C. (1995) Comparison of extracellular tortuosity derived from diffusion of 3 kDa dextran and TMA+ in rat cortical slices. *Soc. Neurosci. Abstr.*, 21: 604.

Thomas, L.B. and Steindler, D.A. (1995) Glial boundaries and scars: programs for normal development and wound healing in the brain. *Neuroscientist*, 1: 142–154.

Van der Toorn, A., Syková, E., Dijkhuizen, R.M., Vořišek, I., Vargová, L., Škobisová, E., Van Lookeren Campagne, M., Reese, T. and Nicolay, K. (1996) Dynamic changes in water ADC, energy metabolism, extracellular space volume, and tortuosity in neonatal rat brain during global ischemia. *Magn. Reson. Med.*, 36: 52–60.

Van Harreveld, A., Dafny, N. and Khattab, F.I. (1971) Effects of calcium on electrical resistance and the extracellular space of cerebral cortex. *Exp. Neurol.*, 31: 358–367.

Vargová, L., Tao, L., Syková, E., Ulbrich, K., Šubr, V. and Nicholson, C. (1998) Diffusion of large polymers in rat cortical slices measured by integrative optical imaging. *J. Physiol. (Lond.)*, 511: 16P.

Vargová, L., Prokopová, Š., Chvátal, A. and Syková, E. (1999) Are the changes in intrinsic optical signals a tool to measure changes in extracellular space volume? *Soc. Neurosci. Abstr.*, 25: 741.

Vořišek, I. and Syková, E. (1997a) Evolution of anisotropic diffusion in the developing rat corpus callosum. *J. Neurophysiol.*, 78: 912–919.

Vořišek, I. and Syková, E. (1997b) Ischemia-induced changes in the extracellular space diffusion parameters, K^+ and pH in the developing rat cortex and corpus callosum. *J. Cereb. Blood Flow Metab.*, 17: 191–203.

Vořišek, I., Roitbak, T., Nicolay, K. and Syková, E. (1999) Water ADC, extracellular space volume and tortuosity in the rat cortex during astrogliosis. *Physiol. Res.*, 48: S136.

Walz, W. (1989) Role of glial cells in the regulation of the brain ion microenvironment. *Prog. Neurobiol.*, 33: 309–333.

West, M.J. (1993) Regionally specific loss of neurons in the aging human hippocampus. *Neurobiol. Aging*, 14: 287–293.

Xie, Y., Zacharias, E., Hoff, P. and Tegtmeier, F. (1995) Ion channel involvment in anoxic depolarisation indused by cardiac arrest in rat brain. *J. Cereb. Blood Flow Metab.*, 15: 587–594.

Yang, P., Yin, X. and Rutishauser, U. (1992) Intercellular space is affected by the polysialic acid content of NCAM. *J. Cell Biol.*, 116: 1487–1496.

Zimmermann, S.B. and Minton, A.P. (1993) Macromolecular crowding: Biochemical, biophysical and physiological consequences. *Annu. Rev. Biophys. Biomol. Struct.*, 22: 27–65.

Zoli, M., Jansson, A., Syková, E., Agnati, L.F. and Fuxe, K. (1999) Intercellular communication in the central nervous system. The emergence of the volume transmission concept and its relevance for neuropsychopharmacology. *Trends Pharmacol. Sci.*, 20: 142–150.

L. F. Agnati, K. Fuxe, C. Nicholson and E. Syková (Eds.)
Progress in Brain Research, Vol 125

CHAPTER 7

Diffusion of radiolabeled dopamine, its metabolites and mannitol in the rat striatum studied by dual-probe microdialysis

Jan Kehr*, Malin Höistad, and Kjell Fuxe

Department of Neuroscience, Karolinska Institute, 171 77 Stockholm, Sweden

Introduction

One of the often studied aspects of volume transmission (VT) involves its spatio-temporal characterization described as a long distance and a long-lasting spread of chemical signals within the brain microenvironment (Agnati et al., 1986; Fuxe and Agnati, 1991). A number of morphological and physiological techniques have shown that neurotransmitters, traditionally believed to act only within the limited space of a synaptic cleft, can diffuse over distances exceeding the volume of the synaptic space. Thus, immunocytochemical analysis at both light and electron microscopic levels has revealed differences between neuroanatomical positions of release sites and their respective receptors or transporter proteins, as well as, the existence of extra-synaptic release and release from non-junctional varicosities of transmitters such as monoamines or acetylcholine and neuromodulators such as neuropeptides or adenosine (for review, see Fuxe and Agnati, 1991; Descarries and Umbriaco, 1995; Zoli et al., 1998; 1999; Vizi and Kiss, 1998). Similarly, using neurotransplantation techniques it was shown that neurotransmitters such as dopamine (DA) can diffuse over long distances out from the grafted tissue (Strömberg et al., 1984).

*Corresponding author. Tel.: +46 8 728 7084; Fax: +46 8 30 28 75; e-mail: Jan.Kehr@neuro.ki.se

Another approach to study VT is based on the use of invasive techniques which allow continuous in vitro or in vivo monitoring of substances involved in chemical neuronal signaling and cellular metabolism (for review, see Kehr, 1999). The principle of such measurements is the same for all types of sensing or sampling devices: a sensor/probe is stereotaxically implanted into the brain parenchyma, fixed and left to stabilize for a certain period of time (hours to days). Once the steady-state conditions are achieved, the device can generate physical (electrical, optical) signals which reflect the extracellular concentrations of a given analyte, first at 'basal' conditions, followed by a period of pharmacological or physiological stimulation. A major advantage of using the invasive techniques to study VT is a possibility to elucidate mechanisms regulating the properties of the extracellular space in health and disease and the role of extracellular molecular transport for design of novel drug therapies.

Microdialysis (Ungerstedt, 1984) has been proven to serve as an efficient tool in studies of brain circuitry and mechanisms of drug actions in vivo. Several mathematical models of diffusion and empirical methods were proposed for estimation of absolute concentrations of extracellular substances (for review, see Kehr, 1993). The principle of microdialysis sampling is based on the diffusion of substances across the concentration gradient in the extracellular space and as such, the technique indirectly implies the concept of VT. In spite of that fact, microdialysis has only sparsely been applied in the studies of VT, as compared to the techniques based directly on sensing devices such as voltammetry on carbon fiber electrodes or potentiometry using the ion-selective microelectrodes. A relatively large surface of the microdialysis probe makes it difficult to combine microdialysis with pressure-ejection technique, which is often used with the other two methods as a tool to administer diffusable substances. On the other hand, microdialysis offers a unique possibility to measure several compounds at a time using analytical techniques, such as high-performance liquid chromatography (HPLC), radioimmunoassay or mass spectrometry. Today, microdialysis is probably the most efficient in vivo technique to study brain chemical signaling and to correlate neurochemical analysis with behavioral methods (Zetterström et al., 1986; Young, 1993; Ögren et al., 1996).

Concentration gradients within the extracellular space

The extracellular space (ECS) under normal circumstances comprises about 20% of the total brain tissue volume. The extracellular fluid (ECF) contains ions, the concentrations of which are very similar to those found in the cerebrospinal fluid (CSF), as well as a number of long-chain glycosaminoglycans, proteoglycans and glycoproteins tethered to membranes. Neurotransmitters and neuromodulators and also neurotrophic factors and other cytokines in glial and neuronal cells, once released into the ECF, all have to traverse the extracellular space on their way to the target receptors. However, the extracellular concentrations of these chemical messengers are far away from being homogeneously distributed within the ECS compartment. For example, for neurotransmitters such as DA, there is a steep gradient of concentrations being up to 6–7 orders of magnitude higher at a release site (a synaptic cleft) than in the outer interstitial space. It was estimated that striatal DA vesicles with a radius of 25 nm contain 25 mM DA, while the concentration of DA in the synaptic cleft, which is about 15 nm wide and 300 nm long, was around 1.6 mM (Garris et al., 1994). However, the actual extracellular DA concentrations measured by in vivo voltammetry using a thin (5 μm O.D.) carbon fiber electrode was not higher than 20 nM in pargyline-treated rats (Gonon and Buda, 1985) or 0.1–2 μM during electrical stimulation (Kawagoe et al., 1992). Using the devices of even larger diameter such as a 200 μm O.D. microdialysis probe the calculated basal extracellular DA concentration was as low as 5 nM (Parsons and Justice, 1992). Thus, implantable sensors can measure only an 'echo' of the actual release event, whether defined as a spill-over from the synaptic cleft (a leaking synapse) or as a release from a non-junctional bouton, varicosity or as extra-synaptic release. All these definitions are based on measurements of extracellular chemical signals by invasive means and have a common prerequisite: the overflow of the substance diffusing to the sensing device must be higher than its reuptake, sequestration, clearance to blood or CSF, and enzymatic or other inactivation mechanisms.

Invasive techniques for monitoring volume transmission: role of microdialysis

The spatial and temporal resolution of any implantable monitoring device is dependent both on its geometry and the applied detection principle. The invasive techniques can be divided into two main groups: (1) intracorporeal biosensors providing immediate measures of a given analyte and (2) continuous sampling devices generating samples in a form of extracts/perfusates which are further analyzed by an appropriate analytical technique. The response time can be as low as 0.1–1 s for directly detecting voltammetric electrodes, while a relatively laborious and time-consuming analysis of neurotransmitters or neuropeptides sampled by microdialysis requires the fractions to be collected in 5–30 min intervals. The antibody-coated microprobes as described by Duggan and Hendry, 1986, provide excellent spatial resolution but only one time point of about 30 min.

The sensing devices and related detection techniques within the first group are highly selective towards one analyte, which is detected directly at the surface of the biosensor implanted into the brain tissue. In the simplest case, the selectivity for a detected endogenous substance is assured by its intrinsic chemical properties such as a relatively specific oxidation potential for DA. A more common way is to utilize the highly specific molecular interactions such as those between ions-ionophores, enzymes-substrates and antibodies-antigens, which all lead, either directly or via an intermediate product, to the changes in physical (electrical or optical) signals.

Contrary to most of the biosensors, the implantable sampling devices allow relatively easy intracerebral monitoring of conscious freely moving animals. The sampling device, for example a microdialysis probe is continuously perfused at a low flow-rate (0.1–2 μl/min) with a physiological solution, typically a Ringer solution or an artificial cerebrospinal fluid (aCSF). In vivo microdialysis on awake animals eliminates the negative effects of general anesthetics on chemical neurotransmission and cell metabolism. A complete recovery of physiological functions occurs at about 5–7 days after the surgery as revealed by telemetry (Drijfhout et al., 1995).

A very useful way to study functional neuroanatomy and the role of specific brain circuits in various behaviors is the so-called dual-probe. Here, one probe is implanted at the cell body level, whereas the second probe is implanted in the terminal area. The first probe is used for chemical stimulation by infusing the drugs while the second probe is used to measure the neurotransmitter release (for review, see Westerink et al., 1998). Here, we describe a newly developed method for studies of long-distance diffusion in the extracellular space. The technique combines dual-probe microdialysis sampling which allows simultaneous infusion and recovery of labeled molecules. The collected samples are separated by HPLC and individual fractions measured by a liquid scintillation counter. Inert markers such as ^{3}H-mannitol can be used to study direct changes of extracellular volume characteristics. Endogenous compounds (e.g. ^{3}H-DA) provide measures of brain permeability, as well as on the rate of cellular uptake and metabolism.

As already mentioned in the Introduction, microdialysis sampling builds on the ability of molecules to diffuse through the ECS; the length of the diffusion path for various molecules has been a matter of intensive research. For example, using voltammetric techniques it was calculated that the half-life and diffusion distance of stimulated DA release from its release site is only 25 ms and 7–10 μm respectively (for review, see Gonon et al., this book). On the contrary, long distance DA diffusion paths were proposed on the basis of immunohistochemical (Fos-like), electrophysiological and microdialysis studies conducted in unilaterally DA denervated rats, following e.g. *d*-amphetamine treatment (Bjelke et al., 1994; Schneider et al., 1994). A possibility of long-distance diffusion arising from non-stimulated DA in the rat striatum could be elegantly investigated by use of a dual-probe microdialysis approach as schematically depicted in Fig. 1.

Briefly, two microdialysis probes (CMA/11, cuprophane membrane: cut-off 6000 Da, length 4 mm and 240 μm O.D.) were implanted stereotaxically into the lateral and medial striatum (Paxinos and Watson, 1986) of the halothane anaesthetized

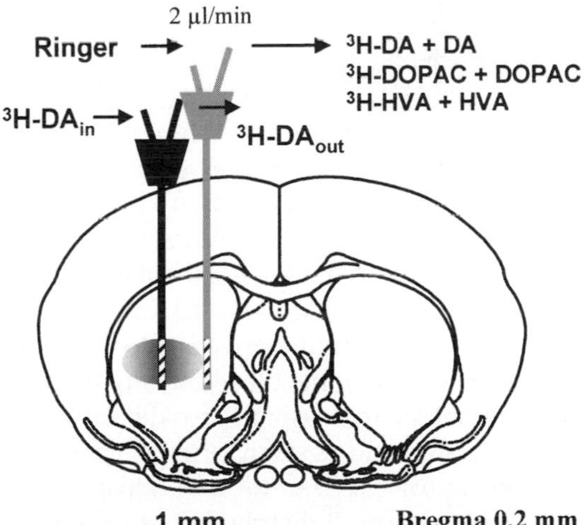

Fig. 1. Principal scheme of a dual-probe microdialysis used for in vivo studies of long-distance diffusion and metabolism of radiolabeled molecules. The first (lateral) microdialysis probe was used for the infusion of radioactive label (e.g. ^{3}H-DA or ^{3}H-mannitol), whereas the second probe, implanted 1 mm apart, recovered all the low-molecular weight substances present in the extracellular fluid including the diffused ^{3}H-label or its metabolites. The samples were separated by microbore LCEC and fractions corresponding to the peaks of DA and metabolites were collected and measured in a liquid scintillation counter.

rat. The distance between the membrane centers was 1 mm. The lateral probe was used for the infusion of a radioactive label: ^{3}H-DA or D-^{3}H-mannitol diluted in Ringer solution, at a flow-rate of 2 μl/min. The medial probe, perfused with Ringer solution at the same flow-rate was used for recovery of endogenous (cold) DA and its metabolites, as well as the tritiated labels. Samples were collected at 30-min intervals. Since ^{3}H-mannitol was used as an inert marker, which did not metabolize during its passage through the brain, its radioactivity could be measured directly by use of a liquid scintillation counter. On the contrary, infusion of ^{3}H-DA caused its immediate uptake and enzymatic cleavage to its respective acidic metabolites ^{3}H-DOPAC and ^{3}H-HVA. Thus, measuring the total tritium in samples of ^{3}H-DA infused animals could provide an in vivo index of: (a) DA metabolism, (b) clearance by cerebral blood vessels and (c) uptake into the nerve terminals. To investigate the partition of radioactivity between individual substances, the microdialysis samples were first separated by microbore column liquid chromatography with electrochemical detection (LCEC), as described elsewhere (Kehr, 1999). Fractions corresponding to the individual peaks of DA and its metabolites were collected and measured by a scintillation counter. The clearance of tritiated labels by cerebral microcirculation can be studied by measuring the radioactivity in the blood samples. However, in our experiments we could not detect any radioactivity in the blood collected from the tail vein or intracardially, most likely due to a massive dilution of 'brain-derived' blood at this sampling site. Sampling the blood directly from the jugular vein and/or increasing the radioactivity of the infused label could possibly provide detectable levels. Uptake of ^{3}H-DA to the dopaminergic terminals and the overall distribution of radioactivity within the brain tissue can be studied by using autoradiography of respective brain slices. In fact, the concept of tissue pre-loading, well known from in vitro superfusion techniques, was already demonstrated in the first report on dopamine microdialysis by Ungerstedt and Pycock in 1974. A single probe was used to load ^{3}H-DA into the dopaminergic terminals and following a washout period, the effect of amphetamine on the release of DA-related radioactivity could be observed. Here, the density of ^{3}H-DA labeling is evaluated histologically for the first time. The diffusion profile of ^{3}H-DA infused via the microdialysis probe in the rat striatum and the analysis of relative tritium density as a function of distance from the probe is shown in Fig. 2. As seen, following 5 hours of continuous infusion, the radioactivity profile shows a steep decay within the first 500 μm and being less than 5% of the initial level at a distance of 1 mm, i.e. at the site of the second (detector) probe (not seen in the phospho-imager print of the 10 μm section).

Interestingly, the infusion of ^{3}H-mannitol at similar tritium activity concentrations and experimental conditions as for ^{3}H-DA resulted in an almost undetectable radioactive trace of mannitol following 5 days' exposure of the brain slice in a phospho-imager (data not shown). Omitting the brain fixation procedure by intracardial perfusion did not increase the tissue radioactivity of infused ^{3}H-mannitol. This led to the assumption that infused ^{3}H-mannitol is a truly extracellular marker of brain microenvironment being rapidly cleared by the cerebral microcirculation lacking any significant incorporation to the intracellular compartments.

^{3}H-mannitol as a marker of extracellular space

The usefulness of dual-probe microdialysis method to study diffusion of neurotransmitters within the ECS was initially evaluated by use of the inert low-molecular weight markers such as ^{3}H-mannitol. Such labels should exhibit fast kinetics of diffusion and equilibration within a typical time frame of an acute microdialysis experiment (6–8 hours including the initial stabilization phase of 90 min). Indeed, as shown in Fig. 3, the recovery of ^{3}H-mannitol (1.5 μM; specific activity 736.3 GBq/mmol, NEN, USA) reached the steady state within 90 min after start of the infusion, with a corresponding half-time ($t_{50\%}$) of 42 ± 1.6 min (mean ± SD, $n = 6$).

From the total ^{3}H-mannitol activity of $820\,000 \pm 13\,000$ dpm/10 μl only about 3% was delivered (lost) into the brain tissue. The curves were calculated using linear regression analysis or Boltzmann sigmoidal nonlinear regression algo-

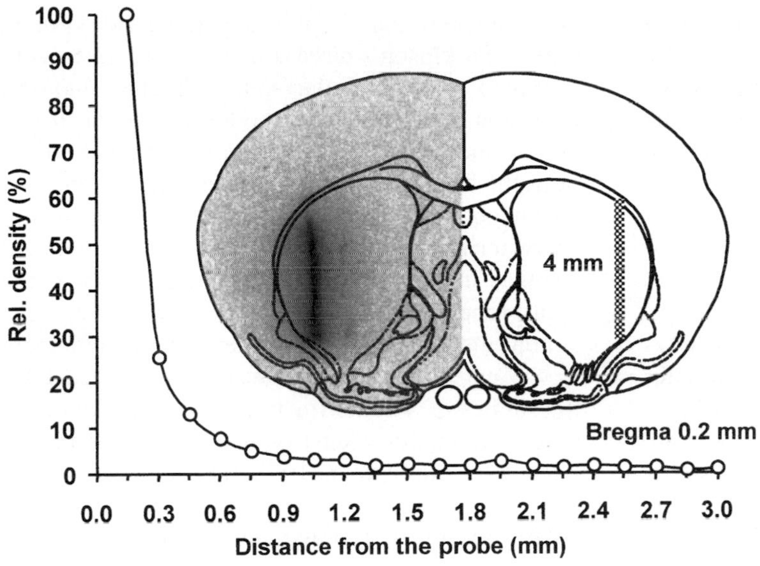

Fig. 2. Diffusion profile of ^{3}H-DA infused for five hours through the microdialysis probe into the rat striatum as a function of a distance from the probe surface. The placement of the infusion probe is schematically depicted on the contralateral side. The detector (medial) probe is not shown.

rithm (Prism™, GraphPad Software, USA), according to the following formula:

$$c = c_o + (c_o - c_{ss})/\{1 + \exp[(t_{50\%} - t)/k_{ss}]\}$$

where c, c_{ss} and c_o are the concentrations of ^{3}H-mannitol at any time (t) after start of the infusion, at steady state and the background counts, respectively. The coefficient $t_{50\%}$ represent the time needed to reach half of the steady state level and and k_{ss} is the empirical constant inversely related to the equilibration rate of diffusion (the steepness of the curve), which was calculated as 14 ± 2 min. A

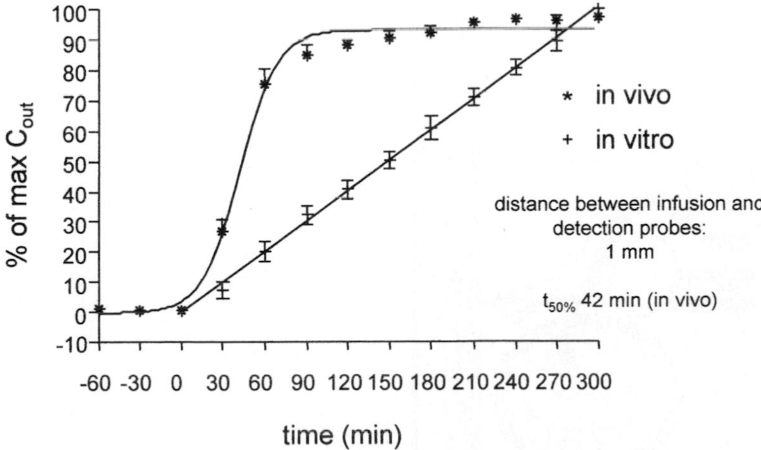

Fig. 3. Diffusion profiles of ^{3}H-mannitol in vivo (rat striatum) and in vitro (Ringer solution) using the dual-probe microdialysis. Under in vivo conditions, the steady state was achieved within 90 min after start of the infusion, as demonstrated by the regression curve and the calculated time required to reach half of the steady state ($t_{50\%}$) which was 42 ± 1.6 min. This indicates the presence of an active clearance processes of ^{3}H-mannitol in the ECF. No such equilibrium was seen in a quiescent Ringer solution. Values expressed as percentage of maximal recovered ^{3}H-mannitol concentration at steady state (max c_{out}), means ± S.E.M., $n = 6$ animals.

corresponding dual-probe in vitro experiment in Ringer solution at +24°C and 2 μl/min showed a constant linear increase of recovered radioactivity, indicating non-saturable conditions during the study period (5 hours). The fast equilibrium of mannitol diffusion in the brain together with low background radioactivity suggest that the clearance by blood brain capillaries is the major inactivating mechanism for this molecule. This situation is schematically depicted in Fig. 4.

The diffusion curve of ^{3}H-mannitol in 6-hydroxydopamine (6-OHDA) lesioned rats 5 weeks after the unilateral lesions was much faster showing a significantly shorter half-time $t_{50\%}$ (33 ± 1 min, $P<0.03$) and a significant change in curve steepness (rate constant k_{ss} 6 ± 2 min, $P<0.05$) vs. the control group. The maximal outflow levels at steady state conditions were about the same in both groups (3600 dpm/10 μl). This is in agreement with histochemical data, showing that the disappearance of dopaminergic terminals has only moderate effects on gliosis as evaluated by GFAP immunostaining at this time-interval (Strömberg et al., 1986).

Our data on facilitated diffusion of ^{3}H-mannitol in DA denervated striatum of a rat suggests that the neurodegeneration of the dopaminergic system, seen in Parkinson's disease may be associated with a change of extracellular volume fraction and/or a reduction in tortuosity of the basal ganglia. It can be concluded that under given experimental conditions the kinetic profile of recovered ^{3}H-mannitol is dependent only on the volume fraction/tortuosity characteristics of the ECS and with no significant changes in the brain-blood clearance. Finally, the dual-probe microdialysis using ^{3}H-mannitol or a similar marker could be an interesting complement to the existing electrochemical methods used for the studies of volume fraction and tortuosity of the brain extracellular environment (for review, see Nicholson and Syková, 1998; this volume).

Diffusion and metabolism of ^{3}H-dopamine

Infusion of ^{3}H-DA (specific activity 2187.1 GBq/mmol, NEN, USA) at a concentration of 500 nM (corresponds to the mean value of 710 000 ± 20 000 dpm/10 μl) through the microdialysis probe resulted in an average delivery of about 6% which corresponds to in vivo delivery of 30 nM DA. This is only 5–6 times higher than the basal extracellular DA levels estimated by zero-flow and no-net-flux

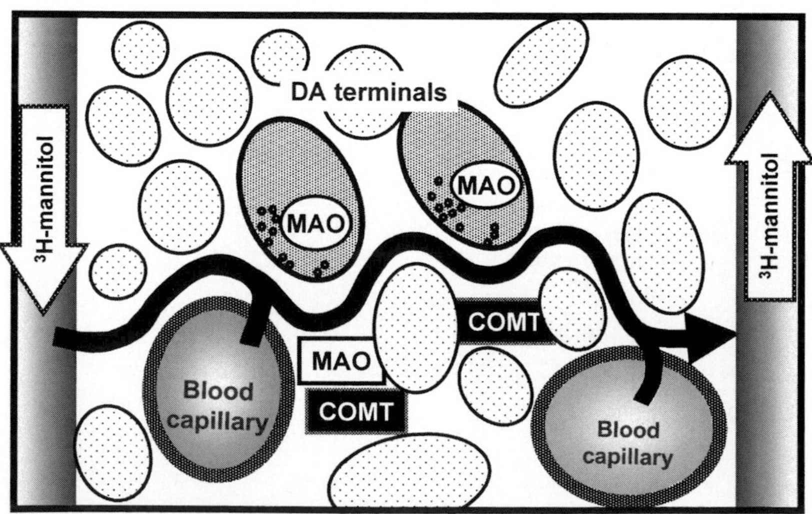

Fig. 4. A scheme of the diffusion path of ^{3}H-mannitol, in the rat striatum. At steady state, ^{3}H-mannitol diffusing between the two probes, was distributed only within the ECS compartment and cleared by cerebral microvessel circuitry. Signifcantly faster diffusion of ^{3}H-mannitol was observed in rats with loss of DA terminals induced by 6-OHDA injections into the substantia nigra. Thus, ^{3}H-mannitol may serve as an inert marker of extracellular volume fraction and tortuosity.

microdialysis techniques (Parsons and Justice, 1992). In addition, this concentration of DA should still lie within the physiological range of stimulated DA release and as such should not affect the normal function of DA release, autoreceptor feedback and uptake machinery. Under these conditions, the total tritium radioactivity recovered by the second, detector probe was only about 5000 dpm/10 μl in control animals, which means that only 12.5% of delivered ^{3}H-DA was recovered by the second probe. However, this value is not corrected for the in vivo recovery of the second microdialysis probe. Calculating the $t_{50\%}$ and k_{ss} parameters for the total recovered radioactivity gave the average values of 100 ± 4 min and 64 ± 3 min, respectively, i.e. the levels are about twice higher than those obtained with ^{3}H-mannitol. Separation of microdialysis samples by microbore LCEC and subsequent counting of fractions representing DA, DOPAC, HVA and an unidentified compound eluted in the front peak revealed that DOPAC was the fastest diffusing compound ($t_{50\%}$ 85 ± 4 min, k_{ss} 20 ± 1 min; means ± S.E.M., $n = 6$) and contained the largest portion of radioactivity (1634 ± 234 dpm/peak) followed by HVA (1343 ± 159 dpm/peak) and the unidentified front peak (850 ± 115 dpm/peak), as shown in Fig. 5. Again, a relatively fast steady state could be achieved for all the substances within 120–180 min.

The initial hypothesis of long-distance diffusion of striatal DA at basal conditions was not confirmed in this experimental model. The levels of recovered ^{3}H-DA peak at steady-state were only slightly elevated to 45.2 ± 1.4 dpm/10 μl (mean ± SEM, range 42–48 dpm/10 μl) over the threshold values which were 31.3 ± 1.3 dpm/10 μl (range 21–41 dpm/10 μl). On the contrary, the basal levels of endogenous (cold) DA in control animals were easily detectable with a mean value of 6.2 nM in 10 μl injected onto the HPLC column. Neither, could ^{3}H-DA be detected in 6-OHDA lesioned animals where the intraneuronal pools of DA terminal monoamine oxidase (MAO) and the dopamine uptake transporter were reduced to a minimum due to an almost complete disappearance of DA-ergic striatal terminals. Here, the counts in fractions of DA peaks were in a similar range as in the non-lesioned animals, i.e. about 35 and 43 dpm/10 μl in pre-infused and ^{3}H-DA infused periods. Surprisingly, the overall recovered ^{3}H-DA-derived radioactivity in the lesioned group was significantly reduced to 53% of the control group, from 4919 ± 530 dpm/10 μl ($n = 6$) to 2609 ± 205 dpm/10 μl ($n = 5$) respectively. The distribution of DOPAC, HVA and the unidentified front peak in control and 6-OHDA lesioned groups is schematically depicted in Fig. 6A. The absolute level of radioactivity in the front peak was unaffected by

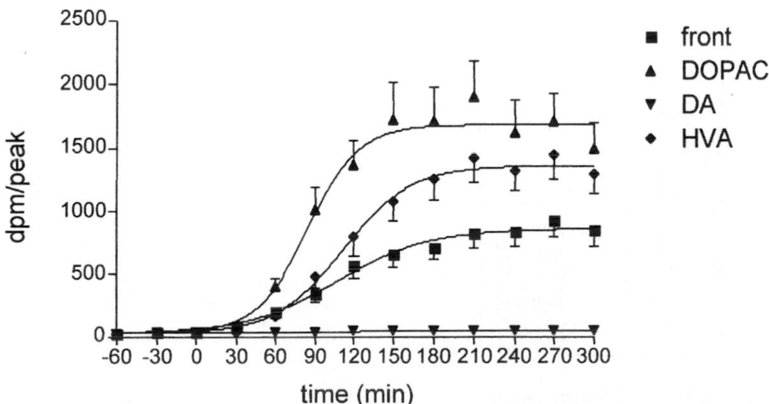

Fig. 5. Distribution, kinetics and steady state levels of infused ^{3}H-DA, its metabolites ^{3}H-DOPAC, ^{3}H-HVA and an unidentified ^{3}H-front peak in the striatum of control anesthetized rats. 30 min samples were separated by LCEC and fractions corresponding to the individual peaks were collected and their radioactivity was measured in a liquid scintillation counter. The diffusion curves were calculated using Boltzmanns sigmoidal non-linear regression analysis (Prism™, GraphPad). Values expressed as means ± S.E.M., n = 6 animals.

the 6-OHDA lesion, similar to the case of the inert marker ^{3}H-mannitol as discussed above. These data indicate that the in vivo delivery and recovery of the two probes was not affected by the 6-OHDA lesion. However, the steady-state levels of metabolites ^{3}H-DOPAC and ^{3}H-HVA were significantly

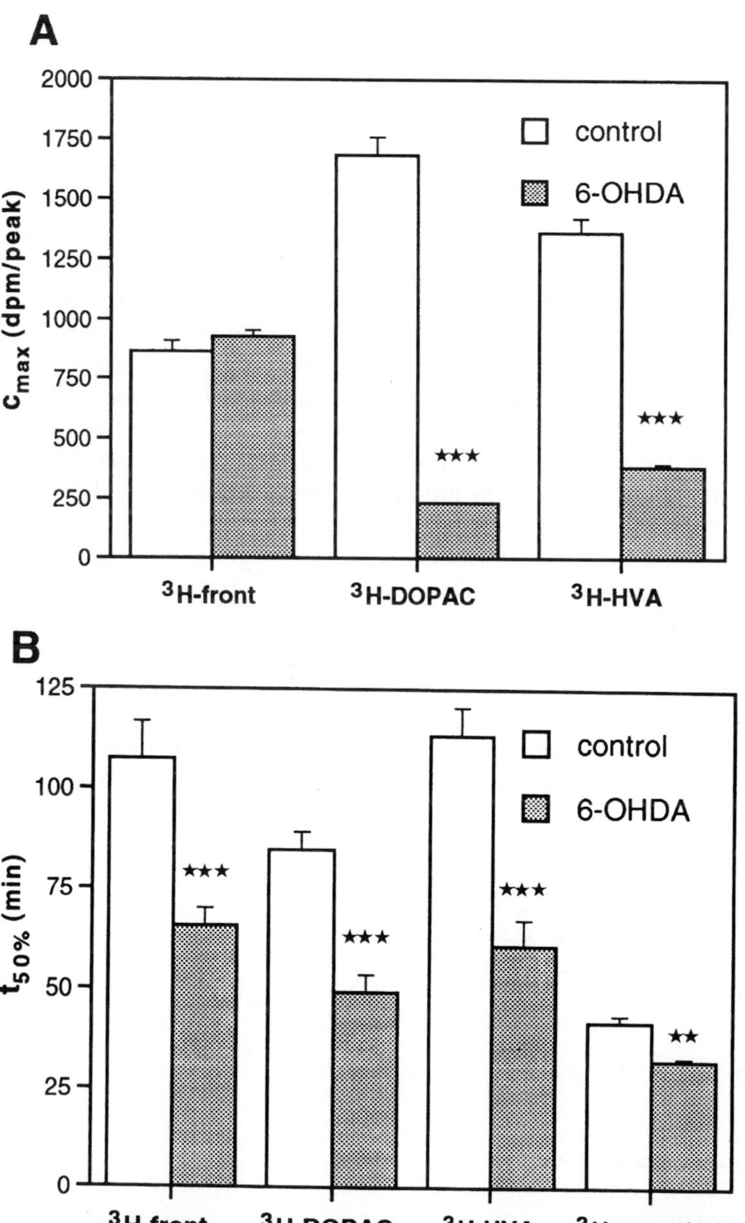

Fig. 6. (A) Maximal levels (c_{max}) of tritiated molecules recovered by a second microdialysis probe at steady state in control ($n = 6$) and 6-OHDA lesioned ($n = 5$) rats. (B) Calculated half-time values ($t_{50\%}$) of tritiated DA metabolites and ^{3}H-mannitol in control and 6-OHDA lesioned rats. (C) Calculated diffusion slope values ($t_{50\%}$) of tritiated DA metabolites and ^{3}H-mannitol in control and 6-OHDA lesioned rats. Statistical analysis was done by Studentś *t*-test, comparing control vs. lesioned groups, values are expressed as means ± S.E.M., *: $P < 0.05$; **: $P < 0.001$; ***: $P < 0.0003$.

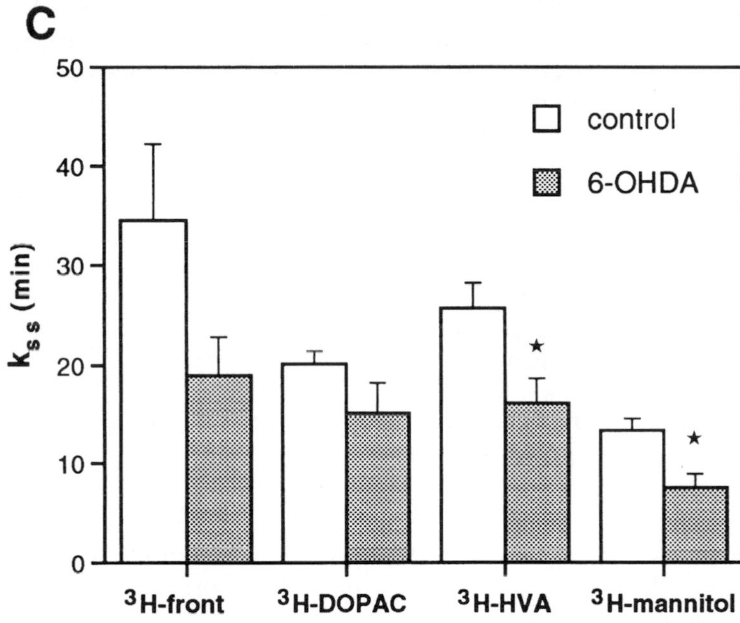

Fig. 6. Continued

reduced in the 6-OHDA lesioned animals, with maximal reduction for DOPAC down to 14% of control, followed by HVA (28% of controls). This indicates that the striatal MAO activity which is localized predominantly in the dopaminergic terminals (Oreland, 1991) was considerably reduced as a consequence of the 6-OHDA lesion. On the other hand, the slightly less pronounced reduction of HVA suggests that this reduction was caused mostly as a consequence of a diminished pool of DOPAC rather than from reduced activity of extracellular COMT.

The calculated kinetic data $t_{50\%}$ and k_{ss} for ^{3}H-DOPAC, ^{3}H-HVA, the unidentified ^{3}H-DA-derived front peak and ^{3}H-mannitol in control and 6-OHDA lesioned groups are shown in Fig. 6B, C.

As seen in Fig. 6B the diffusion half times were significantly reduced by about 40% for all ^{3}H-DA-related peaks, whereas for ^{3}H-mannitol, the corresponding reduction was only 23%. It is notable that ^{3}H-mannitol is cleared much faster under normal conditions than any of the ^{3}H-DA-related compounds in the 6-OHDA lesioned animals. Similarly, the diffusion curves were steeper in the lesioned animals as expressed mathematically by calculating the diffusion rate k_{ss} values using the Boltzmanns non-linear regression (Fig. 6C). Here, again the k_{ss} levels were about 40% lower in the 6-OHDA lesioned group, although only the values for ^{3}H-mannitol (see discussion to Figs. 3, 4) and ^{3}H-HVA were significantly different (controls: 25.7 ± 2.5 min, $n=5$, 6-OHDA: 16.1 ± 2.6 min, $n=6$, $P<0.05$). A relatively large variation of the mean levels could be explained by poor temporal resolution, i.e. too long (30 min) sampling intervals. Interestingly, there was no difference in maximal radioactivity counts between the control and lesioned animals for ^{3}H-mannitol and the ^{3}H-front peak, whereas ^{3}H-DOPAC and ^{3}H-HVA levels were strongly reduced in the lesioned group (see Fig. 6A). This suggests that in the lesioned animals, infused ^{3}H-DA was, in spite of the diminished MAO and DA reuptake activity, unable to diffuse over the longer distances, most likely due to the activation of compensatory mechanisms or the increased brain-blood clearance. Also, it should be considered that ^{3}H-DA itself may exhibit an increased clearance over the brain-blood barrier after the DA denervation. Limited amounts of data are available on the effects of 6-OHDA lesions on the cerebral blood flow (CBF) and on the cerebromicrovascular permeability. Earlier studies have shown a two fold increase of CBF in DA denervated rats following stimulation with apomorphine (Ingvar et al., 1983) probably as a consequence of dopamine receptor supersensitivity. The existence of both D1 and D2 receptors was demonstrated on rat cerebral blood vessels (Amenta et al., 1991). These receptors mediate DA-induced vasodilatation associated with increases in CBF (Sharkey and McCulloch, 1986). In agreement with these data, a reduction of basal CBF was observed in DA denervated caudate-putamen (Mraovitch et al., 1993). Recently, it was reported that DA may act as a vasoconstrictor on cortical microvessels (Krimer et al., 1998). However, this effect is probably mediated via agonist activity of DA at adrenergic and serotonergic receptors as discussed by Iadecola, 1998. Also DA terminals contact cerebral endothelial cells indicating a modulation of brain-blood clearance processes of extracellular chemical signals through DA receptors (Krimer et al., 1998).

In view of these reports, we hypothesize that the infusion of ^{3}H-DA into the DA denervated rat striatum may cause an immediate increase of local CBF and capillary permeability augmenting the clearance processes of extacellular ^{3}H-DA and its metabolites as manifested by their reduced extracellular levels and the inability to measure the long-distance diffusion of ^{3}H-DA. Infusion of ^{3}H-mannitol, on the other hand, may not affect CBF nor capillary permeability and consequently no differences at steady state were observed in control and lesioned animals. These results provide new insights into DA VT in the DA denervated striatum and thus in the striatum of Parkinsonś disease patients.

Dual-probe microdialysis – a new tool for diffusion studies

A scheme shown in Fig. 7 summarizes the proposed diffusion path and metabolism of ^{3}H-DA as studied by the dual-probe microdialysis technique.

In control non-lesioned rats, infused ^{3}H-DA is removed from the ECS compartment by three major mechanisms: (a) uptake into the DA-ergic terminals, (b) brain-blood clearance, and (c) enzymatic cleavage by extracellular COMT and mainly DA terminal MAO. Hence, the appearance of

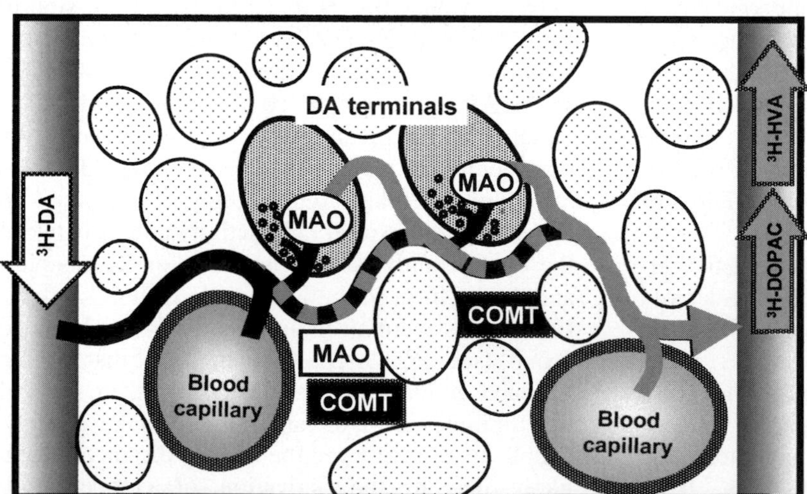

Fig. 7. Scheme of the diffusion path of ^{3}H-DA and formation of its metabolites ^{3}H-DOPAC and ^{3}H-HVA in control rat striatum. At steady state, ^{3}H-DA was cleared by blood capillaries, uptake into the dopaminergic terminals and metabolism by both intracellular and extracellular enzymes. In DA denervated animals, the formation of ^{3}H-DOPAC and ^{3}H-HVA was strongly attenuated due to the absence of DA terminals being the major source of MAO activity for DA metabolism. Consequently, the portion of recovered ^{3}H-HVA formed through extracellular COMT was now higher than ^{3}H-DOPAC (see Fig. 6A).

^{3}H-DOPAC and ^{3}H-HVA metabolites and ultimately of ^{3}H-DA in the perfusates sampled by the second microdialysis probe should reflect the ratio between these three factors. This combined metabolic-diffusion process is much slower than the diffusion of a labeled inert marker shown in Fig. 4, as evidenced by significantly longer diffusion half times of ^{3}H-DOPAC and ^{3}H-HVA compared to ^{3}H-mannitol. It is suggested that for the DA metabolites there exist in addition to the diffusion characteristics through the ECS compartment also contributions of trans-cellular transport involving, e.g. the brain-blood barrier and the activity of the corresponding metabolizing enzymes. Removal of the DA terminals by lesions with 6-OHDA resulted in a dramatic reduction of the formation of ^{3}H-DA metabolites, but only marginally affected their diffusion speed with no detectable facilitation of ^{3}H-DA diffusion. On the contrary, the inert marker ^{3}H-mannitol diffused significantly faster in the lesioned animals than in the controls. In summary, the dual-probe microdialysis approach shows a clear difference in diffusion properties of the neutral molecules used only as markers of ECS and those endogenously synthesized or metabolized by the brain tissue. In a pathological state, the marker molecules seem to mirror only the 'geometrical' changes of the brain microenvironment (tortuosity, volume fraction), whereas diffusion of the endogenously-derived molecules is affected by all the compensatory mechanisms occurring in the damaged tissue including the brain-blood barrier and cerebral blood flow.

Conclusions

The intracerebral dual-probe microdialysis technique allows the study of diffusion of labeled molecules in vivo in anaesthetized or awake animals. The labeled markers can be biologically inert molecules or endogenous substances delivered and sampled without any volume changes or other disturbances to brain homeostasis.

Diffusion kinetics and steady-state levels of DA metabolites could be studied over long (≥ 1 mm) inter-probe distances, whereas studies of DA diffusion should be conducted at shorter (~ 0.1 mm) distances. The dual-probe microdialysis technique can provide an in vivo index of a complex interplay of clearance and metabolic processes of extracellular labels at basal, non-stimulated conditions and as such, the technique can be a complement to the existing voltammetric and potentiometric methods. Finally, the dual-probe technique can be used to study molecular transport within the brain microenvironment in various neuropathological models of brain diseases or following pharmacological stimuli, helping to understand the role of VT at such conditions.

References

Amenta, F., Ricci, A. and Vega, J.A. (1991) Autoradiographic localization of dopamine receptors in rat cerebral blood vessels. *Eur.J. Pharmacol.*, 192: 123–132.

Agnati, L.F., Fuxe, K., Zoli, M., Zini, I., Toffano, G. and Ferraguti, A. (1986) A correlation analysis of the regional distribution of central enkephalin and beta-endorphin immunoreactive terminals and of opiate receptors in adult and old male rats. Evidence for the existence of two main types of communication in the central nervous system: the volume transmission and the wiring transmission. *Acta Physiol. Scand.*, 128: 201–207.

Bjelke, B., Strömberg, I., O'Connor W.T., Andbjer, B., Agnati, L.F. and Fuxe, K. (1994) Evidence for volume transmission in the dopamine denervated neostriatum of the rat after a unilateral nigral 6-OHDA microinjection. Studies with systemic *d*-amphetamine treatment. *Brain Res.*, 662: 11–24.

Descarries, L., Beaudet, A. and Watkins, K.C. (1975) Serotonin nerve terminals in adult rat neocortex. *Brain Res.*, 100: 563–588.

Descarries, L. and Umbriaco, D. (1995) Ultrastructural basis of monoamine and acetylcholine function in CNS. *Sem. Neurosci.*, 7: 309–318.

Drijfhout, W.J., Kemper, R.H., Meerlo, P., Koolhaas, J.M., Grol, C.J. and Westerink, B.H. (1995) A telemetry study on the chronic effects of microdialysis probe implantation on the activity pattern and temperature rhythm of the rat. *J. Neurosci. Meth.*, 61: 191–196.

Duggan, A.W. and Hendry, I.A. (1986) Laminar localization of the sites of release of immunoreactive substance P in the dorsal horn with antibody coated microelectrodes. *Neurosci. Lett.*, 68: 134–140.

Fuxe, K. and Agnati, L.F. (1991) Two principal modes of electro-chemical communication in the brain: volume versus wiring transmission. In: K. Fuxe and L.F. Agnati (Eds), *Volume Transmission in the Brain. Novel Mechanisms for Neural Transmission. Advances in Neuroscience*, Vol. 1, Raven Press, New York, pp. 1–11.

Garris, P.A., Ciolkowski, E.L., Pastore, P. and Wightman, R.M. (1994) Efflux of dopamine from the synaptic cleft in the nucleus accumbens of the rat brain. *J. Neurosci.*, 14: 6084–6093.

Gonon, F. and Buda, M. (1985) Regulation of dopamine release by impulse flow and by autoreceptors as studied by in vivo voltammetry in the rat striatum. *Neuroscience*, 14: 765–774.

Iadecola, C. (1998) Neurogenic control of the cerebral microcirculation: is dopamine minding the store? *Nature*, 1: 263–265.

Ingvar, M., Lindvall, O. and Stenevi, U. (1983) Apomorphine-induced changes in local cerebral blood flow in normal rats and after lesions of the dopaminergic nigrostriatal bundle. *Brain Res.*, 262: 259–265.

Jansson, A., O'Connor, W.T., Strömberg, I., Förander, P., Rimondini, R., Zoli, M., Agnati, L.F. and Fuxe, K. (1999) Actions of cerebroventricular dopamine injections on the unilaterally dopamine denervated striatum. Focus on periventricular striatum. Manuscript.

Kawagoe, K.T., Garris, P.A. and Wiedemann, D.J. and Wightman, R.M. (1992) Regulation of transient dopamine concentration gradients in the microenvironment surrounding nerve terminals in the rat striatum. *Neuroscience*, 51: 55–64.

Kehr, J. (1993) A survey on quantitative microdialysis: theoretical models and practical implications. *J. Neurosci. Meth.*, 48: 251–261.

Kehr, J. (1999) Monitoring chemistry of brain microenvironment: biosensors, microdialysis and related techniques. In: U. Windhorst and H. Johansson (Eds), *Modern Techniques in Neuroscience Research*. Springer-Verlag, Heidelberg, pp. 1149–1198.

Krimer, L.S., Mully, III, E.C., Williams, G.V. and Goldman-Rakic, P. (1998) Dopaminergic regulation of cerebral cortical microcirculation. *Nature*, 1: 286–289.

Mraovitch, S., Calando, Y., Onteniente, B., Peschanski, M. and Seylaz, J. (1993) Cerebrovascular and metabolic uncoupling in the caudate-putamen following unilateral lesion of the mesencephalic dopaminergic neurons in the rat. *Neurosci. Lett.*, 157: 140–144.

Nicholson, C. and Syková, E. (1998) Extracellular space structure revealed by diffusion analysis. *Trends Neurosci.*, 21: 207–215.

Ögren, S.O., Kehr, J. and Schött, P.A. (1996) Effects of ventral hippocampal galanin on spatial learning and on in vivo acetylcholine release in the rat. *Neuroscience*, 75: 1127–1140.

Oreland, L. (1991) Monoamine oxidase, dopamine and Parkinsonś disease. *Acta Neurol. Scand.*, 84: 60–65.

Parsons, L.H. and Justice, J.B., Jr. (1992) Extracellular concentration and in vivo recovery of dopamine in the nucleus accumbens using microdialysis. *J. Neurochem.*, 58: 212–218.

Paxinos, G. and Watson, C. (1986) *The Rat Brain in Stereotaxic Coordinates*. Academic Press, Sydney.

Schneider, J.S., Rothblat, D.S. and DiStefano, L. (1994) Volume transmission of dopamine over large distances may contribute to recovery from experimental Parkinsonism. *Brain Res.*, 643: 86–91.

Sharkey, J. and McCulloch, J. (1986) In: C. Owman and J.E. Hardebo (Eds), *Neuronal Regulation of Brain Circulation*, Elsevier, New York, pp. 111–127.

Strömberg, I., Björklund, H., Dahl, D., Jonsson, G., Sundström, E. and Olson, L. (1986) Astrocytic responses to dopaminergic denervations by 6-hydroxydopamine and 1-methyl-4-phenyl-1,2,3,6-tetrahydropyridine as evidenced by glial fibrillary acidic protein immunohistochemistry. *Brain Res. Bull.*, 17: 225–236.

Strömberg, I., Herrera-Marschitz, M., Hultgren, L., Ungerstedt, U. and Olson L. (1984) Adrenal medullary implants in the dopamine denervated striatum. I. Acute catecholamine levels in grafts and host caudate as determined by HPLC-electrochemistry and fluorescence histochemical image analysis. *Brain Res.*, 297: 41–51.

Ungerstedt, U. (1984) Measurement of neurotransmitter release by intracranial dialysis. In: C.A. Marsden (Ed.), *Measurement of Neurotransmitter Release In Vivo. Methods in Neurosciences*, Vol. 6. Wiley, New York, pp. 81–105.

Ungerstedt, U. and Pycock, C. (1974) Functional correlates of dopamine neurotransmission. *Bull. Schweiz. Akad. Med. Wis.*, 30: 44–55.

Vizi, E.S. and Kiss, J.P. (1998) Neurochemistry and pharmacology of the major hippocampal transmitter systems: synaptic and non-synaptic interactions. *Hippocampus*, 8: 566–607.

Westerink, B.H., Drijfhout, W.J., van Galen, M., Kawahara, Y. and Kawahara, H. (1998) The use of dual-probe microdialysis for the study of catecholamine release in the brain and pineal gland. *Adv. Pharmacol.*, 42: 136–140.

Young, A.M. (1993) Intracerebral microdialysis in the study of physiology and behaviour. *Rev. Neurosci.*, 4: 373–395.

Zetterström, T., Herrera-Marschitz, M. and Ungerstedt, U. (1986) Simultaneous measurement of dopamine release and rotational behaviour in 6-hydroxydopamine denervated rats using intracerebral dialysis. *Brain Res.*, 376: 1–7.

Zoli, M., Torri, C., Ferrari, R., Jansson, A., Zini, I., Fuxe, K. and Agnati, L.F. (1998) The emergence of the volume transmission concept. *Brain Res. Rev.*, 26: 136–147.

Zoli, M., Jansson, A., Syková, E., Agnati, L.F. and Fuxe, K. (1999) Volume transmission in the brain and its relevance for neuropsychopharmacology. *Trends Pharmacol. Sci.*, 20: 142–150.

SECTION III

Glia-neuronal signaling

L. F. Agnati, K. Fuxe, C. Nicholson and E. Syková (Eds.)
Progress in Brain Research, Vol 125

CHAPTER 8

Relationship between glia and the perineuronal nets of extracellular matrix in the rat cerebral cortex: importance for volume transmission in the brain

Davide Viggiano[1,2], Merdol Ibrahim[2] and Marco R. Celio[2,*]

[1] *Istituto di Anatomia Umana, Seconda Università, Via L. Armanni 5, 80138 Napoli, Italy*
[2] *Institute of Histology and General Embryology, University of Fribourg, CH-1705 Fribourg, Switzerland*

Introduction

That the central nervous system elaborates a distinct extracellular matrix went unrecognized for many years. But the existence of such a substance is now undisputed, its molecular composition being similar to that present in the connective tissue (Celio and Blümcke, 1994). The molecules are not homogeneously distributed throughout the intercellular space of the central nervous system but form macromolecular aggregates around specific neuronal populations (Carlson and Hockfield, 1996). For example, using lectins such as *Wisteria floribunda* agglutinin (WFA), that recognize *N*-acetylgalactosamine, or specific antibodies (VC1.1), it has proved possible to selectively label net-like entities around parvalbumin-immunoreactive neurones (Kosaka and Heizmann, 1989; Härtig, 1992; Härtig et al., 1992, 1994; Celio and Chiquet-Ehrismann, 1993a; Celio and Blümcke, 1994). Such structures were first observed more than a century ago by Golgi around the soma, proximal dendrites and axonal cones of some neurones (Golgi 1898). Using his silver impregnation technique other authors have since observed an apparent continuity between nets and glial end-feet. On the basis of these findings, it has been surmised that glial processes represent the main constituents of 'perineuronal nets', which accounts for their being referred to by some authors, as 'perineuronal nets of glia' (Brauer et al., 1982, 1984; Lüth et al., 1992). Although it is now clear that perineuronal nets are extracellular entities, electron microscopic evidence indicates that they are very intimately associated with glial processes (Derouiche et al., 1996). Perineuronal nets are generally considered to consist of a network of extracellular matrix molecules, the windows of the mesh being occupied by synaptic boutons (Carlson and Hockfield, 1996).

However, 3-dimensional reconstruction of macerated brain tissue by scanning electron microscopy reveals the situation to be more complex than was previously supposed (Takahashi-Iwanaga et al., 1998). Indeed, very little is known about the spatial relationship existing between oligodendrocytes and perineuronal nets, albeit that this cell type is known to synthesize some of the components of these structures (Wintergerst et al., 1993). The aim of the present study was to unravel the nature of the 3-dimensional relationship existing between extracellular matrix proteoglycans and glial processes. To this end, astroglia, oligodendrocytes and WFA-immunoreactive proteoglycans within the rat cerebral cortex were labelled with fluorescent

*Corresponding author. Tel.: (41) (26) 300 84 90; Fax: (41) (26) 300 97 32; e-mail: marco-celio@uniFr.ch

markers emitting at different wavelengths, and the images thereby obtained in the laser-scanning confocal microscope reconstructed using the IMARIS-programme (Bitplane, Zürich) on a Silicon Graphics.

Materials and methods

Adult male Wistar rats (250–300g) were asphyxiated with carbon dioxide and then perfused via the ascending aorta with 4% (w/v) paraformaldehyde for 5 minutes at 4°C The brain was removed, postfixed in the same fixative at 4°C for 1 hour and then transferred to phosphate-buffered saline (PBS: NaCl, 137 mM; Na_2PO_4, 8.1 mM; KH_2PO_4, 1.8 mM; KCl: 2.7 mM (pH 7.4), maintained at the same temperature for a similar period. Frontal sections, 100 μm in thickness, were prepared from the cerebral cortex using a vibratome. Free-floating slices were incubated overnight at 4°C with three primary markers: (i) biotin-conjugated WFA ((Sigma) diluted 1:100) (ii) mouse monoclonal anti-Rip (diluted 1:20) and (iii) rabbit polyclonal anti-S100 ((Swant, Bellinzona, Switzerland) diluted 1:300). Primary antibodies were diluted in PBS containing 10% (v/v) bovine serum and 0.1% (v/v) Triton X–100 (Sigma). After incubating overnight, sections were washed in PBS and then floated at ambient temperature on a solution containing three fluorescent secondary antibodies: FITC-conjugated anti-mouse IgG (Sigma), Avidin-Texas Red™ (Molecular Probes) and CY5™-conjugated and anti-rabbit IgG (Jackson Laboratories), each diluted 1:100 in PBS containing 10% bovine serum. After rinsing in PBS, the sections were laid on untreated glass slides and permitted to dry before mounting in Slow-fade™ (Molecular Probes). They were then scanned in a MRC 1024 Bio-Rad krypton/argon laser microscope (LSCM, Bio-Rad Microsciences Division Ltd) all lines of excitation being activated. Fluorescence emission was detected at 520 nm, 615 nm and 670 nm for FITC-, Texas Red- and CY5-labelled structures, respectively. The possibility of fluorescence bleed from one fluorochrome to another was excluded by initially visualizing individual fluorochromes in the single-channel mode using a 494-nm, 596 nm, and 650 nm excitation filters for FITC-Rip, Texas-Red-WFA and CY5-S100 labelled structures, respectively. Specimens were examined using a $\times 60$ oil-immersion objective (Nikon CF Apochromat, numerical aperture: 1.4) and zoom factors up to 3, the pinhole aperture being kept as small as possible. Images of single optical planes were Kalman-averaged. Z-series were taken at intervals of 0.2 μm, this distance being chosen to avoid the interference known to occur when using greater optical-section thicknesses (Salisbury, 1994). 3-dimensional reconstruction from series of images were then computed using the Isosurface module of Imaris (Bitplane, Zurich, Switzerland) in conjunction with a Silicon Graphics workstation. Cells labelled with FITC-conjugated mouse monoclonal Rip-antibody (oligodendrocytes) appeared green, those marked with CY5-conjugated rabbit polyclonal antibody (astrocytes) were rendered blue, and structures labelled with Texas Red-conjugated avidine (perineuronal nets) appeared red. In order to avoid the possibility of threshold artefacts, the images obtained after 3-dimensional reconstruction were compared with those observed directly on the confocal microscope.

Results

The fluorophores used in this study were chosen with the view of maximally separating the excitation/emission spectra using the available Kr/Ar laser confocal system. Within serial, 0.2-μm-thick optical sections, the windows of the perineuronal meshwork, as well as the spatial relationship existing between these elements and glial end-feet, were well resolved. The lectin WFA delineated net-like structures around a subpopulation of neurones predominantly within layers III-V of the cerebral cortex; these nets ensheathed the soma as well as the proximal dendrites of the target cells. After immunodetection of the S100 protein, the perikarya as well as the processes of protoplasmic astrocytes were intensely stained; their surfaces appeared prickly, and fine processes broke up the neuropil into a mosaic of small islands. Astrocyte processes were so intermingled and numerous that it was difficult to trace them back to the parent cell body

(Fig. 1A). They completely ensheathed neuronal surfaces, thereby delimiting their profiles. A single neurone often received contributions from more than one astrocyte, and a single astrocyte, in turn, was observed to extend processes to more than one neuronal soma (Fig. 1A). A given astrocyte could simultaneously ensheath both neurones with and without WFA-labelled nets (PNEMs) (Fig. 1A).

Although the reticular structures built up by astrocytic processes (PNGs) were closely associated with PNEMs, they did not coincide with the latter but formed independent networks external to these. Astrocytic end-feet were frequently observed to pass through the windows of PNEMs, by which means they came into closer contact with the surfaces of neurones (Fig. 1B–E); scattered astrocytes were also seen in a perineuronal position relative to cells mantled by PNEMs. After immunodetection of the Rip-antigen, the soma and processes of oligodendrocytes were occasionally observed in close contact with the surfaces of WFA-labelled neurones (Fig. 1A), but were also associated with unlabelled ones.

Discussion

The first methodological procedure employed to demonstrate perineuronal nets, namely Golgi's silver impregnation technique, revealed a continuity between these structures and glial end-feet processes. This observation led to the conclusion that glial processes themselves form the perineuronal nets (Brauer et al., 1982; Lafarga et al., 1984; Lüth et al., 1992). This hypothesis seemed to be confirmed by the net-like structures rendered visible around neurones when using certain glial markers. Although it is now clear that perineuronal nets consist of extracellular molecules, electron microscopic evidence indicates that they are nonetheless intimately associated with glial processes (Derouiche et al., 1996). However, it is still not known whether these entities are produced by glia (Brauer et al., 1984) or whether they are the product of different kind of cells (Lüth et al., 1992). In the electron microscope, glial processes are seen to be separated from the neuronal cell body by the extracellular matrix, whereas axon terminals apparently approach the perikarya of postsynaptic neurones without intervening substance (Lüth 1992; Brückner et al., 1996; Derouiche et al., 1996). However, a more recent study using NaOH-macerated brain tissue indicates that a fine proteoglycans meshwork is present even under the synaptic boutons (Takahashi-Iwanaga et al., 1998).

Confocal microscopy permits the differential labelling of extracellular matrix molecules and different types of glia, and the 3-dimensional reconstruction of these. Using this technique, we have demonstrated that astrocytic processes form reticular-like structures around WFA-labelled perineuronal nets. The fine astrocytic processes were revealed to break up the neuropil into a mosaic of small islands, each of which is believed to contain a synaptic field (Varon and Somjen, 1979). This observation supports the contention that astrocytes form tracks and barriers for neurite outgrowth (McKeon and Silver, 1995) – possibly mediated by the secretion of axonal guidance molecules such as chondroitin sulphate and tenascin, which represent two components of perineuronal nets (Celio, 1993; Härtig et al., 1994). Indeed, extracellularly located chondroitin sulphate – to which WFA most probably binds (Brückner et al., 1993, 1996) – typically forms net-like arrays (Fig. 1A–E). Proteoglycans and glycoproteins organized in a tridimensional network may influence the diffusion of transmitters released from nerve cells. It could be envisaged that local differences in the composition of the PNEM modulate the rate and distance of diffusion of substances released for volume transmission (Nicholson and Syková, 1998).

The relationship revealed between WFA-immunoreactive structures and S–100-immunoreactive glial processes suggests the existence of two reticular entities around neuronal cell bodies: an outer one – the perineuronal net of glia (PNG), and an inner one – the perineuronal net of extracellular matrix (PNEM). A third, intracellular and concentric net-like structure – the membrane cytoskeleton, composed of spectrin and ankyrin-, has been recently described (Celio et al., 1998). According to our previous conception (Celio et al, 1998), these three almost concentric nets were deemed to be in register, the holes in the meshworks representing sites of synaptic contact between axons and the

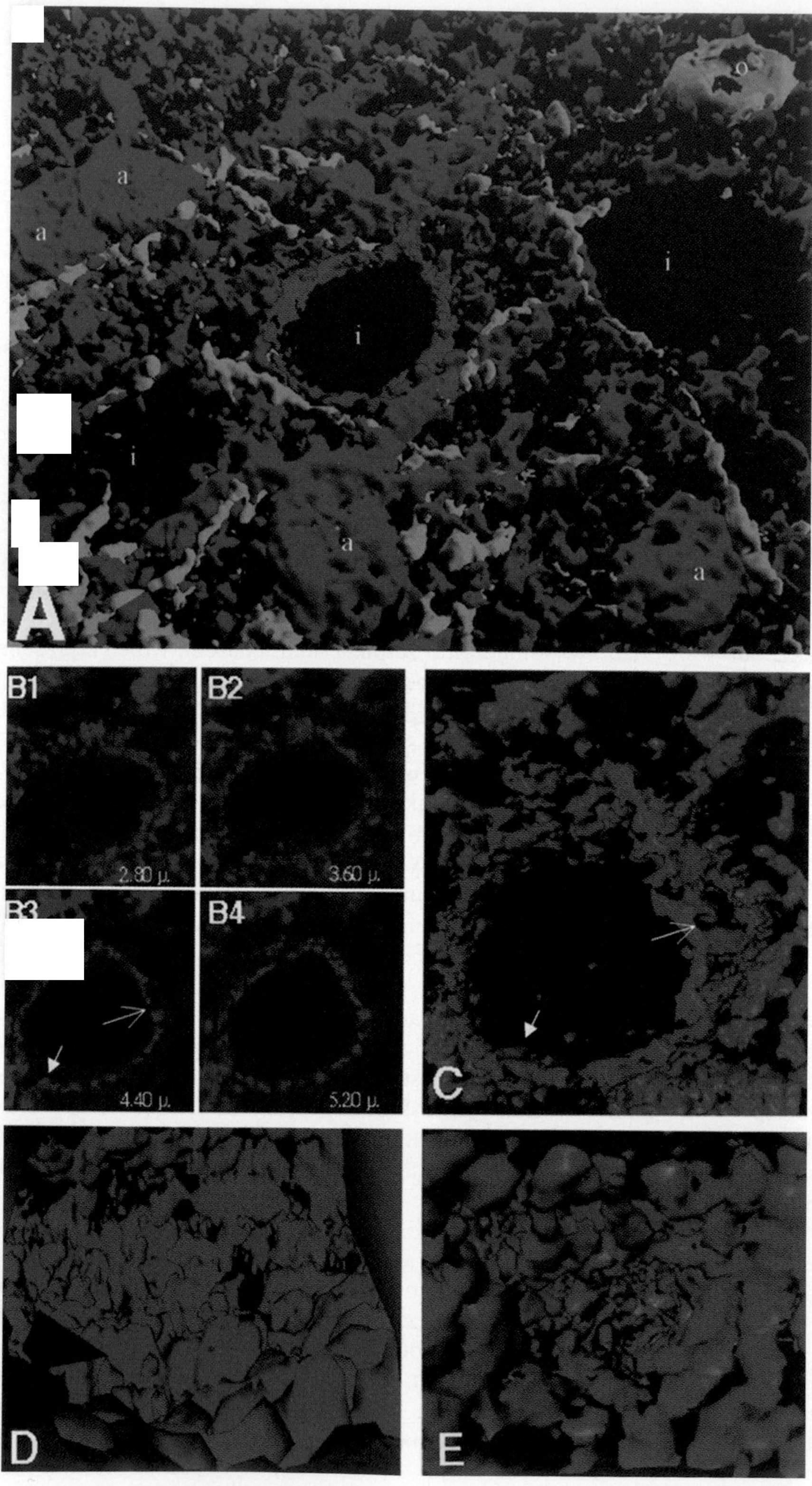
A
a
a
a
a
i
i
i
o
B1
2.80 μ.
B2
3.60 μ.
B3
4.40 μ.
B4
5.20 μ.
C
D
E

neuronal soma. The findings of the present study reveal a slightly different picture: although PRGs and PNEMs are both reticular structures, they are not perfectly superimposed (Fig. 1A–E). Astrocytic processes can and do pass through the windows of the PNEMs and thereby come into closer contact with neuronal cell bodies (Fig. 1B, C). This feature had already been suspected by Brückner et al. (1993), who described fine, S 100-immunoreactive cell processes intermingled among the axosomatic synapses reaching the surfaces of neurones. It may be speculated that the area of contact established between astroglial processes (that have insinuated themselves between the windows of PNEMs) and neuronal cell bodies represents sites of privileged information- exchange beetween the two (King, 1968; Peters et al., 1991).

Our findings further suggest that no specific subclass of astrocytes is involved in the formation of PNGs and PNEMs. The same astrocyte can simultaneously ensheath both neurones with or without PNEMs.

Conclusions

In this work, we have disclosed the relationship pertaining between WFA-labelled perineuronal nets and astroglial end-feet. Two concentric reticular structures exist: (i) an external one consisting of astroglial end-feet (PNG) and (ii) an inner one composed of extracellular matrix molecules (PNEM). The two entities are not superimposed, and glial end-feet are sometimes observed to pass through the windows of PNEMs. PNEMs recognized by WFA are associated with only a fraction of the neurones surrounded by PNGs.

References

Brauer, K., Brückner, G., Leibnitz, L. and Werner, L. (1984) Structural and cytochemical features of perineuronal glial nets in the rat brain. *Acta Histochem.*, 74: 53–60.

Brauer, K., Werner, L. and Leibnitz, L. (1982) Perineuronal nets of glia. *J. Hirnforsch.*, 23: 701–708.

Brückner, G., Brauer, K., Härtig, W., Wolff, J.R., Rickmann, M.J., Derouiche, A., Delpech, B., Girard, N., Oertel, W.H. and Reichenbach, A. (1993) Perineuronal nets provide a polyanionic, glia-associated form of microenvironment around certain neurones in many parts of the rat brain. *Glia*, 8: 183–200.

Brückner, G., Härtig, W., Kacza, J., Seeger, J., Welt, K. and Brauer, K. (1996) Extracellular matrix organization in various regions of rat brain gray matter. *J. Neurocytol.*, 25: 333–346.

Carlson, S.S. and Hockfield, S. (1996) Central nervous system. In: W.D. Comper (Ed.), *Extracellular Matrix*, Vol.1, Cap.1, Harwood Academic Publisher GmbH, The Nederlands, pp. 1–22.

Celio, M.R. (1993) Perineuronal nets of extracellular matrix around parvalbumin-containing neurones of the hippocampus. *Hippocampus*, 3 (special issue): 55–60.

Celio, M.R. and Chiquet-Ehrismann, R. (1993) 'Perineuronal nets' around cortical interneurones expressing parvalbumin are rich in tenascin. *Neurosci. Lett.*, 162: 137–140.

Celio, M.R. and Blümcke, I. (1994) Perineuronal nets – a specialized form of extracellular matrix in the adult nervous system. *Brain Res. Rev.*, 19: 128–145.

Celio, M.R., Spreafico, R., De Biasi, S. and Vitellaro-Zuccarello, L. (1998) Perineuronal nets: past and present. *Trends Neurosci.*, 21: 510–515.

Derouiche, A., Härtig, W., Brauer, K. and Brückner, G. (1996) Spatial relationship of lectin labelled extracellular matrix and glutamine synthetase-immunoreactive astrocytes in rat cortical forebrain regions. *J. Anat.*, 189: 363–372.

Golgi C. (1898) *Intorno alla struttura delle cellule nervose.* Comunicazione alla società Medico-chirurgica di Pavia.

Härtig, W., Brauer, K. and Brückner, G. (1992) Wisteria floribunda agglutinin-labelled nets surround parvalbumin-containing neurones. *NeuroReport*, 3: 869–872.

Härtig, W., Brauer, K., Bigl, V. and Brückner, G. (1994) Chondroitin sulfate proteoglycan immunoreactivity of lectin-labelled perineuronal nets around parvalbumin-containing neurones. *Brain Res.*, 635: 307–311

Fig. 1A. 3-dimensional reconstruction of the relationships existing between astrocytes (a: blue), oligodendrocytes (o: green) and a perineuronal net surrounding an interneurone (i: red) in the cerebral cortex of a rat. Note the presence of oligodendrocytic processes passing close to the interneurone and of astrocytic processes appearing in the windows of the perineuronal net. B. B1-B4 represent four consecutive sections (0.2 μm in thickness) of astrocytic end-feet penetrating a window of the perineuronal net (solid arrow). Another, unoccupied hole is indicated (open arrow). C. 3-dimensional reconstruction of the four images represented in B. D. View from within the cytoplasm of a nerve cell surrounded by a perineuronal net (red). Some of the windows of the mesh are occupied by astrocytic processes (blue); others are empty. The latter probably harbour axonal terminals. E. View through an hole of a perineuronal net (red) within the cytoplasm of the nerve cell surrounded by the net itself.

King, J.S. (1968) A light and electron microscopic study of perineuronal glial cells and processes in the rabbit neocortex. *Anat. Rec.*, 161: 111–124.

Kosaka, T. and Heizmann, C.W. (1989) Selective staining of a population of parvalbumin containing GABAergic neurones in the rat cerebral cortex by lectins with specific affinity for terminal N-acetylgalactosamine. *Brain Res.*, 483: 158–163.

Lafarga M., Berciano M.T. and Blanco M. (1984) The perineuronal net in the fastigial nucleus of the cerebellum. *Anat. Embryol.*, 170: 79–85.

Lüth, H.-J., Fischer, J. and Celio, M.R. (1992) Soyebean lectin binding neurones in the visual cortex of the rat contain parvalbumin and are covered by glial nets. *J. Neurocytol.*, 21: 211–221.

McKeon R.J. and Silver J. (1995) Functional significance of glial-derived matrix during development and regeneration. In: H. Kettenmann and B.R. Ransom (Eds), *Neuroglia, Cap.*, 27, Oxford University Press, New York Oxford, pp. 398–410.

Nicholson C. and Sykova E. (1998) Extracellular space structure revealed by diffusion analysis. *Trends Neurosci.*, 51: 207–215.

Peters, A., Palay, S.L. and Webster, H.deF. (1991). *The fine structure of the nervous system.* Cap. 5, Oxford University Press, USA, pp. 210–211.

Salisbury, J.R. (1994) Three-dimensional reconstruction in microscopical morphology. *Hist. Histopath.*, 9: 773–780.

Takahashi-Iwanaga H., Murakami T. and Abe K. (1998) Three-dimensional microanatomy of perineuronal proteoglycan nets enveloping motor neurones in the rat spinal cord. *J. Neurocytol.*, 27: 817–827.

Varon, S.S. and Somjen, G.G. (1979) Neuron-glia interactions. In: G. Adelman (Ed.), *Neuroscience Research Program Bulletin*, Vol. 17, n.1, MIT Press, Cambridge, pp. 117–128.

Wintergerst, E.S., Fuss, B. and Bartsch, U. (1993) Localization of janusin mRNA in the central nervous system of the developing and adult mouse. *Eur. J. Neurosci.*, 5 (4): 299–310.

L. F. Agnati, K. Fuxe, C. Nicholson and E. Syková (Eds.)
Progress in Brain Research, Vol 125

CHAPTER 9

Glial influence on neuronal signaling

Alexandr Chvátal* and Eva Syková

Department of Neuroscience, 2nd Medical Faculty, Charles University and Institute of Experimental Medicine, Academy of Sciences of the Czech Republic, Videňská 1083, 142 20 Prague 4, Czech Republic

Introduction

Glial cells, non-excitable cells in the central nervous system (CNS), have been the focus of interest for many investigators since the last century, when they were first observed and described morphologically. Systematic study of the electrophysiological properties of glial cells, astrocytes and oligodendrocytes, began during the 1960s, when Kuffler et al. (1966) first observed that glial cells are depolarized by neuronal activity and thus formulated a possible role for glial cells in the buffering of K^+, which accumulates in the extracellular space of the CNS. Astrocytes, oligodendrocytes and neurons develop from the same type of progenitor cell, and extensive experimental research done over more than 30 years has revealed that glial cells do not play only a passive supporting role in the CNS but have distinct electrophysiological properties, which are determined by the presence of membrane proteins, i.e. by different types of ionic channels and transporters in the cell membrane. Activation of ion channels on glial cells may also lead to the activation of second messengers and intracellular metabolic pathways, and to changes in their volume, particularly swelling and rearrangement of processes, accompanied by dynamic variations in the ECS volume (Syková, 1997). Glial cells may thus, by regulating their volume, influence extracellular pathways for neuroactive substances and affect extrasynaptic 'volume' transmission in the CNS. There is increasing evidence that another type of glial cell found in the CNS, microglial cells, also plays an important role in the normal as well as the pathological brain and can have both neurotrophic and neurotoxic properties (Eder, 1998). In contrast to neurons, astrocytes and oligodendrocytes, microglial cells develop from bone marrow-derived monocytes and invade the brain during embryonic development. Since a number of comprehensive reviews describing the membrane properties of glial cells have already been published (Walz, 1989; Chesler, 1990; Barres, 1991; Sontheimer, 1994; Gallo and Russell, 1995; Steinhäuser and Gallo, 1996; Oh, 1997; Porter and McCarthy, 1997; Verkhratsky et al., 1998; Eder, 1998), the present chapter focuses on the relation between the electrophysiological properties of glial cells in the brain in situ, the properties of the extracellular space and synaptic and volume transmission in the CNS.

Membrane properties of glial cells in situ

The first studies of the electrophysiological properties of glial cells were performed in tissue cultures (Hild and Tasaki, 1962). Since it was soon apparent that cultured glial cells, that develop in an environment lacking neurons, may express a wide variety of voltage- and ligand-activated currents, the question arose whether the properties of the glial membrane in culture match the properties of glial

*Corresponding author. Tel.: (+ 420 2) 475 2204; Fax: (+ 420 2) 475 2783; e-mail: chvatal@biomed.cas.cz

membrane in vivo (for review see Porter and McCarthy, 1997). The first attempt to avoid cell culture artifacts was made by developing the technique of freshly isolating glial cells from brain tissue, when the membrane properties of glial cells were studied from different regions of the nervous tissue, e.g. optic nerve or hippocamus (Barres et al., 1990; Tse et al., 1992). By applying the patch-clamp technique to thin brain slices (Edwards et al., 1989), it became possible to investigate glial cells in their normal cellular environment and to re-evaluate the results obtained in tissue culture and by other techniques. Furthermore, brain slices permit the study of glial cells during development, since slices from animals of different postnatal ages can be compared. During the past decade of extensive research on the electrophysiological properties of glial cells in situ by means of patch-clamp technique, the interest of investigators has been predominantly focused on the following areas of the mammalian CNS: corpus callosum, hippocampus, cerebellum and spinal cord. It was found that glial cells studied in brain slices may express, like neurons, a number of voltage- and ligand-activated channels. In contrast to neurons, a large variability in the current patterns among glial cell types is observed in response to depolarizing and hyperpolarizing voltage steps (Fig. 1). Such stimulation may activate Na^+ channels, delayed-rectifier K^+ channels (K_{DR}), A-type K^+ channels

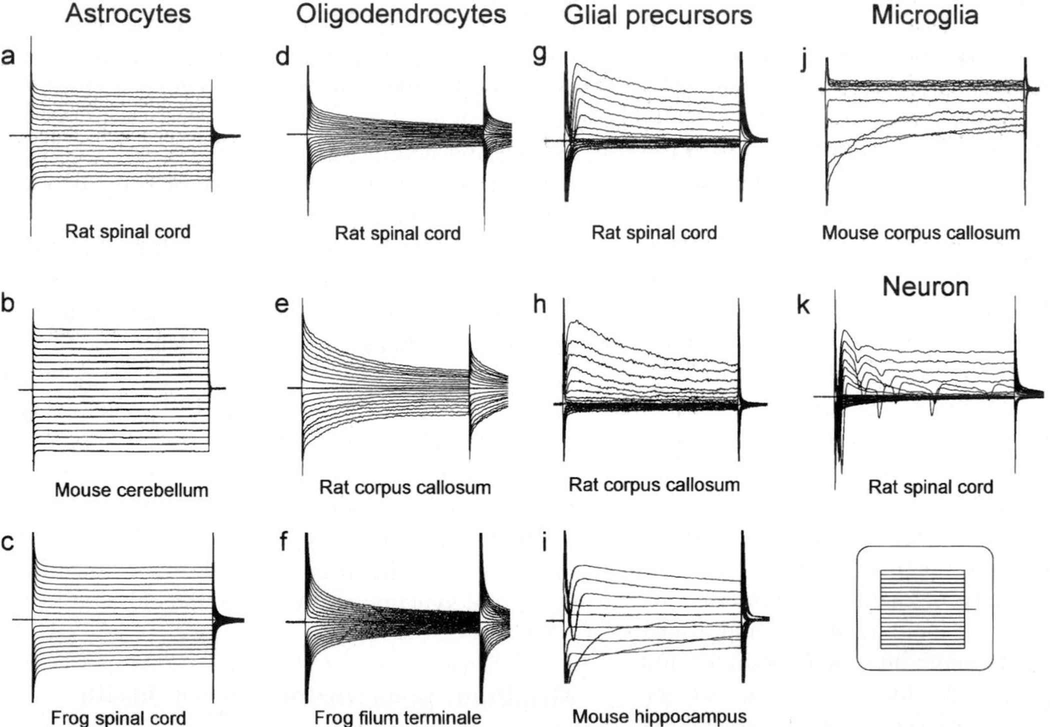

Fig. 1. Typical membrane current patterns of astrocytes (a, b, c), oligodendrocytes (d, e, f), glial precursor cells (g, h, i), microglial cells (j) and neurons (k) evoked by depolarizing and hyperpolarizing voltage steps in different areas of the brain in situ. Membrane currents were recorded while the membrane potential was clamped at values close to –70 mV. To activate voltage-gated currents, the membrane was clamped for 50 ms to increasing de- and hyperpolarizing potentials (pattern of voltage commands in inset) ranging from –160 mV to +20 mV with 10 mV increment. Current traces are not corrected for leakage and capacitance currents. Mature astrocytes and oligodendrocytes express passive currents, while glial precursors, neurons and microglia cells express various types of voltage-activated currents. Current traces in (b) adapted from Kirischuk et al., 1996; current traces in (e, h) adapted from Chvátal et al., 1997; current traces in (i) adapted from Steinhäuser et al., 1992; current traces at (j) adapted from Brockhaus et al., 1993; current traces in (a, c, d, f, g, k) are unpublished data.

(K_A), inward rectifier K^+ channels (K_{IR}) and Ca^{2+} channels. A combined approach of patch-clamp technique, Ca^{2+} imaging and confocal microscopy revealed that glial cells in situ respond to the application of glycine, GABA, glutamate, ATP, histamine, norepinephrine, serotonin, angiotensin II, bradykinin, and substance P (Fig. 2) and that the ionotropic effect of some ligands in glial cells is different from that described in neurons (Fig. 3).

Astrocytes and oligodendrocytes in the spinal cord gray matter

In the gray matter of rat spinal cord slices, astrocytes, oligodendrocytes and their respective precursors can be distinguished at postnatal days 1 to 19 (P1–19), based on their pattern of membrane currents and their morphological and immunocytochemical features (Chvátal et al., 1995). Astrocytes are identified by positive staining for glial fibrillary acidic protein (GFAP), while oligodendrocytes are identified by a typical oligodendrocyte-like morphology, a lack of GFAP staining, and positive labeling with O1 or O4 antibodies – markers of the oligodendrocyte lineage. Electrophysiologically, mature astrocytes are characterized by symmetrical, non-decaying K^+-selective currents. The membrane of mature oligodendrocytes exhibits symmetrical passive, but decaying K^+ currents with prominent tail currents (I_{tail}) after the offset of the voltage command (see also Chvátal et al., 1999a). Astrocyte precursor cells express a complex pattern of voltage gated channels, namely Na^+, K_{DR}, K_A and K_{IR} channels. In comparison to those of neurons, the amplitude of these Na^+ currents is at least one order of magnitude less; none of these cells show the ability to generate action potentials in the current clamp mode. Oligodendrocyte precursors are distinguished by the presence of K_{IR}, K_{DR}, and K_A. Studies performed on glial cells in spinal cord slices from GFAP-positive and GFAP-negative mice revealed a similar distribution of voltage-activated currents in astrocytes, oligodendrocytes and glial precursor cells as was described in the rat spinal cord (Anděrová et al., 1999). The same pattern of voltage-activated membrane currents was also observed in preliminary studies performed on glial cells in frog spinal cord slices (Anděrová, Chvátal, Syková, unpublished results).

Astrocytes, glial precursor cells, and a subpopulation of oligodendrocytes in the gray matter of the rat spinal cord respond to glycine, GABA and glutamate (Pastor et al., 1995). The involvement of glycine receptors is inferred from the

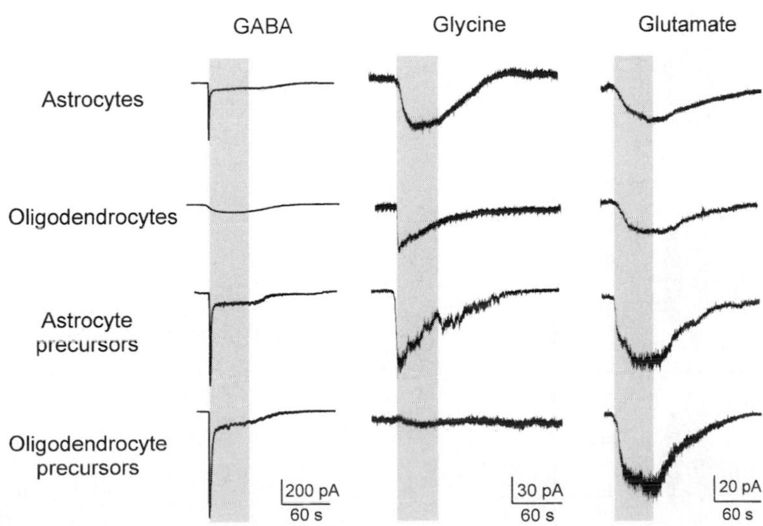

Fig. 2. Membrane currents evoked by GABA (10^{-3} M), glycine (10^{-3} M) and glutamate (10^{-3} M) shown for the four types of glial cells in gray matter of the rat spinal cord slice. Gray bars indicate the time of application of the test substance. Note the difference in the calibrations for the ligand-activated currents. Adapted from Pastor et al., 1995 and Žiak et al., 1998.

observation that the response to glycine is blocked by strychnine and that the induced current reverses close to the Cl^- equilibrium potential. Also, in experiments in which patch-clamp technique was combined with reverse transcription-mediated PCR analysis, the expression of glycine receptor subunits was detected in glial cells in situ (Kirchhoff et al., 1996). GABA-induced membrane currents are mimicked by muscimol and blocked by bicuculline and are large in astrocytes and precursor cells, while oligodendrocytes show only small responses. $GABA_A$ receptor activation also decreases the resting K^+ conductance. All glial cells in the spinal cord gray matter are sensitive to glutamate, kainate, and NMDA (Žiak et al., 1998). Kainate evokes larger currents in glial precursors than in astrocytes and oligodendrocytes, while NMDA induces larger currents in astrocytes and oligodendrocytes than in precursors. Kainate-evoked currents are blocked by the AMPA/kainate receptor antagonist CNQX and are, with the exception of precursors, larger in the dorsal than in the ventral horn, as are NMDA-evoked currents. Currents evoked by NMDA are unaffected by CNQX, are not sensitive to Mg^{2+} and are present when synaptic transmission is blocked in a Ca^{2+}-free solution. NMDA-evoked currents are not abolished during the block of K^+ inward currents in glial cells by Ba^{2+}; thus they are unlikely to be mediated by an increase in extracellular K^+ during neuronal activity.

Oligodendrocytes and microglial cells in corpus callosum

In the mouse corpus callosum, where the first tight-seal patch-clamp recordings from glial cells in situ

receptor	region of CNS	effect
AMPA /kainate	hippocampus cerebellum spinal cord retina	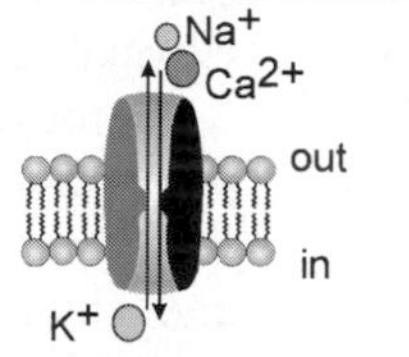 Na^+ Ca^{2+} out in K^+
NMDA	hippocampus cerebellum spinal cord retina cortex	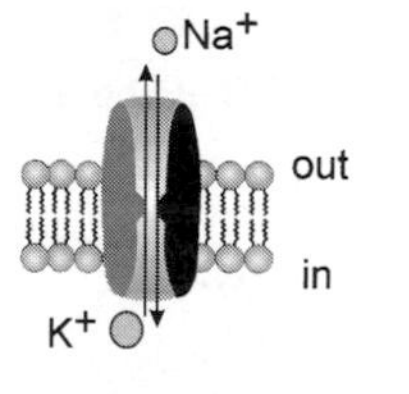 Na^+ out in K^+
$GABA_A$	hippocampus cerebellum spinal cord retina	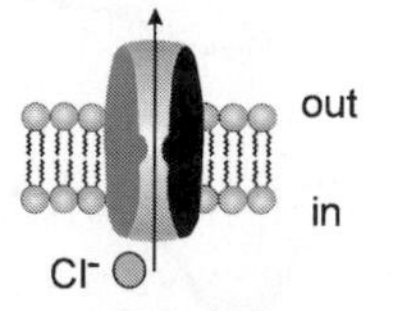 out in Cl^-
glycine	spinal cord	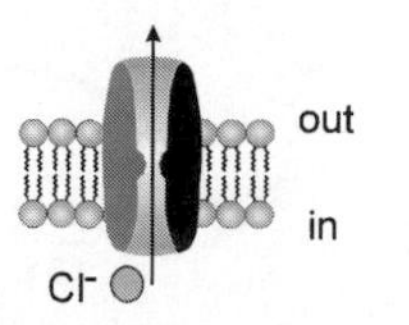 out in Cl^-

Fig. 3. Effects of the activation of AMPA/kainate, NMDA, GABA and glycine receptors in situ as described in different regions of the CNS. Partially based on the data in Porter and McCarthy, 1997.

were performed by Berger et al. (1991), more than 99% of all perikarya belong to glial cells during the early postnatal period (Sturrock, 1980). The morphological investigation of these cells, as studied by Lucifer yellow injection, revealed that at P5 numerous thin processes extend radially from the perikaryon of these cells, while at P10–13 most processes extend parallel to each other, thus showing the typical oligodendrocyte-like morphology. In slices from P6–8, oligodendrocyte precursors were found that predominantly expressed K_{DR}, while K_A or Na^+ currents, described in oligodendrocyte precursors in culture, were either absent or only rarely observed. On the other hand, oligodendrocyte precursors observed in a rat corpus callosum preparation (Chvátal et al., 1997) were characterized by the presence of K_A, K_{DR} and K_{IR}. Mature oligodendrocytes in both species are characterized by large passive currents (with an almost linear current-voltage relationship), which decay during the voltage command, and by large I_{tail} after the offset of the voltage pulse.

Oligodendrocyte precursor cells, as well as mature oligodendrocytes, express GABA and glutamate receptors (Berger et al., 1992). GABA-evoked currents show similarities to those described for $GABA_A$ receptors, i.e. they are mimicked by muscimol, blocked by bicuculine and enhanced by pentobarbital. Glutamate-evoked currents show similarities to those of the kainate/quisqalate receptor subtype; oligodendrocyte precursors in the rat corpus callosum express weakly Ca^{2+}-permeable AMPA receptors (Berger, 1995). Kainate in these cells induces two effects: the activation of a cationic current and the block of K^+ conductance. It has also been shown in studies, in which fluorescent dyes were used concomitantly with confocal microscopy to measure intracellular Ca^{2+} transients, that glial cells from mouse corpus callosum slices respond to a variety of neuroligands (Bernstein et al., 1996). Subpopulations of glial cells obtained at P3–7 responded with an intracellular Ca^{2+} signal to the application of ATP, glutamate, histamine, GABA, norepinephrine, serotonin, angiotensin II, bradykinin and substance P, while glial cells obtained from animals at P11–18 responded only to glutamate, ATP and norepinephrine.

Microglial cells in corpus callosum slices from P6–9 mice express very small currents during hyperpolarizing voltage steps in situ, while hyperpolarization induces large inward currents (Brockhaus et al., 1993). Since the reversal potential of these currents depends on the transmembrane K^+ gradient, inactivation time constants decrease with hyperpolarization, and the currents are blocked by tetraethylammonium, the inward currents observed in microglial cells exhibit properties of the inwardly rectifying K^+ channel.

Glial cells in hippocampus

The first patch-clamp recordings from glial cells in the stratum radiatum of the hippocampus were done by Steinhäuser et al. (1992) in brain slices from P10–12 mice. The authors did not distinguish between astrocytes and oligodendrocytes; nevertheless, based on the current pattern, glial cells were grouped into four types. The first is characterized by passive and symmetrical K^+ currents activated both in depolarizing and hyperpolarizing directions. The second population of glial cells expresses a similar current pattern, but with a marked decay of the current during voltage jumps and with a prominent I_{tail} after the offset of the voltage jump. In the third population the decaying passive currents are superimposed with a K_{DR} and, in some cases, with T-type and L-type Ca^{2+} channels, while the fourth population of glial cells expresses K_{DR}, K_{IR} and, in some cases, K_A and Na^+ currents (see also Kressin et al., 1995; Akopian et al., 1996). In another study performed on hippocampal slices of P5–24 rats, glial cells were identified as astrocytes and oligodendrocytes by staining for GFAP (Sontheimer and Waxman, 1993). Astrocytes express K_{DR}, K_{IR}, K_A and Na^+ currents (see also Bordey and Sontheimer, 1997), while oligodendrocytes express either time-independent currents with linear current/voltage properties or time- and voltage-dependent K^+ currents characterized by inward rectification.

In mouse hippocampal slices, the application of GABA evokes inward currents in all types of glial cells (Steinhäuser et al., 1994). There is no apparent difference in the amplitude and time course of GABA-activated currents between GFAP-positive and GFAP-negative cells. The specific $GABA_A$ receptor agonist muscimol evokes inward currents with similar kinetics and amplitudes as does GABA. In addition, the GABA response in glial cells is reversibly blocked by the $GABA_A$ receptor antagonist bicuculline. The application of glutamate and kainate induces inward currents in all glial cell types in the CA1 stratum pyramidale of the mouse hippocampus, while the application of NMDA induces inward currents only in cells with passive K^+ currents (Steinhäuser et al., 1994). Glutamate increases cation conductance and the blockade of the resting K^+ conductance. It was shown in a study performed in the CA1 stratum radiatum of the mouse hippocampal slice that application of kainate or AMPA also leads to the activation of cationic conductance and to a long-lasting blockade of voltage-gated K^+ channels and, in addition, to an elevation of cytosolic Ca^{2+} (Jabs et al., 1994; Seifert and Steinhäuser, 1995). Similarly, in the hilus of the dentate gyrus of the rat hippocampus, Backus and Berger (1995) showed that glutamate, kainate and AMPA evoked inward currents in astrocytes and presumed glial precursor cells. AMPA receptors in this study were permeable to Ca^{2+}. The electrophysiological data were confirmed by experiments in which confocal microscopy and fluorescent dyes were used in rat hippocampal slices to show that GFAP-positive astrocytes located in the stratum oriens or stratum radiatum from CA1 respond to the perfusion of glutamate, kainate and NMDA (Porter and McCarthy, 1995) or to glutamate released from synaptic terminals (Porter and McCarthy, 1996).

Bergmann glial cells in cerebellum

Bergmann glial cells, an astrocyte-type glial cell in the cerebellum, are closely associated with neurons. During development they provide guiding structures for migrating granule cells; in the adult cerebellum they display intimate interactions with Purkinje cells (Berger et al., 1995; see also Grosche et al., 1999). In Bergmann glial cells from P5–7 mice, K_{DR} and K_{IR} are present, while the same cells in older animals are characterized by large, voltage and time-independent K^+ currents (Müller et al., 1994).

The application of GABA induces a rapid activation of Cl^- conductance and a long-lasting decrease in the resting K^+ conductance. Both effects are mediated by benzodiazepine-insenstitive $GABA_A$ receptors. GABA-evoked currents in cells of P5–7 mice are large compared to the small or even undetectable currents in P20–30 cells. Glutamate, kainate and NMDA induce inward currents in Bergmann glial cells (Müller et al., 1992, 1993). The AMPA/kainate receptor is characterized by a high Ca^{2+} permeability and a sigmoidal current-voltage relationship. The influx of Ca^{2+} blocks the resting K^+ current and the junctional conductances of the cells (Müller et al., 1996). NMDA-induced inward currents are not accompanied by a Ca^{2+} influx and are not affected by the presence of Mg^{2+} or potentiated by glycine.

It was also shown using a combined patch-clamp and confocal microscopy approach, that Bergmann glial cells in situ express metabotropic glutamate receptors, purinergic receptors, α_1-adrenoreceptors and H_1 histamine receptors (Kirischuk et al., 1995, 1996).

Membrane properties of glial cells are determined by their developmental stage

During the last decade of electrophysiological investigations of glial currents in brain slices in situ from different areas of the brain of different species, it has become evident that the membrane properties of glial cells are not uniform, but depend on the developmental stage of the cell (for review see also Berger et al., 1995). For example, astrocyte precursors in spinal cord and in hippocampus are characterized by the presence of Na^+ currents, while in astrocyte-like Bergmann glia these currents have never been detected. Similarly, in oligodendrocyte precursors in the rat spinal cord and corpus callosum, Na^+ currents are not detected, but they are present in mouse corpus callosum slices. On the other hand, all immunohistochemically identified mature astrocytes and oligodendrocytes are characterized by large passive K^+ currents.

In studies of the developmental regulation of ligand-activated currents in glial cells during the first 2–3 postnatal weeks, two approaches have been used. The first compares precursor vs. mature glial cells. In comparison to mature astrocytes and oligodendrocytes, glial precursor cells are as a rule characterized by significantly larger currents evoked by GABA, kainate and glycine (astrocyte lineage only in the spinal cord). Surprisingly, NMDA-evoked currents are larger in mature astrocytes and oligodendrocytes in the spinal cord (Žiak et al., 1998), in astrocyte-like glial cells in the hippocampus (Steinhäuser et al., 1994), and in Bergmann glial cells in the cerebellum than in glial precursor cells (Müller et al., 1993). The second approach, which so far has been applied only in spinal cord slices, follows the developmental regulation of the ligand-evoked currents within distinct populations of glial cells, i.e. glial precursor cells, mature astrocytes and mature oligodendrocytes (Pastor et al., 1995; Žiak et al., 1998). GABA-evoked currents are down-regulated only in astrocytes, while in the other cell populations they do not significantly change. Glycine-evoked currents significantly decrease only in glial precursor cells, while kainate-evoked currents do not change during development in any glial population and NMDA-evoked currents significantly decrease in glial precursor cells as well as mature astrocytes and oligodendrocytes.

It is possible to conclude that during the first 2–3 weeks of postnatal development, the occurrence of voltage-activated currents in glial cells decreases, while the passive K^+ conductance rapidly increases. Similarly, ligand-activated currents either do not change or decrease during the transition from glial precursor to mature glial cell as well as during early postnatal development within each glial cell population.

Glial membrane properties are affected by regional differences in ECS volume

Glial cells as well as neurons in the CNS are surrounded by the extracellular space (ECS), which creates their natural microenvironment. The composition and properties of the ECS are established during development, but the most dramatic changes in the composition of the ECS are observed during neuronal activity, when a number of neuroactive substances are released from neurons into the ECS, e.g. ions, transmitters, peptides, neurohormones and metabolites, which diffuse via the ECS to their targets located on nerve as well as glial cells, distant from the release sites (for reviews see Syková, 1997; Nicholson and Syková, 1998; Zoli et al., 1999). The ECS serves as a communication channel between cellular elements in the CNS. In addition to classical synaptic transmission, substances move by diffusion through the volume of the ECS and this type of signal transmission is called extrasynaptic or volume transmission (Agnati et al., 1995; Syková, 1997; Nicholson and Syková, 1998; Zoli et al., 1999). Properties of the ECS and their changes during neuronal activity, development, aging and many pathological states, affect the diffusion of substances in the ECS and may thus affect neuronal signaling and neuron-glia communication (see Chapter 6 by E. Syková et al.). Extrasynaptic transmission is altered by diffusion barriers that may be formed by fine glial and neuronal processes as well as by adhesion molecules and molecules of the extracellular matrix. It was also shown in recent studies that such diffusion barriers exist around oligodendrocytes and that these regional differences in the ECS affect oligodendrocyte membrane currents (Chvátal et al., 1997, 1999a).

Tail currents in oligodendrocytes arise from the accumulation of K^+ in the ECS

Mature glial cells are almost exclusively permeable for K^+, thus depolarization of the glial membrane causes a massive efflux of K^+ out of the cell. Depolarization or hyperpolarization of oligodendrocytes in culture evokes currents that match the shape of the de- or hyperopolarizing pulse (Sontheimer and Kettenmann, 1988; Sontheimer et al., 1989). On the other hand, oligodendrocytes in corpus callosum or in spinal cord or oligodendrocyte-like cells in hippocampus are characterized by passive but decaying K^+ currents with prominent I_{tail} after the offset of the voltage jump (Fig. 4; Berger et al., 1991; Steinhäuser et al., 1992; Chvátal et al., 1995). Further analysis of oligodendrocyte currents revealed that in contrast to cultured cells where the ECS is almost infinite, current decays during voltage jumps and I_{tail}

observed in oligodendrocytes in brain slice preparations are produced by the extensive accumulation of K^+ in the vicinity of the oligodendrocyte membrane and by the shift of K^+ across the cell membrane either to or from the ECS. This conclusion is based on the following observations:

(1) Depolarizing or hyperpolarizing voltage jumps produce a more positive or negative shift of reversal potential (V_{rev}) of I_{tail}, respectively. This behavior is compatible with an efflux of K^+ during depolarization and an influx during hyperpolarization (Berger et al., 1991; Steinhäuser et al., 1992; Chvátal et al., 1997; Chvátal et al., 1999a).
(2) The time decay during, as well as after, the voltage jump is independent of voltage, but varies markedly from cell to cell. For any given cell the time constant of decay is voltage independent (Berger et al., 1991; Chvátal et al., 1999a).
(3) The glial membrane potential is strongly dependent on $[K^+]_e$. The superfusion of rat spinal cord slices with 55 mM K^+ shifts the reversal potential of all glial cell types, including oligodendrocytes, to –16 mV, which is

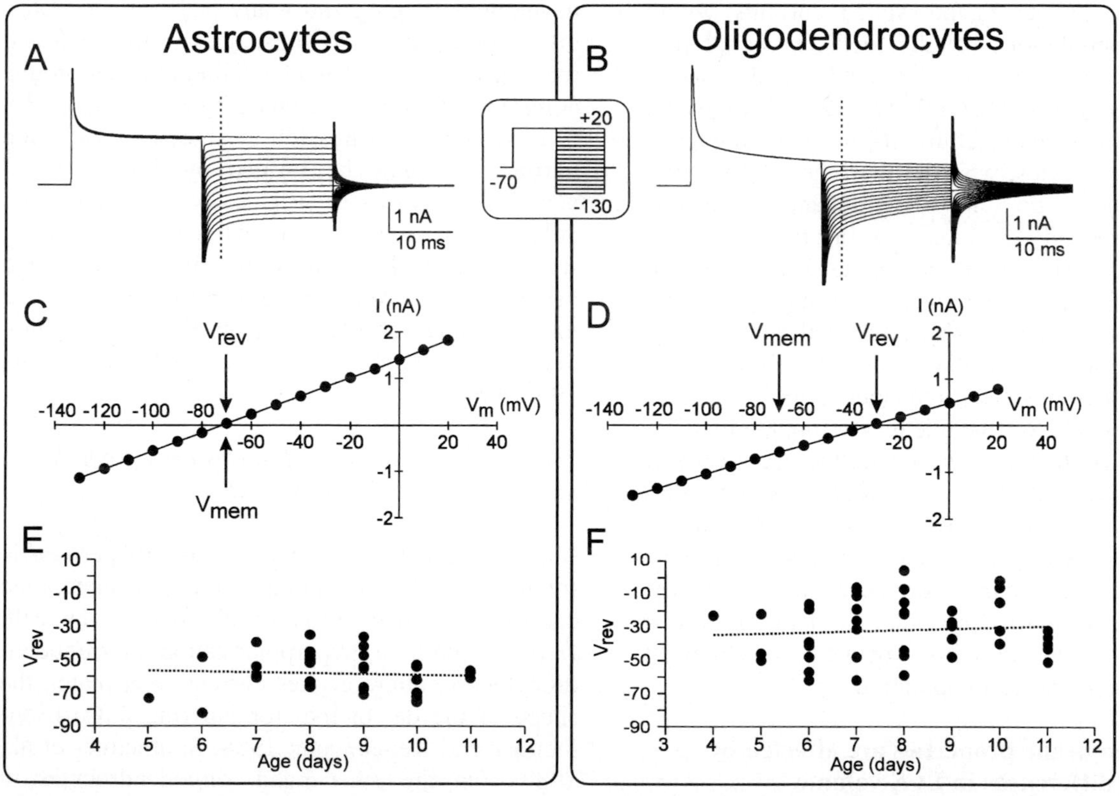

Fig. 4. Tail current analysis of the currents evoked in astrocytes and oligodendrocytes in the gray matter of spinal cord slices and the reversal potential of the tail currents during postnatal development. A, B: The membranes of astrocytes and oligodendrocytes were clamped from a holding potential of –70 mV to +20 mV for 20 ms. After this prepulse, the membrane was clamped for 20 ms to increasing de- and hyperpolarizing potentials (pattern of voltage commands in inset) ranging from –130 mV to +20 mV, at 10 mV increments. C, D: From traces as shown in A and B, currents (I) were measured five ms after the onset of the de- and hyperpolarizing pulses (dashed lines) and plotted as a function of the membrane potential (V_m). Glial membrane potentials (V_{mem}) and the reversal potentials (V_{rev}) after the depolarizing prepulse are indicated in the graphs by the arrows. In oligodendrocytes, in contrast to astrocytes, the depolarizing prepulse shifted the reversal potential from –70 mV to –31 mV. E, F: Reversal potential of the tail currents in astrocytes and oligodendrocytes during postnatal development. The values of reversal potential (V_{rev}) as a function of animal age and the corresponding linear regressions are shown for astrocytes and oligodendrocytes. Adapted from Chvátal et al., 1999.

close to the estimated equilibrium potential (–23 mV) as calculated from the Nernst equation (Chvátal et al., 1995).

(4) In mouse corpus callosum as well as in rat spinal cord gray matter, the application of Ba^{2+}, a K^+ channel blocker, inhibits I_{tail} after de- and hyperpolarizing voltage steps (Berger et al., 1991; Chvátal et al., 1995).

In summary, the I_{tail}, observed in oligodendrocytes, in white as well as in gray matter, do not arise from changes in the capacitance of the cell, but represent a rapid shift of K^+ caused by a change in the K^+ gradient across the cell membrane during the voltage step. Such a rapid shift of K^+ is mediated through K^+ channels or by K^+/Cl^- uptake (Kettenmann, 1986) but not by Na^+/K^+ ATPase activity, since the time-course of the tail currents is in the range of milliseconds, while the uptake of K^+ mediated via the activity of Na^+/K^+ ATPase may last several minutes (Kettenmann et al., 1987) and, in addition, requires an increased intracellular Na^+ concentration (Tang et al., 1980).

Oligodendrocyte tail currents reveal the inhomogeneity of the ECS

In experiments performed in the hippocampal slice or in the gray matter of the spinal cord slice, i.e. in tissue containing neurons as well as mature astrocytes and oligodendrocytes and their respective precursors (Steinhäuser et al., 1992; Chvátal et al., 1995), the occurrence of large I_{tail} was observed in oligodendrocytes or oligodendrocyte-like cells but not in other cell types (Fig. 4). Since the glial membrane is exclusively permeable for K^+, the Nernst equation $V_{rev} = (RT/F)\ \ln\ ([K^+]_e/[K^+]_i)$ can be used to calculate the extracellular K^+ concentration ($[K^+]_e$) from the values of V_{rev} of I_{tail} in the vicinity of the cell membrane. In oligodendrocyte-like cells in hippocampal slices, a +20 mV prepulse evokes a shift of V_{rev} of I_{tail} to –19 mV (Steinhäuser et al., 1992), which corresponds to 61 mM $[K^+]_e$. In the rat corpus callosum slice preparation, a +20 mV prepulse evokes in oligodendrocyte precursors an increase of $[K^+]_e$ up to 12 mM, while in mature oligodendrocytes it is 37 mM (Chvátal et al., 1997). In the rat spinal cord slice a +20 mV prepulse evokes in oligodendrocytes an increase of $[K^+]_e$ up to 47 mM, while in astrocytes it is 12 mM, in astrocyte precursors 15 mM and in oligodendrocyte precursors 22 mM (Chvátal et al., 1999a). These data indicate that the ECS is not homogeneous, i.e. it is more 'condensed' or 'compact' around mature oligodendrocytes than around precursor cells or mature astrocytes.

ECS inhomogeneities represented by the differences in K^+ accumulation around different cell types are also revealed by inducing cell swelling or shrinkage, i.e. changing the size of the ECS volume in spinal cord slices (Chvátal et al., 1999a, b). Cell swelling induced by the application of 50 mM K^+ or by hypotonic solution produces in the vicinity of astrocytes an increase in $[K^+]_e$ in the range of 200–240%, while in oligodendrocytes such an increase is only 22–30%, apparently because the ECS is already small around oligodendrocytes and cannot shrink below a certain level. Cell shrinkage evoked by hypertonic solutions has no effect on $[K^+]_e$ evoked by a depolarizing prepulse in astrocytes, while in oligodendrocytes a rapid decrease of $[K^+]_e$ is observed. These results also indicate that the swelling is more pronounced in astrocytes than in oligodendrocytes. It is possible to speculate that astrocytes are responsible for the majority of the cell volume changes in nervous tissue.

A model has been proposed that explains the differences in the K^+ shift across astrocytic and oligodendrocytic membranes on a cellular level (Fig. 5A; Chvátal et al., 1999a). The model is based on the assumption that during the depolarizing pulse, K^+ is extruded from the cell to the extracellular space. In astrocytes, extruded K^+ freely moves away from the membrane, and within several milliseconds a new steady-state is established, i.e. the same amount of K^+ that is leaving the cell is freely redistributed in the ECS. After the offset of the voltage command, only a small amount of K^+ re-enters the cell, creating a very small I_{tail}. In oligodendrocytes, presumed diffusion barriers (Fig. 5A, dashed ellipse) prevents K^+ from moving away freely and thus K^+ accumulates in the vicinity of the cell membrane. During this accumulation, a new K^+ equilibrium is established, which is reflected by the decreasing current during the voltage step. After the offset of the depolarizing prepulse, K^+ moves back into the cell and produces

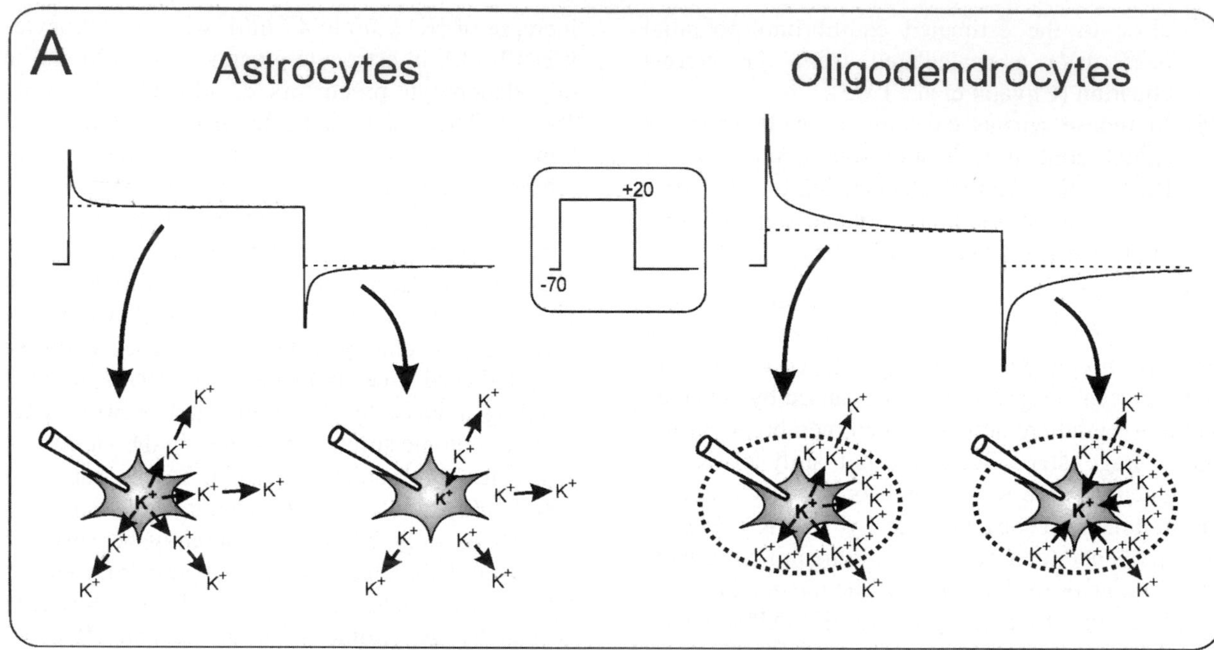

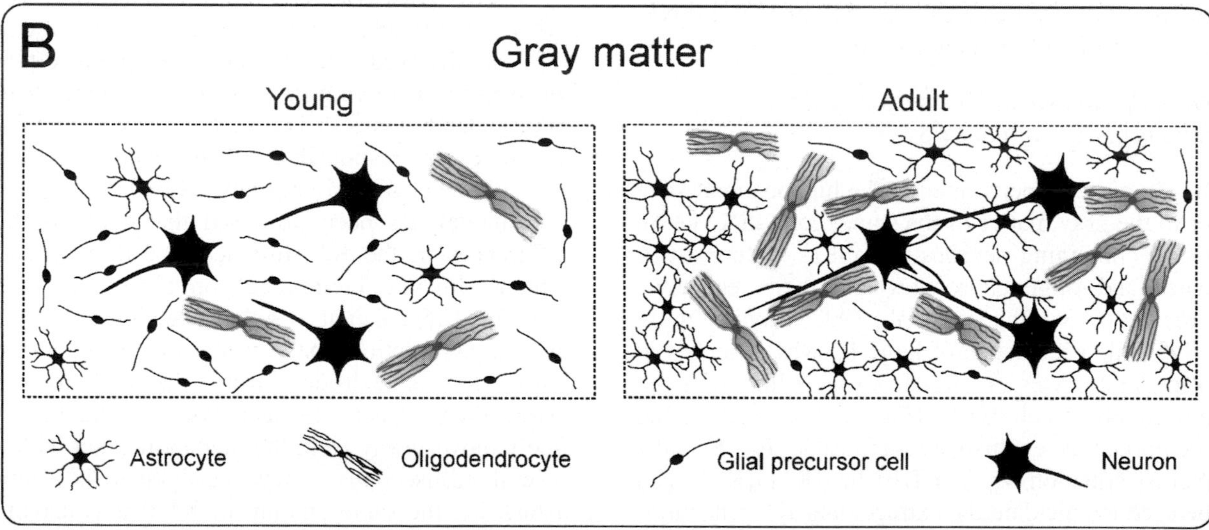

Fig. 5. A model of K^+ movements across the astrocytic and oligodendrocytic membrane (A) and a hypotheses of the developmental changes in the spinal cord gray matter (B) A: Hypothesis of the K^+ shift across astrocytic and oligodendrocytic membranes during and after the depolarizing pulse. In astrocytes, a small amount of K^+ re-enters the cell after the offset of the voltage command, creating a very small tail current. In oligodendrocytes, K^+ accumulates in the vicinity of the cell membrane due to presumed diffusion barriers (dashed ellipse) and thus produces prominent tail currents after the offset of the voltage command. B: Drawing of the developmental changes in the spinal cord gray matter. In the nervous tissue of young animals the majority of the cells are glial precursors and the ECS volume remains relatively large. In adulthood there is a substantial increase in the number of mature astrocytes and oligodendrocytes together with the elongation and branching of glial processes, axonal and dendritic outgrowth and a decrease of the ECS volume. The gray shading between oligodendrocyte processes indicates clusters of the 'compact' ECS in the vicinity of the oligodendrocyte membrane. Adapted from Chvátal et al., 1999.

prominent I_{tail}. The time-course of the K^+ shifts is in the range of milliseconds, and thus they are likely not mediated by ATP-based uptake mechanisms.

Tail currents and ECS volume during the maturation of the CNS

The possible link between I_{tail}, i.e. large and fast transmembrane shifts of K^+, and the ECS size was also suggested in experiments in which the membrane properties of cells of the oligodendrocyte lineage were compared with the ECS diffusion parameters of rat corpus callosum during development, at the period of extensive myelination (Chvátal et al., 1997). The real-time iontophoretic method was used to determine the absolute values of ECS diffusion parameters and their dynamic changes in nervous tissue in situ (Syková, 1997; Nicholson and Syková, 1998). At P10, when the majority of cells in the rat brain white matter are glial precursors, no I_{tail} were observed in these cells and the ECS volume fraction was $\alpha = 0.36$, i.e. the size of the ECS was 36%. In contrast, in white matter after myelination at P20, characterized by the presence of mature oligodendrocytes with decaying currents and large I_{tail} after the offset of a voltage jump, the ECS volume fraction was significantly smaller, $\alpha = 0.25$. These studies demonstrate that the appearance of I_{tail} in oligodendrocytes during early postnatal development coincides with the decrease of the ECS volume (Chvátal et al., 1997).

The above studies raise the question whether the appearance and increase of I_{tail} in oligodendrocytes develop concomitantly with the 'maturation' of the nervous tissue, i.e. the outgrowth and prolongation of cell processes and changes in the extracellular matrix composition, or whether the presence of large I_{tail} in individual cells is independent of the overall decrease of the ECS volume during early postnatal development. It was shown in the study of Prokopová et al. (1997) that the extracellular volume fraction, measured over a volume on the order of 10^{-3} mm^3, significantly decreases in spinal cord gray matter during postnatal development between P4–5 and P7–8. On the other hand, values of V_{rev} of I_{tail} in patch-clamp experiments, which represent the changes in $[K^+]_e$ not in the bulk of the tissue but in the immediate vicinity of the glial membrane, do not significantly change among all glial cell types studied between P5 and P11 (Fig. 4). The appearance of I_{tail} in oligodendrocytes seems not to be related to the average ECS volume fraction of the brain tissue but rather to the increasing number of mature oligodendrocytes surrounded by the 'compact' ECS (Fig. 5B). Such an increased number of mature oligodendrocytes during postnatal development has been found in spinal cord gray matter (Chvátal et al., 1995). An increased number of oligodendrocytes surrounded by a 'compact' ECS might be, together with the elongation and branching of glial processes (Takahashi et al., 1990) and axonal and dendritic outgrowth (Bicknell and Beal, 1984), an additional factor responsible for the decrease in the overall ECS volume during postnatal development (Chvátal et al., 1999a).

The questions remain, do inhomogenities of the ECS affect neuronal signaling, what are the obstacles or diffusion barriers which are present around oligodendrocytes and prevent K^+ from moving away from the oligodendrocyte membrane, and how might these barriers affect neuronal function? The finding that tail currents in oligodendrocytes are observed in brain slices but not in tissue culture with an infinite ECS volume (Sontheimer and Kettenmann, 1988; Sontheimer et al., 1989), indicates that the diffusion barriers arise, at least partially, from the compartments of the nervous tissue. The larger accumulation of K^+ around oligodendrocytes is most likely caused by a smaller ECS around this type of cell, but regional differences in molecules of the extracellular matrix, which have been shown to be produced by glial cells, cannot be excluded (Maleski and Hockfield, 1997; for review see Celio et al., 1998). The unique electrophysiological behavior of oligodendrocytes could also be correlated with their specific morphology, i.e. long and parallel processes, not seen in other types of glial cells, which probably corresponds to their tight contact with axons.

Patch-clamp studies of the electrophysiological properties of glial cells in different regions of the CNS and in different species have revealed that mature oligodendrocytes are surrounded by clusters of compact ECS. Since the number of oligodendrocytes increases during early postnatal development, such clusters in the CNS may contribute to the overall decrease of the ECS volume fraction in the brain.

Role of glial cells in ionic homeostasis and in neuronal signaling

It is evident that during development of the ECS, the membrane properties of glial cells change concomitantly with the maturation of the nervous tissue, i.e. axonal outgrowth, branching of the neuronal processes and formation of the synapses. During the first 2–3 weeks after birth the cellular composition of the nervous tissue changes: the number of glial precursor cells decreases and the number of mature astrocytes and oligodendrocytes increases (Chvátal et al., 1995). It was shown that the ability of mature nervous tissue to efficiently buffer K^+ and pH changes is closely related to gliogenesis (Jendelová and Syková, 1991) and that impairment of glial cell maturation by early postnatal X-irradiation alters K^+ and pH homeostasis (Syková et al., 1992; Syková and Chvátal, 1993) and blocks the normal pattern of volume fraction decrease during postnatal development, resulting in a significant increase in the ECS volume (Syková et al., 1996). Glial cells do not form synapses, however they interact with neurons and with other distant cells by the diffusion of ions and other neuroactive substances through the volume of ECS, i.e. by 'volume transmission' (Agnati et al., 1995; Syková, 1997; Nicholson and Syková, 1998; Zoli et al., 1999). The presence of voltage- and ligand-activated channels on glial precursor cells and on mature astrocytes and oligodendrocytes are very important prerequisites for the volume transmission and are involved in the maintenance of ionic homeostasis and brain energy metabolism, neuron-to-glia and glia-to-neuron signaling, synaptic transmission and neuronal function, and the changes of ECS volume.

Maintenance of ionic homeostasis

It was already shown in the study of Orkand et al. (1966) that during neuronal activity, K^+ increases in the relatively small ECS. It was therefore proposed that to prevent extensive accumulation of K^+, which may impair neuronal signaling, glial cells regulate extracellular K^+ concentration by spatial K^+ buffering. Since the glial membrane is permeable to K^+ and glial cells create a syncitium, this buffering mechanism is maintained by the K^+ influx into the glial cell in the region of increased K^+, which is then released in the distant regions of the CNS. Some studies indicate that K_{IR} are also important for spatial buffer currents (Sontheimer, 1994). In addition to spatial K^+ buffering, K^+ is removed from the ECS by Na^+, K^+-ATPase. It has been suggested in a Na^+ recycling model that astrocyte Na^+ channels are involved in extracellular K^+ homeostasis by providing a 'return pathway' for Na^+ that is required for Na^+, K^+-ATPase operation during neuronal firing (Sontheimer et al., 1993; Sontheimer, 1994). Glial cells may be also involved in the maintenance of Cl^- homeostasis during neuronal activity (Fig. 6B; Bormann and Kettenmann, 1988). The release of GABA from presynaptic nerve terminals results in an influx of Cl^- from the ECS into the postsynaptic neuron and $[Cl^-]_e$ decreases. GABA receptors on glial cells that are activated in the vicinity of the synaptic cleft could cause an efflux of Cl^- and K^+ from glial cells and could be operative in the rapid restoration of $[Cl^-]_e$.

Glial cells also play an important role in buffering activity-related alkaline changes in extracellular pH (Chesler, 1990; Syková, 1992; Deitmer and Rose, 1996). Some of the membrane transport processes regulating intra- and extracellular pH, such as Na^+/H^+ exchange and $Na^+/H^+/Cl^-/HCO_3^-$ cotransport, are present in both neurons and glia, while others are specific either for neurons (e.g. H^+ channels, H^+ or HCO_3^- permeability of the ionic channels activated by GABA or glutamate) or for glia (e.g. voltage-dependent Na^+-HCO_3^- cotransport and lactate extrusion). The glial cell membrane is also readily permeable to CO_2, which reacts with H_2O to form carbonic acid, which in turn quickly dissociates into water and protons. This reaction is enhanced by the catalytic action of the enzyme carbonic anhydrase, which is almost exclusively present in glial cells.

Brain energy metabolism

Recent studies indicate that neurotransmitters such as noradrenaline, vasoactive intestinal peptide, adenosine or ATP may regulate the glucose fluxes and glycogen metabolism in astrocytes (Pellerin et al., 1997, for reviews see Magistretti and Pellerin, 1996; Tsacopoulos and Magistretti, 1996; Magistretti et al., 1999; see also Chapter 11 by Magistretti and Pellerin). Glutamate released from neurons is taken up by astrocytes, where it is converted into glutamine. Glutamine is released by astrocytes and taken up by neuronal terminals, where it is converted back to glutamate to re-establish its physiological levels (for review see Westergaard et al., 1995). Glutamine delivery to neurons is maintained by the volume transmission through the ECS. In addition, glutamate released from neurons stimulates aerobic glycolysis in astrocytes and thus provides lactate to neurons to fulfill their energy needs. This process is activated in several steps: glutamate uptake via Na^+-dependent transporters, activation of Na^+, K^+-ATPase, activation of glucose uptake, lactate release and finally, utilization of lactate by neurons. It was shown that neurotransmitters stimulate glucose utilization by the enhanced formation of glycogen, the major energy reserve in the brain.

Neuron-to-glia and glia-to-neuron signaling

During neuronal activity neurons release not only neurotransmitters, which evoke responses in postsynaptic cells and in glial cells located in the

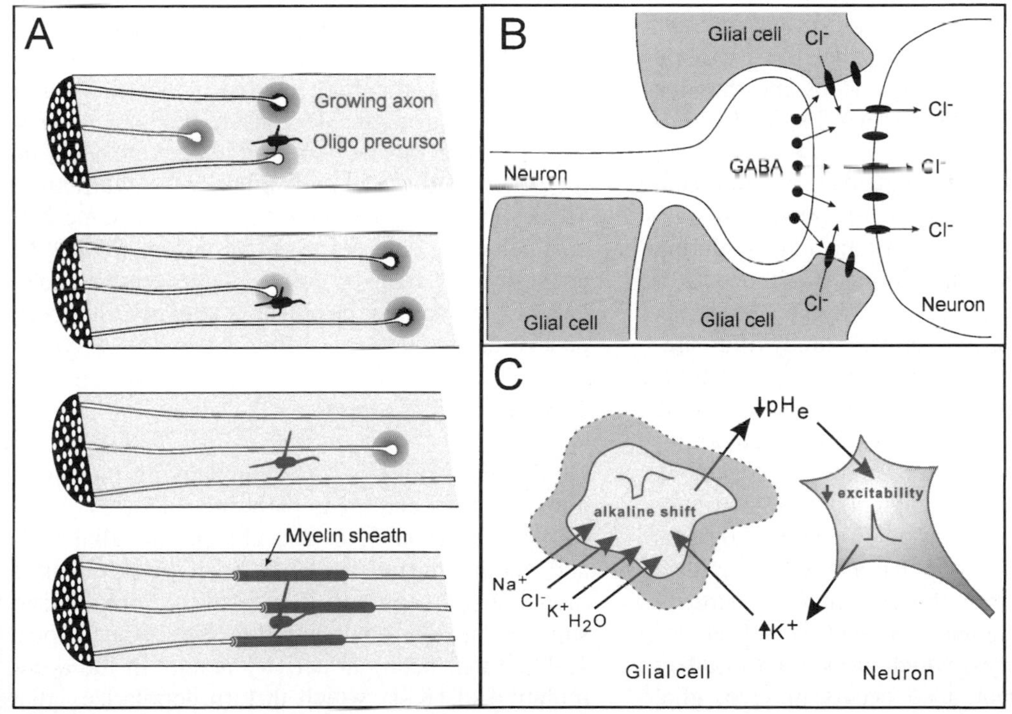

Fig. 6. Some of the proposed mechanisms of neuron-to-glia and glia-to-neuron interactions. A: Oligodendrocyte precursors may detect neurotransmitters released from the growth cones of migrating neurites. Neurotransmitter release thus may initiate oligodendrocyte maturation and start the ensheathment of the axons (adapted from Kettenmann et al., 1992). B: Maintenance of Cl^- homeostasis by glial cells during neuronal activity. The release of GABA into the synaptic cleft results in an influx of Cl^- into the postsynaptic neuron and $[Cl^-]_e$ decreases. Activation of GABA receptors on glial cells may in turn cause an efflux of Cl^- from glial cells and thus restore $[Cl^-]_e$ (adapted from Bormann and Kettenmann, 1988). C: A model of a non-specific feedback mechanism suppressing neuronal activity. Firing neurons release K^+, which accumulates in the extracellular space and thus depolarizes glial cells. This results in an increase of glial intracellular pH, a decrease of extracellular pH and the suppression of neuronal activity. Ionic transmembrane shifts also result in cell swelling, a decrease of the ECS volume and the larger accumulation of ions and neuroactive substances in the ECS (adapted from Syková, 1997).

vicinity of the synapse, but they also release a variety of other neuroactive substances, e.g. ions, transmitters, neurohormones, peptides and metabolites, which diffuse via the ECS to adjacent neurons and glial cells. The effect of such a release may be either short-term, e.g. ion regulation, release of substrates, transmitter clearance, or long-term, e.g. mitosis, regeneration, neuronal guidance during development or the onset of myelination, and is often associated with an increase of intracellular Ca^{2+} (for review see Verkhratsky et al., 1998). It has been proposed, for example, that the release of K^+ or other neuroactive substances, e.g. GABA, from neurons may regulate glial cell differentiation and myelin formation in white matter (Fig. 6A; Kettenmann et al., 1992). Oligodendrocyte precursors may detect the releasing transmitter from the growth cones of the migrating neurites and start the ensheathment of the axons. When the migration ends, the release of transmitter stops and thus the development and maturation of oligodendrocytes may halt.

Glial cells, in response to excitatory amino acid receptor activation, may also release growth factors, neuropeptides or neurotransmitters during development as well as during adulthood (for review see Gallo and Russell, 1995). For example, in astrocyte-neurone co-cultures bradykinin significantly increased calcium levels in neurons co-cultured with astrocytes, but not in single neurons (Parpura et al., 1994). The bradykinin-induced neuronal calcium elevation was blocked by the glutamate receptor antagonists D-2-amino-5-phosphonopentaonic acid or D-glutamylglycine. It was therefore suggested that astrocytes regulate neuronal calcium levels through the Ca^{2+}-dependent release of glutamate. If such a release of neuroactive substances, which are known to affect neuronal development, also occurs in vivo, glial cells may participate in activity-dependent plasticity in the adult brain (Gallo and Russell, 1995).

Synaptic transmission and neuronal function

It was shown in a number of studies that in the mature CNS, glial cell processes encapsulate synapses, thus several functions have been suggested for glial cells at the synaptic site (for reviews see Barres, 1991; Porter and Mccarthy, 1997). The glial cells associated with synapses are thought to take up transmitters, such as glutamate, to terminate neurotransmitter action; they may provide synaptic insulation to prevent neurotransmitter spillover to nearby synapses; they may accumulate Ca^{2+} from the synaptic cleft and reduce Ca^{2+}-dependent transmitter release; and they may even participate in long-term potentiation by synthesizing and releasing arachidonic acid, which induces a long-term, activity-dependent enhancement of synaptic transmission in the hippocampus.

Ritchie and colleagues have proposed in a Na^+ channel transfer model that neuroglial cells may support neuronal function by donating Na^+ channels to nearby axons to reduce the biosynthetic load of neurons, since the half-life of Na^+ channels has been estimated to be about 2 days (Bevan et al., 1985; Ritchie, 1988; for review see Oh, 1997). There is no compelling evidence that this transference exists in the nervous tissue, but at least three observations may support it: Na^+ channels are localized at the node of Ranvier as well as in astrocytes in the perinodal regions; the same Na^+ channel genes are expressed in neurons and glial cells; and the translocation of large protein molecules from axons to the adjacent glial cells has been observed.

Changes of ECS volume

Glial cell swelling is an accompanying phenomenon of normal and many pathological states; a link between ionic and volume changes in glial cells and signal transmission has been proposed in a model of a non-specific feedback mechanism suppressing neuronal activity (Fig. 6C; Syková, 1997). First, neuronal activity results in the accumulation of $[K^+]_e$, which in turn depolarizes glial cells, and this depolarization induces an alkaline shift in glial pH_i. Second, the glial cells extrude acid and the resulting acid shift causes a decrease in neuronal excitability. The ionic transmembrane shifts are accompanied by water movement, and thus this feedback mechanism may be amplified by activity-related glial swelling compensated for by ECS volume shrinkage and by increased tortuosity, presumably by the crowding of molecules of the

ECS matrix and/or by the swelling of fine glial processes. This, in turn, results in a larger accumulation of ions and other neuroactive substances in the nervous tissue due to more hindered diffusion in the ECS and thus affects volume transmission (Syková, 1997). In addition to $[K^+]_e$ increase, the activation of different neurotransmitter receptors on glial cells may also change astroglial morphology and volume and affect the excitability of neurons (for review see Porter and McCarthy, 1997). Electrically induced seizures in the hippocampal slice, for example, can be prevented by furosemide, which is known to block astrocytic swelling (Hochman et al., 1995; Syková et al., 1999). Such receptor-mediated communication between neurons and astrocytes may provide a feedback of the astrocytic regulation of neuronal excitability (Porter and McCarthy, 1997). In addition, clusters of compact ECS around oligodendrocytes revealed by tail current analysis in the corpus callosum as well as in the gray matter of the spinal cord (Chvátal et al., 1997, 1999a), may facilitate spatial K^+ buffering and selectively affect the diffusion of neuroactive substances in the CNS, i.e. volume transmission, in specific areas and/or directions.

Concluding remarks

Studies of glial cells during a period of more than 100 years have shown that they have distinct electrophysiological and morphological properties and that they play an important role in different brain functions, a role that was for many years under-estimated. Astrocytes and oligodendrocytes in the ECS express voltage- as well as ligand-activated currents that are involved in the maintainance of ionic and volume homeostasis as well as neuronal functions, in the release of neuroactive substances, in energy metabolism, in synaptic transmission and in volume transmission (see also Chapter 6 by Syková et al.). However, to fully understand how glial cells participate in neuronal signaling, recent studies of glial cells have focused on their role during pathological states such as anoxia and ischemia, during neurodegenerative diseases, in brain tumors and during aging and regeneration of the nervous tissue. Such research should help to elucidate the complex relationship between neurons and glial cells in normal as well as pathologically altered nervous tissue and to establish new transplantation and regeneration procedures in the treatment of CNS diseases.

Acknowledgements

Supported by grants VS 96 130, GAČR 307/96/K226, GAČR 309/97/K048, GAČR 309/99/0657 and GAČR 305/99/0655.

List of abbreviations

AMPA	a-amino-3-hydroxy-5-methyl-4-isoxazolepropionic acid
ATP	adenosinetriphosphate
CNS	central nervous system
CNQX	6-cyano-7-nitroqinoxaline-2,3-dione
ECS	extracellular space
GABA	γ-aminobutyric acid
GFAP	glial fibrillar acidic protein
I_{tail}	tail current
$[K^+]_e$	extracellular concentration of potassium ions
K_{DR}	potassium delayed outwardly rectifying currents
K_A	A-type potassium current
K_{IR}	potassium inwardly rectifying currents
NMDA	N-methyl-D-aspartate
P	postnatal day
PCR	polymerase chain reaction
V_{rev}	reversal potential
V_{mem}	membrane potential

References

Agnati, L.F., Zoli, M., Strömberg, I. and Fuxe, K. (1995) Intercellular communication in the brain: wiring vs. volume transmission. *Neuroscience*, 69: 711–726.

Akopian, G., Kressin, K., Derouiche, A. and Steinhäuser, C. (1996) Identified glial cells in the early postnatal mouse hippocampus display different types of Ca^{2+} currents. *Glia*, 17: 181–194.

Anděrová, M., Chvátal, A., Eliasson, C., Pěkný, M., Syková, E. (1999) Membrane properties and swelling of astrocytes in GFAP-/- mice. *Physiol. Res.*, Suppl. 1., 48: 52.

Backus, K.H. and Berger, T. (1995) Developmental variation of the permeability to Ca^{2+} of AMPA receptors in presumed hilar glial precursor cells. *Pflügers Arch.*, 431: 244–252.

Barres, B.A. (1991) New roles for glia. *J. Neurosci.*, 11: 3685–3694.

Barres, B.A., Koroshetz, W.J., Chun, L.L. Y. and Corey, D.P. (1990) Ion channel expression by white matter glia: the type–1 astrocyte. *Neuron.*, 5: 527–544.

Berger, T. (1995) AMPA-type glutamate receptors in glial precursor cells of the rat corpus callosum: ionic and pharmacological properties. *Glia*, 14: 101–114.

Berger, T., Schnitzer, J. and Kettenmann, H. (1991) Developmental changes in the membrane current pattern, K^+ buffer capacity, and morphology of glial cells in the corpus callosum slice. *J. Neurosci.*, 11: 3008–3024.

Berger, T., Walz, W., Schnitzer, J. and Kettenmann, H. (1992) GABA- and glutamate-activated currents in glial cells of the mouse corpus callosum slice. *J. Neurosci. Res.*, 31: 21–27.

Berger, T., Müller, T. and Kettenmann, H. (1995) Developmental regulation of ion channels and receptors on glial cells. *Persp. Dev. Neurobiol.*, 2: 347–356.

Bernstein, M., Lyons, S.A., Möller, T. and Kettenmann, H. (1996) Receptor-mediated calcium signalling in glial cells from mouse corpus callosum slices. *J. Neurosci. Res.*, 15: 152–163.

Bevan, S., Chiu, S.Y., Gray, P.T.A. and Ritchie, J.M. (1985) The presence of voltage-gated sodium, potassium and chloride channels in rat cultured astrocytes. *Proc. R. Soc. (Lond.)*, 225: 229–313.

Bicknell, H.R. and Beal, J.A. (1984) Axonal and dendritic development of substantia gelatinosa neurons in the lumbosacral spinal cord of the rat. *J. Comp. Neurol.*, 226: 508–522.

Bordey, A. and Sontheimer, H. (1997) Postnatal development of ionic currents in rat hippocampal astrocytes in situ. *J. Neurophysiol.*, 78: 461–477.

Bormann, J. and Kettenmann, H. (1988) Patch-clamp study of γ-aminobutyric acid receptor Cl^- channels in cultured astrocytes. *Proc. Natl Acad. Sci. USA.*, 85: 9336–9340.

Brockhaus, J., Illschner, S., Banati, R.B. and Kettenmann, H. (1993) Membrane properties of ameboid microglial cells in the corpus callosum slice from early postnatal mice. *J. Neurosci.*, 13: 4412–4421.

Celio, M.R., Spreafico, R., De Biasi, S. and Vitellaro-Zuccarello, L. (1998) Perineuronal nets: past and present. *Trends Neurosci.*, 21: 510–515.

Chesler, M. (1990) The regulation and modulation of pH in the nervous system. *Prog. Neurobiol.*, 34: 401–427.

Chvátal, A., Pastor, A., Mauch, M., Syková, E. and Kettenmann, H. (1995) Distinct populations of identified glial cells in the developing rat spinal cord slice: ion channel properties and cell morphology. *Eur.J. Neurosci.*, 7: 129–142.

Chvátal, A., Berger, T., Voříšek, I., Orkand, R.K., Kettenmann, H. and Syková, E. (1997) Changes in glial K^+ currents with decreased extracellular volume in developing rat white matter. *J. Neurosci. Res.*, 49: 98–106.

Chvátal, A., Anděrová, M., Žiak, D. and Syková, E. (1999a) Glial depolarization evokes a larger potassium accumulation around oligodendrocytes than around astrocytes in gray matter of rat spinal cord slices. *J. Neurosci. Res.*, 56: 493–505.

Chvátal, A., Anděrová, M., Vargová, L., Prokopová, Š., Žiak, D. and Syková, E. (1999b) Effect of osmotic stress on glial potassium tail currents and extracellular space diffusion parameters in the rat spinal cord slice. *Physiol. Res.*, Suppl. 1., 48: 61.

Deitmer, J.W. and Rose, C.R. (1996) pH regulation and proton signaling by glial cells. *Prog. Neurobiol.*, 48: 73–103.

Eder, C. (1998) Ion channels in microglia (brain macrophages). *Am. J. Physiol.*, 275: C327-C342.

Edwards, F.A., Konnerth, A., Sakmann, B. and Takahashi, T. (1989) A thin slice preparation for patch clamp recordings from neurones of the mammalian central nervous system. *Pflügers Arch.*, 414: 600–612.

Gallo, V. and Russell, J.T. (1995) Excitatory amino acid receptors in glia: different subtypes for distinct function? *J. Neurosci. Res.*, 42: 1–8.

Grosche, J., Matyash, V., Möller, T., Verkhratsky, A., Reichenbach, A. and Kettenmann, H. (1999) Microdomains for neuron-glia interaction: parallel fiber signalling to Bergmann glial cells. *Nature Neurosci.*, 2: 139–143.

Hild, W. and Tasaki, I. (1962) Morphological and physiological properties of neurons and glial cells in tissue culture. *J. Neurophysiol.*, 25: 277–304.

Hochman, D.W., Baraban, S.C. Owens, J.W.M. and Schwartzkroin, P.A. (1995) Dissociation of synchronization and excitability in furosemide blockade of epileptiform activity. *Science*, 270: 99–102.

Jabs, R., Kirchhoff, F., Kettenmann, H. and Steinhäuser, C. (1994) Kainate activates Ca^{2+}-permeable glutamate receptors and blocks voltage-gated K^+ currents in glial cells of mouse hippocampal slices. *Pflügers Arch.*, 426: 310–319.

Jendelová, P. and Syková, E. (1991) Role of glia in K^+ and pH homeostasis in the neonatal rat spinal cord. *Glia*, 4: 56–63.

Kettenmann, H. (1986) Oligodendrocytes control extracellular potassium by active uptake and spatial buffering. *Adv. Biosci.*, 61: 155–163.

Kettenmann, H., Syková, E., Orkand, R.K. and Schachner, M. (1987) Glial potassium uptake following depletion by intracellular ionophoresis. *Pflügers Arch.*, 410: 1–6.

Kettenmann, H., Blankenfeld, G.V. and Trotter, J. (1992) Physiological properties of oligodendrocytes during development. *Ann. NY Acad. Sci. USA*, 633: 64–77.

Kirchhoff, F., Mülhardt, C., Pastor, A., Becker, C.-M. and Kettenmann, H. (1996) Expression of glycine receptor subunits in glial cells of the rat spinal cord. *J. Neurochem.*, 66: 1383–1390.

Kirischuk, S., Möller, T., Voitenko, N., Kettenmann, H. and Verkhratsky, A. (1995) ATP-induced cytoplasmatic calcium mobilisation in Bergmann glial cells. *J. Neurosci.*, 15: 7861–7871.

Kirischuk, S., Tuschick, S., Verkhratsky, A. and Kettenmann, H. (1996) Calcium signalling in mouse Bergmann glial cells mediated by α_1-adrenoreceptors and H_1 histamine receptors. *Eur. J. Neurosci.*, 8: 1198–1208.

Kressin, K., Kuprijanova, E., Jabs, R., Seifert, G. and Steinhäuser, C. (1995) Developmental regulation of Na^+ and K^+ conductances in glial cells of mouse hippocampal brain slices. *Gli*, 15: 173–187.

Kuffler, S.W., Nicholls, J.G. and Orkand, R.K. (1966) Physiological properties of glial cells in the central nervous system of amphibia. *J. Neurophysiol.*, 29: 768–787.

Magistretti, P.J. and Pellerin, L. (1996) Cellular bases of brain energy metabolism and their relevance to functional brain imaging: Evidence for a prominent role of astrocytes. *Cereb. Cort.*, 6: 50–61.

Magistretti, P.J., Pellerin, L., Rothman, D.L. and Shulman, R.G. (1999) Energy on demand. *Science*, 283: 496–497.

Maleski, M. and Hockfield, S. (1997) Glial cells assemble hyaluronan-based pericellular matrices in vitro. *Glia*, 20: 193–202.

Müller, T., Möller, T., Berger, T., Schnitzer, J. and Kettenmann, H. (1992) Calcium entry through kainate receptors and resulting potassium-channel blockade in Bergmann glial cells. *Science*, 256: 1563–1566.

Müller, T., Grosche, J., Ohlemeyer, C. and Kettenmann, H. (1993) NMDA-activated currents in Bergmann glial cells. *NeuroReport*, 4: 671–674.

Müller, T., Fritschy, J.M., Grosche, J., Pratt, G.D., Mohler, H. and Kettenmann, H. (1994) Developmental regulation of voltage-gated K^+ channel and $GABA_A$ receptor expression in Bergmann glial cells. *J. Neurosci.*, 14: 2503–2514.

Müller, T., Möller, T., Neuhaus, J. and Kettenmann, H. (1996) Electrical coupling among Bergmann glial cells and its modulation by glutamate receptor activation. *Glia*, 17: 274–284.

Nicholson, C. and Syková, E. (1998) Extracellular space structure revealed by diffusion analysis. *Trends Neurosci.*, 21, 207–215.

Oh, Y. (1997) Ion channels in neuroglial cells. *Kaohsiung. J. Med. Sci.*, 13: 1–9.

Orkand, R.K., Nicholls, J.G. and Kuffler, S.W. (1966) Effect of nerve impulses on the membrane potential of glial cells in the central nervous system of amphibia. *J. Neurophysiol.*, 29: 788–806.

Parpura, V., Basarsky, T.A., Liu, F., Jeftinija, K., Jeftinia, S. and Haydon, P. (1994) Glutamate-mediated astrocyte-neuron signaling. *Nature*, 369: 744–747.

Pastor, A., Chvátal, A., Syková, E. and Kettenmann, H. (1995) Glycine- and GABA-activated currents in identified glial cells of the developing rat spinal cord slice. *Eur. J. Neurosci.*, 7: 1188–1198.

Pellerin, L., Stolz, M., Sorg, O., Martin, J.-L., Deschepper, C.F. and Magistretti, P.J. (1997) Regulation of energy metabolism by neurotransmitters in astrocytes in primary culture and in an immortalized cell line. *Glia*, 21: 74–83.

Porter, J.T. and McCarthy, K.D. (1995) GFAP-positive hippocampal astrocytes in situ respond to glutamatergic neuroligands with increases in $[Ca^{2+}]_i$. *Glia*, 13: 101–112.

Porter, J.T. and McCarthy, K.D. (1996) Hippocampal astrocytes in situ respond to glutamate released from synaptic terminals. *J. Neurosci.*, 16: 5073–5081.

Porter, J.T. and McCarthy, K.D. (1997) Astrocytic neurotransmitter receptors in situ and in vivo. *Prog. Neurobiol.*, 51: 439–455.

Prokopová, Š., Vargová, J. and Syková, E. (1997) Heterogenous and anisotropic diffusion in the developing rat spinal cord. *NeuroReport*, 8: 3527–3532.

Ritchie, J.M. (1988) Sodium-channel turnover in rabbit cultured Schwann cells. *Proc. R. Soc. (Lond.)*, 223: 423–430.

Seifert, G. and Steinhäuser, C. (1995) Glial cells in the mouse hippocampus express AMPA receptors with an intermediate Ca^{2+} permeability. *Eur. J. Neurosci.*, 7: 1872–1881.

Sontheimer, H. (1994) Voltage-dependent ion channels in glial cells. *Glia*, 11: 156–172.

Sontheimer, H. and Kettenmann, H. (1988) Heterogeneity of potassium currents in cultured oligodendrocytes. *Glia*, 1: 415–420.

Sontheimer, H. and Waxman, S.G. (1993) Expression of voltage-activated ion channels by astrocytes and oligodendrocytes in the hippocampal slice. *J. Neurophysiol.*, 70: 1863–1873.

Sontheimer, H., Trotter, J., Schachner, M. and Kettenmann, H. (1989) Channel expression correlates with differentiation stage during the development of oligodendrocytes from their precursor cells in culture. *Neuron*, 2: 1135–1145.

Sontheimer, H., Fernandez-Marques, E., Ullrich, N., Pappas, C.A. and Waxman, S.G. (1993) Astrocyte Na^+ channels are required for maintenance of Na^+/K^+-ATPase activity. *J. Neurosci.*, 14: 2464–2475.

Steinhäuser, C. and Gallo, V. (1996) News on glutamate receptors in glial cells. *Trends Neurosci.*, 19: 339–345.

Steinhäuser, C., Berger, T., Frotscher, M. and Kettenmann, H. (1992) Heterogeneity in the membrane current pattern of identified glial cells in the hippocampal slice. *Eur.J. Neurosci.*, 4: 472–484.

Steinhäuser, C., Jabs, R. and Kettenmann, H. (1994) Properties of GABA and glutamate responses in identified glial cells of the mouse hippocampal slice. *Hippocampus*, 4: 19–36.

Sturrock, R.R. (1980) Myelination of the mouse corpus callosum. *Neuropathol. Appl. Neurobiol.*, 6: 415–420.

Syková, E. (1992) Ionic and volume changes in the microenvironment of nerve and receptor cells. In: D. Ottoson, (Ed.), *Progress in Sensory Physiology*, Springer-Verlag, Heidelberg, pp. 1–167.

Syková, E. (1997) The extracellular space in the CNS: its regulation, volume and geometry in normal and pathological neuronal function. *Neuroscientist*, 3: 28–41.

Syková, E. and Chvátal, A. (1993) Extracellular ionic and volume changes: the role in glia-neuron interaction. *J. Chem. Neuroanatom.*, 6: 247–260.

Syková, E., Jendelová, P., Šimonová, Z. and Chvátal, A. (1992) K^+ and pH homeostasis in the developing rat spinal cord is impaired by early postnatal X-irradiation. *Brain Res.*, 594: 19–30.

Syková, E., Svoboda, J., Šimonová, Z., Lehmenkühler, A. and Lassmann, H., (1996) X-irradiation-induced changes in the diffusion parameters of the developing rat brain. *Neuroscience*, 70: 597–612.

Syková, E., Vargová, L., Prokopová, Š. and Šimonová, Z. (1999) Glial swelling and astrogliosis produce diffusion barriers in the rat spinal cord. *Glia*, 25: 56–70.

Takahashi, T., Misson, J.-P. and Caviness, V.S., Jr. (1990) Glial process elongation and branching in the developing murine neocortex: a qualitative and quantitative immunohistochemical analysis. *J. Comp. Neurol.*, 302: 15–28.

Tang, C.-M., Cohen, M.W. and Orkand, R.K. (1980) Electrogenic pumps in axons and neuroglia and extracellular potassium homeostasis. *Brain Res.*, 194: 283–286.

Tsacopoulos, M. and Magistretti, P.J. (1996) Metabolic coupling between glia and neurons. *J. Neurosci.*, 16: 877–885.

Tse, F.W., Fraser, D.D., Duffy, S. and MacVicar, B.A. (1992) Voltage-activated K^+ currents in acutely isolated hippocampal astrocytes. *J. Neurosci.*, 12: 1781–1788.

Verkhratsky, A., Orkand, R.K. and Kettenmann, H. (1998) Glial calcium: Homeostasis and signaling function. *Physiol. Rev.*, 78: 99–141.

Walz, W. (1989) Role of glial cells in the regulation of the brain ion microenvironment. *Prog. Neurobiol.*, 33: 309–333.

Westergaard, N., Sonnewald, U. and Schousboe, A. (1995) Metabolic trafficking between neurons and astrocytes: the glutamate/glutamine cycle revisited. *Dev. Neurosci.*, 17: 203–211.

Žiak, D., Chvátal, A. and Syková, E. (1998) Glutamate-, kainate- and NMDA-evoked membrane currants in identified glial cells in rat spinal cord slice. *Physiol. Res.*, 47: 365–375.

Zoli, M., Jansson, A., Syková, E., Agnati, L.F. and Fuxe, K. (1999) Volume transmission in the CNS and its relevance for neuropsychopharmacology. *TIPS*, 20: 142–150.

L. F. Agnati, K. Fuxe, C. Nicholson and E. Syková (Eds.)
Progress in Brain Research, Vol 125

CHAPTER 10

Glial modulation of neural excitability mediated by extracellular pH: a hypothesis revisited

Bruce R. Ransom

University of Washington School of Medicine, Department of Neurology, 1959 NE Pacific, Room RA 650, Box 356465, Seattle, WA 98195, USA

Introduction

By its nature, volume transmission (VT) focuses attention on the extracellular space (ECS) as a communication corridor. The essence of the process can be described as follows; information molecules are released by cells into the ECS where they travel by diffusion to signal cells at a distance. The concept of VT has mainly been employed for non-synaptic communication between neurons. Glial cells were not usually thought of as active participants in this process. If they were involved at all, it was mainly as structural elements that helped determine ECS dimensions.

Research discoveries in the last fifteen years, however, have established that signaling between neurons and glial cells, or between glial cells themselves, via VT is feasible; a vast array of established and potential information molecules can be released by glial cells. Glial cells also express receptors for most, if not all, of these same molecules (Kettenmann and Ransom, 1995). Given that glial cells can release and respond to a diverse collection of communication substances, a complex array of physiological interactions is possible. Perhaps the most fascinating to contemplate are the potential interactions between glia and neurons. While few details are presently known, I predict that numerous concrete examples of meaningful glial-neuronal signaling will emerge in the next five to ten years. Most important is the implication that brain function is not just the sum total of interactions between neurons, but also between glia and neurons. The latter, of course, must be mediated exclusively by VT because synapses are not found (to any appreciable extent (Peters et al., 1991)) between glia and neurons.

About 10 years ago, I proposed a model of an interaction between astrocytes and neurons based on activity-induced changes in extracellular ion concentrations, K^+ and H^+, to be specific (Ransom, 1992). The proposed interaction represented an example of VT mediated by ions. It was viewed as a negative feedback loop whereby active neurons would signal to nearby astrocytes the extent of their current activity, in the form of graded changes in extracellular K^+ concentration ($[K^+]_o$), and astrocytes would respond by producing graded degrees of extracellular acidification (Fig. 1). The acidification would, in turn, dampen neuronal excitability. I argued that this feedback system could function to adjust local brain excitability in relationship to ongoing activity; the more intense the activity, the greater the resulting acidification and reduction in excitability. This chapter will update certain aspects of the model that were not well understood 10 years ago, mainly, the nature of the astrocyte-induced extracellular acidification and the manner in which pH changes can alter neural excitability.

*Corresponding author. Tel.: (206) 543-2340; Fax: (206) 685-8100; e-mail: bransom@u.washington.edu

The model: Glial modulation of neural excitability mediated by extracellular pH

The initial model of how glia might modulate neural excitability via extracellular pH (pH_o) is shown in Fig. 1. Parts of this scheme were well established when it was published in 1992. Work on squid axons had established that K^+ released with action potential discharge would briefly accumulate within the surrounding ECS (Frankenhaeuser and Hodgkin, 1956). It was clear that axons and neurons released K^+ into the ECS in proportion to their activity, both action and synaptic potentials (Heinemann and Lux, 1975; Connors et al., 1982). In fact, Kuffler had speculated that increases in $[K^+]_o$ so nicely reflected the intensity of ongoing neural activity that this change could serve as a useful signal (Kuffler and Nicholls, 1966). In other words, changes in $[K^+]_o$ 'encode' the level of nearby neuronal activity, the larger the increase in $[K^+]_o$, the greater the activity. This $[K^+]_o$ signal, of course, represents a regional integral of neuronal activity, not highly specific information about the behavior of individual neurons.

Only a few studies have attempted to determine the magnitude of $[K^+]_o$ fluctuations during physiological activity. In most areas of the brain, $[K^+]_o$ seems to increase by no more than 1 to 3 mM during routine activity (Kelly and Van Essen, 1974; Somjen, 1979). Some areas may experience considerably greater activity-related increases in $[K^+]_o$, such as the retina (Coles et al., 1986). These estimates may be on the low side for technical reasons; the K^+-selective microelectrodes used to record $[K^+]_o$ fluctuations are very large compared to the diameter of the ECS. They create a small extracellular pool at the recording site because of local tissue destruction and this may 'dilute' the true $[K^+]_o$ increase (Ransom et al., 1987). It has also been claimed, based on indirect data, that there are restricted spaces in the brain, such as those between oligodendrocytes and axons, where K^+ discharged from neurons accumulates briefly to very high levels (i.e. in excess of 20 mM) (Lev-Ram and Grinvald, 1986; Chvátal et al., 1997). High levels of $[K^+]_o$ increase are seen with abnormal levels of neural discharge, epitomized by epilepsy. During interictal epileptiform bursts, for example, $[K^+]_o$ can reach levels of 10 to 12 mM (Futamachi and Pedley, 1976), close to the so-called ceiling level of $[K^+]_o$ accumulation that is not exceeded except during anoxia, hypoglycemia or spreading depression (Heinemann and Lux 1975; Connors et al., 1982). $[K^+]_o$ would accumulate to much higher levels were it not for active regulation by glial cells, especially astrocytes (Ballanyi et al., 1987; Newman, 1995; Ransom et al., 2000).

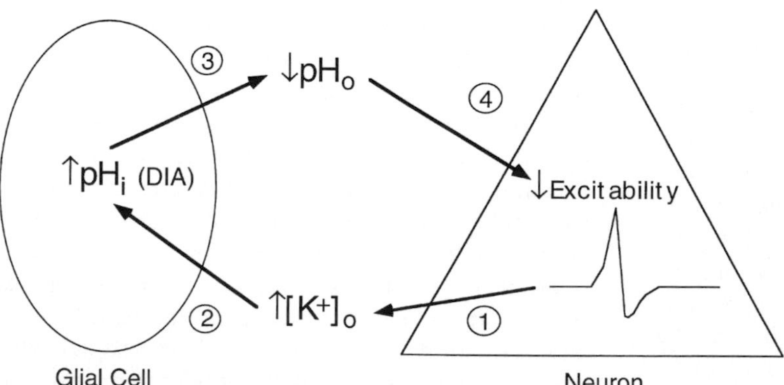

Fig. 1. Diagram of the hypothesized neuronal-glial interaction mediated by pH_o. Active neurons cause an increase in $[K^+]_o$ (1) which causes nearby glial cells to depolarize (2) leading to an intracellular alkaline shift known as depolarization-induced alkalinization (DIA). The same mechanism that causes DIA in glia also leads to an acid shift in pH_o (3) that decreases the excitability of neighboring neurons (4). Modified from Ransom (Ransom, 1992).

Kuffler and colleagues were also the first to show that glia are uniquely suited for registering increases in $[K^+]_o$ (Kuffler and Nicholls, 1966; Ransom and Carlini, 1986). Glial processes tightly invest neurons and their membranes are selectively permeable to K^+ making them, in essence, K^+ electrodes. Increases in $[K^+]_o$ instantly contact glial membranes and steeply depolarize them. What was unclear then, and remains unclear today, is what exactly glial cells make of these 'K^+ signals'.

One established consequence of K^+-induced depolarization of glia (both oligodendrocytes and astrocytes) is development of an alkaline shift in pH_i (Kraig, Ferreira-Filho et al., 1983; Chesler and Kraig, 1987; Kettenmann and Schlue, 1988; Pappas and Ransom, 1994; Rose and Ransom, 1998). This was first suspected based on a study in rat cortex by Chesler and Kraig (Chesler and Kraig, 1987). Stimulation of the cortical surface resulted in graded levels of intracellular alkaline shift in glial cells impaled by pH-sensitive microelectrodes. The alkaline shift mirrored the time course of membrane depolarization and increased as the magnitude of depolarization increased (Chesler and Kraig, 1989).

Because astrocyte membrane depolarization under these circumstances is due to the increase in $[K^+]_o$ (Ransom and Carlini, 1986), it was reasonable to assume that the proximate cause of depolarization-induced alkalinization (DIA) was the stimulated increase in $[K^+]_o$. Cortical stimulation, however, has multiple effects; it causes the release of neurotransmitters and transmembrane fluctuations of other ions, leaving open the possibility that DIA might have alternative explanations.

Neural stimulation also causes changes in pH_o (Kraig et al., 1983; Ransom et al., 1986; Deitmer, 1998). Activity-dependent changes in pH_o are especially marked with intense neural activity, and occur in a reliable but region-specific manner (Carlini and Ransom, 1986; Chesler, 1990; Syková, 1998). Some regions, like the CA1 region of the hippocampus, exhibit a large and rapid alkaline shift with stimulation, which gives way to a slower and more prolonged acid shift as the stimulation is continued. Stimulation in many other regions, such as the cingulate cortex and cerebellum, produces small alkaline shifts followed by acid shifts. Even regions that exhibit a prominent early extracellular alkaline shift (e.g. CA1 region of the hippocampus), develop an extracellular acid shift with continued intense neural activity (Carlini and Ransom, 1986, unpublished observations). The mechanisms of these activity-dependent changes in pH_o, both acidification and alkalinization, were not well understood 10 years ago. Circumstantial evidence strongly suggested that DIA importantly contributed to ECS acidification (Ransom, 1992). Using the enucleated rat optic nerve, which contains only glial cells (mainly astrocytes; (Ransom and Yamate, 1984)), increases in $[K^+]_o$ produced a concentration-dependent extracellular acidification that could be as large as 0.3 pH units (Ransom et al., 1986). Assuming that DIA was produced in optic nerve glial cells during elevation of $[K^+]_o$, it seemed reasonable to conclude that this mechanism alone was capable of robust extracellular acidification.

The idea that pH_o has the capacity to modulate neural activity has a long history (Tombaugh and Somjen, 1998). Hyperventilation has been known to trigger seizures in epileptic patients for three-quarters of a century. Hyperventilation, of course, lowers CO_2 producing a respiratory alkalosis within the brain. While it remains controversial if molecular CO_2, within the physiological range, can directly alter brain function, it is firmly established that the accompanying changes in pH_o can have powerful effects on neural excitability. For example, the evoked population spike amplitude in the CA1 area of the hippocampus decreases by 15–30% per 0.1 change in pH_o in the acid direction (Balestrino and Somjen, 1988). These authors felt that the effect of pH_o was on post-synaptic electrical excitability and not on transmitter release or action (see also (Church and McLennan, 1989)). The exact mechanism of these effects was not clear, however.

The above discussion briefly summarizes the arguments in favor of the model shown in Fig. 1. Steps 1 and 2 of this model, neural activity increasing $[K^+]_o$ and leading to DIA, were reasonably well established in 1992; the subsequent steps were more hypothetical. Advances in our understanding of DIA and its ability to produce extracellular acidification are discussed below.

Likewise, I will discuss new insights about how pH_o can modulate neural activity. While the model in question focuses on a negative feedback loop mediated by glial production of an extracellular acid shift, changes in pH_o with neural activity are complex and include alkaline shifts in pH_o. It is probable, therefore, that the feedback loop described here works in concert with other processes that alter pH_o.

Depolarization-induced alkalinization (DIA)

Intracellular pH regulation in neurons and glial cells has been intensively studied over the last ten years (Kaila and Ransom, 1998). Both neurons and glial cells (I will focus primarily on astrocytes) actively regulate intracellular pH; astrocytes maintain pH_i at a level 0.1–0.2 pH unit more acidic than pH_o. Because astrocytic membrane potential is about –80 mV, however, glial pH_i is still significantly more alkaline than would be expected from a passive distribution of H^+ across the cell membrane. This means that astrocytes actively extrude acid by way of plasma membrane transporters, creating an inwardly directed electrochemical gradient for H^+. Essentially the same situation exists for neurons. Mammalian neurons maintain their intracellular pH at a level which is much more alkaline than would be expected if H^+ were at electrochemical equilibrium. For example, if a neuron had a membrane potential of –60mV in a solution of pH 7.3, pH_i would be about 6.3 if H^+ equilibrated freely across the membrane. Active processes, however, keep pH_i of typical neurons above 7.0 (Bevensee and Boron, 1998).

Several membrane transport mechanisms are involved in regulating intracellular pH (Fig. 2A). Many of these are common to both neurons and glia. Both cell types express the Na^+-H^+ exchanger. This powerful mechanism harnesses the electrochemical gradient of Na^+ to directly extrude H^+ from cells. The other net acid extruder common to both cell types is the Na^+-dependent Cl^-/HCO_3^- exchanger. This mechanism again uses the energy stored in the Na^+ gradient to drive HCO_3^- uptake, an event equivalent to H^+ efflux. The Na^+-H^+ exchanger is the predominant acid extruder in neurons. Glial cells alone express Na^+/HCO_3^- cotransport, and this is the main acid extruder in these cells. Both cell types express Cl^-/HCO_3^- exchange which acts, under normal conditions, as an acid loader. Details about these transport mechanisms should be sought in recent reviews (Deitmer and Rose, 1996; Bevensee and Boron, 1998; Rose and Ransom, 1998).

DIA is seen in glial cells but not neurons (Rose and Ransom, 1998). This phenomenon was first described in kidney epithelial cells (Siebens and Boron, 1989). The term DIA is somewhat misleading because it suggests that only membrane potential changes in the depolarizing direction affect glial pH_i. In fact, pH_i is a continuous function of membrane potential in glial cells under physiological conditions, such that depolarization causes an alkaline shift in pH_i and hyperpolarization causes an acid shift in pH_i (Pappas and Ransom, 1994; Deitmer, 1998; Rose and Ransom, 1998). DIA, therefore, is a mechanism whereby depolarizing or hyperpolarizing events can predictably affect glial pH_i. Important pH-sensitive glial properties can therefore be influenced by changes in membrane potential, including gap junction communication (Ransom, 1995), glucose metabolism (Salem et al., 1975), and glutamate metabolism (Brookes, 1997).

Theoretically, DIA can be caused by activation of Na^+/HCO_3^- cotransport, H^+/lactate cotransport, or the H^+ pump. While evidence exists in support of each of these mechanisms under certain conditions, the preponderance of data strongly indicate that under physiological conditions depolarization-activation of electrogenic Na^+/HCO_3^- cotransport is the most important mechanism (Rose and Ransom, 1998).

The glial Na^+/HCO_3^- cotransporter has been studied most exhaustively in giant neuropil glial cells of the leech (Deitmer, 1998). The established stoichiometry of the transporter in these cells is $1Na^+$ and $2HCO_3^-$. Inward cotransport via this mechanism, therefore, results in membrane hyperpolarization. Like most Na^+/HCO_3 cotransporters, but not all (Kettenmann and Schlue, 1988; Grichtchenko and Chesler, 1994), this mechanism is strongly inhibited in leech glial cells by the stilbene 4,4′-diisothiocyanostilbene-2,2′-disulphonic acid (DIDS). Because of its electrogenic nature, inward cotransport is stimulated by depolarization and outward cotransport by hyperpolarization. In the presence of CO_2/HCO_3^-, voltage clamped leech glial cells showed robust and rapid intracellular pH shifts with changes in membrane potential; alkaline shifts with depolarization and acid shifts with hyperpolarization (Deitmer, 1998).

Several groups have studied DIA in mammalian astrocytes. Cultured rat hippocampal astrocytes depolarized by increases in $[K^+]_o$ developed a largely CO_2/HCO_3^--dependent alkalinization that is markedly reduced by DIDS (O'Connor et al., 1994; Pappas and Ransom, 1994). Similar findings were reported for cultured cerebral astrocytes (Brookes and Turner, 1994). Even small membrane depolarizations, e.g. increasing $[K^+]_o$ from 3 to 6 mM, cause a significant alkaline shift (about 0.1 pH unit). Reactive astrocytes within slices of gliotic

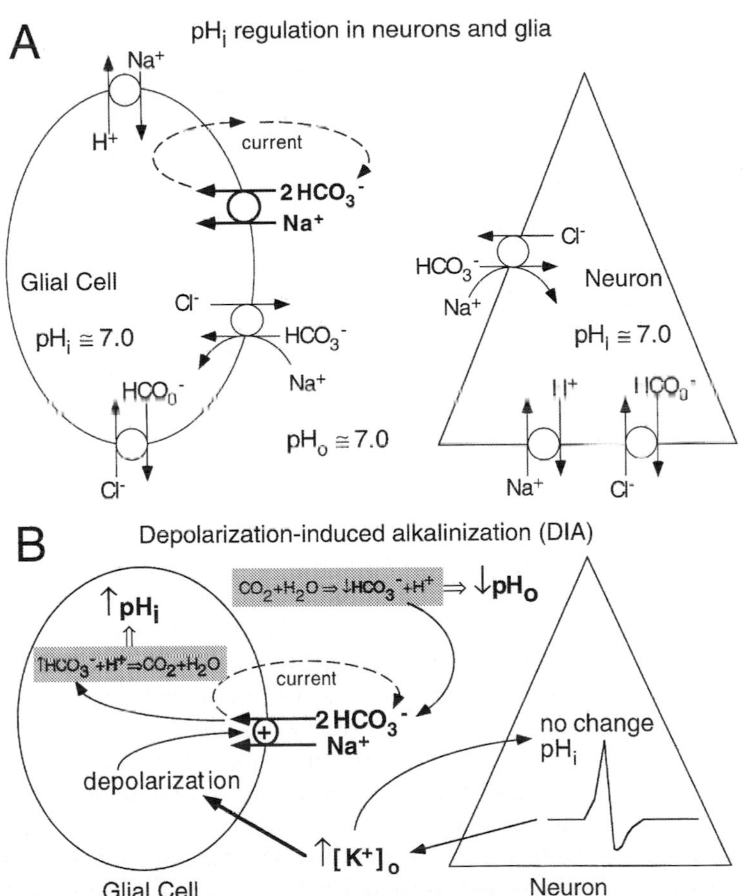

Fig. 2. Intracellular pH regulation and depolarization-induced alkalinization (DIA). A. The major mechanisms of pH_i regulation in neurons and astrocytes are shown. Both neurons and astrocytes exhibit the following three pH_i regulating exchangers: H^+-Na^+ exchange, Na^+-dependent Cl^- – HCO_3^- exchange, and Cl^- – HCO_3^- exchange. Under physiological conditions, H^+-Na^+ exchange and Na^+-dependent Cl^- – HCO_3^- exchange act as acid extruders. The Cl^- – HCO_3^- exchanger functions as an acid loader. Only glial cells (both astrocytes and oligodendrocytes) exhibit Na^+- HCO_3^- cotransport (stoichiometry of $2HCO_3^-$: Na^+). Operation of this transport mechanism, therefore, generates a current making it electrogenic and subject to influence by membrane potential. The Na^+- HCO_3^- cotransporter usually acts as a net acid extruder. B. DIA is primarily mediated by activation of Na^+- HCO_3^- cotransport (see text). Depolarization of glial cells due to increased $[K^+]_o$ causes DIA because it thermodynamically favors inward flux of HCO_3^- and Na^+ by the cotransporter. As a consequence, the intracellular concentration of HCO_3^- increases leading to a decrease in $[H^+]_i$ (i.e. an increase in pH_i) via the bicarbonate buffer system. DIA also leads to a decrease in pH_o: the $[HCO_3^-]_o$ falls as the cotransporter operates and this increases $[H^+]_o$ (i.e. a decrease in pH_o).

hippocampus exhibit DIA which can be largely accounted for by Na^+/HCO_3^- cotransport. DIA, however, was not reduced by DIDS and was partially Na^+-independent (Grichtchenko and Chesler, 1994).

DIA and extracellular acidification

It is important to consider how DIA is coupled to extracellular acidification. Kaila and Chesler (Kaila and Chesler, 1998) point out that "acid-base equivalents are neither generated de novo, nor consumed, within the ECS, which implies that all pH_o shifts are caused by net movements of acid or base into or out of the interstitium." DIA represents net movement of HCO_3^- into glial cells. Of necessity, this net movement of base out of the ECS must produce an extracellular acidification (Fig. 2B). The actual magnitude of the shift in pH_o associated with movement of acid-base equivalents across cell membranes depends upon extracellular buffering capacity; the higher the buffering capacity, the smaller will be the change in pH_o for a given flux of acid or base equivalents. It is now believed that the buffering power of the ECS may be only about one-fourth of its theoretical value in a CO_2/HCO_3^- system (Chesler et al., 1994). The expression of rapid changes in pH_o also depends upon sufficient carbonic anhydrase (CA) activity in the ECS. In the absence of CA the hydration and dehydration of CO_2 proceeds very slowly delaying changes in H^+ concentration. Whether CA inhibition increases or decreases a given change in pH_o can be complex and depends in part on the type of instantaneous acid or alkaline load imposed on the ECS (Voipio, 1998).

The coupling of DIA to changes in pH_o has been carefully studied in gliotic hippocampal slices. Astrocytes in these slices exhibit typical DIA, due at least in part, to accelerated Na^+/HCO_3^- cotransport (Grichtchenko and Chesler, 1994). Graded increases in $[K^+]_o$ produced graded extracellular acid shifts. These shifts were entirely blocked in Na^+-free bathing solution as would be expected if this solution completely abolished Na^+/HCO_3^- cotransport. Further definitive evidence linking inward Na^+/HCO_3^- cotransport with extracellular acidification has been obtained in studies of specialized glial cells in the retina called Muller cells (Newman, 1996). Freshly dissociated individual Muller cells were exposed to increases in $[K^+]_o$ while recording both pH_i and pH_o. Extracellular pH was measured using an indicator dye attached to the surface on which the Muller cell lay; increasing $[K^+]_o$ caused DIA and produced a rapid extracellular acidification immediately under the cell. The magnitude of the extracellular acidification was greatest beneath the cell endfeet where cotransporter sites are preferentially localized (Newman, 1996). The leech nervous system has also proved useful in linking DIA and interstitial acidification (Rose and Deitmer, 1994). If the principal ganglionic glial cell was prevented by voltage clamp from depolarizing with repetitive stimulation, this blocked extracellular acidification and actually unmasked an interstitial alkaline shift.

Other mechanisms may contribute to activity-dependent acidification. The rat optic nerve, a representative white matter tract, exhibits activity-dependent extracellular acid shifts that are partly due to direct acid secretion (Ransom et al., 1994; Kaila and Chesler, 1998). A likely candidate is lactate/H^+ transport from astrocytes into the ECS. Astrocytes are known to secrete far more lactate than neurons and this is increased with K^+ application ((Walz and Mukerji, 1988); see also (Kraig et al., 1983)). Astrocytes are metabolically well adapted for this; they have higher glucose consumption than neurons (Hertz and Peng, 1992) and are the only cells in the brain with significant glycogen stores (Ransom and Fern, 1997). Although lactate/H^+ transport is not explicitly shown in Fig. 2B, it may play an important role in glial-mediated extracellular acidification, perhaps with regional variability. The participation of this mechanism, it should be noted, does not change the basic tenet of the scheme under discussion (i.e. Fig. 1). Astrocyte lactate/H^+ transport would be activated in a graded fashion by increases in $[K^+]_o$ and would lead to an acid shift in pH_o.

Effects of pH_o on neural excitability

A number of pH-sensitive targets on neurons could mediate pH_o effects on excitability. *N*-methyl-D-aspartate (NMDA)-type glutamate receptors are powerfully modulated by pH_o (Tang et al., 1990; Traynelis and Cull-Candy, 1990; Vyklicky et al., 1990) (Fig. 3). The mechanism of this modulation is hypothesized to be a pH sensor located on one of the NMDA receptor subunits (Traynelis, 1998). Certain splice variants of this subunit are proton-insensitive (Traynelis et al., 1995). Neurons with pH-insensitive NMDA receptors may be more likely to mediate damage during ischemia or seizures because they will not be inhibited by the falling pH associated with these conditions (Traynelis, 1998). AMPA-type glutamate receptors are also inhibited by extracellular protons but at much lower pH_o values than NMDA receptors. Nevertheless, this may have some importance under extreme pathological conditions.

GABA is the dominant inhibitory neurotransmitter in the brain (Kaila, 1994). GABA-mediated inhibition is potentiated by acid shifts in pH_o. The pH sensitivity of the GABA receptor is dependent upon its subunit composition; receptors containing the delta subunit have substantially increased pH-sensitivity. It is obvious that the actions of pH_o on glutamate and GABA receptors has a similar net effect on neural excitability: acid shifts in pH_o would tend to decrease excitability by diminishing excitatory synaptic transmission mediated by glutamate, especially NMDA-type, receptors and augmenting inhibitory synaptic transmission mediated by GABA receptors. Other ligand-gated ion channels exhibit variable degrees of pH-sensitivity (Traynelis, 1998). For example, protons potentiate ATP-activated channels; at this time, however, the consequences of this modulation on aggregate neural activity remain uncertain.

Voltage-gated ion channels are also variably sensitive to pH_o (Tombaugh and Somjen, 1998). In general, acidification tends to depress voltage-gated ion currents. Na^+ and Ca^{2+} channels are considerably more sensitive to pH than are most K^+ channels (Fig. 3). An exception to this is the inward rectifier potassium channel HIR which is strongly depressed by protons and exhibits a pK near 7.4 (Coulter et al., 1995). High-voltage activated Ca^{2+} channels are especially sensitive to pH (Tombaugh and Somjen, 1998). The pK of the peak high-voltage activated Ca^{2+} current in mammalian neurons is close to physiological pH_o. Ca^{2+} channel conductance could be substantially modulated bidirectionally by pH shifts as small as 0.1 pH unit.

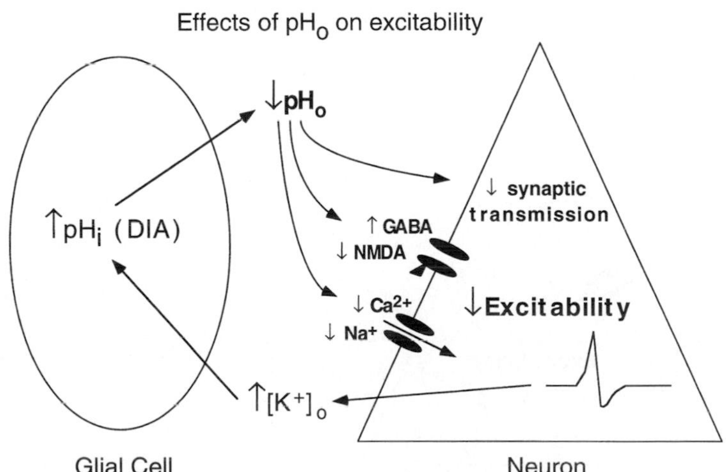

Fig. 3. Effects of pH_o on excitability. This diagram illustrates some of the presumed mechanisms whereby pH_o influences neural excitability. In the context of DIA, pH_o shifts in the acid direction. This decreases neural excitability by: 1. increasing $GABA_A$–mediated inhibition, 2. decreasing excitation mediated by NMDA-type glutamate receptors, 3. blocking Ca^{2+} channels (present also at presynaptic terminals where they mediate transmitter release), 4. blocking Na^+ channels, 5. decreasing synaptic transmission. The effects of an acid shift in pH_o are illustrated but it should be emphasized that an alkaline shift in pH_o would have exactly the opposite effects on excitability.

In general, these data provide strong support for the idea that acid shifts in pH_o depress and alkaline shifts increase neuronal excitability. Modulation of Ca^{2+} channels by pH_o may mediate especially complex effects on excitability because of the unique role of these channels in Ca^{2+} entry at synapses. Still, the net effect of Ca^{2+} channel blockade, at both synaptic and non-synaptic sites, would be a reduction in excitability. While I have primarily summarized information about the effects of pH_o on neural excitability, there is reason to expect that excitability may also be influenced by changes in pH_i. The effects of changing pCO_2 on excitatory synaptic transmission for example, are thought to be mediated at least in part by changes in pH_i (Lee et al., 1996). Figure 3 schematically summarizes some of the ways in which acid shifts in pH_o might depress the excitability of neural ensembles. The magnitude of pH_o effects may vary from region to region within the brain as a consequence of regional variability in the expression of pH-sensitive voltage-gated and ligand-gated channels. It is also likely that some kinds of neural activity may be more susceptible to pH modulation than others. We have found, for example, that the frequency of epileptiform discharges in hyperactive hippocampal brain slices is extremely sensitive to changes in pH_o; on average, spike frequency changed ~25% for each 0.05 pH unit change in pH_o (Lee et al., 1993).

The model revisited

Global changes in brain pH have marked effects on neural excitability; alkaline shifts in brain pH are excitatory and pro-convulsant while acid shifts are inhibitory and anti-convulsant (Tombaugh and Somjen, 1998). Undoubtedly these imposed alterations in pH, for example inhalation of CO_2, exert their influence by acting on several of the proton-sensitive targets discussed above. The theory under discussion, however, requires that brain generated changes in pH_o influence neural activity; specifically, that glial-generated extracellular acidification downmodulate excitability. In the strictest sense, there is no direct experimental evidence to support this. In fact, in most areas of the brain, brief periods of neural activity produce alkaline, not acid changes in pH_o.

It is important to keep in mind, however, that the acidification generated by DIA in glial cells will be integrated with other activity-evoked changes in pH_o (Fig. 4). In the hippocampus, for example, neural activity produces extremely rapid and large

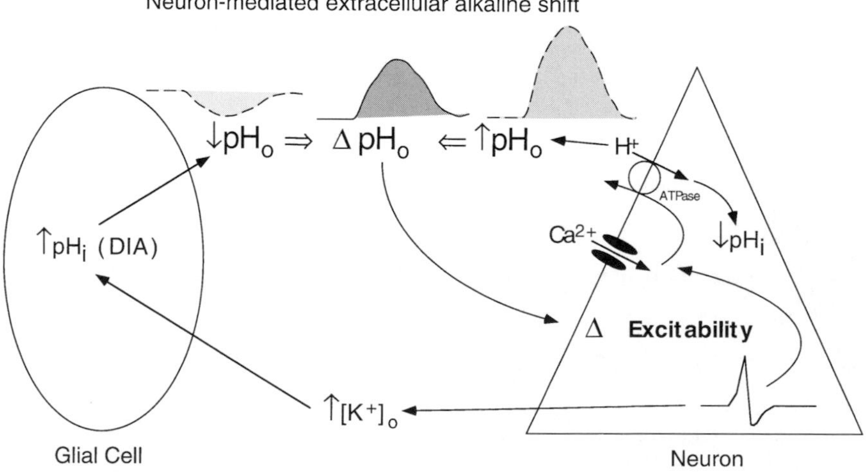

Fig. 4. Neuron-mediated extracellular alkaline shift. The mechanism of neuron-mediated extracellular alkalinization is illustrated; it is believed that Ca^{2+} enters neurons through activated Ca^{2+} channels and is then rapidly removed by Ca^{2+}-ATPase which countertransports H^+. This produces an acid shift within the neuron and an extracellular alkaline shift. The diagram also emphasizes that the extracellular alkaline shift mediated by neurons will be modified by the extracellular acid shift mediated by glial cells. In the case illustrated, the neuron-mediated alkaline shift is reduced in size.

extracellular alkaline shifts. These are believed to be a consequence of Ca^{2+} entry into neurons followed by rapid extrusion by the Ca^{2+}/H^+ exchanger (i.e. Ca^{2+}-ATPase) (Schwiening et al., 1993, Trapp, 1996 p. 548). At the same time that this large alkaline shift is being generated by neurons, nearby glial cells are experiencing DIA and act to modulate the extent of the alkaline shift by acid excretion (i.e. HCO_3^- removal from the ECS). Viewed in this manner, the *observed* alkaline transient represents the integral of the neuron-mediated alkalinizing event and the glial-mediated acidifying event; in the hippocampus, the sum of these two events is a substantial alkaline shift (Fig. 4). Without DIA, however, the alkaline shift would be larger still. The exact changes in pH_o that will result from neural activity will vary from region to region but are likely to represent in all cases the integral of glial and neuronal influences.

Considering that even alkaline shifts harbor a masked contribution from glial DIA, there is evidence that these 'intrinsic' pH_o signals can affect neural activity. Stimulus evoked net alkaline shifts in the hippocampus can clearly modulate excitatory synaptic transmission (Taira et al., 1993; Gottfried and Chesler, 1994). When these alkaline shifts were magnified by blocking CA, NMDA-mediated synaptic responses were enhanced. This represents an important proof of principle: Alterations of pH_o secondary to neural activity can have local and very rapid effects on excitability. In brain regions exhibiting powerful alkaline shifts, glial DIA probably functions primarily to reduce the magnitude of these shifts (Figs 4 and 5). Of course, as mentioned earlier, with very intense and prolonged activity an acid shift in pH_o gradually emerges and is likely to be mainly glial in origin (Carlini and Ransom, 1986).

The elaborated version of the hypothesized proton-mediated interaction between glia and neurons is shown schematically in Fig. 5. The initial steps are unchanged from the original proposal

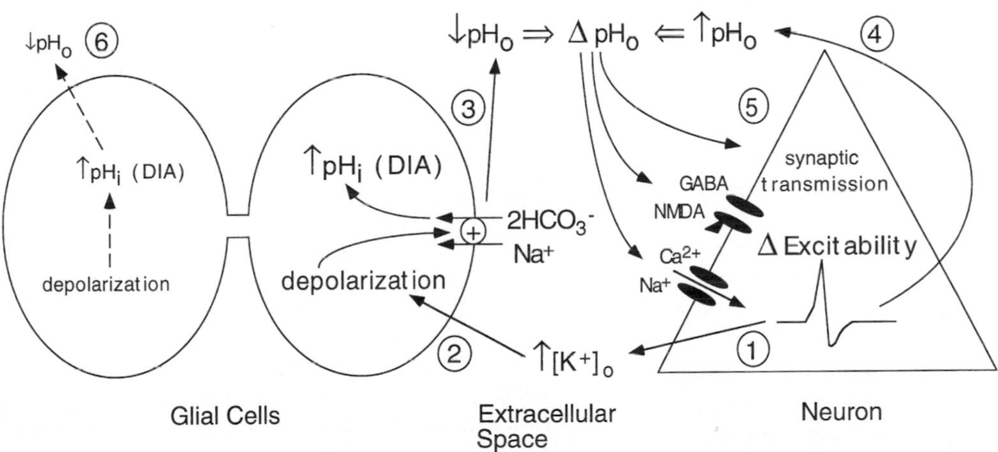

Fig. 5. Proton-mediated VT between glia and neurons. Schematic illustration of the steps in the proposed negative feedback loop between glia and neurons. Discharging neurons increase $[K^+]_o$ in proportion to their activity (1). Nearby glial cells are depolarized and this enhances Na^+–HCO_3^- cotransport causing DIA of graded magnitude (2). DIA leads to extracellular acidification, again in proportion to the inciting increase in $[K^+]_o$ (3). Activity in neurons can also cause extracellular alkaline shifts (4) (see Fig. 4). The final change in pH_o (Δ pH_o) is the integral of glial acidification and neuronal alkalinization. Finally, the change in pH_o, in a feedback fashion, alters the excitability of nearby neurons by modulation of ligand-gated and voltage-gated channels, and by affecting synaptic transmission (5). Neuronal excitability may be increased or decreased depending on whether the pH_o transient is alkaline or acid, respectively. According to the hypothesis, however, the glial contribution to pH_o is always acid excretion (i.e. always in the direction of decreasing excitability). Glial cells, especially astrocytes, are connected by gap junctions, which would mediate depolarization of cells some distance away from the site of focal increased $[K^+]_o$. These distal glia would experience weak DIA producing weak extracellular acidification (6); this could function as a special form of surround inhibition.

((Ransom, 1992), Fig. 1). Neural activity increases $[K^+]_o$ which signals the extent of this activity to adjacent glial cells in the form of graded depolarization; this, in turn, generates a proportional extracellular acidification by way of DIA (perhaps assisted by lactate/H^+ efflux; see above). At the same time, however, active neurons often import acid equivalents which tend to alkalinize the ECS (step 4). The net change in pH_o will be the integral of these opposite influences (see also Fig. 4). The changing level of pH_o will exercise an ongoing effect on neural excitability through actions on neurotransmitter receptors, voltage-gated ion channels and synaptic transmission. The polarity of this effect is clear: extracellular acidosis depresses and extracellular alkalosis heightens neural activity. The most marked changes in pH_o are likely to occur in the immediate vicinity of active neuronal ensembles but glial cells are well suited to extend the consequences of this interaction to distant sites. Astrocytes are strongly connected to one another by gap junctions allowing depolarization to spread to distant cells (Ransom, 1995). If the inciting depolarization is intense, as with epileptic activity, this could produce significant DIA in the distant cells and lead to an interesting variant of surround inhibition mediated by extracellular acidification (step 6).

While the concept discussed here remains a hypothesis, there is growing conviction that pH is an important mediator of brain function (Kaila and Ransom, 1998). The complexity of studying this variable has delayed recognition of its importance and will continue to challenge those of us interested in understanding how pH fits into the larger picture of brain function and dysfunction. I have attempted to be clear in stating one scenario about how things may work, hoping that this will sharpen our collective aim at key targets which are experimentally testable.

References

Balestrino, M. and Somjen, G.G. (1988) Concentration of carbon dioxide, interstitial pH and synaptic transmission in hippocampal formation of the rat. *J. Physiol (Lond.)*, 396: 247–266.

Ballanyi, K., Grafe, P. and ten Bruggencate, G. (1987) Ion activities and potassium uptake mechanisms of glial cells in guinea-pig olfactory cortex slices. *J. Physiol. (Lond.)*, 382: 159–174.

Bevensee, M.O. and Boron, W.F. (1998) pH regulation in mammalian neurons. In: K. Kaila and B.R. Ransom (Eds), *pH and Brain Function*. New York, Wiley-Liss: 211–231.

Brookes, N. (1997) Intracellular pH as a regulatory signal in astrocyte metabolism. *Glia*, 21: 64–73.

Brookes, N. and Turner, R.J. (1994) K^+-induced alkalinization in mouse cerebral astrocytes mediated by reversal of electrogenic Na^+-HCO_3^- cotransport. *Am. J. Physiol.*, 267: C1633–1640.

Carlini, W.J. and Ransom, B.R. (1986) Regional variation in stimulated extracellular pH transients in the mammalian CNS. *Soc. Neurosci. Abstr.*, 12: 452.

Chesler, M. (1990). The regulation and modulation of pH in the nervous system. *Prog. Neurobiol.*, 34: 401–427.

Chesler, M., Chen, J.C. and Kraig, R.P. (1994) Determination of extracellular bicarbonate and carbon dioxide concentrations in brain slices using carbonate and pH-selective microelectrodes. *J. Neurosci. Meth.*, 53: 129–136.

Chesler, M. and Kraig, R.P. (1987) Intracellular pH of astrocytes increases rapidly with cortical stimulation. *Am. J. Physiol.*, 253: R666–670.

Chesler, M. and Kraig, R.P. (1989) Intracellular pH transients of mammalian astrocytes. *J. Neurosci.*, 9: 2011–2019.

Church, J. and McLennan, H. (1989). Electrophysiological properties of rat CA1 pyramidal neurones in vitro modified by changes in extracellular bicarbonate. *J. Physiol. (Lond.)*, 415: 85–108.

Chvátal, A., Berger, T., Vorisek, I., Orkand, R.K., Kettenmann, H. and Sykova, E. (1997) Changes in glial K^+ currents with decreased extracellular volume in developing rat white matter. *J. Neurosci. Res.*, 49: 98–106.

Coles, J.A., Orkand, R.K., Yamate, C.L. and Tsacopoulos, M. (1986) Free concentrations of Na, K, and Cl in the retina of the honeybee drone: stimulus-induced redistribution and homeostasis. *Ann. NY Acad. Sci.*, 481: 303–317.

Connors, B.W., Ransom, B.R., Kunis, D.M. and Gutnick, M.J. (1982) Activity-dependent K^+ accumulation in the developing rat optic nerve. *Science*, 216: 1341–1343.

Coulter, K.L., Perier, F., Radeke, C.M. and Vandenberg, C.A. (1995) Identification and molecular localization of a pH-sensing domain for the inward rectifier potassium channel HIR. *Neuron*, 15: 1157–1168.

Deitmer, J.W. (1998) pH Regulation in Invertebrate Glia. In: K. Kaila and B.R. Ransom (Eds), *pH and Brain Function*. New York, Wiley-Liss: 233–252.

Deitmer, J.W. and Rose, C.R. (1996) pH regulation and proton signalling by glial cells. *Prog. Neurobiol.*, 48: 73–103.

Frankenhaeuser, B. and Hodgkin, A.L. (1956) The after-effects of impulses in the giant nerve fibers of Loligo. *J. Physiol. (Lond.)*, 131: 341–376.

Futamachi, K.J. and Pedley, T.A. (1976) Glial cells and extracellular potassium: their relationship in mammalian cortex. *Brain Res.*, 109: 311–322.

Gottfried, J.A. and Chesler, M. (1994) Endogenous H^+ modulation of NMDA receptor-mediated EPSCs revealed by carbonic anhydrase inhibition in rat hippocampus. *J. Physiol. (Lond.)*, 478.3: 373–378.

Grichtchenko, II. and Chesler, M. (1994) Depolarization-induced acid secretion in gliotic hippocampal slices. *Neuroscience*, 62: 1057–1070.

Grichtchenko, II. and Chesler, M. (1994) Depolarization-induced alkalinization of astrocytes in gliotic hippocampal slices. *Neuroscience*, 62: 1071–1078.

Heinemann, U. and Lux, H.D. (1975) Undershoots following stimulus-induced rises of extracellular potassium concentration in cerebral cortex of cat. *Brain Res.*, 93: 63–76.

Hertz, L. and Peng, L. (1992) Energy metabolism at the cellular level of the CNS. *Can. J. Physiol. Pharmacol.*, 70: S145–157.

Kaila, K. (1994) Ionic basis of GABAA receptor channel function in the nervous system. *Prog. Neurobiol.*, 42: 489–537.

Kaila, K. and Chesler, M. (1998) Activity-evoked changes in extracellular pH. In: K. Kaila and B.R. Ransom (Eds), *pH and Brain Function*. New York, Wiley-Liss: 309–338.

Kaila, K. and Ransom, B.R. (1998) *pH and Brain Function*. New York, Wiley-Liss.

Kelly, J.P. and Van Essen, D.C. (1974) Cell structure and function in the visual cortex of the cat. *J. Physiol. (Lond.)*, 238: 515–547.

Kettenmann, H. and Ransom, B.R. (1995) *Neuroglia*. New York, Oxford University Press.

Kettenmann, H. and Schlue, W.R. (1988) Intracellular pH regulation in cultured mouse oligodendrocytes. *J. Physiol. (Lond.)*, 406: 147–162.

Kraig, R.P., Ferreira-Filho, C.R. and Nicholson, C. (1983) Alkaline and acid transients in cerebellar microenvironment. *J. Neurophysiol.*, 49: 831–850.

Kuffler, S.W. and Nicholls, J.G. (1966) The physiology of neuroglial cells. *Ergeb. Physiol.*, 57: 1–90.

Lee, J., Taira, T., Pihlaja, P., Ransom, B.R. and Kaila, K. (1996) Effects of CO_2 on excitatory transmission apparently caused by changes in intracellular pH in the rat hippocampal slice. *Brain Res.*, 706: 210–216.

Lee, J.H., Vollmer, T.L. and Ransom, B.R. (1993) Extracellular pH and epileptiform activity in rat hippocampal slices. *Soc. Neurosci. Abstr.*, 19: 1870.

Lev-Ram, V. and Grinvald, A. (1986) Ca^{2+}- and K^+-dependent communication between central nervous system myelinated axons and oligodendrocytes revealed by voltage-sensitive dyes. *Proc. Natl Acad. Sci. USA*, 83: 6651–6655.

Newman, E.A. (1995) Glial cell regulation of extracellular potassium. In: H. Kettenmann and B.R. Ransom (Eds), *Neuroglia*. New York, Oxford University Press: 717–731.

Newman, E.A. (1996) Acid efflux from retinal glial cells generated by sodium bicarbonate cotransport. *J. Neurosci.*, 16: 159–168.

O'Connor, E.R., Sontheimer, H. and Ransom, B.R. (1994) Rat hippocampal astrocytes exhibit electrogenic sodium-bicarbonate co-transport. *J. Neurophysiol.*, 72: 2580–2589.

Pappas, C.A. and Ransom, B.R. (1994) Depolarization-induced alkalinization (DIA) in rat hippocampal astrocytes. *J. Neurophysiol.*, 72: 2816–2826.

Peters, A., Palay, S.L. and Webster, H.D. (1991) *The Fine Structure of the Nervous System*. New York, Oxford Press.

Ransom, B., Walz, W., Carlini, C. and Davis, P. (1989) Ionic alterations during anoxia in mammalian white matter. *Acta Physiol. Scand. Suppl.*, 582: 56.

Ransom, B.R. (1992) Glial modulation of neural excitability mediated by extracellular pH: A hypothesis. In: A. Yu, E. Sykova, L. Hertz, M. Norenberg and S.G. Waxman (Eds), *Prog. Brain Res.* Amsterdam, Elsevier. 94: 37–46.

Ransom, B.R. (1995) Gap Junctions. In: H. Kettenmann and B.R. Ransom (Eds), *Neuroglia*. New York, Oxford University Press: 299–318.

Ransom, B.R. and Carlini, W.G. (1986) Electrophysiological properties of astrocytes. In: S. Fedoroff and A. Vernadakis (Eds), *Astrocytes: Biochemistry, Physiology and Pharmacology of Astrocytes*. Orlando, FL, Academic Press. 2: 1–49.

Ransom, B.R., Carlini, W.G. and Connors, B.W. (1986) Brain extracellular space: developmental studies in rat optic nerve. *Ann. NY Acad. Sci.*, 481: 87–105.

Ransom, B.R., Carlini, W.G. and Yamate, C.L. (1987) Tip size of ion-exchanger based K^+-selective microelectrodes. II. Effects on measurement of evoked $[K^+]_o$ transients. *Can. J. Physiol. Pharmacol.*, 65: 894–897.

Ransom, B.R. and Fern, R. (1997) Does astrocytic glycogen benefit axon function and survival in CNS white matter during glucose deprivation? *Glia*, 21: 134–141.

Ransom, B.R., Voipio, J., Kaila, K. and Paalasmaa, P. (1994) Mechanisms of activity-dependent extracellular acid shifts in white matter of the mammalian brain. *Soc. Neurosci. Abstr.*, 20: 575.

Ransom, B.R. and Yamate, C.L. (1984) The rat optic nerve following enucleation: a pure preparation of mammalian glia. *Soc. Neurosci. Abstr.*, 10: 949.

Ransom, C.B., Ransom, B.R. and Sontheimer, H. (2000) Activity-dependent extracellular K^+ accumulation in rat optic nerve: the role of glial and axonal Na^+ pumps. *J. Physiol.*, 522.3: 427–442.

Rose, C.R. and Deitmer, J.W. (1994) Evidence that glial cells modulate extracellular pH transients induced by neuronal activity in the leech central nervous system. *J. Physiol. (Lond.)*, 481: 1–5.

Rose, C.R. and Ransom, B.R. (1998) pH regulation in mammalian glia. In: K. Kaila and B. Ransom (Eds), *pH and Brain Function*. New York, Wiley-Liss: 253–275.

Salem, R.D., Hammerschlag, R., Brancho, H. and Orkand, R.K. (1975) Influence of potassium ions on accumulation and metabolism of (^{14}C)glucose by glial cells. *Brain Res.*, 86: 499–503.

Schwiening, C.J., Kennedy, H.J. and Thomas, R.C. (1993) Calcium-hydrogen exchange by the plasma membrane Ca-

ATPase of voltage-clamped snail neurons. *Proc. R. Soc. Lond. Biol.*, 253: 285–289.

Siebens, A.W. and Boron, W.F. (1989) Depolarization-induced alkalinization in proximal tubules. II. Effects of lactate and SITS. *Am. J. Physiol.*, 256: F354–365.

Somjen, G.G. (1979). Extracellular potassium in the mammalian central nervous system. *Annu. Rev. Physiol.*, 41: 159–177.

Sykova, E. (1998) Extracellular pH and ionic shifts associated with electrical activity and pathological states in the spinal cord. In: K. Kaila and B.R. Ransom (Eds), *pH and Brain Function*. New York, Wiley-Liss: 339–358.

Taira, T., Smirnov, S., Voipio, J. and Kaila, K. (1993) Intrinsic proton modulation of excitatory transmission in rat hippocampal slices. *Neuroreport*, 4: 93–96.

Tang, C.M., Dichter, M. and Morad, M. (1990) Modulation of the *N*-methyl-D-aspartate channel by extracellular H^+. *Proc. Natl Acad. Sci. USA*, 87: 6445–6449.

Tombaugh, G.C. and Somjen, G.G. (1998) pH Modulation of Voltage-Gated Ion Channels. In: K. Kaila and B.R. Ransom (Eds), *pH and Brain Function*. New York, Wiley-Liss: 395–416.

Trapp, S., Lückermann, M., Kaila, K. and Ballanyi, K. (1996) Acidoric of hippocampal neurons mediated by a plasmalemmal Ca^{2+}/H^+ sump. *Neuroreport*, 7: 2000–2004.

Traynelis, S.F. (1998) pH Modulation of Ligand-Gated Ion Channels. In: K. Kaila and B.R. Ransom (Eds), *pH and Brain Function*. New York, Wiley-Liss: 417–446.

Traynelis, S.F. and Cull-Candy, S.G. (1990) Proton inhibition of *N*-methyl-D-aspartate receptors in cerebellar neurons. *Nature*, 345: 347–350.

Traynelis, S.F., Hartley, M. and Heinemann, S.F. (1995) Control of proton sensitivity of the NMDA receptor by RNA splicing and polyamines. *Science*, 268: 873–876.

Voipio, J. (1998) Diffusion and Buffering aspects of H^+, HCO_3^- and CO_2 movements in brain tissue. In: K. Kaila and B.R. Ransom (Eds), *pH and Brain Function*. New York, Wiley-Liss: 45–66.

Vyklicky, L., Jr., Vlachova, V. and Krusek, J. (1990) The effect of external pH changes on responses to excitatory amino acids in mouse hippocampal neurones. *J. Physiol. (Lond.)*, 430: 497–517.

Walz, W. and Mukerji, S. (1988) Lactate release from cultured astrocytes and neurons: a comparison. *Glia*, 1: 366–370.

L. F. Agnati, K. Fuxe, C. Nicholson and E. Syková (Eds.)
Progress in Brain Research, Vol 125

CHAPTER 11

The astrocyte-mediated coupling between synaptic activity and energy metabolism operates through volume transmission

Pierre J. Magistretti* and Luc Pellerin

Institute of Physiology, Lausanne University School of Medicine, 7 Rue du Bugnon, 1005 Lausanne, Switzerland

Volume transmission and the trafficking of metabolites in the extracellular space: role in neurometabolic coupling

An important principle in brain physiology is the tight coupling of neuronal activity with both energy metabolism and blood flow. This principle was already formulated over a hundred years ago by Charles Sherrington who, in association with Roy, published in the Journal of Physiology (Roy and Sherrington, 1890) the following : "... we conclude then that the chemical products of cerebral metabolism ... of the brain can cause variations of the caliber of the cerebral vessels ... that in this reaction the brain possesses an intrinsic mechanism by which the vascular supply can be varied locally in correspondence with local variations of functional activity." This physiological principle – which has been validated also for the coupling between activity and energy metabolism with the determination of local rates of glucose utilization with the 2-deoxyglucose technique (Sokoloff et al., 1977) – has provided the basis for all functional imaging techniques, since local increases in brain activity produce blood flow and metabolism signals that can be detected with various imaging techniques. Indeed, positron emission tomography (PET) can monitor the increase in cerebral blood flow (CBF), in glucose utilization and oxygen consumption (Phelps et al., 1979; Frackowiak et al., 1980); the degree of blood oxygenation yields the signals that are detected with functional magnetic resonance imaging (fMRI) (Ogawa et al., 1992). However, PET and fMRI do not detect synaptic activity directly but measure signals that reflect energy consumption by activity-dependent neural processes. Despite the remarkable advance in imaging techniques and their extensive applications to neurology, neuropsychology and psychiatry, the precise understanding of the cellular and molecular mechanisms that underlie the signals detected with these imaging techniques – which rely on the coupling of synaptic activity with energy metabolism and blood flow – is still quite fragmentary.

Over the last decade, the research efforts of our laboratory have been focused on the understanding of the mechanisms that underlie such coupling. In particular, we have been interested in determining the nature of the effects of neurotransmitters on glucose metabolism. In the present article, we will briefly review the salient features that have emerged from such an analysis, which clearly indicates that astrocytes play a key role in coupling synaptic activity to glucose utilization and metabolism. Considering the fact that communication

*Corresponding author. Tel.: (41) 21 692 55 42;
Fax: (41) 21 692 55 95;
e-mail: Pierre.Magistretti@iphysiol.unil.ch

between astrocytes and neurons operates extra-synaptically, through metabolites exchanged across the extracellular space, neuron-glia metabolic coupling may therefore illustrate a modality for the concept of volume transmission.

Structural and functional features of astrocytes

Before reviewing the recent functional evidences indicating such a role of astrocytes, some structural features of these cells should be mentioned. Astrocytes represent a significant portion of brain volume, 25 to 50% depending on the regions (Wolff, 1968). An astrocyte:neuron ratio of up to 10:1 is a feature of most brain regions (Bignami, 1991). Astrocytes are stellate cells with multiple fine processes. However, evidence obtained from quantitative morphometric studies indicates that astrocytes are in fact polarized cells, with one process contacting a cell of mesodermal origin (most frequently an endothelial cell of the capillary) and a multitude of process that are intertwined within the neuropil and ensheath synaptic contacts (Rohlmann and Wolff, 1996). The processes in close apposition to capillary walls, called astrocytic end-feet, cover the entire surface of intraparenchymal capillaries (Rohlmann and Wolff, 1996). The other astrocytic processes which ensheath virtually every synapse of the CNS possess receptors for a variety of neurotransmitters as well as reuptake sites, in particular for glutamate (Barres, 1991). Indeed the reuptake of synaptically-released glutamate is one of the well-established functions of astrocytes. These synaptically-targeted lamellar process represent approximately 80% of an astrocyte membrane surface and are devoid of organelles (Rohlmann and Wolff, 1996), notably mitochondria. This latter feature provides an indirect indication that glucose, in such peri-synaptic lamellar processes, is processed glycolytically. The cytoarchitectural arrangements and morphological features of astrocytes suggest that indeed these cells are in a position to sense and modulate synaptic transmission through the pervasive lamellar processes that surround synaptic contacts; in addition, the pericapillary end-feet, on which glucose transporters of the GLUT1 type are expressed (Morgello et al., 1995) are a likely site of glucose uptake (Fig. 1).

Central to understanding the emerging role of astrocytes in coupling neuronal activity to energy metabolism and in modulating synaptic transmission is the clear demonstration that synaptically-released glutamate exerts profound effects on astrocytes.

Astrocytes sense glutamate released during synaptic activity

Glutamate is the predominant excitatory neurotransmitter in the brain. In the cerebral cortex, activation of afferent pathways by specific modalities (e.g. somatosensory, visual, auditory) or of cortico-cortical association circuits results in a spatially and temporally defined local release of glutamate from the activated synaptic terminals (Fonnum, 1984). The released glutamate exerts profound effects on the excitability of target neurons, which are mediated by specific subtypes of glutamate receptors. The action of glutamate on postsynaptic neurons is rapidly terminated by an avid reuptake system present on astrocyte processes, which ensheath synaptic contacts (Rothstein et al., 1994). This removal of glutamate from the synaptic cleft is operated through specific glutamate transporters.

Five main types of glutamate transporters have been described: GLAST (EAAT 1), GLT-1 (EAAT 2), EAAC 1 (EAAT 3), EAAT 4 and EAAT 5 (Robinson and Dowd, 1997; Trotti et al., 1998). The latter two appear to be predominantly localized in cerebellum and retina, respectively (Trotti et al., 1998). All have been cloned, functionally characterized and their localization and distribution at the regional, cellular and subcellular levels in the CNS has been performed (Robinson and Dowd, 1997; Trotti et al., 1998). A detailed description of this fertile field of research is beyond the scope of this review; only the main points will therefore be underlined. EAAC 1 transporters are neuronal, mostly localized on the cell body and dendrites, while GLAST and GLT–1 are glial (Robinson and Dowd, 1997; Trotti et al., 1998). Glutamate uptake into astrocytes is driven by the electrochemical gradient of Na^+, with a stoichiometry of one glutamate for three Na^+ (Robinson and Dowd, 1997). In fact, the coupling between synaptic glutamate release and its reuptake into astrocytes is so tight that determination of the Na^+ current generated in astrocytes by the co-transport of glutamate and Na^+ through the glutamate transporter provides an accurate reflection of glutamate release from the synapse with a millisecond temporal resolution (Bergles and Jahr, 1997) (Fig. 1). Once in astrocytes glutamate is predominantly converted to glutamine through an ATP-requiring reaction catalyzed by the astrocyte-specific enzyme glutamine synthase. Glutamine is then released by astrocytes and taken up by neurons to replenish the neurotransmitter pool of glutamate.

Glutamate reuptake into astrocytes appears to be critical for preventing neuronal excitotoxic damage. Indeed, a dysfunction of astroglial glutamate transporters appears to be involved in the pathogenesis of motorneuron degeneration observed in Amyotrophic Lateral Sclerosis (Aoki et al., 1998).

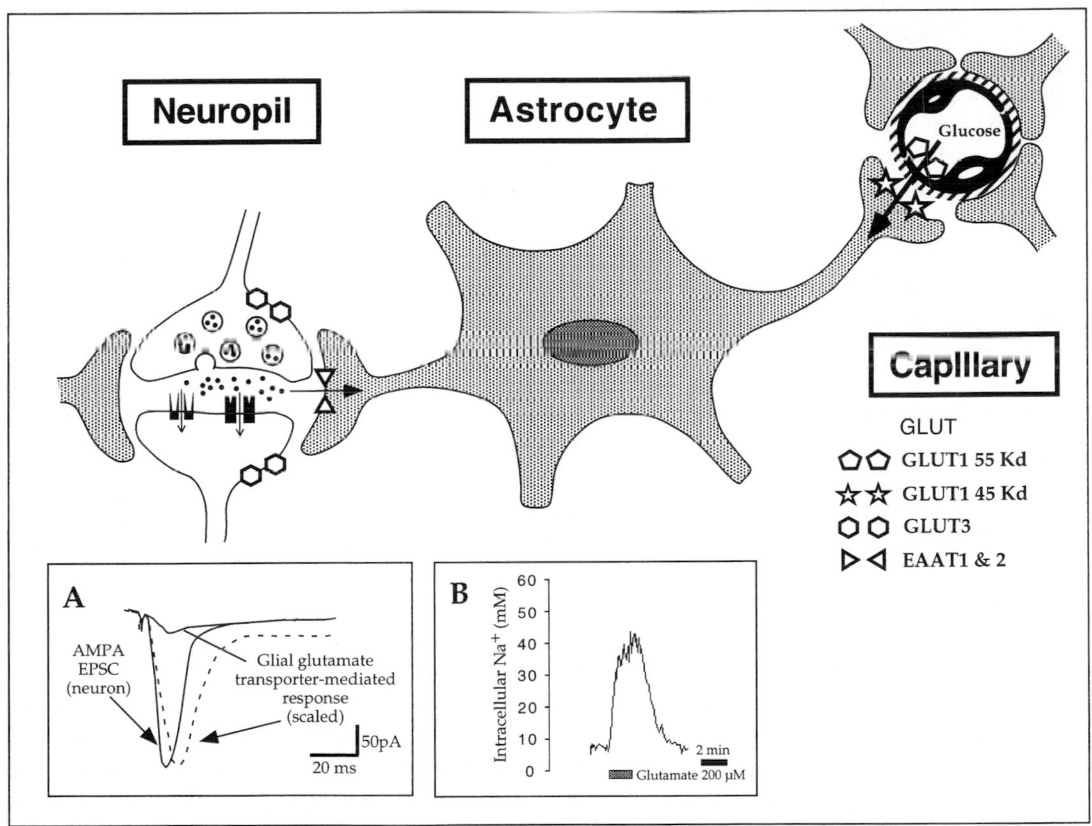

Fig. 1. Cytological relationships of astrocytes with neurons and capillaries. Specialized astrocytic processes, the end-feet, cover virtually all capillary endothelial walls. GLUT1 glucose transporters are present both on endothelial cells (55 kDa form) and on astrocytic end-feet (45 kDa form). Other astrocytic processes ensheath synaptic contacts. These processes are enriched in glutamate transporters (EAAT1 and 2). Neurons also express glucose transporters (GLUT3). A. Tight coupling between glutamate-mediated synaptic responses and glutamate reuptake into hippocampal astrocytes. The AMPA receptor-mediated excitatory post-synaptic current (EPSC) monitored in the post-synaptic neuron following glutamate release is tightly coupled to a sodium current generated by the co-transport of glutamate and sodium through the glial glutamate transporter. The glial glutamate transporter-mediated current (dotted line) has been scaled to the EPSC. Note that both the neuronal and glial currents begin at the same time, while the glial response shows a slower kinetics (Panel A : Redrawn and adapted from Bergles and Jahr, 1997). B. Glutamate-mediated increase in intracellular sodium in cultured astrocytes. The intracellular sodium concentration was determined by fluorescence microscopy using the specific fluorescent dye SBFI (sodium-binding benzofuran isophthalate). Note that following glutamate application (200 μM) the intracellular sodium concentration rapidly rises from ≈ 8 to 40 mM. (Results for panel FB, kindly provided by Dr Jean-Yves Chatton, Institute of Physiology, University of Lausanne). (Taken from Magistretti and Pellerin, 1999).

The neuron-astrocyte communication mediated by glutamate may in fact function in a bidirectional fashion: indeed neuronally-released glutamate, by evoking Ca^{++} waves which propagate through the astrocytic syncitium, can in turn cause glutamate release from astrocytes at distant sites (Pasti et al., 1997). Ca^{++} elevation in astrocytes increases the frequency of miniature synaptic currents in hippocampal neurons (Araque et al., 1998). The modulation of synaptic activity by astrocytes clearly involves soluble molecules, since as elegantly shown by Pfrieger and Barres, the occurrence of spontaneous synaptic potentials in cultured optic nerve neurons is markedly increased by astrocyte-conditioned medium (Pfrieger and Barres, 1997).

Finally, a series of observations from our and other laboratories (to be reviewed in the next sections), has revealed a key role of the astrocytic glutamate transporters in coupling synaptic activity with glucose utilization and provision of energy substrates from astrocytes to active neurons.

Astrocyte couple synaptic activity to glucose utilization

The central question around which the research projects of our laboratory are organized is understanding the molecular mechanisms that mediate the tight coupling of neuronal activity with energy metabolism in the brain. Over the years this question has led us to focus on the effects of neuronally-derived signals, such as certain neurotransmitters, on astrocytes, the cells which appear to play a key role in neuro-metabolic coupling.

Directed by this biological question, we have explored the effect of glutamate, the main excitatory neurotransmitter released by activated circuits, on metabolic indices in astrocytes. We have demonstrated that glutamate is a potent stimulator of glycolysis, i.e. of glucose uptake and lactate production, in primary astrocyte cultures (Pellerin and Magistretti, 1994; Magistretti et al., 1999). These observations have since been confirmed by several laboratories (Takahashi et al., 1995; Keller et al., 1996; Hamai et al., 1999). In such a purified preparation, the molecular steps involved in neurometabolic coupling can readily be addressed. This level of cellular resolution allowed us to show that the metabolic effect of glutamate is not mediated by receptors, but rather by glutamate transporters, selectively expressed in astrocytes, in particular GLAST (Pellerin and Magistretti, 1994). As noted earlier, one of the well established functions of astrocytes is the rapid removal of synaptically-released glutamate. Given their cytoarchitectural characteristics, with processes around synaptic contacts and others covering the capillary walls (see Section 2), this observation suggested that astrocytes could critically contribute to the uptake of glucose into the brain parenchyma in register with synaptic activity and that the glutamate-mediated glycolysis could provide a simple mechanism to couple neuronal activity to glucose utilization (Pellerin and Magistretti, 1994). Following this first observation, we set out to identify the molecular mechanisms that would allow this coupling to occur through astrocytes. We first demonstrated that the NaK-ATPase is critical for this coupling. Thus, ouabain, a specific inhibitor of the NaK-ATPase, completely inhibits the glutamate-evoked glycolysis in astrocytes (Pellerin and Magistretti, 1994). Complementary to this observation was the demonstration that glutamate directly stimulates the NaK-ATPase (Pellerin and Magistretti, 1997). A pharmacological analysis of this effect of glutamate, suggested a mobilization of the $\alpha 2$ subunit of the NaK-ATPase, which is highly sensitive to nanomolar concentrations of ouabain (Pellerin and Magistretti, 1997). Activation of the NaK-ATPase by glutamate is a consequence of the increased intracellular Na^+ concentration triggered by the co-transport of Na^+ with glutamate through the glutamate transporters. Based on the experimental results obtained we have concluded that the overall stoichiometry of the molecular steps involved in the coupling between glutamate uptake and glucose utilization should be as follows : one glutamate is taken up with 3 sodium ions while one glucose consumed through glycolysis produces 2 ATPs. One ATP is used by the NaK-ATPase for the extrusion of 3 sodium ions; the other ATP is used for the synthesis of glutamine from glutamate by the glutamine synthase (Magistretti and Pellerin, 1996; Magistretti et al., 1999) (Figs. 2 and 4). Experimentally, we have observed that the glutamate-stimulated glycolytic processing of glucose results in approximately 2 lactate molecules produced per 1 glucose molecule, i.e. an expected stoichiometrical relationship between glucose and lactate (Pellerin and Magistretti, 1994).

We have also provided initial evidence that these mechanisms, which mediate glucose uptake by astrocytes, can undergo strong plastic adaptations. Indeed the neuroactive cytokines Tumor Necrosis Factor α (TNFα) and Interleukin-1 (IL-1) upregulate glucose uptake by astrocytes in a time- (several hours) and transcription-dependent manner (Yu et al., 1995). One of the genes that appears to be upregulated is phospholipase A_2, a fact which results in increased arachidonic acid formation by astrocytes (Yu et al., 1995). Interestingly, we had previously shown that arachidonic acid is a key modulator in neurotransmitter-mediated glucose uptake by astrocytes (Yu et al., 1995).

Lactate produced by astrocytes fuels neuronal activity

Having demonstrated that glutamate stimulates glucose uptake and lactate production by astrocytes through a mechanism involving a specific subunit

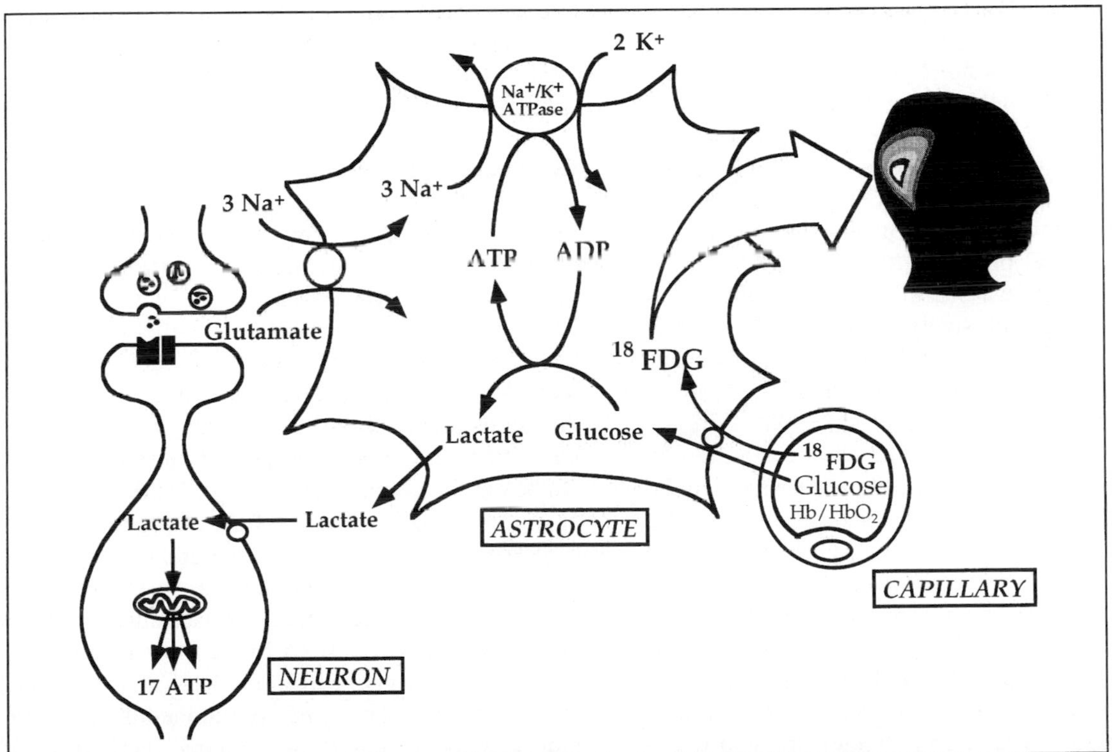

Fig. 2. Schematic representation of the mechanism for glutamate-induced glycolysis in astrocytes during physiological activation and its implication for functional brain imaging. At glutamatergic synapses, presynaptically released glutamate depolarizes postsynaptic neurons by acting at specific receptor subtypes. The action of glutamate is terminated by an efficient glutamate uptake system located primarily in astrocytes (see Fig. 1). Glutamate is cotransported with Na^+, resulting in an increase in the intra-astrocytic concentration of Na^+, leading to an activation of the astrocyte NaK-ATPase. Activation of the NaK-ATPase stimulates glycolysis, i.e. glucose utilization and lactate production. Lactate, once released by astrocytes, can be taken up by neurons and serve them as an adequate energy substrate. The proposed model of glutamate-induced glycolysis in astrocytes implies that the activity-linked uptake of ^{18}FDG monitored with PET, reflects primarily an astrocyte-based signal. Since neuronally-released glutamate triggers the cascade of events that leads to glucose uptake, the ^{18}FDG-PET signal will faithfully reflect activation of neuronal circuits. The model is also consistent with the transient lactate peak detected during activation by MRS in humans and by microdialysis and electrochemical detection in animals (see text for details) (Modified from Magistretti and Pellerin, 1996).

of the NaK-ATPase, we then focused on the issue of the mechanisms of lactate production and export by astrocytes, as well as import and utilization by neurons. Antibodies were raised against two isoforms of lactate dehydrogenase (LDH), one, LDH5, enriched in lactate-producing glycolytic tissues such as skeletal muscle, and the other, LDH1, expressed in highly oxidative tissues such as the heart, which utilize lactate as a substrate. We found a selective distribution of LDH isoforms in the human brain, with LDH5 localized only in astrocytes while neurons were exclusively labeled by LDH1 antibodies (Bittar et al., 1996); these observations are consistent with an intercellular exchange of lactate produced by astrocytes and consumed by neurons.

Since lactate does not cross cellular membranes readily, the next issue that we addressed was that of the presence and nature of specific transporters for monocarboxylates (MCT) in neural cells. Based on published sequences of MCTs in peripheral tissues, we demonstrated by Northen blot the presence of mRNA encoding one isoform of MCT, MCT1 in cultured astrocytes, while primary cultures of neurons almost exclusively express MCT2 (Bröer et al., 1997). By in situ hybridization we have provided a description of the distribution of MCTs in the mouse brain (Pellerin et al., 1998). With the level of cellular resolution that can be achieved by in situ hybridization, we could not provide definite conclusions as to the cellular localization of MCTs. However, by coupling in situ hybridization with histochemistry we have shown a developmentally-regulated transient overexpression of MCT1 in intraparenchymal capillaries from post-natal days 1 to 30 (Pellerin et al., 1998). This period precisely corresponds to the ketogenic diet associated with breastfeeding. Since ketone bodies are also transported through MCTs, these observations suggest that MCTs can undergo marked plastic adaptations dictated, among other possible factors, by the nature of the transported substrates.

Results obtained in a variety of in vivo paradigms both in laboratory animals and in humans, support the existence of a transient lactate production during activation. Thus, microdialysis studies in rats indicate a marked increase in the concentration of lactate in the dialysate, in striatum and hippocampus during physiological sensory stimulation (Fellows et al., 1993). Interestingly, this activity-linked lactate peak is completely inhibited when the glutamate uptake inhibitor THA is present in the perfusate, thus providing further supporting evidence for the existence of glutamate stimulated glycolysis during activation (Fray et al., 1996). A transient lactate production during physiological activity has also been demonstrated by electrochemical detection (Hu and Wilson, 1997). Magnetic resonance spectroscopy (MRS) in humans has also revealed that during physiological activation of the visual system, a transient lactate peak is observed in primary visual cortex (Prichard et al., 1991; Sappey-Marinier et al., 1992). Thus, microdialysis, electrochemical and MRS data in vivo would support the notion of a transient glycolytic processing of glucose during activation.

Relevance to functional brain imaging

The mechanisms of glutamate-mediated neuron-glia metabolic interactions that we have demonstrated over the last five years by using combined functional in vitro and structural ex vivo localization approaches have provided an initial basis to better understand the cellular and molecular steps involved in neuro-metabolic coupling. In particular, the model proposed on the basis of studies at the cellular level (Figs. 2 and 4) suggests an initial glycolytic processing of glucose occurring in astrocytes during neuronal activation, resulting in a transient lactate overproduction, followed by a recoupling phase during which lactate would be oxidized by neurons (Magistretti et al., 1999). In vivo MRS data lend strong support to this model. Thus the simultaneous measurements, over a range of synaptic activities, of tricarboxylic acid cycle and the cycling of glutamate to glutamine (a process which occurs exclusively in astrocytes) using ^{13}C MRS, has revealed a striking stoichiometric relationship of 1:1 between glutamate cycling (a reflection of synaptic activity) and glucose utilization (Sibson et al., 1998). According to these data, for each glutamate released from active terminals and taken up by astrocytes one glucose would be oxidized, a result fully consistent with the stoichiometry that we have proposed on the basis of the experimental data obtained in vitro (Magistretti et al., 1999).

Certain PET studies have indicated that oxygen consumption does not increase commensurately with blood flow and glucose utilization in activated brain areas (Fox et al., 1988), suggesting the occurrence of an activity-dependent glycolytic processing of glucose. This issue has been the object of debate since other studies have found that the degree of uncoupling between glucose utilization and oxygen consumption during activation may be minimal, or may even not occur, depending on the stimulations used (Marrett et al., 1995). In addition, using ^{13}C-glucose MRS, recently reported data are consistent with a significant increase in oxygen utilization during activation (Hyder et al., 1996).

The model that we have proposed (Pellerin and Magistretti, 1994; Magistretti et al., 1999) and its in vivo support (Sibson et al., 1998) could provide a substrate to reconcile these contrasting views by suggesting that glucose imported into the brain parenchyma during activation undergoes a transient glycolysis in astrocytes, resulting in the production of lactate which is then oxidized by neurons (Figs. 2 and 4). This latter process would imply a metabolic recoupling with increase oxygen consumption. The spatial and temporal 'window' during which a transient glycolysis occurs and a lactate peak can be detected, may depend on the rapidity and degree of recoupling existing between astrocytic glycolysis and neuronal oxidative phosphorylation. The existence of a transient non-oxidative glucose utilization in face of increased blood flow, would provide a change in the oxy-/deoxy-hemoglobin ratio sufficient to generate a signal detectable by fMRI.

This operational model for coupling is consistent with the notion that the signals detected during physiological activation in humans with FDG PET and fMRI may result from signalling and metabolic exchanges between neurons (glutamate release) and astrocytes (glycolysis) (Pellerin and Magistretti, 1994; Magistretti and Pellerin, 1996; Magistretti et al., 1999). This conclusion does not question the validity of these techniques for monitoring neuronal function since the triggering event is neuronal glutamate release; rather this conclusion provides a cellular and molecular basis for these functional brain imaging techniques (Fig. 2). In addition, functional brain imaging may reflect the operation of volume transmission at the neuron-glia metabolic unit.

Supporting the glial origin of the FDG PET signal, a series of recent functional imaging and neuropathological studies has revealed that in subjects affected by familial forms of depressive disorders or by bipolar disorders, a significant decrease in glucose utilization revealed by FDG PET is observed in the subgenual prefrontal cortex (Drevets et al., 1997); this decrease in glucose metabolism is accompanied by a specific reduction in the number of glial cells in this area while neuronal counts are normal (Ongür et al., 1998).

Regulation of glycogen metabolism by neurotransmitters

Another neuron-glia metabolic interaction that we have examined, is the regulation by certain neurotransmitters of glycogen metabolism in astrocytes. Following up on our early work indicating that neurotransmitters whose receptors are coupled to cAMP-generating systems, such as vasoactive intestinal peptide (VIP) and noradrenaline (NA), resulted in a rapid (within seconds) glycogenolysis (Magistretti et al., 1981; Sorg and Magistretti, 1991; Magistretti et al., 1993), we had shown that the same neurotransmitters could induce a long-lasting plastic response resulting in massive glycogen resynthesis (Sorg and Magistretti, 1992). This effect is expressed after several hours and is transcriptionally-regulated, involving the expression of new genes (Sorg and Magistretti, 1992) (Fig. 3). We have now identified three genes related to glycogen metabolism, two immediate-early and a late one, that are induced in a cAMP-dependent manner by VIP and NA in astrocytes. The immediate-early genes, C/EBP β and δ, are members of a family of transcription factors called CCAAT/Enhancer binding protein (Cardinaux and Magistretti, 1996). This family of transcription factors is predominantly involved in two types of physiological responses : inflammation, through the control of expression of several acute phase response genes, and energy metabolism, in particular through the regulation of cAMP-sensitive genes controlling glucose metabolism in peripheral tissues (Darlington et al., 1995). We demonstrated the presence of this family of transcription factors in the mammalian CNS, their localization to astrocytes and the modulation of their expression by two neuronally-derived signals, the neurotransmitters VIP and NA (Cardinaux and Magistretti, 1996). Through transfection experiments with vectors that we designed, we also provided evidence for the involvement of C/EBP β and δ in the upregulation of glycogen resynthesis by VIP and NA in astrocytes (Cardinaux and Magistretti, 1996). The third gene whose expression is regulated by VIP and NA is, not unexpectedly, glycogen synthase (Pellegri et al., 1996). We have cloned and sequenced the cDNA encoding astrocyte-specific glycogen synthase (Pellegri et al., 1996), and showed that its mRNA levels are potently induced by VIP and NA with a time-course that is compatible with the glycogen resynthesis induced by both neurotransmitters in astrocytes (Sorg and Magistretti, 1992; Pellegri et al., 1996). Glycogen metabolism in astrocytes appears therefore to be under the dynamic control of at least two neurotransmitters, VIP and NA, with the balance between short term (glycogenolysis) and transcriptionally-regulated long term effects (glycogen resynthesis) setting the intracellular levels of glycogen (Fig. 3).

The circuits that contain VIP and NA in the cerebral cortex, the area where we have examined the metabolic effect of the two neurotransmitters, differ in a striking manner. Thus, VIP is contained in a homogeneous population of bipolar, radially oriented neurons (Magistretti and Morrison, 1988) while for the noradrenergic system the cell bodies of NA-containing neurons are localized in the locus coeruleus in the brain stem from where axons project to various brain areas including the cerebral cortex; here, they enter the rostral end and progress caudally with a predominantly horizontal trajectory, across a vast rostrocaudal expanse of cortex (Morrison et al., 1978). Given these morphological features we suggested that the VIP-containing intracortical neurons could regulate glycogen metabolism locally, within cortical columns, while the noradrenergic system could regulate energy homeostasis globally, spanning across functionally

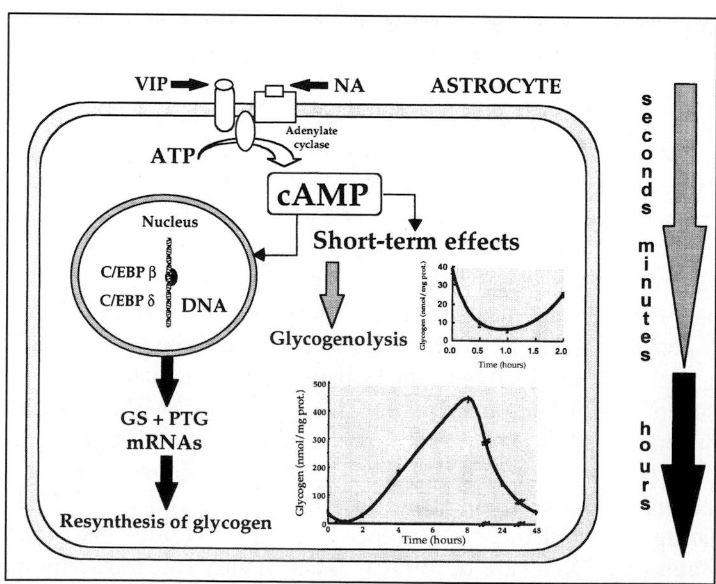

Fig. 3. Short-term (sec-min) and long-term (hour) control of glycogen metabolism of astrocytes by VIP and NA. Within minutes of application, VIP and NA promote glycogenolysis (short-term effect). This effect is due to cAMP-dependent phosphorylation of preexisting proteins. Within a few hours of application of VIP and NA, glycogen levels are increased 6–10 times above control levels (long-term effect). This effect involves the cAMP-dependent induction of C/EBP β and C/EBP δ immediately-early genes as well as of Glycogen Synthase (GS) and Protein Targeting to Glycogen (PTG) mRNAs.

distinct cortical areas (Magistretti et al., 1981; Magistretti and Morrison, 1988).

Recently, we have identified a fourth gene, Protein Targeting to Glycogen (PTG), which could play a critical role in the dynamic adaptations that characterize glycogen levels in astrocytes (Allaman et al., 2000). PTG has been recently discovered and cloned in the liver (Printen et al., 1997). The physiological function of PTG appears to be that of a chaperone protein coordinating the activity and compartmentalization of glycogen-synthesizing enzymes (Berman et al., 1998). Using the published sequence, we have found PTG mRNA to be expressed in the mouse brain and have evidence that its level of expression are strongly induced in astrocytes by VIP and NA (Allaman et al., 2000).

The regulation of genes involved in glycogen metabolism in astrocytes is of particular interest to us for two main reasons; first, it provides a very favourable opportunity to study a complete cascade of signalling events from the receptor (VIP and β-adrenergic) to the gene, with a particular emphasis at this stage on transcriptional regulation of at least three genes involved in glycogen synthesis (Fig. 3). Second, from a purely physiological point of view, the role of glycogen in the brain is still largely unknown, except for its intuitively deduced function (for which there begins to be evidence) of providing an additional energy substrate in register with synaptic activity. However the fact that in vivo, glycogen levels appear to undergo rapid and rather drastic dynamic adaptations, e.g. during the

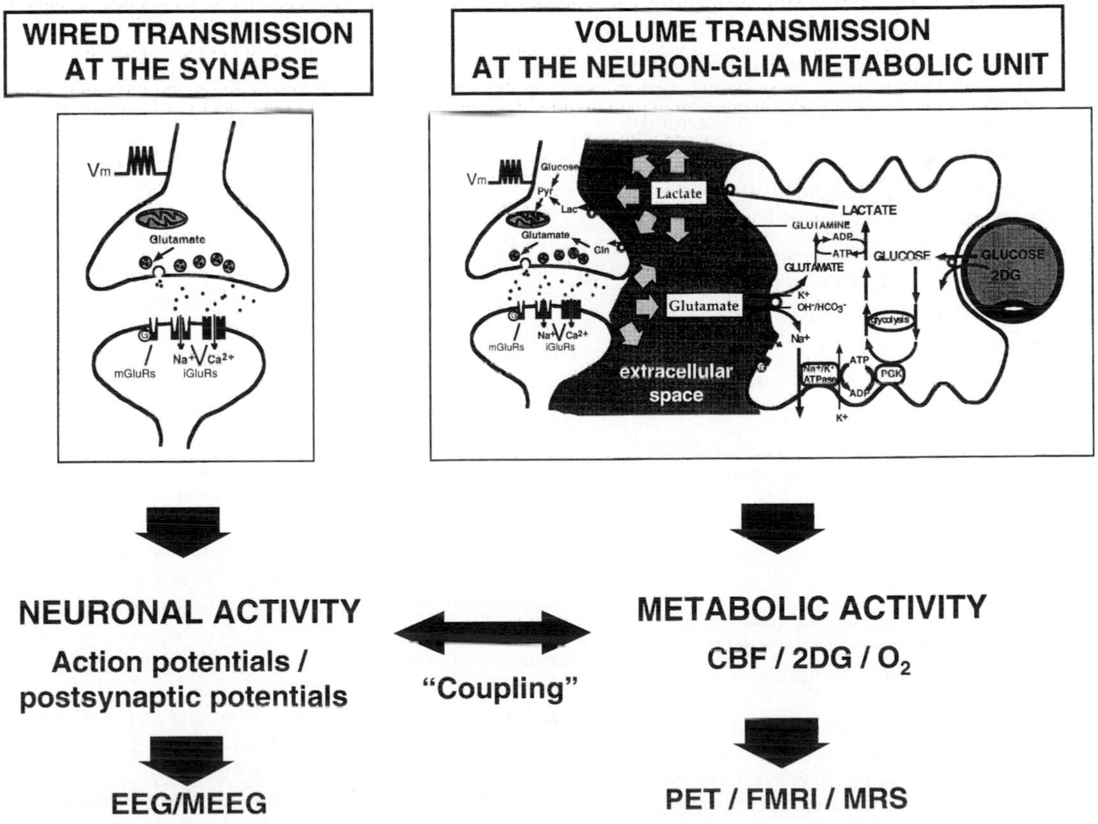

Fig. 4. Functional brain imaging may reflect the operation of volume transmission at the neuron-glia metabolic unit. Left: Action potentials and synaptic potentials generate signals at the synapse that are detected by EEG or MEEG. Right: Functional brain imaging techniques detect metabolic signals generated through the coupling of neuronal activity with energy metabolism. Communication at the neuron-glia metabolic unit operates extrasynaptically, through signals (e.g. glutamate, lactate) exchanged across the extracellular space and may therefore illustrate a modality for the concept of volume transmission.

sleep-wake cycle (Benington and Heller, 1995; Petit et al., 1999), suggest that by better understanding the factors that regulate the expression of the genes involved in its metabolism, it may be possible to gather new insights into its physiological role(s).

Neuron-glia metabolic coupling and volume transmission

The observations reviewed herein provide strong support for the existence of a tight regulation of energy metabolism by the activity-dependent release of certain neurotransmitters. The site of action of these neurotransmitters is the astrocyte which therefore is the central operator of such a neurometabolic coupling. Since neuron-glia signalling operates through extra-synaptic modes, the concept of volume transmission applies to this mode of cell-cell communication in the nervous system. Neuronal activity and *synaptic transmission* (*or wired transmission*) through the associated *electrochemical events* (action potentials and synaptic potentials) generate the signals detected in vivo by electroencephalography and magnetoelectroencephalography. In contrast, as noted in the first paragraph of this review, functional imaging techniques, such as PET or fMRI, detect the *metabolic changes* that are associated with synaptic activity, which appear to rely on neuron-glia interaction. This could suggest that functional brain imaging methods reflect the operation of *volume transmission* at the neuron-glia metabolic unit (Fig. 4).

References

Allaman, I., Pellerin, L. and Magistretti, P.J. (2000). Protein targeting to glycogen mRNA expression is stimulated by noradrenaline in mouse cortical astrocytes. *Glia*, 30: 382–391.

Aoki, M., Lin, C.-L.G., Rothstein, J.D., Geller, B.A., Hosler, B.A., Munsat, T.L., Horvitz, H.R. and Brown, R.H., Jr. (1998) Mutations in the glutamate transporter EAAT2 gene do not cause abnormal EAAT2 transcripts in amyotrophic lateral sclerosis. *Ann. Neurol.*, 43: 645–653.

Araque, A., Sanzgiri, R.P., Parpura, V. and Haydon, P.G. (1998) Calcium elevation in astrocytes causes an NMDA receptor-dependent increase in the frequency of miniature synaptic currents in cultured hippocampal neurons. *J. Neurosci.*, 18: 6822–6829.

Barres, B.A. (1991) New roles for glia. *J. Neurosci.*, 11: 3685–3694.

Benington, J.H. and Heller, H.C. (1995) Restoration of brain energy metabolism as the function of sleep. *Prog. Neurobiol.*, 45: 347–360.

Bergles, D.E. and Jahr, C.E. (1997) Synaptic activation of glutamate transporters in hippocampal astrocytes. *Neuron*, 19: 1297–1308.

Berman, H.K., O'Doherty, R.M., Anderson, P. and Newgard, C.B. (1998) Overexpression of protein targeting to glycogen (PTG) in rat hepatocytes causes profound activation of glycogen synthesis independent of normal hormone- and substrate-mediated regulatory mechanisms. *J. Biol. Chem.*, 273: 26421–26425.

Bignami, A. (1991) *Glial cells in the central nervous system. Discussions in neuroscience*, Vol. 8, Elsevier, Amsterdam, pp. 1–45.

Bittar, P., Charnay, Y., Pellerin, L., Bouras, C. and Magistretti, P.J. (1996) Selective distribution of lactate dehydrogenase isoenzymes in neurons and astrocytes of the human brain. *J. Cereb. Blood Flow Metab.*, 16: 1079–1089.

Bröer, S., Rahman, B., Pellegri, G., Pellerin, L., Martin, J.-L., Verleysdonk, S., Hamprecht, B. and Magistretti, P.J. (1997) Comparison of lactate transport in astroglial cells and monocarbosylate transporter 1 (MCT1) expressing xenopus laevis oocytes. *J. Biol. Chem.*, 272: 30096–30102.

Cardinaux, J.-R. and Magistretti, P.J. (1996) Vasoactive intestinal peptide, pituitary adenylate cyclase-activating peptide, and noradrenaline induce the transcription factors CCAAT/enhancer binding protein (C/EBP)-β and C/EBPδ in mouse cortical astrocytes: involvement in cAMP-regulated glycogen metabolism. *J. Neurosci.*, 16: 919–929.

Darlington, G.J., Wang, N. and Hanson, R.W. (1995) C/EBPα: a critical regulator of genes governing integrative metabolic processes. *Curr. Opinion Genet. Dev.*, 5: 565–570.

Drevets, W.C., Price, J.L., Simpson, J.R.J., Todd, R.D., Reich, T., Vannier, M. and Raichle, M.E. (1997) Subgenual prefrontal cortex abnormalities in mood disorders. *Nature*, 386: 824–827.

Fellows, L.K., Boutelle, M.G. and Fillenz, M. (1993) Physiological stimulation increases nonoxidative glucose metabolism in the brain of the freely moving rat. *J. Neurochem.*, 60: 1258–1263.

Fonnum, F. (1984) Glutamate: a neurotransmitter in mammalian brain. *J. Neurochem.*, 42: 1–11.

Fox, P.T., Raichle, M.E., Mintun, M.A. and Dence, C. (1988) Nonoxidative glucose consumption during focal physiologic neural activity. *Science*, 241: 462–464.

Frackowiak, R.S.J., Lenzi, G.L., Jones, T. and Heather, J.D. (1980) Quantitative measurement of regional cerebral blood flow and oxygen metabolism in man using ^{15}O and positron emission tomography: theory, procedure and normal values. *J. Comput. Assist. Tomogr.*, 4: 727–736.

Fray, A.E., Forsyth, R.J., Boutelle, M.G. and Fillenz, M. (1996) The mechanisms controlling physiologically stimulated changes in rat brain glucose and lactate – a microdialysis study. *J. Physiol.*, 496: 49–57.

Hamai, M., Minokoshi, Y. and Shimazu, T. (1999) L-glutamate and insulin enhance glycogen synthesis in cultured astrocytes from the rat brain through different intracellular mechanisms. *J. Neurochem.*, 73: 400–407.

Hu, Y. and Wilson, G.S. (1997) Rapid changes in local extracellular rat brain glucose observed with an in vivo glucose sensor. *J. Neurochem.*, 68: 1745–1752.

Hyder, F., Chase, J.R., Behar, K.L., Mason, G.F., Siddeek, M., Rothman, D.L. and Shulman, R.G. (1996) Increased tricarboxylic acid cycle flux in rat brain during forepaw stimulation detected with ^{1}H [^{13}C] NMR. *Proc. Natl Acad. Sci. USA*, 93: 7612–7617.

Keller, J.N., Steiner, M.R., Mattson, M.P. and Steiner, S.M. (1996) Lysophosphatidic acid decreases glutamate and glucose uptake by astrocytes. *J. Neurochem.*, 67: 2300–2305.

Magistretti, P.J. and Morrison, J.H. (1988) Noradrenaline- and vasoactive intestinal peptide-containing neuronal systems in neocortex: functional convergence with contrasting morphology. *Neuroscience*, 24: 367–378.

Magistretti, P.J., Morrison, J.H., Shoemaker, W.J., Sapin, V. and Bloom, F.E. (1981) Vasoactive intestinal polypeptide induced glycogenolysis in mouse cortical slices: a possible regulatory mechanism for the local control of energy metabolism. *Proc. Natl Acad. Sci. USA*, 78: 6535–6539.

Magistretti, P.J. and Pellerin, L. (1996a) Cellular bases of brain energy metabolism and their relevance to functional brain imaging: evidence for a prominent role of astrocytes. *Cereb. Cortex*, 6: 50–61.

Magistretti, P.J. and Pellerin, L. (1996b) The contribution of astrocytes to the ^{18}F-2-deoxyglucose signal in PET activation studies. *Mol. Psychiatry*, 1 : 445–452.

Magistretti, P.J. and Pellerin, L. (1999) Cellular mechanisms of brain energy metabolism and their relevance to functional brain imaging. *Phil. Trans. R. Soc. (Lond.)*, B, 354: 1155–1163.

Magistretti, P.J., Pellerin, L., Rothman, D.L. and Shulman, R.G. (1999) Energy on demand. *Science*, 283: 496–497.

Magistretti, P.J., Sorg, O. and Martin, J.-L. (1993) Regulation of glycogen metabolism in astrocytes: physiological, pharmacological, and pathological aspects. In: S. Murphy (Ed.), *Astrocytes: Pharmacology and Function*, Academic Press, pp. 243–265.

Marrett, S., Meyer, E., Kuwabara, H., Evans, A. and Gjedde, A. (1995) Differential increases of oxygen metabolism in visual cortex. *J. Cereb. Blood Flow Metab.*, 15: S80.

Morgello, S., Uson, R.R., Schwartz, E.J. and Haber, R.S. (1995) The human blood-brain barrier glucose transporter (GLUT1) is a glucose transporter of gray matter astrocytes. *Glia*, 14: 43–54.

Morrison, J.H., Grzanna, R., Molliver, M. and Coyle, J.T. (1978) The distribution and orientation of noradrenergic fibers in neocortex of the rat: an immunofluorescence study. *J. Comp. Neurol.*, 181: 17–40.

Ogawa, S., Tank, D.W., Menon, R., Ellermann, J.M., Kim, S.G., Merkle, H. and Ugurbil, K. (1992) Intrinsic signal changes accompanying sensory stimulation: functional brain mapping with magnetic resonance imaging. *Proc. Natl. Acad. Sci., USA*, 89: 5951–5955.

Ongür, D., Drevets, W.C. and Price, J.L. (1998) Glial reduction in the subgenual prefrontal cortex in mood disorders. *Proc. Natl. Acad. Sci., USA*, 95: 13290–13295.

Pasti, L., Volterra, A., Pozza, T. and Carmignoto, G. (1997) Intracellular calcium oscillations in astrocytes: a highly plastic, bidirectional form of communications between neurons and astrocytes in situ. *J. Neurosci.*, 17: 7817–7830.

Pellegri, G., Rossier, C., Magistretti, P.J. and Martin, J.L. (1996) Cloning, localization and induction of mouse brain glycogen synthase. *Mol. Brain Res.*, 38: 191–199.

Pellerin, L. and Magistretti, P.J. (1994) Glutamate uptake into astrocytes stimulates aerobic glycolysis: a mechanism coupling neuronal activity to glucose utilization. *Proc. Natl. Acad. Sci., USA*, 91: 10625–10629.

Pellerin, L. and Magistretti, P.J. (1997) Glutamate uptake stimulates Na^+/K^+-ATPase activity in astrocytes via activation of a distinct subunit highly sensitive to ouabain. *J. Neurochem.*, 69: 2132–2137.

Pellerin, L., Pellegri, G., Martin, J.-L. and Magistretti, P.J. (1998) Expression of monocarboxylate transporter mRNA in mouse brain: support for a distinct role of lactate as an energy substrate for the neonatal vs adult brain. *Proc. Natl. Acad. Sci., USA*, 95: 3990–3995.

Petit, J.-M., Tobler, I., Allaman, I., Borbély, A.A. and Magistretti, P.J. (1999) Effects of a 6 hour sleep deprivation on transcriptional regulation of proteins involved in glycogen metabolism in the mouse cerebral cortex. In: H. Schulz, P.L. Parmeggiani and M.H. Chase (Eds), *Sleep Research Online*, 2 (Suppl. 1), 549 (Abstract).

Pfrieger, F.W. and Barres, B.A. (1997) Synaptic efficacy enhanced by glial cells in vitro. *Science*, 277: 1684–1687.

Phelps, M.E., Huang, S.C., Hoffman, E.J., Selin, C., Sokoloff, L. and Kuhl, D.E. (1979) Tomographic measurement of local cerebral glucose metabolic rate in humans with (F-18)2-fluoro-2-deoxy-d-glucose: validation of method. *Ann. Neurol.*, 6: 371–388.

Prichard, D., Rothman, D., Novotny, E., Petroff, O., Kuwabara, T., Avison, M., Howseman, A., Hanstock, C. and Shulman, R. (1991) Lactate rise detected by ^{1}H NMR in human visual cortex during physiologic stimulation. *Proc. Natl. Acad. Sci., USA*, 88: 5829–5831.

Printen, J.A., Brady, M.J. and Saltiel, A.R. (1997) PTG, a protein phosphatase 1-binding protein with a role in glycogen metabolism. *Science*, 275: 1475–1478.

Robinson, M.B. and Dowd, L.A. (1997) Heterogeneity and functional properties of subtypes of sodium-dependentglutamate transporters in the mammalian central nervous system. *Advances in Pharmacology*, 37: 69–115.

Rohlmann, A. and Wolff, J.R. (1996) Subcellular topography and plasticity of gap junction distribution on astrocytes. In:

D.C. Spray and R. Dermietzel (Eds), *Gap Junctions in the Nervous System*, R.G. Landes Company, pp. 175–192.

Rothstein, J.D., Martin, L. and Levey, A.I. (1994) Localization of neuronal and glial glutamate transporters. *Neuron*, 13: 713–725.

Roy, C.S. and Sherrington, C.S. (1890) On the regulation of the blood supply of the brain. *J. Physiol.*, 11: 85–108.

Sappey-Marinier, D., Calabrese, G., Fein, G., Hugg, J.W., Biggins, C. and Weiner, M.W. (1992) Effect of photic stimulation on human visual cortex lactate and phosphates using ^1H and ^{31}P magnetic resonance spectroscopy. *J. Cereb. Blood Flow Metab.*, 12: 584–592.

Sibson, N.R., Dhankhar, A., Mason, G.F., Rothman, D.L., Behar, K.L. and Shulman, R.G. (1998) Stoichiometric coupling of brain glucose metabolism and glutamatergic neuronal activity. *Proc. Natl. Acad. Sci., USA*, 95: 316–321.

Sokoloff, L., Reivich, M., Kennedy, C., Des Rosiers, M.H., Patlak, C.S., Pettigrew, K.D., Sakurada, O. and Shinohara, M. (1977) The [^{14}C]deoxyglucose method for the measurement of local cerebral glucose utilization: theory, procedure, and normal values in the conscious and anesthetized albino rat. *J. Neurochem.*, 28: 897–916.

Sorg, O. and Magistretti, P.J. (1991) Characterization of the glycogenolysis elicited by vasoactive intestinal peptide, noradrenaline and adenosine in primary cultures of mouse cerebral cortical astrocytes. *Brain Res.*, 563: 227–233.

Sorg, O. and Magistretti, P.J. (1992) Vasoactive intestinal peptide and noradrenaline exert long-term control on glycogen levels in astrocytes: blockade by protein synthesis inhibition. *J. Neurosci.*, 12: 4923–4931.

Takahashi, S., Driscoll, B.F., Law, M.J. and Sokoloff, L. (1995) Role of sodium and potassium ions in regulation of glucose metabolism in cultured astroglia. *Proc. Natl. Acad. Sci., USA*, 92: 4616–4620.

Trotti, D., Danbolt, N.C. and Volterra, A. (1998) Glutamate transporters are oxidant-vulnerable: a molecular link between oxidative and excitotoxic neurodegeneration? *TIPS*, 19: 328–334.

Wolff, J. (1968) Die Astroglia im Gewebsverband des Gehirns. *Acta Neuropathol.*, Suppl. IV: 33–39.

Yu, N., Maciejewski-Lenoir, D., Bloom, F.E. and Magistretti, P.J. (1995) Tumor necrosis factor-α and interleukin-1α enhance glucose utilization by astrocytes: involvement of phospholipase A2. *Mol. Pharmacol.*, 48: 550–558.

L. F. Agnati, K. Fuxe, C. Nicholson and E. Syková (Eds.)
Progress in Brain Research, Vol 125

CHAPTER 12

Metabolic trafficking between cells in nervous tissue

Jonathan A. Coles[1,*], Céline Véga[2,**] and Païkan Marcaggi[2,***]

[1] *INSERM U438, NMR Bioclinique, CHU, Pavillon B, BP 217, 38043 Grenoble Cedex 09, France*
[2] *INSERM U394, Institute François Magendie, 1, rue Camille Saint-Saëns, 33077 Bordeaux Cedex, France*

Introduction

There is evidence that two pathways of intercellular transfer of molecules between neurons and glial cells are quantitatively predominant in mammalian brain. The first is the route, still unproven in the brain itself, but compatible with many experimental results, that transfers carbon fuel, starting as glucose in the capillaries, through astrocytes to neurons. We will briefly review some of the main arguments in favor of its existence, most of which concern the role of lactate in aerobic metabolism and its possible transfer from astrocytes to neurons. The second pathway is the transfer of glutamate from neurons to astrocytes, and the transfer of glutamine in the opposite direction. For this pathway, the 'glutamate–glutamine shuttle', the evidence is solid, although little is known about the detailed mechanisms. Lactate and glutamine during their transfer from astrocytes to neurons are probably not functioning primarily as 'informational substances' (Schmitt, 1984) and the reader may wish to exclude them from the field of 'volume transmission'. But the *control* of the transfer involves diffusion of molecules that could appropriately be described by this term. For the case of lactate transfer, the question of the control is dealt with more fully in the article by Magistretti in Chapter 11. For the glutamate–glutamine shuttle we will concentrate here on an apparent corollary of it, transfer of ammonium from neurons to astrocytes, which clearly falls within the concept of volume transmission. We will also give attention to our own recent results.

Transfer of lactate from astrocytes to neurons

NMR spectroscopy is now capable of detecting the incorporation of ^{13}C into metabolites in the rat brain in vivo. Figure 1A, from Pfeuffer et al., 1999, shows the time course of appearance of ^{13}C in lactate, aspartate and GABA after infusion of [1-^{13}C]glucose into the femoral vein of an anesthetized, ventilated, rat. The infusion was rapid for the first 5 min, then slow, to maintain the blood concentration roughly constant at about 16 mM. Despite the uncertainty in the data points at short times, it is clear that the incorporation into lactate was very rapid, well before aspartate, for example. This incorporation occurred despite the fact that hyperglycemia does not increase the brain lactate concentration (Ljunggren et al., 1974) so that the incorporation of ^{13}C was not the result of net synthesis of lactate. Since a steady state was reached so rapidly, the flux of carbon passing through lactate must have been large. Where was the lactate produced and where did it go?

*Corresponding author. Tel.. 00334 76 63 74 64/00334 76 76 59 20; Fax: 00334 76 76 58 96; e-mail: jonathan.coles@ujf-grenoble.fr
**Present address: Institut de Physiologie, Université de Lausanne, Lausanne, Switzerland.
***Present address: Department of Physiology, University College London, Gower Street, London, WC1E 6BT, UK.

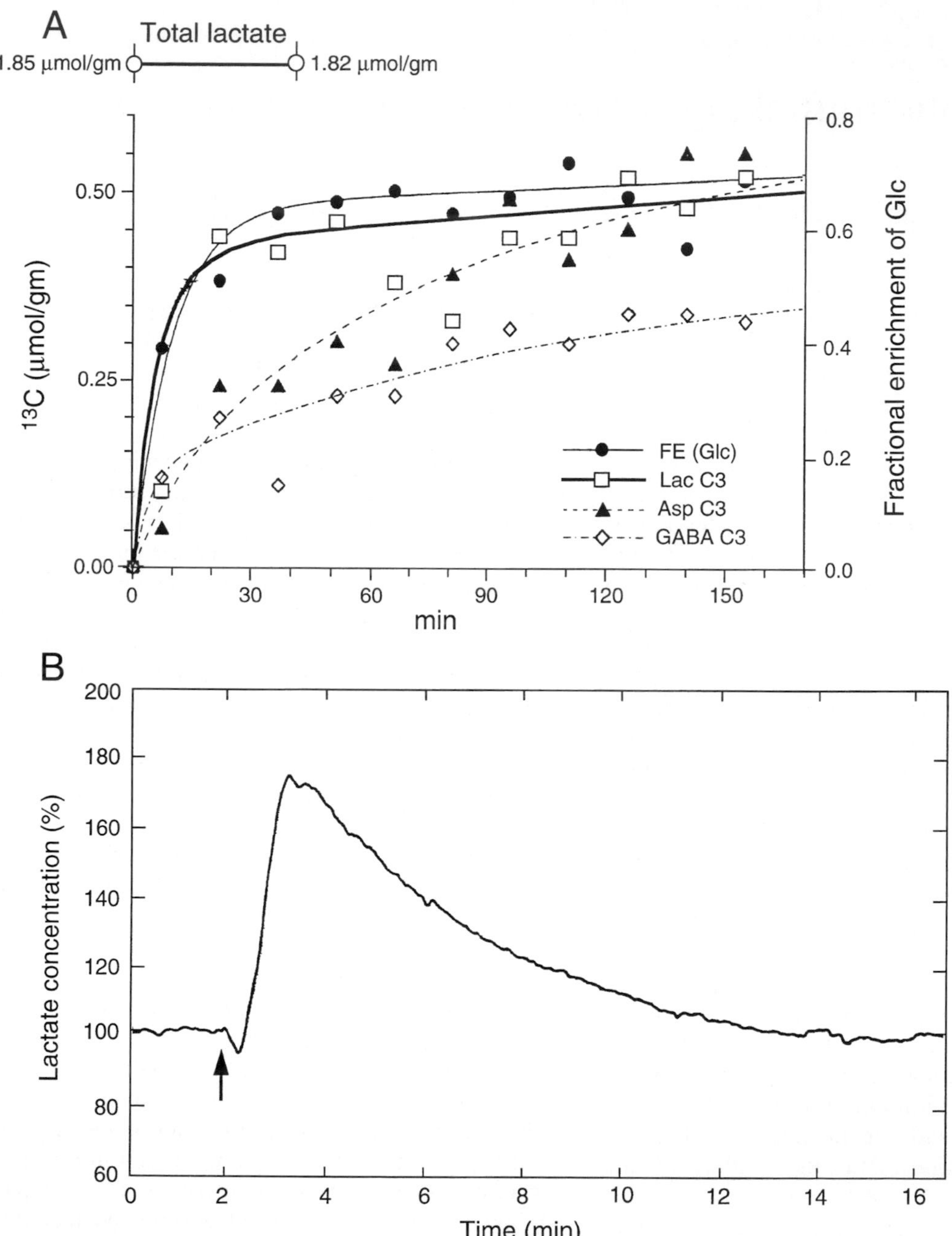

Fig. 1. Brain lactate in vivo. A. Incorporation of ^{13}C from blood glucose into lactate and other compounds in the rat brain (l.h. ordinate scale). From time zero, glucose 70% enriched in [1–^{13}C] glucose was infused at a rate such that it reached an approximately steady concentration in the blood of 16 mM within 5 min. From Pfeuffer et al., 1999. For comparison, the stability of total brain lactate in normoglycemia (7.7 μmol/gm in the blood) and a hyperglycemia of 28.4 μmol/gm (even greater than the 16 mM of Pfeuffer et al., 1999) is shown (from Ljunggren et al., 1974). B. Changes in local extracellular lactate concentration in the dentate gyrus of the rat brain in response to electrical stimulation of the perforant pathway. The measurement was made with a cylindrical lactate sensor 110 μm o.d. and 0.5 mm in length. From Hu and Wilson, 1997.

A specific role of astrocytes in brain metabolism was hinted at by Golgi (1886), whose staining method was the first to reveal the astrocyte endfeet that extend over the surface of the capillaries. He laconically suggested that the astrocyte might provide a route for transfer of substances from the capillary to the neuron.* The question has been raised many times since (see review by Coles, 1995) but it is only recently that a pathway through astrocytes for the supply of metabolic substrate to neurons has become widely accepted. A major factor in this acceptance (see Tsacopoulos and Magistretti, 1996), was a series of studies on a model preparation, the retina of the honey bee drone. In this nervous tissue, in which energy metabolism is extremely compartmented between neurons and glial cells, it has been shown quantitatively that transfer of metabolic substrate from glial cells to neurons can occur (Tsacopoulos et al., 1987) and, more strikingly, that it appears to be the obligatory route (Tsacopoulos et al., 1988). The drone retina will be discussed in more detail in a later section.

Concerning mammalian brain, the basic experimental result suggesting that lactate is produced in astrocytes, is that astrocytes in culture in aerobic conditions produce lactate from glucose and release it to the extracellular medium (Dringen et al., 1993a; Pellerin and Magistretti, 1994; Walz and Mukerji, 1988a, b). Neurons in culture produce much less lactate (Walz and Mukerji, 1988b). Both astrocytes and neurons in culture take up both glucose and lactate and the two substrates are both metabolized, at least to amino acids (Dringen et al., 1993b; Waagepetersen et al., 1998).

The cultured cells used in these experiments were taken from animals at, or just before, birth, a time when glial neuron interactions have not developed (Ransom et al., 1985; Pellerin et al., 1998). However, Poitry-Yamate et al. (1995) showed that Müller glial cells acutely isolated from retinas of 5–12 day old guinea pigs also produced and released lactate. The accumulation of lactate in the bath was less when the Müller cells were present in complexes with photoreceptor neurons. In the presence of these complexes, the accumulation in the bath of radioactive lactate formed from [U–^{14}C] glucose was increased by the addition of unlabelled lactate, which suggests that it was not the release of lactate that had decreased, but the uptake that had increased. The authors' conclusion that lactate was transferred from the Müller cells to the photoreceptors in these complexes is compelling.

Whether lactate transfer occurs in nervous tissue in vivo under physiological conditions, and in particular in the brain itself, is difficult to assess. It has long been remarked that lactate is released from brain slices supplied with oxygen (McIlwain et al., 1951). Conversely, in brain slices and in isolated sympathetic ganglia, neuronal activity is maintained when bath glucose is replaced by lactate (Elliott and Sutherland, 1952; McIlwain, 1953a; McIlwain, 1953b; Larrabee and Horowicz, 1956; Schurr et al., 1988). However, other compounds besides glucose and lactate, notably pyruvate, will also support nervous tissue function to some extent (McIlwain, 1953b; Greene and Winegrad, 1979). The concentration of lactate in the extracellular space of brain ($[\text{lact}]_e$) has been measured by means of implanted microdialysis probes as about 0.35 mM (Demestre et al., 1997) compared to a total brain concentration of 1.3 μmol/g (Mies et al., 1991)†.

---

* The astuteness of this idea is illustrated by the fact that another anatomist, Andriezen (1893), drew exactly the opposite conclusion and suggested that astrocytes provided an open mesh that would allow substances to diffuse freely through the extracellular space.

† Similar microdialysis measurements of glucose from the same laboratory gave values <0.45 mM in awake rats (Fellows et al., 1992, 1993), as did measurements with an enzyme electrode (Lowry et al., 1998). These values are lower than that found by Silver and Erecinska (1994) for extracellular glucose, also measured with an enzyme electrode (2.4 mM), and still less than the mean concentration of total glucose measured in rapidly frozen brain (about 3 mM, corresponding to 2.5 μmol/g, e.g., Mies et al., 1991). Most brain glucose transporters are of the facilitative type (Vannucci et al., 1998), so that glucose can only cross cell membranes down a concentration gradient. Hence, it is clear that at present we are unable to give a coherent description of the distribution in the brain even of a compound as fundamental as glucose.

Although concentration in itself gives no indication of a flux, changes in concentration are potentially more informative. Total tissue lactate concentration in vivo has been shown to increase with neuronal activation (Pritchard et al., 1991) and in Fig. 1B, from an article by Hu and Wilson (1997) it is seen that a 5 sec stimulation of the perforant pathway produced first a slight decrease in extracellular lactate concentration in the dentate gyrus of the hippocampus followed by an increase that reached its maximum in about 90 s. The rapidity and the amplitude of the response confirms that there must be considerable transmembrane fluxes, and one might speculate that the initial decrease is the consequence of increased uptake by neurons while the subsequent increase results from increased release from astrocytes. But, of course, without other information, it is equally possible that the response was the consequence of an initial decrease in lactate release followed by a decrease in uptake. Changes in $[\text{lact}]_e$ occur not only in response to unnaturally intense stimulation but also during physiological activity: Demestre et al. (1997) measured a mean peak increase in $[\text{lact}]_e$ in the striatum of freely moving rats of 58% during grooming.

Although lactate, being a weak acid, can diffuse through lipid membranes in its neutral form (Thomas, 1984), it can also be cotransported across membranes with protons on the monocarboxylic acid transporter (Poole et al., 1989). Two isoforms of this transporter, MCT1 and MCT2, which can also transport certain other metabolically important substances, such as pyruvate, are present within the parenchyme of the adult brain (Pellerin et al., 1998). Their distribution, and also the distributions of glucose transporters and lactate dehydrogenase are described in Chapter 11 by Magistretti.

Our assessment is that transfer of lactate from glial cells to neurons as an important energy substrate under physiological conditions is a most promising hypothesis. What is lacking is a convincing proof, confirmed in more than one laboratory. Poitry-Yamate et al. (1995) come very close, but their preparation of dissociated cells is technically so demanding that there are limitations to the kinds of experiments that can reasonably be undertaken to analyze the system in more detail. A further question is whether lactate is a neuronal substrate at all times (Poitry-Yamate et al., 1995), uniquely during increased neuronal electrical activity (Pellerin and Magistretti, 1994; Schurr et al., 1999) and/or during recovery after pathological events such as transient hypoxia (Schurr et al., 1997). In the case of the retinal rod photoreceptor neurons, part of the cell (the outer segment) appears to take up glucose while the inner segments take up lactate (see Poitry et al., 2000). Nor has the complexity of the control of astrocyte metabolism been unraveled: systems of regional control by neurons releasing neuropeptides or monoamines and local control by glutamate (Pellerin and Magistretti, 1994; Demestre et al., 1997) are discussed by Magistretti in Chap. 11. Work on the vertebrate retina has led Poitry et al. (2000) to suggest that ammonium also regulates glial cell metabolism. Because so many questions like these remain to be answered, we have started to study a different preparation, the isolated rat vagus nerve, which we believe offers certain advantages.

Vagus nerve

Since the putative glial-neuron transfer of lactate in brain is so little understood, we thought it would be useful to look at a preparation of mammalian nervous tissue that was convenient to work with and in which the in vivo relation of the glial cells and neurons was maintained. We chose the vagus nerve for reasons that include those of Ritchie and Straub (1956), namely that most of the axons, being unmyelinated and of small diameter, have a high specific metabolic rate, especially when stimulated (see also Ritchie, 1967). All the axons, myelinated and non-myelinated, are surrounded by Schwann cells, and, like astrocytes of the brain, the Schwann cells are situated between the neurons and endothelial cells rich in the GLUT-1 isoform of the glucose transporter (Orte et al., 1999). Véga et al. (1998) did the classic metabolic experiment that consists of incubating tissue in the presence of radioactive substrate and analyzing the radioactive products. When rabbit vagus nerve, stripped of its epineurium, was incubated in [U–^{14}C] D-glucose, of the products that accumulated in the nerve or in the bath, lactate was overwhelmingly predominant.

This result shows that lactate is produced from glucose and released from cells, but, since lactate is so freely released, it tends to argue against lactate being taken up avidly by axons and metabolized. As is usually the case, the intermediates of glycolysis were undetectable (apart from glucose–6P which precedes the rate-limiting step catalyzed by phosphofructokinase). Véga et al. (1998) took care to show that lactate did not come from a possible anoxic core of the nerve: decreasing bath PO_2 from 1 atm to 0.2 atm, which should have enlarged any putative anoxic core, did not increase lactate release (Fig. 2A). Exogenous lactate was readily oxidized to CO_2, more rapidly than was exogenous glucose, even when both were present (Fig. 2B).

We do not yet have evidence that the release and uptake of lactate in vagus nerve is associated with its transfer from glial cells to neurons, and can, at present, only extrapolate from the results on mammalian retinal cells and the circumstantial evidence for brain. Making this extrapolation, we

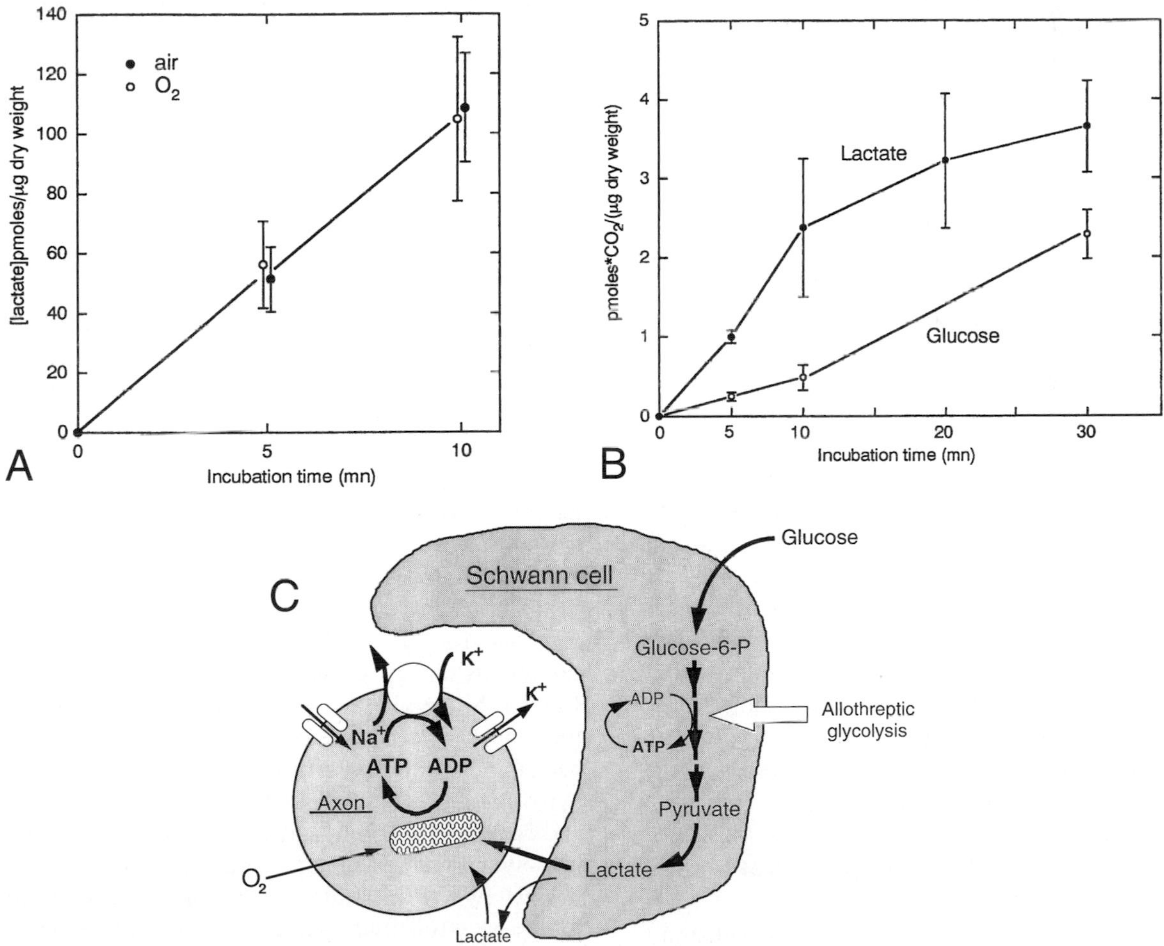

Fig. 2A. Lactate release from vagus nerve at different PO_2s. B. Production of $^{14}CO_2$ from exogenous [U–^{14}C] glucose (o) and from exogenous [U–^{14}C] lactate (●). In all experiments, the nerve was electrically stimulated at 2 Hz. From Véga et al., 1998. In A and B the bars indicate the S.E.M.s. In B the difference between the values at 5 min is significant with $P < 0.0005$. C. Working hypothesis for release and uptake of lactate in vagus nerve. The scheme shows the extreme case in which glucose is taken up only by the Schwann glial cells which supply lactate as a substrate for the intense metabolism necessary to maintain transmembrane ion gradients in the axon.

take as a working hypothesis the core of the schemes proposed by Poitry-Yamate et al. (1995) and Tsacopoulos and Magistretti (1996) as shown in Fig. 2C. Since, in this scheme, the main role of the glycolysis in the glial cells (in this case Schwann cells) is to produce lactate for the neurons we call it 'allothreptic glycolysis' – 'glycolysis to nourish another (cell).'‡

Apart from lactate, the main substance formed from exogenous glucose and released to the bath was glutamate, which raises the possibility that glutamate may be involved in intercellular trafficking or signaling. Kriegler and Chiu (1993) have reported that the intracellular free Ca^{2+} concentration in Schwann cells of rabbit sciatic nerve is increased by glutamate (see Chiu and Kriegler, 1994, for a general review of axon to Schwann cell signaling). Release of glutamate has been described for other peripheral nerves as well as the vagus (see Weinreich and Hammerschlag, 1975); in the vagus, it is formed from exogenous lactate in the presence of exogenous glucose (Véga et al., 1998), so according to the lactate transport theory, it would be formed in the axons. To date, no significant changes in the net release of lactate to the bath have been observed when incubation conditions were changed by omitting the nerve stimulation, or by adding 1 mM α-hydroxy 4 cyano cinnamate (an inhibitor of monocarboxylic acid transport (Poole et al., 1989; Bröer et al., 1999) or 5 mM glutamate (C. Véga, unpublished results). This lack of effect may reflect tight coupling between the metabolism of the axons and that of the Schwann cells. The convenient size, defined geometry and intense and readily-stimulated energy metabolism of the vagus nerve make it a promising preparation for further investigation of these and other aspects of metabolic trafficking in an intact tissue.

Detection of metabolic exchanges in real time

The tissue in which transfer of metabolic substrate from glial cells to neurons is best established is the retina of the honey bee drone. The metabolic compartmentation of this tissue is extreme: the only neurons, the photoreceptor cells, contain large numbers of mitochondria while the principal glial cells (the 'outer pigment cells') contain almost none, but are packed with glycogen (Perrelet, 1970; Dimitracos and Tsacopoulos, 1985). It has been shown that not only is transfer of substrate possible, since the neurons can draw on the glial glycogen reserves (Coles and Tsacopoulos, 1981; Tsacopoulos et al., 1987), but transfer is obligatory. The arguments are that autoradiographs made after incubation in [U–^{3}H] 2-deoxyglucose (2DG), which is taken up on glucose transporters, appear to show uptake only in the glial cells (Tsacopoulos et al., 1988); and 30 min superfusion of a slice with 10 mM 2DG and no glucose has no effect on the oxygen consumption, showing that 2DG does not interfere with photoreceptor metabolism, and further supporting the conclusion that glucose is not taken up by the photoreceptors (Tsacopoulos et al., 1987). A corollary of this conclusion is that the substance transferred cannot be glucose. Alanine is the strongest candidate. After incubation of slices in [U–^{14}C] glucose, alanine is the compound which has the greatest radioactivity and it is both present in the slices and released to the bath (Tsacopoulos et al., 1994). In vivo it is present at a surprisingly high concentration (31 mM) in the interstitial fluid (Cardinaud et al., 1994). The enzyme that catalyses the interconversion of alanine and pyruvate (alanine amino transferase) is present in both the glial cells and the neurons (Tsacopoulos et al., 1994). Pyruvate is taken up by neurons, is an immediate substrate for the TCA cycle, and is an effective exogenous energy substrate for neurons (McIlwain, 1953b). Since alanine, in bee retina, as suggested for lactate in vertebrate nervous tissue, is converted back to pyruvate in neurons, one naturally asks why pyruvate is not transferred directly. Tsacopoulos and Magistretti (1996) have pointed out that while glycolysis results in the reduction of NAD to NADH (which in neurons and muscle is oxidized by the electron transport chain) the conversion of pyruvate to alanine or lactate oxidizes NADH and the stoichiometry is such as to maintain the redox potential constant. This could therefore be a reason for the evolution of the metabolic detour.

If pyruvate is aminated to form alanine in the glial cells and alanine is converted back to pyruvate

‡ From αλο = other, θρεπτηρ = wet-nurse.

in the neurons, then fixed N must be returned to the glial cells. There is good evidence that ammonium fills this role. Ammonium is released from slices (Tsacopoulos et al., 1997b) and, when applied to isolated glial cells, it increases the formation and release of alanine and reduces nicotinamide adenine dinucleotides (Fig. 3A; Tsacopoulos et al., 1997a). Bath application of ammonium to a retinal slice also makes pH in the extracellular clefts (pH_e) more alkaline (Coles et al., 1996). This observation was made during investigation of the effect on pH_e of light stimulation of the photoreceptors. After brief (10s) light stimulation of the photoreceptors, pH_e becomes more alkaline, reaching a peak after about 30 sec, and recovering to baseline within 3–4 min (Fig. 3B; Coles et al., 1988; Coles et al., 1996). An alkalization of similar amplitude and time course can be induced either by stimulating the energy metabolism of the photoreceptors, e.g. by uncoupling mitochondrial respiration from ATP production by application of dinitrophenol (Fig. 3C) or by bath application of ammonium at concentrations up to about 2 mM (Fig. 3B).

The effect of ammonium was studied further by measuring changes in the intracellular pH (pH_i) of isolated retinulae (the clusters of photoreceptor cells that lie behind each corneal facet) and of isolated bundles of glial cells. pH_i was measured with the fluorescent indicator BCECF (Marcaggi et al., 1999). Ammonium produced marked acidification of the glial cells, normally preceded by a small, transient alkalization. The acidification was blocked by removal of external Cl^- (Fig. 4A), or by application of the loop diuretic bumetanide, and the acidification saturated with increasing concentrations of ammonium (Marcaggi et al., 1999). In contrast, Ba^{2+}, which blocks the voltage response of the glial cells to increased K^+ (Marcaggi and Coles, 1998), presumably by blocking a K^+ conductance, had no effect on the ammonium-induced acidification. It was concluded that NH_4^+ was transported into the cell on a cotransporter with functional similarities to transporters of the K, Cl cotransporter family (Haas and Forbush, 1998). In contrast to its effect on the pH_i of glial cells, ammonium had relatively little effect on the pH_i of isolated photoreceptor neurons. The small changes observed were in the alkaline direction, as is usually the case for neurons and muscle cells (Thomas, 1984). This is because the membranes are permeable to the uncharged form of ammonium, but not to NH_4^+; within the neuron, nearly all the NH_3 combines with H^+ ions to form NH_4^+, thereby making the pH alkaline.

The way experimental application of ammonium to isolated glial cells induces changes in pH_i is probably non-physiological, as the acidification depends not only on the entry of NH_4^+ but also on the efflux of NH_3 and is associated with the accumulation of NH_4^+ within the cells (Fig. 4B right; Boron and de Weer, 1976; Marcaggi et al., 1999). We suggest that what happens in the intact tissue is that NH_3 leaves the photoreceptors and picks up an H^+ ion in the extracellular space, so making pH_e alkaline. The departure of NH_4^+ from the extracellular clefts by movement into the glial cells will have almost no effect on pH_e, and, presumably, the ammonium within the glial cells is consumed to form alanine (Fig. 4B, left). The results of two kinds of experiment support this interpretation. First, when extracellular Cl^- was replaced by the large anion gluconate, the alkaline ΔpH_e induced by ammonium application was reduced while the light-induced ΔpH_e was not reduced (Marcaggi et al., 1999). Second, the ΔpH_e induced either by light stimulation or by application of ammonium was modified in the same way when the pH gradient across the cell membranes was changed. When pH_e was increased and pH_i decreased by bath application of a weak acid, such as propionate, ΔpH_e was greatly decreased. Conversely, when pH_e was decreased and pH_i increased by bath application of a weak base, such as 2-(diethylamino)-ethyl-amine, light-induced ΔpH_e was increased (Coles et al., 1996). These results can be explained by the changes in the ratio $[NH_4^+]/[NH_3]$ in the intra- and extracellular spaces when pH was changed.

At present, the main interest of this interpretation of the origin of the stimulus-induced increase in pH_e is that (provided its origin is confirmed) it gives a measure not just of an extracellular concentration but of a *flux*. This should make it possible to investigate, for example, the regulation of the transformation of alanine to pyruvate in the photoreceptors in slices of retina.

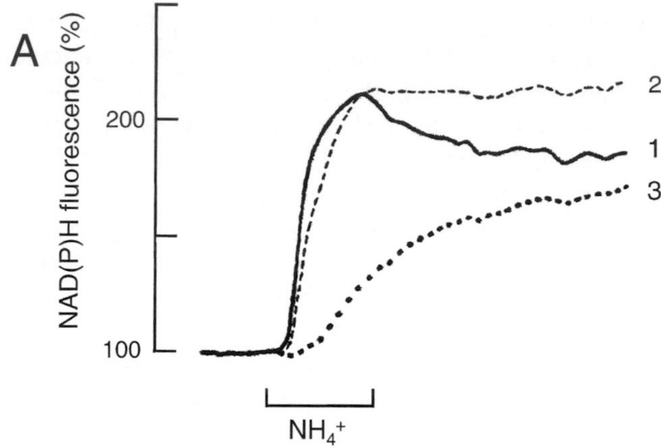
A
NAD(P)H fluorescence (%)
200
100
2
1
3
NH4+

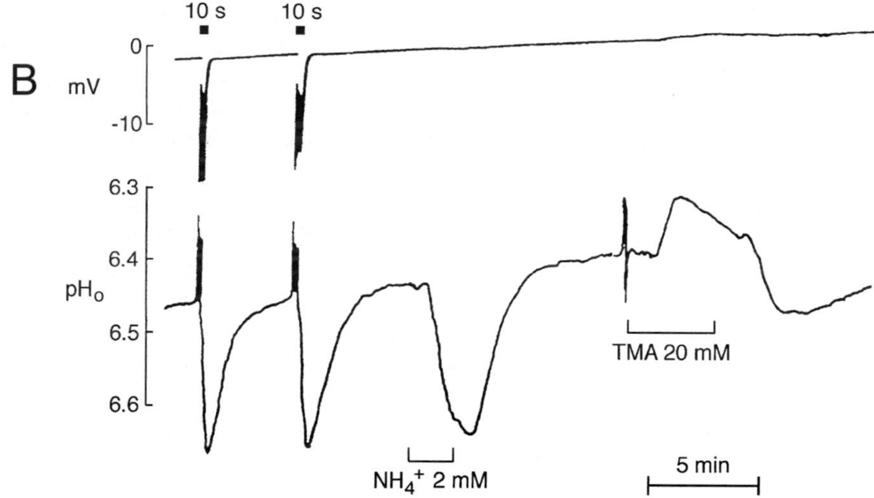
10 s
10 s
0
B
mV
-10
6.3
6.4
pHo
6.5
6.6
TMA 20 mM
NH4+ 2 mM
5 min

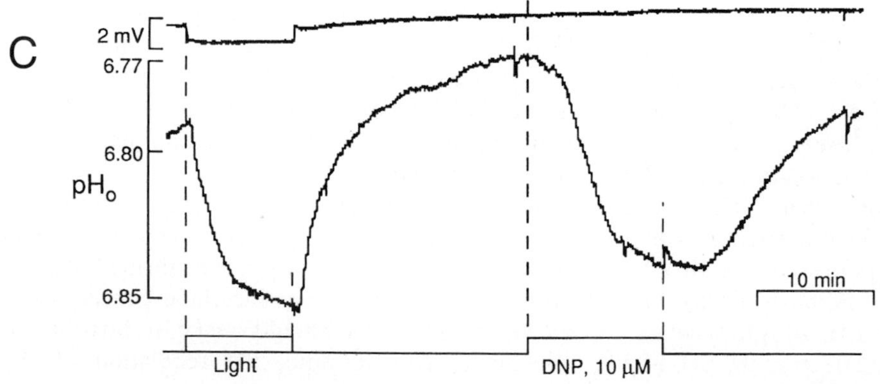
2 mV
C
6.77
6.80
pHo
6.85
10 min
Light
DNP, 10 μM

Fig. 3A. Ammonium reduces nicotinamide adenine dinucleotides in bundles of glial cells isolated from drone retina. Application of a puff of ammonium solution 1 mM increased fluorescence from NADH (and/or NADPH). The curves are from parts of the bundle progressively further from the puffer tip, in the order 1,2,3. The time scale is given by the ammonium application, which lasted 20 s From Tsacopoulos et al., 1997a. B. pH in the extracellular clefts (pH_e) of a superfused slice of bee retina measured with a double-barreled pH microelectrode. Stimulation of the photoreceptor neurons with trains of light flashes, 1 per s for 10s, produced a negative-going field potential recorded by the reference barrel (top trace). After each stimulation, pH_e became transitorily more alkaline (downward deflexion, preceded by electrical artifacts). Inclusion of 2 mM ammonium in the superfusate caused an alkalization of similar amplitude. In contrast, an amine structurally close to ammonium, trimethyl amine, caused an acidification. C. pH_e, measured as in B. Stimulation of the photoreceptors with continuous light caused a sustained alkalization, as did application of 10 μM dinitrophenol (DNP). Both light and DNP increase oxygen consumption. B and C are from Coles et al., 1996.

The glutamate/glutamine shuttle in vertebrate brain

Berl et al. (1962) infused ammonium labelled with ^{15}N into the blood stream of cats and analysed the incorporation of the ^{15}N into glutamate and glutamine. Of the two nitrogens of glutamine, that of the amide group had a much higher specific activity than that of the α-amino group confirming that most of the glutamine was formed from glutamate. What interested them more was that the specific labelling of the α-amino group of glutamine was higher than that of glutamate. On the basis of this, they concluded that glutamate freshly formed from

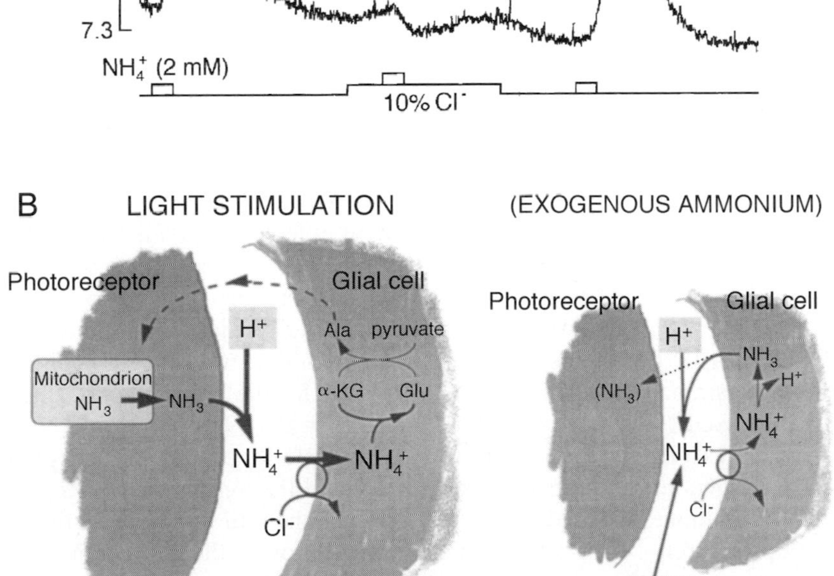

Fig. 4A. Changes in intracellular pH in an acutely isolated, superfused bundle of bee retinal glial cells. Application of ammonium caused an acidification which was abolished when [Cl^-] was reduced to 10% of its normal value. From Marcaggi et al., 1999. B. Schemes showing how light stimulation (left hand scheme) or application of exogenous ammonium to a retinal slice (right hand scheme) are thought to cause changes in pH in the extracellular clefts of of bee retina.

α-ketoglutarate and ammonium in a 'small' compartment was more likely to be converted to glutamine than was glutamate in a larger, more stable pool. They further made the prescient suggestion that the 'small' pool corresponded to astrocytes. Their basic result was confirmed with refined methods by Cooper et al. (1979), and their suggestion that glutamine was formed in astrocytes was also confirmed when Norenberg and Martinez-Hernandez (1979) showed by immunocytochemistry that glutamine synthetase is localized predominantly in astrocytes. It was therefore suggested that glutamine was transferred from astrocytes to neurons, as a precursor for glutamate as a neurotransmitter. Glutamate transporters on astrocytes are now well-described (Billups et al., 1998) and their inactivation by chronic administration of antisense nucleotides leads to neurodegeneration, suggesting that they make an important contribution to homeostasis of extracellular glutamate (Rothstein et al., 1996). Although the localization of glutaminase is not clear, it does appear to be mainly in neurons, although not in all neurons (Kaneko et al., 1995). Finally, Sibson et al. (1998) using a mathematical model (Mason et al., 1995) to interpret data from ^{13}C NMR spectroscopy estimated that, in the brain as a whole, the flux of carbon from glutamate to glutamine was approximately equal half that through the TCA cycle. Taken together these results provide compelling evidence for the existence of a major flux of glutamine from astrocytes to neurons.

Uptake of ammonium by astrocytes

As pointed out by Benjamin and Quastel (1975), amination of glutamate in astrocytes and deamination of glutamine in neurons implies that fixed N, probably in the form of ammonium, must be returned to the astrocytes. In vivo, exogenous ammonium is incorporated rapidly into glutamine in the brain, so it is presumably taken up preferentially into astrocytes rather than neurons (Berl et al., 1962; Cooper et al., 1979). Nagaraja and Brookes (1998) have studied the effect of ammonium on the intracellular pH of cultured astrocytes. They found, as for the bee retinal glial cells, that the cells acidified after only a very small and brief alkalization. However, in striking contrast to the bee glial cells, the acidification was almost entirely blocked by 50 μM Ba^{2+}, which suggests that NH_4^+ entered through K^+ channels of the inwardly-rectifying type. The remaining component was blocked by bumetanide, implying some transport on a member of the K, Cl cotransporter family. These conclusions were supported in independent work by P. Marcaggi and P. Sartor (personal communication) who found that removal of Cl^- had no effect on the ammonium-induced acidification. Concerning adult brain, mean tissue pH has been measured in brain slices by NMR and found to decrease 9 min after application of 1 mM ammonium (Brooks et al., 1989). Since ammonium normally increases pH in neurons (Thomas, 1984), this result may reflect preferential uptake of ammonium by astrocytes as NH_4^+. A result in the same sense was obtained by Chesler (1986) who made simultaneous measurements of pH_i in a neuron and pH_e in the brainstem of the lamprey. When he applied 10 mM NH_4Cl he observed that pH_e first decreased and then increased and, after 5 min, was more than 0.1 pH units more alkaline than the baseline. At this time, pH_i in the neuron he was sampling had returned to baseline; this result suggests that more NH_4^+ was going into astrocytes than into neurons.

Possible effects of NH_4^+ on astrocytes

Irrespective of how it enters, NH_4^+ is likely to have at least two regulatory effects on astrocyte metabolism. First, the expectation that exogenous ammonium will stimulate formation of glutamine has been confirmed for astrocytes in culture (Waniewski, 1992) and for the brain in vivo (Shen et al., 1998; Tsukada et al., 1998). Second, NH_4^+ has a stimulatory allosteric effect on phosphofructokinase (Lowry and Passoneau, 1966; Sugden and Newsholme, 1975) and will therefore favor the glycolysis necessary to provide the ATP for, among other reactions, the amination of glutamate. Pellerin and Magistretti (1994) have shown that uptake of glutamate by cultured astrocytes stimulates uptake of glucose and release of lactate, and suggested that this contributes to coupling of neuronal activity to supply of energy substrate to the neurons. We therefore support the suggestion of Poitry et al. (2000) that since glutamatergic synaptic transmission is associated with entry of ammonium into astrocytes (Benjamin and Quastel, 1975), ammonium will also contribute to this coupling.

Interaction of NH_4^+ and K^+ on glial cells

When the hydrogen ions of NH_4^+ interact with the chemical environment, the remaining N species has a diameter close to that of K^+ (Hille, 1992) which will explain why NH_4^+ will to some extent permeate K^+ ion channels and be transported on K^+ transporters (Choe et al., 2000). The Ba^{2+}-sensitive K^+ conductances that conduct K^+ at resting potentials on glial cells are (at least on astrocytes) of the inwardly-rectifying type (K_{IR}). On both bee retinal glial cells and on astrocytes they subserve both the membrane depolarization in response to elevation of extracellular $[K^+]$ ($[K^+]_e$), and the movement of K^+ into astrocytes in regions where $[K^+]_e$ is elevated by neuronal activity, and out of astrocytes in regions where $[K^+]_e$ is not so high (Karwoski et al., 1989, see Amédée et al., 1997). The NH_4^+/K^+ permeability ratio of vertebrate K_{IR} appears to be variable; it is low (0.03–0.04) on starfish eggs (Hagiwara and Takahashi, 1974). Irrespective of the permeability ratio, K^+ presumably competes with NH_4^+ for access to the K_{IR} channel. We can speculate, therefore, that if $[K^+]_e$ rises to pathologically high levels, it will inhibit entry of NH_4^+ into the astrocytes and reduce its stimulatory effects on glutamine synthesis and glycolysis.

Conclusion

There are fluxes between neurons and glial cells of several compounds involved directly or indirectly in the energy metabolism of nervous tissue. These compounds include, in mammals: glutamate, glutamine, probably lactate, and ammonium. All four will diffuse through extracellular space over distances determined in part by the rates of uptake by cells. In species as different as the guinea pig and the honey bee it appears that allothreptic glycolysis in the glial cells results in the production of respectively lactate and alanine that is transferred to neurons. In both mammalian brain and bee retina, ammonium is transferred from neurons to glial cells. We cannot claim to understand nervous tissue metabolism in either species until we can propose plausible scenarios for the evolution of such similarities and differences.

Acknowledgements

We thank Mme Monique Lavielle for technical help. Financial support was received from the Conseil Régional d'Aquitaine (grant no. 97–0301208)

References

Amédée, T., Robert, A. and Coles, J.A. (1997) Potassium homeostasis and glial energy metabolism. *Glia*, 21: 46–55.

Andriezen, W.L. (1893) On the system of fibre-like cells surrounding the blood vessels of the brain of man and mammals, and its physiological significance. *Int. Monatsschr. Anat. Physiol.*, 10: 532–540.

Benjamin, A. and Quastel, J. (1975) Metabolism of amino acids and ammonia in rat brain cortex slices in vitro: a possible role of ammonia in brain function. *J. Neurochem.*, 25: 197–206.

Berl, S., Takagaki, G., Clarke, D. and Waelsch, H. (1962) Metabolic compartments in vivo. Ammonia and glutamic acid metabolism in brain and liver. *J. Biol. Chem.*, 237: 2562–2569.

Billups, B., Rossi, D., Warr, O., Takahashi, M., Sarantis, M., Szatkowski, M. and Attwell, D. (1998) Physiological and pathological operation of glutamate transporters. *Prog. Brain Res.*, 116: 45–57.

Boron, W.F. and de Weer, P. (1976) Intracellular pH transients in squid giant axon caused by CO_2, NH_3 and metabolic inhibitors. *J. Gen. Physiol.*, 67: 91–112.

Bröer, S., Bröer, A., Schneider, H.P., Stegen, C., Halestrup, A.P. and Deitmer, J.W. (1999) Characterization of the high-affinity monocarboxylate transporter MCT2 in xenopus laevis oocytes. *Biochem. J.*, 341: 529–535.

Brooks, K.J., Kauppinen, R.A., Williams, S.R., Bachelard, H.S., Bates, T.E. and Gadian, D.G. (1989) Ammonia causes a drop in intracellular pH in metabolizing cortical brain slices. A $[^{31}P]$- and $[^1H]$-nuclear magnetic resonance study. *Neuroscience*, 33: 185–192.

Cardinaud, B., Coles, J.A., Perrottet, P., Spencer, A.J., Osborne, A.J. and Tsacopoulos, M. (1994) The composition of the interstitial fluid in the retina of the honeybee drone: implications for the supply of substrates of energy metabolism from blood to neurons. *Proc. Roy. Soc. (Lond.) B*, 257: 49–58.

Chesler, M. (1986) Regulation of intracellular pH in reticulospinal neurones of the lamprey, *Petromyzan marinus*, *J. Physiol. (Lond.)*, 381: 241–261.

Chiu, S.Y. and Kriegler, S. (1994) Neurotransmitter-mediated signaling between axons and glial cells. *Glia*, 11: 191–200.

Choe, H., Sackin, B. and Palmer, L.G. (2000) Permeation properties of inward-rectifier potassium channels and their molecular determinants. *J. Gen. Physiol.*, 115: 391–404.

Coles, J.A. (1995) Glial cells and the supply of substrates of energy metabolism to neurons. In: H. Kettenmann and B.R. Ransom (Eds), *Neuroglia*, pp. 793–814. Oxford University Press: New York.

Coles, J.A., Giovannini, P. and Thomas, R.C. (1988) Changes in extracellular pH induced by light stimulation in slices of honeybee drone retina. *J. Physiol. (Lond.)*, 398: 57P.

Coles, J.A., Marcaggi, P., Véga, C. and Cotillon, N. (1996) Effects of photoreceptor metabolism on interstitial and glial cell pH in bee retina: evidence of a role for NH_4^+. *J. Physiol. (Lond.)*, 495.2: 305–318.

Coles, J.A. and Tsacopoulos, M. (1981) Ionic and possible metabolic interactions between sensory neurones and glial cells in the retina of the honeybee drone. *J. Exp. Biol.*, 95: 75–92.

Cooper, A.J.L., McDonald, J.M., Gelbard, A.S., Gledhill, R.F. and Duffy, T.E. (1979) The metabolic fate of ^{13}N-labeled ammonia in rat brain. *J. Biol. Chem.*, 254: 4982–4992.

Demestre, M., Boutelle, M.G. and Fillenz, M. (1997) Stimulated release of lactate in freely moving rats is dependent on the uptake of glutamate. *J. Physiol. (Lond.)*, 499: 825–832.

Dimitracos, S.A. and Tsacopoulos, M. (1985) The recovery from a transient inhibition of the oxidative metabolism of photoreceptors of the drone (*Apis mellifera*). *J. Expt. Biol.*, 19: 165–181.

Dringen, R., Gebhardt, R. and Hamprecht, B. (1993a) Glycogen in astrocytes: possible function as lactate supply for neighbouring cells. *Brain Res.*, 623: 208–214.

Dringen, R., Wiesinger, H. and Hamprecht, B. (1993b) Uptake of L-lactate by cultured rat brain neurons. *Neurosci. Lett.*, 163: 5–7.

Elliott, H.W. and Sutherland, V.C. (1952) The oxygen uptake of human cerebral cortex slices and the effects of some inhibitors. *J. Cell. Comp. Physiol.*, 40: 221–241.

Fellows, L.K., Boutelle, M.G. and Fillenz, M. (1992) Extracellular brain glucose levels reflect local neuronal activity: a microdialysis study in awake, freely moving rats. *J. Neurochem.*, 59: 2141–2147.

Fellows, L.K., Boutelle, M.G. and Fillenz, M. (1993) Physiological stimulation increases non-oxidative glucose metabolism in the brain of the freely moving rat. *J. Neurochem.*, 60: 1258–1263.

Golgi, C. (1886) *Sulla fina anatomia degli organi centrali del sistema nervosa*. Hoepli: Milan.

Greene, D.A. and Winegrad, A.I. (1979) In vitro studies of the substrates for energy production and the effects of insulin on glucose utilization in the neural components of peripheral nerve. *Diabetes*, 28: 878–887.

Haas, M. and Forbush, B.R. (1998) The Na-K-Cl cotransporters. *J. Bioenerg. Biomembr.*, 30, 161–172.

Hagiwara, S. and Takahashi, K. (1974) The anomalous rectification and cation selectivity of the membrane of a starfish egg cell. *J. Membr. Biol.*, 18: 61–80.

Hille, B. (1992) *Ion channels in excitable membranes*. Sinauer: Sunderland.

Hu, Y. and Wilson, G.S. (1997) A temporary local energy pool coupled to neuronal activity: fluctuation of extracellular lactate levels in rat brain monitored with rapid-response enzyme based sensor. *J. Neurochem.*, 69: 1484–1490.

Kaneko, T., Kang, Y. and Mizuno, N. (1995) Glutaminase-positive and glutaminase-negative pyramidal cells in layer VI of the primary motor and somatosensory cortices: a combined analysis by intracellular staining and immunocytochelmistry in the rat. *J. Neurosci.*, 15: 8362–8377.

Karwoski, C.J., Lu, H.K. and Newman, E.A. (1989) Spatial buffering of light-evoked increases in extracellular K^+ in the plexiform layers of amphibian retinas. *Science*, 244: 578–580.

Kriegler, S. and Chiu, S.Y. (1993) Calcium signalling of glial cells along mammalian axons. *J. Neurosci.*, 13: 4229–4245.

Larrabee, M.G. and Horowicz, P. (1956) Glucose and oxygen utilization in sympathetic ganglia. I. Effects of anesthetics. II Substrates for oxidation at rest and in activity. In: *Molecular Structure and Functional Activity of Nerve Cells*, pp. 84–107. American Institute of Biological Science: Washington.

Ljunggren, B., Norberg, K. and Siesjö, B.K. (1974) Influence of tissue acidosis upon restitution of brain energy metabolism following total ischemia. *Brain Res.*, 77: 173–186.

Lowry, J.P., O'Neill, R.D., Boutelle, M.G. and Fillenz, M. (1998) Continuous monitoring of extracellular glucose concentrations in the striatum of freely moving rats with an implanted glucose sensor. *J. Neurochem.*, 70: 391–396.

Lowry, O.H. and Passoneau, J.V. (1966) Kinetic evidence for multiple binding sites on phosphofructokinase. *J. Biol. Chem.*, 241: 2268–2279.

Marcaggi, P. and Coles, J.A. (1998) The major routes of entry of NH_4^+ and K^+ into bee retinal glial cells are independent. *J. Physiol. (Lond.)*, 513P: 15–16P

Marcaggi, P., Thwaites, D.T., Deitmer, J.W. and Coles, J.A. (1999) Chloride-dependent transport of NH_4^+ into bee retinal glial cells. *Eur. J. Neurosci.*, 11: 167–177.

Mason, G.F., Gruetter, R., Rothman, D.L., Behar, K.L., Shulman, R.G. and Novotny, E.J. (1995) Simultaneous determination of the rates of the TCA cycle, glucose utilization, a-ketoglutarate/glutamate exchange, and glutamine synthesis in human brain by NMR. *J. Cerebr. Blood Flow Metab.*, 15: 12–25.

McIlwain, H. (1953a) Glucose level, metabolism, and response to electrical impulses in cerebral tissues from man and laboratory animals. *Biochem. J.*, 55: 618–624.

McIlwain, H. (1953b) Substances which support respiration and metabolic response to electrical impulses in human cerebral tissues. *J. Neurol. Neurosurg. Psychiatr.*, 16: 257–266.

McIlwain, H., Anguiano, G. and Cheshire, J.D. (1951) Electrical stimulation in vitro of the metabolism of glucose by mammalian cerebral cortex. *Biochem. J.*, 50: 12–18.

Mies, G., Cruz, N. and Sokoloff, L. (1991) Comparison of freeze-blowing and funnel-freezing of rat brain for the measurement of cerebral glucose concentration in vivo. *J. Neurochem.*, 56:, 1673–1676.

Nagaraja, T.N. and Brookes, N. (1998) Intracellular acidification induced by passive and active transport of ammonium ions in astrocytes. *Am. J. Physiol.*, 274: C883-C891.

Norenberg, M.D. and Martinez-Hernandez, A. (1979) The fine structural localization of glutamine synthetase in astrocytes of rat brain. *Brain Res.*, 161: 303–310.

Orte, C., Lawrenson, J.G., Finn, T.M., Reid, A.R. and Allt, G. (1999) A comparison of blood-brain barrier and blood-nerve barrier endothelial markers. *Anat. Embryol. (Berl.)*, 199: 509–517.

Pellerin, L. and Magistretti, P.J. (1994) Glutamate uptake into astrocytes stimulates aerobic glycolysis: a mechanism coupling neuronal activity to glucose utilization. *Proc. Natl Acad. Sci., USA*, 91: 10625–10629.

Pellerin, L., Pellegri, G., Martin, J. and Magistretti, P.J. (1998) Expression of monocarboxylate transporter mRNAs in mouse brain: support for a distinct role of lactate as an energy substrate for the neonatal vs. adult brain. *Proc. Natl Acad. Sci., USA*, 95: 3990–3995.

Perrelet, A. (1970) The fine structure of the retina of the honey bee drone. *Z. Zellforsch. Mikrosk. Anat.*, 108: 530–562.

Pfeuffer, J., Tkác, I., Choi, I.-Y., Merkle, H., Ugurbil, K., Garwood, M. and Gruetter, R. (1999) Localized in vivo ^{1}H NMR detection of neurotransmitter labeling in rat brain during infusion of [1–^{13}C] D-glucose. *Mag. Res. Med.*, 41: 1077–1083.

Poitry, S., Poitry-Yamate, C.L., Ueberfeld, J., MacLeish, P.R. and Tsacopoulos, M. (2000) Mechanisms of glutamate metabolic signaling in retinal glial (Müller) cells. *J. Neurosci.*, 20: 1809–1821.

Poitry-Yamate, C.L., Poitry, S. and Tsacopoulos, M. (1995) Lactate released by Müller cells is metabolized by photoreceptors from mammalian retina. *J. Neurosci.*, 15: 5179–5191.

Poole, R., Halestrup, A.P., Price, S.J. and Levi, A.J. (1989) The kinetics of transport of lactate and pyruvate into isolated cardiac myocytes from guinea pig. *Biochem. J.*, 264: 409–418.

Pritchard, J., Rothman, D., Novotny, E., Petroff, O., Kuwabara, T., Avison, M., Howseman, A., Hanstock, C. and Shulman, R. (1991) Lactate rise detected by ^{1}H NMR in human visual cortex during physiological stimulation. *Proc. Natl Acad. Sci., USA*, 88: 5829–5831.

Ransom, B., Yamate, C. and Connors, B. (1985) Activity-dependent shrinkage of extracellular space in rat optic nerve: a developmental study. *J. Neurosci.*, 5: 532–535.

Ritchie, J.M. (1967) The oxygen consumption of mammalian non-myelinated nerve fibres at rest and during activity. *J. Physiol. (Lond.)*, 188: 309–329.

Ritchie, J.M. and Straub, R.W. (1956) The after-effects of repetitive stimulation on mammalian non-medulated fibres. *J. Physiol. (Lond.)*, 137: 698–711.

Rothstein, J.D., Dykes-Hoberg, M., Pardo, C.A., Bristol, L.A., Jin, L., Kuncl, R.W., Kanai, Y., Hediger, M.A., Wang, Y., Schielke, J.P. and Welty, D.F. (1996) Knockout of glutamate transporters reveals a major role for astroglial transport in excitotoxicity and clearance of glutamate. *Neuron*, 16: 675–686.

Schmitt, F. (1984) Molecular regulators of brain function: a new view. *Neuroscience*, 13: 991–1001.

Schurr, A., Miller, J.J., Payne, R.S. and Rigor, B.M. (1999) An increase in lactate output by brain tissue serves to meet the energy needs of glutamate-activated neurons. *J. Neurosci.*, 19: 34–39.

Schurr, A., Payne, R.S., Miller, J.J. and Rigor, B.M. (1997) Brain lactate is an obligatory aerobic energy substrate for functional recovery after hypoxia: further in vitro validation. *J. Neurochem.*, 69: 423–426.

Schurr, A., West, C.A. and Rigor, B.M. (1988) Lactate-supported synaptic function in the rat hippocampal slice preparation. *Science*, 240: 1326–1328.

Shen, J., Sibson, N.R., Cline, G., Behar, K.L., Rothman, D.L. and Shulman, R.G. (1998) 15N-NMR spectroscopy studies of ammonia transport and glutamine synthesis in the hyperammonemic rat brain. *Dev. Neurosci.*, 20: 434–443.

Sibson, N.R., Dhankhar, A., Mason, G.F., Rothman, D.L., Behar, K.L. and Shulman, R.G. (1998) Stoichiometric coupling of brain glucose metabolism and glutamatergic neuronal activity. *Proc. Natl Acad. Sci., USA*, 95: 316–321.

Silver, I.A. and Erecinska, M. (1994) Extracellular glucose concentration in mammalian brain: continuous monitoring of changes during increased neuronal activity and upon limitation in oxygen supply in normo-, hypo-, and hyperglycemic animals. *J. Neurosci.*, 14: 5068–5076.

Sugden, P. and Newsholme, E. (1975) The effect of ammonium, inorganic phosphate and potassium ions on the activity of phosphofructokinases from muscle and nervous tissues of vertebrates and invertebrates. *Biochem. J.*, 150: 113–122.

Thomas, R.C. (1984) Experimental displacement of intracellular pH and the mechanism of its subsequent recovery. *J. Physiol. (Lond.)*, 354: 3P–22P.

Tsacopoulos, M., Coles, J.A. and Van de Werve, G. (1987) The supply of metabolic substrate from glia to photoreceptors in the retina of the honeybee drone. *J. Physiol. (Paris)*, 82: 279–287.

Tsacopoulos, M., Evêquoz-Mercier, V., Perrottet, P. and Buchner, E. (1988) Honeybee retinal glial cells transform glucose and supply the neurons with metabolic substrate. *Proc. Natl Acad. Sci., USA*, 85: 8727–8731.

Tsacopoulos, M. and Magistretti, P.J. (1996) Metabolic coupling between glia and neurons. *J. Neurosci.*, 16: 877–885.

Tsacopoulos, M., Veuthey, A.-L., Saravelos, S.G., Perrottet, P. and Tsoupras, G. (1994) Glial cells transform glucose to alanine, which fuels the neurons in the honeybee retina. *J. Neurosci.*, 14: 1339–1351.

Tsacopoulos, M., Poitry-Yamate, C.L. and Poitry, S. (1997a) Ammonium and glutamate released by neurons are signals regulating the nutritive function of a glial cell. *J. Neurosci.*, 17: 2383–2390.

Tsacopoulos, M., Poitry-Yamate, C.L., Poitry, S., Perrottet, P. and Veuthey, A.L. (1997b) The nutritive function of glia is regulated by signals released by neurons. *Glia*, 21: 84–91.

Tsukada, Y., Kanamatsu, T., Watanabe, H. and Okamoto, K. (1998) In vivo investigation of glutamate-glutamine metabolism in hyperammonemic monkey brain using ^{13}C-magnetic resonance spectroscopy. *Dev. Neurosci.*, 20: 427–433.

Vannucci, S.J., Clark, R.R., Koehler-Stec, E., Li, K., Smith, C.B., Davies, P., Maher, F. and Simpson, I.A. (1998) Glucose transporter expression in brain: relationship to cerebral glucose utilization. *Dev. Neurosci.*, 20: 369–379.

Véga, C., Poitry-Yamate, C.L., Jirounek, P., Tsacopoulos, M. and Coles, J.A. (1998) Lactate is released and taken up by isolated rabbit vagus nerve during aerobic metabolism. *J. Neurochem.*, 71: 330–337.

Waagepetersen, H.S., Bakken, I., Larsson, O.M., Sonnewald, U. and Schousboe, A. (1998) Comparison of lactate and glucose metabolism in cultured neocortical neurons and astrocytes using ^{13}C-NMR spectroscopy. *Dev. Neurosci.*, 20: 310–320.

Walz, W. and Mukerji, S. (1988a) Lactate production and release in cultured astrocytes. *Neurosci. Lett.*, 86: 296–300.

Walz, W. and Mukerji, S. (1988b) Lactate release from cultured astrocytes and neurons: a comparison. *Glia*, 1: 366–370.

Waniewski, R.A. (1992) Physiological levels of ammonia regulate glutamine synthesis from extracellular glutamate in astrocyte cultures. *J. Neurochem.*, 58: 167–174.

Weinreich, D. and Hammerschlag, R. (1975) Nerve impulse-enhanced release of amino acids from non-synaptic regions of peripheral and central nerve trunks of bullfrog. *Brain Res.*, 84: 137–142.

L. F. Agnati, K. Fuxe, C. Nicholson and E. Syková (Eds.)
Progress in Brain Research, Vol 125

CHAPTER 13

Cell volume and water exchange in neural cells monitored by diffusion weighted ^{1}H NMR spectroscopy

D. Leibfritz*, J. Pfeuffer, U. Flögel, C. Meier and S. Bröer

Fachbereich Biologie/Chemie, Universität Bremen, 28334 Bremen, Germany

Introduction

The structure of brain tissue consists of a filigree network built primarily by glia cells and neurons; neurons, and especially astrocytes, form innumerable branched processes with delicate filamentary processes. As a consequence of its highly differentiated functions, neural tissue has a very high energy-consumption to maintain neural transmission and metabolic trafficking between neurons and astrocytes (see Chapter 12 by Coles et al. and Chapter 11 by Magistretti et al. in this volume). The primary fuel components are in particular glucose, but also branched chain amino acids and keton bodies; in addition, end products of oxidation (bicarbonate, glutamine etc.) have to be removed from the brain. As these compounds are osmotically active, this high energy consumption causes a net flux of more than hundred milliliters of water per hour through the resting brain (van Gelder 1983a, b, 1990) associated with transport of energy substrates, primarily glucose, to the brain and the waste product from the brain.

Both the manifold metabolic fluxes among neural cells and the net flux of water across the brain require a very active control of osmotic pressure and cell volumes. Without this, volume transmission in particular would be compromised because there would be continual fluctuation in the extracellular space. Consequently, active and passive transport of ions and metabolites through membranes must be accompanied by a sizeable water transport to preserve all the homeostatic conditions needed for neural functionality. In addition, information about local deviations of the osmotic pressure and cell volumes must be communicated to induce mechanisms and processes of regulatorial volume control.

In order to monitor the dynamics of tissue water non-invasively and to study mechanisms of volume control one must use a method that is able to

- differentiate intracellular from extracellular water,
- monitor the exchange of water across the cell membrane,
- register anisotropic motions of water.

The method of diffusion weighted NMR spectroscopy (Stejskal and Tanner, 1965, 1968) meets all these requirements and will be described briefly in the next section.

Diffusion weighted NMR spectroscopy

Diffusion weighted NMR spectroscopy (DW-MRS) uses water as a reporter molecule; this molecule carrries two protons. Their nuclear spins are polarized in a static external magnetic field; in addition, they can be magnetically labelled as a function of their spatial localization. The principle is briefly described below.

*Corresponding author.
e-mail: Dieter.Leibfritz@uni-bremen.de

If an ensemble of nuclear spins (e.g. nuclear spins of water protons) is excited with their resonance frequency in a homogeneous magnetic field B_o, all spins within a volume will precess coherently about the axis of the vector B_o (Fig. 1a). A magnetic field gradient applied subsequently to the excitation along the z-axis will linearily increase the precession frequency of water spins along the z-axis. By this means the frequency coherence of precessing nuclear spins within the x, y-plane will vanish (Fig. 1b) and a MR signal is no longer observable in the x, y-plane. A second magnetic field gradient of opposite sign and equal strength as the previous one will restore a coherent precession frequency in all nuclear spins within this volume (Fig. 1c). A spin echo of refocussed spins will induce a detectable signal in a receiver coil in the x, y-plane. However, proton spins, e.g. water molecules, changing their location within this volume between the application of the first field gradient and the subsequent refocussing gradient will not be refocussed properly. As a consequence of water diffusion, the signal intensity of water is decreased in the refocussed gradient echo. In practice, a 180° spin inversion pulse is applied in between the two magnetic field gradients to avoid B_o inhomogeneity effects. In this case both field gradients will have the same sign.

The observable signal intensity will depend on the following parameters:

$$\ln (S/S_o) = -\gamma^2\delta^2G^2(\Delta - \delta/3)D = b \cdot D \qquad (1)$$

S and S_o are the intensities of the water signal with and without diffusion weighting, respectively, γ is the gyromagnetic ratio of protons, which is an isotope specific constant; δ the duration for which the field gradient is switched on; G the strength of magnetic field gradient; $(\Delta - \delta/3) = t_D$ the diffusion time; D the diffusion constant of water; b represents all experimental parameters affecting the diffusion weighting.

In practice, there are two ways to measure diffusion-sensitized signal intensity loss: either by varying the diffusion time t_D, while the gradient strength is kept constant ('constant gradient' experiment, cg) or varying the gradient strength G and keeping the diffusion time constant ('constant time' experiment, ct).

The just-described sequence of a spin-echo excitation in the presence of magnetic field gradients is called 'pulsed-field-gradient nuclear magnetic resonance' (PFG–NMR) and represents an experimental tool to assess the water diffusion (Stejskal and Tanner, 1965, 1968). The method outlined above will not recognize different locations within an object, but has to be combined with an additional tool for spatial localization, called diffusion weighted imaging (DWI) or diffusion weighted spectroscopy (DW–MRS). To follow parameters of importance in volume transmission, such as osmotic disorder and its regulatorial correction will require spatially selective recording of translational water dynamics.

Diffusion weighted MR imaging

Diffusion weighted preparation of the water signal in the PFG experiment is now combined with any of the spatial encoding techniques of water detection (magnetic resonance imaging) to obtain an image contrast, which depends upon the diffusion behaviour of the observed water within individual volume elements (voxels) of an object. This combination is termed: diffusion weighted imaging (DWI) (Moseley et al.,1995). The diffusion contrast is based on variations of the measured, 'apparent' diffusion coefficient (ADC) D_{app} of water in tissues.

Pulsed-field-gradient magnetic resonance imaging (PFG–MRI) is well suited to assess non-invasively the dynamics of water behaviour in biological tissues under normal and pathological conditions. Extracellular water molecules will diffuse away faster, e.g. their signal intensity will be lost more quickly, than intracellular water, which is restricted by the cell wall. Usually a mixed diffusion contribution of both compartments is recorded in vivo. However, if the compartment with slower or restricted diffusion increases in volume, as in the case of a cell swelling process, areas with disordered volume control will show up with a 'brighter' water signal intensity and smaller D_{app} than control. By this means, diffusion weighted imaging is able to monitor changes in the normal volume control occurring concomitantly with various neuropathologies e.g. ischemic lesions and

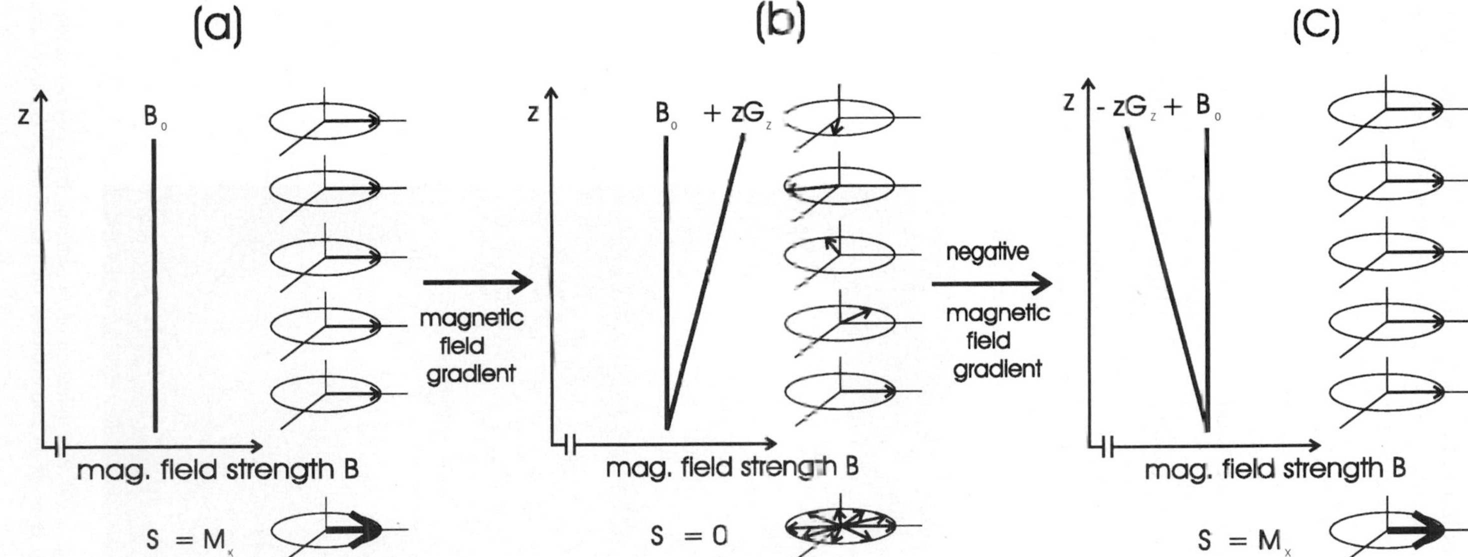

Fig. 1. Schematic of a diffusion weighted NMR experiment. After radio frequency excitation all nuclear spins will precess coherently along the z-axis, as indicated by the magnetization vector in y direction (a), a magnetic field gradient in z-direction will linearily increase the precession frequency in z-direction and destroy the coherence in the x,y plane (b), application of a second field gradient of the same strength, but of opposite sign, will restore coherent magnetization (c). Spins leaving their location after the first gradient pulse and before the second will not be refocussed properly and no longer contribute to the observable refocussed spin echo.

early stage of cytotoxic and vascular edemas or neoplastic malformations in brain (Moseley et al., 1995). Changes or loss of volume control after an ischemic insult can be detected typically a few minutes after onset of the insult and and can be used to follow up the time course of the lesion development (Davis et al., 1994; Hoehn-Berlage et al., 1995). Figure 2 shows a diffusion weighted MR image of rat brain. An ischemic insult is induced in the right hemisphere by occlusion of the middle (MCAO) cerebral artery. Figure 2A, B and 2C, D, respectively, show two examples of a control and a MCAO lesion.

Cytotoxic cell swelling caused by an influx of interstitial water into cells is one of the reasons for the ADC change. However, the change of the extracellular volume fraction and extracellular tortuosity as measured by the real-time iontophoresis (TMA^+) method cannot account quantitatively for the ADC decrease entirely (van der Torn et al., et al., 1996). Besides the net change of the ratio of extracellular to intracellular volume it is of interest to understand the mechanisms involved in 'cell volume control', such as cell membrane permeability, active transport phenomena, variation of relaxation times etc.

Although the usefulness of DWI contrast is well accepted, the physiological and molecular mechanisms of ADC changes are still under discussion, as the time and spatial resolution of whole body or animal scanners is insufficient to elucidate cellular mechanisms. For this reason, cellular models must often be used to study individual contributions to the cell volume regulation to simplify the rather

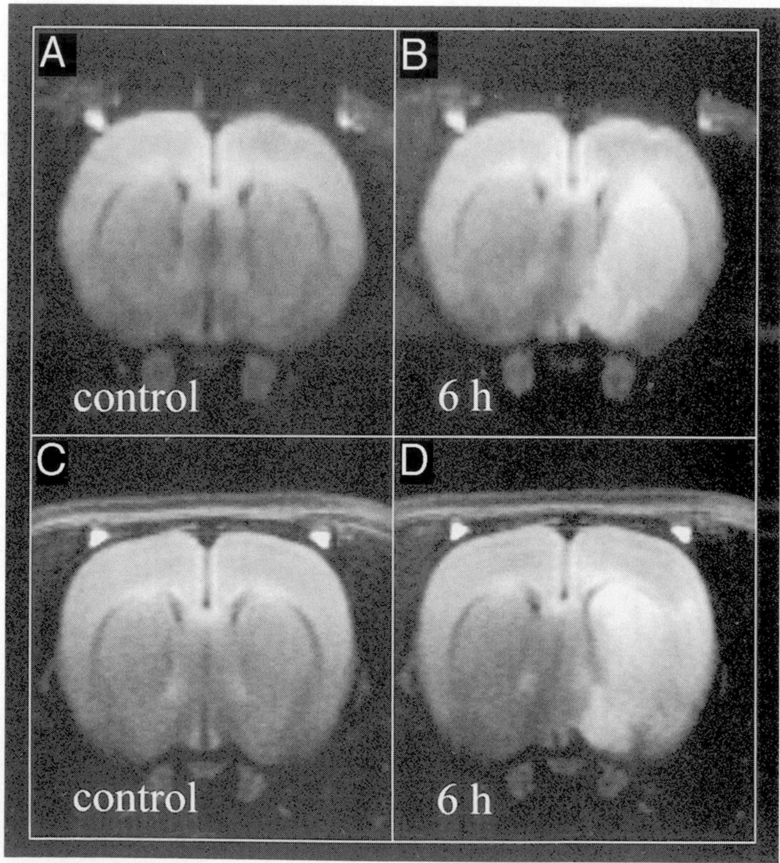

Fig. 2. Comparison of control (A, C) rat brains and after induction of an ischemic lesion by occlusion of the middle cerebral artery, using MCAO model, (B, D). Comparison of B, gives an impression of the individual deviations of this stroke model. DW $b = 1500$ s mm^{-2}.

complex situation in living tissue. Regardless of the drawback of an in vitro model, viable cells can be subjected to controlled volume changes while they still have control of membrane transport and permeability.

Cell model

Clonal glioma cells (F98) or primary rat astrocytes were cultivated in Dulbecco's modified Eagle medium (DMEM) incl. 5% fetal calf serum and embedded in basal membrane gel (BMG) threads. BMG threads with approx. 10^8 cells were transferred under sterile conditions into 8 mm NMR tubes with a perfusion insert and perfused at 310 K with culture medium (1 ml/min) oxygenated with 95% O_2/5% CO_2. Experimental details have been described before (Flögel et al., 1994, 1995). The intracellular volume fraction was calculated to be 0.05 based on an average cell diameter of 6–8 μm. The osmolarity of the medium was modified by changing the NaCl concentration (hypotonic: 180 $mosm \cdot l^{-1}$; isotonic: 300 $mosm \cdot l^{-1}$; hypertonic: 420 $mosm \cdot l^{-1}$).

Diffusion-weighted MRS

Diffusion weighted NMR measurements were performed on a Bruker AMX 360 MHz (8.4 T) system with a maximal gradient strength of 333 mT. A 90°-τ-180°-τ-spin-echo sequence with unipolar gradients during the τ intervals was used. Diffusion weighting was achieved by varying the gradient strength G (constant time experiments) or by varying the separation time Δ between the leading edges of the field gradients (constant gradient experiments). The gradient duration was δ. This defines a diffusion time $t_D = \Delta - \delta/3$ and the q value $q^2 = (\gamma\delta G)^2$ and the b-value $b = q^2\ t_D$. When the interpulse τ is kept constant, T_2 effects also remain constant. For further experimental details see (Pfeuffer et al., 1998a, b).

Intracellular volume

In order to get a measure of cell volume and cell volume control it is necessary to measure the intracellular volume independently from the extracellular compartment and acurately enough. The signal intensity of an immobilized cell culture of primary astrocytes is shown in Fig. 3a under various osmotic conditions. Note the biexponential type behaviour of the slope, which reflects the contributions of extracellular water (steep slope) and intracellular water (flat slope). The signal intensity for an individual b-value at any point (e.g. $b = 15000\ s \cdot mm^{-2}$) within the linear part of the flat slope (intracellular water) can be used as a measure of the cell volume. Figure 3b shows the intracellular volume changes as a function of different osmotic conditions. This experimental set up can be used to monitor cell volume changes after various exogenous or endogenous interferences like necrotic cell swelling, cell shrinkage upon induction of apoptosis, volume effects after drug treatment, etc. (Pfeuffer et al., 1998b,; Engelmann et al., 1996; Henke et al., 1998; Zwingmann et al., 1998). The time resolution of cell volume measurements is about one second, if approximately 10^8 cells immobilized in basal membrane gel threads were used.

Exchange time

If no water exchange from the extracellular to the intracellular space occurs within the time frame of a diffusion weighted experiment, the signal intensity of the intracellular water intensity would remain independent of the diffusion time for long diffusion times. That means that the flat slope (intracellular water) of Fig. 3 would have to be parallel to the abscissa, as the water could not move beyond the cell membrane. However, the signal intensity of the intracellular contribution in a 'constant time' experiment is not parallel to the abscissa for long diffusion times, but decreases with increasing diffusion weighting as a function of the q-value (see section on *Diffusion MRS*) (Fig. 4). This means water leaves the cell within the allowed diffusion time and will diffuse away with the extracellular diffusion constant; an exchange time of approximately 50 ms was calculated from the constant slope for astrocytes and glioma cells. For further experimental details see Pfeuffer et al., (1998a).

To assess changes in brain tissue that would influence volume transmission it is of interest to know whether the membrane structure of the cell is still intact or partially disordered, which would affect the membrane permeability for water.

Mechanism of water exchange

Astrocytes, in particular, are very sensitive to osmotic stress because of their thin, filamentous processes (Rohlmann and Wolff, 1996). Additionally, they are responsible for a highly active metabolic trafficking with neurons, in particular in the thin neuronal processes within the vicinity of synapses (Coles et al., Chapter 12 and Magistretti et al., Chapter 11 this book). Astrocytes, therefore require an efficient water regulation system to maintain their morphology and function. Furthermore, one has to keep in mind that a net flux of more than 100 ml$\cdot$h^{-1} water passes the brain, even at rest (van Gelder 1983a). Maintenance of a

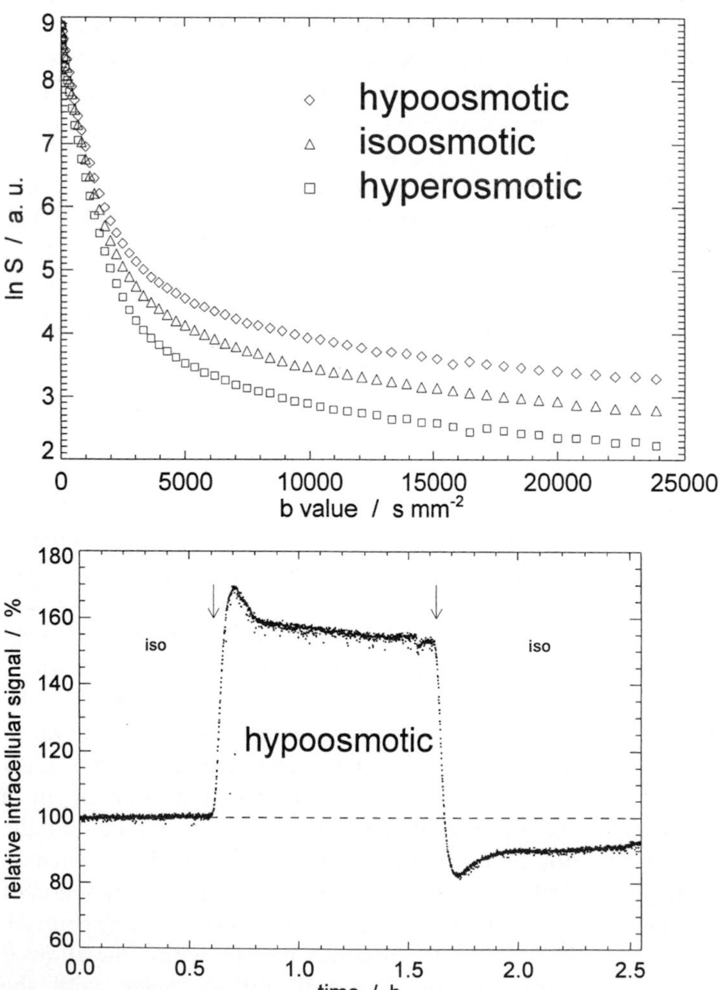

Fig. 3. The semilogarithmic plot of the water signal intensity shows a first order biexponential behaviour as function of increasing diffusion weighting. The first linear part of the curve represents extracellular water, and the second part intracellular water. The intracellular part of the curve will be shifted in parallel with varying osmotic conditions indicating swelling or shrinking cell volume. Monitoring the intracellular water intensity at a fixed b value ($b = 15000$ s$\cdot$mm^{-2}) is proportional to the cell volume (B). This set up can be used to measure cell volume changes with a time resolution of 3–5 s.

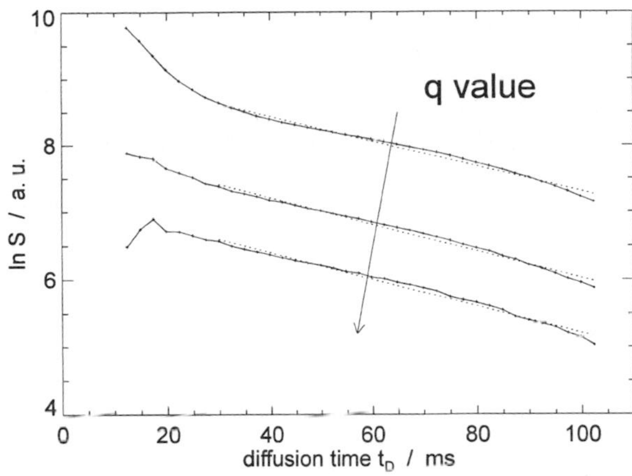

Fig. 4. Semilogarithmic plot of the water signal intensity from the intracellular water as function of the diffusion time for different gradient strength. The parallel slope of all curves indicates a constant exchange time of approx. 50 ms of intracellular water accross the cell membrane.

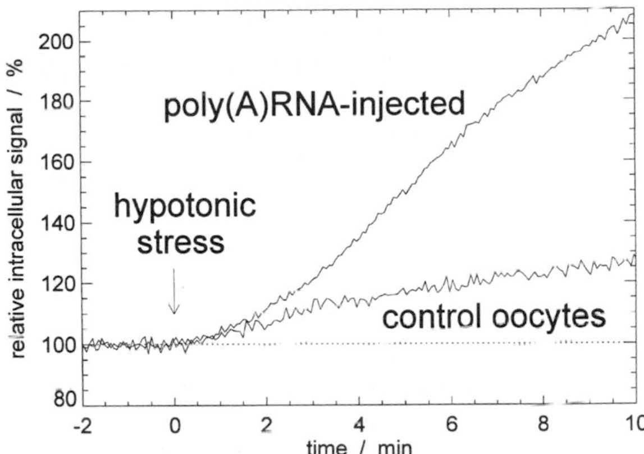

Fig. 5. Swelling of poly(A) RNA injected oocytes, which express aquaporin AQ1 of C6 glioma cells, is detected upon changing from isotonic buffer to hypotonic buffer by diffusion weighted proton NMR. Successive signals of intracellular water are detected with a diffusion weighting with $b = 5400\ s\ mm^{-2}$. Control oocytes show a minor swelling only.

constant cell volume in the presence of a dynamic influx and efflux of metabolites might either occur simply by physical diffusion or though the control of an efficient and water regulation system. Protein-formed water channels (aquaporins, AQP) are one selective pathway to move water rapidly across cell membranes (Nielsen et al., 1993a). Five different aquaporins have been found in mammalian cells (Agre et al., 1995). In brain, the mercury-sensitive AQP1 and the mercury-insensitive AQP4 have been identifed so far (Nielsen et al., 1993b, 1997, Frigeri et al., 1995).

To test the hypothesis of whether aquaporins can account for the water transport dynamics measured in the previous section, oocytes from *Xenopus laevis* have been used as a cellular model system to express astrocytic AQP1 upon induction with the corresponding mRNA from C6 glioma cells (Pfeuffer et al., 1998c). If poly(A) RNA-injected oocytes are exposed to hypotonic stress (Fig. 5), they swell rapidly without regulatory volume control. The swelling is inhibited upon the addition of mercury (II) chloride and restimulated after the addition of mercaptoethanol (Pfeuffer et al., 1998c). Cell swelling of C6-glioma poly(A) RNA treated oocytes can also be prevented after treatment with antisense oligonucleotide against the AQP1 mRNA (Pfeuffer et al., 1998c).

Conclusion

Diffusion weighted NMR spectroscopy is sensitive to translational movements of water molecules in living tissue. By this means it is feasible to measure cell volumes and to reveal restricted intracellular diffusion. However, the barrier for restricted diffusion e.g. the cellular membrane is partially permeable, which accounts for why intracellular water leaves cells and extracellular water enters cells within the experimentally allowed diffusion time. Aquaporins represent the major pathway for water exchange in cells, which are normally exposed to extreme osmotic situations, like red blood cells and kidney cells; they are also present in neural cells. If astrocytic aquaporins are expressed in model oocytes, their volume behaviour is similar to clonal astrocytes under osmotic stress. Whether results obtained on cell cultures are applicable to the in vivo situation in tissues, or to what extent, is still open to question. However, diffusion weighted spectroscopy in rat brain has been shown to be also multiexponential (Norris et al., 1995, Pfeuffer et al., 1999). In addition to the restricted intracellular water diffusion and in combination with water permeable membranes (c.f. the model of 'restricted intracellular water diffusion at permeable boundaries' (Pfeuffer et al., 1998a)), extracellular water diffusion is hindered by tortuosity effects imposed by macromolecular structures on the cell surface and within the interstitial space.

A further complexity arises from different cell size among neural cells and their anisotropic cell shape. The branched and often filamentous processes are necessary to maintain intercellular volume transmission though gap junctions and for sustaining various homeostatic conditions and allowing an essential metabolic trafficking among astrocytes and neurons. It should be emphasized, that the major part of the astrocytic cell volume consists of thin processes (Rohlmann and Wolff, 1996). A water molecule in such a filigree cellular process will reach the cell membrane after a very short intracellular diffusion length. The water channels in these processes will allow a fast water export/import, thus small cells, or regions of small cells, will have a smaller ADC (Meier et al., 1999). The contribution of the cell boundaries to the intracellular diffusion will dominate therefore for very long diffusion times and constant gradient experiments. Preliminary in vivo DWI measurements at high field and strong field gradients indicate this complex diffusion behaviour of tissue and cellular water. Such results are better represented by 'diffusograms', e.g. ensemble of diffusion constants, rather than individual diffusion constants (Pfeuffer et al., 1999).

Diffusion weighted magnetic resonance spectroscopy and imaging are sensitive methods to monitor non-invasively intracellular and extracellular water diffusion and by this means also cell volumes and volume regulation under normal and pathological conditions. Aquaporins can account for the observable membrane permeability; controlled cell volume and membrane permeability are essential prerequisites for metabolic trafficking and other signalling via volume transmission in neural tissue.

References

Agre, P., Brown, D. and Nielsen, S. (1995) Aquaporin water channels: unanswered questions and unresolved controversies. *Curr. Opin. Cell Biol.*, 7: 472–483.

Coles, J.A., Véga, C. and Marcaggi, P. (2000) Metabolic traficking between cells in nervous tissue, 241–254, this volume.

Davis, D., Ulatowski, J., Eleff, S., Izuta, M., Mori, S., Shungu, D. and van Zijl, P.C.M. (1994) Rapid monitoring of changes in water diffusion coefficients during reversible ischemia in cat and rat brain. *Mag. Res. Med.*, 31: 454–460.

Engelmann, J., Henke, J., Willker, W., Kutscher, B., Nößner G. and Leibfritz, D. (1996) Early stage monitoring of miltefosine induced apoptosis in KB cells by multinuclear NMR spectroscopy. *Anticanc. Res.*, 16: 1429–1440.

Flögel, U., Willker, W. and Leibfritz, D. (1994) Regulation of intracellular pH in neuronal and glial cells, studied by multinuclear NMR spectroscopy. *NMR in Biomed.*, 7: 157–166.

Flögel, U., Niendorf, T., Serkova, N., Brand, A. and Leibfritz, D. (1995) Changes in organic solutes, volume, energy state and metabolism associated with osmotic stress in a glial cell line: a multinuclear NMR study. *Neurochem. Res.*, 20: 793–802.

Frigeri, A., Gropper, M.A., Umenishi, F., Kawashima, M., Brown, D. and Verkman, A.S. (1995) Localization of MIWC and GLIP water channel homologs in neuromuscular, epithelial and glandular tissues. *J. Cell Sci.*, 108: 2993–3002.

Henke, J., Engelmann, J., Flögel, U., Pfeuffer, J., Kutscher, B., Nößner, G., Engel, J., Voegeli, R. and Leibfritz, D. (1998) Apoptotic effects of hexadecylphosphocholine on resistant and non-resistant cells monitored by NMR spectroscopy. *Drugs Today* 34 (Suppl. F): 37–50.

Hoehn-Berlage, M., Eis, M., Back, T., Kohno, K. and Yamashita, K. (1995) Changes of relaxation times (T_1, T_2) and apparent diffusion coefficient after permanent MCAO in the rat. Temporal evolution, regional extent and comparison with histology, *Mag. Res. Med.*, 34: 824–838.

Magistretti, P.J. and Pellerin, L. (2000) The astrocyte-mediated coupling between synaptic activity and energy metabolism operates through volume transmission, 230–240, this volume.

Meier, C., Dreher, W. and Leibfritz, D. (1999) Water diffusion and exchange in rat brain measured in-vivo at very high *b*-values. *Ann. Meet. Int. Soc. Mag. Res. Med.*, 563.

Moseley, M.E., Butts, K., Yenari, M.A., Marks, M. and de Crespigny, A. (1995) Clinical aspects of DWI. *NMR in Biomed.*, 8: 387–396.

Nielsen, S., Smith, B.L., Christensen, E.I., Knepper, M.A. and Agre, P. (1993a) CHIP28 water channels are localized in constitutively water-permeable segments of the nephron. *J. Cell Biol.*, 120: 371–383.

Nielsen, S., Smith, B.L., Christensen, E.I. and Agre, P. (1993) Distribution of the aquaporin CHIP in secretory and resorptive epithelia and capillary endothelia. *Proc. Natl. Acad. Sci., USA*, 90: 7275–7279.

Nielsen, S., Nagelhus, E.A., Amiry-Moghaddam, M., Bourque, C., Agre, P. and Ottersen, O.P. (1997) Specialized membrane domains for water transport in glial cells: high-resolution immunogold cytochemistry of aquaporin-4 in rat brain. *J. Neurosci.*, 17: 171–180.

Norris, D.G., Niendorf, T. and Leibfritz, D. (1995) Healthy and infarcted brain tissues studied at short diffusion times: the origin of apparent restriction and the reduction in apparent diffusion coefficient. *NMR in Biomed.*, 7: 304–310.

Pfeuffer, J., Flögel, U., Dreher, W. and Leibfritz, D. (1998a) Restricted diffusion and exchange of intracellular water: theoretical modelling and diffusion time dependence of ^{1}H-NMR measurements on perfused glial cells. NMR in Biomed., 11: 19–31.

Pfeuffer, J., Flögel, U. and Leibfritz, D. (1998b) Monitoring of cell volume and water exchange time in perfused cells by diffusion weighted ^{1}H-NMR spectroscopy. *NMR in Biomed.*, 11: 11–18.

Pfeuffer, J., Bröer, S., Bröer, A,, Lechte, M., Flögel, U. and Leibfritz, D. (1998c) Expression of aquaporins in *Xenopus laevis* oocytes and glial cells as detected by diffusion-weighted ^{1}H NMR spectroscopy and photometric swelling assay. *BBA*, 1448: 27–36.

Pfeuffer, J., Provencher, S.W. and Gruetter, R. (1999) Water diffusion in rat brain in-vivo as detected at very large *b* values is multicompartmental. *MAGMA*, 8: 98–108.

Rohlmann, A. and Wolff, J.R. (1996) Subcellular topography and plasticity of gap junction distribution on astrocytes. In: D.C. Spray (Ed.), *Gap Junctions in the Nervous System.* Landous Company.

Stejskal, E.O. and Tanner, J.E. (1965) Spin diffusion measurements. Spin echos in the presence of a time-dependent field gradient. *J. Chem. Phys.*, 42: 288–292.

Stejskal, E.O. and Tanner, J.E. (1968) Restricted diffusion of protons in colloidal systems by the pulsed gradient spin echo method. *J. Chem. Phys.*, 49: 1768–1777.

van der Torn, A,. Syková, E., Dijkhuisen, R.M., Vorsek, I., Vargová, L., Skobisova, E., van Campagne, M., Reese, T. and Nicolay, K. (1996) Dynamic changes in water ADC, energy metabolism extracellular space volume and tortuosity in neonatal rat brain during global ischemia. *Mag. Reson. Med.*, 36: 52–60.

van Gelder, N.M. (1983a) Metabolic interactions between neurons and astroglia: glutamine synthetase,carbonic anhydrase, and water balance. In: H.H. Jasper and N.M. van Gelder (Eds), *Basic Mechanisms of Neuronal Hyperexcitability* pp. 5–29. A. R. Liss Inc. N.Y.

van Gelder. N.M. (1990) Neuronal discharge hypersynchrony and the intracranial water balance in relation to glutamic acid and taurine redistribution: migraine and epilepsy. *Prog. Clin. Biol. Res.*, 351: 1–20.

van Gelder, N.M.A. (1983b) Central mechanism of action for taurine: osmoregulation, bivalent cations, and excitation threshold. *Neurochem. Res.*, 8: 687–699.

Zwingmann, C., Brand, A., Richter-Landsberg, C. and Leibfritz, D. (1998) Multinuclear NMR studies on NH_4Cl-induced metabolic alterations and detoxification processes in primary astrocytes and glioma cells. *Dev. Neurosci.*, 20: 417–426.

SECTION IV

Monoamines and VT

L. F. Agnati, K. Fuxe, C. Nicholson and E. Syková (Eds.)
Progress in Brain Research, Vol 125

CHAPTER 14

Extrasynaptic distribution of monoamine transporters and receptors

Virginia M. Pickel

Department of Neurology and Neuroscience, Weill Medical College of Cornell University, New York, NY 10021, USA

Introduction

Monoamines are released from vesicular storage sites in axon terminals to activate selective receptors in target neurons (Reith and Selmeci, 1992; Moore, 1993). Transmission is terminated largely by plasmalemmal reuptake into the axon terminals through specific transporters (for recent review see Beckman and Quick, 1998; Blakely et al., 1998; Ase et al., 1999). Electron microscopic studies showing that monoaminergic terminals form symmetric synapses with thin pre-and post-synaptic membrane specializations, as well as asymmetric synapses having thick postsynaptic densities suggest that these neurons use the 'synaptic mode' of transmission, and may produce either inhibitory or excitatory responses that are typically associated with these types of junctions (see Pickel et al., 1997 for review). This concept is challenged by studies showing that dopamine is released from both axons and dendrites (Geffen et al., 1976), and that the sites of release do not necessarily correlate with the morphologically identified synaptic contacts (Descarries et al., 1977; Beaudet and Descarries, 1978). In addition, monoamine storage vesicles include not only small synaptic vesicles (SSVs), but large dense core vesicles (DCVs) that are often located at a distance from the synaptic junctions (see Pickel et al., 1997 for review). Monoamines are also known to diffuse through the neuropil, and to bind to receptors that are located at nonsynaptic sites on the plasma membrane (Nicholson and Rice, 1991). Together, these observations suggest that monoaminergic transmission occurs at conventional synaptic junctions (Bloom, 1991), but also may occur at extrasynaptic sites, which has been termed 'parasynaptic' (Schmitt, 1984) or 'volume' transmission (Zoli et al., 1998).

Extrasynaptic, as well as synaptic sites for monoaminergic transmission have been elucidated by using electron microscopic immunocytochemistry for the localization of sequence specific antipeptide antisera against recently cloned vesicular and plasmalemmal monoamine transporters and receptors (Pickel et al.,1997; Nirenberg and Pickel, 1997). The ultrastructural evidence for extrasynaptic monoaminergic transmission is described in the present review, focusing primarily on research conducted in striatum: caudate-putamen nucleus (CPN) and core or shell of the nucleus accumbens (NAc). These striatal compartments show regional differences in extracellular monoamine concentrations that are attributed, in part, to transporter availability. The extracellular availability of monoamines markedly influences the expression and distribution of receptors controlling motor (CPN and NAc core) and limbic (NAc shell) functions (Koob and Swerdlow, 1988; Koob, 1992; Amalric and Koob, 1993; Wise, 1998).

*Corresponding author. Tel.: (212) 570-2900; Fax: (212) 988-3672; e-mail: vpickel@med.cornell.edu

Vesicular monoamine transporter

The reserpine-sensitive vesicular monoamine transporter2 (VMAT2) (Liu et al., 1992; Erickson et al., 1992) uses the proton electrochemical gradient to mediate the uptake of monoamines into vesicles or other acidic intracellular organelles in central and peripheral monoaminergic neurons (reviewed in Edwards, 1992; Schulinder, 1994). Antisera directed against VMAT2 are selectively localized to monoaminergic neurons by both light (Peter et al., 1995) and electron microscopy (Nirenberg et al., 1995; Nirenberg and Pickel, 1997; Nirenberg et al., 1998).

We have used antisera against a synthetic peptide sequence at the C-terminus of VMAT2 within brainstem monoaminergic neurons, as well as nigrostriatal dopaminergic neurons targeting the CPN and mesolimbic dopaminergic neurons terminating in the NAc (Nirenberg et al., 1995; Nirenberg and Pickel, 1997; Nirenberg, et al., 1998). Immunogold labeling of VMAT2 is mainly localized to membranes of SSVs and DCVs within these monoaminergic axons and terminals. In the brainstem nucleus of the solitary tract, the VMAT2-containing terminals, which are mainly noradrenergic and serotonergic, rarely form synaptic junctions, but when observed, are either symmetric or asymmetric (Nirenberg et al., 1995). Furthermore, when present in these terminals, DCVs are intensely immunoreactive, and often located at sites distant from synaptic specializations, suggesting that monoamines are released from DCVs at extrasynaptic sites serving neuromodulatory functions (Matteoli et al., 1991).

In midbrain dopaminergic neurons, VMAT2 is most often seen in dendrites, where the labeling is localized to a few DCVs as well as clusters of SSVs, but more often is associated with membranes of tubulovesicular organelles near nonsynaptic portions of the plasma membranes (Nirenberg et al., 1996a). Furthermore, the immunoreactive organelles are significantly less prevalent in dopaminergic neurons within the substantia nigra, the origin of nigrostriatal projections, than in the ventral tegmental area (VTA) giving rise to mesolimbic projections (Bjorklund and Lindvall, 1984). The results suggest greater potential for storage and release of dopamine in the mesolimbic pathway that may have direct relevance to the reward and motivational functions ascribed to dopamine in this system (Koob, 1992; Wise, 1998).

In the dorsolateral striatum, VMAT2 immunogold labeling is prominently seen in small varicose axons, many of which contain the catecholamine synthesizing enzyme, tyrosine hydroxylase (TH) (Nirenberg et al., 1997a). The VMAT2 labeling in these axons is mainly localized to membranes of SSVs, whereas in terminals without detectable TH, DCVs are more prevalent, and these vesicles, like those in the brainstem, are often intensely VMAT2 immunoreactive. The absence of TH immunoreactivity in these terminals suggests that they are non-catecholaminergic and possibly contain serotonin, which is the major monoamine, other than catecholamines in this region (Cumming et al., 1992).

Plasmalemmal monoamine transporters

The dopamine transporter (DAT) and serotonin transporter (SERT) are plasmalemmal monoamine transporters that are responsible for reuptake of dopamine and 5-HT, respectively. As reviewed in this report, these transporters are mainly localized by electron microscopic immunocytochemistry to sites on axonal plasma membrane at a distance from synaptic specializations (Nirenberg et al., 1996b, 1997b; Nirenberg and Pickel, 1997; Pickel et al., 1997; Zhou et al., 1998; Pickel and Chan, 1999).

Dopamine transporter

The sodium- and chloride-dependent rat DAT (Ciliax et al., 1995; Freed et al., 1995) has been identified and cloned (Giros et al., 1991; Kilty et al., 1991), thus providing a useful tool for the identification of dopamine reuptake sites. In situ hybridization has shown that the DAT mRNA is enriched in midbrain dopaminergic neurons (Shimada et al., 1992; Cerruti et al, 1993) comprising the mesostriatal dopaminergic projection system. The DAT protein has also been identified by light microscopic immunocytochemistry in these midbrain neurons and their striatal terminals (Ciliax et al., 1995; Freed et al., 1995). Furthermore, electron microscopic immunocytochemistry has shown that DAT antiserum (Freed et al., 1995) has a precise subcellular localization within dopaminergic terminals in the dorsolateral striatum (Nirenberg et al., 1996b, 1997b; Nirenberg and Pickel 1997). In dopaminergic terminals identified by their content of tyrosine hydroxylase (TH), immunogold labeling for DAT is seen along varicose and intervaricose segments of the plasma membrane (Fig. 1; Nirenberg, et al., 1997b). The gold particles identifying DAT immunoreactivity are usually seen at a distance from the punctate symmetric synapses that are typical of striatal dopaminergic terminals (Fig. 2).

Serotonin transporter

The cloning of SERT has permitted the use of in situ hybridization for detection of SERT mRNA,

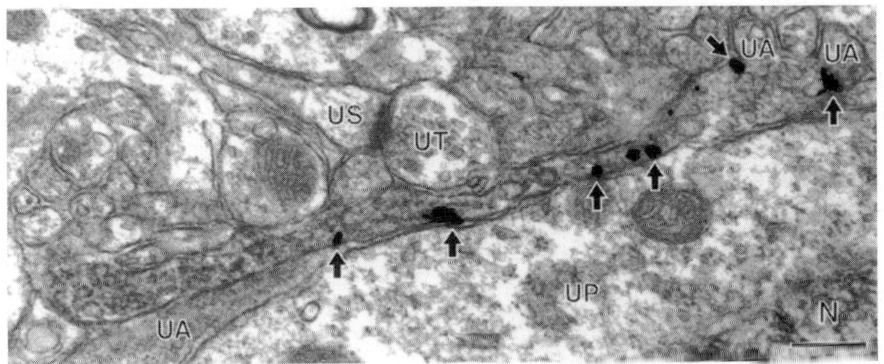

Fig. 1. DAT labelling on the plasma membrane of a TH-immunoreactive axon in the NAc core. Electron micrograph showing colocalization of immunogold labeling for DAT and immunoperoxidase for TH in a varicose axon in the NAc core. The gold particles (arrows) are located near an unlabeled perikaryon (UP) and several small unlabeled axons (UA). The DAT labeled axon is apposed to an unlabeled axon terminal (UT) that forms an asymmetric (excitatory-type) input onto an unlabeled dendritic spine (US). N, Nucleus of unlabeled perikaryon. Scale bar, 0.25 μm. From Nirenberg et al., 1997b.

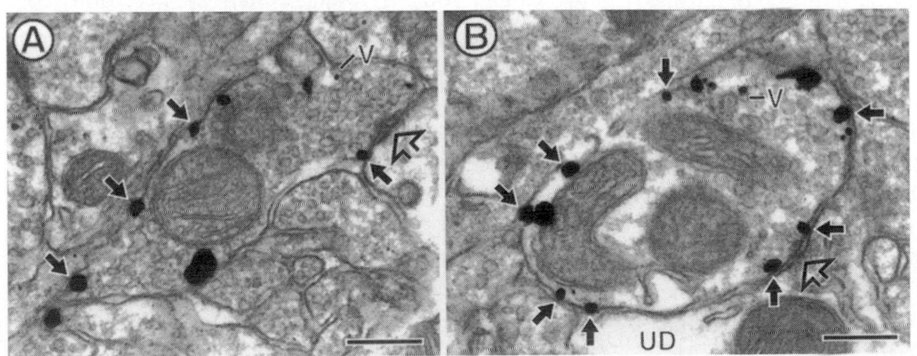

Fig. 2. DAT labeling on extrasynaptic plasma membranes of axon terminals. Electron micrographs show tissue that has been singly labeled with immunogold for DAT, derived from the NAc core (A) and shell (B). In A, immunogold labeling for DAT (closed arrows) is located on extrasynaptic portions of the plasma membrane of an axon terminal in the NAc core. The labeled terminal forms a symmetric synapse (open arrow) onto an unlabeled dendrite. In B, the primary DAT antiserum has been used at twice the concentration as in A. Immunogold labeling for DAT (closed arrows) is located on extrasynaptic portions of the plasma membrane of an axon terminal in the NAc shell. The labeled terminal forms a symmetric synapse (open arrow) onto the spine of an unlabeled dendrite (UD). In A and B, several plasmalemmal DAT gold particles are seen just lateral to the synaptic junctions, whereas others contact electron-lucent intracellular vesicles (V). Scale bars, 0.25 μm. From (Nirenberg et al., 1997b).

which is prominently localized within serotonergic raphe neurons (Austin et al., 1994; Charnay et al., 1996). Immunolabeling of sequence specific SERT antisera is also seen by light microscopy in these raphe neurons, as well as in axonal processes throughout many forebrain regions that receive serotonergic innervation (Sur et al., 1996; Zhou et al., 1998). Electron microscopic immunocytochemistry shows a prominent SERT distribution in small unmyelinated axons, suggesting that serotonin reuptake is also largely extrasynaptic (Zhou et al., 1998). We have extensively examined the subcellular distribution of SERT immunolabeling in the NAc shell and core (Pickel and Chan, 1999), both of which receive afferent input from serotonergic neurons in the dorsal raphe (Van Bockstaele et al., 1993). In each region, SERT immunogold labeling is intensely localized to small unmyelinated axons, and more diffusely distributed along extrasynaptic plasma membrane of axon terminals (Fig. 3). In axon terminals, SERT immunoreactivity is often seen along plasma membranes near 5-HT containing DCVs (Fig. 3). In comparison with the NAc core, SERT-labeled axon terminals in the shell are larger and contain more numerous DCVs, suggesting greater potential for tonic activity and transmitter release in the shell versus core region (Lnenicka et al., 1991).

In addition to axon terminals, we unexpectedly observed SERT immunolabeling in a few dendrites of non-serotonergic neurons and in select glial processes in the NAc core and shell (Pickel and Chan, 1999). In the dendrites, SERT was targeted mainly to extrasynaptic plasma membranes, some of which were apposed to other dendrites showing gap-like specializations. The results suggest that dendritic SERT may play a role in the regulation of membrane excitability through voltage-sensitive channels as is known to occur in certain other systems (see Beckman and Quick, 1998 for review). The ultrastructural localization of SERT within selective glial processes in the striatum is consistent with the known expression of SERT mRNA and monoamine oxidase-A (MAO-A), the enzyme that is mainly responsible for serotonin degradation, in cultured astrocytes (Fitzgerald et al., 1990; Hirst et al., 1998). Pharmacologically, the astrocytic uptake of serotonin appears to be similar, if not identical to that seen in serotonergic neurons (Dave and Kimelberg, 1994; Bal et al., 1997). Together, these observations suggest that SERT mediated transport of 5-HT in astrocytes may play a role in uptake and release or metabolism of 5-HT (Hirst et al., 1998).

Monoamine receptors

Monoaminergic (adrenergic, dopaminergic, and serotonergic) receptors were first identified based on their relative affinities for the respective amines (Harrison et al., 1991; Bahouth et al., 1991). The genes encoding pharmacologically identified receptor subtypes within these major groups, as well as those encoding many other receptor subclasses within these subtypes have been cloned and sequenced (Monsma, Jr. et al., 1989; 1990; Mansour et al., 1990; Harrison et al., 1991; Kroeze and Roth, 1998; Pasqualetti et al., 1999). These receptors have seven membrane-spanning domains, as characteristic of G-protein coupled receptors (Bahouth et al., 1991; Kroeze and Roth, 1998). The cloning of monoaminergic receptors has allowed the production of antipeptide antibodies specific for particular receptor subtypes, thereby permitting their discrete cellular and subcellular immunocytochemical localization (Sesack et al., 1994; Morales and Bloom, 1997; Backstrom and Sanders-Bush, 1997; Cornea-Hébert et al., 1999). We review the striatal localization of two monoamine receptor subtypes, the dopamine D2 receptor and the 5-HT2A receptor, both of which are known to play an important role in normal motor function and in the beneficial effects of antipsychotic drugs (see reviews by Burnet et al., 1995; Sibley, 1999; Vezina and Kim, 1999).

D2 dopamine receptors

Dopamine D2 receptors are present in midbrain dopaminergic neurons and their striatal terminals (Sesack et al., 1994; Pickel and Sesack, 1995; Delle Donne et al., 1997). This distribution is consistent with their known role in the autoregulation of dopamine release (Mercuri et al., 1997; Ekesbo et al., 1999). Within striatal axon terminals, the receptor is localized to membranes of SSVs, and to segments of the plasma membrane (Fig. 4) (Sesack et al., 1994; Pickel and Sesack, 1995). These dopamine receptors are present in axons showing extrasynaptic plasmalemmal distributions of DAT (Hersch et al., 1997). Together, the results suggest that the spatial buffering of dopamine is not necessarily restricted to the region of the synaptic cleft, but instead may occur extensively at extra-synaptic sites including those involved in D2 mediated autoregulation of dopamine release.

The dopaminergic axons containing D2-receptors are often apposed to, or form symmetric synapses with the necks of dendritic spines, many of which also contain D2 receptor labeling (Sesack

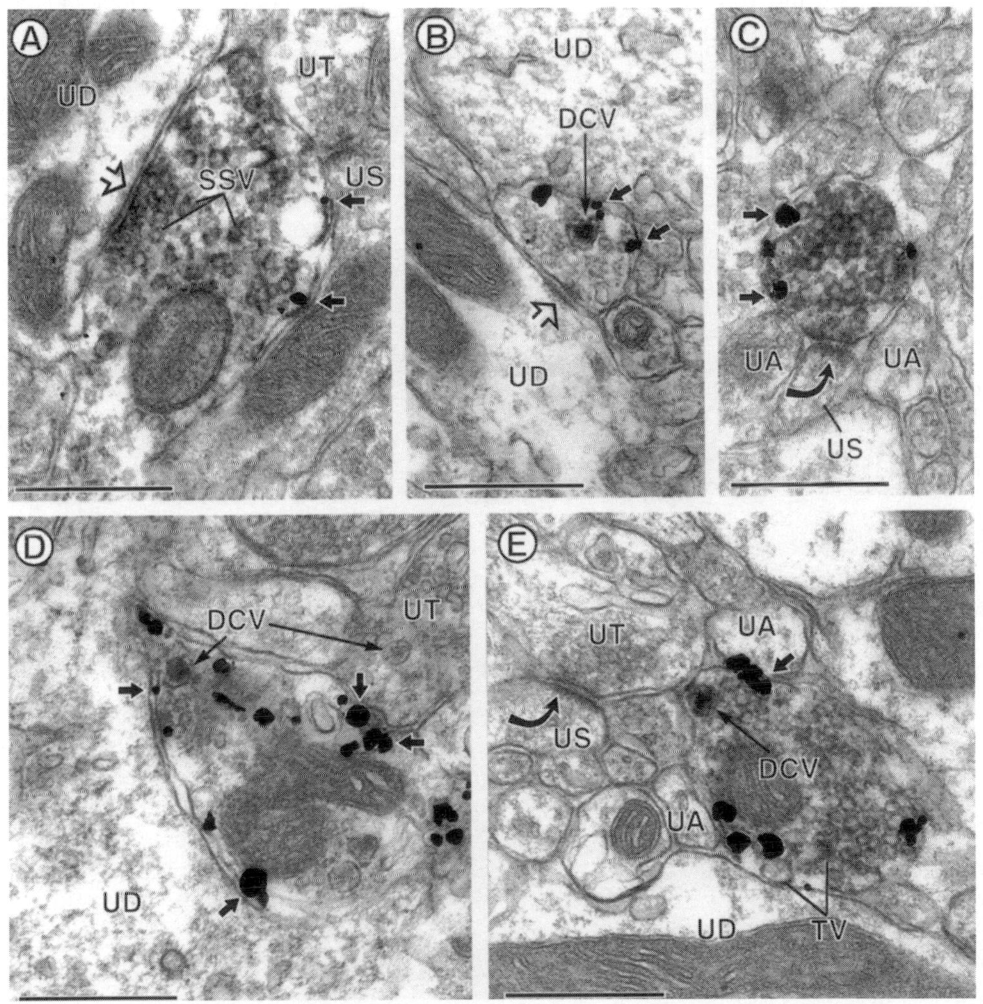

Fig. 3. Dual labeling for SERT and 5-HT in axon terminals within the NAc from brain tissue that was fixed by using glutaraldehyde (A, C) or acrolein (B, D, E). Immunogold labeling for SERT (small arrows) is seen along portions of the plasma membrane away from symmetric axodendritic (open arrowhead in A from the shell and B from the core) and asymmetric axospinous (curved filled arrow inC from the shell) synapses. The labeling is also seen near appositional contacts with unlabeled dendrites (UD) and unlabeled terminals (UT) in the shell (D, E). The peroxidase labeling appears as a diffuse precipitate surrounding small synaptic vesicles (SSV) as seen in A, or more discretely localized to membranes of large dense core vesicles (DCV) as seen in B,D and E. The gold particles are prevalent near DCV that contact the plasma membrane at appositional contacts with unlabeled dendrites and terminals. In E, the apposed unlabeled terminal forms an asymmetric synapse (curved arrow) with an unlabeled spine (US). Tubulovesicular organelles (TV) are also seen near the plasma membrane and contacted by gold particles. Scale bars = 0.5 μm. From Pickel and Chan, 1999.

et al., 1994). When processed for dual detection of the dopamine D2 receptor and γ-amino butyric acid (GABA), many of the dendrites containing D2 receptors are also GABA immunoreactive (Fig. 4). In addition, the receptors are sometimes present in axon terminals, and these terminals are more prevalent in the NAc shell than in the dorsal CPN, suggesting greater involvement of presynaptic D2 receptors in GABA release within the limbic as compared with striatal motor regions (Delle Donne et al., 1997). In each region, several axon terminals forming asymmetric synapses with spine heads also contain D2 receptor labeling. In axon terminals, immunoreactivity is mainly associated either with the plasma membrane or membranes of nearby vesicles. Together, the results support the known involvement of dopamine D2 receptors in modulation of the postsynaptic responses of GABAergic striatal spiny neurons. In addition, they suggest involvement of these receptors not only in the presynaptic release of dopamine, but also in the release of both inhibitory and excitatory amino acids (Gerfen, 1988; Hsu et al., 1995; Delle Donne et al., 1997).

5-HT2A receptors

5-HT2A mRNA and binding sites are widely distributed in forebrain regions that receive ser-

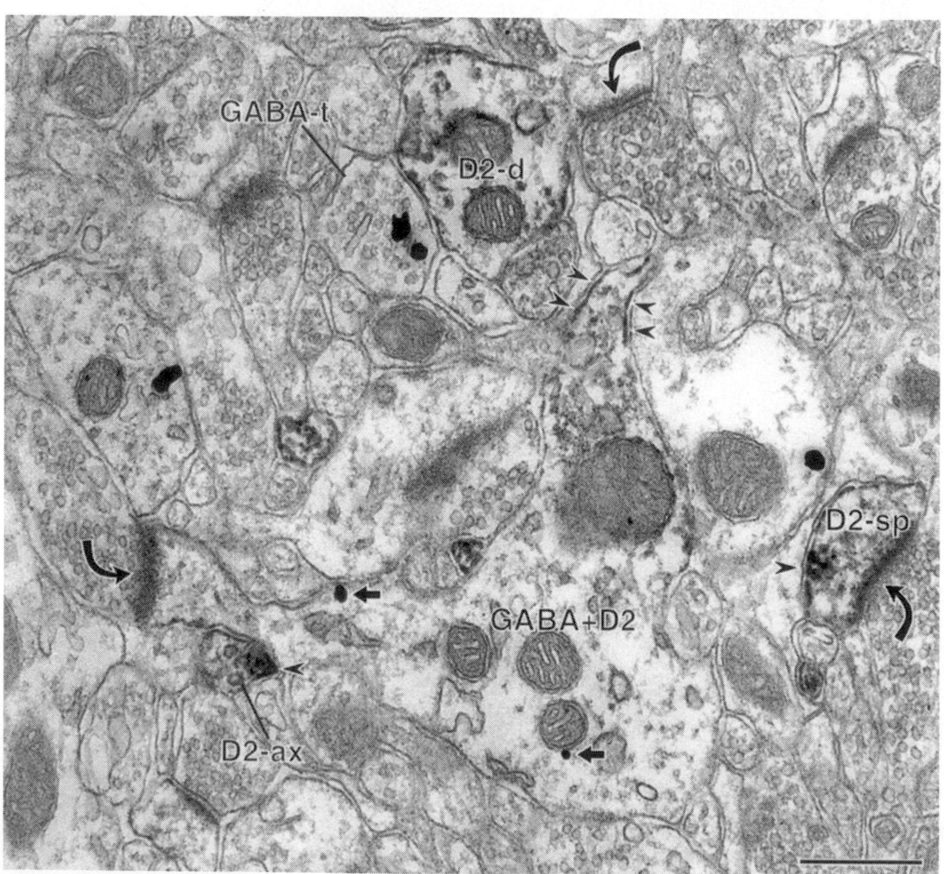

Fig. 4. Axonal and dendritic immunoperoxidase labeling for the dopamine D2 receptor. Intense plasmalemmal D2-immunoreactity (arrowheads) is seen in a small axon (D2-ax) and apposes the neck of a dendritic spine. This spine, as well as several other spines is lightly D2-immunoreactive. Several axon terminals form asymmetric synapses (curved arrows) with labeled and unlabeled dendritic spines within the neuropil. The tissue shown in this micrograph was processed also for immunogold labeling of GABA, and the gold particles (arrows) are seen within one of the D2-labeled spiny dendrites (GABA + D2). More abundant GABA gold particles are seen in a nearby terminal (GABA-t). Scale bar, 0.05 μm.

otonergic innervation (Burnet et al., 1995, 1996). The generation of sequence specific antipeptide antisera has permitted the immunolabeling of these receptors (Backstrom and Sanders-Bush, 1997). In regions that have been examined by electron microscopy, the receptor protein is prominently associated with the cytoplasmic compartment of dendrites (Cornea-Hébert et al., 1999), but is also present at synaptic and non-synaptic sites on their plasma membranes (Rodríguez et al., 1999). We comparatively examined the localization of 5-HT2A receptors with that of NMDA receptors in the CPN and NAc shell, thereby showing that in each region the receptors are often contained in single neurons, but have largely non-overlapping cytoplasmic distributions (Rodríguez, et al., 1999). Whereas NMDA receptors are seen frequently at asymmetric postsynaptic densities in dendritic spines consistent with the known termination of glutamatergic cortical afferents (Gerfen, 1988), 5-HT2A receptors are most commonly seen in dendritic shafts where the labeling is only occassionally detected at perisynaptic sites near excitatory type terminals (Rodríguez, et al., 1999). Dendrites containing NMDA and/or 5-HT2A receptors are more prevalent in the CPN than in the NAc shell (Rodríguez et al., 1999), which is consistent with the known higher levels of 5-HT2A receptor mRNA in the CPN (Mijnster et al., 1997). These differences may reflect regional variations in serotonergic innervation and/or extracellular availability of serotonin, which is, in part, dictated by SERT expression (Rioux et al., 1999).

In addition to dendrites, many small axons, but few axon terminals show immunolabeling for 5-HT2A receptors in brain regions that have been examined by electron microscopy (Cornea-Hébert, et al., 1999; Rodríguez et al., 1999). This distribution is interesting in view of the fact that SERT immunoreactivity is also more prominently localized in small axons than in axon terminals (Zhou et al., 1998; Pickel and Chan, 1999). Furthermore, in these axons SERT is often seen along portions of the plasma membrane near 5-HT containing large DCVs, suggesting that exocytotic release in axons may be closely linked with reuptake. Alternatively, SERT reverse transport in axons may play a role in non-vesicular release of 5-HT (Attwell et al., 1993), or 5-HT may diffuse along fiber bundles following release from axon terminals (Bjelke et al., 1995).

Conclusions

Electron microscopic studies identifying reuptake and receptor sites provide increasing support for the idea that central neurons can release monoamines in a paracrine manner to affect diverse targets (Bunin and Wightman, 1999). While monoaminergic axon terminals sometimes form recognizable synapses, the subcellular distributions of plasmalemmal transporters suggest that reuptake occurs largely outside the synaptic cleft thus facilitating the activation of distant receptors. This conclusion is consistent with the localization of subtype selective dopamine and serotonin receptors on synaptic and non-synaptic plasma membranes of dendrites, as well as small axons and glia in regions that receive dense monoaminergic innervation.

Acknowledgments

VMP receives salary support from the National Institute of Mental Health (NIMH; Grant MH00078), and research support for this work from NIMH (Grant MH40342) and the National Institute on Drug Abuse (NIDA: Grant DA04600).

References

Amalric, M. and Koob, G.F. (1993) Functionally selective neurochemical afferents and efferents of the mesocorticolimbic and nigrostriatal dopamine system. [Review]. *Prog. Brain Res.*, 99: 209–226.

Ase, A.R., Amdiss, F., Hébert , C., Huang, N., Van Gelder, N.M. and Reader, T.A. (1999) Effects of antipsychotic drugs on dopamine and serotonin contents and metabolites, dopamine and serotonin transporters and serotonin1A receptors. *J. Neural Trans.*, 106: 75–105.

Attwell, D., Barbour, B. and Szatkowski, M. (1993) Non-vesicular release of neurotransmitter. [Review]. *Neuron*, 11: 401–407.

Austin, M.C., Bradley, C.C., Mann, J.J. and Blakely, R.D. (1994) Expression of serotonin transporter messenger RNA in the human brain. *J. Neurochem.*, 62: 2362–2367.

Backstrom, J.R. and Sanders-Bush, E. (1997) Generation of anti-peptide antibodies against serotonin 5-HT_{2A} and 5-HT_{2C} receptors. *J. Neurosci. Met.*, 77: 109–117.

Bahouth, S.W., Wang, H. and Malborn, C.C. (1991) Immunological approaches for probing receptor structure and function. *TINS*, 12: 338–343.

Bal, N., Figueras, G., Vilaro, M.T., Sunol, C. and Artigas, F. (1997) Antidepressant drugs inhibit a glial 5-hydroxytryptamine transporter in rat brain. *Eur. J. Neurosci.*, 9: 1728–1738.

Beaudet, A. and Descarries, L. (1978) The monoamine innervation of rat cerebral cortex: Synaptic nonsynaptic axon terminals. *Neuroscience*, 3: 851–860.

Beckman, M.L. and Quick, M.W. (1998) Neurotransmitter transporters: regulators of function and functional regulation. *J. Memb. Bio.*, 164: 1–10.

Bjelke, B., England, R., Nicholson, C., Rice, M.E., Lindberg, J., Zoli, M., Agnati, L.F. and Fuxe, K. (1995) Long distance pathways of diffusion for dextran along fibre bundles in brain. Relevance for volume transmission. *NeuroReport*, 6: 1005–1009.

Bjorklund, A. and Lindvall, O. (1984) Dopamine-containing systems in the CNS. In: A. Bjorklund and T. Hokfelt (Eds), *Handbook of Chemical Neuroanatomy*. Vol. 2, *Classical Transmitters in the CNS*, Part I. Elsevier Science Publishers, Amsterdam, pp. 55–121.

Blakely, R.D., Ramamoorthy, S., Schroeter, S., Qian, Y., Apparsundaram, S., Galli, A. and DeFelice, L.J. (1998) Regulated phosphorylation and trafficking of antidepressant-sensitive serotonin transporter proteins. *Biol. Psych.*, 44: 169–178.

Bloom, F.E. (1991) An integrative view of information handling in the CNS. In: K. Fuxe and L.F. Agnati (Eds), *Volume Transmission in the Brain: Novel Mechanisms for Neural Transmission*. Raven Press, Ltd., New York, pp. 11–23.

Bunin, M.A. and Wightman, R.M. (1999) Paracrine neurotransmission in the CNS involvement of 5-HT. *TINS*, 22: 377–382.

Burnet, P.W., Eastwood, S.L. and Harrison, P.J. (1996) 5-HT1A and 5-HT2A receptor mRNAs and binding site densities are differentially altered in schizophrenia. *Neuropsychopharmacology*, 15: 442–455.

Burnet, P.W., Eastwood, S.L., Lacey, K. and Harrison, P.J. (1995) The distribution of 5-HT1A and 5-HT2A receptor mRNA in human brain. *Brain Res.*, 676: 157-168.

Cerruti, C., Walther, D.M., Kuhar, M.J. and Uhl, G.R. (1993) Dopamine transporter mRNA expression is intense in rat midbrain neurons and modest outside midbrain. *Brain Res. Mol. Brain Res.*, 18: 181–186.

Charnay, Y., Léger, L., Vallet, P.G., Greggio, B., Hof, P.R., Jouvet, M. and Bouras, C. (1996) Mapping of serotonin transporter messenger RNA-containing nerve cell populations in the cat brainstem. *J. Chem. Neuroanat.*, 10: 93–100.

Ciliax, B.J., Heilman, C., Demchyshyn, L.L., Pristupa, Z.B., Ince, E., Hersch, S.M., Niznik, H.B. and Levey, A.I. (1995) The dopamine transporter: Immunochemicalcharacterization and localization in brain. *J. Neurosci.*, 15: 1714–1723.

Cornea-Hébert, V., Riad, M., Wu, C., Singh, S.K. and Descarries, L. (1999) Cellular and subcellular distribution of the serotonin 5-HT2A receptor in the central nervous system of adult rat. *J. Comp. Neurol.*, 409: 187–209.

Cumming, P., Brown, E., Damsma, G. and Fibiger, H. (1992) Formation and clearance of intestinal metabolites of dopamine and serotonin in the rat striatum: an in vivo microdialysis study. *J. Neurochem.*, 59: 1905–1914.

Dave, V. and Kimelberg, H.K. (1994) Na(+)-dependent, fluoxetine-sensitive serotonin uptake by astrocytes tissue-printed from rat cerebral cortex. *J. Neurosci.*, 14: 4972–4986.

Delle Donne, K.T., Sesack, S.R. and Pickel, V.M. (1997) Ultrastructural immunocytochemical localization of the dopamine D_2 receptor within GABAergic neurons of the rat striatum. *Brain Res.*, 746: 239–255.

Descarries, L., Watkins, K.C. and Lapierre, Y. (1977) Noradrenergic axon terminals in the cerebral cortex of rat III: Topometric ultrastructural analysis. *Brain Res.*, 133: 197–222.

Edwards, R.H. (1992) The transport of neurotransmitters into synaptic vesicles. *Curr. Opin. Neurobiol.*, 2: 586–594.

Ekesbo, A., Rydin, E., Torstenson, R., Sydow, O., Låengström, B. and Tedroff, J. (1999) Dopamine autoreceptor function is lost in advanced Parkinson's disease. *Neurology*, 52: 120–125.

Erickson, J.D., Eiden, L. and Hoffman, B. (1992) Expression cloning of a reserpine-sensitive vesicular monoamine transporter. *Proc. Natl. Acad. Sci. USA*, 89: 10993–10997.

Fitzgerald, L.W., Kaplinsky, L. and Kimelberg, H.K. (1990) Serotonin metabolism by monoamine oxidase in rat primary astrocyte cultures. *J. Neurochem.*, 55: 2008-2014.

Freed, C., Revay, R., Vaughan, R.A., Kriek, E., Grant, S., Uhl, G.R. and Kuhar, M.J. (1995) Dopamine transporter immunoreactivity in rat brain. *J. Comp. Neurol.*, 359: 340–349.

Geffen, L.B., Jessel, T.M., Cuello, A.C. and Iversen, L.L. (1976) Release of dopamine from dendrites in rat substantia nigra. *Nature (Lond.)*, 260: 258–260.

Gerfen, C.R. (1988) Synaptic organization of the striatum. *J. Electron. Microsc. Tech.*, 10: 265–281.

Giros, B., el Mestikawy, S., Bertrand, L. and Caron, M.G. (1991) Cloning and functional characterization of a cocaine-sensitive dopamine transporter. *FEBS. Lett.*, 295: 149–154.

Harrison, J.K., Rearson, W.R. and Lynch, K.R. (1991) Molecular characterization of alpha1- and alpha-2-adrenoreceptors. *Trends Pharmacol. Sci.*, 12: 62–67.

Hersch, S.M., Yi, H., Heilman, C.J., Edwards, R.H. and Levey, A.I. (1997) Subcellular localization and molecular topology of the dopamine transporter in the striatum and substantia nigra. *J. Comp. Neurol.*, 388: 211–227.

Hirst, W.D., Price, G.W., Rattray, M. and Wilkin, G.P. (1998) Serotonin transporters in adult rat brain astrocytes revealed by $[^3H]$5-HT uptake into glial plasmalemmal vesicles. *Neurochem. Int.*, 1, 33: 11–22.

Hsu, K.S., Huang, C.C., Yang, C.H. and Gean, P.W. (1995) Presynaptic D_2 dopaminergic receptors mediate inhibition of excitatory synaptic transmission in rat neostriatum. *Brain Res.*, 690: 264–268.

Kilty, J.E., Lorang, D. and Amara, S.G. (1991) Cloning and expression of a cocaine-sensitive rat dopamine transporter. *Science*, 254: 578–579.

Koob, G.F. (1992) Neural mechanisms of drug reinforcement. *Ann. NY Acad. Sci.*, 654: 171–191.

Koob, G.F. and Swerdlow, N.R. (1988) The functional output of the mesolimbicdopamine system. *Ann. NY Acad. Sci.*, 537: 216–227.

Kroeze, W.K. and Roth, B.L. (1998) The molecular biology of serotonin receptors: therapeutic implications for the interface of mood and psychosis. *Biol. Psych.*, 44: 1128–1142.

Liu, Y., Peter, D., Roghani, A., Schuldiner, S., Prive, G.G., Eisenberg, D., Brecha, N. and Edwards, R.H. (1992) A cDNA that suppresses MPP+ toxicity encodes a vesicular amine transporter. *Cell*, 70: 539–551.

Lnenicka, G.A., Hong, S.J., Combatti, M. and LePage, S. (1991) Activity-dependent development of synaptic varicosities at crayfish motor terminals. *J. Neurosci.*, 11: 1040–1048.

Mansour, A., Meador-Woodruff, J.H., Bunzow, J.R., Civelli, O., Akil, H. and Watson, S.J. (1990) Localization of dopamine D2 receptor mRNA and D1 and D2 receptor binding in the rat brain and pituitary: an in situ hybridization-receptor autoradiographic analysis. *J. Neurosci.*, 10: 2587–2600.

Matteoli, M., Reetz, T. and DeCamilli, P. (1991) Small synaptic vesicles and large dense-core vesicles: secretory organelles involved in two modes of neuronal signalling. In: K. Fuxe and L.F. Agnati (Eds), *Volume Transmission in the Brain: Novel Mechanisms for Neural Transmission*. Raven Press, Ltd., New York, pp. 181–193.

Mercuri, N.B., Saiardi, A., Bonci, A., Picetti, R., Calabresi, P., Bernardi, G. and Borrelli, E. (1997) Loss of autoreceptor function in dopaminergic neurons from dopamine D_2 receptor deficient mice. *Neuroscience*, 79: 323–327.

Mijnster, M.J., Raimundo, A.G.V., Koskuba, K., Klop, H., Docter, G.J., Groenewegen, H.J. and Voorn, P. (1997) Regional and cellular distribution of serotonin 5- hydroxytryptamine$_{2a}$ receptor mRNA in the nucleus accumbens, olfactory tubercle and caudate putamen of the rat. *J. Comp. Neurol.*, 389: 1–11.

Monsma, F.J., Jr., Mahan, L.C., McVittie, L.D., Gerfen, C.R. and Sibley, D.R. (1990) Molecular cloning and expression of a D1 dopamine receptor linked to adenylyl cyclase activation. *Proc. Natl Acad. Sci. USA*, 87: 6723–6727.

Monsma, F.J. Jr., McVittie, L.D., Gerfen, C.R., Mahan, L.C. and Sibley, D.R. (1989) Multiple D2 dopamine receptors produced by alternative RNA splicing. *Nature*, 342: 926–929.

Moore, R.Y. (1993) Principles of synaptic transmission. [Review]. *Ann. NY Acad. Sci.*, 695: 1–9.

Morales, M. and Bloom, F.E. (1997) The 5-HT$_3$ receptor is present in different subpopulations of GABAergic neurons in the rat telencephalon. *J. Neurosci.*, 17: 3157–3167.

Nicholson, C. and Rice, M.E. (1991) Diffusion of ions and transmitters in the brain cell microenvironment. In: K. Fuxe and L.F. Agnati (Eds), *Volume Transmission in the Brain: Novel Mechanisms for Neural Transmission*. Raven Press, Ltd., New York, pp. 279–294.

Nirenberg, M.J., Chan, J., Liu, Y., Edwards, R.H. and Pickel, V.M. (1998) Ultrastructural localization of the vesicular monoamine transporter 2 in mesolimbic and nigrostriatal dopaminergic neurons. *Adv. Pharmacol.*, 42: 240-243.

Nirenberg, M.J., Chan, J., Liu, Y.J., Edwards, R.H. and Pickel, V.M. (1996a) Ultrastructural localization of the vesicular monoamine transporter-2 in midbrain dopaminergic neurons: Potential sites for somatodendritic storage and release of dopamine. *J. Neurosci.*, 16: 4135–4145.

Nirenberg, M.J., Chan, J., Liu, Y.J., Edwards, R.H. and Pickel, V.M. (1997a) Vesicular monoamine transporter-2: Immunogold localization in striatal axons and terminals. *Synapse*, 26: 194–198.

Nirenberg, M.J., Chan, J., Pohorille, A., Vaughan, R.A., Uhl, G.R., Kuhar, M.J. and Pickel, V.M. (1997b) The dopamine transporter: Comparative ultrastructure of dopaminergic axons in limbic and motor compartments of the nucleus accumbens. *J. Neurosci.*, 17: 6899–6907.

Nirenberg, M.J., Liu, Y.J., Peter, D., Edwards, R.H. and Pickel, V.M. (1995) The vesicular monoamine transporter 2 is present in small synaptic vesicles and preferentially localizes to large dense core vesicles in rat solitary tract nuclei. *Proc. Natl Acad. Sci. USA*, 92: 8773–8777.

Nirenberg, M.J. and Pickel, V.M. (1997) The vesicular and plasmalemmal dopamine transporters: ultrastructural localization in nigrostriatal dopaminergic neurons. In: *Proceedings of the NATO Advanced Research Workshop on Neurotransmitter Release and Uptake*. Springer-Verlang.

Nirenberg, M.J., Vaughan, R.A., Uhl, G.R., Kuhar, M.J. and Pickel, V.M. (1996b) The dopamine transporter is localized to dendritic and axonal plasma membranes of nigrostriatal dopaminergic neurons. *J. Neurosci.*, 16: 436–447.

Pasqualetti, M., Ori, M., Castagna, M., Marazziti, D., Cassano, G.B. and Nardi, I. (1999) Distribution and cellular localization of the serotonin type 2C receptor messenger RNA in human brain. *Neuroscience*, 92: 601–611.

Peter, D., Liu, Y.J., Sternini, C., De Giorgio, R., Brecha, N. and Edwards, R.H. (1995) Differential expression of two vesicular monoamine transporters. *J. Neurosci.*, 15: 6179–6188.

Pickel, V.M. and Chan, J. (1999) Ultrastructural localization of the serotonin transporter in limbic and motor compartments of the nucleus accumbens. *J. Neurosci.*, 19: 7356–7366.

Pickel, V.M., Nirenberg, M.J. and Milner, T.A. (1997) Ultrastructural view of central catecholaminergic transmission: immunocytochemical localization of synthesizing enzymes, transporters and receptors. *J. Neurocytol.*, 25: 843–856.

Pickel, V.M. and Sesack, S.R. (1995) Electron microscopy of central dopamine systems. In: F.E. Bloom and D.J. Kupfer (Eds.), *Psychopharmacology.* Raven Press, Ltd., New York, pp. 257–268.

Reith, M.E. and Selmeci, G. (1992) Radiolabeling of dopamine uptake sites in mouse striatum: comparison of binding sites for cocaine, mazindol and GBR 12935. *Naunyn. Schmiedebergs. Arch. Pharmacol.*, 345: 309–318.

Rioux, A., Fabre, V., Lesch, K.P., Moessner, R., Murphy, D.L., Lanfumey, L., Hamon, M. and Martres, M.P. (1999) Adaptive changes of serotonin 5-HT$_{2A}$ receptors in mice lacking the serotonin transporter. *Neurosci. Lett.*, 262: 113–116.

Rodríguez, J.J., Garcia, D.R. and Pickel, V.M. (1999) Subcellular distribution of 5-hydroxytryptamine 2A and *N*-methyl-D-aspartate receptors within single neurons in rat motor and limbic striatum. *J. Comp Neurol.*, 413: 219–231.

Schmitt, F.O. (1984) Molecular regulators of brain function: a new view. *Neuroscience*, 13: 991–1001.

Schulinder, S. (1994) A molecular glimpse of vesicular monoamine transporters. *J. Neurochem.*, 62: 2067–2078.

Sesack, S.R., Aoki, C. and Pickel, V.M. (1994) Ultrastructural localization of D_2 receptor-like immunoreactivity in midbrain dopamine neurons and their striatal targets. *J. Neurosci.*, 14: 88–106.

Shimada, S., Kitayama, S., Walther, D. and Uhl, G. (1992) Dopamine transporter mRNA: dense expression in ventral midbrain neurons. *Brain Res. Mol. Brain Res.*, 13: 359–362.

Sibley, D.R. (1999) New insights into dopaminergic receptor function using antisense and genetically altered animals. *Annu. Rev. Pharmacol. Toxicol.*, 39: 313–341.

Sur, C., Betz, H. and Schloss, P. (1996) Immunocytochemical detection of the serotonin transporter in rat brain. *Neuroscience*, 73: 217–231.

Van Bockstaele, E.J., Biswas, A. and Pickel, V.M. (1993) Topography of serotonin neurons in the dorsal raphe nucleus that send axon collaterals to the rat prefrontal cortex and nucleus accumbens. *Brain Res.*, 624: 188–198.

Vezina, P. and Kim, J.H. (1999) Metabotropic glutamate receptors and the generation of locomotor activity: interactions with midbrain dopamine. *Neurosci. Biobehav. Rev.*, 23: 577–589.

Wise, R.A. (1998) Drug-activation of brain reward pathways. *Drug Alc. Depend.*, 51: 13–22.

Zhou, F.C., Tao-Cheng, J.H., Segu, L., Patel, T. and Wang, Y. (1998) Serotonin transporters are located on the axons beyond the synaptic junctions: Anatomical and functional evidence. *Brain Res.*, 805: 241–254.

Zoli, M., Torri, C., Ferrari, R., Jansson, A., Zini, I., Fuxe, K. and Agnati, L.F. (1998) The emergence of the volume transmission concept. *Brain Res. Rev.*, 26: 136-147.

L. F. Agnati, K. Fuxe, C. Nicholson and E. Syková (Eds.)
Progress in Brain Research, Vol 125

CHAPTER 15

Distinct regional differences in dopamine-mediated volume transmission

Margaret E. Rice

New York University School of Medicine, Departments of Neurosurgery and Physiology and Neuroscience, NYU School of Medicine, 550 First Avenue, New York, NY 10016, USA

Introduction

Dopamine is an essential neurotransmitter in motor and emotive pathways of the brain. Dysfunction of dopaminergic neurotransmission plays a major role in a variety of brain disorders including the extrapyramidal motor disturbances of Parkinson's and Huntington's disease, the functional psychoses of schizophrenia, and the pathogenesis of cocaine addiction. DA is also critical for light-dark adaptation in the retina. In each of these CNS systems, DA acts as a neuromodulator to set the tone of excitability. As such, its actions and regulation differ in range and time course from those of conventional neurotransmitters, including glutamate and GABA. To mediate fast synaptic transmission, glutamate and GABA act via their ionotropic receptors to activate excitatory or inhibitory postsynaptic potentials (EPSPs and IPSPs) that last a few milliseconds (Misgeld et al., 1995; Michaelis, 1998). Both glutamate and GABA can act through metabotropic and G-protein-coupled receptors, as well, to modulate neurotransmission on a slower time scale (Misgeld et al., 1995; Michaelis, 1998). By contrast, DA acts only via G-protein coupled receptors, with response times exceeding 100 ms (Robinson and Caron, 1997; Grenhoff and Johnson, 1997; Cragg and Greenfield, 1997).

*Corresponding author. Tel.: 212-263-5438; Fax: 212-689-0334; e-mail: margaret.rice@nyu.edu

With the inherently slower kinetics of DA receptor activation, DA is an ideal candidate to mediate volume transmission (see Fuxe and Agnati, 1991). Whether DA acts synaptically or extrasynaptically, of course, depends on where its receptors are located. In addition, whether it can travel far enough from a site of release to reach extrasynaptic receptors depends on the local distribution of DA uptake transporters. This chapter will address the potential for DA-mediated volume transmission in three distinct regions of the CNS: striatum; substantia nigra; and retina. In each region, voltammetric microelectrodes were used with fast-scan cyclic voltammetry to monitor the behavior of extracellular DA in diffusion studies or during stimulated release. These voltammetric data, taken together with known patterns of receptor and transporter localization in each region, indicate that synaptically released DA is most constrained in the striatum, but that somatodendritically release DA in the substantia nigra and DA released from amacrine cells in the retina act primarily by volume transmission.

Fast-scan cyclic voltammetry

In the experimental studies described here, DA was monitored with carbon fiber microelectrodes used with fast-scan cyclic voltammetry (Armstrong-James et al., 1981; Millar et al., 1985). The electrodes were made from 8 μm carbon fibers with an active surface that was either a plane that was flush with the glass insulation (Rice and Nicholson, 1989) or spark-etched conical tip that extended 30–50 μm beyond the insulation (Millar, 1992). Fast-scan cyclic voltammetric records were obtained using either an EI-400 potentiostat (currently available through Cypress Systems, Lawrence, KS, USA) or a Millar Voltammeter (PD Systems International, West Molesey, Surrey, KT8 ORN, UK). Scans rates were 800–900 V s^{-1}.

Fast-scan cyclic voltammetry has been used to detect DA release from CNS tissue, both in vivo and in vitro (Millar et al., 1985; Bull et al., 1990; Garris and Wightman, 1995; Rice et al., 1997; Cragg et al., 1997). This method is well-suited to monitor DA for two main reasons. First, sampling time with fast-scan cyclic voltammetry is commensurate with physiological time scales (Armstrong-James et al., 1981; Kuhr and Wightman 1986; Stamford et al., 1986; Rice and Nicholson, 1989). A voltammetric scan applied as a triangle wave over a potential range of –0.5 to + 1.1 V (vs. Ag/AgCl reference) has a duration of less than 4 ms at a scan rate of 800 V s^{-1}. Sampling is usually repeated at 100–250 ms intervals. Second, the use of a triangle waveform permits identification of a monitored substance by both its oxidation and reduction peak potentials (Fig. 1). On most carbon fiber microelectrodes, DA oxidizes at +0.5–0.7 V vs. Ag/AgCl, with a subsequent reduction peak at about –0.2 V. Like all voltammetric methods, fast-scan cyclic voltammetry is also used to quantify concentration changes in electroactive substances, similar to DA. Current output at the oxidation potential for DA is proportional to the concentration of DA at the electrode tip, such that calibration responses are linear with concentration (Fig. 1). Current at the DA oxidation potential can therefore be plotted against time to indicate the time course of a change in extracellular DA concentration ($[DA]_o$).

The measurements described here were all obtained from in vitro preparations; specific methods for each have been described in detail elsewhere and will therefore be mentioned only briefly. It is relevant to note that fast-scan cyclic voltammetry is used most successfully to measure dynamic changes in $[DA]_o$ or other electroactive species over a few seconds or minutes. We have used this method to monitor diffusion of DA introduced by pressure ejection or iontophoresis, as well as electrically evoked release of endogenous DA. This method is generally less well-suited to assess basal levels on longer time scales because background currents are sensitive to shifts in the ionic environment, including those from Ca^{2+} and pH (Rice and Nicholson, 1989, 1995; Jones et al., 1994, 1995). One exception, however, is detection of basal $[DA]_o$ in *Xenopus* retina (Witkovsky et al., 1993), which will be discussed further below.

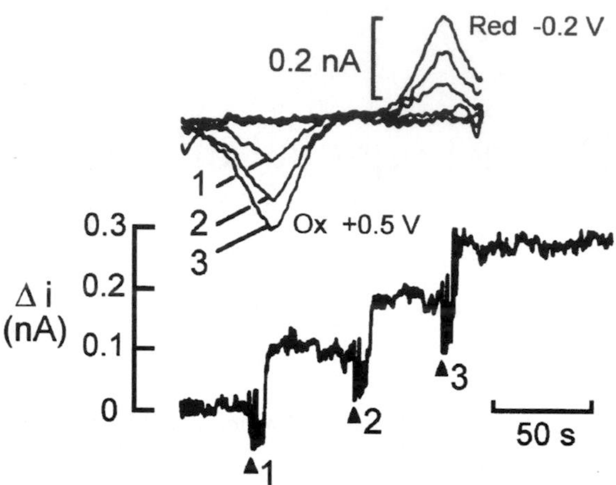

Fig. 1. Dopamine calibration at a carbon-fiber microelectrode used with fast-scan cyclic voltammetry. Inset is a series of background-subtracted voltammograms recorded for 200 nM increments of DA in phosphate-buffered saline, pH 7.4; oxidation peak potential (Ox) was +0.5 V vs. Ag/AgCl with a reduction peak potential (Red) of –0.2 V. Scan rate was 900 V s^{-1}. Peak oxidation current was also plotted against time; DA increments were added at points 1–3, which correspond to the numbering of the voltammograms in the inset. Characteristic oxidation and reduction peaks are use to confirm that the monitored substance in brain tissue is DA; plots of peak current vs. time indicate the timecourse of changes in DA concentration (modified from Rice and Nicholson, 1989).

Striatum

The striatum is a central component of basal ganglia circuitry. In the striatum, dopaminergic input from the substantia nigra pars compacta (SNc) modulates excitatory glutamatergic input from the cortex to the primary output cells of the striatum, which are GABAergic medium spiny neurons (Freund et al., 1984; Smith and Bolam, 1990). Medium spiny neurons project to the globus pallidus and substantia nigra pars reticulata (SNr) via two pathways (Albin et al., 1989): the direct pathway, which activates movement, and the indirect pathway, which inhibits movement. In this model of basal ganglia function, DA has been proposed to facilitate movement by acting at D_1 receptors that enhance cortical activation of the direct pathway and at D_2 receptors that inhibit cortical activation of the inhibitory, indirect pathway (Gerfen et al., 1990). Although specific details of this circuitry are still being refined (e.g. Wichmann and DeLong, 1996; Waszczak et al., 1998), there is no question that loss of DA innervation to in this system causes the motor deficits characteristic of Parkinson's disease and that Parkinsonian symptoms can be alleviated by the DA precursor, L-DOPA.

The structural features of DA input to the striatum are well defined. The nigrostriatal DA pathway forms a dense network of synaptic innervation to the striatum, with an estimated 10^8 DA synapses per mm^3 (Pickel et al., 1981; Doucet et al., 1986). These synapses are well positioned to modulate glutamatergic input to the striatum, with primary localization on the dendrites and necks of dendritic spines of medium spiny neurons that receive cortical excitation (Fig. 2; Freund et al., 1984; Smith and Bolam, 1990; Hersch et al., 1995). Relevant for the issue of volume transmission, non-synaptic DA release sites in striatum have also been reported (Groves et al., 1994; Descarries et al., 1996; Descarries and Mechawar, Chapter 3, this volume). Despite the predominant point-to-point synaptic organization of the striatum, however, both pre- and postsynaptic DA receptors in this region are predominantly extrasynaptic (Sesack et al., 1994; Yung et al., 1995; Hersch et al., 1995; Pickel, Chapter 14, this volume).

Of the two major classes of DA receptors, D_1 receptors are located primarily on medium spiny neurons (Hersch et al., 1995; LeMoine and Bloch, 1995; Yung et al., 1995; Caillé et al., 1996; Dumartin et al., 1998), whereas D_2 receptors are expressed on presynaptic DA terminals, where they act as autoreceptors (Cragg and Greenfield, 1997), as well as on a distinct population of medium spiny cells where they are found on dendrites and spines (Hersch et al., 1995; Yung et al., 1995; Delle Donne et al., 1997; Khan et al., 1998). The D_2 receptors of the striatum have recently been shown to be expressed in two isoforms, D_2 short and D_2 long, with D_2 short found primarily in pre-synaptic terminals in striatum and D_2 long on postsynaptic striatal cells (Khan et al., 1998). In addition, medium spiny neurons appear to express either D_1 or D_2 receptors, with only few cells expressing both subtypes (Hersch et al., 1995; LeMoine and Bloch, 1995).

The DA transporter (DAT) is also densely expressed in the striatum. It is localized presynaptically on nigrostriatal DA terminals, with current data suggesting that the transporters are found adjacent to, but not within DA synapses (Ciliax et al., 1995; Nirenberg et al., 1996a; Hersch et al., 1997). This localization would allow the DAT to 'gate' the overflow of synaptically released DA

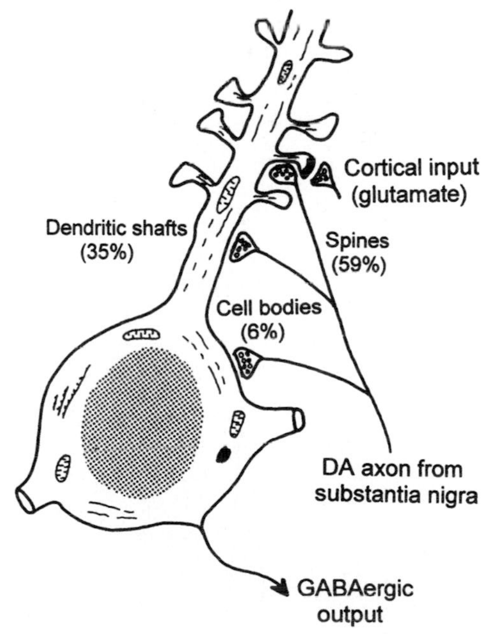

Fig. 2. Distribution of dopamine terminals forming synapses on medium spiny neurons of the rat striatum. The data were obtained from analysis of tyrosine hydroxylase immunoreactive boutons in contact with identified striatonigral neurons. These symmetrical contacts are ideally situated to modulate incoming glutamatergic input from cortex, which makes asymmetric synapses on the tips of the spines of these cells (modified from Smith and Bolam, 1990).

to adjacent cells. Moreover, Dumartin and colleagues (1998) recently provided evidence that increases in $[DA]_o$ can indeed affect extrasynaptic receptors. In that study, D_1 receptor internalization, which follows agonist activation, was monitored after local injection of a D_1 agonist or after systemic administration of the same agonist or amphetamine. As expected, receptor internalization was induced by the D_1 agonist, but more importantly, D_1 internalization was also seen after amphetamine-induced release of endogenous DA (Dumartin et al., 1998). Gonon (1997) has further reported physiological effects of stimulated increases in $[DA]_o$ in striatum (see also Gonon et al., Chapter 16, this volume).

In the striatum, therefore, the three-dimensional density of DA release sites is identically matched by the density of uptake sites. The effect of this density of uptake sites on $[DA]_o$ can be readily seen in DA diffusion curves recorded in slices of rat striatum taken from animals with neonatal, unilateral 6-OHDA lesions (Fig. 3). For these

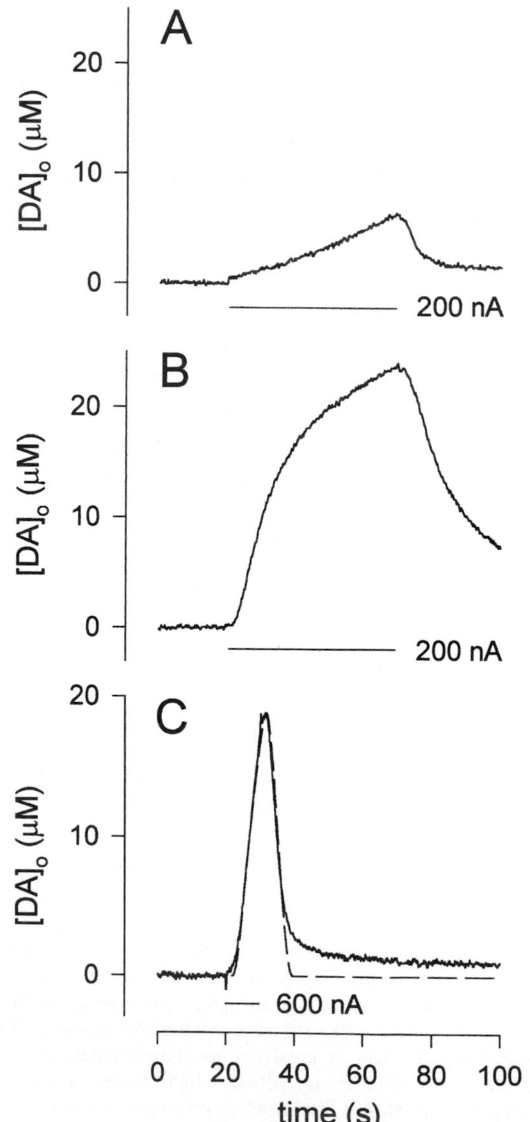

Fig. 3. DA diffusion profiles in slices of intact and 6-OHDA-lesioned rat striatum. (A) DA concentration time profile in an intact rat striatal slice, recorded at a distance of 123 μm from an iontophoretic pipette during a 200 nA current pulse. The slow increase in extracellular DA concentration ($[DA]_o$) reflected avid DA uptake in normal striatal tissue and could not be fitted using standard diffusion equations. (B) DA diffusion profile in the striatum after unilateral 6-OHDA lesion of the substantia nigra. This record was obtained in the contralateral hemisphere of the same slice in which the curve in (A) was recorded. In the 6-OHDA lesioned striatum, DA diffusion was similar to that of the extracellular marker tetramethylammonium (TMA+) (see Rice and Nicholson, 1991). The record in (B) could be fitted to the diffusion equation to give diffusion parameters of $\alpha = 0.26$, $\lambda = 1.50$, and $k' = 2.5 \times 10^{-3}$. (C) DA diffusion profile in the intact striatum during application of a higher iontophoretic current (600 nA) to overcome DA uptake and give quantifiable increases in $[DA]_o$. This curve could be fitted after incorporation of Michaelis-Menten uptake kinetic parameters in the diffusion equation (Nicholson, 1995). The dashed line represents the fitted solution for this record, using Michaelis-Menten parameters $V_{max} = 0.9$ μM s^{-1} and $K_m = 0.15$ μM, where V_{max} is the maximum rate of uptake and K_m is the Michaelis-Menten constant. The theoretical curve was calculated using standard diffusion parameters for striatum: $\alpha = 0.21$, $\lambda = 1.54$ (Rice and Nicholson, 1991). In all records, DA was monitored using fast-scan cyclic voltammetry at a carbon fiber microelectrode; scan rate was 900 V s^{-1}; iontophoresis backfill contained 1 mM DA in 150 mM TMA+; transport number for DA was 0.021; recording chamber temperature was 32°C; slice thickness was 400 μm (Rice and Nicholson, unpublished data).

experiments, coronal slices containing both the lesioned and non-lesioned striatum were prepared using conventional brain slice methods (e.g. Rice and Nicholson, 1991). When DA was introduced by iontophoresis into intact striatal tissue, a carbon fiber microelectrode positioned about 120 μm away detected very little DA (Fig. 3A). On the 6-OHDA lesioned side (Fig. 3B), however, the same iontophoresis parameters and diffusion distance produced DA diffusion records that were indistinguishable from those obtained using the extracellular marker TMA^+ (Rice and Nicholson, 1991). Diffusion of DA in lesioned striatum, like TMA^+ in any CNS region, is governed by the geometric parameters extracellular volume fraction (α), tortuosity (λ) and the non-specific uptake term k' (Nicholson and Phillips, 1981; Rice and Nicholson, 1991; Nicholson and Syková, 1998; Nicholson et al., Chapter 5, this volume). By contrast, diffusion of DA in intact striatum is further subject to Michelis-Menten uptake kinetics and can therefore be considered to be 'uptake-limited' (Nicholson, 1995; Rice and Nicholson, 1995).

This avid DA uptake system effectively 'clamps' $[DA]_o$ at low basal levels, with estimates of 4–20 nM in the striatum in vivo (Gonon and Buda, 1985; Parsons and Justice, 1992). Under conditions of stimulated DA release (Wightman et al., 1988; Garris and Wightman, 1995; Luthman et al., 1993), however, or when higher levels of DA are introduced by iontophoresis or pressure ejection (Fig. 3C), the uptake system can be overwhelmed and $[DA]_o$ can reach micromolar levels. Continued activity of the DAT under these conditions, however, is indicated by the fact that clearance of DA after such elevation is much faster than that possible by diffusion alone. Indeed, such curves can be used to extract Michaelis-Menten uptake parameters (Nicholson, 1995). The record in Fig. 3C was fitted with standard diffusion parameters coupled with an assumed value for K_m of 0.15 μM (Nicholson, 1995) and a best-fit value for V_{max} of 0.9 $\mu M\ s^{-1}$.

Substantia nigra

A special characteristic of DA neurons in the SNc is that they release transmitter from their dendrites, in addition to classical release from their axon terminals in the striatum. Somatodendritic release of DA from neurons of the SN was postulated in the mid-1970s by several groups working independently. Björklund and Lindvall (1975) proposed somatodendritic release based on anatomical evidence that DA is localized in SN dendrites, whereas Groves et al. (1975) concluded that DA release from dendrites was the only possible explanation for autoinhibition of cell firing by locally released DA in the SNc. Release of DA in the SN was confirmed shortly afterwards, both in vitro (Geffen et al., 1976) and in vivo (Nieoullon et al., 1977).

DA release in SNc is often referred to as 'somatodendritic' rather than simply 'dendritic'. This reflects the juxtaposition of soma and lateral dendrites in the SNc (Fig. 4), as well as the limited spatial resolution of techniques that have been used to monitor release. Recent evidence, however, suggests that DA release can be elicited from both dendrites and cell bodies in the SN. The substantia nigra pars reticulata (SNr) contains ventral-projecting dendrites from DA cells in the SNc (Fig. 4), but few cell bodies (Fallon and Moore, 1978; Fallon et al., 1978). Release from dendrites has been confirmed by studies of depolarization-induced release of ^3H-DA from midbrain slices that contained only

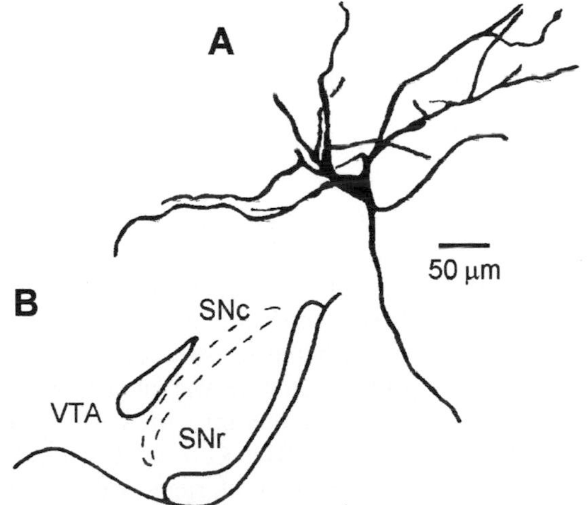

Fig. 4. Midbrain dopamine cells and cell body regions. (A) Ventral tier cell of the substantia nigra pars compacta (SNc). The ventrally projecting dendrite extends into the SN pars reticulata (SNr). (B) Relative location of SNc, SNr and ventral tegmental area (VTA) in coronal view of midbrain (midline is to the left of the VTA) (modified from Fallon et al., 1978).

the SNr (Geffen et al., 1976) and by direct detection of stimulated DA release monitored with carbon fiber microelectrodes positioned in this region (Rice et al., 1994). On the other hand, Jaffe et al. (1998), using amperometry in midbrain slices, described what appears to be quantal release of DA from SNc cell bodies as well. Taken together, these data suggest that non-classical release of DA can be elicited from soma and dendrites, such that 'somatodendritic' is indeed an accurate term.

To date, no characteristics of somatodendritic DA release contradict the original hypothesis of Geffen et al. (1976) that release is vesicular and mediated by exocytosis, as it is in axon terminals. Somatodendritic release is Ca^{2+}-and depolarization-dependent in a variety of paradigms (Geffen et al., 1976; Cheramy et al., 1981; Rice et al., 1994, 1997) and sensitive to depletion by reserpine (which irreversibly inhibits the vesicular monoamine transporter, VMAT2) (Elverfors and Nissbrandt, 1991; Rice et al., 1994; Heeringa and Abercrombie, 1995). These features are consistent with characteristics of exocytosis. Moreover, other manipulations that alter DA release in striatum generally affect $[DA]_o$ in SNc in a parallel manner, offering further support for similar mechanisms of release in terminals and dendrites (Santiago and Westerink 1992; Heeringa and Abercrombie 1995).

On the other hand, the number of vesicles in SNc dendrites is small, which would imply a limited source for dendritic release (Wilson et al., 1977; Groves and Linder, 1983; Nirenberg et al., 1996a). In addition, although dendrites contain DA (Björklund and Lindvall, 1975; Cuello and Kelly, 1977), the primary storage site remains uncertain. Wilson et al. (1977) first proposed that storage was primarily in vesicles, however, subsequent studies suggested that primary storage was in saccules of smooth endoplasmic reticulum (Mercer et al., 1978; Wassef et al., 1981). Groves and Linder (1983) later proposed that storage occurred in vesicles *and* endoplasmic reticulum, which is consistent with more recent data showing expression of VMAT2 in both organelles (Nirenberg et al., 1996b).

Another reason that exocytotic release in the SN has been questioned is that synaptic sites available for vesicle fusion are rare. Although dendrodendritic synapses do occur in the SNc (Wilson et al., 1977), these are largely absent in the SNr, so that they comprise less than 1% of synaptic input to DA dendrites (Groves and Linder, 1983). Since depolarization-induced DA release can be elicited from the SNr in isolation (Geffen et al., 1976; Rice et al., 1994), this further suggests that dendrodendritic synapses are not required for release. It is relevant to note that vesicular release of catecholamines from adrenal chromaffin cells, for example (Wightman et al., 1991), occurs in the absence of synapses.

The substantia nigra expresses both D_1 and D_2 receptors. D_2 receptors are expressed by DA cells, especially on dendrites (Morell et al., 1988; Mansour et al., 1990; Sesack et al., 1994; Khan et al., 1998), whereas D_1 receptors are on afferents to DA cells (Beckstead 1988; Yung et al., 1995). Based on this localization, D_2 receptors are considered autoreceptors, and have been shown to regulate basal DA firing rate (for reviews see Kalivas, 1993; Lacey, 1993) and to modulate stimulated increases in $[DA]_o$ (Cragg and Greenfield, 1997).

An important finding about D_1 and D_2 receptor localization in the substantia nigra is that expression of both subtypes is exclusively extrasynaptic (Yung et al., 1995). This indicates that volume transmission is the predominant form of intercellular communication mediated by DA in this region. Further, it underscores the importance of defining factors that regulate the $[DA]_o$ in the SNc and SNr because the effect of DA at its extrasynaptic DA receptors will be concentration-dependent. In addition to the D_2-mediated autoreceptor effects noted above, functional volume transmission in the midbrain has been demonstrated for D_1-mediated actions in the ventral tegmental area (VTA; see Fig. 4). Cameron and Williams (1993) found that DA regulates $GABA_B$-mediated inhibition via D_1 receptors in VTA, despite the absence of DA synapses onto GABAergic cells in this region.

We have addressed regulation of $[DA]_o$ in the SN by investigating electrically stimulated DA release in guinea pig midbrain slices (Rice et al., 1997; Cragg et al., 1997). Local electrical stimulation elicited Ca^{2+}-dependent release that was also dependent on stimulation frequency. In these studies, 10 s pulse trains of 1–10 Hz frequency elicited increasing $[DA]_o$ from 0.2–0.5 μM (Rice et al., 1997), which is the usual range of firing frequency for SNc DA cells (Grace and Bunney, 1983, 1984). Frequencies above 10 Hz did not elicit further increases in $[DA]_o$; this optimal 10 Hz stimulation frequency was used in all subsequent studies of the ionic and pharmacologic dependence of somatodendritic release.

More relevant for the issue of volume transmission, we then investigated the effect of DA uptake inhibition on stimulated $[DA]_o$ in midbrain and in striatum (Fig. 5A; Cragg et al., 1997). In caudal SNc, stimulated $[DA]_o$ increased to roughly 200% of control levels in the presence of the selective DA uptake inhibitor GBR 12909 (300 nM). By contrast, uptake inhibition in striatum caused an increase in stimulated $[DA]_o$ to 400% of control levels (Fig. 5A). These data are consistent with the higher density of catecholamine uptake sites in DA terminal regions compared to somatodendritic regions in midbrain (Donnan et al., 1991; Ciliax et al., 1995; Freed et al., 1995). Moreover, the greater enhancement of stimulated release from DA axon terminals in striatum compared to somatodendritic release in SNc indicate that DAT has a greater role in the regulation of $[DA]_o$ in striatum than in SN.

In contrast to the potentiating effect of DA uptake inhibition on evoked $[DA]_o$ in caudal SNc, GBR12909 had no effect in two other midbrain regions: rostral SNc and VTA (Fig. 5A; Cragg et al., 1997). The DA cells of these regions differ from those in caudal SNc. On the basis of anatomical localization, DA neurons have been classified as 'dorsal tier' and 'ventral tier', with caudal SNc comprised of ventral tier cells and rostral SNc and VTA comprised of dorsal tier cells (Fallon and Moore, 1978; Gerfen et al., 1987). These cells can also be distinguished on the basis of morphological and biochemical differences, including expression of DA transporter mRNA, with greater expression in the dorsal tier cells of caudal SNc compared to ventral tier cells in rostral SNc and VTA (Blanchard et al., 1994; Hurd et al., 1994; Sanghera et al., 1994). The much lower effect of DAT uptake inhibition on stimulated $[DA]_o$ in rostral SNc and VTA than in caudal SNc demonstrated for the first time a functional consequence of these differences in mRNA levels (Cragg et al., 1997).

Importantly, the enhancement of evoked $[DA]_o$ in caudal SNc and lack of effect in rostral SNc and VTA argue against reversal of the DAT as a mechanism of somatodendritic release (Groves and Linder, 1983; Nirenberg et al., 1996a; Elverfors et al., 1997). If transporter reversal were the primary mechanism of DA release, DAT inhibition would be expected to decrease evoked $[DA]_o$ rather than

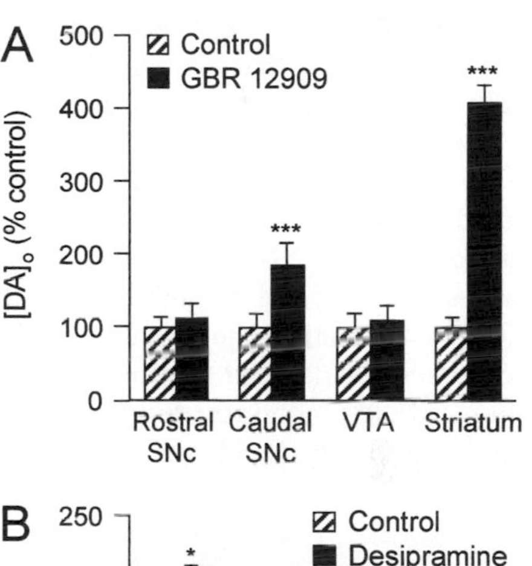

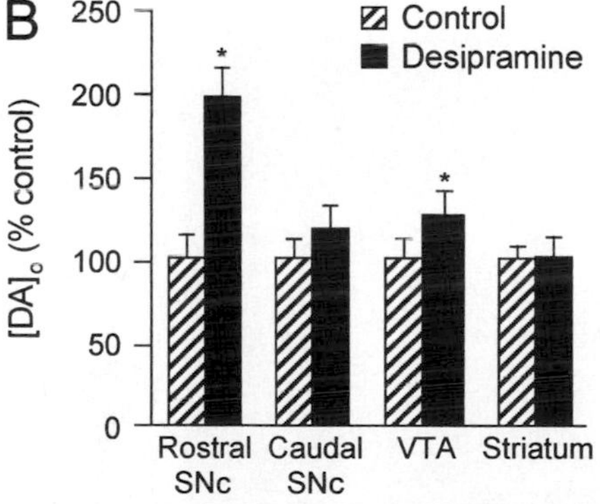

Fig. 5. Comparison of the effects of the DA and NE uptake inhibitors GBR 12909 and desipramine (300 nM) on electrically evoked increases $[DA]_o$ in midbrain and striatum. (A) GBR 12909 significantly increased evoked $[DA]_o$ compared to controls in both caudal SNc and dorsal CPu, but had no effect in rostral SNc or VTA. (B) The NE uptake inhibitor desipramine significantly increased evoked $[DA]_o$ compared to controls in both rostral SNc ($n = 11$) and VTA ($n = 15$) but had no effect in caudal SNc or CPu. SNc, substantia nigra pars compacta; VTA, ventral tegmental area; CPu, caudate-putamen (striatum). Data are means ± s.e.m; $*p < 0.05$, $***p < 0.001$ compared to control (from Cragg et al. 1997).

increase it or leave it unaffected, as is actually seen (Fig. 5A; Cragg et al., 1997).

A novel role for volume transmission in these midbrain DA regions was suggested by the regionally dependent effect of inhibition of the norepinephrine (NE) transporter on stimulated $[DA]_o$. Immunoreactivity to the NE synthesizing enzyme, dopamine-β-hydroxylase (DβH), is found in a subset of NE fibers passing throughout SN and VTA, with a greater density in rostral SNc and VTA than in caudal SNc (Cragg et al., 1997). Importantly, these DβH-immunoreactive (DβH-ir) fibers are also dense in sections rostral to SNc, where no catecholamine signals were detected during stimulation. This indicated that NE release from these fibers did not interfere with measurements of evoked $[DA]_o$ in midbrain.

On the other hand, DβH-ir fibers apparently contributed to DA uptake in rostral SNc and VTA, although not in caudal SNc or in striatum (Fig. 5B). Desipramine, a NE uptake inhibitor, caused a 2-fold increase in stimulated levels of $[DA]_o$ in rostral SNc and a small but significant increase in VTA (Fig. 5B). By contrast, desipramine had no effect on stimulated $[DA]_o$ in caudal SNc or striatum. Uptake of DA in rostral SNc and VTA by the NE transporter would provide an alternative or supplemental mechanism for DA clearance in these regions (Carboni et al., 1990; Cragg et al., 1997). Intriguingly, the DβH-ir fibers in these regions are immunonegative for tyrosine hydroxylase, which suggests that DA taken up into NE fibers might serve as a precursor for NE synthesis by DβH. The necessary diffusion of DA from a somatodendritic site of release to a NE fiber might therefore represent a novel role for volume transmission in the midbrain.

Retina

Of the CNS regions considered in this review, the retina provides the clearest example of DA-mediated volume transmission. In the retina, DA is released from is the dopaminergic amacrine or interplexiform cells located in a plane defined by the inner plexiform layer (Fig. 6) (Dowling and Ehinger, 1978; Witkovsky and Schütte, 1991). These cells make synapses with other amacrine cells and their processes within this layer (Dowling and Ehinger, 1978; Witkovsky and Schütte, 1991),

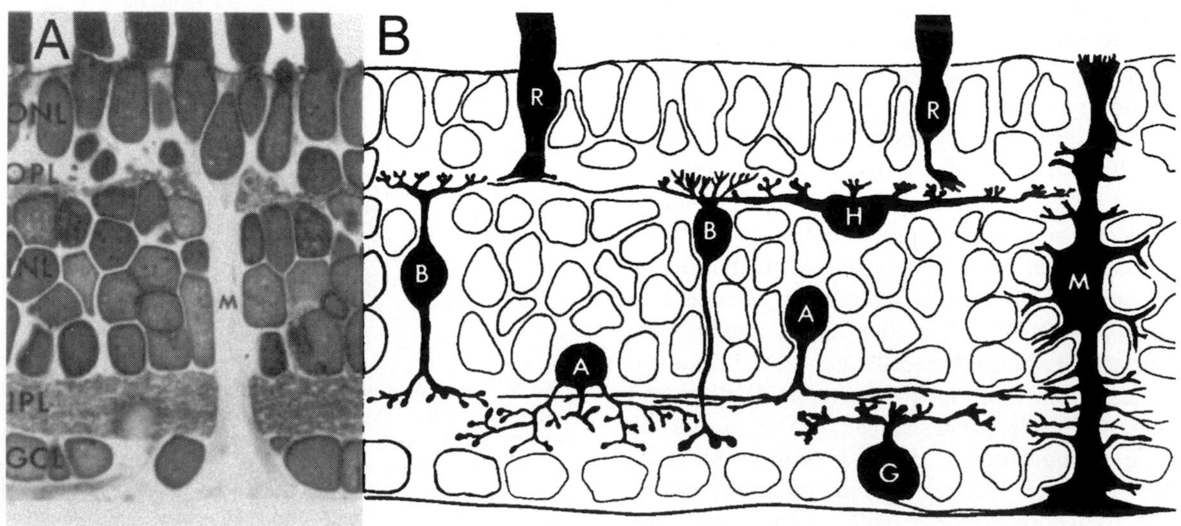

Fig. 6. Retinal circuitry indicating amacrine cell processes in the inner plexiform layer. (A) Light micrograph of mudpuppy retina. Three nuclear and two plexiform layers are see: GLC, ganglion cell layer; IPL, inner plexiform layer; INL, outer nuclear layer; OPL, outer plexiform layer; ONL, outer nuclear layer. A prominent Müller (glial) cell (M) is also seen. Magnification is 345 ×. (B) Schematic of major cell types in vertebrate retina: G, ganglion cells; A, amacrine cells: B, bipolar cells; H, horizontal cells; R, receptors (from Dowling and Dubin, 1984).

so that DA release is presumably synaptic in origin. Both D_1 and D_2 receptors are found in the inner plexiform layer (Muresan and Besharse, 1993; Behrens and Wagner, 1995; Bjelke et al., 1996; Mora-Ferrer et al., 1999), including $D_{2/3}$ receptor expression on dopaminergic amacrine cells, which are likely to function as autoreceptors (Veruki, 1997). In addition, however, D_1 and D_2 receptors are expressed by horizontal cells, photoreceptors, and pigment epithelium cells that are in the outer retina, at least 10–100 μm away from the source of DA in the inner plexiform layer (Muresan and Besharse, 1993; Wagner et al., 1993; Behrens and Wagner, 1995; Bjelke et al., 1996; Veruki and Wässle, 1996; Mora-Ferrer et al., 1999). This source/receptor mismatch has led a number of groups to suggest that volume transmission is the major form of intercellular communication by DA in the retina (Wagner et al., 1993; Witkovsky et al., 1993; Behrens and Wagner, 1995; Bjelke et al., 1996; Veruki and Wässle, 1996; Mora-Ferrer et al., 1999).

Importantly, the DA receptors expressed in the outer retina are functional: retinal DA exerts its principal actions on horizontal cells and photoreceptors. Actions of DA include disc shedding in outer rod segments, changes in receptive field size of horizontal cells, suppression of melatonin production, and light-dark adaptation by altering the balance of rod-to-cone input to horizontal cells (see Witkovsky et al., 1993; Behrens and Wagner, 1995; Križaj et al., 1998).

The geometric orientation of DA release and uptake sites in the inner plexiform layer facilitates volume transmission. In vertebrate retina, DA cells are distributed more or less evenly and at a low density (10–60 cells per mm^2), with overlapping dendritic fields of adjacent DA cells (Witkovsky and Schütte, 1991). Release of DA, therefore, can be modeled as being uniform from a thin sheet located in the inner plexiform layer of DA; critically, sites of DA uptake are also localized in this monolayer (Witkovsky et al., 1993; Bjelke et al., 1996). Consequently, as DA diffuses away from release sites, its three-dimensional diffusion path will permit it to escape the sheet of amacrine cells and uptake sites and thus reach the outer retina, which is bounded by the pigment epithelium. It will also diffuse into the vitreous body located only 25 μm from the inner plexiform layer (Witkovsky et al., 1993). This geometric organization of release and uptake sites in the retina contrasts markedly with that in the striatum, where the dense three-dimensional network of uptake sites are positioned to recapture released DA, as discussed above.

A striking consequence of the minimal influence of reuptake on released DA in the retina is that $[DA]_o$ in this tissue (Witkovsky et al., 1993) can be as much as two orders of magnitude higher than that in the uptake-limited striatum. The lowest DA concentrations in the vitreous of *Xenopus laevis* are found during dark adaptation, yet even then, the concentration exceeds 150 nM (Fig. 7; Witkovsky et al., 1993). Indeed, DA concentrations in the vitreous approach micromolar levels after light exposure to a dark-adapted eye (Fig. 7). These levels, assessed using HPLC, gradually fall back to dark adapted levels during daylight hours. Impres-

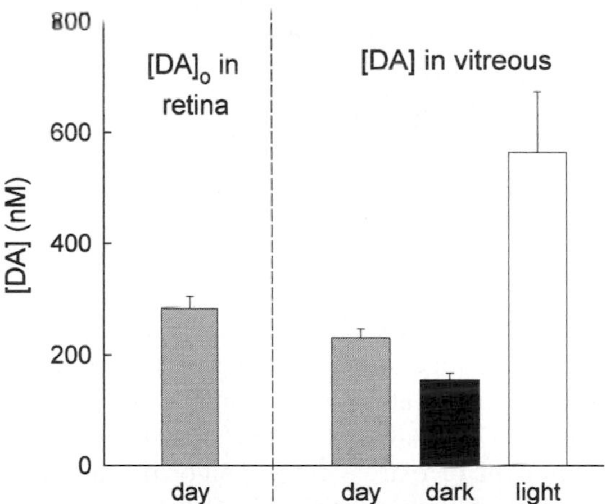

Fig. 7. Basal extracellular dopamine concentration ($[DA]_o$) in the isolated *Xenopus* retina and dopamine concentration in the vitreous body of *Xenopus* as a function of time of day and lighting conditions. Basal extracellular dopamine concentration ($[DA]_o$) in the isolated retina was monitored with fast-scan cyclic voltammetry after several hours of light adaptation (day; 10 a.m.–4 p.m.). Average $[DA]_o$ was 283 ± 22 nM (mean $\pm$ SEM; $n = 24$). This was similar to [DA] in the vitreous sampled during the same period (day) and analyzed using HPLC. Vitreous [DA], and by extension $[DA]_o$ in the retina where DA is released, was lowest during dark adaptation, but increased markedly in the period following exposure of dark-adapted eyes to light (light; 6 a.m.–9 a.m.) (modified from Witkovsky et al., 1993).

sively, $[DA]_o$ in the isolated retina also proved to be several hundred nanomolar, when detected using carbon-fiber microelectrodes and fast-scan cyclic voltammetry (Fig. 7). In contrast to striatum, in which basal levels of $[DA]_o$ are lower than detection limits of fast-scan cyclic voltammetry, clearly recognizable voltammograms for DA were observed in the outer retina (Witkovsky et al., 1993). The average $[DA]_o$ after several hours of light adaptation was about 280 nM, which was somewhat higher than that in the vitreous (230 nM) during the same period (Fig. 7). Indeed, it is likely that our voltammetric measurements underestimated $[DA]_o$ since DA would have continued to diffuse from the superfused retina during the brief period (3–5 min) between removal of the retina and the time of the voltammetric measurement (Witkovsky et al., 1993).

It is important to note that the 100–1000 nM range of retinal $[DA]_o$, indicated by levels in the vitreous as well as by direct detection in the retina, is consistent with concentrations of DA found to be effective in mediating local physiological responses (see Križaj and Witkovsky, 1993; Witkovsky et al., 1993). Taken together, these findings demonstrate that dopaminergic neurotransmission/neuromodulation in the retina is a well-characterized example of volume transmission.

Conclusions

The regions of the CNS examined in this paper exhibit a range of DA release and reuptake characteristics that will differentially influence the nature of DA-mediated volume transmission. The requirement of extrasynaptic and/or distant receptor sites is met in all three areas: the striatum, the substantia nigra, and the retina. The range of influence of DA in each region, however, is regulated primarily by local DA reuptake processes. The effectiveness of each uptake process, in turn, is governed by the geometry of DA uptake transporter distribution: in striatum, a dense plexus of DA uptake sites limits DA diffusion from sites of release; in SNc and SNr, less dense networks apparently permit a greater radius of influence; in retina, the planar orientation of release and reuptake sites readily allow DA to diffuse to adjacent cell layers to mediate volume transmission.

The variation in $[DA]_o$ regulation among these regions illustrates two key issues that need to be incorporated into further refinements in our understanding of volume transmission. The first issue is one of dimensionality; is there a minimal distance over which a substance must diffuse to reach a receptor for the interaction to be considered 'volume transmission'? In the retina, for example, there is no question that when DA diffuses a distance of 100 μm from an amacrine cell to a photoreceptor expressing DA receptors, this is an example of volume transmission. The question of scale becomes more nebulous, however, when considering DA-mediated volume transmission in striatum, where the avid uptake system constrains DA to remain within a few μm of a release site. Data from Garris and Wightman (1995) indicate that the 'half life' of stimulated $[DA]_o$ in the striatum, that is the time required for 50% of a DA release signal to be removed by uptake, is 40 ms. In 40 ms (time, t), the mean diffusion radius (r) for DA would be 8.8 μm, calculated using the expression $r = (6D^*t)^{0.5}$ for diffusion in three dimensions, where the *in vivo* diffusion coefficient $D^* = D\lambda^{-2}$. The diffusion coefficient (D) for DA at 37°C is 7.6×10^{-6} cm^2 s^{-1} and the tortuosity factor (λ) for striatum is 1.54 (Rice and Nicholson, 1991), to yield a D^* of 3.2×10 cm^2 s^{-1}. The volume of the sphere defined by a radius of 8.8 μm is roughly 680 μm^3. Given the density of DA synapses in the striatum (10^8 synapses per mm^3 or 0.1 synapses per μm^3; Pickel et al., 1981; Doucet et al., 1986), this volume of tissue would contain an average of 68 DA synapses. Moreover, DA terminals comprise only 10–20% of striatal synapses (Pickel et al., 1981; Doucet et al., 1986). Even with uptake-limited diffusion of DA in striatum, therefore, an increase in $[DA]_o$ could have considerable influence on synaptic transmission within distances of only a few micrometers from a release site. Consequently, the process of volume transmission would be best defined as a *functionally significant association of release and receptor sites via extrasynaptic diffusion*, rather than by stricter definitions based on absolute distances between such sites.

The second issue raised by the regional differences described here is the futility of trying to categorize a given substance *either* as a mediator of volume transmission *or* as a synaptic transmitter. These patterns of DA regulation illustrate only a few of the possible behaviors and regulatory processes that might govern the extent of extracellular communication in the CNS. We have gained an understanding of these processes for DA (and other electroactive species, like 5-HT; see Bunin and Wightman, 1999) because we have tools, carbon-fiber microelectrodes, that allow us to monitor $[DA]_o$ under a variety of conditions (for reviews, see Adams, 1990; Boulton et al., 1995). These electrodes have permitted measurement of the time course of exogenously introduced DA, stimulated release of endogenous DA, and even detection of basal $[DA]_o$ under certain conditions. As new probes become available for other substances, understanding of the range of possible behaviors will surely increase. For example, the novel process of somatodendritic release, increasingly well-characterized for DA, may prove less unique than it now appears. At the present time, however, the extensive data available for DA provide the best indication of how neuroactive substances can behave in the extracellular microenvironment to mediate volume transmission.

Acknowledgements

These studies were supported by NINDS grants NS–28642 and NS–36362.

References

Adams, R.N. (1990) In vivo electrochemical measurements in the CNS. *Prog. Neurobiol.* 35: 297–311.

Albin, R.L., Young, A.B. and Penney, J.B. (1989) The functional anatomy of basal ganglia disorders. *Trends Neurosci.*, 12: 366–75.

Armstrong-James, M., Fox, K., Kruk, Z.L. and Millar, J. (1981) Quantitative iontophoresis of catecholamines using multibarrel carbon fibre electrodes, *J. Neurosci. Meth.*, 4: 385–406.

Beckstead, R.M. (1988) Association of dopamine D_1 and D_2 receptors with specific cellular elements in the basal ganglia of the cat: the uneven topography of dopamine receptors in the striatum is determined by intrinsic striatal cells not nigrostriatal axons. *Neuroscience*, 27: 851–863.

Behrens, U.D. and Wagner, H.-J. (1995) Localization of dopamine D_1-receptors in vertebrate retinae. *Neurochem. Int.*, 27: 497–507.

Bjelke, B., Goldstein, M., Tinner, B., Andersson, C., Sesack, S.R., Steinbusch, H.W.M., Lew, J.Y., He, X., Watson, S., Tengroth, B. and Fuxe, K. (1996) Dopaminergic transmission in the rat retina: evidence for volume transmission. *J. Chem. Neuroanat.*, 12: 37–50.

Björklund, A. and Lindvall, O. (1975) Dopamine in dendrites of substantia nigra neurons: suggestions for a role in dendritic terminals. *Brain Res.*, 83: 531–537, 1975.

Blanchard, V., Raisman-Vozari, R., Vyas, S., Michel, P.P., Javoy-Agid, F., Uhl, G. and Agid, Y. (1994) Differential expression of tyrosine hydroxylase and membrane dopamine transporter genes in subpopulations of dopaminergic neurons of the rat mesencephalon. *Mol. Brain Res.*, 22: 29–40.

Boulton, A.A., Baker, G.B. and Adams, R.N., Eds. (1995) *Voltammetric Measurements in Brain Systems. Neuromethods.* Vol. 27, Humana Press, Totowa, NJ.

Bull, D.R., Palij, P., Sheehan, M.J., Millar, J., Stamford, J.A., Kruk, Z.L. and Humphrey, P.P.A. (1990) Application of fast-scan cyclic voltammetry to measurement of electrically evoked dopamine overflow from brain slices in vitro. *J. Neurosci. Meth.*, 32: 37–44.

Bunin, M.A. and Wightman, R.M. (1999) Paracrine neurotransmission in the CNS: involvement of 5-HT. *Trends Neurosci.*, 22: 377–382.

Caillé, I., Dumartin, B. and Bloch, B (1996) Ultrastructural localization of D_1 dopamine receptor immunoreactivity in rat striatonigral neurons and its relation with dopaminergic innervation. *Brain Res.*, 730: 17–31.

Cameron, D.L. and Williams, J.T. (1993) Dopamine D_1 receptors facilitate transmitter release. *Nature*, 366: 344–347, 1993.

Carboni, E., Tanda, G.L., Frau, R. and Di Chiara, G. (1990) Blockade of the noradrenaline carrier increases extracellular dopamine concentrations in the prefrontal cortex: evidence that dopamine is taken up in vivo by noradrenergic terminals. *J. Neurochem.*, 55: 1067–1070.

Cheramy, A., Leviel, V. and Glowinski, J. (1981) Dendritic release of dopamine in the substantia nigra. *Nature*, 289: 537–542, 1981.

Ciliax, B.J., Heilman, C., Demchyshyn, L.L. Pristupa, Z.B., Ince, E., Hersch, S.M., Niznik, H.B. and Levey, A. (1995) The dopamine transporter: immunochemical characterization and localization in brain. *J. Neurosci.*, 15: 1714–1723.

Cragg, S.J. and Greenfield, S.A. (1997) Differential autoreceptor control of somatodendritic and axon terminal dopamine release in substantia nigra, ventral tegmental area and striatum. *J. Neurosci.*, 17: 5738–5746.

Cragg, S.J., Rice, M.E. and Greenfield, S.A. (1997) Heterogeneity of electrically-evoked dopamine release and uptake between substantia nigra, ventral tegmental area and striatum. *J. Neurophysiol.*, 77: 863–873.

Cuello, A.C. and Kelly, J.S. (1977) Electron microscopic autoradiographic localization of [^{3}H]-dopamine in the dendrites of the dopaminergic neurones of the rat substantia nigra in vivo. *Br. J. Pharmacol.*, 59: P527–528.

Descarries, L., Watkins, K.C., Garcia, S., Bosler, O. and Doucet, G. (1996) Dual character, asynaptic and synaptic, of the dopamine innervation in adult rat neostriatum: a quantitative autoradiographic and immunocytochemical analysis. *J. Comp. Neurol.*, 375: 167–186.

Delle Donne, K.T., Sesack, S.R. and Pickel, V.M. (1997) Ultrastructural immunocytochemical localization of the D_2 receptor within GABAergic neurons of the rat striatum. *Brain Res.*, 746: 239–255.

Donnan, G.A., Kaczmarczyk, S.J., Paxinos, G., Chilco, P.J., Kalnins, R.M. Woodhouse, D.G. and Mendelsohn, F.A. (1991) Distribution of catecholamine uptake sites in human brain as determined by quantitative [^{3}H] mazindol autoradiography. *J. Comp. Neurol.*, 304: 4, 19–34.

Doucet, G., Descarries, L. and Garcia, S. (1986) Quantification of the dopamine innervation in adult rat neostriatum. *Neuroscience*, 19: 427–445.

Dowling, J.E. and Ehinger, B. (1978) The interplexiform cell system: I. Synapses of the dopaminergic neurons of the goldfish retina. *Proc. R. Soc. (Lond.)*, B 201: 7–26.

Dowling, J.E. and Dubin, M.W. (1984) The vertebrate retina. In: I. Darian-Smith (Ed.), *Handbook of Physiology: The Nervous System III*, Oxford Univ. Press, pp. 317–339. .

Dumartin, B., Caillé, I., Gonon, F. and Bloch, B. (1998) Internalization of D_1 dopamine receptor in striatal neurons *in vivo* as evidence of activation by dopamine agonists. *J. Neurosci.*, 18: 1650–1661.

Elverfors, A. and Nissbrandt, H. (1991) Reserpine-insensitive dopamine release in the substantia nigra? *Brain Res.*, 557: 5–12.

Elverfors, A., Jonason, J., Jonason, G. and Nissbrandt, H. (1997) Effects of drugs interfering with sodium channels and calcium channels on the release of endogenous dopamine from superfused substantia nigra slices. *Synapse*, 26: 359–369.

Fallon, J.H. and Moore, R.Y. (1978) Catecholamine innervation of the basal forebrain. IV. Topography of the dopamine projection to the basal forebrain and neostriatum. *J. Comp. Neurol.*, 180: 545–580, 1978.

Fallon, J.H., Riley, J.N. and Moore, R.Y. (1978) Substantia nigra dopamine neurons: separate populations project to neostriatum and allocortex. *Neurosci. Lett.*, 7: 157–162.

Freed, C., Revay, R., Vaughan, R.A., Kriek, E. Grant, S., Uhl, G.R. and Kuhar, M.J. (1995) Dopamine transporter uptake immunoreactivity in rat brain. *J. Comp. Neurol.*, 359: 340–349.

Freund, T.F., Powell, J.F. and Smith, A.D. (1984) Tyrosine hydroxylase-immunoreactive boutons in synaptic contact with identified striatonigral neurons, with particular reference to dendritic spines. *Neuroscience*, 13: 1189–1215.

Fuxe, K. and Agnati, L.F. (Eds) (1991) *Volume Transmission in the Brain.* Raven Press, New York.

Garris, P.A. and Wightman, R.M. (1995) Regional differences in dopamine release, uptake and diffusion measured by fast-scan cyclic voltammetry. In: A.A. Boulton, G.B. Baker and R.N. Adams (Eds), *Neuromethods Vol. 27, Voltammetric Methods in Brain Systems*, Humana Press, Totowa, NJ, pp 179–220.

Geffen, L.B., Jessell, T.M., Cuello, A.C. and Iversen, L.L. (1976) Release of DA from dendrites in rat substantia nigra. *Nature*, 260: 258–260.

Gerfen, C.R., Herkenham, M. and Thibault, J. (1987) The neostriatal mosaic. II. Patch- and matrix-directed mesostriatal dopaminergic and non-dopaminergic systems. *J. Neurosci.*, 7: 3915–3934.

Gerfen, C.R., Engber, T.M., Mahan, L.C., Susel, Z., Chase, T.N., Monsma, F.J. Jr. and Sibley, D.R. (1990) D1 and D2 dopamine receptor-regulated gene expression of striatonigral and striatopallidal neurons. *Science*, 250: 1429–1432.

Gonon, F. (1997) Prolonged and extrasynaptic excitatory action of dopamine mediated by D_1 receptors in the rat striatum in vivo. *J. Neurosci.*, 17: 5972–5978.

Gonon, F.G. and Buda, M.J. (1985) Regulation of dopamine release by impulse flow and by autoreceptors as studied by in vivo voltammetry in the rat striatum. *Neuroscience*, 14: 765–774.

Grace, A.A. and Bunney, B.S. (1983) Intracellular and extracellular electrophysiology of nigral dopaminergic neurons. 1. Identification and characterization. *Neuroscience*, 10: 301–315.

Grace, A.A. and Bunney, B.S. (1984) The control of firing pattern in nigral dopamine neurons: burst firing. *J. Neurosci.*, 4: 2877–2890.

Grenhoff, J. and Johnson, S.W. (1997) Electrophysiological effects of dopamine receptor stimulation. In: K.A. Neve and R.L. Neve (Eds) *The Dopamine Receptors*, Humana Press, Totowa, NJ, pp. 267–304.

Groves, P.M. and Linder, J.C. (1983) Dendro-dendritic synapses in substantia nigra: descriptions based on analysis of serial sections. *Exp. Brain. Res.*, 49: 209–217.

Groves, P.M., Linder, J.C. and Young, S.J. (1994) 5-hydroxydopamine-labeled dopaminergic axons: three-dimensional reconstructions of axons, synapses and postsynaptic targets in rat neostriatum. *Neuroscience*, 58: 593–604.

Groves, P.M., Wilson, C.J., Young, S.J. and Rebec, G.V. (1975) Self-inhibition by dopaminergic neurons. *Science*, 190: 522–529.

Heeringa, M.J. and Abercrombie, E.D. (1995) Biochemistry of somatodendritic dopamine release in the substantia nigra: an in vivo comparison with striatal dopamine release. *J. Neurochem.*, 65: 192–200.

Hersch, S.M., Ciliax, B.J., Gutekunst, C.-A., Rees, H.D., Heilman, C.J., Yung, K.K.L., Bolam, J.P., Ince, E., Yi, H. and Levey, A.I. (1995) Electron microscope analysis of D_1 and D_2 dopamine receptors proteins in the dorsal striatum and their synaptic relationships with motor corticostriatal afferents. *J. Neurosci.*, 15: 5222–5237.

Hersch, S.M., Yi, H., Heilman, C.J., Edwards, R.H. and Levey, A.I. (1997) Subcellular localization and molecular topology of the dopamine transporter in the striatum and substantia nigra. *J. Comp. Neurol.*, 388: 211–227.

Hurd, Y.L., Pristupa, Z.B., Herman, M.M., Niznik, H.B. and Kleinman, J.E. (1994) The dopamine transporter and dopamine D_2 receptor messenger RNAs are differentially expressed in limbic- and motor-related subpopulations of human mesencephalic neurons. *Neuroscience*, 63: 357–362.

Jaffe, E.H., Marty, A., Schulte, A. and Chow, R.H. (1998) Extrasynaptic vesicular transmitter release from the somata of substantia nigra neurons in rat midbrain slices. *J. Neurosci.*, 18: 3548–353.

Jones, S.R., Mickelson, G.E., Collins, L.B., Kawagoe, K.T. and Wightman, R.M. (1994) Interference by pH and Ca^{2+} ions during measurements of catecholamine release in slices of rat amygdala with fast-scan cyclic voltammetry. *J. Neurosci. Meth.*, 52: 1–10.

Kalivas, P.W. (1993) Neurotransmitter regulation of dopamine neurons in the ventral tegmental area. *Brain Res. Rev.*, 18: 75–113.

Khan, Z.U., Mrzljak, L., Gutticrrrez, A., de la Calle, A. and Goldman-Rakic, P.S. (1998) Prominence of the D_2 short isoform in dopaminergic pathways. *Proc. Natl Acad. Sci. USA*, 95: 7731–7736.

Križaj, D. and Witkovsky, P. (1993) effects of submicromolar concentrations of dopamine on photoreceptor to horizontal cell communication. *Brain Res.*, 627: 122–128.

Križaj, D., Gábriel, R., Owen, W.G. and Witkovsky, P. (1998) Dopamine D_2 receptor-mediated modulation. *J. Comp. Neurol.*, 398: 529–538.

Kuhr, W. and Wightman R.M. (1986) Real-time measurement of dopamine release in rat brain. *Brain Res.*, 81: 168–171.

Lacey. M.G. (1993) Neurotransmitter receptors and ionic conductances regulating thc activity of neurones in substantia nigra pars compacta and ventral tegmental area. *Prog. Brain Res.*, 99: 251–276.

LeMoine, C. and Bloch, B. (1995) D_1 and D_2 dopamine receptor gene expression in the rat striatum: sensitive cRNA probes demonstrate prominent segregation of D_1 and D_2 mRNAs in distinct neuronal populations of the dorsal and ventral striatum. *J. Comp. Neurol.*, 355: 418–426.

Luthman, J., Friedemann, M., Bickford, P., Olson, L., Hoffer, B.J. and Gerhardt, G.A. (1993) In vivo electrochemical measurements and electrophysiological studies of rat striatum following neonatal 6-hydroxydopamine treatment. *Neuroscience*, 52: 677–687.

Mansour, A., Meador-Woodruff, J.H., Bunzow, J.R., Civelli, O., Akil, H. and Watson, S.J. (1990) Localization of dopamine D_2 receptor mRNA and D_1 and D_2 receptor binding in the rat brain and pituitary: an in situ hybridization-receptor autoradiographic analysis. *J. Neurosci.*, 10: 2587–2600.

Mercer, L., del Fiacco, M. and Cuello, A.C. (1978) The smooth endoplasmic reticulum as a possible storage site for dendritic dopamine in substantia nigra neurones. *Experientia*, 35: 101–103.

Michaelis, E.K. (1998) Molecular biology of glutamate receptors in the central nervous system and their role in excitotoxicity, oxidative stress and aging. *Prog. Neurobiol.*, 54: 369–415.

Millar, J. (1992) Extracellular single and multiple unit recording with microelectrodes. In: J.A. Stamford (Ed.), *Monitoring Neuronal Activity*, Oxford Univ. Press, pp. 1–27.

Millar, J., Stamford, J.A., Kruk, Z.L. and Wightman, R.M. (1985) Electrochemical, pharmacological and electrophysiological evidence of rapid dopamine release and removal in rat caudate nucleus following electrical stimulation of the median forebrain bundle. *Eur. J Pharmacol.*, 109: 341–348.

Misgeld, U., Bijak, M. and Jarolimek, W.A. (1995) A physiological role for $GABA_B$ receptors and the effects of baclofen in the mammalian central nervous system. *Prog. Neurobiol.*, 46: 423–462.

Mora-Ferrer, C., Yazulla, S., Studholme, K.M. and Haak-Frendscho, M. (1999) Dopamine D_1-receptor immunolocalization in goldfish retina. *J. Comp. Neurol.*, 411: 705–714.

Morell, M., Menning, T. and DiChiara, G. (1988) Nigral dopamine autoreceptors are exclusively of the D_2 type: quantitative autoradiography of [^{125}I]iodosulpiride and [^{125}I]SCH 23982 in adjacent brain sections. *Neuroscience*, 27: 865–870.

Muresan, Z. and Besharse, J.C. (1993) D_2-like dopamine receptors in amphibian retina: localization with fluorescent ligands. *J. Comp. Neurol.*, 331: 149–160.

Nicholson, C. (1995) Interaction between diffusion and Michaelis-Menten uptake of dopamine following iontophoresis in striatum. *Biophys. J.*, 68: 1699–1715.

Nicholson, C. and Phillips, J.M. (1981) Ion diffusion modified by tortuosity and volume fraction in the extracellular microenvironment of the rat cerebellum. *J. Physiol. (Lond.)*, 321: 225–257.

Nicholson, C. and Syková, E. (1998) Extracellular space structure revealed by diffusion analysis. *Trends Neurosci.*, 21: 207–215.

Nieoullon, A., Cheramy, A. and Glowinski, J. (1977) Release of DA in vivo from cat SN. *Nature*, 266: 375–377.

Nirenberg, M.J., Vaughan, R.A., Uhl, G.R., Kuhar, M.J. and Pickel, V.M. (1996a) The dopamine transporter is localized to dendritic and axonal plasma membranes of nigrostriatal dopaminergic neurons. *J. Neurosci.*, 16: 436–447.

Nirenberg, M.J., Chan, J., Liu, Y., Edwards, R.H. and Pickel, V.M. (1996b) Ultrastructural localization of the vesicular monoamine transporter–2 in midbrain dopaminergic neurons: potential sites for somatodendritic storage and release of dopamine. *J. Neurosci.*, 16: 4135–4145.

Parsons, L.H. and Justice, J.B. Jr. (1992) Extracellular concentration and in vivo recovery of dopamine in the nucleus accumbens using microdialysis. *J. Neurochem.*, 58: 212–218.

Pickel, V.M., Beckley, S.C., Joh, T.H. and Reis, D.J. (1981) Ultrastructural immunocytochemical localization of tyrosine hydroxylase in the neostriatum. *Brain Res.*, 225: 373–385.

Rice, M.E. and Nicholson, C. (1989) Measurement of nanomolar dopamine diffusion using low-noise perfluorinated ionomer coated carbon fiber microelectrodes and high-speed cyclic voltammetry. *Analyt. Chem.*, 61: 1805–1810.

Rice, M.E. and Nicholson, C. (1991) Diffusion characteristics and extracellular volume fraction during normoxia and hypoxia in slices of rat neostriatum. *J. Neurophysiol.*, 65: 264–272.

Rice, M.E., Nicholson, C. (1995) Diffusion and ion shifts in the brain extracellular microenvironment and their relevance for voltammetric measurements: the brain is not a beaker. In: A.A. Boulton, G.B. Baker and R.N. Adams (Eds), *Neuromethods*, Vol. 27, *Voltammetric Methods in Brain Systems*, Humana Press, Totowa, NJ, pp 27–79.

Rice, M.E., Cragg, S.J. and Greenfield, S.A. (1997) Characteristics of electrically-evoked somatodendritic dopamine release in substantia nigra and ventral tegmental area in vitro. *J. Neurophysiol.* 77: 853–862.

Rice, M.E., Richards, C.D., Nedergaard, S., Hounsgaard, J., Nicholson, C. and Greenfield, S.A. (1994) Direct monitoring of dopamine and 5-HT release in substantia nigra and ventral tegmental area in vitro. *Exp. Brain Res.*, 100: 395–406.

Robinson, S.E. and Caron, M.G. (1997) Interactions of dopamine receptors with G proteins. In: K.A. Neve and R.L. Neve (Eds), *The Dopamine Receptors*, Humana Press, Totowa, NJ, pp. 137–165.

Sanghera, M.K., Manaye, K.F., Liang, C-L., Iacopino, A.M., Bannon, M.J. and German, D.C. (1994) Low dopamine transporter mRNA levels in mid-brain regions containing calbindin. *NeuroReport*, 5: 1641–1644.

Santiago, M. and Westerink, B.H.C. (1992) Simultaneous recording of the release of nigral and striatal dopamine in the awake rat. *Neurochem. Int.*, 20: 107S–110S.

Sesack, S.R., Aoki, C. and Pickel, V.M. (1994) Ultrastructural localization of D_2 receptor-like immunoreactivity in midbrain dopamine neurons and their striatal targets. *J. Neurosci.*, 14: 88–106. .

Smith, A.D. and Bolam, J.P. (1990) The neural artwork of the basal ganglia as revealed by the study of synaptic connections of identified neurons. *Trends Neurosci.*, 13: 259–265.

Stamford, J.A., Kruk, Z.L. and Millar, J. (1986) Sub-second striatal dopamine release measured by in vivo voltammetry. *Brain Res.*, 381: 351–355.

Veruki, M.L. (1997) Dopaminergic neurons in the rat retina express dopamine D_2/D_3 receptors. *Eur.J. Neurosci.*, 9: 1096–1997.

Veruki, M.L. and Wässle, H. (1996) Immunohistochemical localization of dopamine D_1 receptors in rat retina. *Eur.J. Neurosci.*, 8: 2286–2297.

Wagner, H.-J., Luo, B.-G., Ariano, M.A., Sibley, D.R. and Stell, W.K. (1993) Localization of D_2 dopamine receptors in vertebrate retinae with anti-peptide antibodies. *J. Comp. Neurol.*, 331: 469–481.

Wassef, M., Berod, A. and Sotelo, C. (1981) Dopaminergic dendrites in the pars reticulata of the rat substantia nigra and their striatal input. Combined immunocytochemical localization of tyrosine hydroxylase and anterograde degeneration. *Neuroscience*, 6: 2125–2139.

Waszczak, B.L., Martin, L.P., Greif, G.J. and Freedman, J.E. (1998) Expression of a dopamine D_2 receptor-activated K^+ channel on identified striatopallidal and striatonigral neurons. *Proc. Natl Acad. Sci. USA*, 95: 11440–11444.

Wichmann, T. and DeLong, M.R. (1996) Functional and pathophysiological models of the basal ganglia. *Curr. Opin. Neurobiol.*, 6: 751–758.

Wightman, R.M. and Zimmerman, J.B. (1990) Control of dopamine extracellular concentration in rat striatum by impulse flow and uptake. *Brain Res. Revs.*, 15: 135–144.

Wightman, R.M., Amatore, C., Engstrom, R.C., Hale, P.D., Kristensen, E.W., Kuhr, W.G. and May, L.J. (1988) Real-time characterization of dopamine overflow and uptake in the rat striatum. *Neuroscience*, 25: 513–523.

Wightman, R.M., Jankowski, J.A., Kennedy, R.T., Kawagoe, K.T., Schroeder, T.J., Leszczyszyn, D.L., Near, J.A., Diliberto, E.J. Jr. and Viveros, O.H. (1991) Temporally resolved catecholamine spikes correspond to single vesicle release from individual chromaffin cells. *Proc. Natl Acad. Sci. USA*, 88: 10754–10758.

Wilson, C.J., Groves, P.M. and Fifková, E. (1977) Monoaminergic synapses, including dendro-dendritic synapses in the rat substantia nigra. *Exp. Brain Res.*, 30: 161–174.

Witkovsky, P. and Schütte, M. (1991) The organization of dopaminergic neurons in vertebrate retinas. *Vis. Neurosci.*, 7: 113–124.

Witkovsky, P., Nicholson, C., Rice, M.E., Bohmaker, K. and Meller, E. (1993) Extracellular dopamine concentration in the retina of the clawed frog, *Xenopus laevis*. *PNAS*, 90: 5667–5671.

Yung, K.K.L., Bolam, J.P., Smith, A.D., Hersch, S.M., Ciliax, B.J. and Levey, A.I. (1995) Immunocytochemical localization of D_1 and D_2 dopamine receptors in the basal ganglia of the rat: light and electron microscopy. *Neuroscience*, 65: 709–730.

L. F. Agnati, K. Fuxe, C. Nicholson and E. Syková (Eds.)
Progress in Brain Research, Vol 125

CHAPTER 16

Geometry and kinetics of dopaminergic transmission in the rat striatum and in mice lacking the dopamine transporter

F. Gonon[1,*], J.B. Buric[2], M. Jaber, M. Benoit-Marand[1], B. Dumartin[1] and B. Bloch[1]

[1] *CNRS UMR 5541, and* [2] *CNRS UMR 5466 UFR M.I. 2S, Université Victor Segalen Bordeaux 2, France*

Introduction

In the mid 80s Agnati et al. (1986) proposed that intercellular communication in the central nervous system can be grouped into two broad classes based on spatio-temporal characteristics: wiring transmission (WT) and volume transmission (VT) (for a recent review of this concept see Zoli et al., 1998). "WT was defined as a mode for intercellular communication which occurs via a relatively constrained cellular chain (wire), while VT was defined as the three-dimensional diffusion of a signal in the extracellular fluid volume for a distance larger than the synaptic cleft" (Zoli et al., 1998).

Dopaminergic transmission in the striatum (including the ventral part, i.e. the nucleus accumbens) has been extensively studied since it is involved in major pathologies (Parkinson, schizophrenia, drug addiction). During the first symposium devoted to 'Volume Transmission in the Brain' in 1989 we favored the view that dopaminergic transmission in the striatum is of the WT type. In fact, in the light of the data available at that time, we proposed that the direct dopaminergic transmission in the striatum occurs inside the synaptic cleft (Gonon et al., 1991). The main functional argument supporting this thesis was that the extrasynaptic extracellular dopamine concentration, as measured by microdialysis or with electrochemical techniques, was too low to stimulate dopaminergic receptors of the D1 type (Gonon and Buda, 1985; Gonon, 1988; Gonon et al., 1991). However, we also pointed out that some other aspects of dopaminergic transmission such as autoregulation of dopamine release via presynaptic D2 autoreceptors exhibit the characteristics of VT (Gonon et al., 1991; Suaud-Chagny et al., 1991).

Since this first symposium, several important observations have been reported concerning dopaminergic transmission in the striatum; all these data support the view that this transmission occurs outside the synaptic cleft. Therefore, according to the definition given by Zoli et al. (1998) dopaminergic transmission must be considered of the VT type. In this chapter these data are reviewed and the spatiotemporal characteristics of dopaminergic transmission in the striatum are discussed in the light of a comparison between normal animals and mice lacking the dopamine transporter.

Geometry of dopaminergic transmission in normal rodents

Localization of dopaminergic receptors

D1 receptors

In the striatum the most abundant dopaminergic receptors are D1 and D2 receptors (Missale et al.,

*Corresponding author. Tel.: 33 557 57 15 40;
Fax: 33 556 98 61 83;
e-mail: Francois.Gonon@umr5541.u-bordeaux2.fr

1998). The immunocytochemical localization of these receptors has been studied at the subcellular level by several distinct groups. Most striatal D1 receptors are located in dendrites and "D1 immunoreactive axon terminals are exceedingly rare" (Hersch et al., 1995). In the striatum most dopaminergic terminals form symmetrical synaptic contact on the neck of the dendritic spine (Pickel et al., 1981; Freund et al., 1984). The vast majority of the postsynaptic D1 receptors are not located in front of these dopaminergic synapses but are distributed along the dendritic membranes with a higher density in the peripheral zone of asymmetrical synapses formed by glutamatergic terminals on the head of dendritic spines (Fig. 1) (Levey et al., 1993; Yung et al., 1995; Hersh et al., 1995; Caillé et al., 1996).

D2 receptors

In contrast to D1 receptors, presynaptic D2 receptors have been identified in many axon terminals in the striatum. Most of the terminals labeled with D2 immunoreactivity have been identified as dopaminergic and these observations provide an anatomical basis regarding the role of presynaptic D2 receptors on autoregulation of dopamine release (Levey et al., 1993; Sesack et al., 1994; Yung et al., 1995; Hersh et al., 1995). The subcellular localization of postsynaptic D2 receptors on striatal dendrites has been less firmly established than that of D1 receptors. Nevertheless, similar to D1 receptors, most D2 receptors located on the internal surface of the dendritic membranes were not found in front of dopaminergic terminals but outside these synapses (Levey et al., 1993; Yung et al., 1995; Hersh et al., 1995). These observations strongly suggest that dopamine released at the level of the symmetric contacts formed by dopaminergic terminals must diffuse in the extracellular space to reach the postsynaptic striatal receptors (Levey et al., 1993; Yung et al., 1995; Hersh et al., 1995; Caillé et al., 1996; Pickel et al., 1996).

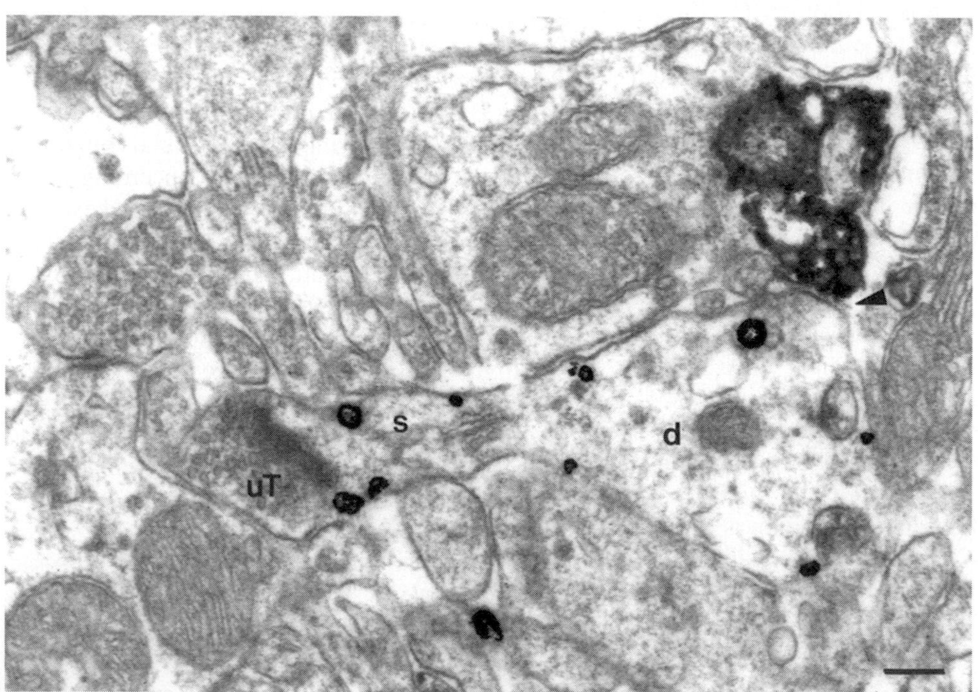

Fig. 1. Double detection of D1 receptors (by immunogold) and tyrosine hydroxylase immunoreactivity (by immunoperoxidase) at the electron microscopic level in the rat striatum. A D1 positive dendrite (d) gives rise to a D1 positive spine (s). A tyrosine hydroxylase positive profile is apposed to the dendritic shaft (arrow head). An unlabelled terminal (uT) forms an asymmetrical synapse on the head of the spine. Scale bar : 0.2 μm (from Caillé et al., 1996).

Extrasynaptic dopaminergic receptors are functional

Extrasynaptic D2 receptors are functional

According to in vivo estimates by microdialysis (Justice, 1993) or by electrochemical monitoring (Gonon and Buda, 1985), the basal extracelluar dopamine concentration in the rat striatum is in the range of 5 to 20 nM. Since D2 receptors are thought to exhibit a nanomolar affinity for dopamine (Kebabian and Calne, 1979; Creese, 1982; Missale et al., 1998), extrasynaptic D2 receptors can be stimulated by dopamine diffusing in the extracellular fluid. Indeed, striatopallidal neurons specifically express D2 receptors and the expression of immediate early genes (e.g. *c-fos*) is selectively enhanced in these neurons when the tonic influence of the basal extracellular level is blocked by D2 antagonists (Robertson et al., 1992) or by an acute decrease in the extracellular dopamine level (Cole and Difiglia, 1994; Svenningsson et al., 1999).

Extrasynaptic D1 receptors are functional

D1 receptors exhibit at least a 10 times lower affinity for dopamine than D2 receptors (Kebabian and Calne, 1979; Creese, 1982; Missale et al., 1998). Indeed, stimulation of adenylate cyclase activity by dopamine in the striatum is mediated by D1 receptors and requires a dopamine concentration of at least 0.3 μM (Kebabian et al., 1972; Kelly and Nahorski, 1987). This high level of extracellular dopamine concentration can be reached either after stimulation of dopamine release by amphetamine (Kuczenski et al., 1991) or, according to earlier electrochemical studies, by electrical stimulation of the dopaminergic pathway at frequencies which exceed the physiological discharge rate of dopaminergic neurons (Ewing et al., 1983; Gonon and Buda, 1985). Therefore, we proposed that the stimulation of extrasynaptic D1 receptors by dopamine in physiological conditions was unlikely (Gonon and Buda, 1985; Gonon et al., 1991).

More recent measurement of the rapid variations in the extracellular dopamine concentration by means of fast electrochemical techniques showed that even electrical stimulation with a single pulse evoked a transient increase in the extracellular dopamine concentration which reached 0.2 μM (Garris et al., 1994a; Dugast et al., 1994; Gonon, 1997). In the light of these recent data it appears now that extrasynaptic D1 receptors can actually be stimulated by dopamine in physiological conditions. Additional arguments supporting this view have been put forward by comparing the transient changes in the extracellular dopamine concentration and the postsynaptic electrophysiological responses mediated by D1 receptors (Gonon, 1997). We showed a positive correlation between the amplitude of the evoked dopamine overflow in the extracellular fluid and that of the postsynaptic response. Moreover, the postsynaptic response triggered by the evoked dopamine release was compared to that induced in the same neurons by the administration of a D1 agonist, SKF 82958. Since we were able to estimate the extracellular concentration of this drug, we compared its in vivo potency to that of dopamine and we found a potency ratio consistent with the much better affinity of this drug for D1 receptors (Andersen and Jansen, 1990). These data strongly suggest that extrasynaptic D1 receptors can be stimulated by the released dopamine in physiological conditions.

Diffusion of dopamine in the extracellular fluid

Dopamine can diffuse outside the synaptic cleft

The released dopamine is cleared from the extracellular fluid by reuptake via the dopamine transporter (DAT) (Graefe and Bönish, 1988). The DAT was initially thought to be localized at the level of dopaminergic synapses but recent studies demonstrated DAT labeling on the plasma membrane of dopaminergic terminal fibers near and distant from symmetrical contact formed by dopaminergic terminals (Nirenberg et al., 1996; Pickel et al., 1996). This localization suggests that the released dopamine can diffuse outside the synaptic cleft before clearance by DAT (Nirenberg et al., 1996; Pickel et al., 1996).

This view was further supported by functional studies showing that inhibition of DAT drastically slowed down the dopamine clearance but only moderately enhanced the maximal amplitude of the dopamine overflow evoked in the nucleus accumbens by a 'pseudo-one-pulse' stimulation (Garris et al., 1994a) or in the striatum by a single pulse stimulation (Gonon, 1997). These observations suggest that the released dopamine can invade the whole extracellular space to reach the electrode surface before major elimination.

Half-life of the released dopamine in the extracellular fluid

The maximal distance to which dopamine can diffuse from release sites is limited by the kinetics of the dopamine clearance by DAT. Fast electrochemical monitoring of the dopamine overflow evoked by brief stimulation provides an estimate of the dopamine half-life in the striatal extracellular space. This half-life was found to be in the range of 60 to 100 ms (Garris and Wightman, 1994; Dugast et al., 1994; Suaud-Chagny et al., 1995; Gonon, 1997). However, as discussed in more detail in Fig. 2, these experimental values represent an overestimate of the dopamine half-life. In fact, we hypothesize that in the living tissue the kinetics of the dopamine overflow are faster than those observed by carbon fiber electrodes. This is not due to the time resolution of the electrochemical technique per se (carbon fiber electrode coupled to continuous amperometry) since this time resolution is below 1 ms as exemplified by the electrochemical monitoring of excytotic events (Chow et al., 1996). Our hypothesis is based on the assumption that there is a diffusion zone between the living tissue and the electrode surface (Fig. 2). Diffusion of dopamine through this zone delays and slows down the kinetics observed by the electrode. We attempted to take into account this distortion and to calculate the kinetics of the evoked dopamine overflow in the living tissue by deconvolution following a procedure already used by Kawagoe et al. (1992). According to this calculation the dopamine half-life might be in the range of 20–30 ms in the dorsal striatum (Fig. 2).

Another approach to estimate the dopamine half-life has been proposed by Garris et al. (1994a): assuming that dopamine clearance by DAT can be described by Michaelis–Menten's kinetics, in vivo estimate of the corresponding parameters (Km and Vm) can be used to calculate the dopamine half-life. These authors proposed a value of 37 ms for the dopamine half-life in the nucleus accumbens.

Fig. 2. Experimental and calculated dopamine overflows evoked in the striatum of one mice before (***c***. and ***d***.) and 20 min after (***b.***) inhibition of dopamine uptake by nomifensine (20 mg/kg, s.c.). The release of dopamine was evoked by electrical stimulation of the medial forebrain bundle and the resulting dopamine overflow was continuously monitored by amperometry at +0.4 V with a carbon fiber electrode (Dugast et al., 1994; Benoit-Marand et al., 2000). Rough experimental recordings are shown as dotted lines. They show a delay of 50 to 70 ms between the stimulation and the maximum of the dopamine overflow. This delay cannot be accounted for by a diffusion delay inside the living striatal tissue since, due to the high density of dopaminergic terminals, the released dopamine can invade the whole extracellular fluid within 3 ms (see text). As illustrated in ***a***. we hypothesize that the delay we observed on experimental curves is due to a diffusion process in a diffusion zone between the living tissue and the carbon fiber electrode. The temporal distortion caused by this diffusion can be removed by deconvolution if the response function of this diffusion process to a concentration step is known (Press et al., 1991). We assume that the overflow evoked by a single pulse in the presence of uptake inhibition represents a suitable experimental response to a concentration step. In ***b***. this experimental curve was fitted with a function C (t, τ) describing the concentration at the electrode surface after diffusion through a film (thickness: l) in response to a concentration step according to equation 4.17 in Crank (1992):

$$C(t,\tau)/C_{max} = 1 - (4/\pi)\sum_{n=0}^{\infty}[(-1)^n/(2n+1)] \times exp[-(2n+1)^2 t/\tau]$$

with $\tau = 4l^2/D\pi^2$ (D is the diffusion coefficient for dopamine)

The best fit of the experimental curve was obtained for $\tau = 43$ ms and $C_{max} = 230$ nM. In ***c***. and ***d***. the experimental curve was smoothed (continuous line) using cubic spline interpolation. The calculated dopamine overflow in the living tissue corresponds to the part of the curve for $t > 0$ which was obtained by deconvolution of the smoothed data using the derivative of the function C (t, τ) normalized to 1 and with $\tau = 43$ ms. Notice that the maximum of the calculated overflow evoked by a single pulse is close to that of the experimental overflow measured after uptake inhibition.

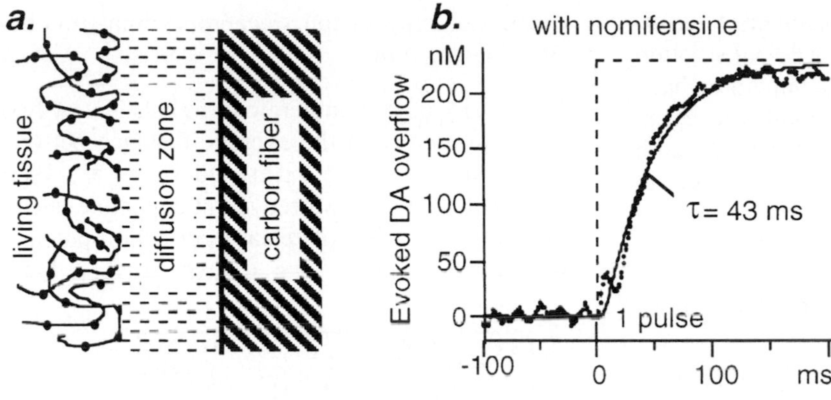
a.
living tissue
diffusion zone
carbon fiber
b.
with nomifensine
nM
Evoked DA overflow
200
150
100
50
0
τ= 43 ms
1 pulse
-100
0
100
ms

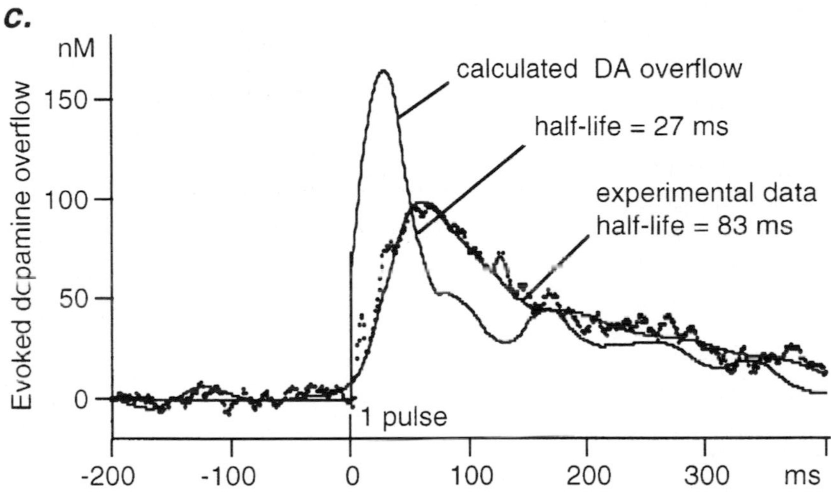
c.
nM
Evoked dopamine overflow
150
100
50
0
calculated DA overflow
half-life = 27 ms
experimental data
half-life = 83 ms
1 pulse
-200
-100
0
100
200
300
ms

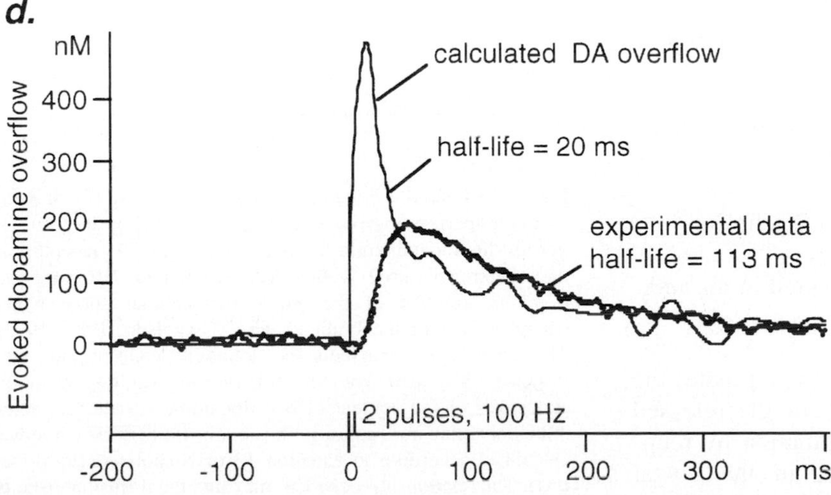
d.
nM
Evoked dopamine overflow
400
300
200
100
0
calculated DA overflow
half-life = 20 ms
experimental data
half-life = 113 ms
2 pulses, 100 Hz
-200
-100
0
100
200
300
ms

Since the density of DAT per dopaminergic terminals is about two times higher in the dorsal striatum than in the nucleus accumbens, the dopamine half-life is briefer in the former than in the latter region (Suaud-Chagny et al., 1995). Therefore, both estimates of the real dopamine half-life are in excellent agreement.

Diffusion distance of dopamine from release sites

The diffusion of dopamine in the extracellular fluid has been carefully studied (Nicholson, 1985, 1995; Rice and Nicholson, 1991). The case of dopamine diffusion in the extracellular fluid following its quantal release by exocytosis of a single vesicle in a synaptic cleft, can be considered to be equivalent to the diffusion resulting from the instantaneous deposition at time zero of a very small amount of dopamine in a point. Nicholson (1985) proposed a simple solution to this problem: the dopamine concentration (C) at a distance r from the release site varies with time (t) according to the following relationship.

$C = K\, t^{-3/2} \exp(-r^2\lambda^2/4Dt)$ (from equation 13 in Nicholson, (1985))
K is a constant term (see Nicholson, 1985)
λ is the tortuosity coefficient ($\lambda = 1.54$ in the striatum, Nicholson, (1995))
D is the diffusion coefficient for dopamine ($D = 7.6 \times 10^{-6}$ cm^2/sec, value from Nicholson (1995) at a temperature of 37°C)

As illustrated in Fig. 3, this concentration reaches a maximum at a time t_{max} given by:

$t_{max} = (r^2/6)(\lambda^2/D)$ (equation 14 in Nicholson (1985))
$t_{max} = 0.52 \times r^2$ when t_{max} is expressed in ms and r in μm

This relationship can be used to estimate the distance r at which at least 50% of the released dopamine can diffuse before elimination by reuptake. With a dopamine half-life in the dorsal striatum of 25 ms this distance is about 7 μm (Fig. 3).

Kinetics of extrasynaptic receptors stimulation by diffusing dopamine

The density of dopaminergic terminals is very high in the striatum and the average distance between release sites is about 4 μm (Doucet et al., 1986; Groves et al., 1994). Moreover, at dopamine release sites the probability of release per action potential seems very high in vivo (Garris et al., 1994a).

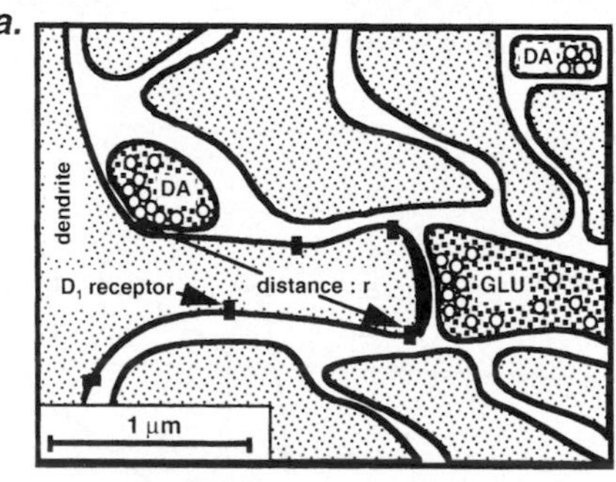

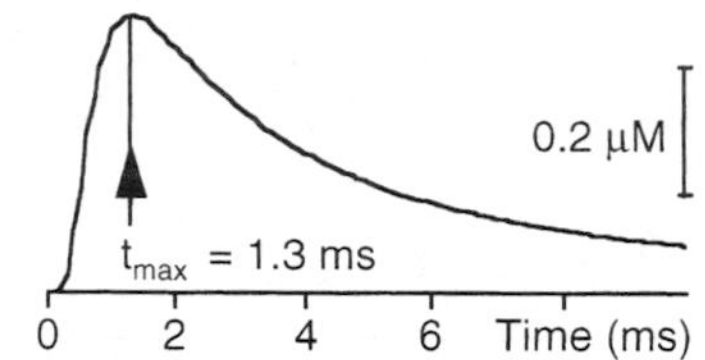

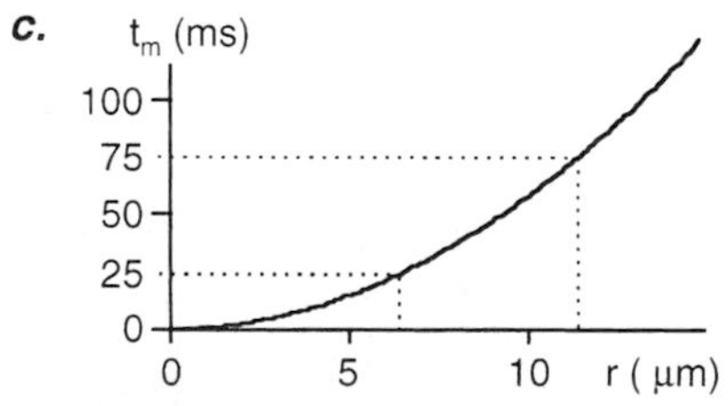

Fig. 3. Diffusion of dopamine from release sites. The drawing in ***a***. summarizes morphological data concerning the localization of the dopaminergic terminals and of the D1 receptors in the striatum (see text). Notice that the synaptic cleft represents a minute fraction of the whole extracellular space which occupies 21% of the brain volume (Nicholson, 1985, 1995). The curve in ***b***. represents the dopamine concentration at a distance r = 1.5 μm from the release site resulting from the exocytosis of the content of one dopamine vesicle (i.e. about 3000 dopamine molecules, Pothos et al., 1998). This curve was calculated according to equation 13 of Nicholson (1985) (see text). The relationship between tmax and the diffusion distance r is shown in ***c***. and has been calculated according to equation 14 of Nicholson (1985) (see text).

Therefore, if dopamine is released simultaneously from all release sites, the whole extracellular fluid is invaded by dopamine within less than 3 ms (as estimated from Equation (14) of Nicholson (1985), see above). The kinetics of dopamine diffusion outside the synaptic cleft has been studied in more details by Garris et al. (1994a) and they also found that, 3 ms after release, the dopamine concentration can be considered as uniform throughout the extracellular space. These considerations have two important consequences. First, dopamine diffusion up to a locally uniform extracellular dopamine concentration is much faster than the elimination of dopamine by reuptake. Second, the delay due to diffusion process before complete stimulation of extrasynaptic dopaminergic receptors is short (below 3 ms) and the duration of this stimulation is regulated by reuptake.

Conclusions

(1) The striatal dopaminergic transmission is extrasynaptic but local

In the striatum dopaminergic transmission undoubtedly occurs outside synaptic clefts; the released dopamine must diffuse in the extracellular space to reach its target but this diffusion process is fast (<3 ms). Since dopamine is rapidly cleared by reuptake, the maximal diffusion distance is in the range of 7 to 10 μm. Therefore, regional variations regarding dopaminergic innervation (density, probability of release per action potential) would result in heterogeneous dopamine overflows with well defined boundaries. Such regional variations have been actually reported (Garris et al., 1994b). In conclusion, the striatal dopaminergic transmission is extrasynaptic but local.

(2) Dopaminergic transmission in the striatum must be classified as a VT

All recent data discussed above inescapably lead to the conclusion that, contrary to previous proposals (Gonon and Buda, 1985; Gonon et al., 1991; Agnati et al., 1995), dopaminergic transmission in the striatum of normal rodents must be classified as VT. In fact, according to Agnati and Fuxe's definition (Agnati et al., 1995; Zoli et al., 1998), dopamine fulfills all the criteria of a substance to be recognized as a VT signal: (i) it is released by a cell in a regulated fashion; (ii) it diffuses at relevant concentration in the extracellular fluid for a distance larger than the synaptic cleft; (iii) it is able to activate selective receptors in a number of target cells; (iv) it triggers physiological responses in target cells.

Kinetics of dopaminergic transmission

Discharge activity of dopaminergic neurons

In rats and mice, dopaminergic neuronal cell bodies exhibit two kinds of discharge activity: single spikes and bursts of two to six action potentials (Grace and Bunney, 1984; Sanghera et al., 1984). Individual dopaminergic neurons can switch from one pattern to another and sensory stimuli favor the bursting pattern in unrestrained rats (Freeman and Bunney, 1987). In monkeys, dopaminergic neurons respond to an appetitive stimulus by a burst (Mirenowicz and Schultz, 1996). Grace and Bunney (1984) hypothesized that "bursting may be as important as firing frequency in affecting DA release" and this hypothesis was supported by our early studies (Gonon and Buda, 1985; Gonon, 1988). More recently, we showed however that the main mechanism subserving the high extracellular DA level evoked by stimulations mimicking the bursting activity is not facilitation of the DA release per se, but accumulation of the released DA as a result of overcoming of DA uptake (Chergui et al., 1994a).

The single spike pattern of activity is due to the intrinsic properties of dopaminergic neurons since it is observed in slices (Sanghera et al., 1984; Grace and Onn, 1989). In contrast, the bursting activity, which is not observed in vitro (Grace and Onn, 1989), is due to excitatory afferents and is mediated by NMDA receptors (Chergui et al., 1993, 1994b). In anesthetized animals the incidence of bursting is low in the vast majority of dopaminergic neurons (Grace and Bunney, 1984; Sanghera et al., 1984). About 80% of the basal extracellular dopamine concentration is due to the release of dopamine triggered by the tonic activity of dopaminergic neurons (Gonon and Buda, 1985; Gonon, 1988).

Transmission mediated by D1 receptors

In anesthetized rats, when dopaminergic neurons of the ventral tegmental area (VTA) are forced to fire in bursts by local injection of NMDA, the extracellular dopamine concentration in the nucleus accumbens is enhanced by almost one order of magnitude (i.e. up to 150 nM) (Suaud–Chagny et al., 1992). This enhanced dopamine concentration induces a prominent increase in the discharge activity of a subpopulation of target neurons and this excitation is mediated by D1 receptors (Gonon and Sundstrom, 1996). When the dopamine release is evoked by an electrical stimulation mimicking one burst (i.e. 4 pulses at 15 Hz, whole duration: 200 ms), an excitatory response can be recorded in a subpopulation of striatal neurons (Gonon, 1997). This response is mediated by D1 receptors, lasts for 0.5 s and is delayed as compared to the evoked dopamine overflow: it starts 0.3 s after the beginning of the stimulation train, i.e. when the released dopamine has been almost completely eliminated by reuptake (Gonon, 1997). This delayed excitatory response fully develops when D1 receptors are stimulated by dopamine for a sufficient duration (i.e. by a burst) since single pulse stimulations are much less effective (Gonon, 1997). All the characteristics of this delayed excitatory response are very similar to those observed in vitro concerning effects mediated by receptors which are coupled with G-proteins (Cole and Nicoll, 1984; Surprenant and Williams, 1987; Isaacson et al., 1993; Batchelor et al., 1994).

These data and other studies (Chergui et al., 1996, 1997) strongly suggest that stimulation of D1 receptors by the dopamine release which is evoked by bursts, facilitates the discharge activity of target neurons. The time course of this postsynaptic response is delayed and prolonged as compared to that of the presynaptic signal (i.e. the dopamine overflow) and these kinetics are governed by that of the intracellular G-protein-mediated messenger systems in the target neurons. The diffusion delay which is due to the extrasynaptic localization of D1 receptors plays a negligible role in the whole time course of dopaminergic transmission.

Transmission mediated by D2 receptors

Several studies showed that the expression of early genes (e.g. *c-fos*) in striatopallidal neurons is tonically inhibited, via D2 receptors, by the basal extracellular dopamine level (Robertson et al., 1992; Cole and Difiglia, 1994; Svenningsson et al., 1999). This view is in line with the tonic D2-mediated attenuation of cortical excitation in nucleus accumbens reported by O'Donnell and Grace (1994). Transient inhibition of the tonic discharge activity of dopaminergic neurons induces a parallel decrease in the extracellular dopamine concentration (Suaud–Chagny et al., 1992). Unfortunately, the postsynaptic response to transient changes in the extracellular dopamine concentration mediated by D2 receptors in striatopallidal neurons has not yet been investigated. However, such phasic responses are likely to occur. In fact, the tonic activity of dopaminergic neurons is interrupted by brief (200 ms) silent periods which seem to code errors in the temporal prediction of reward (Hollerman and Schultz, 1998). These silent periods might represent the relevant phasic signal as regards the transmission mediated by D2 receptors.

Conclusion: the temporal properties of the dopamine signal are those of a WT signal

The classification of intercellular communication in the brain either as WT or VT has been associated with differential properties. The WT was associated with a brief transmission delay (ms), a time scale in the range of ms to s and a phasic biological effect whereas the VT was associated with a prolonged delay (s to min.), a time scale in the range of s to min. and with a tonic biological effect (Zoli et al., 1998). The data discussed above clearly show that dopaminergic transmission in the striatum exhibits properties associated with WT rather than with VT. Indeed, the fact that dopamine is a VT signal does not imply a prolonged delay: the delay due to dopamine diffusion outside the synaptic cleft is below 3 ms. In other words the onset of the dopamine signal is below 3 ms and its offset (i.e. dopamine clearance by reuptake) occurs within less than 50 ms. Considering together both pre- and postsynaptic aspects of dopaminergic transmission the whole time scale is still below one second and it must be emphasized again that this duration is governed by the kinetics of the intracellular G-protein-mediated messenger system in target neurons. Thus, the spatial characteristics, which lead to the classification of dopamine as a VT signal, play a negligible role in the time scale of dopaminergic transmission.

Last, but not least, the dopaminergic transmission mediated by D1 receptors is solely activated by brief bursts. Although the postsynaptic response is delayed and prolonged, it is closely timely locked to the presynaptic signal. This exclusively phasic transmission might play a major role in learning since it is involved in the coding of salient stimuli. These characteristics might be also those of the transmission mediated by D2 receptors. In conclusion, the temporal properties of dopaminergic transmission in the striatum are those which have been associated with a WT by Zoli et al. (1998).

Mice lacking the dopamine transporter

Spatiotemporal characteristics of dopaminergic transmission

Mice lacking the DAT have been generated by Giros et al. (1996). In these mice the extracellular dopamine level measured in vivo by microdialysis is five times higher than in normal mice despite a whole tissue content of dopamine reduced to 5% of the normal level (Jones et al., 1998). In vitro and in vivo electrochemical studies have shown that the dopamine half-life was prolonged up to the range of several seconds (Giros et al., 1996; Jones et al., 1998; Benoit–Marand et al., 2000) and that the amplitude of the DA release per pulse was reduced to 7% in the striatum (Benoit–Marand et al., 2000). From this prolonged half-life it can be inferred that the maximal diffusion distance in the striatum of DAT –/– mice is about 100 μm. Therefore, the topographical organization of dopaminergic transmission is preserved at the level of the cerebral region (Benoit–Marand et al., 2000).

Functional consequences

The bursting activity of dopaminergic neurons is specifically involved in positively reinforced learning (Mirenowicz and Schultz, 1996). The DAT plays a crucial role in favoring the expression of a burst over the tonic activity. In fact, the dopamine release is completely cleared between pulses at frequency mimicking the tonic activity (4 Hz) whereas it accumulates at burst frequency (15 Hz) (Chergui et al., 1994a). In DAT –/– mice the released dopamine accumulates in the extracellular fluid even at low frequencies (Benoit–Marand et al., 2000). Therefore, despite a very low release per pulse, the basal dopamine level sustained by the tonic activity is five times higher than in normal mice (Jones et al., 1998) but the dopamine overflow evoked by a burst is hardly detectable above this high basal level (Benoit–Marand et al., 2000). Thus, DAT –/– mice are not able to express the bursting activity in terms of phasic changes in extracellular DA and this deficit might be involved in the difficulties these mice experience in spatial learning (Gainetdinov, 1999).

Conclusion: in DAT –/– mice dopaminergic transmission is a pure VT

In mice lacking DAT, the striatal dopamine signal exhibits all the spatial and temporal characteristics associated with a VT signal. In particular, the tonic transmission due to the basal extracellular dopamine level is exacerbated and this is in line with the hyperactive phenotype of these mice, but the phasic dopaminergic transmission triggered by bursts cannot be expressed. The disruption of this phasic function might be involved in the learning deficit of these mice. The comparison between normal and DAT –/– mice underlines the functional importance of the temporal characteristics of dopaminergic transmission in the striatum and the major role of the dopamine transporter in the control of these temporal characteristics. Interestingly, the amplitude and time course of the dopamine overflow in the striatum of mice lacking the dopamine transporter are close to those of the dopamine overflow evoked in the medial prefrontal cortex and in the basal amygdaloid nucleus of normal rats (Garris and Wightman, 1994). These authors pointed out that in both latter regions “extracellular dopamine has a greater possibility of extrasynaptic sites of action” than in the striatum. In other words the dopamine signal might be considered as a pure VT signal in these two less densely innervated regions.

General conclusion

In summary dopaminergic transmission in the striatum must be classified as VT, according to Agnati and Fuxe's definition, but exhibits temporal properties of a WT. This is certainly not a unique case; the same classification is also valid regarding the sympathetic transmission mediated by noradrenaline (Gonon et al., 1993; Stjärne et al., 1994). In more general terms it is likely that many neurotransmissions mediated by G-protein-coupled receptors exhibit spatiotemporal characteristics similar to both aforementioned types of transmission (Hille, 1992). In our opinion, the classification of this type of neurotransmission as a VT might have negative consequences since it might be associated with an inaccurate view of its temporal characteristics.

Therefore, we feel that the VT concept should be reconsidered according to one of the three following suggestions. First, in the framework of the present VT definition it must be recognized that many examples of VT are fast and can exert phasic functions. Second, the definition of VT vs. WT might be modified to include perisynaptic neurotransmission inside the WT class. In fact, in our opinion, the characteristics of dopaminergic transmission in the striatum (and similar types of neurotransmission) fit well with the definition given for a WT: "a mode for intercellular communication which occurs via a relatively constrained cellular chain" (Zoli et al., 1998). Or third, intercellular communication in the brain might be classified in three rather than only two classes by considering an intermediate class termed perisynaptic WT.

Acknowledgments

This work was supported by CNRS, the Université Victor Segalen Bordeaux 2 and La Région Aquitaine.

References

Agnati, L.F., Fuxe, K., Zoli, M., Zini, I., Toffano, G. and Ferraguti, F. (1986) A correlation analysis of the regional distribution of central enkephalin and beta-endorphin immunoreactive terminals and opiate receptors in adult and old male rats. Evidence for the existence of two main types of communication in the central nervous system: the volume transmission and the wiring transmission. *Acta Physiol. Scand.*, 128: 201–207.

Agnati, L.F., Zoli, M., Srtömberg, I. and Fuxe, K. (1995) Intercellular communication in the brain: wiring versus volume transmission. *Neuroscience*, 69: 711–726.

Andersen, P.H. and Jansen, J.A. (1990) Dopamine receptor agonists: selectivity and dopamine D1 receptor efficacy. *Eur. J. Pharmacol.*, 188: 335–347.

Batchelor, A.M., Madge, D.J. and Garthwaite, J. (1994) Synaptic activation of metabotropic glutamate receptors in the parallel fibre-Purkinje cell pathway in rat cerebellar slices. *Neuroscience*, 63: 911–915.

Benoit–Marand, M., Jaber, M. and Gonon, F. (2000) Release and elimination of dopamine in vivo in mice lacking the dopamine transporter: functional consequences. *Eur. J. Neurosci.*, in press.

Caillé, I., Dumartin, B. and Bloch, B. (1996) Ultrastructural localization of D1 dopamine receptor immunoreactivity in the rat striatonigral neurons and its relation with dopaminergic innervation. *Brain Res.*, 730: 17–31.

Chergui, K., Charlety, P.J., Akaoka, H., Saunier, C.F., Brunet, J.L., Buda, M., Svensson, T.H. and Chouvet, G. (1993) Tonic activation of NMDA receptors causes spontaneous burst discharge of rat midbrain dopamine neurons in vivo. *Eur. J. Neurosci.*, 5: 137–144.

Chergui, K., Suaud–Chagny, M.F. and Gonon, F. (1994a) Nonlinear relationship between impulse flow, dopamine release and dopamine elimination in the rat brain in vivo. *Neuroscience*, 62: 641–645.

Chergui, K., Akaoka, H., Charléty, P.J., Saunier, C.F., Buda, M. and Chouvet, G. (1994b) Subthalamic nucleus modulates burst firing of nigral dopamine neurones via NMDA receptors. *NeuroReport*, 5: 1185–1188.

Chergui, K., Nomikos, G.G., Mathé, J.M., Gonon, F. and Svensson, T.H. (1996) Burst stimulation of the medial forebrain bundle selectively increases *Fos*-like immunoreactivity in the limbic forebrain of the rat. *Neuroscience*, 72: 141–156.

Chergui, K., Svenningsson, P., Nomikos, G., Gonon, F., Fredholm, B.B. and Svensson, T.H. (1997) Increased expression of NGFI-A mRNA in the rat striatum following burst stimulation of the medial forebrain bundle. *Eur. J. Neurosci.*, 9: 2370–2382.

Chow, R.H., Klingauf, J., Heinemann, C., Zucker, R.S. and Neher, E. (1996) Mechanisms determining the time course of secretion in neuroendocrine cells. *Neuron*, 16: 369–376.

Cole, A.E. and Nicoll, R.A. (1984) Characterization of a slow cholinergic post-synaptic potential recorded in vitro from rat hippocampal pyramidal cells. *J. Physiol.*, 352: 173–188.

Cole, D.G. and Difiglia, M. (1994) Reserpine increases *Fos* activity in the rat basal ganglia via a quinpirole-sensitive mechanism. *Neuroscience*, 60: 115–123.

Crank, J. (1992) *The mathematics of diffusion*. Oxford University Press, London.

Creese, I. (1982) Dopamine receptors explained. *Trends Neurosci.*, 5: 40–43.

Doucet, G., Descarries, L. and Garcia, S. (1986) Quantification of the dopamine innervation in adult rat neostriatum. *Neuroscience*, 19: 427–445.

Dugast, C., Suaud–Chagny, M.F. and Gonon, F. (1994) Continuous in vivo monitoring of evoked dopamine release in the rat nucleus accumbens by amperometry. *Neuroscience*, 62: 647–654.

Ewing, A.G., Bigelow, J.C. and Wightman, R.M. (1983) Direct in vivo monitoring of dopamine release from two striatal compartments in the rat. *Science*, 221: 169–171.

Freeman, A.S. and Bunney, B.S. (1987) Activity of A9 and A10 dopaminergic neurons in unrestrained rats: further characterization and effects of apomorphine and cholecystokinin. *Brain Res.*, 405: 46–55.

Freund, T.F., Powell, J.F. and Smith, A.D. (1984) Tyrosine hydroxylase-immunoreactive boutons in synaptic contact with identified striatonigral neurons, with particular reference to dendritic spines. *Neuroscience*, 13: 1189–1215.

Gainetdinov, R.R., Wetsel, W.C., Jones, S.R., Levin, E.D., Jaber, M. and Caron, M.G. (1999) Role of serotonin in the paradoxical calming effect of psychostimulants on hyperactivity. *Science*, 283: 397–401.

Garris, P.A. and Wightman, R.M. (1994) Different kinetics govern dopaminergic transmission in the amygdala, prefrontal cortex, and striatum: an in vivo voltammetric study. *J. Neurosci.*, 14: 442–450.

Garris, P.A., Ciolkowski, E.L., Pastore, P. and Wightman, R.M. (1994a) Efflux of dopamine from the synaptic cleft in the nucleus accumbens of the rat brain. *J. Neurosci.*, 14: 6084–6093.

Garris, P.A., Ciolkowski, E.L. and Wightman, R.M. (1994b) Heterogeneity of evoked dopamine overflow within the striatal and striatoamygdaloid regions. *Neuroscience*, 59: 417–427.

Giros, B., Jaber, M., Jones, S.R., Wightman, R.M. and Caron, M.G. (1996) Hyperlocomotion and indifference to cocaine and amphetamine in mice lacking the dopamine transporter. *Nature*, 379: 606–612.

Gonon, F. and Buda, M. (1985) Regulation of dopamine release by impulse flow and by autoreceptors as studied by in vivo voltammetry in the rat striatum. *Neuroscience*, 14: 765–774.

Gonon, F. (1988) Nonlinear relationship between impulse flow and dopamine release by rat midbrain dopaminergic neurons as studied by in vivo electrochemistry. *Neuroscience*, 24: 19–28.

Gonon, F., Suaud–Chagny, M.F., Mermet, C. and Buda,, M. (1991) Relation between impulse flow and extracellular catecholamine levels as studied by in vivo electrochemistry in CNS. In: K. Fuxe and L.F. Agnati (Eds), *Volume Transmission in the Brain: Novel Mechanism for Neural Transmission*. Raven Press, New York, pp. 337–350.

Gonon, F., Msghina, M. and Stjärne,, L. (1993) Kinetics of noradrenaline released by sympathetic nerves. *Neuroscience*, 56: 535–538.

Gonon, F. and Sundström, L. (1996) Excitatory effects of dopamine released by impulse flow in the rat nucleus accumbens in vivo. *Neuroscience*, 75: 13–18.

Gonon,, F. (1997) Prolonged and extrasynaptic excitatory action of dopamine mediated by D1 receptors in the rat striatum in vivo. *J. Neurosci.*, 17: 5972–5978.

Grace, A.A. and Bunney, B.S. (1984) The control of firing pattern in nigral dopamine neurons: burst firing. *J. Neurosci.*, 4: 2877–2890.

Grace, A.A. and Onn, S.P. (1989) Morphology and electrophysiological properties of immunocytochemically identified rat dopamine neurons recorded in vitro. *J. Neurosci.*, 9: 3463–3481.

Graefe, K.H. and Bönish, H. (1988) The transport of amines across the axonal membranes of noradrenergic and dopaminergic neurons. In: U. Trendelenburg and N. Weiner (Eds), *Handbook of Experimental Pharmacology*, Vol. 90, Springer Verlag, Berlin, pp. 193–235.

Groves, P.M., Linder, J.C. and Young,, S.J. (1994) 5-Hydroxydopamine-labeled dopaminergic axons – three-dimensional reconstructions of axons, synapses and postsynaptic targets in rat neostriatum. *Neuroscience*, 58: 593–604.

Hersch, S.M., Ciliax, B.J., Gutekunst, C.A., Rees, H.D., Heilman, C.J., Yung, K.K., Bolam, J.P., Ince, E., Yi, H. and Levey, A.I. (1995) Electron microscopic analysis of D1 and D2 dopamine receptor proteins in the dorsal striatum and their synaptic relationships with motor corticostriatal afferents. *J. Neurosci.*, 15: 5222–5237.

Hille, B. (1992) G-protein-coupled mechanisms and nervous signaling. *Neuron*, 9: 187–195.

Hollerman, J.R. and Schultz, W. (1998) Dopamine neurons report an error in the temporal prediction of reward during learning. *Nat. Neurosci.*, 1: 304–309.

Isaacson, J.S., Solis, J.M. and Nicoll, R.A. (1993) Local and diffuse synaptic actions of GABA in the hippocampus. *Neuron*, 10: 165–175.

Jones, S.R., Gainetdinov, R.R., Jaber, M., Giros, B., Wightman, R.M. and Caron, M.G. (1998) Profound neuronal plasticity in response to inactivation of the dopamine transporter. *Proc. Natl Acad. Sci. USA*, 95: 4029–4034.

Justice, J.B. (1993) Quantitative microdialysis of neurotransmitters. *J. Neurosci. Meth.*, 48: 263–276.

Kawagoe, K.T., Garris, P.A., Wiedemann, D.J. and Wightman, R.M. (1992) Regulation of transient dopamine concentration gradients in the microenvironment surrounding nerve terminals in the rat striatum. *Neuroscience*, 51: 55–64.

Kebabian, J.W., Petzold, G.L. and Greengard, P. (1972) Dopamine-sensitive adenylate cyclase in caudate nucleus of rat brain and its similarity to the 'dopamine receptor'. *Proc. Natl Acad. Sci. USA*, 69: 2145–2149.

Kebabian, J.W. and Calne, D.B. (1979) Multiple receptors for dopamine. *Nature*, 277: 93–96.

Kelly, E. and Nahorski, S.R. (1987) Endogenous dopamine functionally activates D-1 and D-2 receptors in striatum. *J. Neurochem.*, 49: 115–120.

Kuczenski, R., Segal, D.S. and Aizenstein, M.L. (1991) Amphetamine, cocaine, and fencamfamine – relationship between locomotor and stereotypy response profiles and caudate and accumbens-dopamine dynamics. *J. Neurosci.*, 11: 2703–2712.

Levey, A.I., Hersch, S.M., Rye, D.B., Sunahara, R.K., Niznik, H.B., Kitt, C.A., Price, D.L., Maggio, R., Brann, M.R. and Ciliax, B.J. (1993) Localization of D1 and D2 dopamine receptors in brain with subtype-specific antibodies. *Proc. Natl Acad. Sci. USA*, 90: 8861–8865.

Mirenowicz, J. and Schultz, W. (1996) Preferential activation of midbrain dopamine neurons by appetitive rather than aversive stimuli. *Nature*, 379: 449–451.

Missale, C., Nash, S.R., Robinson, S.W., Jaber, M. and Caron, M.G. (1998) Dopamine receptors: from structure to function. *Physiolog. Rev.*, 78: 189–225.

Nicholson, C. (1985) Diffusion from an injected volume of a substance in brain tissue with arbitrary volume fraction and tortuosity. *Brain Res.*, 333: 325–329.

Nicholson, C. (1995) Interaction between diffusion and Michaelis–Menten uptake of dopamine after iontophoresis in striatum. *Biophys. J.*, 68: 1699–1715.

Nirenberg, M.J., Vaughan, R.A., Uhl, G.R., Kuha,r M.J. and Pickel, V.M. (1996) The dopamine transporter is localized to dendritic and axonal plasma membranes of nigrostriatal dopaminergic neurons. *J. Neurosci.*, 16: 436–447.

O'Donnell, P. and Grace, A.A. (1994) Tonic D2-mediated attenuation of cortical excitation in nucleus accumbens neurons recorded in vitro. *Brain Res.*, 634: 105–112.

Pickel, V.M., Beckley, S.C., Joh, T.H. and Reis, D.J. (1981) Ultrastructural immunocytochemical localization of tyrosine hydroxylase in the neostriatum. *Brain Res.*, 225: 373–385.

Pickel, V.M., Nirenberg, M.J. and Milner, T.A. (1996) Ultrastructural view of central catecholaminergic transmission: immunocytochemical localization of synthesizing enzymes, transporters and receptors. *J. Neurocytology*, 25: 843–856.

Pothos, E.N., Davila, V. and Sulzer, D. (1998) Presynaptic recording of quanta from midbrain dopamine neurons and modulation of the quantal size. *J. Neurosci.*, 18: 4106–4118.

Press, W.H., Flannery, B.P., Teukolsky, S.A. and Vetterling, W.T. (1991) *Numerical recipes*. Cambridge University Press, London.

Rice, M.E. and Nicholson, C. (1991) Diffusion characteristics and extracellular volume fraction during normoxia and hypoxia in slices of rat neostriatum. *J. Neurophysiol.*, 65: 264–272.

Robertson, G.S., Vincent, S.R. and Fibiger, H.C. (1992) D1 and D2 dopamine receptors differentially regulate *c-fos* expression in striatonigral and striatopallidal neurons. *Neuroscience*, 49: 285–296.

Sanghera, M.K., Trulson, M.E. and German, D.C. (1984) Electrophysiological properties of mouse dopamine neurons: in vivo and in vitro studies. *Neuroscience*, 12: 793–801.

Sesack, S.R., Aoki, C. and Pickel, V.M. (1994) Ultrastructural localization of d-2 Receptor-Like immunoreactivity in midbrain dopamine neurons and their striatal targets. *J. Neurosci.*, 14: 88–106.

Stjärne, L., Astrand, P., Bao, J.X., Gonon, F., Msghina, M. and Stjärne, E. (1994) Spatiotemporal pattern of quantal release of ATP and noradrenaline from sympathetic nerves: consequences for neuromuscular transmission. In: L. Stjärne, P. Greengard, S. Grillner, T. Hökfelt and D. Ottoson (Eds), *Molecular and Cellular Mechanisms of Neurotransmitter Release*, Raven Press, New York, 1994, pp. 461–496.

Suaud–Chagny, M.F., Ponec, J. and Gonon, F. (1991) Presynaptic autoinhibition of the electrically evoked dopamine release studied in the rat olfactory tubercle by in vivo electrochemistry. *Neuroscience*, 45: 641–652.

Suaud–Chagny, M.F., Chergui, K., Chouvet, G. and Gonon, F. (1992) Relationship between dopamine release in the rat nucleus accumbens and the discharge activity of dopaminergic neurons during local in vivo application of amino-acids in the ventral tegmental area. *Neuroscience*, 49: 63–72.

Suaud–Chagny, M.F., Dugast, C., Chergui, K., Msghina, M. and Gonon, F. (1995) Uptake of dopamine released by impulse flow in the rat mesolimbic and striatal systems in vivo. *J. Neurochem.*, 65: 2603–2611.

Surprenant, A. and Williams, J.T. (1987) Inhibitory synaptic potentials recorded from mammalian neurones prolonged by blockade of noradrenaline uptake. *J. Physiol.*, 382: 87–103.

Svenningsson, P., Fourreau, L., Bloch, B., Fredholm, B.B., Gonon, F. and Le Moine, C. (1999) Opposite tonic modulation by dopamine and adenosine on *c-fos* expression in the striatopallidal neurons. *Neuroscience*, 89: 827–837.

Yung, K.K., Bolam, J.P., Smith, A.D., Hersch, S.M., Ciliax, B.J. and Levey, A.I. (1995) Immunocytochemical localization of D1 and D2 dopamine receptors in the basal ganglia of the rat: light and electron microscopy. *Neuroscience*, 65: 709–730.

Zoli, M., Torri, C., Ferrari, R., Jansson, A., Zini, I., Fuxe, K. and Agnati, L.F. (1998) The emergence of the volume transmission concept. *Brain Res. Rev.*, 26: 136–147.

L. F. Agnati, K. Fuxe, C. Nicholson and E. Syková (Eds.)
Progress in Brain Research, Vol 125

CHAPTER 17

Evidence for the existence of pulses of dopamine in the extracellular space of the rat striatum

L.F. Agnati[1,*], M. Zoli[1], R. Ferrari[1], L. Di Paola[1], C. Torri[1], K. Fuxe[2] and I. Zini[1]

[1] *Department of Biomedical Sciences, section of Physiology, University of Modena and Reggio Emilia, Via Campi 287, I-41100 Modena, Italy*
[2] *Department of Neuroscience, Karolinska Institute, S-171 77 Stockholm, Sweden*

Introduction

Analogy is a logical tool which allows the inference that certain admitted resemblances between two phenomena imply probable further similarities. It is a widely accepted concept that in many ways the endocrine system is similar to the nervous system: in both cases cells communicate via secretion of chemical messages, although peculiar features of chemical transmission can be observed in the two systems. As stated by Keener and Sneyd "hormones are, in a sense, neurotransmitters that are capable of acting on target cells throughout the body, or conversely, neurotransmitters can be thought of as hormones with a localised action" (Keener and Sneyd, 1998). In the last 15 years evidence has been gathered that neurotransmitters can also work *without* a localised action when they are signals involved in the so-called *Volume Transmission* (VT) type of intercellular communication (Agnati et al., 1986, 1992, 1995, 1997; Fuxe and Agnati, 1991; Zoli et al., 1999). Thus, there is a close resemblance between the endocrine intercellular communication and VT.

A prominent feature of the endocrine system is the presence of pulsatile changes in hormonal concentrations ranging in periods from a few minutes to a year (Van Cauter and Aschoff, 1989). The most prevalent endocrine rhythms show:

- an ultradian range (i.e., once every 1 to 4 hours); these rhythms arise from timing systems located in the central nervous system (CNS)
- a circadian range (i.e., once every 24 hours); these rhythms are synchronized by environmental entraining signals (*Zeitgebers*)

On the basis of the analogy between the endocrine and the nervous system and especially between the endocrine and paracrine intercellular communication and the VT, we put forward the hypothesis that at least some VT-signaling display oscillations that have both physiological and, when altered, pathological relevance (Agnati et al., 1999). This hypothesis has been tested for dopamine (DA) release in the basal ganglia of the rat, using microdialysis to monitor the levels of DA in the extracellular space (ECS).

Materials and Methods

Animals

Adult male specific pathogen-free Sprague Dawley rats (250 g b.w.) were used. They were kept under standardized temperature, humidity and lighting conditions (lights on at 8.00 a.m. and off at 8.00 p.m.) and had free access to water and food.

The animals were divided into two groups

*Corresponding author. Tel.: +39 059 428 218; Fax: +39 059 428 236; e-mail: agnati@unimo.it

- Anaesthetised rats with bilateral sampling of the striatal ECS ($n = 7$)
- Awake freely moving rats with unilateral sampling of the striatal ECS ($n = 4$)

Intracerebral microdialysis

Experiments on anaesthetised rats were carried out under halothane anaesthesia, starting the sampling 1 h after probe implantation (500 μm outer diameter, with a 2 mm length dialysing membrane, CMA12, Carnegie Medicine). In the experiments on awake freely moving rats, the animals were deeply anaesthetized with halothane (Fluothane), and a microdialysis probe guide cannula (Carnegie Medicine) was implanted into the striatum and secured to the skull with dental cement. Twenty-four hours after surgery, the awake animal was gently restrained by the experimenter and a microdialysis probe was inserted into the guide cannula; all experiments started at 10: 00 a.m.

In all rats, the probe was continuously perfused at 2 μl/min with Ringer's solution. Starting 1 hr after implantation, dialysate samples were collected every 20 min into 0.5 ml Eppendorf tubes containing 10 μl of mobile phase. DA levels in the dialysate were determined using HPLC with electrochemical detection (464B BAS, sensitivity = 20 fmol/sample). For more details on the microdialysis technique see Ruggeri et al., 1990.

At the end of the microdialysis experiment, the animal was killed, the brain was dissected out and probe localization in striatum was examined.

Statistical analysis

Original data are time series (usually 6 hrs) monitoring DA concentrations in the ECS of the basal ganglia. In order to properly describe the time courses observed, the nomenclature suggested by Urban and coworkers (1988) will be used. In the presence of time series, such as those obtained by monitoring the DA levels in ECS of the basal ganglia, three problems have to be faced:

(1) to give an objective and valid criterion to identify baseline values
(2) to determine if some infrequent value is out of the range of the baseline values and hence can be considered a peak value
(3) to suggest an objective criterion for peak recognition and to try to validate this criterion.

In this chapter we have used the approach proposed by Christian and coworkers (1978) to detect baseline values. The set of data was separated into two groups

(1) one containing the most frequent values which are symmetrically distributed around the mean; these values represent *baseline values*
(2) the second one containing the unfrequent higher values that might represent peak values.

In order to compare different time series, the data obtained in each experiment were expressed as percentage values of the respective median value (central tendency value not affected by possible outliers). Then, the symmetrically distributed values were separated from peak values; with this aim, we used an iterative procedure which, at each step, takes away a class of outliers until the distribution of the remaining values is symmetrical (i.e. the skeweness is ~ 0).

In order to have an overall objective detection of peaks, all the time series recorded for each animal group were set in register using as reference point the first presumptive peak value identified as described above (when this was not possible, the first value above the third quartile was selected). The *Shewhart control charts* were applied to the data in register to detect peaks that, when present, appeared as *values out of control*.

Results

Since in anaestetized rats both striata were monitored through dialysis cannulas, the time series obtained in these animals were used to have an intra-individual test of reliability of the measurements. As was expected, the overall analysis of the data shows a striking similarity between the two sides, so that median (MD) values observed on the right and left side are very close to each other. Therefore, all the data obtained by monitoring the DA levels in the ECS of basal ganglia in the anesthetized rats were pooled together.

The distribution of the entire set of data showed clear departures from the gaussian curve with marked positive skewness and kurtosis (Fig. 1A). Using the iterative procedure described above, a cutoff point was found below which the residual distribution is normal (Fig. 1B). This result indicates that there are baseline values (normally distributed) and peak values (located on the far right tail of the distribution). Using this criterion, almost every time series showed at least a presumptive peak; only for 2 out of 14 time series in the case of the anesthetized rats and for 2 out of 4 time series in the case of the awake freely moving rats presumptive peak values were not observed but only values above the third quartile. This may suggest that DA peaks in awake animals are smaller or less frequent than in anesthetized rats and, therefore, their detection needs a longer sampling.

When the time series from anaesthetized or awake animals were set in register using as reference point the first presumptive peak value, a significant peak was detected in both groups by means of the Shewhart control charts (Figs. 2A, B). In addition, similar to hormone secretion (Urban et al., 1988) a pre-peak nadir was observed in both anesthetized and awake freely moving rats.

Discussion

A pulsatile mode of hormone release has been demonstrated for the vast majority of the endocrine glands. Previous work, especially of Philippu's group, has demonstrated a pulsatile release of catecholamines in several brain areas, such as the nucleus tractus solitarii and the locus coeruleus of

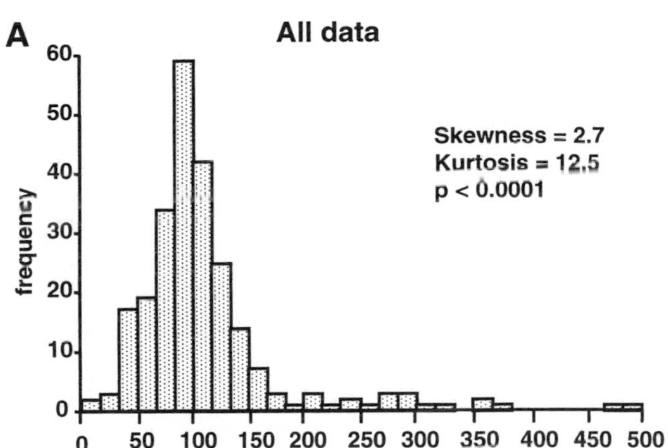

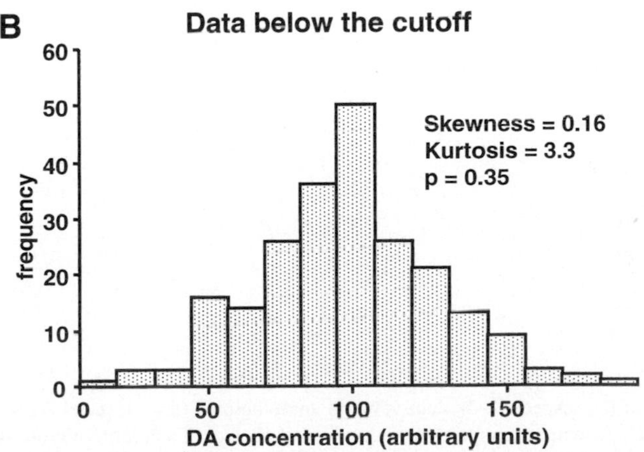

Fig. 1. Frequency distribution of DA content in samples taken from the striata of anaesthetized rats. All samples are shown in panel A, whereas only samples below cutoff are shown in panel B. For further details, see text.

the cat (Lanzinger et al., 1989; Singewald et al., 1994) and the hypothalamus of the rat (Dietl et al., 1993). In this chapter, we provide some evidence that an oscillatory pattern of release may also be present for VT-signals such as DA in the basal ganglia. The present findings show the existence of DA pulses in the ECS of the basal ganglia in anesthetized rats. A similar phenomenon was observed in awake freely moving rats, although, due to the limited number of animals studied, this result needs to be better characterized. Further experiments are needed to detect if some sort of periodicity exists for the interpulse intervals and to try to relate DA peaks with other pulsatile phenomena, such as the secretory patterns of hormones that affect DA neuronal system activity (Piazza and Le Moal, 1997). Indeed, the effect of anaesthesia on DA peaks needs to be further investigated since, although the mean rate of DA cell firing is similar in awake and anaesthetized animals (Kiyatkin, 1995), the pattern of firing may be different. On the other hand, it has to be taken into account that environmental stimuli may have a great impact on DA levels in the awake rats (Kiyatkin and Stein, 1995). In order to solve all these problems, important technical aspects should be improved, such as the temporal resolution and the duration of the sampling procedure (Urban et al., 1988).

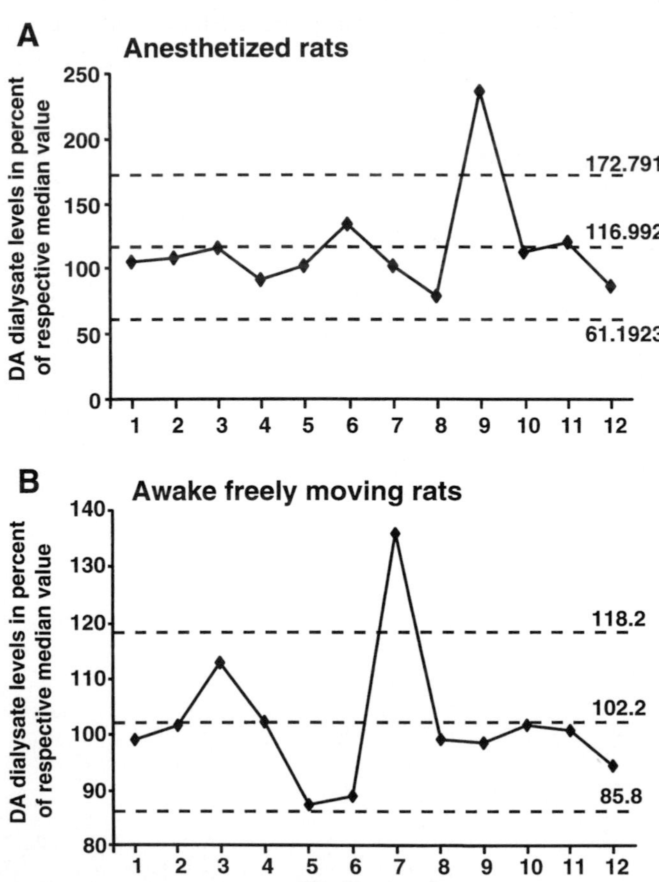

Fig. 2. Time course of DA dialysate levels expressed in percent of the respective median value in anaesthetized ($n = 14$, panel A) and awake freely moving rats ($n = 4$, panel B). The Shewart's control chart, which monitors the values out of control in a repetitive process, was employed. In order to apply this statistical tool, the time course of the striatal DA levels observed in the experiments have been vertically aligned selecting as reference point the first presumptive peak value in each time course. The mean and 95% confidence limits of the DA values are shown.

In the present analysis only one peak of striatal DA could be detected in both awake freely moving and anaesthetized rats. Hence, it may be surmised that an interpulse interval of about 3 hours could be the major ultradian rhythm for DA levels in the basal ganglia, at least in anesthetized rats; this rhythm may show some similarity with catecholamine release in the periphery. With a 20-minute sampling interval, 8–9 catecholamine pulses/24 hour could be detected in the plasma of normal subjects; furthermore it was shown that the magnitude of pulses is greater during the day-time and larger for epinephrine than for norepinephrine (Van Cauter and Aschoff, 1989).

As shown for some endocrine controls, episodic presentation of the VT-signal (first messenger) to the target cells in the CNS may convey trophic information (see, e.g. gonadotropin releasing hormone-regulated lutein hormone secretion and growth hormone effects on liver, muscle, or bone, Friend et al., 1996). In addition to this trophic action, the VT-signal may cause a reset of the functional state of entire far away located compartments that in this way will be properly tuned to interact in the most effective way (Agnati et al., 1990, Agnati and Fuxe, 1997, Agnati et al., this symposium). Thus, oscillations of DA levels in the ECS of the basal ganglia may have a great importance in resetting basal ganglia compartments, i.e. compartments present in a brain region that has a general regulatory role to integrate and control several behaviours necessary for homeostasis (Le Moal and Simon, 1991).

In agreement with a diffuse action of DA on the basal ganglia circuits, it has been reported that DA can decrease responses of postsynaptic cells to various inputs, both excitatory (glutamate, acetylcholine, noxious stimuli) and inhibitory (γ-aminobutyric acid, noxious stimuli) (Kiyatkin, 1995). Thus, the present results suggest that DA oscillations may affect, according to a certain pattern, the activity of the main target cells of the mesostriatal system through the rhythmic modulation of their responses to various transmitters.

As stressed by Keener and Sneyd (1998) for hormone secretion, there is not a completely satisfactory answer as to the question why hormone secretion is pulsatile. Similar to what has been proposed for the endocrine signalling (Li and Goldbeter, 1989), it can be suggested that VT-signal receptors have optimal responses to stimuli of a certain frequency. Actually, different cells in a compartment may respond to various oscillation patterns of the VT-signal in a different way, e.g. according to their WT inputs. Optimal response of target cells to pulsatile signals may stem from different causes. For instance, pulsatile release of transmitters may decrease the chance of transmitter-induced receptor inactivation (desensitization, down-regulation, etc.) compared with continuous transmitter release (see Jones and Westbrook, 1996 and references therein). In addition, it has been shown that pulses of hormones (see, e.g. the case of gonadotropin releasing hormone, Fink 1995) prime the target cells (e.g. increase the responsivity) to further pulses of the same hormone. Future studies will assess whether DA pulsatility subserves similar functions in striatum.

References

Agnati, L.F., Torri, C., Ferrari, R., Fuxe, K., Zini, I. and Zoli, M. (1999) Volume transmission revisited. Volume transmission in a rat model of Parkinson's disease. *Physiol. Res.*, 48: S8.

Agnati, L.F. Fuxe, K. Zoli, M. Zini, I. Toffano, G. and Ferraguti, F. (1986) A correlation analysis of the regional distribution of central enkephalin and beta-endorphin immunoreactive terminals and of opiate receptors in adult and old male rats. Evidence for the existence of two main types of communication in the central nervous system: the volume transmission and the wiring transmission. *Acta Physiol. Scand.*, 128: 201–207.

Agnati, L.F. Zoli, M. Merlo Pich, E. Benfenati, F. and Fuxe, K. (1990) Aspects of neural plasticity in the central nervous system. VII. Theoretical aspects of brain communication and computation. *Neurochem. Int.*, 16: 479–500.

Agnati, L.F. Bjelke, B. and Fuxe, K. (1992) Volume transmission in the brain. Do brain cells communicate solely through synapses? A new theory proposes that information also flows in the extracellular space. *Am. Sci.*, 80: 362–374.

Agnati, L.F. Zoli, M. Strömberg, I. and Fuxe, K. (1995) Intercellular communication in the brain: wiring vs. volume transmission. *Neuroscience*, 69: 711–726.

Agnati, L.F. and Fuxe, K. (1997) The impact of histological techniques in revealing brain function. Volume transmission: from fluorescence histochemistry to confocal laser microscopy. In: K. Fuxe, T. Hökfelt, L. Olson, D. Ottoson, A. Dahlström and A. Björklund (Eds), *Molecular Mechanisms of Neuronal Communication*, pp. 251–277; Pergamon, Oxford.

Christian, L.E. Everson, D.O. and Davis, S.L. (1978) A statistical method for detection of hormone secretory spikes. *J. Anim. Sci.*, 46: 699–706.

Dietl, H. Prast, H. and Philippu, A. (1993) Pulsatile release of catecholamines in the hypothalamus of conscious rats. *Naunyn-Schiemiedeberg's Arch. Pharmacol.*, 347: 28–33.

Fink, G, (1995) The self-priming effect of LHRH: a unique servomechanism and possible cellular model for memory. *Front. Neuroendocrinol.*, 16: 183–190.

Friend, K. Iranmanesh, A. and Veldhuis, J.D. (1996) The orderliness of the growth hormone (GH) release process and the mean mass of GH secreted per burst are highly conserved in individual men on successive days. *J. Clin. Endocrinol. Metab.*, 81: 3746–3753.

Fuxe, K. and Agnati, L.F. (Eds) (1991) *Volume Transmission in the Brain. Novel Mechanisms for Neural Transmission. Advances in Neuroscience*, Vol. 1, Raven Press, New York, 1991.

Jones, M.V. and Westbrook, G.L. (1996) The impact of receptor desensitization on fast synaptic transmission. *Trends Neurosci.*, 19: 96–101.

Keener, J. and Sneyd, J. *Mathematical Physiology*, pp. 579–611; Springer-Verlag, Berlin 1998.

Kiyatkin, E.A. (1995) Functional significance of mesolimbic dopamine. *Neurosci. Biobehav. Rev.*, 19: 573–598.

Kiyatkin, E.A. and Stein, E.A. (1995) Fluctuations in nucleus accumbens dopamine during cocaine self-administration behaviour: an 'in vivo' electrochemical study. *Neuroscience*, 64: 599–617.

Lanzinger, I. Kobilanski, C. and Philuppu, A. (1989) Pattern of catecholamuine release in the nucleus tractus solitarii of the cat. *Naunyn-Schiemiedeberg's Arch. Pharmacol.*, 339: 298–301.

Le Moal, M. and Simon, H. (1991) Mesocortical dopaminergic network: functional and regulatory roles. *Physiol. Rev.*, 71: 155–234.

Li, Y.X. and Goldbeter, A. (1989) Frequency specificity in intercellular communication: influence of patterns of periodic signalling on target cell responsiveness. *Biophys. J.*, 55: 125–145.

Piazza, P.V. and Le Moal, M. (1997) Glucocorticoids as a biological substrate of reward: physiological and pathological implications. *Brain Res. Rev.*, 25: 359–372.

Ruggeri, M. Zoli, M. Grimaldi, R., Ungerstedt, U. Eliasson, A. Agnati, L.F. and Fuxe, K. (1990) Aspects of neural plasticity in the central nervous system. III. Methodological studies on the microdialysis technique. *Neurochem. Int.*, 16: 427–435.

Singewald, N. Schneider, C. Pfitscher, A. and Philippu, A. (1994) In vivo release of catecholamines in the locus coeruleus. *Naunyn-Schiemiedeberg's Arch. Pharmacol.*, 350: 339–345.

Urban, R.J. Evans, W.S. Rogol, A.D. Kaiser, D.L. Johnson, M.L. and Veldhuis, J.D. (1988) Contemporary aspects of discrete peak-detection algorithms. I. The paradigm of the luteinizing hormone pulse signal in men. *End. Rev.*, 9: 3–37.

Van Cauter, E. and Aschoff, J. (1989) Endocrine and other biological rhythms In: L.J. DeGroot (Ed.), *Endocrinology*, W.B. Saunders Company, Philadelphia, pp. 2658–2705.

Zoli, M. Jansson, A. Sykova, E. Agnati, L.F. and Fuxe, K. (1999) Volume transmission in the CNS and its relevance for neuropsychopharmacology. *Trends Pharmacol. Sci.*, 20: 142–150.

L. F. Agnati, K. Fuxe, C. Nicholson and E. Syková (Eds.)
Progress in Brain Research, Vol 125

CHAPTER 18

Restoration of dopamine transmission in graft reinnervated striatum. Evidence for regulation of dopamine D2 receptor function in regions lacking dopamine

Ingrid Strömberg*, Jan Kehr and Kjell Fuxe

Department of Neuroscience, Karolinska Institute, S-17177 Stockholm, Sweden

Introduction to grafting as a tool to restore dopamine deficiency

Grafting of catecholamine-rich tissue into dopamine depleted striatum to restore the loss of dopaminergic input into the striatum has become a useful tool (Björklund and Stenevi, 1979; Perlow et al., 1979; Freed et al., 1981). The first clinical attempts were performed by the grafting of adrenal medullary tissue into the caudate nucleus of patients suffering from Parkinson's disease (Backlund et al., 1985). Although the chromaffin cells store and release adrenaline and noradrenaline (NA) rather than dopamine (DA), it has been shown in animal experiments that implantation of adrenal chromaffin cells to the DA depleted striatum affects drug-induced rotational behavior (Freed et al., 1981, 1986; Strömberg et al., 1985). Studies documenting effects on rotational behavior induced acutely after implantation of the tissue showed that rotations were induced by release of catecholamines from the grafts (Herrera-Marschitz et al., 1984; Strömberg et al., 1984), and consequently, it was suggested that the graft exerted its action via diffusion or volume transmission rather than over synaptic contacts, since no outgrowth was found in this case.

*Corresponding author: Tel.: 46-8-728 7087; Fax: 46-8-728 7437; e-mail: ingrid.stromberg@neuro.ki.se

Later, grafting of fetal ventral mesencephalic tissue was shown to be more powerful to use than chromaffin cells. Robust documentation of graft survival, graft-derived dopaminergic reinnervation of host brain, spontaneous activity of dopaminergic neurons within the transplants, and DA released from the graft-derived outgrowth were shown (Wuerthele et al., 1981; Arbuthnott et al., 1985; Brundin et al., 1985; Rose et al., 1985; Strecker et al., 1987; Zetterström et al., 1986). Furthermore, functional evidence of DA transmission was shown when local application of phencyclidine affected the striatal neuronal activity in graft-reinnervated striatum (Strömberg et al., 1985).

It was suggested that DA transmission in the striatum after reinnervation occurred via synaptic contacts, since reciprocal contacts between graft and host had been found, showing membrane specializations of the same type that had been characterized for a dopaminergic synaptic input to the striatal neurons (Freund et al., 1985; Jaeger, 1985; Mahalik et al., 1985; Bolam et al., 1987). Furthermore, evidence for functional dopaminergic input to the graft-reinnervated striatum was obtained based on studies showing that the increased D2 receptor binding found after a chronic DA depletion (Creese et al., 1977) was normalized after grafting (Freed et al., 1983). Later, it was shown that the D2 receptor binding became normalized in the total volume of dorsal striatum, although the graft-derived dopaminergic reinnervation of dorsal striatum is limited (Dawson et al., 1991; Gagnon et al., 1991; Blunt et al., 1992; Freed et al., 1983). In fact, outgrowth from fetal ventral mesencephalon terminates already approximately 2 weeks after implantation (Barker et al., 1996), and at this time point only 1/3 to 1/2 of the volume of dorsal striatum has become reinnervated.

Conditions of dopamine transmission via volume transmission?

The mismatch between striatal reinnervated areas and normalization of D2 receptor binding raised the question whether the dopamine transmission in graft-reinnervated striatum occurs via volume transmission in addition to synaptic transmission. Indeed it has been shown that clearance time for extracellular dopamine is prolonged distal compared to proximal to the graft (Strömberg et al., 1991). Dopamine reuptake sites are present in TH-positive nerve fibers after grafting (Kordower et al., 1996). Hence, the explanation for the longer clearance time was suggested to be due to the distribution of DA nerve fibers with less density distal than proximal to the graft, resulting in a more efficient DA reuptake close to the graft with higher DA nerve fiber densities and DA reuptake sites than that found distally. The conclusion was drawn that the graft may act over a larger volume of the host brain than that becomes reinnervated, probably via diffusion in the extracellular space (long-distance volume transmission).

Functional dopamine D2 receptors in dopamine reinnervated versus denervated striatum

To further explore the possibility of DA transmission via volume transmission in non-DA-innervated areas of graft-reinnervated striatum, extracellular recordings of graft reinnervated striatum were employed. Recordings were performed in the DA depleted striatum reinnervated by a fetal ventral mesencephalic graft implanted into the lateral ventricle. The host striatum becomes dopaminergically reinnervated as visualized by tyrosine hydroxylase (TH)-positive nerve fibers in a zone close to the ventricle, and thus, reinnervated and non-innervated areas can be localized to medial and lateral striatum respectively (Fig. 1) (Strömberg and Bickford, 1996).

Extracellular recordings in the striatum including local applications of the D2-like agonist quinpirole revealed a dose-dependent reduction in striatal neuronal firing rates. When recording in graft-reinnervated areas of the striatum there was no shift in sensitivity to quinpirole when compared to the control side. Furthermore, there was no difference in sensitivity to quinpirole when recording in noninnervated versus reinnervated regions of graft-reinnervated striatum (Fig. 2). However, a significant supersensitivity was found when recording in 6-hydroxydopamine (6-OHDA) lesioned control animals (Fig. 2). Thus, the D2 receptor supersensitivity seen after an almost complete DA denervation may be functionally removed in graft reinnervated striatum, not only within reinnervated areas, but also lateral to the DA reinnervation. Hence, the normalization of D2 receptor binding in incomplete graft-reinnervated striatum had physiological relevance.

Striatal spontaneous neuronal activity

The striatal neuronal firing rates are significantly upregulated in DA depleted striatum (Siggins et al., 1974; Schultz and Ungerstedt, 1978; Strömberg et al., 1985). After graft-reinnervation the spontaneous discharge rate becomes normalized, but the normalization is found only in areas that are reinnervated (Fig. 3) (Strömberg et al., 1985; Fisher et al., 1991). Since the D2 receptors were shown to be functionally normalized, the upregulated neuronal spontaneous activity might be due to malfunction at the D1 receptor level. It has been suggested that the D1 receptor sensitivity shows either decrease, increase or no change after a 6-OHDA lesion (Buonamici et al., 1986; Marshall et al., 1989; Dawson et al., 1991; Gagnon et al., 1991; Robertson et al., 1991; Blunt et al., 1992; Savasta et al., 1992). Although the reports show divergent results, it has been found that the D1 receptor levels are normalized after grafting and some studies have shown reduced D1 receptor levels (Blunt et al., 1992; Strömberg et al., 1995). Nevertheless, electrophysiological recordings in

DA depleted striatum indicate no significant change of the dose-response curve to the D1 agonist N0437 (Strömberg and Bickford-Wimer, 1991). However, inactivation of the D2 receptor results in a down-regulation of the sensitivity to a D1 agonist, indicating that there is a kind of synergistic action between D1 and D2 receptors in the 6-OHDA lesioned striatum (Strömberg and Bickford-Wimer, 1991). Extracellular recordings using local applications of the D1 agonist SKF 81297 in DA reinnervated striatum revealed no physiological differences of striatal neuronal responses to the D1 agonist in reinnervated versus noninnervated areas of grafted striatum (Strömberg et al., 1999). Thus, the upregulated striatal neuronal discharge rate in non-innervated areas did not seem to be due to a malfunction of the D1 receptor.

Cortical excitatory input to the striatum after grafting

The results showing that the dopaminergic reinnervation normalized the D2 receptor function but not the spontaneous neuronal activity in noninnervated areas of grafted striatum turned our interest to the excitatory input to the striatum, i.e. the cortical glutamatergic innervation of the striatum. It has been shown that a DA depletion enhance extracellular levels of glutamate (Yamamoto and Davy, 1992; Meshul et al., 1999), and accordingly this might be the explanation for the upregulated striatal neuronal activity in noninnervated areas of the graft reinnervated striatum. However, there is a loss of asymmetric synapses after a DA depletion (Ingham et al., 1998; Meshul et al., 1999), and since

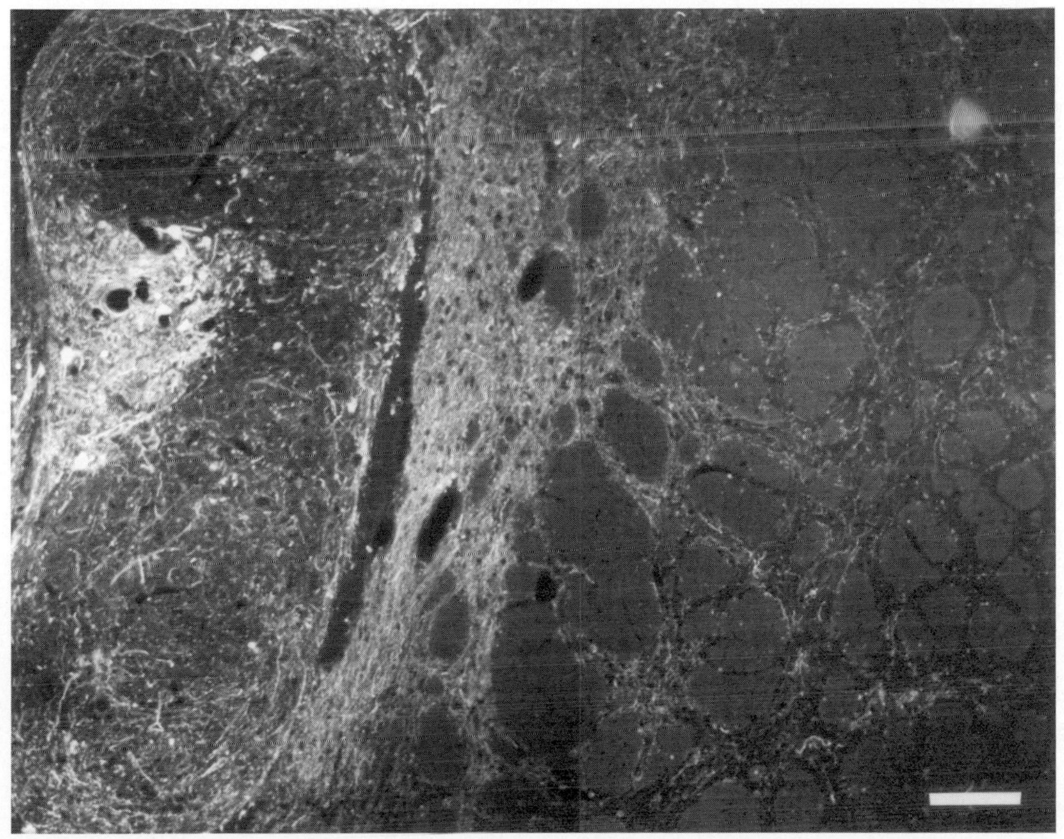

Fig. 1. TH-immunohistochemistry of a fetal ventral mesencephalic graft transplanted to the lateral ventricle of a unilaterally 6-OHDA lesioned rat. Graft outgrowth into the host brain is limited to a zone close to the ventricle. Extracellular electrophysiological recordings were performed in host striatum within the zone of TH-positive nerve fibers and in dopamine denervated regions lateral to TH-immunoreactive areas. Scale bar: 200 μm.

asymmetric synapses are related to the glutamatergic input to striatal dentritic spines (Kemp and Powell, 1971; Somogyi et al., 1981), enhanced glutamate levels are accompanied with loss of glutamate synaptic input.

Microdialysis using the dual-probe approach (Morari et al., 1996) and recording of extracellular levels of glutamate after DA reinnervation showed no differences in glutamate overflow in reinnervated compared to non-innervated areas of graft-derived DA reinnervated striatum. Extracellular levels of potassium-induced glutamate overflow showed approximately 200% increase of baseline levels in all areas measured. Thus, it is not likely that the upregulated neuronal activity in noninnervated areas of graft-reinnervated striatum is due to an increase in extracellular glutamate in these areas.

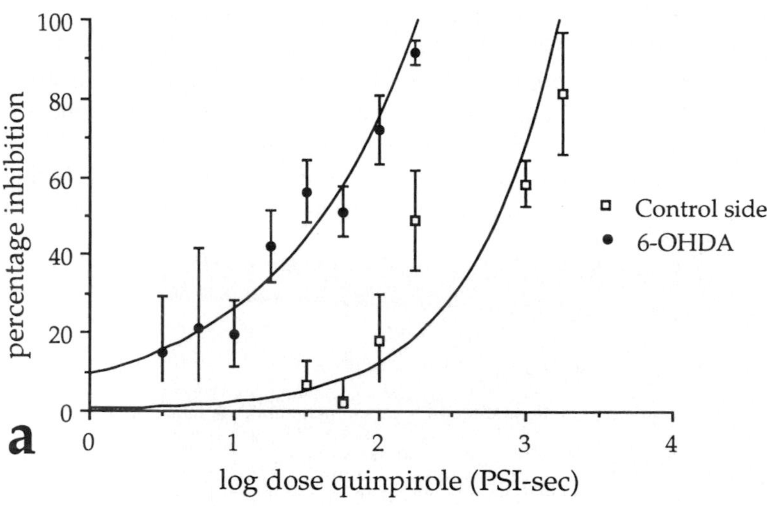

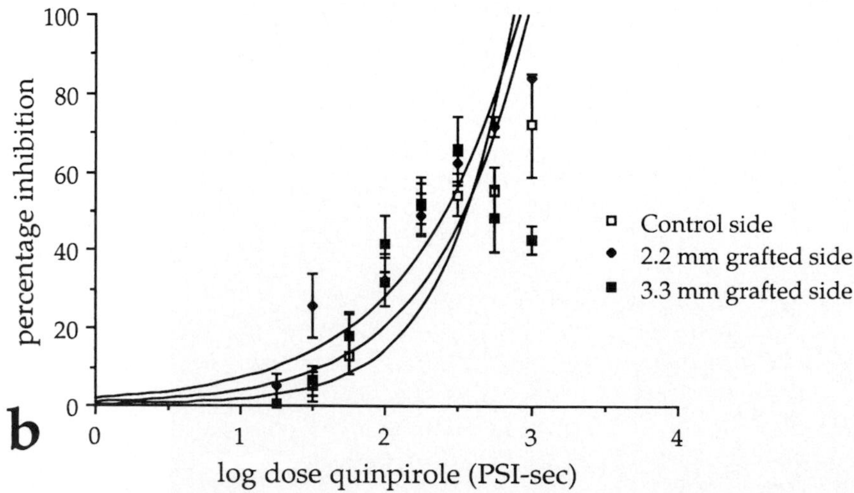

Fig. 2. Dose-response to locally applied quinpirole expressed as percentage change from baseline. Recordings were performed in 6-OHDA lesioned (a) and in graft DA reinnervated striatum (b). In DA depleted striatum there was a shift in dose-response ($p < 0.001$) to quinpirole and the supersensitivity to the D2 agonist was shown, while in graft reinnervated striatum there was no difference in dose response when recording in DA reinnervated vs. noninnervated regions compared to control side. From Strömberg et al., 2000.

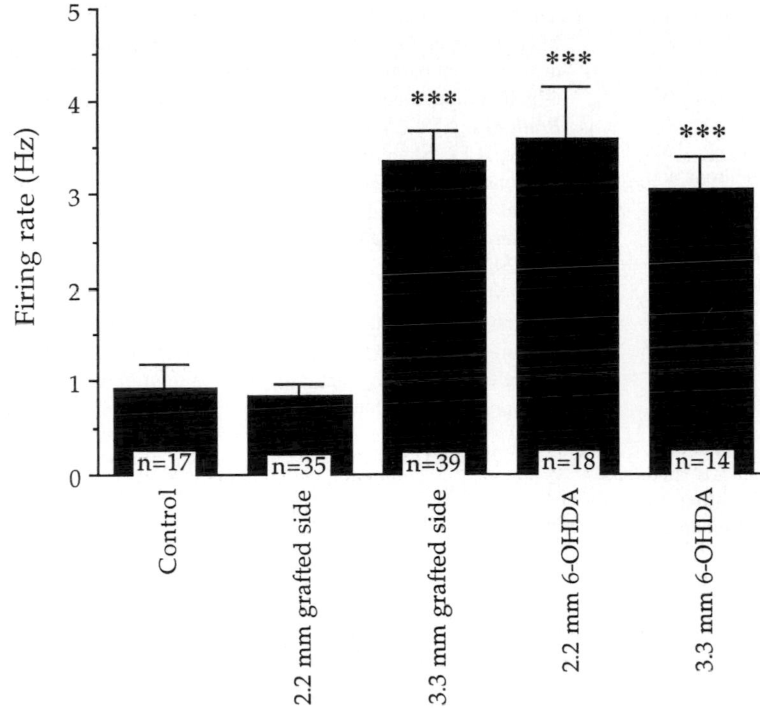

Fig. 3. Extracellular recordings performed at 2.2 and 3.3 mm lateral to bregma in 6-OHDA lesioned rats and in DA-depleted animals followed by grafting of fetal ventral mesencepahlic tissue to the lateral ventricle. Striatal neuronal discharge rates were significantly increased in DA depleted regions compared to normal firing rate. In DA reinnervated regions of grafted striatum (2.2 mm grafted side) the spontaneous discharge rate was normal. $^{***}p<0.0001$.

Conclusion

The results indicate that the functional normalization of the D2 receptors within as well as distal to reinnervated areas of the striatum is regulated by diffusion of DA from graft-derived reinnervated areas to denervated areas. Binding studies have shown a normalization of D2 receptor levels in total volume of dorsal striatum, and it has not been correlated to the exact location of the graft or the reinnervated areas. However, the present data indicate that the distance the volume transmitted DA diffuses to functionally normalize the D2 receptor levels in non-innervated regions lateral to the DA reinnervated areas is rather at the 500 μm levels than at smaller distances, since recordings were performed at a minimum of 500 μm lateral to the zone of DA reinnervation. Trials to measure extracellular levels of DA at 500 μm distance from the reinnervated areas have not been successful, but since the D2 receptors require more than a 90% DA depletion to become upregulated (Heikkila et al., 1981), the DA levels needed to normalize the D2 receptors in denervated areas might be below detection limit. Thus, these data provides evidence that DA action in graft-reinnervated striatum may occur via long distance volume transmission in DA denervated areas.

Acknowledgements

This study was supported by the Swedish Medical Research Council, grant No. 09917, 13233, Tore Nilsson's, Loo and Hans Osterman's, and Karolinska Institutet's foundations.

References

Arbuthnott, G., Dunnett, S. and MacLeod, N. (1985) Electrophysiological properties of single units in dopamine-rich mesencephalic transplants in rat brain. *Neurosci. Lett.*, 57: 205–210.

Backlund, E.-O., Granberg, P.O., Hamberger, B., Knutsson, E., Mårtensson, A., Sedvall, G., Seiger, Å. and Olson, L. (1985) Transplantation of adrenal medullary tissue to striatum in parkinsonism. First clinical trials. *J. Neurosurg.*, 62: 169–173.

Barker, R.A., Dunnett, S.B., Faissner, A. and Fawcett, J.W. (1996) The time course of loss of dopaminergic neurons and the gliotic reaction surrounding grafts of embryonic mesenceohalon to the striatum. *Exp. Neurol.*, 141: 79–93.

Björklund, A. and Stenevi, U. (1979) Reconstruction of the nigrostriatal dopamine pathway by intracerebral nigral transplants. *Brain Res.*, 177: 555–560.

Blunt, S.B., Jenner, P. and Marsden, C.D. (1992) Autoradiographic study of striatal D_1 and D_2 dopamine receptors in 6-OHDA-lesioned rats receiving foetal ventral mesencephalic grafts and chronic treatment with L-DOPA and carbidopa. *Brain Res.*, 582: 299–311.

Bolam, J.P., Freund, T.F., Björklund, A., Dunnett, S.B. and Smith, A.D. (1987) Synaptic input and local output of dopaminergic neurons in grafts that functionally reinnervate the host striatum. *Exp. Brain Res.*, 68: 131–146.

Brundin, P., Isacson, O. and Björklund, A. (1985) Monitoring of cell viability in suspensions of embryonic CNS tissue and its use as a criterion for intracerebral graft survival. *Brain Res.*, 331: 251–259.

Buonamici, M., Caccia, M., Carpenteri, M., Pegrassi, L., Rossi, A.C. and Di Chiara, G. (1986) D-1 receptor supersensitivity in the rat striatum after unilateral 6-hydroxydopamine lesions. *Eur. J. Pharmacol.*, 126: 347–348.

Creese, I., Burt, D.R. and Snyder, S.H. (1977) Dopamine receptor binding enhancement accompanies lesion-induced behavioral sensitivity. *Science*, 17: 596–598.

Dawson, T.M., Dawson, V.L., Gage, F.H., Fisher, L.J., Hunt, M.A. and Wamsley, J.K. (1991) Functional recovery of supersensitive dopamine receptors after intrastriatal grafts of fetal substantia nigra. *Exp. Neurol.*, 111: 282–292.

Fisher, L.J., Young, S.J., Tepper, J.M., Groves, P.M. and Gage, F.H. (1991) Electrophysiological characteristics of cells within mesencephalon suspension grafts. *Neuroscience*, 40: 109–122.

Freed, W., Morihisa, J., Spoor, E., Hoffer, B., Olson, L., Seiger, Å. and Wyatt, R. (1981) Transplanted adrenal chromaffin cells in rat brain reduce lesion-induced rotational behaviour. *Nature*, 292: 351–352.

Freed, W.J., Cannon-Spoor, H.E. and Krauthamer, E. (1986) Intrastriatal adrenal medulla grafts in rats. Long-term survival and behavioral effects. *J. Neurosurg.*, 65: 664–670.

Freed, W.J., Neihoff, D.L., Kuhar, M.J., Hoffer, B.J., Olson, L., Cannon-Spoor, H.E., Morihisa, J.M. and Wyatt, R.J. (1983) Normalization of spiroperidol binding in the denervated rat striatum by homologous grafts of substantia nigra. *Science*, 222: 937–939.

Freund, T.F., Bolam, J.P., Björklund, A., Stenevi, U., Dunnett, S.B., Powell, J.F. and Smith, A.D. (1985) Efferent synaptic connections of grafted dopaminergic neurons reinnervating the host neostriatum: a tyrosine hydroxylase immunocytochemical study. *J. Neurosci.*, 5: 603–616.

Gagnon, C., Bédard, P.J., Rioux, L., Gaudin, D., Martinoli, M.G., Pelletier, G. and Di Paolo, T. (1991) Regional changes of striatal dopamine receptors following denervation by 6-hydroxydopamine and fetal mesencephalic grafts in the rat. *Brain Res.*, 558: 251–263.

Heikkila, R.D., Shapiro, B.S. and Duvoisin, R.C. (1981) The relationship between loss of dopamine nerve terminals, striatal ^{3}H-spiroperidol binding and rotational behavior in unilaterally 6-OHDA lesioned rats. *Brain Res.*, 211: 285–292.

Herrera-Marschitz, M., Strömberg, I., Olsson, D., Olson, L. and Ungerstedt, U. (1984) Adrenal medullary implants in the dopamine-denervated rat striatum. II. Rotational behavior during the first seven hours as a function of graft amount and location and its modulation by neuroleptics. *Brain Res.*, 297: 53–61.

Ingham, C.A., Hood, S.H., Taggart, P. and Arbuthnott, G.W. (1998) Plasticity of synapses in the rat neostriatum after unilateral lesion of the nigrostriatal dopaminergic pathway. *J. Neurosci.*, 18: 4732–4743.

Jaeger, C.B. (1985) Cytoarchitectonics of substantia nigra grafts: a light- and electron-microscopic study of immunocytochemically identified dopaminergic neurons and fibrous astrocytes. *J. Comp. Neurol.*, 231: 121–135.

Kemp JM, Powell TPS (1971) The synaptic organization of the caudate nucleus. *Philos. Trans.R. Soc. (Lond.) B. Biol. Sci.*, 262: 403–412.

Kordower, J.H., Rosenstein, J.M., Collier, T.J., Burke, M.A., Chen, E.-Y., Li, J.M., Martel, L., Levey, A.E., Mufson, E.J., Freeman, T.B. and Olanow, C.W. (1996) Functional fetal nigral grafts in a patient with Parkinson's disease: chemoanatomic, ultrastructural, and metabolic studies. *J. Comp. Neurol.*, 370: 203–230.

Mahalik, T.J., Finger, T.E., Strömberg, I. and Olson, L. (1985) Substantia nigra transplants into denervated striatum of the rat: Ultrastructure of graft and host interconnections. *J. Comp. Neurol.*, 240: 60–70.

Marshall, J.F., Navarrete, R. and Joyce, J.N. (1989) Decreased striatal D1 binding density following mesetelencephalic 6-hydroxydopamine injections: an autoradiographic analysis. *Brain Res.*, 493: 247–257.

Meshul, C.K., Emre, N., Nakamura, C.M., Allen, C., Donohue, M.K. and Buckman, J.F. (1999) Time-dependent changes in striatal glutatmate synapses following 6-hydroxydopamine lesion. *Neuroscience*, 88: 1–16.

Morari, M., O'Connor, W.T., Darvelid, M., Ungerstedt, U., Bianchi, C. and Fuxe, K. (1996) Functional neuroanatomy of the nigrostriatal and the striatonigral pathways as studied with dual probe microdialysis in the awake rat. I. Effects of perfusion with tetrodotoxin and low calcium medium. *Neuroscience*, 72: 79–87.

Perlow, M.J., Freed, W.J., Hoffer, B.J., Seiger, Å., Olson, L. and Wyatt, R.J. (1979) Brain grafts reduce motor abnormalities produced by destruction of nigrostriatal dopamine system. *Science*, 204: 643–647.

Robertson, G.S., Fine, A. and Robertson, H.A. (1991) Dopaminergic grafts in the striatum reduce D1 but not D2 receptor-mediated rotation in 6-OHDA-lesioned rats. *Brain Res.*, 539: 304–311.

Rose, G., Gerhardt, G., Strömberg, I., Olson, L. and Hoffer, B. (1985) Monoamine release from dopamine-depleted rat caudate nucleus reinnervated by substantia nigra transplants: an in vivo electrochemical study. *Brain Res.*, 341: 92–100.

Savasta, M., Mennicken, F., Chritin, M., Abrous, D.N., Feuerstein, C., Le Moal, M. and Herman, J. (1992) Intrastriatal dopamine-rich implants reverse the changes in dopamine D2 receptor densities caused by 6-hydroxydopaine lesion of the nigrostriatal pathway in rats: an autoradiographic study. *Neuroscience*, 46: 729–738.

Schultz, W. and Ungerstedt, U. (1978) Short-term increase and long-term reversion of striatal cell activity after degeneration of the nigrostriatal dopamine system. *Exp. Brain Res.*, 33: 159–171.

Siggins, G., Hoffer, B. and Ungerstedt, U. (1974) Electrophysiological evidence for the involvement of cyclic adenosine monophosphate in dopamine responses of caudate neurons. *Life Sci.*, 15: 779–792.

Somogyi, P., Bolam, J.P. and Smith, A.D. (1981) Monosynaptic cortical input and local axon collaterals of identified striatonigral neurons. a light and electron microscopic study using the Golgi-peroxidase transport-degeneration procedure. *J. Comp. Neurol.*, 195: 567–584.

Strecker, R.E., Sharp, T., Brundin, P., Zetterström, T., Ungerstedt, U. and Björklund, A. (1987) Autoregulation of dopamine release and metabolism by intrastriatal nigral grafts as revealed by intracerebral dialysis. *Neuroscience*, 22: 169–178.

Strömberg, I., Adams, C., Bygdeman, M., Hoffer, B., Boyson, S. and Humpel, C. (1995) Long-term effects of human-to-rat mesencephalic xenografts on rotational behavior, striatal dopamine receptor binding, and mRNA levels. *Brain Res. Bull.*, 38: 221–233.

Strömberg, I. and Bickford, P. (1996) Reduced aging effects of striatal neuronal discharge rate by aged ventral mesencephalic grafts. *NeuroReport*, 7: 693–696.

Strömberg, I. and Bickford-Wimer, P. (1991) Effects of locally applied D_1 and D_2 agonists on striatal neurons with 6-OHDA and pertussis toxin lesions. *Brain Res.*, 564: 279–285.

Strömberg, I., Herrera-Marschitz, M., Hultgren, L., Ungerstedt, U. and Olson, L. (1984) Adrenal medullary implants in the dopamine-denervated rat striatum. I. Acute catecholamine levels in grafts and host caudate as determined by HPLC-electrochemistry and fluorescence histochemical image analysis. *Brain Res.*, 297: 41–51.

Strömberg, I., Herrera-Marschitz, M., Ungerstedt, U., Ebendal, T. and Olson, L. (1985) Chronic implants of chromaffin tissue into the dopamine-denervated striatum. Effects of NGF on graft survival, fiber growth and rotational behavior. *Exp. Brain Res.*, 60: 335–349.

Strömberg, I., Johnson, S., Hoffer, B. and Olson, L. (1985) Reinnervation of dopamine-denervated striatum by substantia nigra transplants. Immunocytochemical and electrophysiological correlates. *Neuroscience*, 14: 981–998.

Strömberg, I., Kehr, J., Andbjer, B. and Fuxe, K. (2000) Fetal ventral mecencephalic grafts functionally reduce the dopamine D2 receptor supersensitivity in partially dopamine reinnervated host striatum. *Exp. Neurol.*, 1648.

Strömberg, I., van Horne, C., Bygdeman, M., Weiner, N. and Gerhardt, G. (1991) Function of intraventricular human mesencephalic xenograft in immunosuppressed rats: an electrophysiological and neurochemical analysis. *Exp. Neurol.*, 112: 140–152.

Wuerthele, S., Freed, W., Olson, L., Morihisa, J., Spoor, L., Wyatt, R. and Hoffer, B. (1981) Effects of dopamine agonists and antagonists on the electrical activity of substantia nigra neurons transplanted into the lateral ventricle of the rat. *Exp. Brain Res.*, 44: 1–10.

Yamamoto, B.K. and Davy, S. (1992) Dopaminergic modulation of glutamate release in striatum as measured by microdialysis. *J. Neurochem.*, 58: 1736–1742.

Zetterström, T., Brundin, P., Gage, F.G., Sharper, T., Isacson, O., Dunnett, S.B., Ungerstedt, U. and Björklund, A. (1986) In vivo measurement of spontaneous release and metabolism of dopamine from intrastriatal nigral grafts using intracerebral dialysis. *Brain Res.*, 362: 344–349.

L. F. Agnati, K. Fuxe, C. Nicholson and E. Syková (Eds.)
Progress in Brain Research, Vol 125

CHAPTER 19

When it comes to communications between neurons, synapses are over-rated: Insights from an animal model of Parkinsonism

Michael J. Zigmond

Department of Neurology, University of Pittsburgh, Pittsburgh, PA 15213, USA

Birth of the concept of a 'synapse'

Where did the 'synapse' come from?

In the middle of the 19th Century, Claude Bernard published his findings on the mechanism of action of curare. He had found that this plant drug, used in South America to paralyze animals during a hunt, blocked the ability of motoneuron stimulation to cause contraction of skeletal muscles. However, the drug worked only when applied to the junction between nerve and muscle. Many other experiments on the relation between a nerve and its target were to follow. Finally, on the basis of extensive electrophysiological studies, Charles Sherrington declared that there must indeed be a 'surface of separation' between neurons. In 1897, Sherrington gave the term *synapse* to this separation.

What did the pioneers have to say about dimensions?

It is likely Bernard, Sherrington, and other early students of the neuron doctrine assumed that the space across which a neuron communicated with its target was small. Certainly, the neurons of the adult central nervous system (CNS) *appeared* to be continuous with their target when examined under the microscope, and only under special conditions could one actually detect any indication of a discontinuity. For example, Cajal, who probably did more than any other person to ensure acceptance of the concept of neurons as separate units, drew his conclusions by examining the *developing* CNS, in which neurons had not yet contacted their targets, or by using special histological techniques such as the Golgi stain, which capriciously stained some neurons and not others. Since the limit of resolution of the light microscope at that time was roughly 250 nm, the space between them was presumably smaller than this, and the communication between neurons would naturally have been assumed to occur here.

Yet, when studies of communication between neurons were initiated in the early 1900s, they offered no insight into the distances involved or even into the role of the synapse in that communication process. Thomas Elliott, generally given credit for the first public presentation of the concept of chemical transmission in 1904, came to the idea of chemical transmission by injecting adrenal extract systemically and measuring its cardiovascular effects. The distances involved in Otto Loewi's experiments nearly two decades later were great still – his *vagusstuff* (ACh) needed to travel all the way from one isolated nerve-muscle preparation to another (see Shepherd, 1991; Zigmond,

*Corresponding author. Tel.: (412) 624-4258; Fax: (412) 624-7327; e-mail: zigmond@pitt.edu

2000). So, whereas it is difficult to be sure what visual image the early pioneers had regarding a cell–cell communication, there is no reason to assume that it was comparable to our current concept of synaptic transmission. Indeed, one of the earliest drawings left behind suggests that for some, at least, the image of neuron-to-neuron communication was far from that of most present-day neuroscientists. It features a neuron releasing transmitter into a large empty space much as a garden hose would release water into a garden (Bacq, 1974).

How did we arrive at the present-day concept of chemical *synaptic* transmission?

The current dogma about chemical transmission between neurons emerged with the application of the electron microscope to the study of the nervous system by Palay and Palade and by others in the 1950s. The electron microscope provided a resolution that was at least two orders of magnitude greater than that of the light microscope and with it investigators were able to see what previous researchers could only surmise – a space between one neuron and the next. More than just a space, they also saw vesicles on the side that appeared to be the nerve terminal as well as structural specializations at the level of the opposing plasma membranes. Quickly labels were applied. The space became the *synapse*, the two opposing membranes were termed the *presynaptic* and *postsynaptic* elements, the vesicles become *transmitter storage sites,* and chemical transmission became synonymous with *synaptic transmission*.

These labels represented important hypotheses and, indeed, the next half-century has shown them to be largely correct. Few if any modern neuroscientist doubts that what Palay, Palade, and others saw was a site across which chemical transmission does act – but was it the only site? Most students of the adult, vertebrate CNS would probably say ‘yes,’ all signaling between neurons within brains takes place across a 20–30 nm space; if so, the adult vertebrate's brain must be very different from other neural systems that have been examined. Indeed, many have noted that communication between a neuron and its target often takes place across distances that greatly exceed 20–30 nm in invertebrates, the vertebrate autonomic nervous system, and even the mammalian CNS (see reviews by Cuello, 1983; Jan and Jan, 1983; Vizi, 1984; Herkenham, 1987; Bach-y-Rita, 1993; Fuxe and Agnati, 1993 and others).

My own realization that conventional concepts of chemical synaptic transmission could not account for all signaling within the CNS came slowly and as a result of studying something that at first seemed unrelated to the issue – the compensations that underlie the preclinical phase of Parkinson's disease. In this chapter, I will briefly summarize some of our studies and discuss their implications for chemical transmission. (For more extensive reviews on these points, see Zigmond et al., 1993; Zigmond, 1994).

Insights from an animal model of Parkinson's disease

Characterization of Parkinson's disease and related animal models

Parkinson's disease typically emerges as a clinical entity in the sixth decade of life with the onset of tremor, rigidity, and bradykinesia. These symptoms appear to be caused primarily from a loss of neurons that utilize dopamine (DA) as their transmitter and project from the substantia nigra to the corpus striatum. Postmortem analysis of the brains of individuals who have been diagnosed with Parkinson's disease, as well as more recent imaging studies of individuals with the condition, indicate that these symptoms do not emerge until the majority of these DA neurons have been lost.

Analogous results can be obtained from animal models in which a portion of the dopaminergic projection has been destroyed. Most of these models utilize one of two neurotoxins that selectively affect dopaminergic neurons, 6-hydroxydopamine (6-OHDA) or 1-methyl-4-phenyl-1,2,3,6-tetra-hydropyridine (MPTP). In these animals, as in patients, gross neurological symptoms appear only after lesions are extensive (Ungerstedt, 1971a; Zigmond and Stricker, 1972, 1973; Marshall and Teitelbaum, 1973; Kopin and

Markey, 1988). Indeed, in these animals, whose lesions are typically abrupt and not progressive as in the case of Parkinson's disease, even the initial neurological deficits gradually dissipate (Ungerstedt, 1971; Zigmond and Stricker, 1973). This capacity of animals to withstand large DA depletions without permanent gross neurological impairments cannot be explained as a transfer of function to other systems, since functional impairments can be induced by the administration of drugs that temporarily disrupt dopaminergic function (Breese et al., 1973; Schoenfeld and Zigmond, 1973; Heffner et al., 1977).

Initial evidence for hyperactive DA neurons

It was Hornykiewicz and his colleagues who first reported on the basis of postmortem analyses that the loss of as much as 80% of the DA in caudate nucleus tissue and 90% of DA in putamen was accompanied by only mild parkinsonian impairments. At that time, Hornykiewicz noted that the concentrations of the DA metabolite homovanillic acid was less affected by the disease than were DA levels themselves, leading to a ratio of homovanillic acid to DA that reached more than 10 times control in some samples (Bernheimer and Hornykiewicz, 1965; Hornykiewicz, 1993). Sourkes and his associates soon made a similar observation in monkeys with surgical lesions of the nigrostriatal projection (Sharman et al., 1967). Likewise, we observed an increase in the ratio of dihydroxyphenylacetic acid to DA in striatal tissue of the rat after DA-depleting lesions produced by 6-OHDA (Zigmond and Stricker, 1977; Zigmond et al., 1984) and this was followed by comparable observations with several animal models of Parkinson's disease (Hefti et al., 1980; Burns et al., 1986; Duvoisin et al., 1986; Altar et al., 1987).

Hornykiewicz proposed a causal relation between the extended preclinical phase of Parkinson's disease and the elevated ratio of DA metabolite to DA in postmortem tissue. He argued that in Parkinson's disease the loss of DA from striatal tissue was proportional to the loss of DA terminals and that the relative increase in DA metabolite levels reflected an increase in DA turnover in the remaining DA neurons, which in turn reflected a compensatory increase in DA release (Hornykiewicz, 1966, 1993). We have suggested a similar hypothesis (Zigmond and Stricker, 1974; Zigmond et al., 1990, 1993).

Increased DA efflux from residual neurons

The assumption that an increase in the ratio of DA metabolites to DA in striatal tissue reflects an increase in DA turnover in residual neurons is supported by numerous observations in animal models. First, there is a strong correlation between tissue DA levels and various other specific indices of DA neurons after partial lesions (Uretsky and Iversen, 1970; Hefti et al., 1980; Zigmond et al., 1984; Onn et al., 1986; Altar et al., 1987). Second, we find an increase in the ratio of tyrosine hydroxylase activity to DA content and an increase in apparent striatal DA synthesis in residual neurons (Zigmond et al., 1984; Onn et al., 1986; see also Acheson and Zigmond, 1981) and comparable data had been reported by several other labs over the years (Bloom et al, 1969; Uretsky and Iversen, 1971; Agid et al., 1973; Altar et al., 1987).

In vivo microdialysis studies performed by us and others indicate that concentration of DA in extracellular fluid is not significantly reduced by 6-OHDA unless the depletion of DA in striatal tissue exceeds 80% (Robinson and Whishaw, 1988; Zhang et al., 1988; Abercrombie et al., 1990). Consequently, there was a large elevation in the ratio of extracellular DA to tissue DA; this ratio provides an estimate of the contribution of each surviving DA terminal to the pool of extracellular DA and is consistent with postmortem findings of increased DA turnover.

What is the basis for dopaminergic hyperactivity after lesions?

There are several possible explanations for the increased DA turnover in residual neurons after partial loss of DA neurons; perhaps the most obvious would be an increase in the firing rate of those neurons. Indeed, we found this to be the case for the noradrenergic neurons of the locus coeruleus – 6-OHDA-induced lesions failed to reduce the extracellular levels of norepinephrine in the hippocampus (Abercrombie et al., 1988) and this was associated with an increase in tyrosine hydroxylase activity (Acheson and Zigmond, 1981) and in the firing rate of locus coeruleus cells (Chiodo et al., 1983). However, when we examined residual DA neurons of the substantia nigra, no such increase in firing rate was observed, nor was there any significant change in the pattern of firing in those cells (Pucak and Grace, 1991; Hollerman and Grace, 1990).

Thus, we were forced to consider an alternative explanation – that there was an increase in the net amount of DA released from residual terminals in response to each action potential. To explore this hypothesis, we prepared striatal slices, subjected them to low frequency stimulation (2–4.5 Hz), and measured the efflux of endogenous DA into the surrounding medium. We observed that although DA efflux was reduced by 6-OHDA-induced lesions, this reduction was significantly smaller than the loss of DA from striatal tissue, resulting in an increase in fractional DA efflux (Stachowiak et al., 1987; Snyder et al., 1990). This observation suggests an increase in the net output of each residual DA terminal per depolarizing stimulus.

The amount of DA present in the superfusate is determined primarily by the rate at which DA is released from DA terminals and the rate at which it is taken back up into those terminals by high affinity transport. Whereas it is often assumed that this process operates primarily to take DA back up into the neuron from which the transmitter had been released, it is also possible that DA diffuses away from the site of release and thus is subject to uptake by neighboring nerve terminals. If so, the increase in DA overflow that we observed after 6-OHDA could have resulted from a reduction in the density of DA terminals in the slice. To test this hypothesis, we repeated our experiment in the presence of an inhibitor of high affinity DA uptake, nomifensine (10 μM). We observed that nomifensine had significantly less impact on overflow from lesioned slices than from control slices. On the other hand, at least in the case of relatively large lesions, fractional DA overflow from 6-OHDA-treated slices was elevated relative to overflow from intact slices even in the presence of nomifensine (Stachowiak et al., 1987; Snyder et al., 1990). Thus, increased ability of DA to diffuse throughout striatal tissue can evidently account for a portion of the increase in DA overflow from lesioned slices, whereas an increase in the actual amount of DA released per terminal must also occur when lesions are extensive.

An increase in DA overflow could occur at the expense of DA stores. However, this did not appear to be the case; the DA content of striatal slices prepared from 6-OHDA-lesioned animals was unaffected by up to 15 min of electrical stimulation under conditions in which overflow was increased several fold (Stachowiak et al., 1987; Snyder et al., 1990). This suggested that a concomitant increase in DA synthesis also had occurred. In support of this assumption, we found that tyrosine hydroxylation and of the activity of the tyrosine hydroxylase enzyme itself were both increased relative to DA content in striatal tissue (Zigmond et al., 1984; Onn et al., 1986; Wolf et al., 1989; Liang et al., 1993).

Mechanism of increased DA release

The increase in DA synthesis and release from residual neurons presumably results from factors operating at the level of the DA terminals. First, the hyperactivity occurred in vivo in the absence of any apparent change in the electrophysiology of DA cells in substantia nigra. Second, this hyperactivity could also be demonstrated in vitro when the firing rate of the DA neurons should have been identical in control and lesioned slices since it was determined by an imposed electrical field. How might such local regulation occur? Several similar influences have been described, including those involving glutamate, acetylcholine, GABA, and nitric oxide, as well as DA itself. Many if not all of these inputs appear to be part of local negative feedback loops (see Chesselet, 1984; Zigmond and Stricker, 1985; Zigmond, 1994). We believe that one or more of these influences are responsible for the homeostatic regulation of extracellular DA and in turn the compensatory hyperactivity that accompanies partial lesions.

One type of feedback loop responsible for the hyperactivity may involve the autoregulation of DA synthesis and release via presynaptic homoreceptors located on the DA terminal. Normally, DA overflow from striatal slices can be inhibited by DA agonists and increased by DA antagonists acting on these sites (Farnebo and Hamberger, 1971; Starke et al., 1978). However, we observed that the effect of the DA receptor antagonist, sulpiride (1 μM), is abolished after 6-OHDA lesions (Snyder et al., 1990). These data suggest that the observed increase in DA efflux per terminal after 6-OHDA lesions is due in part to a reduction in the autoinhibition of DA release.

Why would such a reduction occur? If the presynaptic DA receptors on a given terminal are affected primarily by the DA released from that terminal and the outflow from that terminal is increased as appears to be the case, then autoinhibition should be *increased*, not decreased. One likely explanation for this apparent paradox is that the major task of these autoreceptors is to monitor DA that has diffused away from *neighboring* DA terminals rather than from the terminal on which the receptors actually reside. If so, a reduction in DA terminal density would lead to less DA at any given autoreceptor; with less dopaminergic inhibition to offset, the impact of an autoreceptor inhibitor would be reduced. Additional experiments will be needed to test this hypothesis.

In addition to changes in the extent to which DA release is inhibited by neighboring DA neurons, it is also possible that non-DA neurons play a role in regulating DA release through their action at the terminal. We have not explored the role of these inputs in the response to 6-OHDA. However, we have demonstrated that glutamate can trigger the release of DA via an intrastriatal action that involves a reversal of the high affinity DA transporter (Lonart and Zigmond, 1991). Moreover, it has been reported extracellular glutamate in the striatum is increased by 6-OHDA (Tossman et al., 1986), possibly a reflection of a decrease in the normal inhibitory influence that DA appears to have on glutamate release. Thus, it seems possible that the hyperactivity of residual DA neurons that occurs after 6-OHDA is due in part to an increase in local excitatory input. Further studies are also needed here.

Is dopaminergic hyperactivity 'compensatory'?

Biochemical changes in striatal targets

DA neurons have been shown to have a variety of effects on striatal function. Perhaps the best studied of these is their ability to inhibit acetylcholine (ACh) release from cholinergic interneurons in this region (see Lehman and Langer, 1983). Thus, we employed the capacity of endogenous DA to inhibit ACh release from striatal slices as an index of the influence of DA on striatal function after partial loss of dopaminergic afferents. Previous investigators have shown that when tissue slices are preincubated with [^{3}H]choline and then superfused with Krebs bicarbonate buffer containing hemicholinium, tritium overflow in response to electrical field, stimulation can serve as a measure of ACh release (Richardson and Szerb, 1974; Somogyi and Szerb, 1972). Moreover, when the antagonist of D2 dopamine receptors, sulpiride (1 μM), is added to the superfusate, [^{3}H]choline efflux from striatal slices is increased (Cubeddu and Hoffmann, 1983), further supporting the hypothesis that DA acts at D2 receptors to exert an inhibitory influence over striatal ACh release.

In our lab, we have observed that three days after lesioning striatal DA neurons with 6-OHDA, the effects of sulpiride on ACh release were unaffected so long as at least 40% of the striatal DA content remained. Moreover, two months post-surgery the amount of DA needed to sustain a normal response to sulpiride was only 10% of control (MacKenzie et al., 1989). Still further recovery of the capacity of DA to inhibit ACh efflux can be obtained by transplanting fetal substantia nigra cells into the striatum of a 6-OHDA-lesioned rat (Carder et al., 1989). These findings are consistent with other observations on the impact of 6-OHDA on DA-ACh interactions (Grewaal et al., 1974; Guyenet et al., 1975). They also parallel the results of analogous studies that we have carried out on the impact of sulpiride on GABA efflux (Harsing and Zigmond, 1997a, see also 1997b), the firing rate of striatal cells (Nisenbaum, 1986, 1988; Orr et al.,

1986), and changes in D2 receptor (Zigmond and Stricker, 1980; see also Creese and Snyder, 1979; Heikkila et al., 1981, Staunton et al., 1981). Collectively, these data suggest that after 6-OHDA, the DA from hyperactive terminals finds its way to target neurons within the striatum, i.e. that the hyperactivity is, indeed, compensatory.

Implications for non-conventional forms of chemical transmission

These observations force upon my colleagues and myself the assumption that communication between neurons involves non-conventional influences. First, our studies indicate that after partial lesions produced by 6-OHDA, DA released from residual DA terminals – or from fetal transplants – can 'biochemically reinnervate' striatal targets. This almost surely occurs in the absence of a significant number of conventional synapses between the DA cells and the affected targets. Indeed, in the case of the dopaminergic influence on ACh efflux this must be the case since studies with TTX indicate that this is a terminal-terminal interaction and few if any axo-axonic synapses exist within the striatum (see Lehman and Langer, 1983). Thus, the non-synaptic influences of DA are not limited to instances of partial loss of DA terminals but can occur in intact tissue, as well.

Second, regulation of DA release can occur locally and in the absence of conventional synapses. After partial lesions, there is a marked increase in the DA efflux from terminals that remain, and this cannot be explained in terms of changes in the electrophysiological activity of the residual terminals; thus, local process must be at work. For example, our data suggest that DA released by one neuron can act on autoreceptors of adjacent DA neurons and be taken up by those neurons, as well. Though not an example of non-synaptic transmission, this would be an instance of neurons influencing each other across significant distances; and we have also raised the possibility of alterations in excitatory and inhibitory input from other types of neurons. As in the case of dopaminergic modulation of ACh release, such influences would have to occur in the absence of a synaptic basis.

Characteristics of communication at a distance

If such non-conventional interactions occur, they alter the traditional rules of cell–cell interactions within the CNS. For example, they suggest that transmitter release from a terminal is not only determined by events that occur at the cell body; indeed, transmitter release can be altered in the absence of any electrophysiological change. These results also suggest that one nerve terminal may act on many targets, none of which need be 'post-synaptic' to that terminal; in such a case, the channel of communication is not to be determined by traditional anatomical constraints. Instead, it is governed by the distance over which the transmitter can diffuse and the availability of receptors capable of responding to the transmitter in a highly dilute concentration. The nature of the message being conveyed by non-synaptic transmission must be rather subtle when compared to the type of information normally thought of as being conveyed by neurons; it would only be useful for certain types of communication. Indeed, relatively small changes (20–100% rather than 100-fold) would be signaled and these changes would occur over relatively long time bases (1 sec to several minutes rather than a millisecond).

Clearly, these processes would not subserve many of the most vital processes for which the nervous system has evolved; skeletal muscles will not be provided with precise and rapid instructions by transmitter diffusing slowly over many microns or even millimeters. On the other hand, the general flow of information through the corpus striatum might well be facilitated or inhibited by DA diffusing through this region; and this in the end may well influence skeletal motor function. Indeed, we can be quite sure that it does in the case of Parkinson's disease.

Questions still to be answered

There are still many questions to be answered regarding communication at a distance. For example,

- From where is the transmitter released? Does it merely spill over from conventional synapses?

Or are there separate release sites for transmitter that is to travel more than nanometer distances?
- What is the process by which release occurs? Does exocytosis from vesicles occur or is there some other process, such as reverse transport that operates?
- What is the stimulus that triggers release? Is transmitter always released by action potentials or can some other process be involved, perhaps one triggered by a transmitter acting at a distance on the terminal itself.

Conclusions

In the mid-1950s electron micrographic examination of CNS neurons led to the marriage of the twin concepts of *synapse* and of *chemical transmission*, and in an instant many set aside any further consideration of other forms of neuron-to-neuron communication. True, the conventional synapse can provide speed and a level of privacy that would be difficult to attain when transmitters act across larger distances. Conventional synapses also provide the possibility of greater fidelity between the information flowing through the axon terminal and the signal that reaches the target; and besides there must be *some* good reason for all that careful wiring and those low affinity receptors. Yet, chemical synaptic transmission, with its speed, privacy, and fidelity, may not always be a useful way for neurons to communicate with their targets – sometimes it might even be counter productive. And the evidence for non-synaptic transmission now seems overwhelming; indeed, my own feeling is that in the case of certain transmitters – and DA seems to be one – the burden of proof is no longer on those who believe that non-synaptic transmission exists. On the contrary, the real question is: exactly what are all those synapses for, anyway?

Acknowledgements

Grateful thanks to many colleagues who have participated in these experiments over the years. This work was supported by grants from the National Institutes of Health (NS19608, MH00058).

References

Abercrombie, E.D., Keller, R.W. Jr. and Zigmond, M.J. (1988) Characterization of hippocampal norepinephrine release as measured by microdialysis perfusion: pharmacological and behavioral studies. *Neuroscience*, 27: 897–904.

Abercrombie, E.D., Bonatz, A.E. and Zigmond, M.J. (1990) Effects of L-DOPA on extracellular dopamine in striatum of normal and 6-hydroxydopamine-treated rats. *Brain Res.*, 525: 36–44.

Acheson, A.L. and Zigmond, M.J. (1981) Short and long term changes in tyrosine hydroxylase activity in rat brain after subtotal destruction of central noradrenergic neurons. *J. Neurosci.*, 1: 493–504.

Agid, Y. Javoy, F. and Glowinski, J. (1973) Hyperactivity of remaining dopaminergic neurons after partial destruction of the nigro-striatal dopaminergic system in the rat. *Nature*, 245: 150–151.

Altar, C.A., Marien, M.R. and Marshall, J.F. (1987) Time course of adaptations in dopamine biosynthesis, metabolism and release following nigrostriatal lesions: implications for behavioral recovery from brain injury. *J. Neurochem.*, 48: 390–399.

Bach-y-Rita P. (1993) Non-synaptic diffusion neurotransmission (NDN) in the brain. *Neurochem. Int.*, 23: 297–318.

Bacq, Z.M., (1974) *Chemical transmission of nerve impulses: a historical sketch*. Pergamon Press, Oxford, pp. 10–22.

Bernheimer, H. and Hornykiewicz, O. (1965) Herabgestzte konzentration der homovanillinsäure im gehirn von parkinsonkranken menschen als ausdruck der störung des zentralen dopaminstoffwechsels. *Klim. Wochschr.*, 43: 711–715.

Bloom, F.E., Algeria, S., Groppetti, A., Revuelta, A. and Costa, E. (1969) Lesions of central norepinephrine terminals with 6-hydroxydopamine: biochemistry and fine structure. *Science*, 166: 1284–1286.

Breese, G.R., Smith, R.D., Cooper, B.R. and Grant, L.D. (1973) Alterations in consummatory behavior following intracisternal injection of 6-hydroxydopamine.*Pharmacol. Biochem. Behav.*, 1: 319–328.

Buma, P. and Roubos, E.W. (1986) Ultrastructural demonstration of non-synaptic release sites in the central nervous system of the snail Lymnaea stagnalis, the insect Periplaneta americana and the rat. *Neuroscience*, 17: 867–879.

Burns, R., Chiueh, C., Markey, S., Ebert, M., Jacobowitz, D. and Kopin, I. (1983) A primate model of Parkinsonism: selective destruction of dopaminergic neurons in the pars compacta of the substantia nigra by *N*-methyl-4-phenyl-1,2,3,6-tetrahydropyridine. *Proc. Natl Acad. Sci. USA*, 80: 4546–4550.

Burns, R.S., Chiueh, C.C., Parisi, J., Markey, S. and Kopin, I.J. (1986) In: S. Fahn, C.D. Marsden, P. Jenner and P. Teychenne (Eds), *Recent Developments in Parkinson's Disease*. Raven Press, New York, pp. 127–136.

Carder, R.K., Jackson, D., Morris, H.J., Lund, R.D. and Zigmond, M.J. (1989) Dopamine released from mesencephalic transplants restores modulation of striatal acetylcholine release after neonatal 6-hydroxydopamine: an in vitro analysis. *Exp Neurol.*, 105: 251–259.

Chesselet, M.F. (1984) Presynaptic regulation of neurotransmitter release in the brain. *Neuroscience*, 12: 347–475.

Chiodo, L.A., Acheson, A.L., Zigmond, M.J. and Stricker, E.M. (1983) Subtotal destruction of central noradrenergic projections increases the firing rate of locus coerulleus cells. *Brain Res.*, 264: 123–126.

Creese, I. and Snyder, S.H. (1979) Nigrostriatal lesions enhance striatal [^{3}H]apomorphine and [^{3}H]spiroperidol binding. *Eur. J. Pharmacol.*, 56: 277–281.

Cubeddu, L.X. and Hoffmann, I.S. (1983) Frequency-dependent release of acetylcholine and dopamine from rabbit striatum: its modulation by dopaminergic receptors. *J. Neurochem.*, 41: 94.

Cuello, A.C. (1983) Non-classical neuronal communications. *Fed. Proc.*, 42: 2912–2922.

Duvoisin, R.C., Heikkila, R.E., Nicklas, W.J. and Hess, A. (1986) In: S. Fahn, C.D. Marsden, P. Jenner and P. Teychenne (Eds), *Recent Developments in Parkinson's Disease*. Raven Press, New York, pp. 147–154.

Farnebo, L.O. and Hamberger, B. (1971) Drug-induced changes in the release of ^{3}H-monoamines from field-stimulated rat brain slices. *Acta Physiol. Scand. Suppl.*, 371: 35–44.

Fuxe, K. and Agnati, L.F. (Eds) (1991) Volume transmission in the brain: novel mechanisms for neural transmission. In: *Advances in Neuroscience*. Vol. 1. Raven Press, New York. pp. 1–602.

Grewaal, D., Fibiger, H. and McGeer, E. (1974) 6-Hydroxydopamine and striatal acetylcholine levels. *Brain Res.*, 73: 372–375.

Guyenet, P., Agid, Y., Javoy, F., Beaujouan, J., Rossier, J. and Glowinski, J. (1975) Effects of dopaminergic receptor agonists and antagonists on the activity of the neo-striatal cholinergic system. *Brain Res.*, 84: 227-244.

Harsing, L.G. Jr. and Zigmond, M.J. (1997a) Dopaminergic inhibition of striatal GABA release after 6-hydroxydopamine. *Brain Res Bull.*, 738: 142–145.

Harsing, L.G. Jr. and Zigmond, M.J. (1997b) Influence of dopamine on GABA release in striatum: evidence for D1-D2 interactions and non-synaptic influences. *Neuroscience*, 77: 419–429.

Heffner, T.G., Zigmond, M.J. and Stricker, E.M. (1977) Effects of dopamine agonists and antagonists on feeding in intact and 6-hydroxydopamine treated rats. *J. Pharmacol. Exp. Ther.*, 201: 386–399.

Hefti, F., Melamed, E. and Wurtman, R.J. (1980) Partial lesions of the dopaminergic nigrostriatal system in the brain: biochemical characterization. *Brain Res.*, 195: 95–122.

Heikkila, R.E., Shapiro, B.S. and Duvoisin, R.C. (1981) The relationship between loss of dopamine nerve terminals, striatal [^{3}H]spiroperidol binding and rotational behavior in unilaterally 6-hydroxydopamine lesioned rats. *Brain Res.*, 211: 285–292.

Herkenham, M. (1987) Mismatches between neurotransmitter and receptor localizations in brain: observations and implications. *Neuroscience*, 23: 1–38.

Hertting, G., Zumstein, A., Jackisch, R., Hoffman, I. and Starke, K. (1980) Modulation by endogenous dopamine of the release of acetylcholine in the caudate nucleus of the rabbit. *Naunyn-Schmiedeberg's Arch. Pharmacol.*, 315: 111.

Hollerman, J.R. and Grace, A.A. (1990) The effects of dopamine-depleting brain lesions on the electrophysiological activity of rat substantia nigra dopamine neurons. *Brain Res.*, 19; 533: 203–212.

Hollerman, J.R. and Grace, A.A. (1990) The effects of dopamine-depleting brain lesions on the electrophysiological activity of rat substantia nigra dopamine neurons. *Brain Res.*, 533(2): 203–212.

Hornykiewicz, O. (1993) Parkinson's disease and the adaptive capacity of the nigrostriatal dopamine system: possible neurochemical mechanisms. *Adv. Neurol.*, 60: 140–147.

Hornykiewicz, O. (1966) Dopamine (3-hydroxytyramine) and brain function. *Pharmacol. Rev.*, 18: 925–964.

Jan, Y.N. and Jan, L.Y. (1983) A LHRH-like peptidergic neurotransmitter capable of 'action at a distance' in autonomic ganglia. *Trends Neurosci.*, 6: 320–325.

Kopin, I.J. and Markey, S.P. (1988) MPTP toxicity: implications for research in Parkinson's disease. *Ann. Rev. Neurosci.*, 11: 81–96.

Lehmann, J. and Langer, S.Z. (1983) The striatal cholinergic interneuron: synaptic target of dopaminergic terminals? *Neuroscience*, 10: 1105–1120.

Liang, L.P. and Zigmond, M.J. (1993) Dopamine synthesis in neostriatal slices after intraventricular 6-hydroxydopamine. *Soc. Neurosci. Abstr.*, 19: 401.

Lonart, G. and Zigmond, M.J. (1991) High glutamate concentrations evoke Ca^{++}-independent dopamine release from striatal slices: a possible role of reverse dopamine transport. *J. Pharm. Exp. Ther.*, 256: 1132–1138.

Mackenzie, R.G., Stachowiak, M.K. and Zigmond, M.J. (1989) Dopaminergic inhibition of striatal acetylcholine release after 6-hydroxydopamine. *Eur. J. Pharmacol.*, 168: 43–52.

Marshall, J.F. and Teitelbaum, P. (1973) A comparison of the eating in response to hypothermic and glucoprivic challenges after nigral 6-hydroxydopamine and lateral hypothalamic electrolytic lesions in rats. *Brain Res.*, 55: 229–233.

Mayeri, E., Rothman, B.S., Brownell, P.H., Branton, W.D. and Padgett, L. (1985) Non-synaptic characteristics of neurotransmission mediated by egg-laying hormone in the abdominal ganglion of Aplysia. *J. Neurosci.*, 5: 2060–2077.

Mobley, P. and Greengard, P. (1985) Evidence for widespread effects of noradrenaline on axon terminals in the rat frontal cortex. *Proc. Natl Acad. Sci. USA*, 82: 945–947.

Nisenbaum, E.S., Stricker, E.M., Zigmond, M.J. and Berger, T.W. (1986) Long-term effects of dopamine-depleting brain lesions on spontaneous activity of Type II striatal neurons: relation to behavioral recovery. *Brain Res.*, 398: 221–230.

Nisenbaum, E.S., Stricker, E.M., Zigmond, M.J. and Berger, T.W. (1988b) Spontaneous activity of Type II but not Type I striatal neurons is correlated with recovery of behavioral function after dopamine-depleting brain lesions. *Brain Res.*, 473: 389–393.

Onn, S.P., Berger, T.W., Stricker, E.M. and Zigmond, M.J. (1986) Effects of intraventricular 6-hydroxydopamine on the dopaminergic innervation of striatum: histochemical and neurochemical analysis. *Brain Res.*, 376: 8–19.

Orr, W.B., Gardiner, T.W., Stricker, E.M., Zigmond, M.J. and Berger, T.W. (1986). Short-term effects of dopamine-depleting brain lesions on spontaneous activity of striatal neurons: Relation to local dopamine concentration and behavior. *Brain Res.*, 376: 20–28.

Pucak, M.L. and Grace, A.A. (1991) Partial dopamine depletions result in an enhanced sensitivity of residual dopamine neurons to apomorphine. *Synapse*, 9: 144–155.

Richardson, I.W. and Szerb, J.C. (1974) The release of labelled acetylcholine and choline from cerebral cortical slices stimulated electrically. *Br. J. Pharmacol.*, 52: 499–507.

Robinson, T.E. and Whishaw, I.Q. (1988) Normalization of extracellular dopamine in striatum following recovery from a partial unilateral 6-OHDA lesion of the substantia nigra: a microdialysis study in freely moving rats. *Brain Res.*, 450: 209–224.

Schmitt, F.O. (1984) Molecular regulators of brain function: a new view. *Neuroscience*, 13: 991–1001.

Schoenfeld, R.I. and Zigmond, M.J. (1973) Behavioral pharmacology of 6-hydroxydopamine. In: E. Usdin and S.H. Snyder (Eds), *Frontiers in Catecholamine Research*. Pergamon Press, New York, pp. 695–700.

Schultz, W. and Ungerstedt, U. (1978) Short-term increase and long-term reversion of striatal cell activity after degeneration of the nigrostriatal system. *Exp. Brain Res.*, 33: 159–171.

Sharman, D.F., Poirier, L.J., Murphy, G.F. and Sourkes, T.L. (1967) Homovanillicacid and dihydroxyphenylacetic acid in the striatum of monkeys with brain lesions. *Can. J. Physiol. Pharmacol.*, 45: 57–62.

Shepherd, G.M. (1991) *Foundations of the Neuron Doctrine*. NY: Oxford University Press.

Snyder, G.L., Keller, R.W., Jr. and Zigmond, M.J. (1990) Dopamine efflux from striatal slices after intracerebral 6-hydroxydopamine: evidence for compensatory hyperactivity of residual terminals. *J. Pharmacol. Exp. Therap.*, 253: 867–876.

Somogi, G.T. and Szerb, J.C. (1972) Demonstration of acetylcholine release by measuring efflux of labelled choline from cerebral cortical slices. *J. Neurochem.*, 19: 2667–2677.

Stachowiak, M.K., Keller, R.W., Jr., Stricker, E.M. and Zigmond, M.J. (1987) Increased dopamine efflux from striatal slices during development and after nigrostriatal bundle damage. *J. Neurosci.*, 7: 1648–1654.

Starke, K., Reimann, W., Zumstein, A. and Hertting, G. (1978) Effect of dopamine receptor agonists and antagonists on release of dopamine in the rabbit caudate nucleus in vitro. *Naunyn-Schmied. Arch. Pharmacol.*, 305: 27–36.

Staunton, D.A., Wolff, B.B., Groves, P.M. and Molinoff, P.B. (1981) Dopamine receptor changes following destruction of the nigrostriatal pathway: lack of a relationship to rotational behavior. *Brain Res.*, 211: 315–327.

Tossman, U., Segovia, J. and Ungerstedt, U. (1986) Extracellular levels of amino acids in striatum and globus pallidus of 6 hydroxydopamine-lesioned rats measured with microdialysis. *Acta Physiol Scand.*, 127: 547–551.

Ungerstedt, U. (1971a) Adipsia and aphagia after 6-hydroxydopamine induced degeneration of the nigro-striatal dopamine system. *Acta Physiol. Scand. Suppl.*, 367: 95–122.

Uretsky, N.J. and Iversen, L.L. (1970) Effects of 6-hydroxydopamine on catecholamine containing neurones in rat brain. *J. Neurochem.*, 17: 269–278.

Uretsky, N.J., Simmonds, M.A. and Iversen, L.L. (1971) Changes in the retention and metabolism of 3H-l-norepinephrine in rat brain in vivo after 6-hydroxydopamine pretreatment. *J. Pharmacol. Exp. Ther.*, 176: 489–496.

Vizi, E.S. (1984) *Non-synaptic interactions between neurons: Modulation of neurochemical transmission: Pharmacological and clinical aspects*. Wiley & Sons Ltd., UK, pp.1–260.

Wolf, M.E., Zigmond, M.J. and Kapatos, G. (1989) Tyrosine hydroxylase content of residual striatal dopamine nerve terminals following 6-hydroxydopamine administration: a flow cytometric study. *J. Neurochem.*, 53: 879–885.

Zhang, W.Q., Tilson, H.A., Nanry, K.P., Hudson, P.M. and Hong, J.S. and Stachowiak, M.K. (1988) Increased dopamine release from striata of rats after unilateral nigrostriatal bundle damage. *Brain Res.*, 461: 335–342.

Zigmond, M.J. Chemical transmission in the brain: homeostatic regulation and its functional implications. (1994) In: F. Bloom (Ed.), *Progress in Brain Research Vol. 100*, Elsevier Science B.V., pp. 115–122.

Zigmond, M.J. (1999) Otto Loewi and the demonstration of chemical neurotransmission, *Brain Res. Bull.*, 50: 347–348.

Zigmond, M.J., Abercrombie, E.D., Berger, T.W., Grace, A.A. and Stricker, E.M. (1990) Compensations after lesions of central dopaminergic neurons: some clinical and basic implications. *Trends Neurosci.*, 13: 290–296.

Zigmond, M.J., Abercrombie, E.D., Berger, T.W., Grace, A.A. and Stricker, E.M. (1993) Compensatory responses to partial loss of dopaminergic neurons: studies with 6-hydroxydopamine. In: J.S. Schneider and M. Gupta (Eds), *Current Concepts in Parkinson's Disease Research*, Hogrefe & Huber, Toronto, Ontario, pp. 99–140.

Zigmond, M.J., Acheson. A.L., Stachowiak, M.K. and Stricker, E.M. (1984) Neurochemical compensation after nigrostriatal bundle injury in an animal model of preclinical Parkinsonism. *Arch Neurol.*, 41: 856–861.

Zigmond, M.J. and Stricker, E.M. (1972) Deficits in feeding behavior after intraventricular injection of 6-hydroxydopamine in rats. *Science*, 177: 1211–1214.

Zigmond, M.J. and Stricker, E.M. (1973) Recovery of feeding and drinking by rats after intraventricular 6-hydroxydopamine or lateral hypothalamic lesions. *Science*, 182: 717–720.

Zigmond, M.J. and Stricker, E.M. (1977) Behavioral and neurochemical effects of central catecholamine depletion: a possible model for 'subclinical' brain damage. In: I. Hanin and E. Usdin (Eds), *Animal Models in Psychiatry and*

Neurology, Pergamon Press, New York, pp. 415–429.

Zigmond, M.J. and Stricker, E.M. (1980) Supersensitivity after intraventricular 6-hydroxydopamine: Relation to dopamine depletion. *Experimentia*, 36: 436–437.

Zigmond, M.J. and Stricker, E.M. (1985) Adaptive properties of monoaminergic neurons. In: A. Lajtha (Ed.), *Handbook of Neurochemistry, Alterations of Metabolites in the Nervous System*, Vol. 9, Plenum, New York, pp. 87–102.

SECTION V

The wider world of VT – from ions to peptides

L. F. Agnati, K. Fuxe, C. Nicholson and E. Syková (Eds.)
Progress in Brain Research, Vol 125

CHAPTER 20

GABAergic excitation and K^+-mediated volume transmission in the hippocampus

Juha Voipio and Kai Kaila

Department of Biosciences, Division of Animal Physiology, University of Helsinki, PO Box 17, FIN-00014 Helsinki, Finland

Introduction

Fast synaptic inhibition mediated by GABA (γ-aminobutyric acid) is based on the opening of anion channels which are mainly permeable to chloride ($GABA_A$ receptors). In various kinds of neurons within the brain, $GABA_A$ receptor activation produces a fast inhibitory postsynaptic potential (fIPSP) that is hyperpolarizing and reflects active K^+-dependent extrusion of Cl^- (e.g. Thompson et al., 1988b). Such 'conventional' fIPSPs are, for instance, typically seen in hippocampal pyramidal neurons where the Cl^- driving force is most likely generated by the K^+-Cl^- cotransporter isoform, KCC2 (Rivera et al., 1999). Nevertheless, there are notable and even apparently paradoxical exceptions to this pattern. An important example here is provided by data from neocortical pyramidal neurons, which very often show slightly depolarizing fIPSPs (see Kaila et al., 1993) despite the fact that they have also been unequivocally demonstrated to have a K^+-Cl^- cotransporter in their plasma membrane (Thompson et al., 1988a, b).

Just about a decade ago, we used crayfish preparations to test the idea that fast GABA responses are not solely mediated by Cl^- (Kaila and Voipio, 1987; Kaila et al., 1989; Kaila et al., 1990; Voipio et al., 1991). In brief, using ion-sensitive microelectrodes for measurements of intracellular chloride and bicarbonate in combination with a three-microelectrode voltage clamp, we were able to demonstrate that HCO_3^- can act as a significant carrier of an inward (depolarizing) current component across ionotropic GABA receptor channels, which produces a large deviation of the reversal potential of the anion-mediated GABA response ($E_{GABA\text{-}A}$) from the equilibrium potential of Cl^- (E_{Cl}) to more positive voltages. Thereafter we proceeded to demonstrate that, in fact, the steady-state reversal potential of fIPSPs in neocortical neurons is roughly 10 mV more positive than E_{Cl}, and an alkaline intracellular pH shift will increase this deviation simply because of an increase in the intracellular bicarbonate concentration (Kaila et al., 1993; see also Kaila and Voipio, 1990).

The identification of HCO_3^- as a physiologically relevant current carrier gave a straightforward explanation for the (at that time enigmatic) observations of moderately depolarizing responses in neurons with a hyperpolarizing driving force for purely Cl^- mediated currents. Simple quantitative considerations based on the Goldman–Hodgkin–Katz equation predict that the bicarbonate-dependent positive deviation of $E_{GABA\text{-}A}$ from E_{Cl} is physiologically significant in most kinds of Cl^--extruding neurons. Furthermore, this deviation can lead to 'paradoxical' depolarizing fIPSPs under steady-state conditions particularly in Cl^--extruding neurons that have a distinctly negative resting membrane potential (Kaila and Voipio, 1990), a

Fax +358-9-191 28701
e-mail: juha.voipio@helsinki.fi; kai.kaila@helsinki.fi

situation that holds for neocortical pyramidal neurons *in vitro* (Kaila et al., 1993), and perhaps also for the granule cells in the dentate gyrus (but see Misgeld et al., 1986).

However, much more pronounced GABA-mediated depolarizing responses than those described in the studies cited above can be observed in hippocampal pyramidal neurons under non-steady-state conditions. In the presence of ionotropic glutamatergic antagonists, a bicarbonate-dependent biphasic response with an initial hyperpolarization followed by a long-lasting depolarization is readily evoked in a hippocampal slice preparation by a high-frequency stimulation of GABAergic inputs onto hippocampal CA1 pyramidal neurons (Davies and Collingridge, 1993; Grover et al., 1993; Staley et al., 1995; Kaila et al., 1997; Jackson et al., 1999). For instance, application of a train of 40 pulses at 100 Hz initially leads to a fast hyperpolarizing (h) response that is a result of temporal integration of fIPSPs. Following this hIPSP, it takes about 250 ms until the membrane potential of the postsynaptic cell undergoes a large prolonged depolarization, the GDNSP (see below and Fig. 1A), which is often associated with trains of spikes at its peak (Kaila et al., 1997).

The mechanisms underlying the GDNSP will be in the focus of the present review. The GDNSP is physiologically interesting for a number of reasons: it is generated by stimulation trains of the type that are widely used for the induction of long-term potentiation (LTP) in the CA1 area (e.g. Bliss and Collingridge, 1993); and recent work suggests that the mechanisms generating the GDNSP are also of crucial importance in the generation of high-frequency stimulation-induced synchronous gamma/beta-frequency activity in pyramidal neurons (Taira et al., 1997; Bracci et al., 1999).

In particular, we will provide evidence that the GDNSP (the *G*ABA-mediated *d*epolarizing *n*on-*s*ynaptic *p*otential) is largely due to an activity-induced potassium transient, which has a direct depolarizing effect even in the absence of glutamatergic transmission. This transient increase in the extracellular K^+ concentration ($[K^+]_o$) seems to be tightly linked to the $GABA_A$ receptor-dependent coupling in the local interneuronal network (Michelson and Wong, 1991, 1994) that is

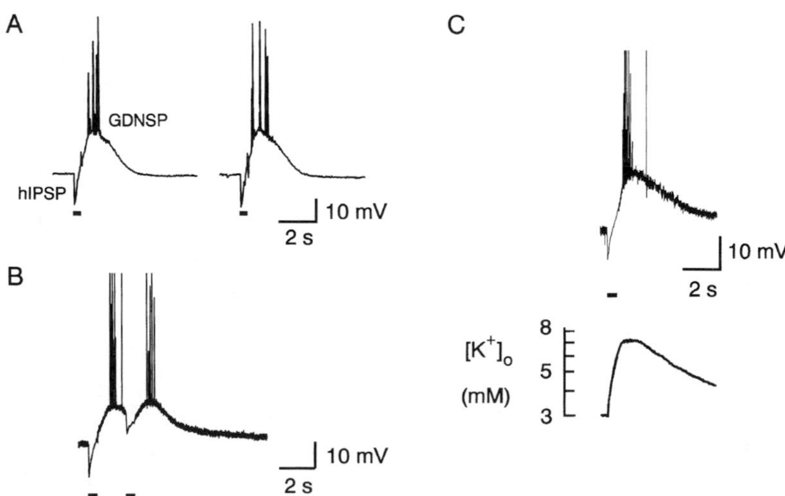

Fig. 1. The peak depolarization during a GDNSP attains a value much more positive than $E_{GABA\text{-}A}$, and is closely linked to an increase in interstitial $[K^+]$. A, two trains of stimuli (40 pulses, 100 Hz; horizontal bars) applied at an interval of a few minutes trigger an hIPSP-GDNSP sequence with constant properties. B, A prompt hyperpolarization is seen if the 2nd train is applied at the peak GDNSP. C, The GDNSP is paralleled by a $[K^+]_o$ transient large enough to account for the depolarization. All recordings (including those depicted in Fig. 2) were obtained in the presence of ionotropic glutamate receptor and $GABA_B$ receptor antagonists, with stimuli applied at a distance of about 0.5 mm to the site of recording in the CA1 stratum pyramidale. (Modified from Kaila et al., 1997).

functionally excitatory in the presence, but not in the absence, of CO_2 and HCO_3^- (Lamsa and Kaila, 1997). In general, it appears that bicarbonate-dependent volume transmission mediated by K^+ plays a salient role in GABAergic neuronal responses during synchronous high-frequency activity.

Another type of ionic shift that is triggered by tonic GABAergic activity is an intraneuronal acidicosis coupled to an extracellular alkaline pH change, which are direct consequences of the $GABA_A$ receptor-mediated net efflux of HCO_3^- (Kaila and Voipio, 1987). In view of the large diversity of both ligand-gated and voltage-gated channels, which are sensitive to perturbations of extracellular pH, both H^+ and HCO_3^- transients generated in the interstitial space are likely to have a volume signaling role of their own. As these mechanisms have been recently reviewed (Kaila and Chesler, 1998; Voipio, 1998), we will not discuss this topic in this chapter.

A brief history of biphasic GABA-mediated responses

During the last two decades, a large number of studies have been devoted to a search of the mechanisms that cause the polarity change from hyperpolarizing to depolarizing during the course of biphasic GABAergic responses. This work has led to the following hypotheses, that will be evaluated below in the light of the data available at present.

Are biphasic responses due to a heterogeneity of $GABA_A$ receptors?

In the early 1980s, Alger and Nicoll suggested that extrasynaptic $GABA_A$ receptors with properties distinct from their synaptic counterparts are activated by transmitter spillover, which leads to the late depolarizing response (Alger and Nicoll, 1982a, b). This idea was based on results from experiments on hippocampal slices with local iontophoretic application of GABA and the $GABA_A$ receptor specific agonist THIP (4,5,6,7-Tetrahydroisoxazolo[5,4-c] pyridin-3-ol). The responses were hyperpolarizations upon somatic agonist injections whereas dendritic injections often gave rise to monophasic depolarizations or biphasic hyperpolarization-depolarization sequences, where the depolarizing responses were enhanced in a dose-dependent manner. In addition to these site- and dose-dependencies, further evidence supporting the suggested model was obtained by pharmacological manipulations involving up-modulation of $GABA_A$ receptor responses (pentobarbital), and inhibition of GABA uptake (cf. Jackson et al., 1999) both resulting in an enhancement of the depolarizations.

But what kind of a receptor heterogeneity might explain the dramatic difference in the voltage responses generated? Recently, Perkins and Wong (Perkins and Wong, 1996; Perkins, 1999) have postulated that long-lasting depolarizing $GABA_A$ responses are generated by $GABA_A$ receptors that show an exceptionally high permeability to HCO_3^-. This implies that the postulated dendritic receptors (e.g. Perkins, 1999) have $GABA_A$ receptors with an ionic filter that is radically different from all those described so far in the literature. A compilation of the pertinent data from invertebrates, as well as from vertebrate species ranging from amphibians to mammals, indicates that the ionic filter appears to be evolutionarily highly conserved (see Fig. 3 in Kaila, 1994). More recent results on the so-called $GABA_C$ receptors support this conclusion (Bormann and Feigenspan, 1995). Taken together, all the permeability estimates so far obtained for this wide diversity of ionotropic GABA receptors yield values within a rather narrow range from about 0.2 to 0.3 for the relative HCO_3^- vs. Cl^- permeability. It appears, therefore, that the rather indirect evidence taken to suggest the existence of aberrant $GABA_A$ receptors with a predominant selectivity to HCO_3^- in pyramidal cell dendrites should be evaluated in the light of the direct biophysical data reviewed below.

Are biphasic responses due to an intraneuronal anion gradient?

The polarity and driving force of a channel-mediated current is not only determined by the ionic permeability properties but also by the transmembrane gradient(s) of the ion(s) involved.

Hence, one might hypothesize that if the GABA-mediated depolarizing responses are generated in dendrites, this situation could perhaps be caused by an intraneuronal anion gradient, where internal [Cl^-] (or internal [HCO_3^-]) is higher in dendrites than in the soma. Indeed, the well-known dual-pump hypothesis (Misgeld et al., 1986; see also Andersen et al., 1980) postulated an intracellular Cl^- gradient as a consequence of active Cl^- uptake in the dendrites and Cl^- extrusion in the soma. This model became a widely-recognized explanation for the qualitative difference between somatic and dendritic GABAergic responses. However, this scheme was based on indirect (pharmacological) evidence and it was not consistent with some key observations, especially with the fact that both synaptic and pharmacological activation of dendritic $GABA_A$ receptors can lead to hyperpolarizing responses (for references, see Kaila, 1994).

In sharp contrast to the original dual-pump hypothesis, a recent electrophysiological study by Misgeld and coworkers (Jarolimek et al., 1999) on cultured midbrain neurons provided convincing evidence for the idea that neuronal Cl^- *extrusion* mainly takes place in the dendrites. Their findings are consistent with novel immunohistochemical results demonstrating a predominantly dendritic localization of the K^+-Cl^- cotransporter, KCC2 (Freund et al., 1999).

Are biphasic responses due to HCO_3^--dependent channel-mediated uptake of Cl^-?

One of the important consequences of the inward current component carried by HCO_3^- is its ability to cause intracellular Cl^- accumulation, i.e. a positive shift in the equilibrium potential of Cl^-. Such a bicarbonate-dependent redistribution of Cl^- was first described in crayfish muscle fibers (Kaila and Voipio, 1987; Kaila et al., 1989). The most convincing evidence was provided by experiments where direct intracellular ion measurements showed that the GABA-induced increase in the intracellular Cl^- concentration was blocked if the simultaneous membrane depolarization was prevented by voltage clamp, while a decrease in intracellular bicarbonate was observed both in the absence and presence of voltage clamp. This is consistent with passive redistribution of Cl^- leading to intracellular chloride accumulation, i.e. a channel-mediated influx of Cl^- down its electrochemical gradient during the HCO_3^- driven depolarization (Kaila et al., 1989). This chain of events leads to a progressive positive shift in $E_{GABA\text{-}A}$.

Using a model incorporating the elements above, Staley and coworkers (Staley et al., 1995; Staley and Proctor, 1999) have proposed that a bicarbonate driven collapse of the postsynaptic transmembrane Cl^- gradient is the cause of the biphasic GABAergic responses in hippocampal pyramidal neurons. There is no doubt that a passive redistribution of Cl^- must take place if HCO_3^- plays a role in a tonic $GABA_A$ receptor-mediated postsynaptic response, thereby contributing to the change in polarity that is characteristic in the biphasic response. However, the crucial question that arises here is whether this type of Cl^- accumulation is sufficient to account for the voltage and current responses that are seen during the late depolarizing (inward current) phase following tetanic activation of the GABAergic afferents.

In what follows, we will provide evidence that indicates an involvement of a potassium-based volume transmission mechanism in the generation of the late depolarizing phase. In other words, following tonic activation of $GABA_A$ receptors, the neuronal responses are composed of a target-specific signal mediated by synaptic transmission, and a more diffuse signal attributable to a GABAergic HCO_3^--dependent increase in the interstitial [K^+]. The relative contributions of these two components depend on the experimental conditions where, for instance, the frequency and duration of the afferent stimulus train plays a decisive role. It should be evident from the data described below, that an increase in the gross activation of the local GABAergic network will lead to a more pronounced depolarizing component (e.g. Grover et al., 1993) that reflects a more prominent (non-synaptic) signalling component mediated by the potassium transient (see below). An important point here is that with high-frequency stimulation paradigms akin to those used to induce LTP and tetanus gamma activity (e.g. 40–100 pulses given at 100 Hz), only a small fraction of the depolarizing phase of the biphasic GABA response can be explained on the basis of purely channel-mediated anion shifts (cf. Kaila et al., 1989) in the target neuron. With briefer depolarizing responses, the non-synaptic mechanism is less dominant, and I–V curves with well-defined reversals can be readily obtained (cf. Cobb et al., 1999). A clear reversal of depolarizing GABA response is also seen in cells internally perfused with a high concentration of QX–314 (cf. Perkins and Wong, 1996) which results in a block of the nonsynaptic potassium mediated input as described below.

Here, it should also be clearly stated that the above findings do not rule out the possibility that under other kinds of experimental conditions, such as those involving cultured neurons (Barker and Ransom, 1978), $GABA_A$ receptor-mediated depolarizations may well show a more pronounced dependence on channel-mediated anionic shifts than those seen in the above studies in slice preparations.

Influence of interstitial K^+ transients on the generation of biphasic GABA-mediated responses

What is the crucial physiological/biophysical evidence for the presence of a non-synaptic, volume-transmission mechanism in the depolarizing part of the biphasic response? A series of key findings were obtained in experiments with two identical trains of stimuli (40 pulses at 100 Hz) given in the presence of antagonists of ionotropic glutamate receptors and $GABA_B$ receptors (Kaila et al., 1997). Here, the 1st train evoked a typical hIPSP-GDNSP sequence. However, if the 2nd train was applied at the *peak depolarization* during the GDNSP, the result was not an accelerated depolarization but a transient hyperpolarization (see Fig. 1B), which clearly indicated that $E_{GABA\text{-}A}$ *must be more negative than the membrane potential at the time of the maximum depolarization*. Therefore, the late depolarization which is dependent on both GABAergic transmission and on the presence of bicarbonate is obviously attributable to something else than just the postsynaptic current mediated by $GABA_A$ receptor channels.

A standard approach in the identification of postsynaptic ionic mechanisms is to examine the current-voltage relationships seen under voltage-clamp conditions. Indeed, in I–V plots taken at various time points starting from the beginning of the hIPSP–GDNSP sequence, a gradual transition was seen from outwardly rectifying currents with reversals typical for $GABA_A$ receptor-mediated IPSPs during the hIPSP time window, to inwardly rectifying currents with progressively more positive reversals. Those currents that corresponded to the peak and late phase of the GDNSP were strongly inwardly rectifying and did not show any clear reversal.

Since currents that are gated by an increase in membrane conductance reverse at predictable membrane potentials, the absence of reversal of the GABAergic late inward current suggested that it could be due to a transient change in ionic concentration on one side of the membrane. Simultaneous measurements (in the presence of inhibitors of ionotropic glutamatergic and $GABA_B$ receptors) of the stimulation-induced GABAergic biphasic responses and extracellular ion concentrations revealed a bicarbonate-dependent transient increase in $[K^+]_o$ with a time course resembling that of the depolarization (Fig. 1C; Kaila et al., 1997). This, together with the sub-Nernstian dependence of the depolarization on $[K^+]_o$, led to the conclusion that the biphasic response is a result of a shift from fused inhibitory postsynaptic potentials to a voltage response evolved by increased $[K^+]_o$. Intracellular glial recordings verified the existence of a fast local increase in $[K^+]_o$ which depolarized the glial cells and resulted in I–V characteristics qualitatively similar (inward rectification; absence of reversal) to those of the late phase of the pyramidal neuron responses (Kaila et al., 1997).

An important aspect of the above findings is that, during tetanic stimulation, the GABAergic signaling mechanisms in a hippocampal slice can undergo a smooth transition from conventional postsynaptic inhibition to excitatory volume transmission mediated by extracellular K^+.

In a recent study (Smirnov et al., 1999a), we used pharmacological tools to isolate the two parts of the biphasic response in a target neuron. The results that are most pertinent with respect to the topic of volume transmission can be summarized as follows: First, the GABAergic transient rise in $[K^+]_o$ is generated by synchronous network activity and causes a depolarization in a given pyramidal neuron even if all the target neuron's $GABA_A$ receptor channels have been blocked by internal perfusion of a high concentration of fluoride (Fig. 2; for the actions of internal F^-, see Bormann et al., 1987; Nelson et al., 1994). This means that the latter phase of the biphasic hyperpolarization-depolarization response will be seen even in the absence of any kind of synaptic input to the cell (hence the attribute 'non-synaptic' in the term GDNSP), a situation reminiscent to that encountered with the glial cells within the same area, as described earlier on. Second, the stimulation-induced rise in $[K^+]_o$ was found to be strongly inhibited (although not fully blocked) by quinine or quinidine. Measurements of interstitial pH transients indicated that in the presence of quini(di)ne, the $GABA_A$ receptor-mediated CO_2/HCO_3^- shuttle was fully intact and it was hence possible to rule out a major role of bicarbonate-driven chloride accumulation in the generation of the GDNSP. (Readers interested in details may refer to Fig. 4D in Smirnov et al., 1999a, for a quantitative comparison of the relative contributions of the shifts in intracellular $[Cl^-]$ and extracellular $[K^+]$.)

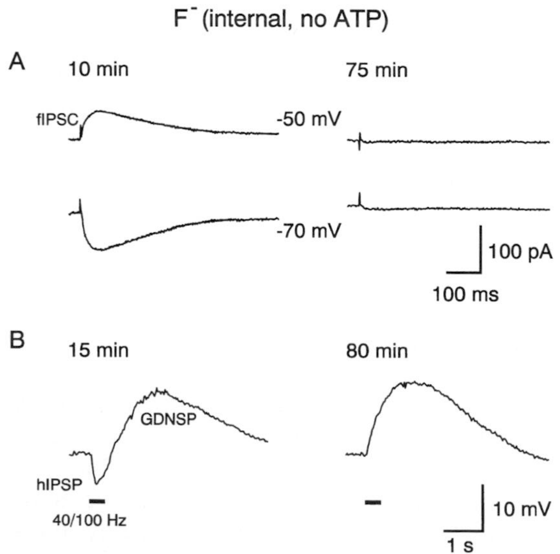

Fig. 2. Direct demonstration of the non-synaptic nature of the GDNSP. Intracellular perfusion for >1 h with fluoride and no ATP blocks $GABA_A$ receptor-mediated fIPSCs (A), and hIPSPs, but does not suppress the GDNSPs recorded in the same CA1 pyramidal neuron (B). (Modified from Smirnov et al., 1999).

A further means to isolate the hyperpolarizing and depolarizing phases of the biphasic response from each other was intracellular perfusion with a high concentration of (5 mM) of the lidocaine derivative QX-314, which selectively blocked the GDNSP while having no effect on the hIPSP. Unlike quini(di)ne, internal QX-314 reduced the sensitivity of the pyramidal neuron to the depolarizing action of brief puffs of K^+, which nicely fits with the non-synaptic, potassium-based model of GDNSP generation (Smirnov et al., 1999a).

GABA and neuronal volume regulation

The $GABA_A$ channel-mediated efflux of HCO_3^- and influx of Cl^- do not produce quantitatively identical changes in the intracellular concentrations of these anions, because the loss of cytoplasmic HCO_3^- is buffered by an influx and intracellular net hydration of CO_2 (see the scheme in Fig. 8 in Kaila et al., 1990). Therefore, during tonic $GABA_A$ receptor activation at an 'electrical' steady-state (i.e. at the time when these fluxes have large and equal values), *the net electric current is zero but yet the intracellular anion content is increasing* (Voipio et al., 1999; note that electroneutrality is maintained by the protons that are released in the net hydration reaction and subsequently taken up by intrinsic buffers). This anion-mediated increase in intraneuronal osmotic pressure leads to swelling unless the cell is equipped with effective means for volume regulation.

Recent data from our laboratory suggest a developmental change, where $GABA_A$ receptor activation leads to neuronal swelling in immature neurons, whereas in more mature neurons the osmotic load is dealt with by an effective volume regulatory response (Voipio et al., 1999). It is intriguing that, in the rat hippocampus, this developmental switch in volume regulation takes place by the end of the first postnatal week, hence coinciding with the expression of KCC2 (cf. Rivera et al., 1999). As to the mechanism(s) underlying the bicarbonate-dependent GABAergic K^+ shifts, it is important to note that *both types* of responses lead to an increase in $[K^+]_o$: in the absence of neuronal volume regulation in response to a GABAergic anion load, neuronal swelling leads to a fall in the fractional volume of the interstitial space, i.e. to a net loss of interstitial water and a consequent increase in the K^+ concentration. On the other hand, if the volume-regulatory response is based on a potassium-dependent mechanism (such as KCC2), the net efflux of K^+ required to maintain a constant neuronal volume results in a net increase in extracellular K^+ and hence, also to an elevated $[K^+]_o$.

Finally, we want to emphasize that tonic activation of $GABA_A$ receptors in the presence of bicarbonate (whether brought about by stimulation of interneuronal networks or using bath-applied agonists; see Barolet and Morris, 1991) will always lead to an increase in $[K^+]_o$, which cannot be ignored in investigations of GABAergic depolarizing responses. This is because any kinds of changes in $[K^+]_o$ do not only affect neuronal membrane potentials, but they also have a powerful effect on intraneuronal Cl^- homeostasis through a direct effect on neuronal K^+–Cl^- cotransport (cf. Thompson et al., 1988b; Jensen et al., 1993).

K^+-mediated volume transmission and the excitability of the interneuronal network

In the light of the evidence described in the work cited so far, the increase in $[K^+]_o$ following high-frequency stimulation is caused by the synchronous activity of GABAergic interneurons, resulting in a long-lasting depolarization of neurons. From a mechanistic point of view, it is of much interest here that a synchronous activation of GABAergic interneurons can be brought about in the absence of stimulation, by exposing a slice preparation to a low concentration (50–100 μM) of 4-aminopyridine (4-AP; see Storm, 1990), in the presence and absence of ionotropic glutamate receptor antagonists (Perreault and Avoli, 1992). This results in spontaneous propagating waves of excitation within the interneuronal network, which are associated with $[K^+]_o$ transients paralleled by long-lasting depolarizations (LLDs) in pyramidal neurons. The LLDs have a number of characteristics similar to the high-frequency stimulation-induced GDNSP, including a sensitivity to quini(di)ne (Smirnov et al., 1999b) and dependence on bicarbonate.

That a postsynaptic $GABA_A$ receptor-mediated excitatory bicarbonate current generated in interneurons is crucially involved in the 4-AP-induced LLDs is supported by a number of observations (Lamsa and Kaila, 1997), including the finding that LLDs are abolished upon wihdrawal of CO_2/bicarbonate. A key observation here was that subsequent addition of formate, a weak-acid anion with a significant permeability in $GABA_A$ receptor channels (Kaila, 1994) and capable of producing a depolarizing current under physiological conditions (Mason et al., 1990), re-established the generation of LLDs and the associated $[K^+]_o$ transients.

The tight link between the $[K^+]_o$ transients and LLDs raises an intriguing possibility (cf. Avoli et al., 1996) that is currently being examined in our laboratory: perhaps the K^+-shifts seen in the presence of 4-AP are volume transmission signals that are largely responsible for the long-lasting excitation within the interneuronal population. This view is fully consistent with the findings by Michelson and Wong (1991, 1994), who reported the presence of paradoxically excitatory $GABA_A$ receptor-mediated postsynaptic responses mediated between interneurons in the presence of 4-AP and, notably, following an elevation of extracellular K^+. Another point worth of attention here is that quini(di)ne has a blocking action both on the 4-AP-induced LLDs and the associated $[K^+]_o$ shifts at exactly the same concentrations that are needed to block the stimulation-induced GDNSPs and the accompanying $[K^+]_o$ transients (Smirnov et al., 1999b).

In the light of the crucial role of extracellular K^+ in the generation and enhancement of excitatory coupling among interneurons (Michelson and Wong, 1991, 1994), the 4-AP model of interneuronal excitation and synchronization may indicate a positive feedback loop, where an initially weak coupling among interneurons results in a rise in $[K^+]_o$, which in turn enhances the excitability of the network. The potassium-induced increase in excitability may be analogous to the tonic depolarizing action of certain types metabotropic glutamate receptors (Miles and Poncer, 1993), and a likely additional factor is that the increase in $[K^+]_o$ acts on interneuronal K^+-Cl^- cotransport, hence leading to a depolarizing shift in E_{GABA-A}.

Conclusions

In summary, it appears that bicarbonate-dependent GABAergic $[K^+]_o$ transients are volume transmission signals that do not affect only those neurons that are postsynaptic with regard to the stimulated ones, but they will have a depolarizing influence on all nearby neurons, glial cells as well as on presynaptic terminals (cf. Poolos et al., 1987; Stasheff et al., 1993). The GABAergic high-frequency stimulation (HFS) evoked $[K^+]_o$ shift described in the present review acts at a time scale fast enough to have a pronounced influence on 'conventional' postsynaptic mechanisms involved in responses generated by trains of HFS. In fact, the idea that $[K^+]_o$ shifts can play a role in the induction of LTP has been repeatedly discussed in the literature (Collingridge, 1992; Bliss and Collingridge, 1993).

It also appears that $[K^+]_o$ transients are able to promote synchronous activity of the interneuronal network, whether triggered by high-frequency stimulation or application of 4-AP. At the pyramidal cell level, the stimulation-induced tonic activation of the interneuronal population and the associated ionic shifts can result in GABAergic excitatory responses that are also observed in the absence of glutamate antagonists (Taira et al., 1997). Recently, it has been postulated (Bracci et al., 1999) that the GABAergic shrinkage of the interstitial space accompanying high-frequency stimulation (Autere et al., 1999) is a key factor involved in the generation of this 'tetanus gamma' (but see also Whittington et al., 1997; Traub et al., 1999), in that it enhances ephaptic interactions between pyramidal neurons.

Acknowledgements

The original work of the authors has been supported by grants from the Academy of Finland (Finnish Centre of Excellence Programme 2000–2005) and from the Sigrid Jusélius Foundation.

References

Alger, B.E. and Nicoll, R.A. (1982a) Feed-forward dendritic inhibition in rat hippocampal pyramidal cells studied in vitro. *J. Physiol. (Lond.)*, 328: 105–123.

Alger, B.E. and Nicoll, R.A. (1982b) Pharmacological evidence for two kinds of GABA receptor on rat hippocampal pyramidal cells studied in vitro. *J. Physiol. (Lond.)*, 328: 125–141.

Andersen, P., Dingledine, R., Gjerstad, L., Langmoen, I.A. and Mosfeldt Laursen, A.M. (1980) Two different responses of hippocampal pyramidal cells to application of γ-amino butyric acid. *J. Physiol. (Lond.)*, 305: 279–296.

Autere, A.-M., Lamsa, K., Kaila, K. and Taira, T. (1999) Synaptic activation of $GABA_A$ receptors induces neuronal uptake of Ca^{2+} in adult rat hippocampal slices. *J. Neurophysiol.*, 81: 811–816.

Avoli, M., Louvel, J., Kurcewicz, I., Pumain, R. and Barbarosie, M. (1996) Extracellular free potassium and calcium during synchronous activity induced by 4-aminopyridine in the juvenile rat hippocampus. *J. Physiol. (Lond.)*, 493: 707–717.

Barker, J.L. and Ransom, B.R. (1978) Amino acid pharmacology of mammalian central neurones grown in tissue culture. *J. Physiol. (Lond.)*, 280: 331–354.

Barolet, A.W. and Morris, M.E. (1991) Changes in extracellular K^+ evoked by GABA, THIP and baclofen in the guinea-pig hippocampal slice. *Exp. Brain Res.*, 84: 591–598.

Bliss, T.V.P. and Collingridge, G.L. (1993) A synaptic model of memory: long-term potentiation on the hippocampus. *Nature*, 361: 31–39.

Bormann, J., Hamill, O.P. and Sakmann, B. (1987) Mechanism of anion permeation through channels gated by glycine and γ-aminobutyric acid in mouse cultured spinal neurones. *J. Physiol. (Lond.)*, 385: 243–286.

Bormann, J. and Feigenspan, A. (1995) $GABA_C$ receptors. *Trends Neurosci.*, 18: 515–519.

Bracci, E., Vreugdenhil, M., Hack, S.P. and Jefferys, J.G. (1999) On the synchronizing mechanism of tetanically induced hippocampal oscillations. *J. Neurosci.*, 19: 8104–8113.

Cobb, S.R., Manuel, N.A., Morton, R.A., Gill, C.H., Collingridge, G.L. and Davies, C.H. (1999) Regulation of depolarizing $GABA_A$ receptor-mediated synaptic potentials by synaptic activation of $GABA_B$ autoreceptors in the rat hippocampus. *Neuropharmacology.*, 38: 1723–1732.

Collingridge, G.L. (1992) The mechanism of induction of NMDA receptor-dependent long-term potentiation in the hippocampus. *Exp. Physiol.*, 77: 771–797.

Davies, C.H. and Collingridge, G.L. (1993) The physiological regulation of synaptic inhibition by $GABA_B$ autoreceptors in rat hippocampus. *J. Physiol. (Lond.)*, 472: 245–265.

Freund, T.F., Gulyás, A.I., Sik, A., Payne, J.A. and Kaila, K. (1999) Distribution of the K-Cl cotransporter (KCC2) parallels glutamatergic input in the developing and adult hippocampus. *Soc. Neurosci. Abstr.*, 25: 881.11.

Grover, L.M., Lambert, N.A., Schwartzkroin, P.A. and Teyler, T.J. (1993) Role of HCO_3^- ions in depolarizing $GABA_A$ receptor-mediated responses in pyramidal cells of rat hippocampus. *J. Neurophysiol.*, 69: 1541–1555.

Jackson, M.F., Esplin, B. and Capek, R. (1999) Inhibitory nature of tiagabine-augmented $GABA_A$ receptor-mediated depolarizing responses in hippocampal pyramidal cells. *J. Neurophysiol.*, 81: 1192–1198.

Jarolimek, W., Lewen, A. and Misgeld, U. (1999) A furosemide sensitive K^+-Cl^- cotransporter counteracts intracellular Cl^- accumulation in cultured rat midbrain neurons. *J. Neurosci.*, 19: 4695–4704.

Jensen, M.S., Cherubini, E. and Yaari, Y. (1993) Opponent effects of potassium on $GABA_A$-mediated postsynaptic inhibition in the rat hippocampus. *J. Neurophysiol.*, 69: 764–771.

Kaila, K. (1994) Ionic basis of $GABA_A$ receptor channel function in the nervous system. *Progr. Neurobiol.*, 42: 489–537.

Kaila, K. and Chesler, M. (1998) Activity-evoked changes in extracellular pH. In: K. Kaila and B.R. Ransom (Eds), *pH and Brain Function*, Chap. 17, Wiley-Liss, New York, pp. 309–337.

Kaila, K. and Voipio, J. (1987) Postsynaptic fall in intracellular pH induced by GABA-activated bicarbonate conductance. *Nature*, 330: 163–165.

Kaila, K. and Voipio, J. (1990) GABA-activated bicarbonate conductance: influence on E_{GABA} and on postsynaptic pH regulation. In: F.J. Alvarez-Leefmans and J.M. Russell (Eds), *Chloride Channels and Carriers in Nerve, Muscle and Glial Cells*, Chapter 12, Plenum Press, New York, pp. 331–352.

Kaila, K., Pasternack, M., Saarikoski, J. and Voipio, J. (1989) Influence of GABA-gated bicarbonate conductance on potential, current and intracellular chloride in crayfish muscle fibres. *J. Physiol. (Lond.)*, 416: 161–181.

Kaila, K., Saarikoski, J. and Voipio, J. (1990) Mechanism of action of GABA on intracellular pH and on surface pH in crayfish muscle fibres. *J. Physiol. (Lond.)*, 427: 241–260.

Kaila, K., Voipio, J., Paalasmaa, P., Pasternack, M. and Deisz, R.A. (1993) The role of bicarbonate in $GABA_A$ receptor-mediated IPSPs of rat neocortical neurones. *J. Physiol. (Lond.)*, 464: 273–289.

Kaila, K., Lamsa, K., Smirnov, S., Taira, T. and Voipio, J. (1997) Long-lasting GABA-mediated depolarization evoked by high-frequency stimulation in pyramidal neurons of rat hippocampal slice is attributable to a network-driven, bicarbonate-dependent K^+ transient. *J. Neurosci.*, 17: 7662–7672.

Lamsa, K. and Kaila, K. (1997) Ionic mechanisms of spontaneous GABAergic events in rat hippocampal slices exposed to 4-aminopyridine. *J. Neurophysiol.*, 78: 2582–2591.

Mason, M.J., Mattsson, K., Pasternack, M., Voipio, J. and Kaila, K. (1990) Postsynaptic fall in intracellular pH and increase in surface pH caused by efflux of formate and acetate anions through GABA-gated channels in crayfish muscle fibres. *Neuroscience*, 34: 359–368.

Michelson, H.B. and Wong, R.K.S. (1991) Excitatory synaptic responses mediated by $GABA_A$ receptors in the hippocampus. *Science*, 253: 1420–1423.

Michelson, H.B. and Wong, R.K.S. (1994) Synchronization of inhibitory neurones in the guinea-pig hippocampus in vitro. *J. Physiol. (Lond.)*, 477: 35–45.

Miles, R. and Poncer, J.C. (1993) Metabotropic glutamate receptors mediate a post-tetanic excitation of guinea-pig hippocampal inhibitory neurones. *J. Physiol. (Lond.)*, 463: 461–473.

Misgeld, U., Deisz, R.A., Dodt, H.U. and Lux, H.D. (1986) The role of chloride transport in postsynaptic inhibition of hippocampal neurons. *Science*, 232: 1413–1415.

Nelson, S., Toth, L., Sheth, B. and Sur, M. (1994) Orientation selectivity of cortical neurons during intracellular blockage of inhibition. *Science*, 265: 774–777.

Perkins, K.L. (1999) Cl^- accumulation does not account for the depolarizing phase of the synaptic GABA response in hippocampal pyramidal cells. *J. Neurophysiol.*, 82: 768–777.

Perkins, K.L. and Wong, R.K.S. (1996) Ionic basis of the postsynaptic depolarizing GABA response in hippocampal pyramidal cells. *J. Neurophysiol.*, 76: 3886–3894.

Perreault, P. and Avoli, M. (1992) 4-aminopyridine induced epileptiform activity and a GABA-mediated long-lasting depolarization in the rat hippocampus. *J. Neurosci.*, 12: 104–115.

Poolos N.P., Mauk M.D. and Kocsis J.D (1987) Activity-evoked increases in extracellular potassium modulate presynaptic excitability in the CA1 region of the hippocampus. *J. Neurophysiol.* 58: 404–416.

Rivera, C., Voipio, J., Payne, J.A., Ruusuvuori, E., Lahtinen, H., Lamsa, K., Pirvola, U., Saarma, M. and Kaila, K. (1999) The K^+/Cl^- co-transporter KCC2 renders GABA hyperpolarizing during neuronal maturation. *Nature*, 397: 251–255.

Smirnov, S., Paalasmaa, P., Uusisaari, M., Voipio, J. and Kaila, K. (1999a) Pharmacological isolation of the synaptic and nonsynaptic components of the GABA-mediated biphasic response in rat CA1 hippocampal pyramidal cells. *J. Neurosci.*, 19: 9252–9260.

Smirnov, S., Paalasmaa, P., Voipio, J. and Kaila, K. (1999b) Pharmacological properties of the potassium-mediated depolarizing $GABA_A$ response evoked by high-frequency stimulation in rat CA1 hippocampal pyramidal cells. *Soc. Neurosci. Abstr.*, 25: 887.2.

Staley, K.J. and Proctor, W.R. (1999) Modulation of mammalian dendritic GABAA receptor function by the kinetics of Cl- and HCO3- transport. *J. Physiol. (Lond.)*, 519: 693–712.

Staley, K.J., Soldo, B.L. and Proctor, W.R. (1995) Ionic mechanism of neuronal excitation by inhibitory $GABA_A$ receptors. *Science*, 269: 977–981.

Stasheff S.F., Mott D.D. and Wilson W.A. (1993) Axon terminal hyperexcitability associated with epileptogenesis in vitro: II. Pharmacological regulation by NMDA and $GABA_A$ receptors. *J. Neurophysiol.*, 70: 976–984.

Storm, J.F. (1990) Potassium currents in hippocampal pyramidal cells. *Progr. Brain Res.*, 83: 161–187.

Taira, T., Lamsa, K. and Kaila, K. (1997) Post-tetanic excitation mediated by $GABA_A$ receptors in rat CA1 pyramidal neurons. *J. Neurophysiol.*, 77: 2213–2218.

Thompson, S.M., Deisz, R.A. and Prince, D.A. (1988a) Relative contributions of passive equilibrium and active transport to the distribution of chloride in mammalian cortical neurons. *J. Neurophysiol.*, 60: 105–124.

Thompson, S.M., Deisz, R.A. and Prince, D.A. (1988b) Outward chloride/cation co transport in mammalian cortical neurons. *Neurosci. Lett.*, 89: 49–54.

Traub, R.D., Whittington, M.A., Buhl, E.H., Jefferys, J.G. and Faulkner, H.J. (1999) On the mechanism of the gamma–beta frequency shift in neuronal oscillations induced in rat hippocampal slices by tetanic stimulation. *J. Neurosci.*, 19: 1088–1105.

Voipio, J. (1998) Diffusion and buffering aspects of H^+, HCO_3^-, and CO_2 movements in brain tissue. In: K. Kaila and B.R. Ransom (Eds), *pH and Brain Function*, Chap. 4, Wiley-Liss, New York, pp. 45–65.

Voipio, J., Pasternack, M., Rydqvist, B. and Kaila, K. (1991) Effect of gamma-aminobutyric acid (GABA) on intracellular pH in the crayfish stretch-receptor neurone. *J. Exp. Biol.*, 156: 349–361.

Voipio, J., Ruusuvuori, E. and Kaila, K. (1999) $GABA_A$ receptor channel-mediated anionic load and the consequent increase in interstitial K^+: a developmental study sheds light on the molecular mechanisms involved. *Soc. Neurosci. Abstr.*, 25: 710.18.

Whittington, M.A., Stanford, I.M., Colling, S.B., Jefferys, J.G. and Traub, R.D. (1997) Spatiotemporal patterns of gamma frequency oscillations tetanically induced in the rat hippocampal slice. *J. Physiol. (Lond.)*, 502: 591–607.

L. F. Agnati, K. Fuxe, C. Nicholson and E. Syková (Eds.)
Progress in Brain Research, Vol 125

CHAPTER 21

Spillover and synaptic cross talk mediated by glutamate and GABA in the mammalian brain

Dimitri M. Kullmann

Department of Clinical Neurology, Institute of Neurology, UCL, Queen Square, London WC1N 3BG, UK

Introduction

Although the concept of 'volume' transmission is well established for monoamines and peptides (Agnati et al., 1995), the amino acids glutamate and GABA have generally been considered as mediating only 'wiring' transmission, that is, private communication between neurons in synaptic apposition. Recently, however, this distinction between transmitters that use volume or wiring transmission has been blurred by several reports that both GABA and glutamate can act relatively remotely from their site of release. This chapter will consider some of this evidence, and attempt to address the physiological and pathological roles of this form of signalling.

Before examining the evidence for remote effects of GABA and glutamate, it is important to bear in mind that neither of these transmitters is normally destroyed in the extracellular space after exocytosis. Although dilution into the extracellular volume will cause the transmitter concentration within the synaptic cleft to drop rapidly to negligible levels, GABA or glutamate molecules must be taken up eventually, or else they will gradually build up and activate and/or desensitise receptors over a large volume of tissue. Considerable effort has gone into characterising the properties of several highly selective transporters that take up either GABA or glutamate (Palacin et al., 1998). However, transmitter molecules within a vesicle probably far outnumber transporter molecules within the synaptic cleft, and transporters have relatively slow kinetics. It therefore follows that a large proportion of the released transmitter molecules must escape into the surrounding extracellular space, with the potential to activate extrasynaptic receptors and/or receptors at neighbouring synapses.

The kinetics of transmitter uptake from the immediate vicinity of the synaptic cleft must be understood in order to determine the spatial and temporal properties of this transmitter 'leakage'. Ultimately, the extent to which remote receptors are activated depends on whether the transporters immediately surrounding the synapse are able to 'soak up' and clear most of the transmitter molecules before they reach them.

Several different types of extrasynaptic transmission can be distinguished conceptually, depending on the location of the target receptors relative to the transmitter release site (Fig. 1). First, transmitter molecules might be released from a presynaptic terminal and activate extrasynaptic receptors on the same neuron (Fig. 1: A). This form of signalling is suggested by the finding that group II metabotropic glutamate receptors tend to be located in the preterminal membrane of hippocampal mossy fibres (Yokoi et al., 1996) relatively far from the synaptic cleft (Takumi et al., 1998). Second, the transmitter leaking out of a synaptic cleft might activate

*Corresponding author. Tel: +44 20 7837 3611; Fax: +44 20 7278 5616; e-mail: d.kullmann@ion.ucl.ac.uk

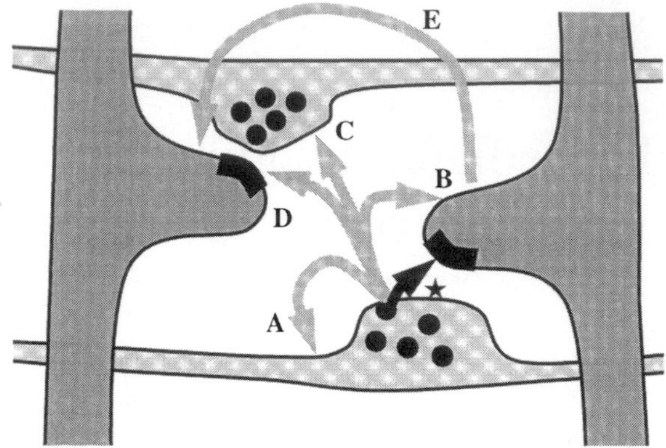

Fig. 1. Transmitter spillover and inter-synaptic cross talk. In addition to mediating wiring transmission (★), neurotransmitter molecules might also activate extrasynaptic receptors on the same terminal where they are released (A) or extrasynaptic postsynaptic receptors (B). 'Cross talk' refers to the possibility that transmitter released at one synapse could activate presynaptic (C) or postsynaptic (D) receptors at another synapse nearby. Where dendritic exocytosis takes place, transmitter might mediate non-synaptic dendro-dendritic communication (E).

postsynaptic receptors outside the boundary of the synaptic apposition (Fig. 1: B). This type of signalling is suggested by the finding that group I metabotropic receptors are preferentially located in a 'halo' around the synaptic specialisation at several cerebellar and hippocampal synapses (Baude et al., 1993; Nusser et al., 1994). The third form of extrasynaptic transmission is where transmitter released from one axonal terminal or varicosity diffuses to and activates presynaptic receptors on another terminal or varicosity (Fig. 1: C). Such signalling is exemplified by heterosynaptic interactions among hippocampal mossy fibre terminals, mediated by group II metabotropic receptors. It is also represented by the presynaptic depression of glutamatergic transmission by GABA, released from interneuron terminals and mediated by presynaptic $GABA_B$ receptors (Isaacson et al., 1993). Fourth, transmitter might escape from one synaptic cleft and activate postsynaptic receptors at a neighbouring synapse, where the pre- and post-synaptic neurons can be quite different from those making up the 'donor' synapse (Fig. 1: D). This has been proposed as an explanation for NMDA receptor activation by glutamate spillover from neighbouring synapses (Kullmann and Asztely, 1998), as well as for $GABA_A$ receptor-mediated slow conductances in cerebellar granule cells (Rossi and Hamann, 1998). Finally, an unusual case is where a transmitter is released from dendrites, activating dendritic receptors at neighbouring neurons (Fig. 1: E). This has been reported for non-synaptic glutamatergic interactions among olfactory mitral cells (Isaacson, 1999).

The first two forms of signalling listed above (pre- and postsynaptic extrasynaptic receptor activation) qualify for the term 'spillover', in the sense that in order for the transmitter to reach the receptors it must spill out of the synaptic cleft and overcome the barrier represented by transporters. The other types of signalling also qualify for the term 'cross talk' because they potentially violate synaptic specificity: that is, intercellular transmission may not be limited to synaptic connections.

GABA spillover

GABA acts at both ionotropic ($GABA_A$) and metabotropic ($GABA_B$) receptors. The latter are well-suited to detect extrasynaptic escape of GABA, because they are present on the presynaptic terminals of a wide range of monoaminergic, GABAergic and glutamatergic neurons, where they mediate the depressant effect of the selective $GABA_B$ agonist baclofen (Bowery et al., 1980; Lanthorn and Cotman, 1981; Dutar and Nicoll, 1988; Edwards et al., 1989). With the development of GABA uptake blockers and $GABA_B$ receptor antagonists, it became apparent that synaptically released GABA can exert an autoinhibitory effect on its own release (Deisz and Prince, 1989; Davies et al., 1990). However, it is not clear from these observations whether synaptically released GABA actually reaches extrasynaptic receptors.

$GABA_B$ receptor-mediated heterosynaptic depression

The first clear demonstration of GABA spillover actually came from studying glutamatergic excitatory postsynaptic currents (EPSCs) in pyramidal neurons in rodent hippocampal slices. Isaacson et al. (1993) examined the interactions between two intermingled but distinct groups of afferent fibres.

They found that a brief train of action potentials in one pathway depressed EPSCs elicited by stimulating the other pathway. This heterosynaptic depression was abolished by applying the $GABA_B$ antagonist CGP35348. Moreover, it was accompanied by an increase in the short-term facilitation seen when two stimuli are delivered in rapid succession, implying that it was mediated by a presynaptic decrease in release probability. Because axo-axonic synapses are not seen in the hippocampus, these observations are most simply explained by extrasynaptic spillover of GABA from inhibitory nerve terminals, diffusing over to the presynaptic $GABA_B$ receptors of neighbouring excitatory axons (Fig. 2). Lending further support to this hypothesis, the heterosynaptic depression was potentiated by inhibiting GABA uptake, suggesting that amino acid transporters normally play an important role in regulating the spatial extent of heterosynaptic interactions. Finally, GABA release into the extracellular space could be detected via an outside-out membrane patch containing $GABA_A$ receptors positioned on the surface of a hippocampal slice. Again, the response monitored by direct observation of channel opening events was potentiated by blocking GABA uptake.

In addition to demonstrating that GABA could exert effects at a distance, this landmark study (Isaacson et al., 1993) also defined several important criteria for subsequent attempts to demonstrate intersynaptic cross-talk: a physiological effect on an anatomically separate synaptic population, blockade of this effect by a selective antagonist, and potentiation by interfering with the mechanisms that normally clear the transmitter from the extracellular space.

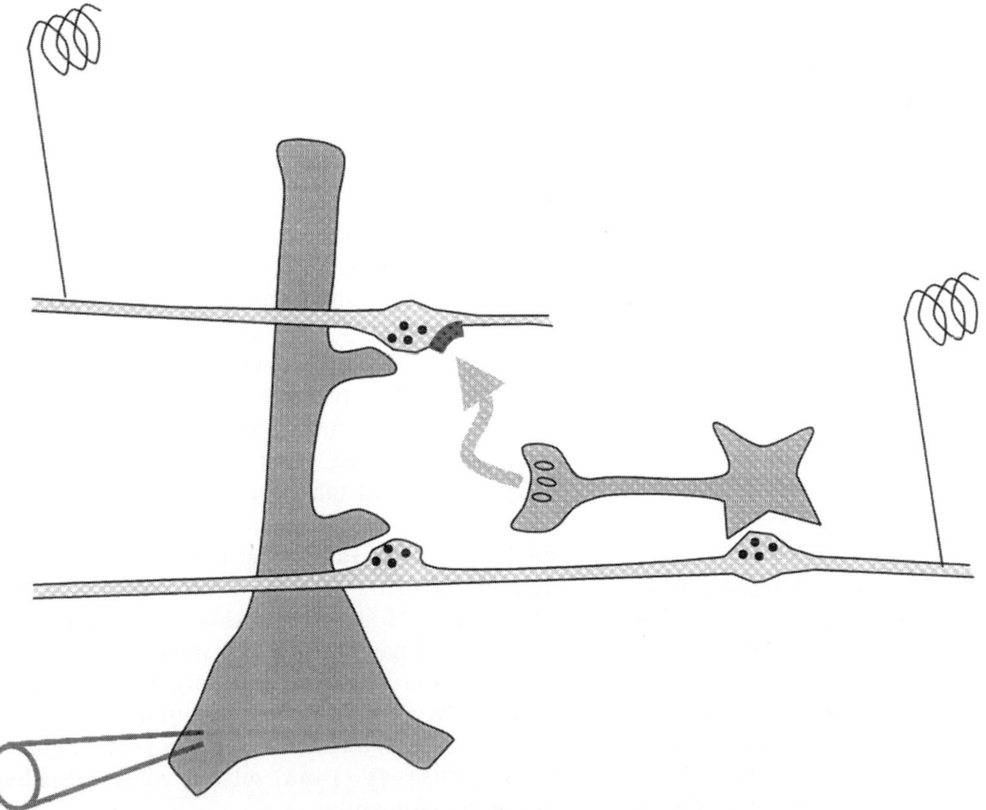

Fig. 2. GABA spillover in the hippocampus. GABA (grey arrow) is released from interneuron terminals, and activates presynaptic $GABA_B$ receptors on neighbouring Schaffer collaterals. This phenomenon explains the heterosynaptic depression seen when a brief train of stimuli is delivered via one electrode (at right) before a test stimulus delivered via another electrode (at left).

GABA spillover onto presynaptic $GABA_B$ receptors of excitatory terminals has also been described in the cerebellum (Dittman and Regehr, 1997), where brief trains of action potentials in parallel fibres caused a transient surround inhibition of parallel fibre-evoked EPSCs in Purkinje cells. The phenomenon was accompanied by a decrease in Ca^{2+} influx into parallel fibre terminals, and was blocked by $GABA_B$ antagonists. The results could be explained by GABA release from interneurons, which were synaptically recruited by the conditioning parallel fibre stimulus. According to this scheme, GABA spillover onto presynaptic $GABA_B$ receptors then reduced glutamate release from the surrounding parallel fibre-Purkinje cell synapses. Somewhat surprisingly, a recent study of the subcellular localisation of $GABA_B$ receptors found them to be present postsynaptically on Purkinje cell spines, in an exclusively extrasynaptic distribution, with no evidence of $GABA_B$ receptors on parallel fibre axons or terminals (Fritschy et al., 1999). Whether another $GABA_B$ receptor exists to explain the presynaptic action of baclofen at these synapses remains to be determined.

Finally, mossy fibre synapses in the hippocampus exhibit a form of heterosynaptic depression that is partly mediated by $GABA_B$ receptors (Min et al., 1998; Vogt and Nicoll, 1999). Although this could be due to recruitment of interneurons by the conditioning stimulus, as proposed above, an alternative possibility is that GABA is actually released synaptically from mossy fibre terminals (Sloviter et al., 1996).

$GABA_A$ receptor-mediated spillover

The GABA-mediated cross-talk phenomena described above are mediated by $GABA_B$ receptors. These receptors have a high affinity for GABA, with an EC_{50} estimated at $\sim 1\ \mu M$ (Jones et al., 1998), which would make them well suited to detect the extracellular diffusion of GABA from neighbouring release sites. $GABA_A$ receptors, on the other hand, generally have a much lower affinity, with an EC_{50} in the range 10–300 μM (Macdonald and Olsen, 1994). They also desensitise rapidly, suggesting that the relatively slow wave of GABA that could reach them from remote release sites might cause a large fraction of receptors to enter a state where they are unable to open. An exception to this description of $GABA_A$ receptors is that receptors containing the α_6 subunit have a much higher affinity ($EC_{50} \sim 0.3$–$2\ \mu M$, comparable to $GABA_B$ receptors), and much less pronounced desensitisation in response to prolonged presence of the agonist (Saxena and Macdonald, 1996). Such receptors are present at both excitatory and inhibitory synapses in the cerebellar cortex (Nusser et al., 1996), and even in extrasynaptic membranes (Nusser et al., 1998b), raising the possibility that they detect extracellular GABA relatively far from the release site. This does indeed appear to occur within cerebellar glomeruli, where glutamatergic mossy fibre terminals, as well as GABAergic Golgi cell axonal varicosities and granule cell dendrites are all surrounded by a glial sheath. The first evidence for spillover onto $GABA_A$ receptors was the finding that granule cells, recorded in rodent cerebellar slices, have a tonic $GABA_A$ receptor-mediated conductance (Brickley et al., 1996; Wall and Usowicz, 1997). This conductance is reduced by blocking Na^+ channels and potentiated by GABA uptake blockers, implying that it results from action potential-dependent exocytosis, which gives rise to a steady background level of GABA, possibly trapped by the glial envelope, and leading to continuous $GABA_A$ receptor activation. Subsequently, Rossi and Hamann (1998) showed that very slow $GABA_A$ receptor-mediated events evoked by electrical stimulation could be resolved, and could be separated from fast synaptic $GABA_A$ receptor-mediated inhibitory postsynaptic currents (IPSCs). The slow responses were potentiated by uptake blockers, suggesting that they were mediated by GABA spillover from relatively distant release sites. Moreover, the slow (spillover) signals and fast (synaptic) signals showed differential sensitivity to furosemide and diazepam, compatible with the hypothesis that high-affinity α_6 subunit-containing receptors detect low concentrations of GABA diffusing from release sites elsewhere within the glomerulus.

The fact that a slow IPSC could be elicited without a fast IPSC suggests that GABA was released not from a terminal in synaptic contact with the postsynaptic granule cell, but from a terminal in synaptic contact with another dendrite (Fig. 3).

Functional considerations

Although the above reports provide compelling evidence for GABA spillover onto both $GABA_A$ and $GABA_B$ receptors, several important questions remain. First, the phenomena have generally been studied under very artificial conditions. Extracellular electrical stimuli might cause synchronous exocytosis from a large number of GABAergic terminals, potentially overwhelming the transporters in the vicinity of the recorded neurons. This would not, however, explain the tonic activation of $GABA_A$ receptors in granule cells in the absence of evoked activity (Brickley et al., 1996; Wall and Usowicz, 1997). In this admittedly unusual anatomical specialisation, therefore, spillover-mediated signalling appears to occur spontaneously. Second, the recordings were made at sub-physiological temperatures, ranging from room temperature to 30–32 °C (Isaacson et al., 1993; Wall and Usowicz, 1997). Again, this aspect of the recording conditions could potentially exacerbate spillover if GABA uptake were impaired. In addition, the dependence on temperature of exocytosis and receptor activation is incompletely documented, making extrapolation to the situation in vivo problematic.

One relatively indirect suggestion that GABA spillover in the cerebellar glomerulus has a physio-

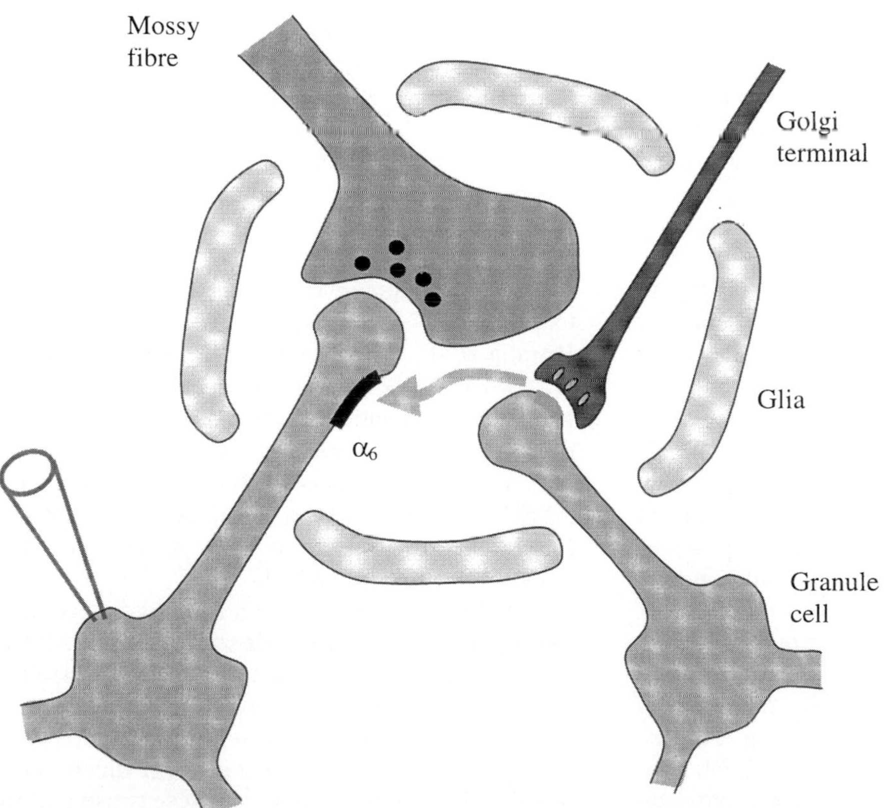

Fig. 3. GABA spillover in cerebellar glomeruli. GABA (arrow) is released from Golgi terminals, and activates extrasynaptic α_6 subunit-containing receptors on neighbouring granule cell dendrites, giving rise to a slow synaptic conductance. Direct activation of synaptic GABA receptors would give rise to a fast $GABA_A$ receptor-mediated IPSC in the granule cell at right. Although the glial cells may play a major role in clearing GABA from the extracellular space, they also trap the neurotransmitter within the glomerulus, explaining the prolonged activation of the α_6 subunit-containing receptors.

logical role is that it appears to increase with development: the tonic $GABA_A$ receptor-mediated current becomes more prominent during the first few weeks of life (Brickley et al., 1996; Wall and Usowicz, 1997). It does not appear to result, therefore, from immaturity of the uptake mechanisms. On the contrary, spillover may increase with development because the glial ensheathment becomes more complete, exacerbating the entrapment of GABA within the glomerulus. In the light of these observations, it is interesting to ask what is the consequence of genetic ablation of the α_6 GABA receptor subunit. Rather surprisingly, α_6 knockout mice have no gross motor deficits (Nusser et al., 1999). It remains to be seen whether spillover transmission is altered (although the interpretation of the results may be complicated by some indirect alterations in the expression of other subunits (Nusser et al., 1999).

Although the unremarkable phenotype of α_6 knockout animals suggests that spillover may not play an important role in motor processing, more subtle functions cannot be ruled out. The escape of GABA from cerebellar glomeruli has not been studied, but one can speculate that the delimiting glial sheath serves to compartmentalise the tonic inhibition. Granule cell dendrites terminating in neighbouring glomeruli might therefore sample a level of GABA that fluctuates independently, albeit slowly. In contrast, the axonal varicosities of hippocampal Schaffer collaterals are relatively free of glial contacts (see below), so the presynaptic inhibition mediated by $GABA_B$ receptors probably represents a spatial integral over an unknown volume. At one extreme of the range of possibilities, if GABA uptake is efficient glutamatergic synapses might sample release from only a few interneuron synapses nearby. At the other extreme, if GABA uptake is relatively slow, the extracellular concentration might represent a very large spatial and temporal integral, increasing only in response to synchronous discharge of a large number of neurons. If so, presynaptic $GABA_B$ receptors on glutamatergic terminals may serve to shut down entire regions of the hippocampus during epileptiform discharges. In order to choose between these alternatives, it will be important to relate the kinetics of GABA uptake, especially under physiological conditions, to the spatial and temporal density of GABA exocytosis and to the properties of $GABA_B$ receptors (Destexhe and Sejnowski, 1995).

Glutamate spillover

Glutamate activates three classes of ionotropic receptors (Dingledine et al., 1999). α-amino–3-hydroxy-5-methyl-4-isoxazolepropionic acid (AMPA) receptors are almost ubiquitous at fast excitatory synapses, and gate a cation-selective ionophore that is relatively insensitive to the postsynaptic membrane potential. In their commonest form, they are relatively impermeable to Ca^{2+}. *N*-methyl-D-aspartic acid (NMDA) receptors, in contrast, allow Ca^{2+} as well as monovalent cations to flow. They act as coincidence detectors, because they only open if two conditions are satisfied: glutamate has to bind, and the membrane has to be relatively depolarised. These properties underlie their crucial role in the induction of an important type of synaptic plasticity (NMDA receptor-dependent long-term potentiation), which may underlie certain forms of learning. Both AMPA and NMDA receptors are localised at synaptic membranes, with much lower densities in extrasynaptic membranes (Benke et al., 1993; Petralia et al., 1994; Baude et al., 1995; Takumi et al., 1998). This suggests that their main function is to subserve synaptic or wiring transmission. However, NMDA receptors have a very high affinity for glutamate, approximately 100-fold higher than AMPA receptors, raising the possibility, explored further below, that they may sense glutamate released from further afield.

The third class, kainic acid receptors, have only recently begun to reveal their secrets (Lerma, 1997; Mody, 1998). Major questions continue to surround the kinetics of kainate receptor-mediated synaptic signals. In particular, there is a discrepancy between their kinetics studied in vitro, which imply that they have a low affinity and desensitise rapidly (Lerma et al., 1993), and the observation that the kainate receptor-mediated component of excitatory synaptic currents is very slow, similar to the time course of the NMDA receptor-mediated component (Castillo et al., 1997; Vignes and Collingridge,

1997). If the slow kinetics are explained by a very low unbinding rate, as for NMDA receptors, then their affinity may be much higher than expected, and they may therefore be candidate target receptors for glutamate escaping from synapses. Their subcellular localisation has not been studied extensively, and although spillover activation of kainate receptors has been suggested (see below), direct evidence is still lacking.

Glutamate also activates metabotropic receptors, which have a much higher affinity and can be classified into three groups, depending on their pharmacology and transduction mechanisms (Pin and Duvoisin, 1995). As a general rule, group I receptors are localised to the postsynaptic membrane, in a ring around the edge of the synaptic cleft (Baude et al., 1993; Lujan et al., 1996). This implies that they may be ideally positioned to detect glutamate spillover, for instance under conditions of high discharge frequency or if the transporters are overwhelmed. Group II receptors, in contrast, are localised to presynaptic membranes, but with a much lower spatial correlation with release sites (Lujan et al., 1997; Yokoi et al., 1996). If anything, this localisation would imply that they may detect glutamate that has diffused a very long distance away from excitatory synapses. Finally, group III receptors tend to have a presynaptic localisation, but within the synaptic cleft (Takumi et al., 1999). Physiological evidence exists for spillover activation of group II metabotropic receptors.

NMDA receptor-mediated synaptic cross talk

A striking feature of several glutamatergic synapses is that NMDA receptors detect a larger number of quanta of glutamate than AMPA receptors. This observation was first made by examining the trial-to-trial variability of excitatory postsynaptic currents (EPSCs) evoked in CA1 pyramidal neurons by Schaffer collateral stimulation: the coefficient of variation of the NMDA receptor-mediated component was consistently smaller than that of the AMPA receptor-mediated component (Kullmann, 1994). Because the coefficient of variation is dimensionless, the discrepancy cannot be due to differences in the absolute amplitude of the quanta mediated by the two receptors. Instead, it can be explained by postulating that NMDA receptors detect a larger number of quanta of glutamate than do AMPA receptors. This conclusion was supported by several studies that used very low intensity stimuli to excite a small number of presynaptic fibres ('minimal stimulation'), so that the postsynaptic response fluctuated between small EPSCs and failures of transmission (Isaac et al., 1995; Liao et al., 1995). The failure rate was consistently higher for the AMPA receptor-mediated component than for the NMDA receptor-mediated component. These two approaches have different inherent weaknesses, but they converge on a view where, at hippocampal CA1 synapses studied at room temperature, the quantal content (average number of quanta) is roughly twice as large for the NMDA receptor-mediated component as for the AMPA receptor-mediated component.

There are two possible explanations for the discrepancy in quantal content. First, there may be a population of synapses with NMDA receptors but where AMPA receptors are non-functional or absent (Kullmann, 1994). This has been termed the 'latent AMPA receptor cluster' hypothesis or the 'postsynaptically silent synapse' hypothesis, and has attracted intense attention because rapid incorporation of functional AMPA receptors could be a major mechanism underlying NMDA receptor-dependent long-term potentiation. The alternative explanation relies on the different pharmacological properties of AMPA and NMDA receptors for glutamate: because NMDA receptors have a much higher affinity, they might respond to low concentrations of glutamate, which are insufficient to activate AMPA receptors. A possible source of a low concentration of glutamate is neighbouring synapses: electron micrographs of the CA1 neuropil frequently show synapses in very close relation, without interposed astrocytic processes (Spacek, 1985; Ventura and Harris, 1999). Glutamate release from an individual terminal would then give rise to a conventional dual-component (AMPA and NMDA receptor-mediated) EPSC at the postsynaptic spine, but in addition could also give rise to a pure NMDA receptor-mediated signal at another spine that happens to have a synaptic cluster of receptors sufficiently close by (Kullmann et al., 1996; Kullmann and Asztely, 1998). Because of the extensive intermingling of different dendrites, such an 'eavesdropping' synapse is most likely to occur on a different postsynaptic neuron, so that the NMDA signal detected via a somatic recording may not be accompanied by an AMPA signal (Fig. 4). The two explanations for 'silent synapses' are not mutually exclusive; indeed, evidence is accumulating for both of them. What follows concerns only the evidence for synaptic cross talk mediated by NMDA receptors.

One observation that argues for spillover as the explanation for much of the discrepancy in quantal signalling mediated by AMPA and NMDA receptors is that the phenomenon was greatly attenuated as the recording temperature was increased (Asztely et al., 1997). That is, rather than a 2 : 1 NMDA : AMPA quantal content ratio, as seen at room temperature, the ratio at physiological temperature was closer to 1.5 : 1, whether it was assessed by comparing the trial-to-trial variability

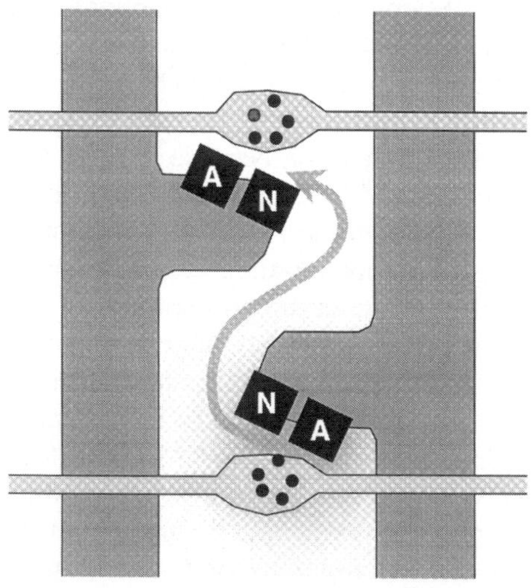

Fig. 4. Synaptic cross talk mediated by NMDA receptors. Two axons synapsing with neighbouring dendritic spines are illustrated. Glutamate released from a presynaptic varicosity activates both NMDA and AMPA receptors (indicated N and A). It may also diffuse to a neighbouring synapse (arrow), where it activates NMDA receptors. The AMPA receptors at the neighbouring synapse fail to detect the spillover of glutamate because their affinity is much lower than that of NMDA receptors. The postsynaptic neuron at left will therefore detect a signal mediated only by NMDA receptors.

of population EPSCs or failure rates with minimal stimulation. This effect of temperature can be explained by postulating that glutamate uptake is enhanced, which would agree with the steep temperature dependence of glutamate transport studied in other systems (Wadiche et al., 1995). In support of this interpretation, inhibiting glutamate uptake with dihydrokainate, which acts on the glutamate transporter Glt-1, partially reversed the effect of increasing the recording temperature: the discrepancy in quantal content mediated by AMPA and NMDA receptors was increased (Asztely et al., 1997). The alternative explanation for the discrepancy in quantal content, that synapses exist without functional AMPA receptors, cannot on its own explain these results. One would have to postulate that increasing the temperature somehow increases the proportion of synapses with functional AMPA receptors. However, this leads to the prediction that exogenous application of AMPA would give a response that grew more steeply with temperature than the response to NMDA. In fact, increasing the temperature had very similar effects on the response to AMPA and NMDA, arguing against recruitment of new clusters of AMPA receptors (Asztely et al., 1997).

The fact that raising the recording temperature did not abolish the discrepancy in quantal signalling (Asztely et al., 1997) does not necessarily imply that there is a residual degree of intersynaptic cross talk under physiological conditions. It might instead reflect absence of AMPA receptors at a proportion of synapses. This interpretation is supported by immunocytochemical studies that have reported that a minority of synapses fail to bind AMPA receptor subunit antibodies (He et al., 1998; Nusser et al., 1998a; Petralia et al., 1999; Takumi et al., 1999). The proportion of synapses that appear to be devoid of AMPA receptors is however insufficient to explain the discrepancy in quantal signalling mediated by AMPA and NMDA receptors at room temperature, although species and age differences confound any direct comparison.

An important aspect of the glutamate spillover hypothesis is that it occurs with single release events, because minimal stimulation evokes AMPA and NMDA receptor-mediated EPSCs with different failure rates. This means that synchronous glutamate release from many terminals is not required for activation of remote NMDA receptors. Deliberate manipulations that enhance the spatial density of glutamate exocytosis have, however, been reported to cause a relatively larger enhancement of the NMDA receptor-mediated component of EPSCs in hippocampal pyramidal neurons (Lozovaya et al., 1999). This can be explained by postulating that neighbouring 'domains' of glutamate released from different terminals might merge, and thereby overwhelm the transporters, further exacerbating the spatial extent of spillover.

Quantitative analysis of glutamate diffusion

An alternative approach to address synaptic cross talk is to estimate the spatial and temporal extent of diffusion of glutamate following exocytosis, and to relate the findings to the anatomical arrangement of synapses (Rusakov and Kullmann, 1998a; Rusakov et al., 1999). In addition to estimating the distance and diffusion obstacles separating synapses, it is necessary to understand the exocytosis of glutamate, and the kinetic properties of glutamate receptors, transporters and other binding sites. Many of these parameters have been estimated, providing biologically plausible ranges of values. A numerical simulation revealed that, following exocytosis, glutamate molecules were highly likely to diffuse over a distance roughly corresponding to the separation between neighbouring synapses (Rusakov and Kullmann, 1998a). As the distance from the site of exocytosis is increased, the wave of glutamate becomes attenuated and smeared out in time. This leads to a precipitous drop in the predicted size of the response mediated by AMPA receptors positioned at various distances. NMDA receptors, however, because of their slower kinetics and higher affinity, tend to be less sensitive to the change in glutamate concentration profile: the predicted response only falls off with distance relatively little.

These simulation results are compatible with the view that glutamate spillover mediates significant inter-synaptic cross talk, detected by NMDA but not AMPA receptors. There is, however, an important parameter that is not known with sufficient precision to be able to derive a precise estimate for the extent of the glutamate wave extending around a typical synapse: the extracellular diffusion coefficient. This is affected by interactions with cell walls and extracellular macromolecules, and is likely to be considerably smaller than in free solution (Rusakov and Kullmann, 1998b). However, it is unclear whether local diffusion obstacles, such as synaptic cleft macromolecules, and glutamate binding sites, can further retard diffusion. A counter-intuitive result from simulations is that retarding diffusion actually enhances the activation of remote NMDA receptors: although glutamate reaches these receptors more slowly, it also persists for longer, and therefore gives the receptors a higher chance of opening (Rusakov and Kullmann, 1998a).

Another phenomenon that has not been explored systematically is that inter-synaptic distances, and the spatial arrangement of astrocytes, which appear to express the overwhelming majority of transporters, are both extremely non-uniform (Lehre and Danbolt, 1998; Ventura and Harris, 1999). There are frequent occurrences of synapses separated by as little as 200 nm without interposed astrocytic processes (Ventura and Harris, 1999). Cross talk among such synapses would appear to be inevitable, even at physiological temperatures. In other cases, synapses are relatively isolated, with good astrocytic coverage. Because astrocytes express a very high concentration of transporters (Lehre and Danbolt, 1998), these synapses are presumably neither donors nor targets for spillover glutamate.

Finally, the simulations have been based on estimates of exocytosis, receptor and transporter kinetics mainly obtained at room temperature. In this context it is notable that NMDA receptor occupancy following exocytosis at near-physiological temperature has recently been reported to be relatively low (Mainen et al., 1999). This finding has yet to be confirmed, but it may imply that glutamate clearance is much faster than expected, which would compromise the relevance of the simulations for the in vivo situation. Alternatively, the NMDA receptor-mediated signals may have been generated by spillover from neighbouring synapses, in which case cross talk may be quite extensive.

In summary, cross talk among glutamatergic synapses is supported by experimental data obtained in hippocampal slices, but the extent of the phenomenon in vivo is unclear.

Synaptic transporter currents

A different insight into the fate of glutamate molecules following exocytosis comes from recording currents elicited by transporters in glial cells. In the presence of ionotropic glutamate receptor blockers, stimulation of afferent fibres in a hippocampal slice elicits a transporter current that can be detected by patch-clamp recordings from astrocytes (Bergles et al., 1999). Much of this current is generated by an anion conductance that opens during the transport cycle, rather than by the stoichiometrically coupled movement of Na^+, K^+ and H^+ down their electrochemical gradients. The main limitation of studying glial transporter currents in brain slices is that they sample over a large volume, and much of the transporter current is shunted by their low membrane resistivity. Nevertheless, some information can be obtained about the diffusion of glutamate out of the synaptic cleft from the onset kinetics of the transporter current. Notably, this current rises more slowly than when brief pulses of glutamate, as might occur at the edge of the synaptic cleft, are applied to isolated astrocytic membrane patches (Bergles and Jahr, 1997). This implies that the wave of glutamate that reaches astrocytes is relatively slow, and therefore that glutamate must diffuse a relatively long distance out of the synaptic cleft.

Metabotropic receptor-mediated heterosynaptic depression

Hippocampal mossy fibres exhibit heterosynaptic depression, which is sensitive to $GABA_B$ receptor antagonists (Min et al., 1998; Vogt and Nicoll, 1999). Because axo-axonic synapses have not been seen in the mossy fibre pathway, this observation argues for extrasynaptic actions of GABA. The interaction between mossy fibres is however not completely abolished by $GABA_B$ antagonists, and another component appears to be mediated by group II metabotropic glutamate receptors (Min et al., 1998; Vogt and Nicoll, 1999). This observation implies that glutamate also exerts remote actions, most likely mediated by mGluR2 receptors that are located in the preterminal membrane of neighbouring mossy fibres (Yokoi et al., 1996). In keeping with this, frequency-dependent facilitation of mossy fibre transmission is under autoreceptor modulation, which can be exacerbated by blocking glutamate uptake and reduced by applying metabotropic glutamate antagonists (Scanziani et al., 1997). Another manipulation which interferes with this form of autoreceptor modulation is application of an enzymatic scavenger, which breaks down extracellular glutamate (Min et al., 1998). Finally, the metabotropic receptor-mediated feedback is exacerbated by an experimental manipulation that slows down diffusion: perfusion of high molecular weight dextran (Min et al., 1998).

A remarkable aspect of cross talk among mossy fibres is that the metabotropic receptors are relatively far from the synaptic cleft (Yokoi et al., 1996). It will be important to carry out a quantitative analysis of the release, diffusion, and uptake of glutamate in this system, in order to understand precisely how far the phenomenon extends. Does it represent a form of surround inhibition, or is it a gain control mechanism, operating over a large volume, which serves to limit the spread of excitation through the hippocampal formation?

Dendro-dendritic glutamate spillover in the olfactory bulb

Perhaps the most surprising form of spillover that has been described for amino acid transmitters is non-synaptic communication between mitral cells in the olfactory bulb. Isaacson (Isaacson, 1999) reported that action potentials in one mitral cell could elicit a slow current mediated by NMDA receptors, not only in the cell that discharged, but also in neighbouring cells. Because mitral cells are not coupled synaptically, this observation leads to the conclusion that glutamate molecules can be released from dendrites, possibly at the same dendritic specialisations that mediate reciprocal interactions with granule cells. Glutamate then activates NMDA receptors over a wide range. Again, this phenomenon has yet to be reproduced in physiological conditions, and one can only speculate about its role in olfactory signal processing. One possibility is that long-range non-synaptic signalling helps to synchronise populations of neurons, which may be a necessary step in odour recognition.

Conclusion

The instances of extrasynaptic signalling, heterosynaptic interactions and cross talk listed above are not necessarily the most important phenomena mediated by amino acid spillover. Instead, they are the examples that have been observed with the available recording techniques. Indeed, the spatial arrangement of group I metabotropic receptors, just outside the edge of the synaptic cleft in many areas, would make them strong candidates for detecting glutamate spillover. However, it is more difficult to unravel activation of extrasynaptic receptors from conventional synaptic phenomena when the same postsynaptic cell detects both types of signal.

Whether spillover-mediated transmission occurs extensively in the normal living brain is an open question; the anatomical evidence suggests that it is highly likely to occur at some synapses that are poorly isolated by glial processes. Because this phenomenon could severely impair the specificity of synaptic communication, it may be necessary to think of amino acid-mediated signalling as a mixture of volume and wiring transmission. The relative importance of the two types of signalling may be determined by such factors as the resting membrane potential of the target cell, its metabolic state and the amount and time course of transmitter release. Even if spillover is not extensive under physiological conditions, it is highly likely that synchronous discharge of many synaptic terminals, for instance in an epileptic seizure, would give rise to a very widespread increase in transmitter concentration. Extrasynaptic receptors may play an important role in detecting such an increase, and in modulating excitability to prevent spread of the seizure.

References

Agnati, L.F., Zoli, M., Stromberg, I. and Fuxe, K. (1995) Intercellular communication in the brain: wiring vs volume transmission. *Neuroscience*, 69: 711–726.

Asztely, F., Erdemli, G. and Kullmann, D.M. (1997) Extrasynaptic glutamate spillover in the hippocampus: dependence on temperature and the role of active glutamate uptake. *Neuron*, 18: 281–293.

Baude, A., Nusser, Z., Molnar, E., McIlhinney, R.A. and Somogyi, P. (1995) High-resolution immunogold localization of AMPA type glutamate receptor subunits at synaptic and non-synaptic sites in rat hippocampus. *Neuroscience* 69: 1031–1055.

Baude, A., Nusser, Z., Roberts, J.D., Mulvihill, E., McIlhinney, R.A. and Somogyi, P. (1993) The metabotropic glutamate receptor (mGluR1 alpha) is concentrated at perisynaptic membrane of neuronal subpopulations as detected by immunogold reaction. *Neuron*, 11: 771–787.

Benke, T.A., Jones, O.T., Collingridge, G.L. and Angelides, K.J. (1993) *N*-Methyl-D-aspartate receptors are clustered and immobilized on dendrites of living cortical neurons. *Proc. Natl. Acad. Sci. USA*, 90: 7819–7823.

Bergles, D.E., Diamond, J.S. and Jahr, C.E. (1999) Clearance of glutamate inside the synapse and beyond. *Curr. Opin. Neurobiol.*, 9: 293–298.

Bergles, D.E. and Jahr, C.E. (1997) Synaptic activation of glutamate transporters in hippocampal astrocytes. *Neuron*, 19: 1297–1308.

Bowery, N.G., Hill, D.R., Hudson, A.L., Doble, A., Middlemiss, D.N., Shaw, J. and Turnbull, M. (1980) (–)Baclofen decreases neurotransmitter release in the mammalian CNS by an action at a novel GABA receptor. *Nature*, 283: 92–94.

Brickley, S.G., Cull-Candy, S.G. and Farrant, M. (1996) Development of a tonic form of synaptic inhibition in rat cerebellar granule cells resulting from persistent activation of $GABA_A$ receptors. *J. Physiol. (Lond.)*, 497: 753–759.

Castillo, P.E., Malenka, R.C. and Nicoll, R.A. (1997) Kainate receptors mediate a slow postsynaptic current in hippocampal CA3 neurons. *Nature*, 388: 182–186.

Davies, C.H., Davies, S.N. and Collingridge, G.L. (1990) Paired-pulse depression of monosynaptic GABA-mediated inhibitory postsynaptic responses in rat hippocampus. *J. Physiol. (Lond)*. 424: 513–531.

Davies, C.H., Starkey, S.J., Pozza, M.F. and Collingridge, G.L. (1991) GABA autoreceptors regulate the induction of LTP. *Nature*, 349: 609–611.

Deisz, R.A. and Prince, D.A. (1989) Frequency-dependent depression of inhibition in guinea-pig neocortex in vitro by $GABA_B$ receptor feed-back on GABA release. *J. Physiol. (Lond.)*, 412: 513–541.

Destexhe, A. and Sejnowski, T.J. (1995) G protein activation kinetics and spillover of gamma-aminobutyric acid may account for differences between inhibitory responses in the hippocampus and thalamus. *Proc. Natl. Acad. Sci. USA*, 92: 9515–9519.

Dingledine, R., Borges, K., Bowie, D. and Traynelis, S.F. (1999) The glutamate receptor ion channels. *Pharmacol. Rev.*, 51: 7–61.

Dittman, J.S. and Regehr, W.G. (1997) Mechanism and kinetics of heterosynaptic depression at a cerebellar synapse. *J. Neurosci.*, 17: 9048–9059.

Dutar, P. and Nicoll, R.A. (1988) Pre- and postsynaptic $GABA_B$ receptors in the hippocampus have different pharmacological properties. *Neuron*, 1: 585–591.

Edwards, F.R., Harrison, P.J., Jack, J.J. and Kullmann, D.M. (1989) Reduction by baclofen of monosynaptic EPSPs in lumbosacral motoneurones of the anaesthetized cat. *J. Physiol. (Lond.)*, 416: 539–556.

Fritschy, J.M., Meskenaite, V., Weinmann, O., Honer, M., Benke, D. and Mohler, H. (1999) $GABA_B$-receptor splice variants GB1a and GB1b in rat brain: developmental regulation, cellular distribution and extrasynaptic localization. *Eur. J. Neurosci.*, 11: 761–768.

He, Y., Janssen, W.G. and Morrison, J.H. (1998) Synaptic coexistence of AMPA and NMDA receptors in the rat hippocampus: a postembedding immunogold study. *J. Neurosci. Res.*, 54: 444–449.

Isaac, J.T., Nicoll, R.A. and Malenka, R.C. (1995) Evidence for silent synapses: implications for the expression of LTP. *Neuron*, 15: 427–434.

Isaacson, J.S. (1999) Glutamate spillover mediates excitatory transmission in the rat olfactory bulb. *Neuron*, 23: 377–384.

Isaacson, J.S., Solis, J.M. and Nicoll, R.A. (1993) Local and diffuse synaptic actions of GABA in the hippocampus. *Neuron*, 10: 165–175.

Jones, K.A., Borowsky, B., Tamm, J.A., Craig, D.A., Durkin, M.M., Dai, M., Yao, W.J., Johnson, M., Gunwaldsen, C., Huang, L.Y., Tang, C., Shen, Q., Salon, J.A., Morse, K., Laz, T., Smith, K.E., Nagarathnam, D., Noble, S.A., Branchek, T.A. and Gerald, C. (1998) GABA(B) receptors function as a heteromeric assembly of the subunits GABA(B)R1 and GABA(B)R2. *Nature*, 396: 674–679.

Kullmann, D.M. (1994) Amplitude fluctuations of dual-component EPSCs in hippocampal pyramidal cells: implications for long-term potentiation. *Neuron*, 12: 1111–1120.

Kullmann, D.M. and Asztely, F. (1998) Extrasynaptic glutamate spillover in the hippocampus: evidence and implications. *Trends Neurosci.*, 21: 8–14.

Kullmann, D.M., Erdemli, G. and Asztely, F. (1996) LTP of AMPA and NMDA receptor-mediated signals: evidence for presynaptic expression and extrasynaptic glutamate spillover. *Neuron*, 17: 461–474.

Lanthorn, T.H. and Cotman, C.W. (1981) Baclofen selectively inhibits excitatory synaptic transmission in the hippocampus. *Brain Res.*, 225: 171–178.

Lehre, K.P. and Danbolt, N.C. (1998) The number of glutamate transporter subtype molecules at glutamatergic synapses: chemical and stereological quantification in young adult rat brain. *J. Neurosci.*, 18: 8751–8757.

Lerma, J. (1997) Kainate reveals its targets. *Neuron*, 19: 1155–1158.

Lerma, J., Paternain, A.V., Naranjo, J.R. and Mellstrom, B. (1993) Functional kainate-selective glutamate receptors in cultured hippocampal neurons. *Proc. Natl. Acad. Sci. USA*, 90: 11688–11692.

Liao, D., Hessler, N.A. and Malinow, R. (1995) Activation of postsynaptically silent synapses during pairing-induced LTP in CA1 region of hippocampal slice. *Nature*, 375: 400–404.

Lozovaya, N.A., Kopanitsa, M.V., Boychuk, Y.A. and Krishtal, O.A. (1999) Enhancement of glutamate release uncovers spillover-mediated transmission by *N*-methyl-D-aspartate receptors in the rat hippocampus. *Neuroscience*, 91: 1321–1330.

Lujan, R., Nusser, Z., Roberts, J.D., Shigemoto, R. and Somogyi, P. (1996) Perisynaptic location of metabotropic glutamate receptors mGluR1 and mGluR5 on dendrites and dendritic spines in the rat hippocampus. *Eur. J. Neurosci.*, 8: 1488–1500.

Lujan, R., Roberts, J.D., Shigemoto, R., Ohishi, H. and Somogyi, P. (1997) Differential plasma membrane distribution of metabotropic glutamate receptors mGluR1 alpha, mGluR2 and mGluR5, relative to neurotransmitter release sites. *J. Chem. Neuroanat.*, 13: 219–241.

Macdonald, R.L. and Olsen, R.W. (1994) GABAA receptor channels. *Annu. Rev. Neurosci.*, 17: 569–602.

Mainen, Z.F., Malinow, R. and Svoboda, K. (1999) Synaptic calcium transients in single spines indicate that NMDA receptors are not saturated [see comments]. *Nature*, 399: 151–155.

Min, M.Y., Rusakov, D.A. and Kullmann, D.M. (1998) Activation of AMPA, kainate, and metabotropic receptors at hippocampal mossy fiber synapses: role of glutamate diffusion. *Neuron*, 21: 561–570.

Mody, I. (1998) Interneurons and the ghost of the sea [news; comment]. *Nat. Neurosci.*, 1: 434–436.

Nusser, Z., Ahmad, Z., Tretter, V., Fuchs, K., Wisden, W., Sieghart, W. and Somogyi, P. (1999) Alterations in the expression of $GABA_A$ receptor subunits in cerebellar granule cells after the disruption of the $\alpha 6$ subunit gene. *Eur. J. Neurosci.*, 11: 1685–1697.

Nusser, Z., Lujan, R., Laube, G., Roberts, J.D., Molnar, E. and Somogyi, P. (1998a) Cell type and pathway dependence of synaptic AMPA receptor number and variability in the hippocampus. *Neuron*, 21: 545–559.

Nusser, Z., Mulvihill, E., Streit, P. and Somogyi, P. (1994) Subsynaptic segregation of metabotropic and ionotropic glutamate receptors as revealed by immunogold localization. *Neuroscience*, 61: 421–427.

Nusser, Z., Sieghart, W. and Somogyi, P. (1998b) Segregation of different $GABA_A$ receptors to synaptic and extrasynaptic membranes of cerebellar granule cells. *J. Neurosci.*, 18: 1693–1703.

Nusser, Z., Sieghart, W., Stephenson, F.A. and Somogyi, P. (1996) The $\alpha 6$ subunit of the $GABA_A$ receptor is concentrated in both inhibitory and excitatory synapses on cerebellar granule cells. *J. Neurosci.*, 16: 103–114.

Palacin, M., Estevez, R., Bertran, J. and Zorzano, A. (1998) Molecular biology of mammalian plasma membrane amino acid transporters. *Physiol. Rev.*, 78: 969–1054.

Petralia, R.S., Esteban, J.A., Wang, Y.X., Partridge, J.G., Zhao, H.M., Wenthold, R.J. and Malinow, R. (1999) Selective acquisition of AMPA receptors over postnatal development suggests a molecular basis for silent synapses. *Nat. Neurosci.*, 2: 31–36.

Petralia, R.S., Yokotani, N. and Wenthold, R.J. (1994) Light and electron microscope distribution of the NMDA receptor subunit NMDAR1 in the rat nervous system using a selective anti-peptide antibody. *J. Neurosci.*, 14: 667–696.

Pin, J.P. and Duvoisin, R. (1995) The metabotropic glutamate receptors: structure and functions. *Neuropharmacology*, 34: 1–26.

Rossi, D.J. and Hamann, M. (1998) Spillover-mediated transmission at inhibitory synapses promoted by high affinity α6 subunit GABA(A) receptors and glomerular geometry [published erratum appears in *Neuron*, 1998 Jul;21(1):527]. *Neuron*, 20: 783–795.

Rusakov, D.A. and Kullmann, D.M. (1998a) Extrasynaptic glutamate diffusion in the hippocampus: ultrastructural constraints, uptake, and receptor activation. *J. Neurosci.*, 18: 3158–3170.

Rusakov, D.A. and Kullmann, D.M. (1998b) Geometric and viscous components of the tortuosity of the extracellular space in the brain. *Proc. Natl. Acad. Sci. USA*, 95: 8975–8980.

Rusakov, D.A., Kullmann, D.M. and Stewart, M.G. (1999) Hippocampal synapses: do they talk to their neighbours? *Trends Neurosci.*, 22: 382–388.

Saxena, N.C. and Macdonald, R.L. (1996) Properties of putative cerebellar gamma-aminobutyric acid A receptor isoforms. *Mol. Pharmacol.*, 49: 567–579.

Scanziani, M., Salin, P.A., Vogt, K.E., Malenka, R.C. and Nicoll, R.A. (1997) Use-dependent increases in glutamate concentration activate presynaptic metabotropic glutamate receptors. *Nature*, 385: 630–634.

Sloviter, R.S., Dichter, M.A., Rachinsky, T.L., Dean, E., Goodman, J.H., Sollas, A.L. and Martin, D.L. (1996) Basal expression and induction of glutamate decarboxylase and GABA in excitatory granule cells of the rat and monkey hippocampal dentate gyrus. *J. Comp. Neurol.*, 373: 593–618.

Spacek, J. (1985) Three-dimensional analysis of dendritic spines. III. Glial sheath. *Anat. Embryol.*, 171: 245–252.

Takumi, Y., Bergersen, L., Landsend, A.S., Rinvik, E. and Ottersen, O.P. (1998) Synaptic arrangement of glutamate receptors. *Prog. Brain Res.*, 116: 105–121.

Takumi, Y., Ramirez-Leon, V., Laake, P., Rinvik, E. and Ottersen, O.P. (1999) Different modes of expression of AMPA and NMDA receptors in hippocampal synapses. *Nat. Neurosci.*, 2: 618–624.

Ventura, R. and Harris, K.M. (1999) Three-dimensional relationships between hippocampal synapses and astrocytes. *J. Neurosci.*, 19: 6897–6906.

Vignes, M. and Collingridge, G.L. (1997) The synaptic activation of kainate receptors. *Nature*, 388: 179–182.

Vogt, K.E. and Nicoll, R.A. (1999) Glutamate and gamma-aminobutyric acid mediate a heterosynaptic depression at mossy fiber synapses in the hippocampus. *Proc. Natl. Acad. Sci. USA*, 96: 1118–1122.

Wadiche, J.I., Arriza, J.L., Amara, S.G. and Kavanaugh, M.P. (1995) Kinetics of a human glutamate transporter. *Neuron*, 14: 1019–1027.

Wall, M.J. and Usowicz, M.M. (1997) Development of action potential-dependent and independent spontaneous $GABA_A$ receptor-mediated currents in granule cells of postnatal rat cerebellum. *Eur. J. Neurosci.*, 9: 533–548.

Yokoi, M., Kobayashi, K., Manabe, T., Takahashi, T., Sakaguchi, I., Katsuura, G., Shigemoto, R., Ohishi, H., Nomura, S., Nakamura, K., Nakao, K., Katsuki, M. and Nakanishi, S. (1996) Impairment of hippocampal mossy fiber LTD in mice lacking mGluR2. *Science*, 273: 645–647.

L. F. Agnati, K. Fuxe, C. Nicholson and E. Syková (Eds.)
Progress in Brain Research, Vol 125

CHAPTER 22

Adenosine as a volume transmission signal. A feedback detector of neuronal activation

Sergi Ferré and Kjell Fuxe

Department of Neuroscience, Karolinska Institute, 171 77 Stockholm, Sweden

Adenosine and adenosine receptors in the CNS

Adenosine is an endogenous nucleoside which functions as a neuromodulator in many areas of the central nervous system (CNS) (for review, see Fredholm, 1995; Fredholm et al., 1999). It is a normal cellular constituent and its intracellular concentration depends mainly on the breakdown and synthesis of ATP, which is metabolized to AMP. Adenosine is then formed from AMP, through the action of a 5′-nucleotidase, and the intracellular and extracellular concentrations are kept in equilibrium by means of equilibrative transporters. The two main metabolic pathways of adenosine removal depend on the enzymes adenosine deaminase (mostly intracellular) and adenosine kinase. Extracellular adenosine concentration depends, therefore, on intracellular adenosine and also on extracellular ATP (released as a neurotransmitter or as an intracellular signal, from neurons or glial cells), which is rapidly hydrolyzed to adenosine and other metabolites. However, the main source of extracellular adenosine seems to be intracellular adenosine and, therefore, extracellular adenosine is mainly released from active cells in response to an increased metabolic demand (Fredholm, 1995; Fredholm et al., 1999). Taking into account its formation and release, together with the demonstration of the existence of extrasynaptic adenosine receptors (Tetzlaff et al., 1987), adenosine seems to represent a strong volume transmission (VT) signal.

Four different subtypes of adenosine receptors have been described and found in the brain: A_1, A_{2A}, A_{2B} and A_3 (for review, see Fredholm et al., 1994). A_1 receptors are widely distributed in the brain, with the highest densities in the hippocampus, cerebellum and neocortical areas. A_{2A} receptors have a much more restricted distribution, with the striatum having the highest density of A_{2A} receptors in the brain, where they are specially concentrated in the GABAergic striopallidal neurons (Schiffmann et al., 1991), colocalized with D_2 receptors (Fink et al., 1992). A_{2B} receptors have a ubiquitous distribution in the brain and most probably play a role under pathological conditions, since they are only activated by high adenosine concentrations, such as those observed with ischemia. On the other hand, the concentration of adenosine in the extracellular space under physiological conditions (submicromolar range, Ballarin et al., 1991) is high enough to assure a partial occupation and a tonic activation of A_1 and A_{2A} receptors (Fredholm, 1995; Fredholm et al., 1999). Furthermore, the affinity of both A_1 and A_{2A} receptors for the non-selective adenosine antagonist caffeine (K_D around 10 micromolar) is in the range of the concentration obtained at doses achieved in normal human consumption. This and other experimental findings strongly suggest that the central effects of caffeine are

*Corresponding author: Tel: 46-8-7287081; Fax: 46-8-337941

mainly produced by the antagonism of the tonic physiological activation of adenosine A_1 and A_{2A} receptors (Fredholm, 1995; Fredholm et al., 1999). Finally, the A_3 receptor is not abundant in the brain, its affinity for caffeine is low and its role under functional and pathological conditions is still poorly understood.

Adenosine as an endogenous VT signal for 'resting' of the neuron

The reasons why excessive activation is deleterious for the neuron are beginning to be understood. Excitotoxic effects of extracellularly raised glutamate concentrations and consecutive intracellular loading with Ca^{2+} are believed to be the major pathogenic process involved in neuronal death (Rudolphi et al., 1992; Schubert et al., 1997). The nervous system displays different mechanisms to overcome this suicidal risk, such as the generation of endogenous neuroprotectants. By using different receptor subtypes differentially localized in specific neuronal systems and different interactions between adenosine receptors and receptors for other neurotransmitters, adenosine seems to have as a major role the protection of the nervous system. This protection involves effects ranging from physiological to pathological conditions; this makes a lot of sense if we take into account the way extracellular adenosine is formed (see above). Thus, adenosine can be used as a very good signal of the degree of neuronal activation. Under physiological conditions, adenosine protects the CNS from the continuous neuronal activation by promoting resting both at the neuronal and at the system levels: by directly inhibiting neuronal activation and by producing motor inactivity and sleep (Fig. 1).

At the neuronal level, adenosine has been shown to be a potent depressant of excitatory neurotransmission, mainly by acting on the widely

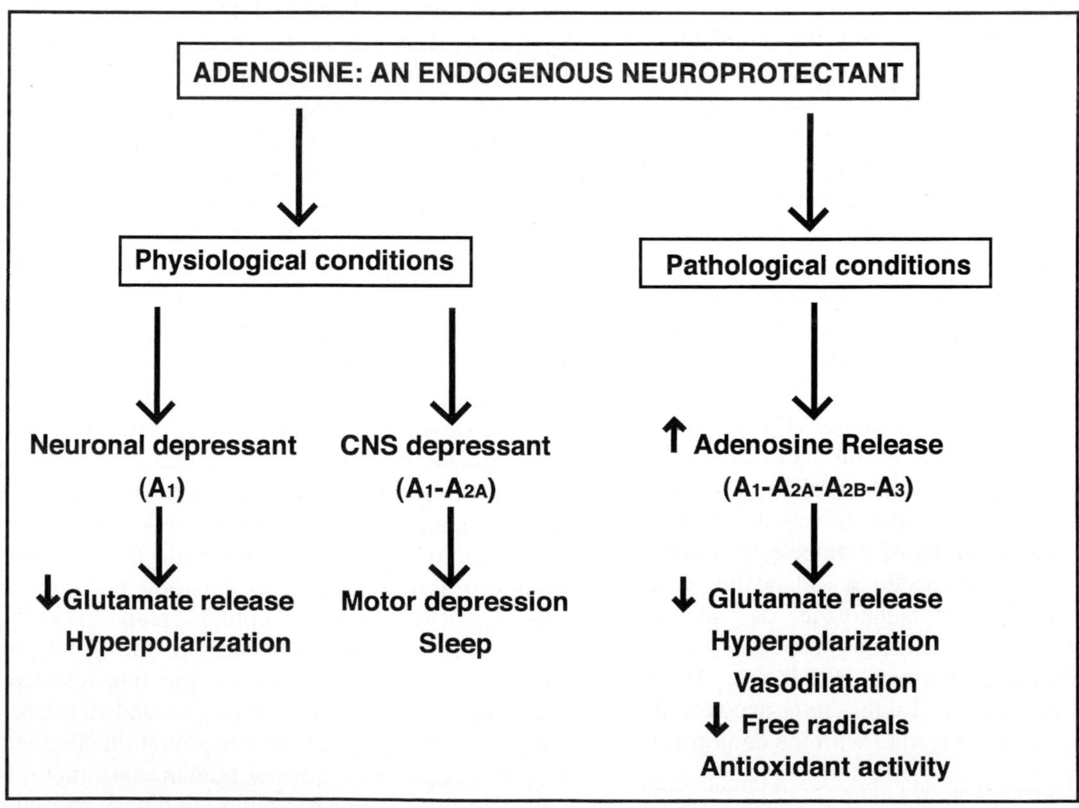

Fig. 1. Scheme of the neuroprotectant effects of adenosine. For further details, see text.

expressed A_1 receptors. Both presynaptic and postsynaptic mechanisms are involved (reviewed in Rudolphi et al., 1992; Schubert et al., 1977). First, stimulation of A_1 receptors localized in glutamatergic nerve terminals inhibits glutamate release. Second, stimulation of A_1 receptors hyperpolarizes the neuronal membrane (by increasing K^+ and Cl^- conductances), which reduces repetitive firing and Ca^{2+} entry through voltage-sensitive and NMDA receptor-operated ion channels. Thus, an A_1 receptor/NMDA receptor interaction exists, by which stimulation of A_1 receptors modulates the excitability of postsynaptic NMDA receptors (de Mendoça et al., 1995). Furthermore, NMDA receptor activation induces further adenosine release (Pedata et al., 1991; Manzoni et al., 1994).

These mechanisms are already in operation under physiological conditions, where submicromolar concentrations of adenosine are present in the extracellular fluid. Under pathological conditions, such as in the ischemic brain, an important increase of the extracellular concentration of adenosine (micromolar range) is produced, which further strengthens the A_1 receptor activation and raises the threshold for a Ca^{2+} influx causing neuronal damage (Rudolphi et al., 1992; Schubert et al., 1977). In addition to A_1 receptors, the other adenosine receptor subtypes (A_{2A}, A_{2B} and A_3) might also play an important role in neuroprotection under ischemic conditions, by inducing local vasodilatation, by decreasing formation and release of free radicals and by increasing antioxidant activity (Rudolphi et al., 1992; Maggirwar et al., 1994; Schubert et al., 1977).

Adenosine as an endogenous VT signal for 'resting' of the central nervous system

Adenosine is formed in the complex cellular networks of the entire brain, made up by neurons, glial cells and ependymal cells (see Agnati and Fuxe, Chapter 1, this volume). As a consequence of the increased metabolic activity, adenosine is released and diffuses during the wakefulness period in the extracellular space of the brain as a powerful VT signal (Huston et al., 1996; Porkka-Heiskanen et al., 1997). By acting on the motor and arousal brain systems and using different adenosine receptor subtypes strategically situated in different neuronal elements of these systems, adenosine acts a local VT signal which promotes motor inactivity and sleep (see below). It can be assumed that the accumulation of extracellular adenosine is functionally related to the subsequent decrease in motor activity and increase in sleep propensity. The decrease in motor activity will not only decrease the activity of the brain circuits directly involved in motor activation, but it will attenuate extero- and interoceptive sensorial stimulation with the concomitant decrease in arousal. Thus, when considering these effects, adenosine can be regarded as a signal to the brain to decrease its activity, in order to avoid the neuronal activity-related damage and to allow some sleep-linked restorative processes to take place.

Basal ganglia circuitry

The basal ganglia are a main target for an important part of the motor depressant and sleep-inducing effects of adenosine and, therefore, some anatomical aspects should be considered. The striatum is the main input structure of the basal ganglia and a key component of the motor system. On the basis of its afferent and efferent connections the striatum is currently subdivided in two parts, the dorsal and the ventral striatum (Heimer et al., 1995; Gerfen and Wilson, 1996). It is generally accepted that the dorsal and the ventral striatum are those parts of the striatal complex which receive information from the motor and the limbic system, respectively. The ventral striatum allows the connection between the motor and the limbic system, between motivationally relevant stimuli into appropriate adaptive behaviours, what it is called the conversion of motivation into action (for review, see Ferré et al., 1997). Accordingly, the dorsal striatum is mainly represented by the dorsolateral part of the nucleus caudate-putamen and the ventral striatum is made of the ventromedial caudate-putamen, the nucleus accumbens (with its two compartments, shell and core) and the olfactory tubercle (Heimer et al., 1995; Gerfen and Wilson, 1996).

The most abundant striatal neuron is the GABAergic spiny projection neuron, which constitutes more than 90% of all the striatal neurons.

The remaining are interneurons, which mainly use either GABA or acetylcholine as neurotransmitter (Heimer et al., 1995; Gerfen and Wilson, 1996). The GABAergic spiny striatal neurons can be subdivided in two different subtypes, which give rise to the two dorsal striatal efferent systems which connect the dorsal striatum with the output structures, the so-called direct and indirect pathways. Stimulation of the direct pathway results in motor activation and stimulation of the indirect pathway produces motor inhibition (Alexander and Crutcher, 1990; Gerfen and Wilson, 1996); the direct pathway is made of strionigral and strioentopeduncular neurons. There are also strionigral neurons which project to the substantia nigra pars compacta, which is the location of the somas of the dopaminergic nigrostriatal neurons. The indirect pathway consists of the striopallidal GABAergic neurons, pallido-subthalamic GABAergic neurons and glutamatergic neurons which connect the subthalamic nucleus with the output structures. Pallidal GABAergic neurons also project directly to the output structures without using the subthalamic nucleus relay. The striopallidal GABAergic neurons contain the peptide enkephalin and dopamine receptors of the D_2 subtype. On the other hand, the strionigral and strioentopeduncular GABAergic neurons contain the peptides dynorphin and substance P and dopamine receptors of the D_1 subtype (Alexander and Crutcher, 1990; Gerfen and Wilson, 1996). Dopamine, will therefore, induce motor activation by activating the direct pathway (acting on stimulatory D_1 receptors) and by depressing the indirect pathway (acting on inhibitory D_2 receptors) (Alexander and Crutcher, 1990).

The same two subtypes of GABAergic neurons are also found in the ventral striatum (Le Moine and Bloch, 1995); however, the organization of the outputs of the ventral parts of the basal ganglia are different from their dorsal counterparts (for review, see Ferré, 1997). First, although the ventral striatum sends projections to the pallidal complex (ventral pallidum), the entopeduncular nucleus and the substantia nigra-ventral tegmental area, it also sends projections to the extended amygdala, the lateral hypothalamus and the lateral mesopontine tegmental nucleus. It is important to point out that some of the targets of the ventral striatum (the ventral pallidum and extended amygdala as part of the basal forebrain cholinergic system and the lateral mesopontine tegmental nucleus) contain corticopetal cholinergic neurons which are part of the ascending cholinergic arousal system (Heimer et al., 1995). In fact, direct efferent projections of the nucleus accumbens to the cholinergic neurons of the ventral pallidum have been described (Záborszky and Cullinan, 1992). Second, unlike the dorsal striatum the substantia nigra pars reticulata is not a main target area for the ventral striatum, which preferentially directs its midbrain efferents to the substantia nigra pars compacta and the ventral tegmental area. Third, similarly to the ventral striatum, the ventral pallidum also projects to the substantia nigra pars compacta and ventral tegmental area. Fourth, the ventral pallidum has characteristics of both the globus pallidus and the entopeduncular nucleus in its afferent and efferent systems. It receives enkephalin-D_2-containing and substance P-D_1-containing neurons and it projects to the subthalamic nucleus and substantia nigra pars reticulata and to the thalamus (mediodorsal nucleus) and the pedunculopontine tegmental nucleus (Heimer et al., 1995; Robertson and Jian 1995). Therefore, the ventral pallidum can also be considered as an output structure of the ventral parts of the basal ganglia. Finally, in addition to the D_1 and D_2 (long and short isoforms) dopamine receptors, the ventral striatum contains the highest densities of dopamine D_3 receptors in the brain (Lévesque et al., 1992), which are localized in a subpopulation of striatal GABAergic efferent neurons, colocalized with either D_1 or D_2 receptors (Le Moine and Bloch, 1996).

Motor depressant effects of adenosine

A relatively high density of A_1 receptors is present in the striatum, where they seem to be located in corticostriatal and dopaminergic afferents. Furthermore, A_1 receptors are also localized in both subtypes of striatal GABAergic efferent neurons and in the striatal cholinergic interneurons (for review, see Ferré et al., 1997). A_{2A} receptors, on the other hand, have a much more restricted distribution, with the striatum having the highest density of A_{2A} receptors in the brain (Jarvis and Williams

1989). They are specially concentrated in the GABAergic striopallidal neurons (Schiffmann et al. 1991), colocalized with D_2 receptors (Fink et al., 1992).

Adenosine agonists induce motor depression and adenosine antagonists, such as caffeine, induce motor activation; many experimental findings indicate that dopaminergic neurotransmission is involved in the motor effects of adenosine agonists and antagonists. Thus, the motor activation induced by adenosine receptor antagonists is counteracted by treatments that cause an acute dopamine depletion or dopamine receptor blockade. Furthermore, adenosine receptor agonists inhibit and adenosine receptor antagonists potentiate the motor activating effects of dopamine receptor agonists (for review, see Ferré et al., 1992). Therefore, a strong antagonistic interaction between adenosine and dopamine seems to exist in the striatum. This seems to be mostly due to the existence of specific antagonistic interactions between specific subtypes of adenosine and dopamine receptors in the basal ganglia, mainly between adenosine A_{2A} and dopamine D_2 and between A_1 and dopamine D_1 receptors. Morphological and functional data suggest that these A_{2A}-D_2 and A_1-D_1 interactions are segregated in the two different subtypes of striatal GABAergic efferent neurons. In this way, by acting on different receptor subtypes adenosine will counteract the effects of dopamine on both the direct and indirect striatal efferent pathways (reviewed in Ferré, 1997; Ferré et al., 1997). Thus, adenosine as a VT signal appears to operate in motor functions mainly by reducing the efficacy at dopamine D_1 and D_2 receptors, through A_1 and A_{2A} receptors, respectively.

At the behavioural level, it has been shown that the administration of A_1 agonists or antagonists specifically antagonizes or potentiates, respectively, the motor effects induced by D_1 agonists (Ferré et al., 1994a,1999a; Popoli et al., 1996a; Rimondini et al., 1998). On the other hand, the administration of A_{2A} agonists or antagonists specifically antagonizes or potentiates, respectively, the motor effects induced by D_2 agonists (Ferré et al., 1994a,1999b; Rimondini et al., 1998; Strömberg et al., 1999). In vivo microdialysis experiments using a dual-probe approach have shown that the decrease in extracellular levels of GABA in the globus pallidus induced by the infusion of a D_2 agonist in the ipsilateral dorsal striatum is antagonized by the previous infusion of an A_{2A} agonist (Ferré et al., 1993), which did not have any effect on its own. Similar experiments performed in the ventral striatum showed a more powerful effect of the A_{2A} agonist, which induced an increase in the extracellular levels of GABA in the ipsilateral ventral pallidum (Ferré et al., 1994b). Finally, the infusion of a selective A_1 agonist in the dorsal striatum inhibited the increase in the extracellular levels of GABA in the ipsilateral entopeduncular nucleus induced by the striatal infusion of a D_1 agonist (Ferré et al., 1996a).

At the biochemical level it was first found that stimulation of A_1 and A_{2A} receptors changes the binding characteristics of D_1 and D_2 receptors, respectively, in membrane preparations from rat striatum. Thus, stimulation of A_{2A} receptors decreases the affinity of D_2 receptors (especially in a high affinity state) for dopamine (Ferré et al., 1991, 1999b) and stimulation of A_1 receptors decreases the proportion of D_1 receptors in a high affinity state (Ferré et al., 1994). Exactly the same kind of specific intramembrane interactions have been demonstrated in mammalian cell lines stably cotransfected with A_{2A} and D_2 receptors (Dasgupta et al., 1996; Kull et al., 1999) or with A_1 and D_1 receptors (Ferré et al., 1998). The use of these cell lines also allowed the demonstration of other levels of A_1-D_1 and A_{2A}-D_2 receptor interactions, namely at the adenylate-cyclase level (Dasgupta et al., 1996; Ferré et al., 1998; Kull et al., 1999). Recent findings in A_1-D_1 cotransfected cells have given a molecular basis for these kind of intramembrane receptor-receptor interactions, by showing heteromerization of D_1 and A_1 receptors (Ginés et al., 1999). These findings show the importance of receptor-receptor subtype specific interactions though heterodimerization in VT to control the activity of target cells.

Sleep-inducing effects of adenosine

The sedative and hypnotic effects induced by the central and peripheral administration of adenosine and adenosine analogues as well as the increase in the degree of wakefulness induced by caffeine have long been recognized (for recent review, see Fredholm et al., 1999). Prolonged wakefulness decreases the degree of arousal (usually measured as EEG activation) and increases the propensity to sleep and adenosine is a very important neural mediator of these effects. The work by McCarley and coworkers has elegantly demonstrated that the basal forebrain and mesopontine tegmental cholinergic neurons, whose discharge activity play a key role in EEG arousal, are a key target for the sleep-promoting effects of adenosine. First, this discharge activity is under a tonic inhibitory control of adenosine, by acting on A_1 receptors (Rainnie et al., 1994). Second, by using in vivo microdialysis measurements in freely behaving cats, local perfusion of adenosine into the basal forebrain area and lateral mesopontine tegmental nucleus dramatically decreases wakefulness (Portas et al., 1997). Finally, the extracellular concentrations of adenosine in the basal forebrain cholinergic region increased during wakefulness and declined during recovery sleep (Porkka-Heiskanen et al., 1997).

In addition to the A_1 receptors localized in the basal forebrain and mesopontine cholinergic neurons, other brain localizations and also the A_{2A} receptors seem to be involved in the sleep-promoting effects of adenosine (Fig. 2). Prostaglandin (PG) D_2 is a very potent endogenous sleep-promoting substance (reviewed in Urade and Hayaishi, 1999). Its main site of action is in the ventral surface of the rostral basal forebrain, where it activates prostanoid DP receptors. Activation of DP receptors in this PGD_2-sensitive chemosensory region results in activation of neurons located in the ventrolateral preoptic area (VLPO). These have been found to be sleep-active neurons which give rise to a GABAergic (and galaninergic) inhibitory pathway which connects with the tuberomammillary nucleus, the source of the ascending histaminergic arousal system (Sherin et al., 1996, 1998). The studies by Sherin et al. (1998) also suggest that VLPO may also provide inhibitory inputs to other components of the ascending monoaminergic arousal system (noradrenergic, serotoninergic and dopaminergic). Satoh et al. (1996, 1999; see also Urade and Hayaishi, 1999) have found that stimulation of A_{2A} receptors localized in the rostral basal forebrain promotes sleep and that those receptors are crucial in the sleep-promoting process triggered by PGD_2. Recent studies suggest that the shell of the nucleus accumbens and the medial portion of the olfactory tubercle are the major sites responsible for these effects mediated

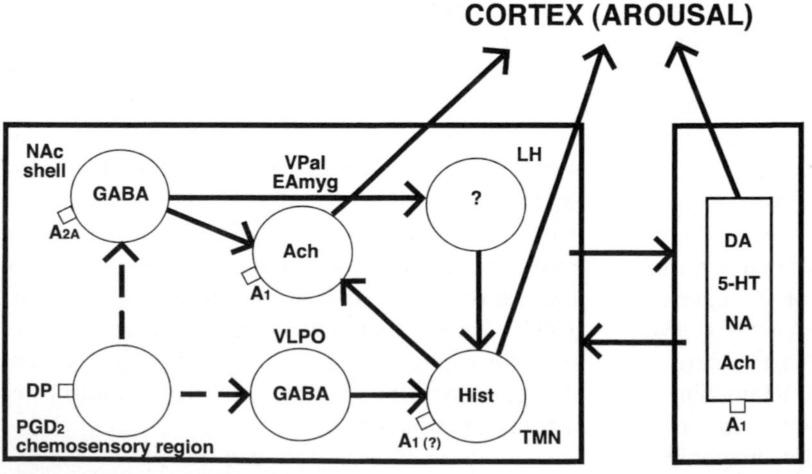

Fig. 2. Scheme of the neuronal circuits most probably involved in the sleep-promoting effects of adenosine. Hist: histaminergic neuron; LH: lateral hypothalamus; NAc shell: shell of the nucleus accumbens; VLPO: ventrolateral preoptic area; VPal, EAmyg: ventral pallidum, extended amygdala. For further details, see text.

by A_{2A} receptors, while VLPO does not seem to be involved. The projections, most probably GABAergic, from the nucleus accumbens to the cholinergic neurons of the basal forebrain (ventral pallidum) and, possibly, to the mesopontine tegmental nucleus and to the mesencephalic dopaminergic neurons (Záborszky and Cullinan, 1992; Heimer et al., 1995; Sarter and Bruno, 1999) are most probably involved in the sleep-promoting effects of adenosine through activation of these striatal A_{2A} receptors. In fact, it has been shown that GABAergic neurotransmission in the basal forebrain regulates the excitability of cortical cholinergic inputs (Sarter and Bruno, 1999). The cholinergic neurons of the ventral pallidum and extended amygdala projecting to the entire cerebral cortex represent an important attentional system, which participates in the gating of cortical information processing. It has therefore been hypothesized that an abnormal regulation of the excitability of these cholinergic neurons by, for instance, alterations in their GABAergic-mediated modulation, might be involved in the symptomatology of neuropsychiatric disorders (Sarter and Bruno, 1999). Finally, stimulation of accumbal A_{2A} receptors could also promote sleep by indirectly influencing the ascending histaminergic arousal system, since afferents from the lateral hypothalamus to the tuberomammillary nucleus have been found (Ericson et al., 1991) (Fig. 2).

Finally, we have obtained experimental evidence for a role of adenosine-dopamine receptor-receptor interactions in the modulation of arousal, through the existence of interacting A_1 and D_1 receptors in the prefrontal cortex. Thus, stimulation of A_1 receptors prevents the EEG arousal due to dopamine D_1 receptor activation in rabbits (Popoli et al., 1996b), an effect which seems to be mediated by D_1 receptors localized in the frontal cortex, and stimulation of A_1 receptors decreases the binding of a labelled D_1 antagonist in the medial prefrontal cortex (Ferré et al., 1996b).

Conclusion

Adenosine is mainly formed and released over the adenosine transporter from complex cellular networks of the brain into the extracellular space, especially upon neronal activation. Here it acts as a strong local VT signal to antagonize activity in the motor and arousal systems. The action at target cells are mediated via high affinity G protein-coupled A_1 and A_{2A} receptors, often by antagonizing the efficacy of other receptors, such as dopamine receptors, via receptor–receptor interactions.

References

Alexander, G.E. and Crutcher, M.D. (1990) Functional architecture of basal ganglia circuits: neural substrates of parallel processing. *Trends Neurosci.*, 13: 266–271.

Ballarin, M., Fredholm, B.B., Ambrosio, S. and Mahy, N. (1991) Extracellular levels of adenosine and its metabolites in the striatum of awake rats: inhibition of uptake and metabolism. *Acta Physiol. Scand.*, 142: 97–103.

Dasgupta, S., Ferré, S., Köll, B., Hendlund, P., Finnman, U.-B., Ahlberg, S. Arenas, E., Fredholm, B.B. and Fuxe, K. (1996) Adenosine A_{2A} receptors modulate the binding characteristics of dopamine D_2 receptors in stably cotransfected fibroblast cells. *Eur. J. Pharmacol.*, 316: 325–331.

Ericson, H., Blomqvist, A. and Köhler, C. (1991) Origin of the neuronal inputs to the region of the tuberomammillary nucleus of the rat brain. *J. Comp. Neurol.*, 311: 45–64.

Ferré, S., von Euler, G., Johansson, B., Fredholm, B.B. and Fuxe K. (1991) Stimulation of high affinity adenosine A_2 receptors decreases the affinity of dopamine D_2 receptors in rat striatal membranes. *Proc. Natl Acad. Sci. USA*, 88: 7238–7241.

Ferré, S., Fuxe, K., von Euler, G., Johansson, B. and Fredholm, B.B. (1992) Adenosine-dopamine interactions in the brain. *Neuroscience*, 51: 501–512.

Ferré, S., O'Connor, W.T., Fuxe, K. and Ungerstedt, U. (1993) The striopallidal neuron: a main locus for adenosine-dopamine interactions in the brain. *J. Neurosci.*, 13: 5402–5406.

Ferré, S., Popoli, P., Giménez-Llort, L., Finnman, U.B., Martnez, E., Scotti de Carolis, A. and Fuxe, K. (1994a) Postsynaptic antagonistic interaction between adenosine A_1 and dopamine D_1 receptors. *NeuroReport*, 6: 73–76.

Ferré, S., O'Connor, W.T., Snaprud, P., Ungerstedt, U. ad Fuxe, K. (1994b) Antagonistic interaction between adenosine A_{2A} and dopamine D_2 receptors in the ventral striopallidal system. Implications for the treatment of schizophrenia. *Neuroscience*, 63: 765–773.

Ferré, S., O'Connor, W.T., Svenningsson, P., Björklund, L., Lindberg, J., Tinner, B., Strömberg, I., Golstein, M., Ögren, S.O., Ungerstedt, U., Fredholm, B. and Fuxe, K. (1996a) Dopamine D_1 receptor-mediated facilitation of GABAergic neurotransmission in the rat strioentopeduncular pathway and its modulation by adenosine A_1 receptor-mediated mechanisms. *Eur. J. Neurosci.*, 8: 1545–1553.

Ferré, S., Popoli, P., Tinner-Staines, B. and Fuxe K. (1996b) Adenosine A_1 receptor-dopamine D_1 receptor interaction in the rat limbic system: modulation of dopamine D_1 receptor antagonist binding sites. *Neurosci. Lett.*, 208: 1–4.

Ferré, S. (1997) Adenosine-dopamine interactions in the ventral striatum. Implications for the treatment of schizophrenia. *Psychopharmacology*, 133: 107–120.

Ferré, S., Fredholm, B.B., Morelli, M., Popoli, P. and Fuxe K. (1997) Adenosine-dopamine receptor-receptor interactions as an integrative mechanism in the basal ganglia. *Trends Neurosci.*, 20: 482–487.

Ferré, S., Torvinen, M., Antoniou, K., Irenius, E., Civelli, O., Arenas, E.,. Fredholm, B.B. and Fuxe, K. (1998) Adenosine A_1 receptor-mediated modulation of dopamine D_1 receptors in stably cotransfected fibroblast cells. *J. Biol. Chem.*, 273: 4718–4724.

Ferré, S., Rimondini, R., Popoli, P., Reggio, R., Pèzzola, A., Hansson, A.C., Andersson, A. and Fuxe K. (1999a) Stimulation of adenosine A_1 receptors attenuates the dopamine D_1 receptor-mediated increase of *NGFI-A*, *c-fos* and *jun B* mRNA levels in the dopamine-denervated striatum and the dopamine D_1 receptor-mediated turning behaviour. *Eur. J. Neurosci.*, 11: 3884–3892.

Ferré, S., Popoli, P., Rimondini, R., Reggio, R., Kehr, J. and Fuxe K. (1999b) Adenosine A_{2A} and group I metabotropic glutamate receptors synergistically modulate the binding characteristics of dopamine D_2 receptors in the rat striatum. *Neuropharmacology*, 38: 129–140.

Fink, J.S., Weaver, D.R., Rivkees, S.A., Peterfreund, R.A., Pollack, A., Adler, E.M. and Reppert, S.M. (1992) Molecular cloning of the rat A_2 adenosine receptor: selective co-expression with D_2 dopamine receptors in rat striatum. *Mol. Brain. Res.*, 14: 186–195.

Fredholm, B.B. and Dunwiddie, T.V. (1988) How does adenosine inhibit transmitter release? *Trends Pharmacol. Sci.*, 9: 130–134.

Fredholm, B.B., Abbracchio, M.P., Burnstock, G., Daly, J.W., Harden, T.K., Jacobson, K.A., Leff, P. and Williams, M. (1994) Nomenclature and classification of purinoceptors. *Pharmacol. Rev.*, 46: 143–156.

Fredholm, B.B. (1995) Adenosine, adenosine receptors and the actions of caffeine. *Pharmacol. Toxicol.*, 76: 93–101.

Fredholm, B.B., Bättig, K., Holmén, J., Nehlig, A. and Zvartau, E.E. (1999) Actions of caffeine in the brain with special reference to factors that contribute to its widespread use. *Pharmacol. Rev.*, 51: 83–133.

Gerfen, C.R. and Wilson, C.J. (1996) The Basal Ganglia. In: Swanson, L.W., Björklund.A., Hökfelt.T. (Eds), *Handbook of Chemical Neuroanatomy*, Vol. 12, Part III. Elsevier, Amsterdam, pp. 371–468.

Ginés, S., Hillion J., Lew, J.Y., Le Crom, S., Vernier, P., Zoli, M., Agnati, L., Lluis, C., Ferré, S., Fuxe, K. and Franco, R. (1999) Dopamine D_1 and adenosine A_1 receptors form functionally interacting heteromeric complexes *Proc. Natl Acad. Sci. USA* (in press).

Heimer, L., Zahm, D.S. and Alheid, G.F. (1995) Basal ganglia. In: G. Paxinos, (Ed.) *The Rat Nervous System*. Academic Press, San Diego, pp. 579–628.

Huston, J.P., Haas, H.L., Boix, F., Pfister, M., Decking, U., Schrader, J. and Schwarting R.K.W. (1996) Extracellular adenosine levels in neostriatum and hippocampus during rest and activity periods of rats. *Neuroscience*, 73: 99–107.

Jarvis, M.F. and Williams, M. (1989) Direct autoradiographic localization of adenosine A_2 receptors in the rat brain using the A_2-selective agonist [^{3}H]CGS 21680. *Eur. J. Pharmacol.*, 168: 243–246.

Kull, B., Ferré, S., Arslan, G., Svenningsson, P., Fuxe, K., Owman, C. and Fredholm, B.B. (1999) Reciprocal interactions between adenosine A_{2A} and dopamine D_2 receptors in CHO cells co-transfected with the two receptors. *Biochem. Pharmacol.*, 58: 1035–1045.

Le Moine, C. and Bloch, B. (1995) D_1 and D_2 dopamine receptor gene expression in rat striatum: Sensitive cRNA probes demonstrate prominent segregation of D_1 and D_2 mRNAs in distinct neuronal populations of the dorsal and ventral striatum. J. Comp. Neur., 355: 418–426.

Le Moine, C. and Bloch, B. (1996) Expression of the D_3 dopamine receptor in peptidergic neurons of the nucleus accumbens: comparison with the D_1 and D_2 dopamine receptors. *Neuroscience*, 73: 131–143.

Lévesque, D., Diaz, J., Pilon, C., Martres, M.P., Giros, B., Souil, E., Schott, D., Morgat, J.L., Schwartz, J.C. and Sokoloff, P. (1992) Identification, characterization, and localization of the dopamine D_3 receptor in rat brain using 7-[^{3}H]hydroxy-N,N-di-n-propyl-2-aminotetralin. *Proc. Natl Acad. Sci. USA*, 89: 8155–8159.

Maggirwar, S.B., Dhanraj, D.N., Somani, S.M. and Ramkumar, V. (1994) Adenosine acts as an endogenous activator of the cellular antioxidant defense system. *Biochem. Biophys. Res. Commun.*, 201: 508–515.

Manzoni, O.J., Manabe, T. and Nicoll, R.A. (1994) Release of adenosine by activation of NMDA receptors in the hippocampus. *Science*, 265: 2098–2101.

de Mendoça, A., Sebastiao, A.M. and Ribeiro, J.A. (1995) Inhibition of NMDA receptor-mediated currents in isolated rat hippocampal neurons by A_1 receptor activation. *NeuroReport*, 6: 1097–1100.

Pedata, F., Pazzagli, M. and Pepeu, G. (1991) Endogenous adenosine release from hippocampal slices: excitatory amino acid agonists stimulate release, antagonists reduce the electrically-evoked release. *Naunyn Schmiedeberg's Arch. Pharmacol.*, 344: 538–543.

Popoli, P., Giménez-Llort, L., Pèzzola, A., Reggio, R., Martnez, E., Fuxe, K. and Ferré, S. (1996a) Adenosine A_1 receptor blockade selectively potentiates the motor effects induced by dopamine D_1 receptor stimulation in rodents. *Neurosci. Lett.*, 218: 209–213.

Popoli, P., Ferré, S., Pèzzola, A., Reggio, R., Scotti de Carolis, A. and Fuxe, K. (1996b) Stimulation of adenosine A_1 receptors prevents the EEG arousal due to dopamine D_1 receptor activation in rabbits. *Eur. J. Pharmacol.*, 305: 123–126.

Porkka-Heiskanen, T., Strecker, R.E., Thakkar, M., Bjorkum, A.A., Greene, R.W. and McCarley, R.W. (1997) Adenosine:

a mediator of sleep-inducing effects of prolonged wakefulness. *Science*, 276: 1265–1268.

Portas, C.M., Thakkar, M., Rainnie, D.G., Greene, R.W. and McCarley, R.W. (1997) Role of adenosine in behavioural state modulation: a microdialysis study in the freely moving cat. *Neuroscience*, 79: 225–235.

Rainnie, D.G., Grunze, H.C.R., McCarley, R.W. and Greene, W. (1994) Adenosine inhibition of mesopontine cholinergic neurons: implications for EEG arousal. *Science*, 263: 689–692.

Rimondini, R., Ferré, S., Giménez-Llort, L., Ögren, S.O. and Fuxe K. (1998) Differential effects of selective adenosine A_1 and A_{2A} receptor agonists on dopamine receptor agonist-induced behavioural responses in rats. *Eur. J. Pharmacol.*, 347: 153–158.

Robertson, G.S. and Jian, M. (1995) D1 and D2 dopamine receptors differentially increase Fos-like immunoreactivity in accumbal projections to the ventral pallidum and midbrain. *Neuroscience*, 64: 1019–1034.

Rudolphi, K.A., Schubert, P., Parkinson, F.E. and Fredholm, B.B. (1992) Neuroprotective role of adenosine in cerebral ischemia. *Trends Pharmacol. Sci.*, 13: 439–445.

Sarter, M. and Bruno, J.P. (1999) Abnormal regulation of corticopetal cholinergic neurons impaired information processing in neuropsychiatric disorders. *Trends Neurosci.*, 22: 67–74.

Satoh, S., Matsumura, H., Suzuki, F. and Hayaishi, O. (1996) Promotion of sleep mediated by the A_{2A}-adenosine receptor and possible involvement of this receptor in the sleep induced by prostaglandin D_2 in rats. *Proc. Natl Acad. Sci.*, 93: 5980–5984.

Satoh, S., Matsumura, H., Koike, N., Tokunaga, Y., Maeda, T. and Hayaishi, O. (1999) Region-dependent difference in the sleep-promoting potency of an A_{2A} receptor agonist. *Eur. J. Neurosci.*, 11: 1587–1597.

Schiffmann, S.N., Jacobs, O. and Vanderhaeghen, J.-J. (1991) Striatal restricted adenosine A_2 receptor (RDC8) is expressed by enkephalin but not by substance P neurons: an in situ hybridization histochemistry study. *J. Neurochem.*, 57: 1062–1067.

Schubert, P., Ogata, T., Marchini, C., Ferroni, S. and Rudolphi, K. (1997) Protective mechanisms of adenosine in neurons and glial cells. *Ann. NY Acad. Sci.*, 825: 1–10.

Sherin, J.E., Shiromani, P.J., McCarley, R.W. and Saper, C.B. (1996) Activation of ventrolateral preoptic neurons during sleep. *Science*, 271: 216–219.

Sherin, J.E., Elmquist, J.K., Torrealba, F. and Saper, C.B. (1998) Innervation of histaminergic tuberomammillary neurons by GABAergic and galaninergic neurons in the ventrolateral preoptic nucleus of the rat. *J. Neurosci.*, 18: 4705–4721.

Strömberg, I., Popoli, P., Müller, C.E., Ferré, S. and Fuxe, K. (1999) Electrophysiological and behavioural evidence for an antagonistic modulatory role of adenosine A_{2A} receptors in dopamine D_2 receptor regulation in the rat dopamine denervated striatum. *Eur. J. Neurosci.*, (submitted).

Tetzlaff, W., Schubert, P. and Kreutzberg, G.W. (1987) Synaptic and extrasynaptic localization of adenosine binding sites in rat hippocampus. *Neuroscience*, 21: 839–875.

Urade, Y. and Hayaishi, O. (1999) Prostaglandin D_2 and sleep regulation. *Biochem. Biophys. Acta*, 1436: 606–615.

Záborszky, L. and Cullinan, W.E. (1992) Projections from the nucleus accumbens to cholinergic neurons of the ventral pallidum: a correlated light and electron microscopic double-immunolabeling study in rat. *Brain Res.*, 570: 92–101.

L. F. Agnati, K. Fuxe, C. Nicholson and E. Syková (Eds.)
Progress in Brain Research, Vol 125

CHAPTER 23

Dynorphins are endogenous opioid peptides released from granule cells to act neurohumorly and inhibit excitatory neurotransmission in the hippocampus

Charles Chavkin*

Department of Pharmacology, University of Washington, Box 357280, Seattle WA 98195-7280, USA

Introduction

Our thinking about how neurons communicate with adjacent cells using chemical messengers has been largely shaped by the early work by Katz, Miledi and others (Katz, 1966). They described in perfect detail the actions of acetylcholine at the frog neuromuscular junction. Every biology undergraduate learns that at this synapse, the neurotransmitter acetylcholine, concentrated in small vesicles, is released from presynaptic nerve terminals following an influx of calcium through voltage gated ion channels activated by the arrival of the sodium channel-mediated action potential. The released acetylcholine diffuses the short distance (20–50 nm) across the synaptic gap to activate specific receptors expressed by the postsynaptic muscle cell. The cholinergic signal is rapidly detected and rapidly terminated.

This archetypal pattern of synaptic transmission is also the principal form of chemical signaling evident in the central nervous system; here the principal chemical messengers: glutamate and GABA, also act as fast neurotransmitters. Fast transmission at the neuromuscular junction and in the central nervous system is controlled by receptors that contain an intrinsic ion channel (known as ligand-gated ion channels or 'ionotropic' receptors). The properties of these synapses dominate our thinking about neurotransmission in the nervous system; however, other forms of neurotransmission are now well established. Glutamate and GABA also regulate ion conductance by activating a specific set of G protein coupled receptors. Because these receptors and ion channels are physically separate and require the activation of an intervening second messenger, the electrical response to this type of 'metabotropic' receptor activation is typically slower (in the order of 100 msec).

We thought we had things pretty well understood until the peptide neurotransmitters seemed to break the rules. Jan and Jan (1982) studied the properties of an LH-RH-like peptide expressed in the bullfrog sympathetic ganglia; they found that intense, high frequency stimulation of the ganglionic afferents caused a late, slow synaptic potential mediated by the neuropeptide. Their results were consistent with the hypothesis that the LH-RH-like peptide was able to diffuse microns in a neurohumoral manner. Similar neurohumoral properties were also found for FMRFamide and related neuropeptides in Aplysia (Cottrell, 1993; Hawkins et al., 1993).

Skeptics might dismiss these observations as aberrations and suggest that the anomalous behav-

*Corresponding author.
e-mail: cchavkin@u.washington.edu.

ior of neuropeptide transmission observed in the ganglia of invertebrates and amphibians had little relevance to understanding the functioning of neuropeptides in the mammalian central nervous system. Unfortunately for the skeptics, three additional observations supported the concept that neuropeptides act differently than fast-transmitters. The first was that peptides typically have very high (nanomolar) affinities for their receptors. Since the concentration of transmitter in a narrow synapse following vesicle fusion is orders of magnitude greater (about 100 micro molar) (Kuffler and Nicholls, 1976), the basis for the very high affinity was a puzzle unless the peptides were diluted in the extracellular space during diffusion to their target sites. Next, the sites of densest receptor concentration did not typically overlap the sites of densest peptide concentration. Third, neurochemical and anatomical studies seemed to suggest that neurotransmitter release occurred from brain regions lacking axon terminals and rich in dendrites (Cheramy et al, 1981; Pow and Morris,1989). Thus, the data suggested that the properties of the neuropeptide synapse in the mammalian brain do not faithfully follow the pattern evident for the neuromuscular junction cholinergic synapse.

Some years ago, we set out to directly explore this issue by studying the actions of the Dynorphin neuropeptides in the hippocampus. We chose these opioid peptides because they were among the first to have well-characterized competitive receptor antagonists. We chose the hippocampus because this mammalian brain structure contained a group of cells (the granule cells of the dentate gyrus) that synthesized the dynorphins and released these peptides in a calcium-dependent manner (Chavkin et al., 1983, 1985). Slices of the hippocampus maintain viability and are also readily studied as an in vitro preparation.

Stimulated release of endogenous dynorphin opioids

Prior immunocytochemical analysis showed that the principal pathway containing dynorphin immunoreactivity in the hippocampus is the mossy fiber axons of the dentate granule cells that project to the CA3 region of the hippocampus (McGinty et al., 1983). We developed an in vitro competition-binding assay using [3H]U69,593, a kappa$_1$-selective radioligand (Wagner et al., 1991). In this assay, focal electrical stimulation of opioid peptide-containing mossy fiber pathways resulted in a calcium-dependent release of Dynorphins as shown by a decrease in [3H]U69,593 binding to the tissue slice. In contrast, stimulation of a pathway lacking opioid immunoreactivity had no effect. Antiserum against dynorphin peptides blocked the reduction in [3H]U69,593 binding; whereas, antisera against other opioid peptides did not. These results indicate that prodynorphin-derived opioids are released under physiological conditions from the mossy fibers to act at kappa1 type opioid receptors in the hippocampus. That conclusion is consistent with previous pharmacological demonstration that Dynorphin is a kappa selective opioid peptide (Charkin et al., 1982).

Electrophysiological recordings showed that kappa1 type opioid receptor activation inhibits excitatory amino acid release presynaptically from mossy fiber and perforant path afferents to the CA3 region and dentate gyrus, respectively (Wagner et al., 1992, 1993; Weisskopf et al., 1993; Simmons et al., 1994). Further ultrastructural analysis of the hippocampus using affinity-purified antisera generated against the cloned kappa opioid receptor also supports the concept that the kappa1 opioid receptors are located presynaptically on excitatory afferent terminals (Drake et al., 1996).

The molecular mechanism responsible for the presynaptic inhibition is not completely clear; however, we found that the inhibitory effects of kappa1 opioid receptor activation could be selectively blocked by dendrotoxin, an inhibitor of Kv1 type delayed rectifying potassium channels (Simmons and Chavkin, 1996). Opioid activation of a delayed rectifying potassium channel in the excitatory nerve terminal would be very effective at reducing calcium-dependent transmitter release. Subsequently, Vaughan and co-workers (1997) confirmed these findings and then further showed that opioid receptor activation may increase the Kv1 potassium channel conductance via an eicosanoid second messenger produced by the activation of phospholipase A2. Important details of this signal transduction mechanism are in the process of being better resolved.

Physiological effects of dynorphins in the hippocampus:

One of the physiological consequences of an opioid-induced inhibition of excitatory amino acid release is the inhibition of certain forms of synaptic plasticity in the hippocampus. Endogenous dynorphins released from the granule cells effectively block long term potentiation (LTP) at the mossy fiber to CA3 pyramidal cell synapse (Weisskopf et al., 1993), The release of the Dynorphins also block LTP induction at both the perforant path to granule cell synapse (Wagner et al., 1993; Terman et al., 1994) and at the mossy fiber collateral pathway to the dentate gyrus (Terman et al., in press). Thus, synaptic plasticity at three different synapses in the hippocampus is regulated by the release of the endogenous Dynorphins. Whether this action normally regulates learning and memory is not clear; however, Gallagher and colleagues (1993) have correlated changes in dynorphin levels in the hippocampus with learning deficits. Their data support the hypothesis that the endogenous dynorphins in the hippocampus normally regulate excitatory neurotransmission to control learning processes.

A second physiological consequence of the Dynorphin-induced inhibition of excitatory amino acid release is an inhibition of the excitotoxicity associated with epilepsy (Tortella and DeCoster 1994; Simonato and Romualdi, 1996; Bausch et al., 1998; Zhang and Houser, 1999). Endogeneous Dynorphins have been demonstrated to be neuroprotective and anticonvulsant, and kappa opioid agonists may have clinical utility as antiepileptic drugs (Przewlocka et al., 1994). After damage to the temporal lobes of humans, a partial complex seizure disorder is common; in that form of epilepsy, the Dynorphin expression in the hippocampus is dramatically increased. The results summarized here support the hypothesis that the increased Dynorphin expression serves to partially compensate and serve as a brake for the increased excitability underlying this disease (Simmons et al, 1997).

Release of dynorphins from dendrites of the dentate granule cells

Dynorphin released from granule cells of the dentate gyrus effectively inhibits excitatory amino acid release at *three* synapses listed above. Release at the mossy fiber terminals in the stratum lucidum of the CA3 pyramidal region and release at the mossy fiber collaterals in the dentate hilus region activates kappa opioid receptors on the same and nearby mossy fiber nerve terminals. This is a form of *autoinhibition* as the released opioid peptides inhibit their own release as well as that of excitatory amino acids. In contrast, the source of the dynorphin responsible for presynaptic inhibition of excitatory amino acid release from the perforant path terminals (the third synapse regulated by kappa opioids) was less evident.

The perforant path terminals are the axon fibers projecting from cells in the entorhinal cortex, and these axons form the predominant, excitatory input to the dentate gyrus. The granule cells of the dentate gyrus are organized in a strikingly polar manner; they have their dendrites projecting in one direction (into the molecular layer), and they send their dynorphin-containing axons into the hilar and CA3 regions of the hippocampus. Autoradiographic analysis of receptor distribution showed that the kappa opioid receptors are concentrated in the molecular layer, whereas the dynorphin containing axons are in the different regions of the hippocampus listed above (Wagner et al 1991).

Stimulation of the granule cells released dynorphins that activate the kappa receptors on the perforant path terminals; however the source of peptide was unlikely to be the distant mossy fiber terminals. While application of dynorphin in the molecular layer did reduce excitatory synaptic transmission from the perforant path, application in the hilus (within the dense dynorphin-containing axon plexus) had no significant effect on perforant path excitatory transmission (Drake et al., 1994). These results suggest that endogenous dynorphin peptides may be released from a local source within the dentate molecular layer and not the mossy fiber axons. Electron microscopic analysis of the molecular layer revealed that dynorphin-immunoreactivity was present in dense-core vesicles in both spiny dendrites and unmyelinated axons with the majority of the dynorphin-immunoreactive dense-core vesicles found in granule cell dendrites.

Additional support for the concept that dynorphins are released from the dendrites of the granule cells came from an analysis of the calcium channels responsible for the peptide release (Simmons et al., 1995). We found that both L- and N-type calcium channel antagonists inhibited dynorphin release in the molecular layer of the dentate gyrus. In contrast, N-type but not L-type calcium channel antagonists inhibited peptide release at the mossy fiber axon terminals. Neither L- nor N-type channel antagonists directly altered the effects of kappa opioid receptor activation; thus L-type calcium channels seem to have a special role in regulating peptide release at dendrites. By inhibiting dynorphin release at the granule cell dendrites, L-type channel antagonists also facilitated the induction of long-term potentiation of the perforant path-granule cell synapse. These studies establish that a single cell type can release a transmitter from three different cellular domains and provide new distinction between axonal and dendritic transmitter release mechanisms.

These results suggest that dynorphins act as retrograde neurotransmitters, released from the dendrites of the granule cells following perforant path activation; retrograde feedback has several conceptual advantages. First, it provides direct feedback regulation of an excitatory input from the target; the active excitatory input sends its signal to the post synaptic cell and the post synaptic cell responds by acknowledging the input and limiting further input. Second, providing the feedback from the dendrite rather than an axon allows a spatially precise regulation of the active input. It would not be anatomically or physically efficient to require that the post synaptic granule cell grow an axon to regulate each perforant path terminal. Third, the diffusion of the dynorphin peptides from the sites of release to the targets enables the peptide to activate kappa receptors within approximately 50–100 μm from the dendritic source (Drake et al., 1994). This large sphere of influence allows dynorphin to inhibit surrounding inputs as well as the primary excitatory afferent; in the retina, this form of surround inhibition sharpens the spatial boundaries of the input and helps this structure detect edges to the signal; surround inhibition by dynorphins may have a similar function in the dentate gyrus. Lastly, the dynorphins produce inhibition that has a much longer duration than that of the standard transmitter; following a short (50 msec) train of activation events, the released dynorphin produces inhibition that is sustained for many seconds to minutes. The long duration of the dynorphin signal has clear advantages for temporal integration of signals.

In summary, the shape of the neuropeptide synapse in the mammalian brain is likely to have very different properties than those of the fast-acting transmitters. We have devoted considerable effort to analyze the properties of one set of neuropeptides acting in one region of the brain in the hope that the insights gained here will have general relevance. Further studies are yet required to establish the full range of possibilities.

Acknowledgments

The author thanks Drs. Gregory Terman, Carrie Drake, Michele Simmons and John Wagner for participating in the studies that are reviewed in this manuscript. The work performed in the author's laboratory was supported by a USPHS grant from the National Institute on Drug Abuse (DA 04123).

References

Bausch, S.B., Esteb, T.M., Terman, G.W. and Chavkin, C. (1998) Administered and endogenously released kappa opioids decrease pilocarpine-induced seizures and seizure-induced histopathology. *J. Pharmacol. Exp. Ther.*, 248: 1147–1155.

Chavkin, C., Bakhit, C., Weber, E. and Bloom, F.E. (1983) Relative contents and concomitant release of prodynorphin/neoendorphin-derived peptides in rat hippocampus. *Proc. Natl Acad. Sci.*, 80: 7669–7673.

Chavkin, C., James, I.F. and Goldstein, A. (1982) Dynorphin is a specific endogenous ligand of the kappa opioid receptor. *Science* 215: 413–415.

Chavkin, C., Shoemaker, W., McGinty, J. Bayon, A. and Bloom, F.E. (1985) Characterization of the prodynorphin and proenkephalin neuropeptides systems in rat hippocampus. *J. Neurosci.*, 5: 808–816.

Cheramy, A., Leviel, V. and Glowinski, J. (1981) Dendritic release of dopamine in the substantia nigra. *Nature*, 289: 537–542.

Cottrell, G.A. (1993) The wide range of actions of the FMRFamide-related peptides and the biological importance of peptidergic messengers. *EXS*, 63: 279–285.

Drake, C., Patterson, T.A., Simmons, M.L., Chavkin, C. and Milner, T. (1996) Kappa opioid receptor-like immunoreactivity in guinea pig brain: ultrastructural localization in presynaptic terminals in hippocampal formation. *J. Comp. Neurol.*, 370: 377–397.

Drake, C., Terman, G.W., Simmons, M.L., Kunkel, D., Milner, T., Schwartzkroin, P.A. and Chavkin, C. (1994) Dynorphin opioids present in dentate granule cells may function as a retrograde inhibitory neurotransmitters, *J. Neurosci.*, 14: 3736–3750.

Gallagher, M. and Nicolle, M.M. (1993) Animal models of normal aging: relationship between cognitive decline and markers in hippocampal circuitry. *Behav. Brain Res.*, 57: 155–162.

Hawkins, R.D., Kandel, E.R. and Siegelbaum, S.A. (1993) Learning to modulate transmitter release: themes and variations in synaptic plasticity. *Annu Rev. Neurosci.*, 16: 625–665.

Jan, L.Y. and Jan, Y.N. (1982). Peptidergic transmission in sympathetic ganglia of the frog. *J. Physiol.*, 327: 219–246.

Katz, B. (1966) *Nerve, Muscle, and Synapse*. McGraw-Hill, New York.

Kieffer, B.L. (1995) Recent advances in molecular recognition and signal transduction of active peptides: receptors for opioid peptides. *Cell Mol. Neurobiol.*, 15: 615–635.

Kuffler, S.W. and Nicholls, J.G. (1976) *From Neuron to Brain*, Sinauer Associates, Sunderland, Massachusetts, pp. 216.

McGinty, J.F., Henriksen, S.J., Goldstein, A., Terenius, L. and Bloom, F.E. (1983) Dynorphin is contained within hippocampal mossy fibers: immunochemical alterations after kainic acid administration and colchicine-induced neurotoxicity. *Proc. Natl Acad. Sci.*, 80: 589–593.

Pow, D.V. and Morris, J.F. (1989) Dendrites of hypothalamic magnocellular neurons release neurohypophysial peptides by exocytosis. *Neuroscience*, 32: 435–439.

Przewlocka, B., Machelska, H. and Lason, W. (1994) Kappa opioid receptor agonists inhibit the pilocarpine-induced seizures and toxicity in the mouse. *Eur. Neuropsychopharmacol.* 4: 527–533.

Simmons, M.L. and Chavkin, C. (1996) Kappa opioid receptor activation of a dendrotoxin-sensitive potassium channel mediates presynaptic inhibition of mossy fiber neurotransmitter release. *Mol. Pharmacol.*, 50: 80–85.

Simmons, M.L., Drake, C.T., Terman, G.W. and Chavkin, C. (1994) Inhibition of glutamate release by presynaptic kappa1 opioid receptors in the guinea pig dentate gyrus. *J. Neurophysiol.*, 72: 1697–1705.

Simmons, M.L., Terman, G.W., Gibbs, S.M. and Chavkin. C. (1995) L-type calcium channel types mediate dynorphin neuropeptide release from dendrites but not axons of hippocampal granule cells. *Neuron* 14: 1265–1272.

Simmons, M.L., Terman, G.W. and Chavkin, C. (1997) Spontaneous excitatory currents and kappa opioid receptor inhibition in dentate gyrus are increased in the rat pilocarpine model of temporal lobe epilepsy. *J. Neurophysiol.*, 78: 1860–1868.

Simonato, M. and Romualdi, P. (1996) Dynorphin and epilepsy. *Prog. Neurobiol.*, 50: 557–583.

Terman, G.W., Wagner, J.J. and Chavkin, C. (1994) Kappa opioids inhibit induction of long-term potentiation in the dentate gyrus of the guinea pig hippocampus. *J. Neurosci.*, 14: 4740–4747.

Terman, G.W., Drake, C.T., Simmons, M.L., Milner, T.A. and Chavkin, C. (2000) Opioid modulation of recurrent excitation in the dentate gyrus of the guinea pig hippocampus. *J. Neuroscience* (in press).

Tortella, F.C. and DeCoster, M.A. (1994) Kappa opioids: therapeutic considerations in epilepsy and CNS injury. *Clin. Neuropharmacol.*, 17: 403–416.

Vaughan, C.W., Ingram, S.L., Connor, M.A. and Christie, M.J. (1997) How opioids inhibit GABA-mediated neurotransmission. *Nature*, 390: 611–614.

Wagner, J.J., Evans, C.J. and Chavkin, C. (1991) Focal stimulation of the mossy fibers releases endogenous dynorphins that bind kappa1 opioid receptors in the guinea pig hippocampus. *J. Neurochem.*, 57: 333–343.

Wagner, J.J., Caudle, R.M. and Chavkin, C. (1992) Kappa opioids decrease excitatory transmission in the dentate gyrus of the guinea pig hippocampus. *J. Neurosci.*, 12: 132–141.

Wagner, J.J., Terman, G.W. and Chavkin, C. (1993) Endogenous dynorphins inhibit excitatory neurotransmission and block long term potentiation in the hippocampus. *Nature*, 363: 451–454.

Weisskopf, M.G., Zalutsky, R.A. and Nicoll, R.A. (1993) The opioid peptide dynorphin mediates heterosynaptic depression of hippocampal mossy fibre synapses and modulates long-term potentiation. *Nature*, 362: 423–427.

Zhang, N. and Houser, C.R.J. (1999) Ultrastructural localization of dynorphin in the dentate gyrus in human temporal lobe epilepsy: a study of reorganized mossy fiber synapses. *Comp. Neurol.*, 405: 472–490.

L. F. Agnati, K. Fuxe, C. Nicholson and E. Syková (Eds.)
Progress in Brain Research, Vol 125

CHAPTER 24

Neuropeptide spread in the brain and spinal cord

A.W. Duggan

Department of Anesthesia and Pain Management, University of Sydney, Royal North Shore Hospital, St. Leonards, NSW 2065, Australia

Introduction

When transmission in the brain and spinal cord was first studied by intracellular methods the time scale of the potentials measured was one of milliseconds. Thus when Eccles (1959) spoke of excitatory or inhibitory post synaptic potentials these typically reached a peak within 2–5 milliseconds and decayed over 10–50 milliseconds. This was comparable to the time scale established in the periphery for acetylcholine acting at the neuromuscular junction; longer lasting potentials were observed with repetitive afferent inputs but this was viewed as temporal summation of short lasting events. For released neurotransmitters to operate within this time scale not only requires a very short diffusion distance from the releasing structure to the relevant receptor but a rapid inactivation process which will effectively prevent significant diffusion away from the site of release. From an operational viewpoint such a system has a minimum of crosstalk between neurones and can encode the onset and offset of afferent information with a good temporal resolution.

Amino acids appear to be the major excitatory and inhibitory transmitters within the mammalian central nervous system (Curtis and Watkins, 1963; Curtis, 1978) and the evidence indicates that they are inactivated predominantly by uptake into neurons and glia adjacent to sites of release (Johnston, 1978; Larsson et al., 1991). There is no evidence for significant enzymic degradation of extracellular amino acids. Since there is abundant evidence that aminoacids mediate the rapid time scale excitatory and inhibitory potentials referred to previously, it is assumed that inactivation of these compounds following release is so rapid that significant diffusion does not occur. The question however, has not been directly investigated experimentally.

The effects of neuropeptides differ in many ways from those just described for amino acids. The post synaptic potentials produced by synaptic events mediated by neuropeptides are, in general, orders of magnitude longer than those produced by aminoacids. For example, when a peripheral nerve is stimulated and intracellular recordings are obtained from a neuron of the dorsal horn, the short latency depolarizing potential blocked by an antagonist of L-glutamate lasts for less than 20 milliseconds whereas the later long-lasting depolarization reduced by a substance P antagonist last for hundreds of milliseconds (Radhakrishnan and Henry, 1995). Uptake systems have not been described for released neuropeptides and their extracellular degradation is presumed to result from membrane bound enzymes (McKelvey and Blumberg, 1986). At the skeletal neuromuscular junction enzymic degradation of acetylcholine is a rapid process and such rapidity probably results from the proximity of cholinesterase to the sites of release. The subject of enzymic degradation of released neuropeptides will be referred to elsewhere in this account but suffice to say at this juncture, that a

*Corresponding author: Tel: 61-2-9926 8420; Fax: 61-2-9906 4079; e-mail: aduggan@med.usyd.edu.au

situation comparable to the neuromuscular junctions appears not to be present in the central nervous system. If a released compound is slowly degraded it follows that not only will its action at the site of release be prolonged but significant diffusion away from the site of release can also occur. This has important implications for function since it means that crosstalk between neurons can occur and relatively global alterations in excitability may also follow. It poses the question what is the message that the neurotransmitter is conveying if it affects the function of many surrounding cells?

For the past decade work in my laboratory has employed antibody microprobes to study release of neuropeptides in the brain and spinal cord. These probes bear antibodies immobilized to their outer surfaces and when placed in the central nervous system will bind a proportion of the molecules recognized by the antibodies from the adjacent extracellular space (Duggan, 1991). The bound molecules are detected on autoradiographs by the failure of binding of a radio labelled form of the peptide under study; the probes are of the same dimensions as glass microelectrodes and hence are relatively atraumatic. Microprobes were designed to detect focal sites of release (100–200 μm) but because they are inserted for several millimetres, they can detect diffusion away from a site of release. As such they can provide direct evidence relevant to the volume transmission hypothesis.

Release of tachykinins

The first real evidence from microprobe experiments which suggested that a neuropeptide was diffusing a relatively large distance from its site of release came from experiments on the release in the spinal cord of the two tachykinins, substance P (SP) and neurokinin A (NKA). Substance P and NKA extensively coexist in primary afferent neurons of the rat (Dalsgaard et al., 1985). Virtually all dorsal root ganglion cells containing NKA also contain SP although the converse is not true. The release of both compounds has been extensively studied in the spinal cord in relation to the function of peripheral nociceptors (Go and Yaksh, 1987; Duggan et al., 1988; Kuraishi et al., 1989; Duggan et al., 1990; Schaible et al., 1997). Since it is the results with antibody microprobes which have drawn attention to differences in the diffusion of SP and NKA following release, these alone will be discussed.

Studies with antibody microprobes have shown that, in the anesthetized spinal cat, electrical stimulation of large diameter primary afferents of the tibial nerve does not produce release of immunoreactive (ir) SP within the spinal cord but that increasing the stimulus strength to include unmyelinated (C) fibres results in release of this peptide in the region of the substantia gelatinosa of the ipsilateral dorsal horn (Duggan and Hendry, 1986). When using noxious peripheral stimuli a release of ir SP was also produced in the region of the substantia gelatinosa by noxious thermal, mechanical and chemical stimuli (Duggan et al., 1988). With noxious heat the hind paw was immersed in a water bath and, although temperatures of 45°–48°C are generally regarded as painful, both in man and cat (Zimmermann, 1976), a bath temperature of 50°C was needed to produce release of irSP in the dorsal horn. The noxious mechanical stimulus was pinching of the skin of digital pads with small alligator clips; non-noxious mechanical stimulation did not result in ir SP release.

Studies of the release of NKA, have shown important differences from SP. In the anesthetized cat, peripheral cutaneous thermal and mechanical stimuli produced a spinal release of ir neurokinin A (Duggan et al., 1990). Unlike release of SP a thermal stimulus at 46°C was adequate to produce detectable release of ir NKA. In marked contrast to the relatively focal release of ir SP in the substantia gelatinosa, ir NKA was detected widely in the dorsal horn and the overlying dorsal columns. These contrasting patterns of release of SP and NKA are illustrated in Fig. 1. This difference is unlikely to have resulted from a wider distribution of NKA-containing primary afferent terminals since, as cited previously, virtually all NKA containing primary afferents also contain SP at least in the rat.

Further experiments (Hope et al., 1990b) showed that ir NKA released by peripheral nerve stimulation in the cat persisted in the dorsal horn in elevated amounts for at least an hour beyond the cessation of the stimulus. This persistence probably accounts for the diffuse presence of ir NKA detected by microprobes when compared with the localized presence of SP.

Neuropeptides appear to be degraded mainly by enzymes but there are often difficulties in determining which enzymes are responsible for the major part of degradation of a particular peptide under physiological conditions and whether these enzymes are located adjacent to the sites of release (McKelvey and Blumberg, 1986; Bunnet, 1987; Csuhai et al., 1995). Whilst a number of enzymes degrade SP, of these only endopeptidase 24.11 significantly degrades NKA and at a slower rate (Hooper et al., 1985; Theodorsson-Norheim et al., 1987). Thus there are reasonable biochemical grounds to support the conclusion of the microprobe experiments that NKA may be slowly degraded when released in the spinal cord.

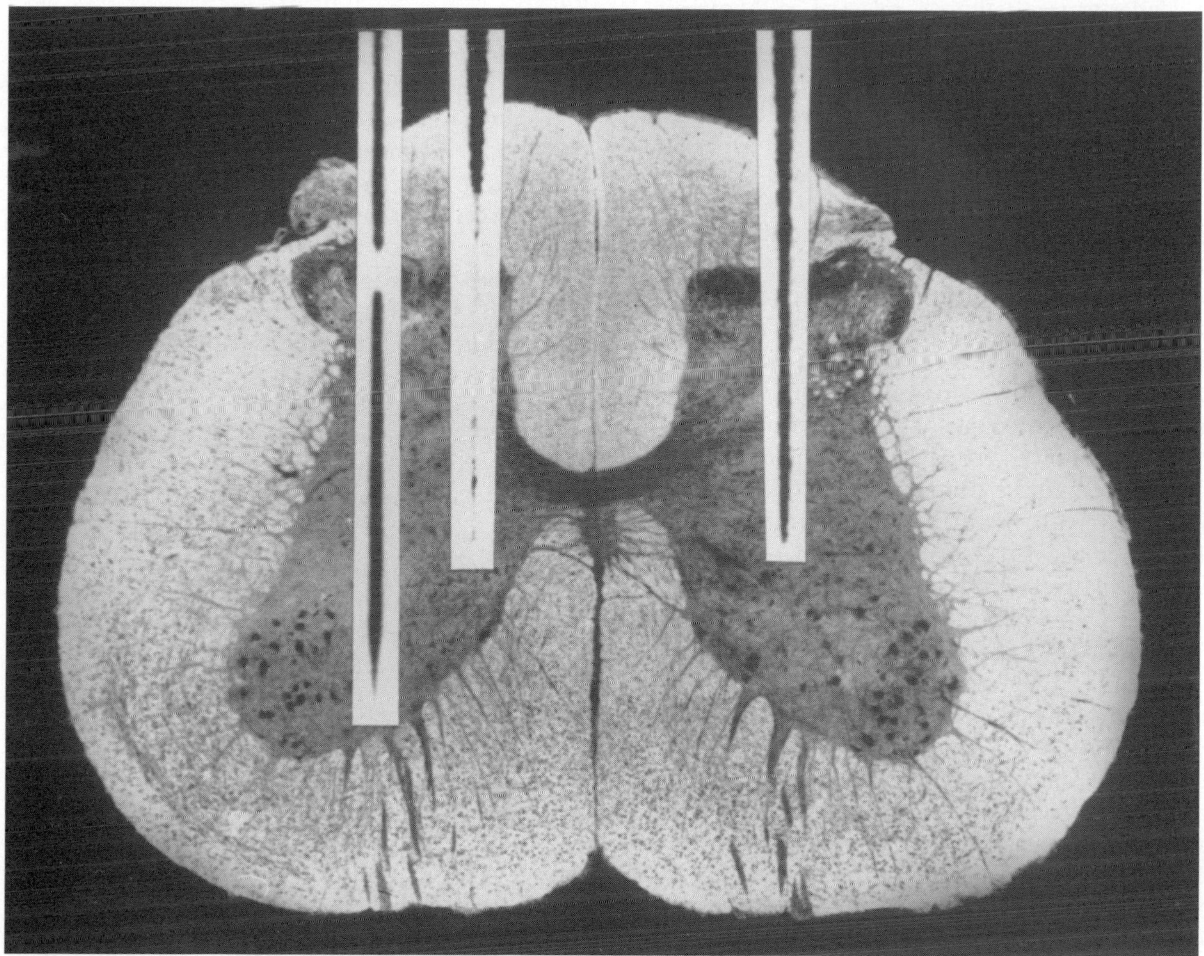

Fig. 1. Contrasting release patterns of substance P and neurokinin A following peripheral noxious mechanical stimuli in the cat. Photographic enlargements of x-ray film images of microprobes have been superimposed on a photograph of a cross section of the L6 spinal cord. The microprobe on the right shows uniform binding of ^{125}I NKA. No peripheral stimulus was applied before and after this microprobe was in the spinal cord. The two microprobes on the left remained for 30 min. in the spinal cord while the digital pads of the ipsilateral hindpaw were pinched. The microprobe on the far left bore antibodies to substance P while the middle microprobe bore antibodies to NKA. Following similar peripheral stimuli, ir-substance P was detected focally in the superficial dorsal horn whereas ir-neurokinin A was found throughout the whole of the dorsal horn and the deeper dorsal columns. (Reproduced, with permission, from Duggan et al. (1990), *Neuroscience*, 35: 195–202).

Substance P release following inhibition of degrading enzymes

An important corollary to these studies on the release of irNKA, is that protecting SP from degradation should result in a release pattern similar to that of NKA. Such an effect was shown by Duggan et al. (1993a) following microinjection of the mixed peptidase inhibitors kelatorphan and enalaprilat into the superficial dorsal horn of the cat. Under these conditions, released irSP not only diffused widely in the spinal cord but also persisted for prolonged periods following release. Whereas the increased amounts in the grey matter could represent protection of previously undetectable amounts at sites of release, this reasoning cannot apply to the widespread presence of irSP in the dorsal columns which must have resulted from diffusion from distal sites of release.

Additional experiments with microinjection of calcitonin gene-related peptide (CGRP) suggested that the sites accessed by synaptically released SP may be under physiological control through a mechanism of varied rate of degradation. Le Greves et al (1985) showed that CGRP is a potent inhibitor of an endopeptidase degrading SP. CGRP is often colocalised with SP in primary afferent neurons (Hökfelt et al., 1992; Galeazza et al., 1995) and hence is likely coreleased. When both neuropeptides were applied topically to the spinal cord, CGRP prolonged facilitation of a flexor reflex by SP (Woolf and Wiesenfeld-Hallin, 1986), a result consistent with an inhibition of SP degradation. Experiments with antibody microprobes have shown that microinjection of CGRP into the superficial dorsal horn gave results similar to those obtained with synthetic peptidase inhibitors: irSP released by noxious stimuli diffused widely in the dorsal horn when compared with release prior to CGRP injection (Schaible et al., 1992). This action is illustrated in Fig. 2.

Arthritis and spinal release of tachykinins

The demonstration that CGRP can protect synaptically released SP raises the possibility that the distribution and hence the sites accessed by SP is under physiological control. An important question is when is this control altered in response to a physiological stimulus? Microprobe experiments of SP release associated with peripheral inflammation suggest that the latter may be such a stimulus. Schaible et al. (1992) studied this directly by inserting antibody microprobes into the spinal cord of the cat to measure release of irSP before and after induction of inflammation in a knee joint by injection of kaolin and carrageenan. Both before, and from 3 to 8 h after joint injection, movement of the injected joint failed to produce a central release of irSP. Beyond that period, joint flexion produced a massive release of SP both in the superficial dorsal horn and deep in spinal laminae VI and VII. Compression of an inflamed joint resulted in an even larger release of SP. Figure 3 illustrates microprobes inserted into the spinal cord at this later phase following induction of joint inflammation and shows the extraordinary patterns of distribution of SP released by simple joint flexion and joint compression.

Spinal release of irNKA was also studied following the induction of peripheral arthritis (Hope et al., 1990a). Unlike substance P, spinal release of irNKA occurred immediately after joint injection and widespread presence in the spinal cord was observed. Indeed, for the duration of each experiment microprobes indicated that irNKA had diffused throughout the whole of the spinal grey matter sampled within 30 min of injecting one knee joint with kaolin and carrageenan.

Functional implications

It is probable that L-glutamate was released together with the neuropeptides in all of the cited studies. Antagonists of glutamate receptors of the AMPA type have been shown to reduce short latency excitation of spinal neurons by peripheral noxious stimuli (Dougherty et al., 1992a; Neugebauer et al., 1993; Procter et al., 1998). Indeed if the onset and offset of a relatively brief noxious peripheral stimulus are to be encoded and transmitted with any accuracy, then a rapidly inactivated transmitter is essential. The motor responses to such a stimulus also need to be of short latency, since nociceptive reflexes (such as flexor withdrawal) normally rapidly withdraw the stimulated area away from the stimulus source.

With peripheral inflammation the situation has changed; there may be a near continuous input from sensitized peripheral nociceptors since they may fire in response to previously innocuous stimuli (Grigg et al., 1986). This heightened awareness is appropriate to the avoidance of furthur damage to the inflamed area and will also act to keep it immobilized. Mechanical allodynia and hyperalgesia typically occur in skin areas outside the actually inflamed site (Schaible and Grubb, 1993) and pain may outlast the duration of an applied stimulus. The process of volume transmission is appropriate to initiating and maintaining these changes since widespread alterations in excitability will influence the responses of the non-inflamed as well the inflamed peripheral areas. The relative abundance of NK1 receptors around but not within the substantia gelatinosa (Bleazard et al., 1994; Liu et al., 1994) supports the proposal that the main actions of SP are exerted at a distance from the initial site of release (Duggan et al., 1979). Thus if a neurotransmitter should normally be thought of as transmitting a particular message, the message when substance P is released in abundance from the central terminals of nociceptors is that real tissue damage has occurred peripherally. The situation with NKA is less certain since this is released by non-damaging noxious stimuli. Because of its persistence, NKA can be likened to a short term memory trace and could contribute to hyperexcitability with repetitive noxious stimuli in the absence of developed inflammation.

Numerous studies have shown that spinal neurons receiving input from inflamed areas have expanded receptive fields (Schaible et al., 1987; Hoheisel et al., 1993) and glutaminergic excitation

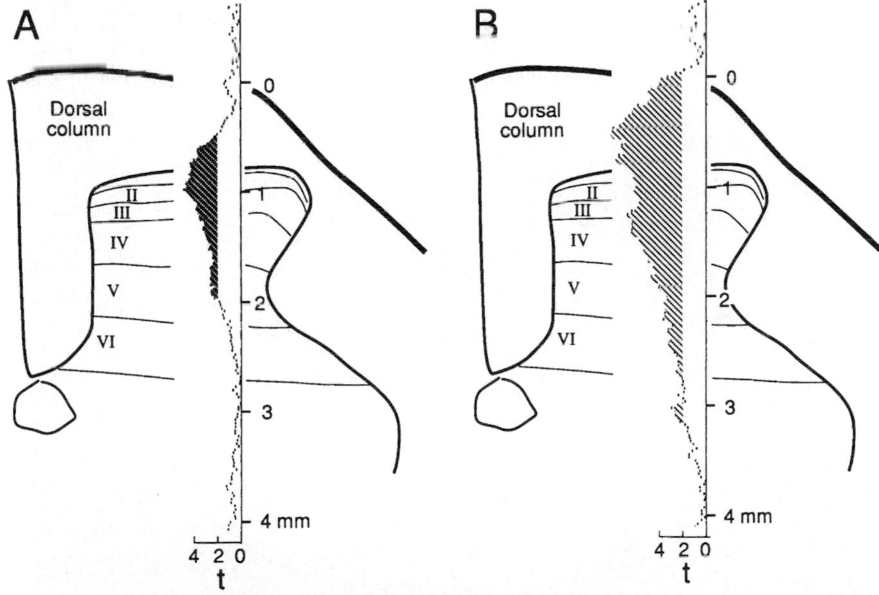

Fig. 2. The effect of injection of peptidase inhibitors of the spread of substance P released by peripheral nerve stimulation in the cat. A. Substance P release prior to injection of peptidase inhibitors. Superimposed on a diagram of the spinal cord is a plot of the *t*-statistics derived from the differences between the mean image analyses of 19 microprobes inserted into the spinal cord in the absence of any stimulation and of 42 microprobes present in the spinal cord during electrical stimulation of the ipsilateral tibial nerve adequate to excite C fibres. The hatched area indicates where these differences are significant with $P < 0.05$. B. Substance P release following microinjection of 0.1 μl of kelatorphan (10^{-3} M) and enalaprilat (10^{-3} M) into the dorsal horn. The format is as in A and 19 microprobes were in the nerve stimulation, post peptidase group. In both A and B microprobes remained in the spinal cord for 10 min. B relates to the period 35 to 45 min after peptidase injection. (Reproduced, with permission, from Duggan et al. (1992) *Brain Res.*, 579: 261–269).

of these cells increasingly involves receptors of the NMDA type (Coderre and Melzack, 1991; Schaible et al., 1991a; Dougherty et al., 1992b). The latter induces firing in bursts (Grubb et al., 1996) which will increase transmitter release at the terminals of these cells which will likely further expand receptive fields. Tachykinins have been shown to potentiate excitation by glutamate at NMDA receptors (Cumberbatch et al., 1995; Randic et al., 1995).

It should not be forgotten that spinal inhibition is also altered with the development of peripheral inflammation (Schaible et al., 1991b) and its function may be to prevent excessive movement or load bearing of an inflamed area. Possibly relevant to this is the enhanced spinal release of dynorphin

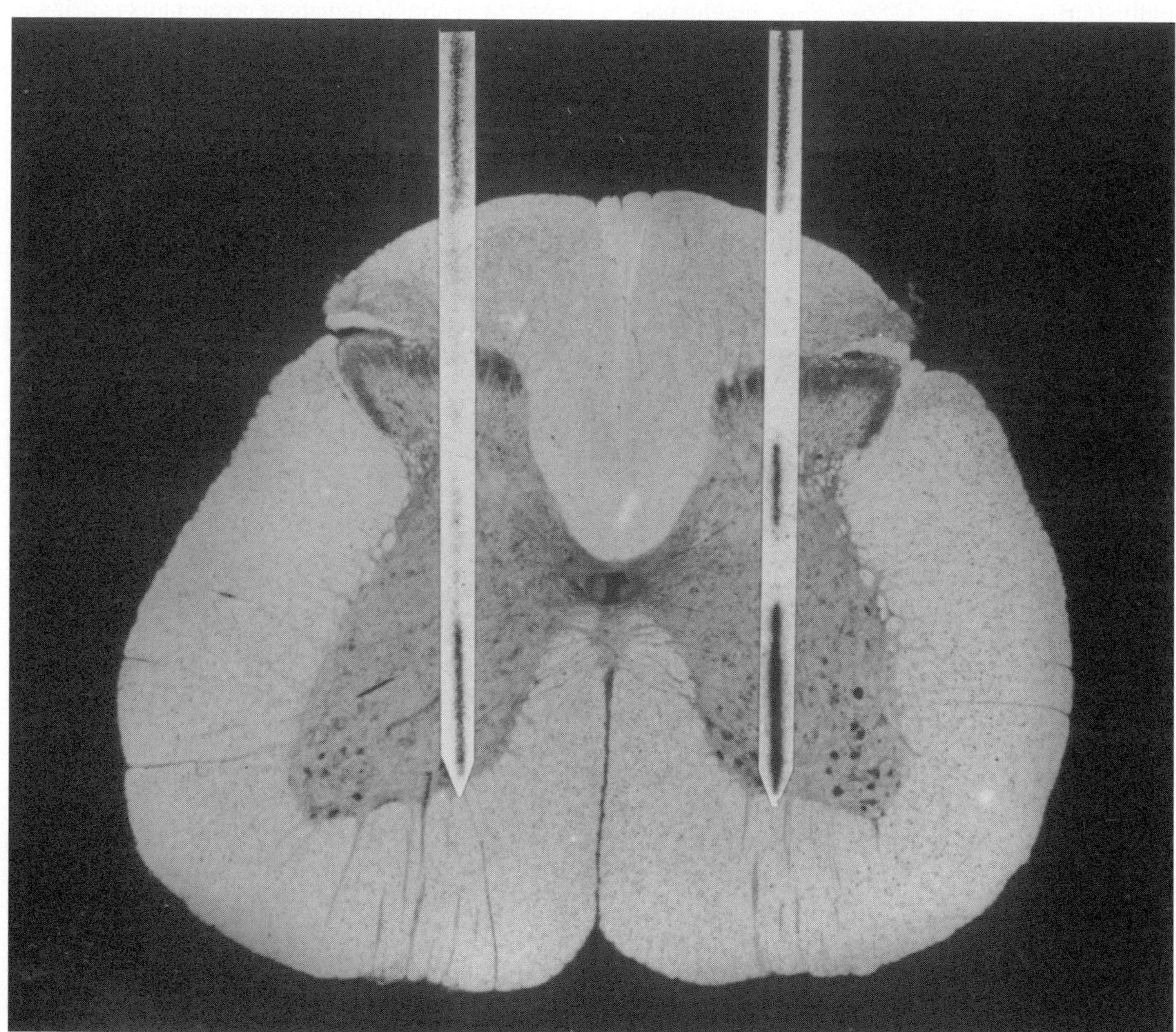

Fig. 3. Release of ir-substance P with acute arthritis in the cat. The mircoprobe on the right was inserted into the spinal cord 560 min. after the ipsilateral knee joint was injected with kaolin and carrageenan. During the 30 min. the probe remained in the spinal cord the inflamed joint was flexed 10 times. Note the two zones of substance P release. The deep one approximately to the laminae VI and VII while the upper one encompasses the whole of the dorsal columns and the superficial dorsal horn. The microprobe on the left was inserted at 640 min. after joint injection and remained in the cord for 5 min. During this time the inflamed joint was twisted. The resultant released substance P was detected over the whole of the dorsal columns, the whole dorsal horn and the upper ventral horn.

(Riley et al., 1996) which again appears to diffuse widely in the spinal cord and operate through a volume transmission process.

Release and spread of β-endorphin

The second series of antibody microprobe experiments which have strongly supported the volume transmission concept dealt with release of β-endorphin following electrical stimulation of the arcuate nucleus. A previous series of experiments had failed to detect β-endorphin release in the periaqueductal grey matter (PAG) of the anaesthetized rat following severe peripheral noxious stimuli (Duggan et al., 1993b). Since such a result may have resulted from an inadequate sensitivity of the microprobes it was decided to directly stimulate the arcuate nucleus, which contains the majority of the cells of origin of the β-endorphin containing fibres of the central nervous system of the rat (Young et al., 1993), and seek release within the PAG (MacMillan et al., 1998). These experiments were also performed in anaesthetized rats. To minimize trauma to nervous tissue, microprobes are inserted into the PAG through the intact cerebral cortex and midbrain colliculi. The sagittal sinus prevented inserting microprobes being introduced vertically and hence they were inclined medially at 15°C and entered the cerebral cortex 2.0 mm from the midline and penetrated for 7.5 mm. Prior to any arcuate stimulation, microprobes detected an extensive basal presence of ir-β-endorphin throughout the whole of the areas sampled (cerebral cortex, colliculi, PAG and midbrain tegmentum). There is evidence in humans for a release of β-endorphin into blood following surgery (Janicki et al., 1993; Acosta et al., 1995), and hence the basal presence of β-endorphin in rat brain may have resulted from the surgery needed for microprobe experiments.

With electrical stimulation of the arcuate nucleus at 2, 5 and 10 Hz there was a small release of ir-β-endorphin within the PAG during the period of stimulation, but after stimulation a peak of the opioid was found in the space between the colliculi and the inferior surface of the cerebral cortex. By 75 to 90 minutes after stimulation, ir-β-endorphin had fully penetrated the areas of cerebral cortex sampled and much of the midbrain.

This sequential spread of ir-β-endorphin is well shown in Fig. 4 which contains scans of single microprobes inserted sequentially in the same experiment. The pre-stimulus and during-stimulus microprobes showed few differences but 45 min following the cessation of stimulation, a peak of

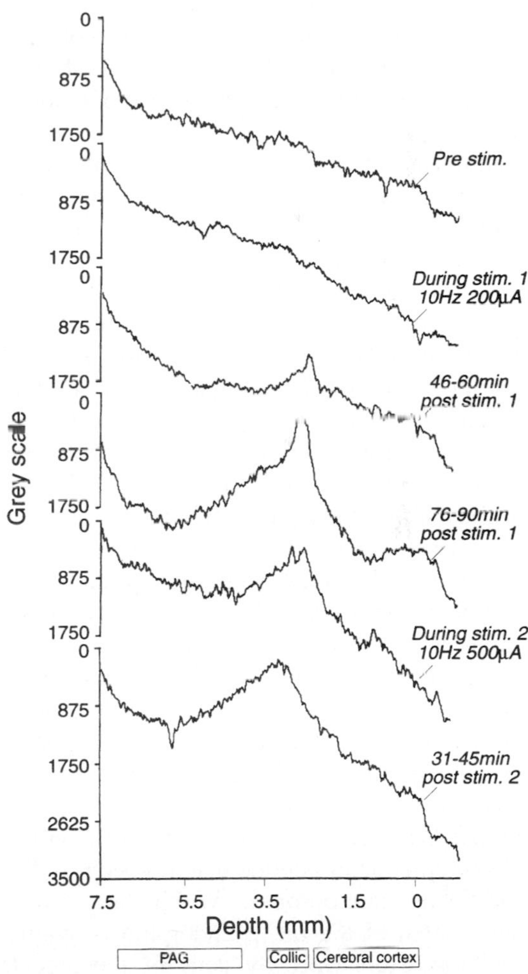

Fig. 4. Release of ir-β-endorphin by electrical stimulation of the arcuate nucleus at 10 Hz, 500 µA. The image analyses are of single microprobes inserted sequentially at 15 min. intervals in the one experiment. Note the peak of ir-β-endorphin appearing at the junction of the inferior surface of the cerebral cortex and the superior colliculus at 46–60 min after stimulation and the sequential spread of the neuropeptide into the brain and midbrain. The second period of stimulation occurred 270 min after the first. (Reproduced, by permission, from MacMillan et al. (1998).

inhibition of binding of radiolabelled β-endorphin appeared 2.25 to 2.5 mm from the surface of the cerebral cortex. From this depth and the location of a marker deposit of pontamine sky blue in this experiment, this peak corresponds to the cerebrospinal fluid (CSF)-containing space between the superior colliculus and the overlying cerebral cortex. In this experiment there was a second period of electrical stimulation at 10 Hz with a strength of 500 μA which may or may not have produced further release but the spread of ir-β-endorphin through the midbrain and cerebral cortex continued.

When release of ir-β-endorphin was produced in these experiments it was associated with peak levels in the subarachnoid space between the lower surface of the cerebral cortex and the superior surface of the superior colliculus. Sequential insertion of microprobes showed a diffusion of ir-β-endorphin bilaterally away from this site into adjacent neural tissue. The simplest explanation of this result is that ir-β-endorphin was released, or rapidly transported following release, into the ventricular system and that transport in CSF was the mechanism of subsequent diffusion into diverse regions of the central nervous system. Active transport to the CSF could occur through tanycytes. In the hypothalamus these glial cells have processes extending from the third ventricle to the median eminence and have been shown to accumulate β-endorphin injected into the third ventricle (Bjelke and Fuxe, 1993) and to possess μ-opiate receptors at sites of apposition with nerve terminals in the hypothalamus (Beauvillain et al., 1992).

This pattern of release of β-endorphin is unlike that of any other neuropeptide that has been studied with antibody microprobes. Within the central nervous system of the rat, β-endorphin is synthesized almost exclusively by cells of the arcuate nucleus and the nucleus of the tractus solitarius (Young et al., 1993). The axons of β-endorphin-containing arcuate neurones pursue an ipsilateral periventricular course both rostrally and caudally with terminations in such structures as the thalamus, hypothalamus and periventricular grey matter (Khachaturian et al., 1985, 1993). It was a surprising finding that release of β-endorphin was not consistently detected within the PAG during electrical stimulation near the arcuate nucleus. This suggests that the relatively dense hypothalamic terminations of β-endorphin-containing fibres released β-endorphin predominantly into the third ventricle initially, and there subsequently occurred diffusion within the CSF and thence into adjacent brain areas.

Based on studies of receptor affinity, β-endorphin is a likely physiological ligand of μ and δ opiate receptors (Corbett et al., 1993; Mansour et al., 1995b). Although μ-receptors are present in the periventricular structures where β-endorphin-containing fibres are found, δ-receptors are relatively sparse at these sites (Mansour and Watson, 1993). In addition high densities of μ-opiate receptors occur in the superficial dorsal horn of the spinal cord, the superior and inferior colliculi, moderate densities occur in some layers of the cerebral cortex and in the basal ganglia (Mansour et al., 1995a) but all of these structures are essentially devoid of β-endorphin-containing fibres (Khachaturian et al., 1985, 1993). Although other opioid peptides may activate these μ-receptors (Corbett et al., 1993; Zadina et al., 1997), the contrasting distributions of β-endorphin-containing fibres and of μ and δ opiate receptors is an example of the so called ligand-receptor mismatch for neuropeptides in the brain (Kuhar, 1985). The experiments just described imply that while β-endorphin-containing fibres are absent in many parts of the brain, this neuropeptide can access receptors in these sites. The results also suggest that an important mechanism of spread of released β-endorphin is through cerebrospinal fluid.

The extensive studies of Bach and Yaksh have shown that β-endorphin is readily found in ventriculo-cisternal CSF following activation of arcuate neurons. This has been shown by both electrical stimulation (Bach and Yaksh, 1995b) and by chemical excitation of the arcuate with *N*-methyl-D-aspartate (Bach and Yaksh, 1995a). There are reports of β-endorphin in human CSF which have attempted to relate alterations in these levels to clinical syndromes (Bach et al., 1987; Myer et al., 1992; Vaeroy et al., 1997). Tseng et al (1986) have proposed that β-endorphin in CSF produces a spinal release of met-enkephalin.

Functional implications

Widespread diffusion implies relatively broad or global alterations in brain function; there is some evidence for this in the case of β-endorphin. It is a popularly held view that release of brain endorphins results in mood elevation and this may occur with exercise in man (Janal et al., 1984). A general inhibition of activity appears to occur in other species. There is evidence that hibernation is opioid dependent (Margules et al., 1979; Beckman and Llados-Eckman, 1985; Bruce et al., 1996) and this may relate to β-endorphin release. This release of β-endorphin with surgery may relate to the need to inhibit body activity following extensive tissue damage as an aid to healing.

Concluding remarks

Microprobe experiments are supportive of the volume transmission for the three neuropeptides considered in this account. Possibly the most surprising result is the extraordinary distance over which diffusion from sites of release was found to occur.

Microprobe experiments of the type described are difficult to quantify and it was not possible to state the concentrations of the tachykinins or of β-endorphin at sites where the neuropeptides were detected. Thus some caution is needed in predicting the functional effects on neurons bearing the appropriate receptors. There are two considerations here – whether the concentration of the peptide is sufficient to activate receptors and whether the receptors are sufficiently abundant to modify cell function when activated. The recently developed technique of direct visualization of receptor internalization (Bowden et al., 1994; Mantyh et al., 1995) offers a partial solution to the first caveat but the second remains largely unexplored; focal administration of antagonists may help. For example it may be possible to show that a μ-opiate antagonist administered microiontophoretically near neurons of the superior colliculus has no effect on firing prior to arcuate stimulation but modifies firing post stimulation. It is important in such experiments that the structure studied should have an absence of the relevant releasing terminals. Although focal administration of an antagonist will help determine whether a compound is affecting the function of single neurons by a process of volume transmission, this will not determine the behavioural consequences of such a mechanism. This is a more difficult task. It can be hinted at, but not proven, by finding that a systemically administered antagonist (which penetrates the blood brain barrier), produces a behavioural change which cannot be reproduced by focally injecting the antagonist at the sites of initial release of the agonist.

Abbreviations

CGRP	calcitonin gene-related peptide
NKA	neurokinin A
SP	substance P

References

Acosta, F., Diaz, J., Moreno, J., Fuente, T., Contreras, R.F., Sansano, T., Bueno, F.S., Robles, R., Ramirez, P. and Parrilla, P. (1995) Relationship between β-endorphin release and surgical stress in orthotopic liver transplantation. *Transplant. Proc.* 27: 2303.

Bach, F.W., Langemark, M., Secher, N.H. and Oleson, J. (1987) Plasma and cerebrospinal fluid beta-endorphin during chronic tension-type headache. *Scand. J. Clin. Lab. Invest*, 47: 751–758.

Bach, F.W. and Yaksh, T.L. (1995a) Release of β-endorphin immunoreactivity from brain by activation of a hypothalamic *N*-methyl-D-aspartate receptor. *Neuroscience*. 65: 775–783.

Bach, F.W. and Yaksh, T.L. (1995b) Release into ventriculo-cisternal perfusate of β-endorphin- and Met-enkephalin-immunoreactivity: effects of electrical stimulation in the arcuate nucleus and periaqueductal gray of the rat. *Brain Res.* 690: 167–176.

Beauvillain, J.C., Moyse, E., Dutriez, I., Mitchell, V., Poulain, P. and Mazzuca, M. (1992) Localization of mu opioid receptors on the membranes of nerve endings and tanycytes in the guinea-pig median eminence by electron microscopic radioautography. *Neuroscience*, 49: 925–936.

Beckman, A.L. and Llados-Eckman, C. (1985) Antagonism of brain opioid peptide action reduces hibernation bout duration. *Brain Res*, 328: 201–205.

Bjelke, B. and Fuxe, K. (1993) Intraventricular beta-endorphin accumulates in DARPP-32 immunoreactive tanycytes. *NeuroReport*, 5: 265–268.

Bleazard, L., Hill, R.G. and Morris, R. (1994) The correlation between the distribution of the NK1 receptor and the actions of tachykinin agonists in the dorsal horn of the rat indicates that substance P does not have a functional role on substantia gelatinosa (lamina II) neurons. *J. Neurosci.*, 14: 7655–7664.

Bowden, J.J., Garland, A.M., Baluk, P., Lefevre, P., Grady, E.F., Vigna, S.R., Bunnett, N.W. and McDonald, D.M. (1994) Direct observation of substance P-induced internalization of neurokinin 1 (NK_1) receptors at sites of inflammation. *Proc. Natl Acad. Sci. USA*, 91: 8964–8968.

Bruce, D.S., Bailey, E.C., Setran, D.P., Tramell, M.S., Jacobson, D., Oeltgen, P.R., Horton, N.D. and Hellgren, E.C. (1996) Cirannual variations in bear plasma albumin and its opioid-like effects on guinea pig ileum. *Pharmacol., Biochem. & Behav.*, 53: 885–889.

Bunnet, N.W. (1987) Postsecretory metabolism of peptides. *Ann. Rev. Respir. Dis.*, 136: 27–34.

Coderre, T.J. and Melzack, R. (1991) Central neural mediators of secondary hyperalgesia following heat injury in rats: neuropeptides and excitatory amino acids. *Neurosci. Lett.*, 131: 71–74.

Corbett, A.D., Paterson, S.J. and Kosterlitz, H.W. (1993) Selectivity of ligands for opioid receptors. In: A. Herz (Ed.), *Handbook of Experimental Pharmacology 104/ I. Opioids I.* 645–679. Berlin, Springer-Verlag.

Csuhai, E., Little, S.S. and Hersh, L.B. (1995) Inactivation of neuropeptides. In: F. Nyberg, H.S. Sharma and Z. Wiesenfeld-Hallin. *Neuropeptides in the Spinal Cord*, 131–142. Amsterdam, Elsevier.

Cumberbatch, M.J., Chinzh, B.A. and Headley, P.M. (1995) Modulation of excitatory amino responses by tachykinins and selective tachykinin receptor agonists in the rat spinal cord. *Br. J. Pharmacol.*, 115: 1005–1012

Curtis, D.R. (1978) Gabergic transmission in the mammalian central nervous system. In: P. Krogsgaard-Larsen, J. Scheel-Kruger and H. Kofod. (Eds), *GABA-Neurotransmitters – Alfred Benzon Symposium*, 12th ed., 17–27. Copenhagen, Munksgaard.

Curtis, D.R. and Watkins, J.C. (1963) Acidic amino acids with strong excitatory actions on mammalian neurons. *J. Physiol.* 166: 1–4.

Dalsgaard, C., Haegerstrand, A., Theodorsson-Norheim, E., Brodin, E. and Hokfelt, T. (1985) Neurokinin-A like immunoreactivity in rat primary sensory neurons: coexistence with substance P. *Histochemistry*, 83: 37–40.

Dougherty, P.M., Palecek, J., Paleckova, V., Sorkin, L.S. and Willis, W.D. (1992a) The role of NMDA and non-NMDA excitatory amino acid receptors in the excitation of primate spinothalamic tract neurons by mechanical, chemical, thermal, and electrical stimuli. *J. Neurosci.*, 12: 3025–3041.

Dougherty, P.M., Sluka, K.A., Sorkin, L.S., Westlund, K.N. and Willis, W.D. (1992b) Neural changes in acute arthritis in monkeys. I. Parallel enhancement of responses of spinothalamic tract neurons to mechanical stimulation and excitatory amino acids. *Brain Res. Rev.*, 17: 1–13.

Duggan, A.W., Griersmith, B.T.,Headley, P.M. and Hall, J.G. (1979) Lack of effect of substance P at sites in the substantia gelatinosa where met-enkephalin reduces the transmission of nociceptive information. *Neurosci. Lett.*, 12: 313–317.

Duggan, A.W., Hendry, I.A., Green, J.L., Morton, C.R. and Zhao, Z.Q. (1988) Cutaneous stimuli releasing immunoreactive substance P in the dorsal horn of the cat. *Brain Res.*, 451: 261–273.

Duggan, A.W., Hope, P.J., Jarrott, B., Schaible, H. and Fleetwood-Walker, S.M. (1990) Release, spread and persistence of immunoreactive neurokinin A in the dorsal horn of the cat following noxious cutaneous stimulation. Studies with antibody microprobes. *Neuroscience*, 35: 195–202.

Duggan, A.W. (1991) Antibody microprobes. In: J. Stamford (Ed.), *Monitoring Neuronal Activity: A Practical Approach*, 181–202. Oxford, Oxford University Press.

Duggan, A.W., Hope, P.J., Lang, C.W. and Bjelke, B. (1993b) Noxious mechanical stimulation of the hind paws of the anaesthetized rat fails to elicit release of immunoreactive beta-endorphin in the periaqueductal grey matter. *Neurosci. Lett.*, 149: 205–208.

Duggan, A.W. and Hendry, I.A. (1986) Laminar localisation of the sites of release of immunoreactive substance P in the dorsal horn with antibody coated microelectrodes. *Neurosci. Lett.*, 68: 134–140.

Eccles, J.C. and Lundberg, A. (1959) Synaptic actions in motoneurones by afferents which may evoke the flexion reflex. *Arch. Ital. Biol.*, 97: 199–221.

Galeazza, M.T., Garry, M.G., Yost, H.J., Strait, K.A., Hargreaves, K.M. and Seybold, V.S. (1995) Plasticity in the synthesis and storage of substance P and calcitonin gene-related peptide in primary afferent neurons during peripheral inflammation. *Neuroscience*, 66: 443–458.

Go, V.L.W. and Yaksh, T.L. (1987) Release of substance P from the cat spinal cord. *J. Physiol.*, 391: 141–167.

Grigg, P., Schaible, H.-G. and Schmidt, R.F. (1986) Mechanical sensitivity of group III and IV afferents from posterior articular nerve in normal and inflamed cat knee. *J Neurophysiol.*, 55: 635–643.

Grubb, B.D., Riley, R.C., Hope, P.J., Pubols, L. and Duggan, A.W. (1996) The burst-like firing of spinal neurons in rats with peripheral inflammation is reduced by an antagonist of *N*-methyl-D-aspartate. *Neuroscience*, 74: 1077–1086.

Hoheisel, U., Mense, S., Simons, D.G. and Yu, X.-M. (1993) Appearance of new receptive fields in rat dorsal horn neurons following noxious stimulation of skeletal muscle: a model for referral of muscle pain. *Neurosci. Lett.*, 153: 9–12.

Hooper, N.M., Kenny, A.J. and Turner, A.J. (1985) Neurokinin A (substance K) is a substrate for endopeptidase–24.11 but not for peptidyl dipeptidase A (angiotensin-converting enzyme). *Biochem. J.*, 231: 357–361.

Hope, P.J., Jarrott, B., Schaible, H., Clarke, R.W. and Duggan, A.W. (1990a) Release and spread of immunoreactive neurokinin A in the cat spinal cord in a model of acute arthritis. *Brain Res.*, 533: 292–299.

Hope, P.J., Lang, C.W. and Duggan, A.W. (1990b) Persistence of immunoreactive neurokinins in the dorsal horn of barbiturate anaesthetised and spinal cats, following release by tibial nerve stimulation. *Neurosci. Lett.*, 118: 25–28.

Hökfelt, T., Arvidsson, U., Ceccatelli, S., Cortés, R., Cullheim, S., Dagerlind, Å., Johnson, H., Orazzo, C., Piehl, F., Pieribone, V., Schalling, M., Terenius, L., Ulfhake, B., Verge, V.M., Villar, M., Wiesenfeld-Hallin, Z., Xu, X.-J. and Xu, Z. (1992) Calcitonin gene-related peptide in the brain, spinal cord, and some peripheral systems. *Ann. NY Acad. Sci.*, 657: 119–134.

Janal, M.N., Colt, E.W.D., Clark, W.C. and Glusman, M. (1984) Pain sensitivity mood and plasma endocrine levels in man following long distance running: effects of naloxone. *Pain*, 19: 13–25.

Janicki, P.K., Erskine, R. and Van der Watt, M.L. (1993) Plasma concentrations of immunoreactive beta-endorphin and substance P in patients undergoing surgery under general vs. spinal anaesthesia. *Horm. Metab. Res.*, 25: 131–133.

Johnston, G.A.R. (1978) Transmitter inactivating processes. *Proc. Aust. Physiol. Pharmac. Soc.*, 9: 94–98.

Khachaturian, H., Lewis, M.E., Kang, T. and Watson, S.J. (1985) Beta-endorphin, alpha-MSH, ACTH and related peptides. In: A. Bjorklund and T. Hokfelt. (Eds), *Handbook of Chemical Neuroanatomy 4. GABA and Neuropeptides in the CNS.* Part I. 216–272. Amsterdam, Elsevier.

Khachaturian, H., Schafer, M.K.-H. and Lewis, M.E. (1993) Anatomy and function of the endogenous opioid systems. In: A. Herz (Ed.), *Handbook of Experimental Pharmacology 104/ I. Opioids I.* 471–495. Berlin, Springer-Verlag.

Kuhar, M.J. (1985) The mismatch problem in receptor mapping studies. *Trends Neurosci.*, 27: 190–191.

Kuraishi, Y., Hirota, N., Sato, Y. and Hanashima, N. (1989) Stimulus specificity of peripherally evoked substance P release from rabbit dorsal horn in situ. *Neuroscience*, 30: 241–250.

Larsson, O.M., Falch, E., Schousboe, A. and Krogsgaard-Larsen, P. (1991) GABA uptake inhibitors: kinetics and molecular pharmacology. *Adv. Biosci.*, 82: 197–200.

Le Greves, P., Nyberg, F., Terenius, L. and Hokfelt, T. (1985) Calcitonin gene-related peptide is a potent inhibitor of substance P degradation. *Europ. J. Pharmacol.*, 115: 309–311.

Liu, H., Brown, J.L., Jasmin, L., Maggio, J.E., Vigna, S.R., Mantyh, P.W. and Basbaum, A.I. (1994) Synaptic relationship between substance P and the substance P receptor: Light and electron microscopic characterization of the mismatch between neuropeptides and their receptors. *Proc. Natl. Acad. Sci. USA*, 91: 1009–1013.

MacMillan, S.J.A., Mark, M.A. and Duggan, A.W. (1998) The release of beta-endorphin and the neuropeptide-receptor mismatch in the brain. *Brain Res.*, 794: 127–136.

Mansour, A., Fox, C.A., Burke, S., Akil, H. and Watson, S.J., (1995a) Immunohistochemical localization of the cloned μ opioid receptor in the rat CNS. *J. Chem. Neuroanat.*, 8: 283–305.

Mansour, A., Hoversten, M.T., Taylor, L.P., Watson, S.J. and Akil, H. (1995b) The cloned μ, delta and kappa receptors and their endogenous ligands: evidence for two opioid peptide recognition cores. *Brain Res.*, 700: 89–98.

Mansour, A. and Watson, S.J. (1993) Anatomical distribution of opioid receptors in mammalians. In: A. Herz (Ed.) *Handbook of Experimental Pharmacology 104/I. Opioids 1.* 79–106. Berlin, Springer-Verlag.

Mantyh, P.W., Allen, C.J., Ghilardi, J.R., Rogers, S.D., Mantyh, C.R., Liu, H., Basbaum, A.I., Vigna, S.R. and Maggio, J.E. (1995) Rapid endocytosis of a G protein-coupled receptor: substance P-evoked internalization of its receptor in the rat striatum in vivo. *Proc. Natl Acad. Sci. USA*, 92: 2622–2626.

Margules, D.L., Goldman, B. and Finck, A. (1979) Hibernation: an opioid-dependent state? *Br. Res. Bull.*, 4: 721–724.

McKelvey, J.F. and Blumberg, S. (1986) Inactivation and metabolism of neuropeptides. *Annu. Rev. Neurosci.*, 9: 415–434.

Myer, E.C., Tripathy, H.L., Brase, D.A. and Dewey, W.L. (1992) Elevated CSF beta-endorphin immunoreactivity in Rett's syndrome:report of 158 cases and comparison with leukemic children. *Neurology*, 42: 357–360.

Neugebauer, V., Lücke, T. and Schaible, H.-G. (1993) Differential effects of *N*-methyl-D-aspartate (NMDA) and non-NMDA receptor antagonists on the responses of rat spinal neurons with joint input. *Neurosci. Lett.*, 155: 29–32.

Procter, M.J., Houghton, A.K., Faber, E.S.L., Chizh, B.A., Ornstein, P.L., Lodge, D. and Headley, P.M. (1998) Actions of kainate and AMPA selective glutamate receptor ligands on nociceptive processing in the spinal cord. *Neuropharmacology*, 37: 1287–1297.

Radhakrishnan, V. and Henry, J.L. (1995) Antagonism of nociceptive responses of cat spinal dorsal horn neurons in vivo by the NK-1 receptor antagonists CP-96,345 and CP-99,994 but not by CP-96,344 *Neuroscience*, 64: 943–958.

Randic, M., Kolaj, M., Kojic, Lj., Cerne, R,. Cheng, G. and Wang, R.A. (1995) Interactions of neuropeptides and excitatory amino acids in the rat superficial spinal dorsal horn. In: F. Nyberg, H.S. Sharma, and Z. Wiesenfeld-Hallin (Eds), *Progress in Brain Research* 104. *Neuropeptides in the Spinal Cord* 225–254. Amsterdam, Elsevier.

Riley, R.C., Zhao, Z.Q. and Duggan, A.W. (1996) Spinal release of immunoreactive dynorphin A(1–8) with the development of peripheral inflammation. *Brain Res.*, 710: 131–142.

Schaible, H.-G., Grubb, B.D., Neugebauer, V. and Oppmann, M. (1991a) The effects of NMDA antagonists on neuronal activity in cat spinal cord evoked by acute inflammation in the knee joint. *Eur.J. Neurosci.*, 3: 981–991.

Schaible, H.-G., Neugebauer, V., Cervero, F. and Schmidt, R.F. (1991b) Changes in tonic descending inhibition of spinal neurons with articular input during the development of acute arthritis in the cat. *J. Neurophysiol.*, 66: 1021–1031.

Schaible, H.-G., Hope, P.J., Lang, C.W. and Duggan, A.W. (1992) Calcitonin gene-related peptide causes intraspinal spreading of substance P released by peripheral stimulation. *Eur. J. Neurosci.*, 4: 750–757.

Schaible, H.-G. and Grubb, B.D. (1993) Afferent and spinal mechanisms of joint pain. *Pain*, 55: 5–54.

Schaible, H.G., Ebersberger, A., Peppel, P., Beck, U. and Messlinger, K. (1997) Release of immunoreactive substance P in the trigeminal brain stem nuclear complex evoked by chemical stimulation of the nasal mucosa and the dura mater encephali – a study with antibody microprobes. *Neuroscience*, 76: 273–284.

Schaible, H-G., Schmidt, R.F. and Willis, W.D. (1987} Enhancement of the responses of ascending tract cells in the cat spinal cord by acute inflammation of the knee joint. *Exp. Brain Res.*, 66: 489–499.

Theodorsson-Norheim, E., Hemsen, A., Brodin, E. and Lundberg, J.M. (1987) Sample handling techniques when analyzing regulatory peptides. *Life Sci.*, 41: 845–848.

Tseng, L.F., Towell, J.F. and Fujimoto, J.M. (1986) Spinal release of immunoreactive Met-enkephalin by intraventricular beta-endorphin and its analogs in anesthetized rats. *J Pharmacol. Exp. Ther. [JP3]*, 237: 65–74.

Vaeroy, H., Helle, R., Forre, O., Kass, E. and Terenius, L. (1997) Cerebrospinal fluid levels of beta-endorphin in patients with fibromyalgia (fibrositis syndrome). *J. Rheum.*, 15: 1804–1806.

Woolf, C.J. and Wiesenfeld-Hallin, Z. (1986) Substance P and calcitonin gene-related peptide synergistically modulate the gain of the nociceptive flexor withdrawal reflex in the rat. *Neurosci. Lett.*, 66: 226–230.

Young, E., Bronstein, D. and Akil, H. (1993) Proopiomelanocortin biosynthesis,processing and secretion: functional implications. In: A. Herz (Ed.), *Handbook of Experimental Pharmacology 104/ I. Opioids I.* 393–422. A. Berlin, Springer-Verlag.

Zadina, J.E., Hackler, L., Ge, L. and Kastin, A.J. (1997) A potent and selective endogenous agonist for the mu opiate receptor. *Nature*, 386: 499–502.

Zimmerman, M. (1976) Neurophysiology of nociception. In: R. Porter (Ed.), *International Review of Physiology* Vol. 10, 179–221. Baltimore, University Park Press.

L. F. Agnati, K. Fuxe, C. Nicholson and E. Syková (Eds.)
Progress in Brain Research, Vol 125

CHAPTER 25

Neuronal mechanisms of synaptic and network plasticity in the lamprey spinal cord

David Parker* and Sten Grillner*

Nobel Institute for Neurophysiology, Department of Neuroscience, Karolinska Institute, SE 171 77 Stockholm, Sweden

Introduction

A major current task for neuroscience is to understand the integrated function of the nervous system in terms of interacting nerve cells. For each function, be it motor, sensory or cognitive, we need to know which nerve cells form the functional module or network under study, their particular molecular and cellular properties, as well as their fast and slow synaptic properties. In a limited number of invertebrate and lower vertebrate motor systems we have reached a level of knowledge that enables us to form a bridge from the molecular and cellular levels to the network and behavioral levels (Kiehn et al., 1998). In most vertebrates, and in particular mammals, however, this is still a major challenge.

The lamprey is an ancient jawless vertebrate that we have used as experimental model (see Grillner et al., 1998a, b for review). Its brainstem – spinal cord – has the same basic components as that of higher vertebrates in terms of general neural organization (cf Nieuwenhuys et al., 1998), but the number of neurons is much lower, and its isolated brainstem – spinal cord – can be maintained in vitro over several days. In this isolated nervous system the networks generating locomotion, respiration, oral motor activity, and a variety of postural and protective reflexes can be activated (Rovainen, 1979; Grillner et al., 1998a, b; Huard et al., 1999; Orlovsky et al., 1999; Ullström et al., 1999). This has allowed us to carry out an extensive analysis of the networks that coordinate locomotor movements with regard to both propulsion and steering functions, as well as the maintenance of body orientation during ongoing movements.

In this chapter we will deal with different, presumably paracrinic, modulatory systems which act on the segmental network that generates and coordinates alternating activity between the left and the right sides of the body during locomotion. This network (Grillner et al., 1998a, b, see Buchanan and Grillner, 1987, 1988) has been studied by making paired intracellular recordings (Fig. 1A, B) between motoneurons and different identified excitatory glutamatergic and inhibitory glycinergic interneurons. It is activated by descending reticulospinal neurons acting via NMDA, AMPA/kainate, and metabotropic glutamate receptors (mGluR; Ohta and Grillner, 1989; Krieger et al., 1998).

To summarize the essential features of the network (Fig. 1):

- Reticulospinal neurons provide (RS in Fig. 1) excitatory glutamatergic drive that determines the level of network activity.

*Corresponding authors: Tel: +46 8 728 6900
Fax: +46 8 349544; e-mail: Sten.Grillner@neuro.ki.se; David.Parker@neuro.ki.se

- Pattern generation is due to a basic glutamatergic (E) – glycinergic (I) interneuronal network.
- Alternation is mediated by reciprocal glycinergic inhibition.
- Burst termination is due to several factors, including the cellular properties of network neurons (e.g. Ca^{2+}, Kca channel subtypes), types of synaptic transmission and connectivity, and finally, in the intact preparation, by sensory input from stretch receptors activated by the locomotor movements.
- Fine tuning of the network occurs through a variety of aminergic and peptidergic modulatory systems acting at both pre- and postsynaptic levels to modulate synaptic transmission and different types of ion channels (see Parker et al., 1998; El Manira et al., 1998).
- Experimentally, the network can be activated in a dose-dependent manner by bath applying glutamatergic agonists (see Fig. 1C).

Modulator systems – action on ion channels manifested on the network – behavioral level

In addition to fast ionotropic glutamatergic, glycinergic, and GABAergic actions on the spinal level, there are a number of control mechanisms that act via G-proteins. These include: (1) descending

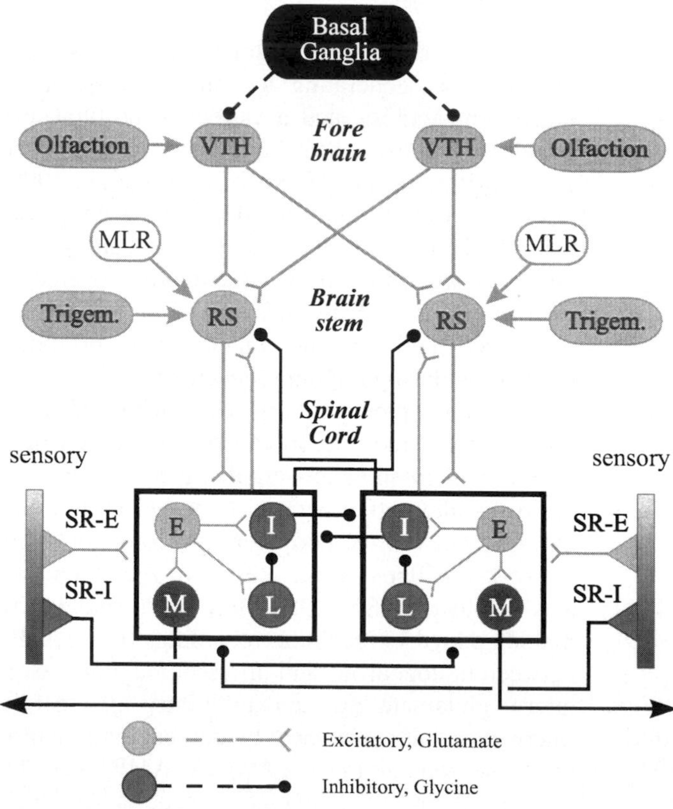

Fig. 1. The lamprey locomotor network; a schematic representation of the forebrain, brain-stem and spinal components of the neural network that generate locomotor burst acitivity. All neuron symbols denote populations of neurons rather than single cells. The reticulospinal (RS) glutamatergic neurons excite all classes of spinal interneurons and motoneurons. The excitatory interneurons (E) excite all types of spinal neurons, that is, the inhibitory glycerinergic interneurons (I) that cross the midline and inhibit neurons on the contralateral side, the lateral interneurons (L), which inhibit ipsilateral I interneurons, and motoneurons (M). The stretch receptor neurons are of an excitatory type (SR-E) that excites neurons on the ipsilateral side, and an inhibitory type (SR-I) that inhibits neurons on the contralateral side. RS neurons receive excitatory synaptic inputs from cutaneous afferents (Trigem.), the mesencephalic locomotor region (MLR), and the ventral thalamus (Vth), which in turn receives inputs from the basal ganglia.

peptidergic (cholecystokinin and peptide YY; Brodin et al., 1988, 1989; Ohta et al., 1988) and 5-HTergic (Brodin et al., 1986) pathways, that originate in the brainstem; (2) dorsal root inputs (calcitonin gene related peptide, bombesin, tachykinins, 5-HT; van Dongen et al., 1985, 1986; Brodin et al., 1988); and finally (3) intraspinal systems, for example, bipolar GABAergic and neuropeptide Y containing neurons (Bongianni et al., 1990; Parker et al., 1998), somatostatin and neurotensin plexa around the lateral edge of the spinal cord (Brodin et al., 1990; Christensson et al., 1991), and intraspinal midline neurons that form a 5-HT, dopamine, and tachykinin-containing plexus (van Dongen et al., 1985, 1986; Schotland et al., 1995, 1996). In addition, fast glutamatergic and GABAergic systems can also act via metabotropic receptors ($GABA_B$, Alford et al., 1991; Bussières and El Manira, 1999; mGluR I-III, Tegnér et al., 1993; Krieger et al., 1996, 1998).

The actions of several of these different modulators have been analyzed in some detail with regard to their molecular, cellular, and overall network effects. Table 1 shows some of the actions mediated by different modulators that often act via different receptor subtypes; the left column shows presynaptic actions. In most cases there is a presynaptic inhibitory effect (I), but with regard to

TABLE 1. Spinal modulation involving G-protein-coupled receptors

Spinal modulation - **G-protein coupled receptors**

	presyn	HVA_{Ca}	LVA_{Ca}	K_{Ca}	K^+	NMDA	network
$GABA_B$	I	↘	↘				↘
$mGluR_I$	0	0	0			↗	↗
$mGluR_{II-III}$	I	0	0				↘
$5\text{-}HT_{1A}$	0			↘			↘
$5\text{-}HT_x$	I	↘					
D_2	I	↘	↘				↘
TK	F				↘	↗	↗
NPY	I						0
Som.		0	0	0	↗		↘
NT		0	0	0			↗

*Metabotropic amino acid, aminergic and peptidergic g-protein-mediated modulation of ion channel, synaptic, cellular and network activity in the lamprey spinal cord. The table summarizes the results of a number of studies (also see text). The effects of different transmitters and receptors on different targets are listed in the columns on the right.

+The presynaptic actions can be targeted to sensory afferents, excitatory or inhibitory interneurons and descending reticulospinal axons (see El Manira et al., 1998) Different transmitters have selective actions on different cellular targets (I indicates presynaptic inhibition and F facilitation). The locomotor network modulates phasically, in each cycle, the synaptic transmission from sensory afferents and interneurons (El Manira et al., 1997a, El Manira et al., 1998). The modulation of HVA_{Ca}, LVA_{Ca}, K_{Ca}, K^+ and NMDA channels is indicated with a downward arrow for depression and an upward arrow for facilitation (cf. Matsushima et al., 1993; Bacskai et al., 1995; Barthe and Grillner 1995; Schotland et al., 1995; El Manira and Bussieres 1997; Parker et al., 1997; Grillner et al., 1998a, b; Parker et al., 1998; El Manira et al., 1998; Krieger et al., 1998). Again, the effects may be specific to particular cell types. Finally, the effects on the network level have been studied on the background of locomotor activity (arrows relate to locomotion burst frequency), and in related modelling experiments (Schotland et al., 1995; Grillner et al., 1998b). 5-HT, 5-hydroxytryptamine (serotonin) receptor; D_2, type 2 dopamine receptor; HVA, high voltage activated; mGluR, metabotropic glutamate receptor; NPY, neuropeptide Y; NT, neurotensin; TK, tachykinin.

tachykinins (Substance P-like peptides) synaptic transmission is instead facilitated (Parker and Grillner, 1998). Presynaptic interactions occur at all levels in the locomotor control system, namely on sensory inputs, inhibitory and excitatory network synapses, and descending reticulospinal axons. Moreover, during locomotor activity there is a phase-dependent gating of synaptic transmission in both interneuronal and sensory axons (Alford et al., 1991; El Manira et al., 1997a), making synaptic transmission more effective in one or other phase of the locomotor cycle. Other modulators (5-HT, dopamine, substance P, mGluR, CCK, PYY) provide a tonic gating of synaptic transmission, for example of glutamatergic synaptic transmission from reticulospinal axons (Buchanan and Grillner 1991; Wikström et al., 1995; Krieger et al., 1996; Parker and Grillner 1998; Parker, 2000).

The five middle columns in Table 1 show modulator/receptor-mediated actions on different types of ion channels (Ca^{2+} channel subtypes, Kca, K^+), and transmitter-gated keep channels (NMDA). It is important to note that the effects are often specific to a given cell type or synapse, and that it is thus not possible to extrapolate from one type of neuron to another. For example, $GABA_B$ receptors mediate powerful presynaptic inhibition of sensory and interneuronal axons but not reticulospinal axons (Christenson and Grillner, 1991; Alford and Grillner, 1991; Alford et al., 1991). The converse is true for dopamine, which acts on the latter synapse (Wikström et al., 1995), but not on the sensory level (El Manira et al., 1997a). In addition to different effects at different levels in the spinal cord, modulatory effects are in many cases also neuron and synapse-specific within the locomotor network itself (see Parker and Grillner, 1998, 1999). It is thus imperative that individual network neurons are identified if a satisfactory understanding of network effects is to be achieved.

The column to the right in Table 1 shows the overall network effects exerted by the different modulator systems; in most cases there is a change in burst frequency when the modulator is applied during ongoing locomotor activity. Intersegmental coordination may also be affected (see for example Matsushima and Grillner, 1992), as well as the amplitude and the duration of the bursts within a given locomotor cycle. Since we have a relatively detailed knowledge of how the network functions, the network effects of a modulator can (to some extent) be predicted from its different cellular actions. Extensive biologically relevant mathematical models of the locomotor network have also been developed, based on knowledge of the properties and synaptic connectivity of nerve cells in the network (Grillner et al., 1988; Hellgren et al., 1992; Tegnér et al., 1998). In these simulations, known specific modulator-mediated changes can be induced to investigate how their different effects contribute to the overall change in the network output. Thus, by combining detailed experimental analyses of modulator effects at the molecular, cellular, synaptic and network levels, and applying this data in computer simulations, we bridge effects from the molecular to behavioral levels.

5-HT, dopamine, and tachykinins

We shall now review the action of three related modulatory systems, 5-HT, dopamine, and tachykinins, on the locomotor network. In the midline below the central canal there is a set of interneurons with bilateral ramifications that form a dense bilateral plexus of varicosities in the ventromedial region of the spinal cord. These neurons and their processes co-store 5-HT and dopamine, and a proportion of them also express immunoreactivity toward tachykinins (van Dongen et al., 1985, 1986; Schotland et al., 1995). The varicosities contain both small and large dense core vesicles; 5-HT is known to be stored in two vesicle fractions thought to correspond to small and dense core vesicles (Frank et al., 1992). Tachykinins are assumed to be stored in dense core vesicles, whereas the storage of dopamine is unknown. In contrast to other synapses in the spinal cord, ventromedial 5-HT/DA/TK plexus varicosities do not contain synaptic specializations (Fig. 2), as studied with electron-microscopy of serial sections (Christensson et al., 1990). This plexus, which with regard to 5-HT, is one of the densest that has been encountered in the vertebrate CNS, thus functions through the release of transmitter onto neurons in the vicinity that have the appropriate receptors. Synaptic effects will thus not depend on point-to-point synaptic transmission, but through paracrinic or volume transmission. The latter expression is broader, however, and is taken to encompass a variety of signals in the extracellular space, some of which have no relation to synaptic action (see companion chapters in this volume). Little is yet known about the release from these types of varicosities, but it can be assumed that the small vesicles may be released by localized and relatively large Ca^{2+} entry following single action potentials (Augustine et al., 1991), whereas the dense core vesicles may depend on an increase in the level of bulk cytosolic Ca^{2+} (see Verhage et al., 1991), which may occur during a relatively higher level of activity.

The ventromedial plexus contains the dendrites of motoneurones and network interneurons; transmitters released from the varicosities can thus influence the activity of network neurons. The large reticulospinal axons also pass very close to the plexus; distances within the thin lamprey spinal cord are small (dorsoventral extent is 200–300 μm). For example, the central canal is less than 100 μm from the ventromedial plexus, and the large reticulospinal axons in the ventral funiculus are much closer. In addition, even large molecules diffuse rapidly within the thin lamprey spinal cord (Wald and Selzer, 1981), thus facilitating paracrinic and volume transmission-mediated effects.

The action of 5-HT and DA have been reviewed several times previously (see Wallén et al., 1989a), and thus will only be summarized briefly here. This

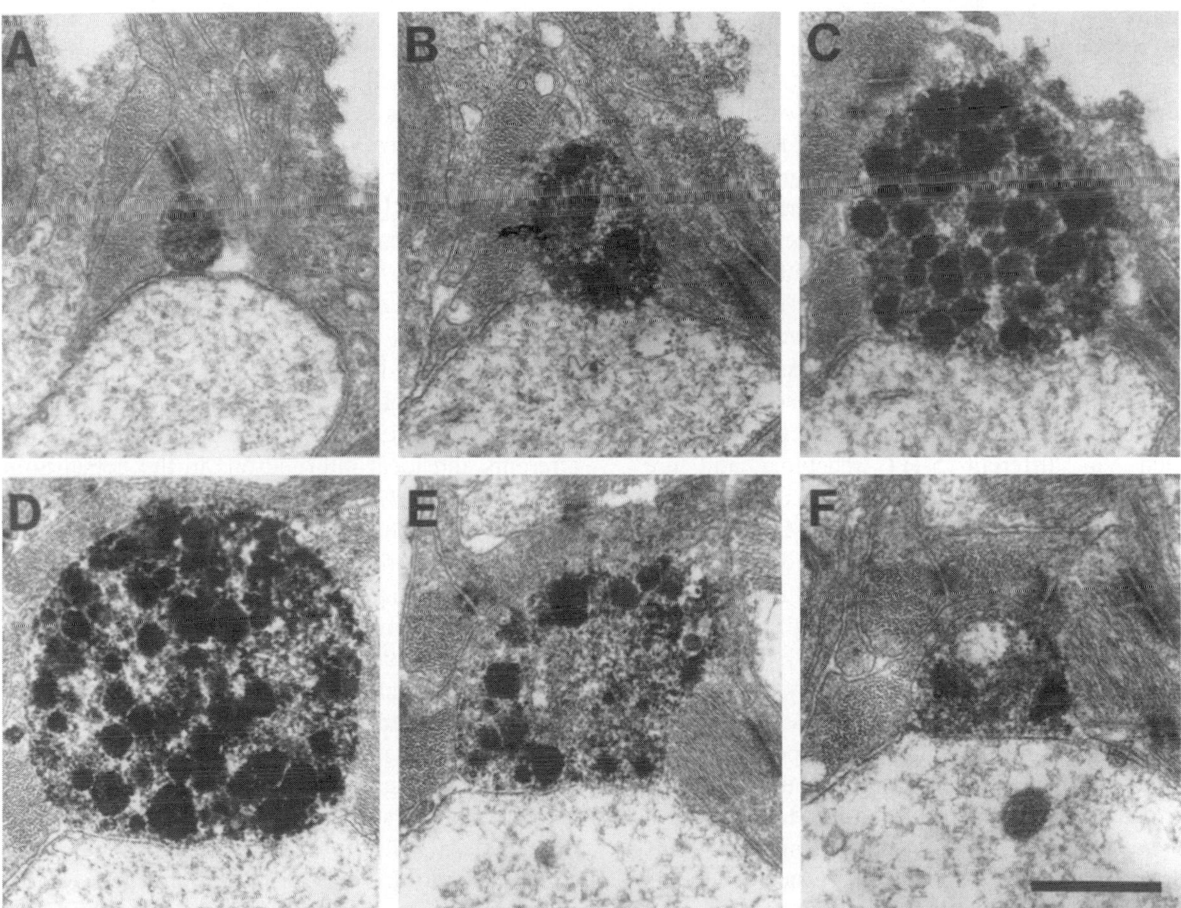

Fig. 2. Electron micrograph showing a serially sectioned 5-HT immunoreactive varicosity. Although a varicosity is in close relation to an unidentified neuronal structure, no synaptic junction is formed. Note the higher content of large granular vesicles in the middle part of the varicosity. Bar = 0.5μm (from Christensson et al., 1990).

review will instead focus on the molecular, cellular, synaptic, and network effects of tachykinins.

The actions of 5-HT and DA are complementary.

5-HT acts via 5-HT_{1a}-like receptors on apamin-sensitive Kca channels (Wallén et al., 1989b; El Manira et al., 1994; Wikström et al., 1995). Since the latter are responsible for generating the relatively slow after-hyperpolarization (AHP) that follows each action potential, it is a main determining factor in the frequency regulation of nerve cells in the locomotor network. The reduction of the Kca-dependent AHP thus reduces spike frequency adaptation, making neurons spike longer and at a higher rate in response to the same level of excitatory synaptic drive. At the network level, 5-HT prolongs ventral root bursts and thus slows the burst rate (Harris-Warrick and Cohen, 1985; Grillner et al., 1988; Christenson et al., 1989; Hellgren et al., 1992). These effects are prominent with even with low concentrations of 5-HT (0.5 μM; Matsushima and Grillner, 1992). In addition to its effects on KCa channels and spike frequency regulation, 5-HT also influences the amplitude and activity-dependent plasticity of network interneuron synaptic transmission (see Parker and Grillner, 1999). 5-HT also depresses glutamatergic synaptic transmission from sensory afferents and reticulospinal axons, although these effects are exerted via different 5-HT receptors and are manifested at higher concentrations (5–10 μM; Buchanan and Grillner, 1991; El Manira et al., 1997b).

DA acts via D2 receptors to inhibit Ca^{2+} channels primarily of the N-type (Fig 3; Schotland et al., 1995, Wikström et al., 1999). Since Kca channels are activated by Ca^{2+} entry through N-type channels, the DA receptor-mediated reduction of Ca^{2+} entry during each action potential will reduce Kca channel activation and thus the amplitude of the AHP (Wikström and El Manira, 1998). The action of DA on Kca channels is thus complementary to that of 5-HT, but it occurs through a different molecular pathway (Fig. 2.). The reduced AHP amplitude will have the same net cellular effect as 5-HT, a reduction of spike frequency adaption, and indeed its network effects are similar, resulting in longer ventral root bursts and a reduced burst frequency. As with 5-HT, DA also has a presynaptic inhibitory effect on reticulospinal glutamatergic synaptic transmission (Wikström et al., 1995), although, unlike 5-HT, it does not affect sensory synaptic transmission (El Manira et al., 1997b).

The effects of tachykinins on the locomotor network

There is an extensive literature on neuropeptides in the spinal cord (see Nyberg et al., 1995), although it primarily relates to their location in the dorsal horn and their roles in sensory processing. However, neuropeptides are also found in ventral regions of the spinal cord involved in motor control. The tachykinins are arguably the most extensively studied family of neuropeptides, principally due to their effects on sensory inputs in relation to nociception. Tachykinins are also found, however, in motor areas; in mammals the tachykinin substance P co-localizes with 5-HT and thyrotrophin releasing hormone in descending bulbospinal neurons (see Cullheim and Arvidsson, 1995). As stated above, in the lamprey tachykinins are located in the ventromedial spinal plexus.

In addition to being located in the ventromedial spinal cord plexus, tachykinin-like peptides (Waugh et al., 1993) are also located in primary afferents and local interneurons in the dorsal horn (Van Dongen et al., 1985, 1986; Svensson and Grillner, unpublished observations). Tachykinin-mediated network modulation and its underlying molecular, cellular, and synaptic mechanisms have been examined in detail. Bath application of the tachykinin substance P for 10 min results in a concentration-dependent increase in the frequency of NMDA-evoked ventral root bursts (Parker et al., 1998); the magnitude of the burst frequency increase is state-dependent, in that it is dependent on the initial burst frequency. With low initial burst frequencies (<1 Hz), the increase can be up to 500% of control, whereas with higher frequencies (>2 Hz) the increase is between 100–200% of control (Parker et al., 1998). In addition to the magnitude of this effect. Its time course is also markedly concentration-dependent; with nM concentrations of substance P (10–100 nM), recovery occurs after washing for between 1–5 hr. However, with 1 μM substance P, which is in the physiological concentration range reported for neuropeptides (Duggan, 1995), no appreciable recovery occurs even after washing for 24 hr (Fig. 3). In addition to increasing the burst frequency, substance P also makes the activity more regular; this is shown quantitatively by the concentration-dependent reduction of the coefficient of variation (SD of cycle duration/mean cycle duration; Parker et al., 1998). This effect also shows no appreciable recovery following the application of 1 μM substance P; both of these network effects are blocked by the general tachykinin receptor antagonist spantide II (Fig. 4A).

The substance P-mediated modulation of the burst frequency is NMDA-dependent; this was shown by the failure of substance P to potentiate the frequency of kainate-evoked ventral root bursts when NMDA receptors were blocked by the NMDA receptor antagonist AP5 (see Parker et al., 1998). However, the modulation of the burst regularity was not affected by AP5, suggesting that it is NMDA-independent, and thus that the two network effects of substance P are due to separate, specific effects on the locomotor network. Since positive insight was available on the burst frequency modulation (i.e. its NMDA-dependence), it is this effect that most is known about, and thus which will be focused on, in the remainder of this chapter (see Parker et al., 1998 for more information on the burst regularity modulation).

There appear to be at least three phases (Fig. 5) to the burst frequency modulation (see Parker and Grillner, 1999a): an initial induction phase that lasts for between 2–3 hr after substance P application; an intermediate phase (2–15 hr) that is blocked by translational, but not transcriptional protein synthesis inhibitors, and which thus appears to require protein but not de novo RNA synthesis; and a final phase (>15–20 hr) that is blocked by

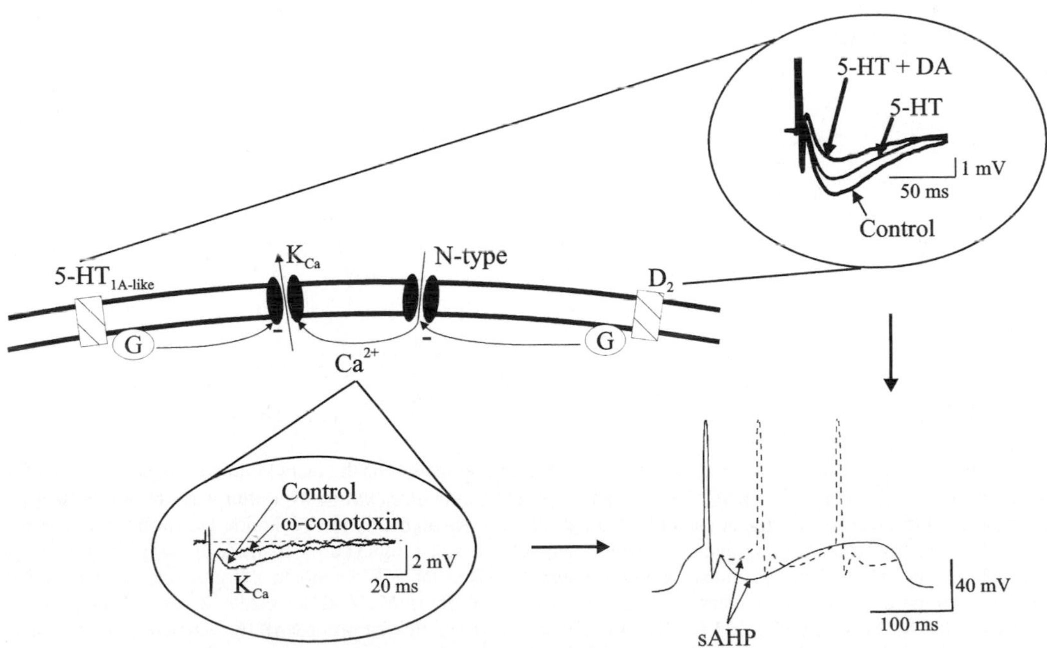

Fig. 3. 5-HT and dopamine (DA) modulate the slow afterpolarization (sAHP) that follows the action potential; this occurs in an additive way. By modifying the sAHP the frequency regulation is affected. DA reduces, via D_2 receptors, the degree of activation of the calcium dependent K^+ channels (K_{Ca}), which generate the sAHP. This occurs by reducing Ca^{2+} entry through N-type channels during the action potential. 5-HT, on the other hand, acts directly on K_{Ca} channels. 5-HT, via 5-$HT_{1a\text{-like}}$, and DA via the two receptors, thus produce the same net effect but through different molecular targets. The inset below in the figure shows that K_{Ca} channels depend on Ca^{2+} entry occurring via N-channels (blocked by μ-conotoxin).

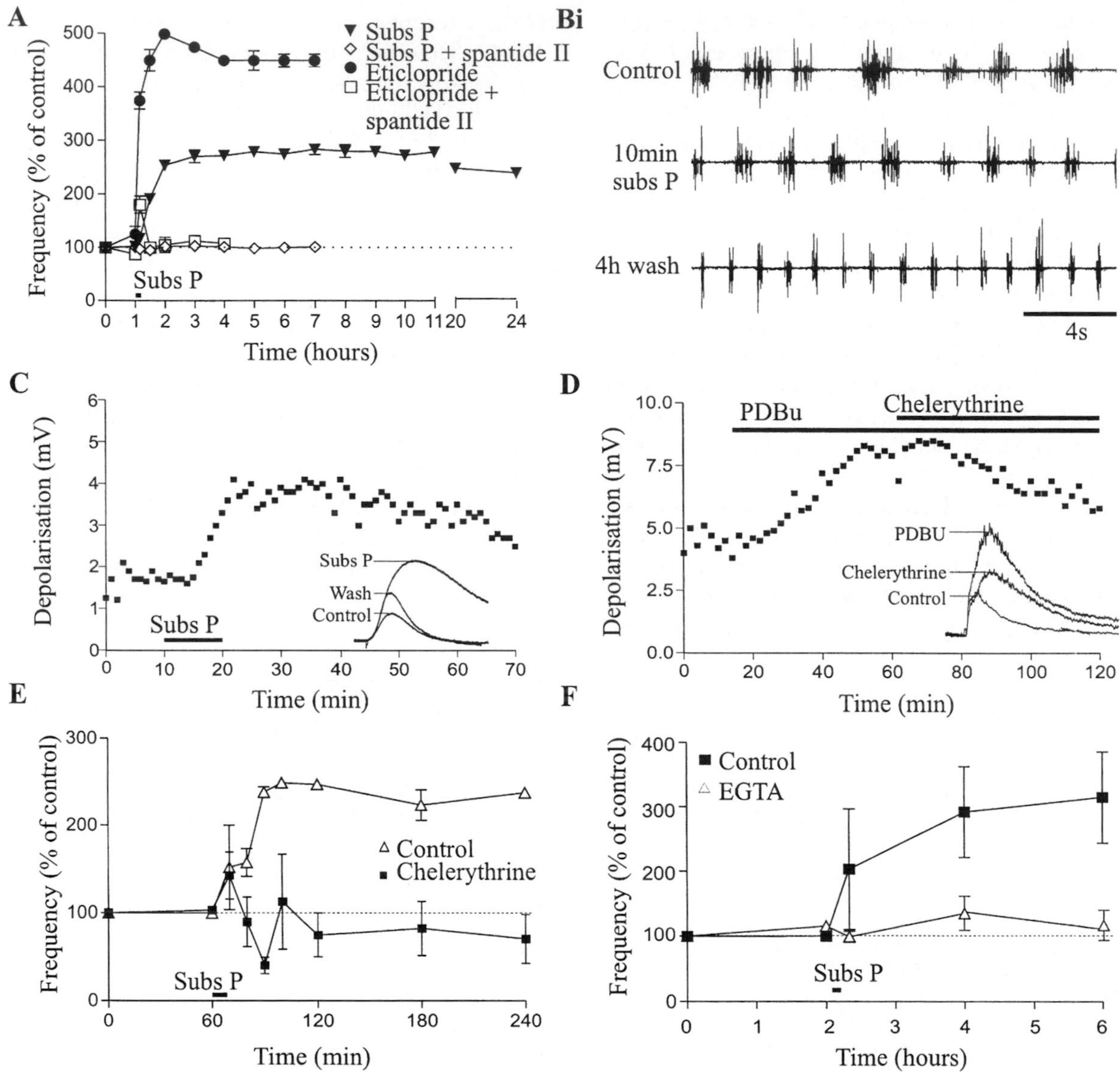

Fig. 4. The effects of tachykinins on the lamprey locomotor network. (A) Bath application of the tachykinin substance P (1 μM) for 10 min results in an increase in the frequency of NMDA-evoked ventral root bursts that does not recover after washing for 24 hr. This effect is blocked by the tachykinin receptor antagonist spantide II. In addition to substance P, the dopamine D2 receptor antagonist eticlopride also evokes a long lasting increase in the burst frequency. This effect is also blocked by spantide II, suggesting that the eticlopride effect could be mediated through endogenoustachykinin release. (C) Substance P potentiates NMDA responses in spinal neurons. The example shown here is of a motoneuron. Pressure application of 1mM NMDA results in an NMDA-evoked depolarization that is potentiated by substance P (1 μM for 10 min). Notice that the potentiation of NMDA responses recovers after washing. (D) The protein kinase C (PKC) activating phorbol ester PDBu potentiates NMDA responses, an effect that is reversed by the PKC antagonist chelerythrine. Cheleythrine also blocks the substance P-mediated potentiation of NMDA responses (data not shown). (E) Chelerythrine blocks the substance P-mediated potentiation of NMDA-evoked ventral root bursts. (F) The intracellular calcium chelator EGTA–AM also blocks the substance P-mediated network modulation. Note that the pre-incubation of chelerythrine or EGTA–AM does not affect ongoing network activity.

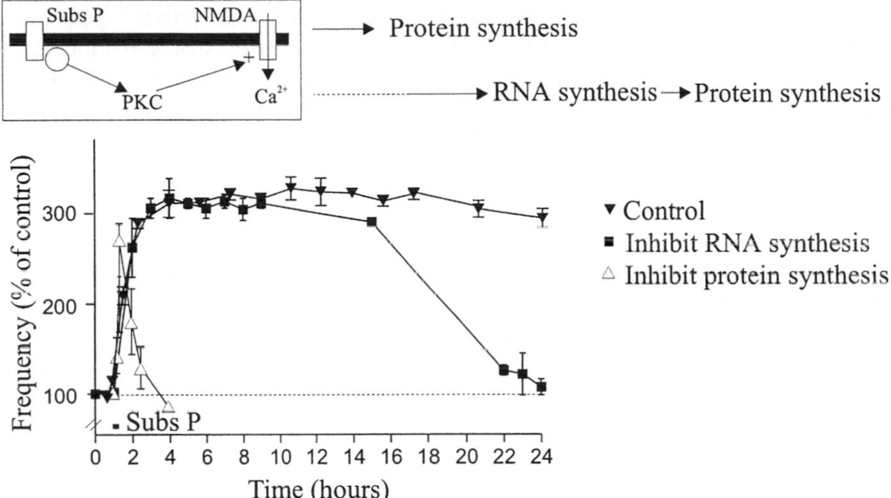

Fig. 5. Summary diagram showing the three phases of the substance P-mediated network modulation. The box shows the putative induction scheme (0–2 hr), in which substance P acts through protein kinase C to potentiate NMDA responses and intracellular calcium levels. This is followed by an intermediate maintenance phase (>2 hr) that is blocked by inhibiting protein synthesis; it is not affected, however, by inhibiting RNA synthesis, suggesting that the protein synthesized at this time does not require de novo RNA synthesis. A final phase develops approximately 15–20 hr after substance P application, which is blocked by inhibiting RNA synthesis.

transcriptional inhibitors, and which thus does require de novo RNA synthesis.

Since the burst frequency modulation was NMDA-dependent, it was obviously important to investigate the effect of substance P on cellular responses to NMDA; this was done by pressure applying NMDA onto the surface of the spinal cord above the neuron being recorded from (Fig. 4C). Since the spinal cord is thin, drugs applied in this way readily gain access to the cell body and dendritic regions of neurons. Substance P potentiated responses to NMDA in motor neurons and one type of network interneuron, the CC interneurons (Parker et al., 1998); however, recovery of this effect usually occurred within 1–1.5 hr after substance P application. In more than 30 experiments, whilst substance P consistently potentiated NMDA responses, the effect never lasted in excess of 2 hr. Thus, tonic NMDA receptor modulation does not underlie the maintenance of the long-term network modulation. The effect of substance P on NMDA responses appears to be due to a specific postsynaptic effect, since it cannot be accounted for by an effect on the postsynaptic input resistance or membrane potential, and substance P also fails to affect responses to the non-NMDA receptor agonist AMPA (Parker et al., 1998).

Since the burst frequency modulation is NMDA-dependent, the potentiation of NMDA responses is consistent with its role as the cellular mechanism underlying this effect. As it stood, however, this was only correlational evidence. Three further aspects suggested a direct link between the potentiation of NMDA responses and the increased network burst frequency. Firstly, as stated above, the potentiation of the frequency of network activity evoked by the non-NMDA receptor agonist kaniate was blocked by the NMDA receptor antagonist AP5, suggesting a requirement for NMDA receptors in the burst frequency modulation. Further support was obtained by examining the intracellular pathway underlying the potentiation of NMDA responses. Protein kinase C (PKC) was implicated, since the effects of substance P were blocked by the specific PKC antagonist chelerythrine, and mimicked and occluded by the PKC activating phorbol ester PDBu (Fig. 4D; Parker et al., 1998). Thus, if the modulation of NMDA receptors was important in the network modulation, the increased burst frequency should be blocked by PKC antagonists; this was found to be the case. Whilst chelerythrine did not usually affect control network activity, it blocked the substance P-mediated potentiation of the burst frequency (Fig. E), again providing a link between the cellular and network levels. A final result that suggested a role of NMDA receptor potentiation in the burst frequency modulation was obtained by examining how the consequences of NMDA receptor modulation contributed to the network modulation. NMDA receptors are permeable to Ca^{2+} (see Mayer and Westbrok, 1987). Thus, potentiating NMDA response should increase Ca^{2+} levels in network neurons. To examine the role of increased Ca^{2+} in the network modulation, the intracellular Ca^{2+} chelator EGTA-AM was used (Fig. 4F); this is a membrane permeable form of EGTA. It cannot buffer calcium extracellularly, but once inside the cell ester groups are removed allowing it to buffer calcium and preventing it from leaving the cell. In most systems the kinetics of Ca^{2+}binding by EGTA are too slow to enable it to buffer the rapid Ca^{2+} transients required for the release of small synaptic vesicles located at active zones (Adler et al., 1991; but see Borst and Sakman, 1996). This also appears to be the case in the lamprey, since EGTA–AM does not affect the amplitude of monosynaptic EPSPs evoked by excitatory network interneurons in motor neurons (Parker, 2000), and EGTA–AM does not affect control network activity up to 2 hr after its application (see Fig. 4F; Parker et al., 1998), which is inconsistent with an effect on network transmitter release. EGTA–AM, however, blocked the effect of substance P on the burst frequency modulation. Thus, increased intracellular Ca^{2+} is required to evoke the burst frequency modulation. Whilst this Ca^{2+}could potentially come from several sources, including voltage-activated calcium channels (but not L-type channels; Parker, unpublished observations) and intracellular calcium stores, the potentiation of NMDA responses is consistent with an increase in Ca^{2+}levels in network neurons, and thus provides further support for the link between the cellular potentiation of NMDA responses and the burst frequency modulation.

The PKC-mediated potentiation of cellular responses to NMDA thus appears to underlie the increased frequency of network activity, providing a direct link between the effects of substance P at the molecular, cellular and network levels. However, it is important to remember that the potentiation of NMDA-receptors does not last in excess of 2 hr, and thus that this effect cannot account for the maintenance of the long-term network modulation. It can thus only contribute to the short-term effects of lower substance P concentrations, or the induction phase of the long-term modulation.

Since the potentiation of NMDA responses could not account for the maintenance of the burst frequency modulation, other potential cellular and synaptic maintenance mechanisms were investigated. At the synaptic level, substance P potentiates low frequency-evoked monosynaptic glutamatergic synaptic transmission, but has not, as yet, been shown to affect GABAergic or glycinergic inputs (Parker and Grillner, 1998). The potentiation of glutamatergic inputs is presumably mediated pre- and postsynaptically, shown by an increase in the frequency and amplitude, respectively, of spontaneous miniature mEPSPs recorded in TTX (Parker and Grillner 1998). As outlined above, the postsynaptic effect is due to the specific potentiation of NMDA receptors (Parker et al., 1998). The postsynaptic potentiation, i.e. the increase in EPSP amplitude, recovers within 2 hr of substance P application, and thus cannot account for the maintenance of the burst frequency modulation. The presynaptic effect, the increased frequency of mEPSPs, also appears to recover within 2 hr (Parker, unpublished observations), and the amplitude of monosynaptic EPSPs evoked in paired recordings by either reticulospinal axons or excitatory network interneurons in motor neurons also recovers within 2 hr of substance P application. There is thus no long-term potentiation of low frequency-evoked glutamatergic synaptic transmission underlying the burst frequency modulation. In addition to its synaptic effects, substance P has several neuron specific effects on network cellular properties, including an increase or decrease in excitability, modulation of the Kca-dependent AHP, and a depolarization and oscillations of the resting membrane potential (Parker and Grillner, 1998; Svensson et al., in preparation). However, all of these effects again recover after washing for between 20 min–2 hr, suggesting that they cannot contribute to the long-term burst frequency modulation; there is thus as yet no cellular or synaptic effect that can account for the long-term maintenance of the burst frequency modulation.

Metaplasticity-the plasticity of network interneuron synaptic plasticity

In addition to the effects of slow-acting modulatory transmitters, network modulation could also be evoked by the activity-dependent plasticity of synaptic transmission between network neurons (Getting, 1989). In recent years, it has become clear that activity-dependent synaptic plasticity is itself plastic, an effect that has been termed 'metaplasticity' (see Abraham and Bear, 1996; Fischer et al., 1997). Metaplastic interactions have been analyzed in the lamprey locomotor network by investigating how endogenous neuromodulators affect the activity-dependent plasticity of network interneuron synaptic transmission (Parker and Grillner, 1999b, Parker, unpublished observations). Glutamatergic inputs from excitatory network interneurons (EIN), and glycinergic inputs from crossed caudal (CC) interneurons have been examined in detail, since these are assumed to form the basis of the locomotor network (Hellgren et al., 1992), the EINs by providing the main source of excitatory drive at the segmental level, and the CC interneurons by mediating reciprocal inhibition between hemisegments on each side of the spinal cord. Substance P and 5-HT were examined, since their opposite network effects (Harris-Warrick and Cohen, 1985; Parker et al., 1998) allow the potential relevance of metaplasticity in the control of the burst frequency to be examined. Under control conditions (Fig. 6), EIN and CC interneuron inputs to motor neurons do not show any significant activity-dependent plasticity in response to physiologically relevant spike trains (Parker and Grillner, 1999b). However, in the presence of substance P or 5-HT physiologically relevant activity-dependent metaplasticity is evoked (Parker and Grillner, 1999b); the effects that are seen depend on the connection and stimulation frequency examined. Both substance P and 5-HT evoke metaplastic facilitation of EIN-evoked EPSPs; their effects differ, however, in that 5-HT reduces the amplitude of the initial EPSP in the train, whereas substance P potentiates it. The resulting net effects are that 5-HT does not significantly affect the EIN input over physiologically relevant spike trains, whereas substance P evokes significant potentiation, an effect that could contribute to its effect on the network burst frequency (Hellgren et al., 1992). In contrast to EIN-evoked EPSPs, substance P and 5-HT have opposite effects on CC interneuron inputs to motor neurons (Fig. 7). In this case, neither modulator affects the amplitude of the initial IPSP in the train, but substance P-evokes significant metaplastic depression, whereas 5-HT evokes significant metaplastic facilitation. Assuming that CC interneurons play a significant role in segmental reciprocal inhibition, substance P and 5-HT will decrease and increase, respectively, the inhibitory drive to the contralateral network. These effects are again consistent with the substance P-mediated increase and the 5-HT-mediated reduction of the burst frequency (see Hellgren et al., 1992). Metaplastic effects of this kind thus provide a further mechanism by which network cellular and synaptic properties, and thus the output, can be modulated.

Interactions between endogenous neuromodulators

In addition to the growing realization that synaptic plasticity is plastic and can change depending on the synapses' history, it has been known for some time that neuromodulatory effects can be altered by interactions between neuromodulators (see Kupfermann, 1991). These interactions can occur either as a result of the simultaneous release of co-localized modulators, or the simultaneous release of modulators contained in separate terminals, that interact through volume transmission. In addition, due to the 'modulatory tone' resulting from the slow onset and duration of their effects, modulators that are temporally independently released can interact. These interactive effects can enhance the subtlety of network modulation (see Kupfermann, 1991), and are particularly relevant in relation to volume transmission, since the diffusion of modulators through the extracellular space provides ample opportunities for these sort of interactive effects to occur. Interactive effects between neuromodulators have recently been examined in detail in the lamprey locomotor network, and will be briefly summarized here.

Interactive presynaptic inhibition of tachykinin release

The co-localization of 5-HT, dopamine, and tachykinins in the ventromedial plexus provides the opportunity for these three modulators to influence each other's effects; the first example of their potential interactions was provided by the finding that blocking D2 receptors with eticlopride could also evoke a long-lasting increase in the network burst frequency (Fig. 4A). Since there does not appear to be any tonic dopaminergic modulation of the locomotor network, and dopamine itself does not have a long-term network effect (McPherson and Kemnitz, 1994; Wikström, 1999), eticlopride could act by influencing or evoking the endogenous effects of another modulator. A possible relationship between the long-term effects of eticlopride and substance P was examined by applying eticlopride with the tachykinin receptor antagonist spantide II, which blocks the network effects of substance P (see above). Spantide II blocked the long, but not short-term effects of eticlopride (Parker et al., 1998), supporting the possibility that its long-term effect was caused by endogenous tachykinin release; this conclusion was further

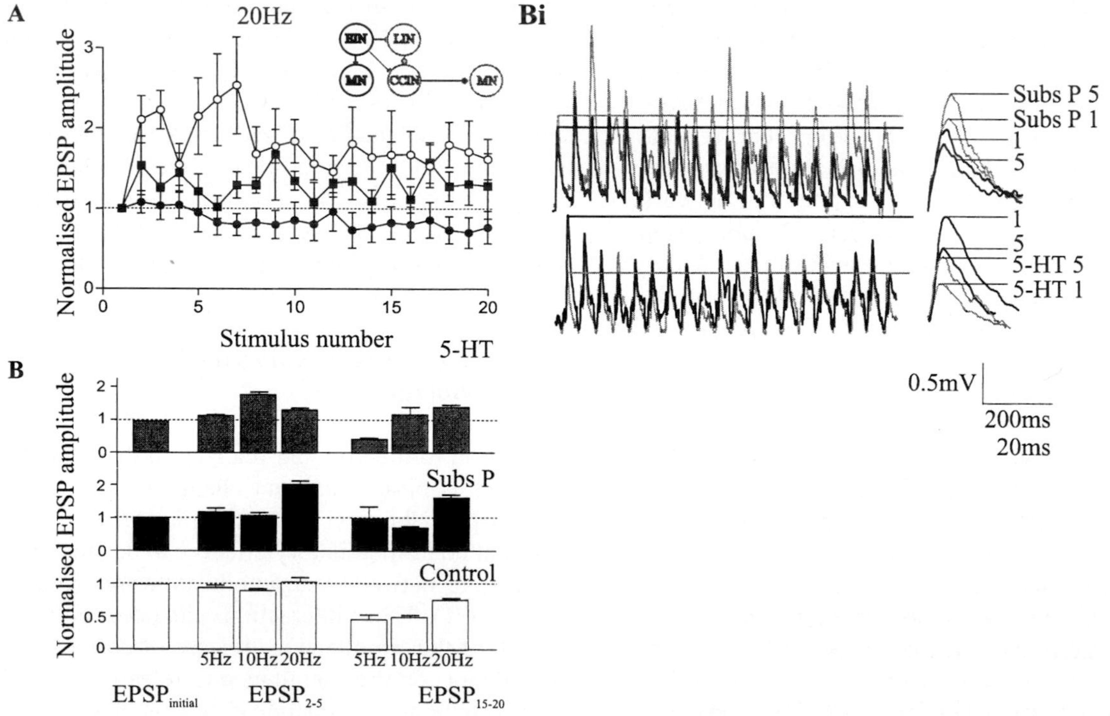

Fig. 6. Substance P and 5-HT evoke activity-dependent metaplastic facilitation of glutamaergic inputs from excitatory network interneurons (EIN). (A) Graph showing the effects of 5-HT and substance P on EIN inputs evoked during a train of 20 spikes delivered at 20 Hz. Both 5-HT and substance P evoke facilitation of the input during the spike train. (B) Traces showing the effects of substance P and 5-HT on spike trains. Note the opposite effects of 5-HT and substance P on the initial EPSP amplitude in the train The substance-mediated facilitation is associated with potentiation of the initial EPSP amplitude, whereas the effect of 5-HT is associated with an initial reduction. (C) Graph showing the effects of 5-HT and substance P on EIN inputs evoked at frequencies of 5 10, and 20 Hz. The train of 20 spikes is divided into four regions, the initial EPSP, the 2nd to 5th EPSPs in the train, the 6th to 10th EPSPs, and the 11th to 20th EPSPs.

supported by the finding that eticlopride mimicked the substance P-mediated improvement of the burst regularity (Parker et al., 1998).

Dopaminergic D2 receptors can inhibit transmitter release by modulating presynaptic calcium channels. By reducing endogenous D2-receptor-mediated inhibition of calcium channels, eticlopride could increase bulk calcium to a level sufficient to evoke endogenous tachykinin release and the long-term spantide II-sensitive increase in the burst frequency. Support for this hypothesis was provided by the dependence of the eticlopride-mediated modulation on ongoing network activity, an effect that is not shared by exogenously applied substance P-evoked modulation (Parker et al., 1998). In the absence of network activity, presynaptic calcium channels would presumably be inactivated, and thus eticlopride should not significantly affect calcium levels in tachykinin-containing terminals. During network activity, however, calcium channels may be activated, and relieving endogenous D2-receptor-mediated inhibition under these conditions could increase bulk calcium levels and evoke tachykinin release. This possibility was investigated using EGTA–AM (see above). In the presence of EGTA–AM, the long-term effects of eticlopride were blocked (Zhang, unpublished observations), suggesting a dependence on increased cytoplasmic Ca^{2+} levels. However, since EGTA–AM blocks the long-term effects of exogenously applied substance P (see above), it is not possible to be certain whether the occlusion of the eticlopride-mediated modulation is due to the inhibition of endogenous tachykinin release, or the direct inhibition of tachykinin-mediated effects. Both possibilities, however, provide further support for endogenous tachykinins in the long-term eticlopride-mediated modulation.

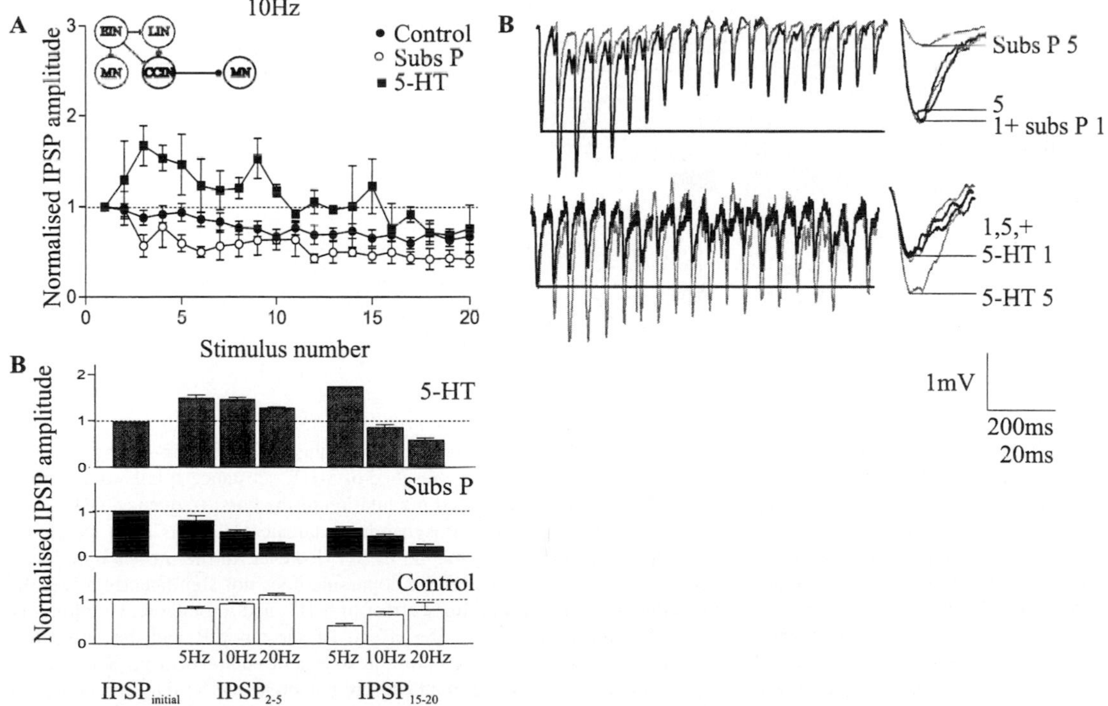

Fig. 7. Substance P and 5-HT evoke activity-dependent metaplastic modulation of glycinergic inputs from inhibitory crossed caudal network interneurons (CC interneurons). (A) Graph showing the effects of 5-HT and substance P on CC interneuron inputs evoked during a train of 20 spikes delivered at 10 Hz. 5-HT evokes metaplastic facilitation of the IPSP during the train, whereas substance P evokes metaplastic depression. (B) Traces showing the effects of 5-HT and substance P on CC interneuron inputs during a 20 spike train. Note that neither 5-HT or substance P affects the initial IPSP in the train. (C) Graph showing the effects of 5-HT and substance P on CC interneuron inputs evoked at frequencies of 5 10, and 20 Hz. The train of 20 spikes is again divided into four regions.

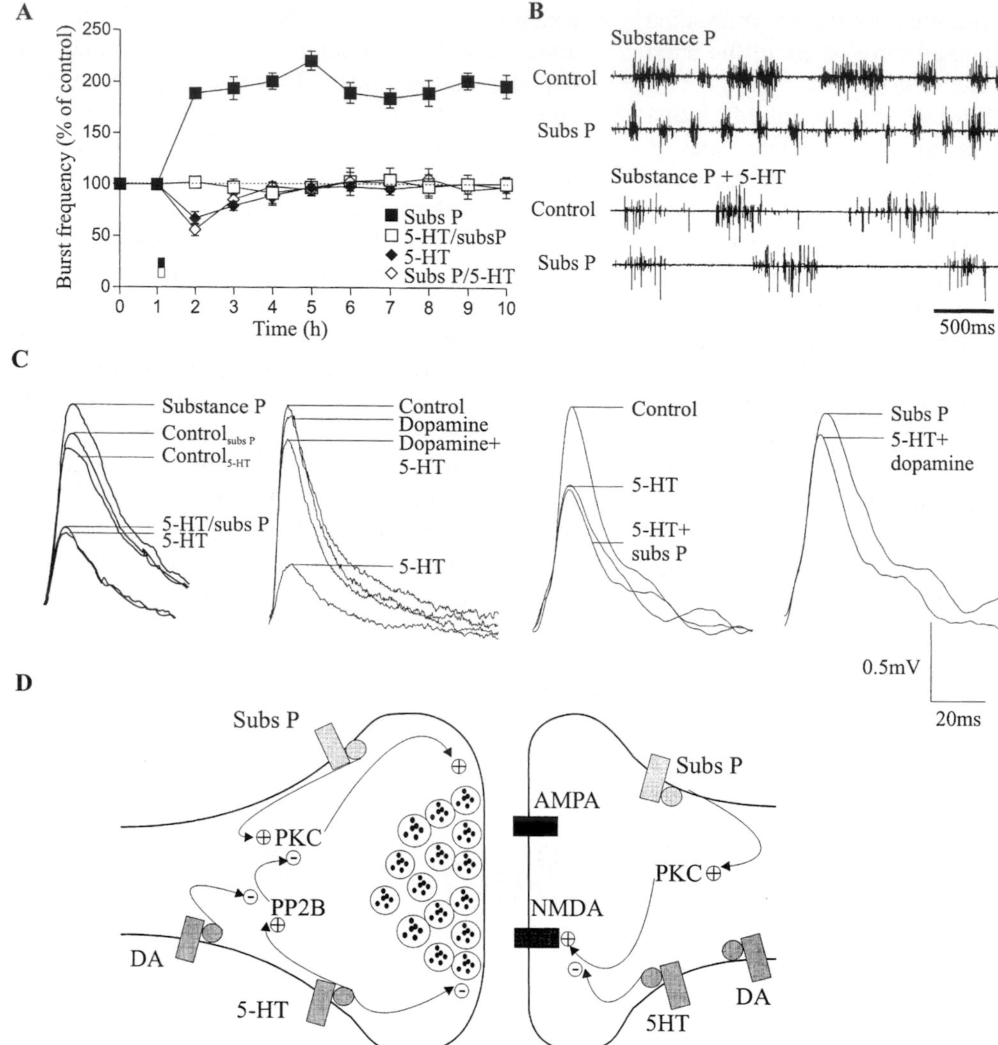

Fig. 8. Interactive effects of co-localized modulators in the ventromedial plexus. (A) Graph showing the usual long-lasting effect of substance P on the frequency of NMDA-evoked ventral root bursts. However, in the presence of 5-HT, substance P fails to affect the burst frequency. Note that substance P does not significantly affect the 5-HT-mediated modulation of the burst frequency. (B) Traces showing the effects of substance P, 5-HT, dopamine, and their interactions on monosynaptic glutamatergic inputs from excitatory network interneurons (EIN). Substance P potentiates the amplitude of the monosynaptic EPSP. However, in the presence of 5-HT, which presynaptically reduces the EPSP amplitude, substance P fails to have an effect; dopamine does not significantly affect the amplitude of EIN-evoked EPSPs. However, it significantly reduces the direct inhibitory effect of 5-HT, and the interactive inhibitory effect of 5-HT on the substance P-evoked potentiation. (C) Summary diagram of the effects of substance P, and the interactive modulation of 5-HT and dopamine; a presynaptic EIN terminal is shown synapsing onto a postsynaptic motor neuron. Substance P presynaptically potentiates glutamatergic transmission from EINs. It also acts postsynaptically to potentiate the NMDA component of glutamatergic inputs, an effect that is crucial for the induction of the long-term network modulation. 5-HT acts presynaptically to prevent the substance P-mediated potentiation of glutamatergic transmission through a protein phosphatase 2B (PP2B)-mediated mechanism. It also inhibits glutamate release. Dopamine does not act directly, but it blocks the presynaptic interactive effect of 5-HT on the substance P-mediated potentiation, and also the 5-HT-mediated inhibition of glutamate release. Postsynaptically, 5-HT acts through an unknown mechanism to prevent or limit the substance P-mediated PKC-dependent potentiation of NMDA responses. Dopamine does not influence the postsynaptic effects of substance P or 5-HT, and thus does not allow the induction of the long-term network modulation.

Gating and braking of modulatory effects by modulator interactions

A single modulator can have several distinct effects. This is exemplified in the lamprey by substance P, which has several specific effects on identified network cellular and synaptic properties (Parker et al., 1998; Parker and Grillner, 1998, 1999b). Depending on the complement of effects elicited, short or long-term network plasticity can be evoked. The information available on tachykinin effects, together with the co-localization of 5-HT and dopamine in plexus neurons, has allowed neuromodulator interactions to be examined in detail by investigating what effects 5-HT and dopamine have on substance P-evoked modulation (Svensson et al., in preparation). 5-HT, but not dopamine, pre- and postsynaptically inhibits tachykinin-mediated cellular and synaptic effects (Fig 8). This includes the presynaptic and NMDA-dependent postsynaptic potentiation of glutamatergic synaptic transmission; thus, the network effects of substance P and the induction of the long-term modulation are blocked by 5-HT. Since substance P has profound network effects, namely a large, apparently irreversible increase in burst frequency, 5-HT may function as a brake on the tachykinin-mediated modulation, ensuring that it only occurs when developmentally or behaviorally appropriate, possibly with respect to migration when the lamprey swims 1,000s of kilometers to spawn.

In contrast to 5-HT, dopamine, which co-localizes with 5-HT in ventromedial plexus neurons, does not directly affect responses to exogenously applied substance P; however, as outlined above, it may act synergistically with co-released 5-HT to inhibit endogenous tachykinin-mediated modulation through its potential inhibitory effect on tachykinin release. Unexpectedly, however, it was found that co-applying dopamine with 5-HT relieved the 5-HT-mediated inhibition of presynaptic potentiation of glutamatergic synaptic transmission by substance P; dopamine disinhibits and can thus gate certain short-term tachykinin-mediated cellular and synaptic effects. Since it does not relieve the 5-HT-mediated occlusion of the NMDA receptor potentiation, dopamine will not allow the long-term network modulation to be induced (see above). Modulation can thus be modulated: by selecting components from a single modulators repertoire, dynamic changes occur in network plasticity.

Conclusions

As outlined in the Introduction, the lamprey provides the opportunity to examine network modulation in detail at the cellular, synaptic, and molecular levels. Several modulatory systems have been studied with respect to their effects on the locomotor network, and its sensory and descending brainstem inputs; of particular interest has been the ventromedial modulatory system. As described in this and other reviews (see Wallén et al., 1989a, b, Grillner et al., 1998a, b), the individual effects of 5-HT, dopamine, and tachykinin peptides contained in this plexus have now been examined in some detail. The terminals that contain these modulators do not appear to form synaptic specializations, and they are thus presumably released paracrinically to act through volume transmission-mediated mechanisms. In addition to the potential for modulator interactions through their co-release, the nature of volume transmission and the relatively long duration of modulatory effects, will also allow modulators whose release is spatially or temporally independent to interact. The analysis of the interactive effects of modulators will thus be a necessary component to analyses of network function.

References

Abraham, W.C. and Bear, M.F. (1996) Metaplasticity: the plasticity of synaptic plasticity. *TINS*, 19: 126–130.

Adler, E.M., Augustine, G.J., Duffy, S.N. and Charlton, M.P. (1991) Alien intracellular calcium chelators attenuate neurotransmitter release at the squid giant synapse. *J. Neurosci.*, 11: 1496–1507.

Alford, S., Christenson, J. and Grillner, S. (1991) Presynaptic $GABA_A$ and $GABA_B$ receptor-mediated phasic modulation in axons of spinal motor interneurons. *Europ. J. Neurosci.*, 3: 107–117.

Alford, S. and Grillner, S. (1991) The involvment of $GABA_B$ receptors and coupled G-proteins in spinal GABAergic presynaptic inhition. *J. Neurosci.*, 11: 3718–3726.

Augustine, G.J., Adler, E.M. and Charlton, M.P. (1991) The calcium signal for transmitter secretion from presynaptic nerve terminals. *Ann. NY Acad. Sci.*, 635: 365–381.

Bacskai, B.J., Wallén, P., Lev-Ram, P., Grillner, S. and Tsien, R.Y. (1995) Activity-related calcium dynamics in lamprey motoneurons as revealed by video-rate confocal microscopy. *Neuron*, 14: 19–28.

Barthe, J.Y., Grillner, S.(1995) Neurotensin-induced modulation of spinal neurons and fictive locomotion in lamprey. *J. Neurophysiol.*, 73: 1308–1312.

Bongianni, F., Christenson, J., Hökfelt, T. and Grillner, S. (1990) Neuropeptide Y-immunoreactive spinal neurons make close appositions on axons of primary sensory afferents. *Brain Res.*, 523: 337–341.

Borst, J.G.G. and Sakmann, B. (1996) Calcium influx and transmitter release in a fast CNS synapse. *Nature*, 383: 431–434.

Brodin, L., Buchanan, J.T., Hökfelt, T., Grillner, S. and Verhofstad, A.A.J. (1986) A spinal projection of 5-hydroxytryptamine neurons in the lamprey brainstem: evidence from combined retrograde tracing and immunohistochemistry. *Neurosci. Lett.*, 67: 53–57.

Brodin, L., Buchanan, J.T., Hökfelt, T., Grillner, S., Rehfeld, J.F., Frey, P., Verhofstad, A.A.J., Dockray, G.J. and Walsh, J.H. (1988) Immuno-histochemical studies of cholecystokininlike peptides and their relation to 5-HT, CGRP, and bombesin immunoreactivities in the brainstem and spinal cord of lampreys. *J. Comp. Neurol.*, 271: 1–18.

Brodin, L., Rawitch, A., Taylor, T., Ohta, Y., Ring, H., Hökfelt, T., Grillner, S. and Terenius, L. (1989) Multiple forms of pancreatic polypeptide related compounds in the lamprey CNS: partial characterization and immunohistochemical localization in the brain stem and spinal cord. *J. Neurosci.*, 9: 3428–3442.

Brodin, L., Theodorsson, E., Christenson, J., Cullheim, S., Hökfelt T., Brown, J.C., Buchanon, A., Panula, P., Verhofsted, A.A.J. and Goldstein, M. (1990) Neurotensin-like peptides in the CNS of lampreys: chromatographic characterization and immunohistochemical localisation with reference to aminergic neurones. *Eur. J. Neurosci.*, 2: 1095–1109.

Buchanan, J.T. (1999) Commissural interneurons in rhythm generation and intersegmental coupling in the lamprey spinal cord. *J. Neurophysiol.*, 81: 2037–2045.

Buchanan, J.T. and Grillner, S. (1987) Newly identified 'glutamate interneurons' and their role in locomotion in the lamprey spinal cord. *Science*, 236: 312–314.

Buchanan, J.T. and Grillner, S. (1988) A new class of small inhibitory interneurones in the lamprey spinal cord. *Brain Res.*, 438: 404–407.

Buchanan, J.T. and Grillner, S. (1991) 5-hydroxytryptamine depresses reticulospinal excitatory postsynaptic potentials in motoneurones of the lamprey. *Neurosci. Lett.*, 112: 71–74.

Buchanan, J.T. and Kasicki, S. (1999) Segmental distribution of common synaptic inputs to spinal motoneurons during fictive swimming n the lamprey. *J. Neurophysiol.*, 82: 1156–1163.

Bussières, N. and El Manira A. (1999) GABA(B) receptor activation inhibits N- and P/Q-type calcium channels in cultured lamprey sensory neurons. *Brain Res.*, 847: 175–185.

Christenson, J., Franck, J. and Grillner, S. (1989) Increase in endogenous 5-hydroxytryptamine modulates the central network underlying locomotion in the lamprey spinal cord. *Neurosci. Lett.*, 100: 188–192.

Christenson, J., Cullheim, S., Grillner, S. and Hökfelt, T. (1990) 5-Hydroxytryptamine immunoreactive varicosities in the lamprey spinal cord have no synaptic specializations – an ultrastructural study. *Brain Res.*, 512: 201–209.

Christenson, J., Alford, S., Grillner, S. and Hökfelt, T. (1991) Co-localized GABA and somatostatin use different ionic mechanisms to hyperpolarize target neurons in the lamprey spinal cord. *Neurosci. Lett.*, 134: 93–97.

Cullheim, S. and Arvidsson, U. (1995) The peptidergic innervation of spinal motoneurons via the bulbospinal 5-hydroxytryptamine pathway. *Progr. Brain Res.*, 104: 21–40.

Duggan, A.W. (1995) Release of neuropeptides in the spinal cord. *Progr. in Brain Res.*, 104: 197–224.

El Manira, A. and Bussieres, N. (1997) Calcium channel subtypes in lamprey sensory and motor neurons. *J. Neurophysiol.*, 78: 1334–1340.

El Manira, A., Tegnér, J. and Grillner, S. (1994) Calcium-dependent potassium channels play a critical role for burst termination in the locomotor network in lamprey. *J. Neurophysiol.*, 72.4: 1852–1861.

El Manira, A., Tegnér, J. and Grillner, S. (1997a) Locomotor-related presynaptic modulation of primary afferents in the lamprey. *Eur. J. Neurosci.*, 9: 696–705.

El Manira, A., Zhang, W., Svensson, E. and Bussières, N. (1997b) 5-HT inhibits calcium current and synaptic transmission from sensory neurons in lamprey. *J. Neurosci.*, 17: 1786–1794.

El Manira, A., Parker, D., Krieger, P., Wikström, M. and Grillner, S. (1998) Presynaptic inhibition of sensory transmission from sensory, interneuronal and supraspinal neurons to spinal target cells in lamprey. In: P. Rudomin, R. Romo, and L. Mendell (Eds), *Presynaptic Inhibition and Neural Control*, Oxford University Press, pp. 329–348.

Fischer, T.M., Blazis, D.E.J., Priver, N.A. and Carew, T.J. (1997) Metaplasticity at identified inhibitory synapses in aplysia. *Nature*, 389: 860–865.

Franck, J., Christenson, J., Fried, G., Grillner, S. and Hökfelt, T. (1992) Subcellular distribution of serotonin in the lamprey spinal cord. *Brain Res.*, 589: 48–54.

Getting, P.A. (1989) Emerging principles governing the operation of neural networks. *Ann. Rev. Neurosci.*, 12: 185–204.

Grillner, S., Buchanan, J.T. and Lansner, A. (1988) Simulation of the segmental burst generating network for locomotion in lamprey. *Neurosci. Lett.*, 89: 31–35.

Grillner, S., Ekeberg, Ö., El Manira, A., Lansner, A., Parker, D., Tegnér, J. and Wallén, P. (1998a) Intrinsic function of a neuronal network – a vertebrate central pattern. *Brain Res. Rev.*, 26: 184–197.

Grillner, S., Parker, D. and El Manira, A. (1998b) Vertebrate locomotion – a lamprey perspective. In: O. Kiehn, R.M. Harris-Warrick, L.M. Jordan, H. Hultborn and N. Kudo (Eds), *Neuronal Mechanisms for Generating Locomotor Activity*, Vol. 860, *Ann. NY Acad. Sci.*, pp. 1–18.

Harris-Warrick, R.M. and Cohen, A.H. (1985) Serotonin modulates the central pattern generator for locomotion in the isolated lamprey spinal cord. *J. Exp. Biol.*, 116: 27–46.

Hellgren, J., Grillner, S. and Lansner, A. (1992) Computer simulation of the segmental neural network generating locomotion in lamprey by using populations of network interneurons. *Biol. Cybern.*, 68: 1–13.

Huard, H., Lund, J.P., Veilleux, D. and Dubuc, R. (1999) An anatomical study of brainstem projections to the trigeminal motor nucleus of lampreys. *Neuroscience*, 91: 363–378.

Kiehn, O., Harris-Warrick, R.M., Jordan, L.M., Hultborn, H. and Kudo, N. (1998) Neuronal Mechanisms for Generating Locomotor Activity. *Ann. NY Acad. Sci.*, Vol. 860; 1–569.

Krieger, P., El Manira, A. and Grillner, S. (1996) Activation of pharmacologically distinct metabotropic glutamate receptors depresses reticulospinal-evoked monosynaptic EPSPs in the lamprey spinal cord. *J. Neurophysiol.*, 76.6: 3834–3841.

Krieger, P., Grillner, S. and El Manira.A. (1998) Endogenous activation of metabotropic glutamate receptors contributes to burst frequency regulation in the lamprey locomotor network. *Eur. J. Neurosci.*, Vol. 10; 3333–3342.

Kupfermann, I. (1991) Functional studies of cotransmission. *Physiol. Rev.*, 71: 683–732.

McPherson, D.R. and Kemnitz, C.P. (1994) Modulation of fictive swimming and motoneuron physiology by dopamine, and its imminocytochemical localization in the spinal cord. *Neurosci. Lett.*, 166: 23–26.

Matsushima, T. and Grillner, S. (1992) Local serotonergic modulation of calcium dependent potassium channels controls intersegmental coordination in the lamprey spinal cord. *J. Neurophysiol.*, 67: 1683–1690.

Matsushima, T., Tegnér, J., Hill, R. and Grillner, S. (1993) $GABA_B$ receptor activation causes a depression of low- and high-voltage-activated Ca^{2+} currents, postinhibitory rebound, and postspike afterhyperpolarisation in lamprey neurons. *J. Neurophysiol.*, 70: 2606–2619.

Mayer, M.L. and Westbrook, G.L. (1987) The physiology of excitatory amino acids in the vertebrate central nervous system. *Prog. Neurobiol.*, 28: 197–276.

Nieuwenhuys, R., Donkelaar, H.J. ten and Nicholson, C. (1998) *The central nervous system of vertebrates*. Springer Verlag, Vol. 1: 1–757.

Nyberg, F. and Sharma, H.S. (1995) In: Z. Wiesenfeld-Hallin, (Eds), *Neuropeptides in the Spinal Cord*, Progress in Brain Research., 104: 1–421.

Ohta, Y. and Grillner, S. (1989) Monosynaptic excitatory amino acid transmission from the posterior rhombencephalic reticular nucleus to spinal neurons involved in the control of locomotion in lamprey. *J. Neurophysiol.*, 62: 1079–1089.

Ohta, Y., Brodin, L., Grillner, S., Hökfelt, T. and Walsh, J.H. (1988) Possible target neurons of the reticulospinal cholecystokinin (CCK) projection to the lamprey spinal cord: immunohistochemistry combined with intracellular staining with Lucifer Yellow. *Brain Res.*, 44: 400–403.

Orlovsky, G.N., Deliagine, T.G. and Grillner, S. (1999) Neuronal control of locomotion. Oxford Univ. Press, 1–322.

Parker, D. (2000) Activity and calcium-dependent mechanisms maintain reliable interneuron synaptic transmission in a rhythmic neural network. *J. Neurosci.*, 20: 1754–1766.

Parker, D. (2000) Presynaptic and interactive peptidergic modulation of reticulospinal synaptic inputs in the lamprey. *J. Neurophysiol.* (in press).

Parker, D. and Grillner, S. (1998) Cellular and synaptic modulation underlying substance P-mediated plasticity of the lamprey locomotor network. *J. Neurosci.*, 18: 8095–8110.

Parker, D. and Grillner, S. (1999a) Long-lasting substance P-mediated modulation of NMDA-induced rhytmic activity in the lamprey locomotor network involves separate protein and RNA synthesis-dependent stages. *Eur. J. Neurosci.*, 11: 1515–1522.

Parker, D. and Grillner, S. (1999b) Activity-dependent metaplasticity of inhibitory and excitatory synaptic transmission in the lamprey spinal cord locomotor network. *J. Neurosci.*, 19: 1647–1656.

Parker, D., Svensson, E. and Grillner, S. (1997) Substance P modulates sensory action potentials in the lamprey via a protein kinase C-mediated reduction of a 4-aminopyridine-sensitive potassium conductance. *Eur. J. Neurosci.*, 9: 2064–2076.

Parker, D., Zhang, W. and Grillner, S. (1998a) Substance P modulates NMDA responses and causes long-term protein synthesis-dependent modulation of the lamprey locomotor network. *J. Neurosci.*, 18: 4800–4813.

Parker, D., Söderberg, C., Zotova, E., Shupliakov, O., Langel, U., Bartfai, T., Larhammar, D., Brodin, L. and Grillner, S. (1998b) Co-localised neuropeptide Y and GABA have complementary presynaptic effects on sensory synaptic transmission. *Eur. J. Neurosci.*, 10: 2856–2870.

Rovainen, C.M. (1979) Neurobiology of lampreys. *Physiol. Rev.*, 59: 1007–1077.

Schotland, J., Shupliakov, O., Wikström, M., Brodin, L., Srinivasan, M., You, Z., Herrera-Marschitz, M., Zhang, W., Hökfelt, T. and Grillner, S. (1995) Control of lamprey locomotor neurons by co-localized monoamine transmitters. *Nature*, 374: 266–268.

Schotland, J.L., Shupliakov, O., Grillner, S. and Brodin, L. (1996) Synaptic and non-synaptic monoaminergic neuron systems in the lamprey spinal cord. *J. Comp. Neurol.*, 372: 229–244.

Tegnér, J., Matsushima, T., El Manira, A. and Grillner, S. (1993) The spinal GABA system modulates burst frequency and intersegmental coordination in the lamprey: differential effects of $GABA_A$ and $GABA_B$ receptors.*J. Neurophysiol.*, 69: 647–657.

Tegnér, J., Lansner, A. and Grillner, S. (1998) Modulation of burst frequency by calcium-dependent potassium channels in the lamprey locomotor system: dependence of the activity level. *J. Comp. Neurosci.*, 5; 121–140.

Ullström, M., Hellgren Kotaleski, J., Tegnér, J., Aurell, E., Grillner, S. and Lansner, A. (1998) Activity-dependent modulation of adaptation produces a constant burst proportion in a model of the lamprey spinal generator. *Biol. Cybern.*, 79: 1–4.

Van Dongen, P.A., Hökfelt, T., Grillner, S., Verhofstad, A.A.J., Steinbusch, H.W.M., Cuello, A.C. and Terenius, L. (1985) Immunohistochemical demonstration of some putative neurotransmitters in the lamprey spinal cord and spinal ganglia: 5-Hydroxytryptamine-, Tachykinin-, and Neuropeptide-Y-immunoreactive neurons and fibers. *J. Comp. Neurol.*, 234: 501–522.

Van Dongen, P.A., Theodorsson-Norheim, E., Brodin, E., Hökfelt, T., Grillner, S. Peters, A., Cuello, A.C., Forssman, W-G. , Reinecke, M., Singer, E.A. and Lazarus, L. (1986) Immunohistochemical and chromatographic studies of peptides with tachykinin-like immunoreactivity in the central nervous system of the lamprey. *Peptides*, 7: 297–313.

Verhage, M., McMahon, H.J., Ghijsen, W.E.J.M., Boomsa, F., Scholten, G., Wiegant, V.M. and Nicholls, D.G. (1991) Differential release of amino acids, neuropeptides and catacholamines from isolated nerve terminals. *Neuron*, 6: 517–524.

Wald, U. and Selzer, M.E. (1981) The inulin space of the lamprey spinal cord. *Brain Res.*, 208: 113–122.

Wallén, P., Christenson, J., Brodin, L., Hill, R., Lansner, A. and Grillner, S. (1989a) Mechanisms underlying the serotonergic modulation of the spinal circuitry for locomotion in the lamprey. *Prog. Brain Res.*, 80: 321–327.

Wallén, P., Buchanan, J.T., Grillner, S., Hill, R.H., Christenson, J. and Hökfelt, T. (1989b) Effects of 5-hydroxytryptamine on the afterhyperpolarization, spike frequency regulation, and oscillatory membrane properties in lamprey spinal cord neurons. *J. Neurophysiol.*, 61: 759–768.

Waugh, D., Wang, Y., Hazon, N., Balment, R.J. and Conlon, J.M. (1993) Primary structures and biological activities of substance P-related peptides from the brain of the dogfish, Scliorhinus canicula. *Eur. J. Biochem.*, 214: 469–474.

Wikström, M. (1999) Dopaminergic and Serotonergic Modulation of Cellular and Locomotor Network Properties in the Lamprey Spinal Cord, thesis, pp 1–66.

Wikström, M. and El Manira, A. (1998) N- and P/Q-type calcium channels activate apamin-sensitive calcium-dependent potassium channels generating the late afterhyperpolarisation in lamprey spinal neurons. *Eur J, Neurosci.*, 10: 1528–1532.

Wikström, M., El Manira, A., Zhang, W., Hill, R.H. and Grillner, S. (1995) Dopamine and 5-HT modulation of synaptic transmission in the lamprey spinal cord. *Soc. Neurosci. Abstract*, 21:1145.

Wikström, M., Hill, R., Hellgren J. and Grillner, S. (1995) The action of 5-HT on calcium-dependent potassium channels and on the spinal locomotor network in lamprey is mediated by 5-HT1A-like receptors. *Brain. Res.*, 678: 191–199.

Wikström, M., Grillner, S. and El Manira, A. (1999) Inhibition of N- and L-type Ca^{2+} currents by dopamine in lamprey spinal motoneurons. *NeuroReport*, 10: 3179–3183.

L. F. Agnati, K. Fuxe, C. Nicholson and E. Syková (Eds.)
Progress in Brain Research, Vol 125

CHAPTER 26

Long distance signalling in volume transmission. Focus on clearance mechanisms

Anders Jansson[1]*, Andrea Lippoldt[2], Tomas Mazel[3], Tamas Bartfai[4], Sven-Ove Ögren[1], Eva Syková[3], Luigi F. Agnati[5] and K. Fuxe[1]

[1]*Department of Neuroscience, Karolinska Institutet, 171 77 Stockholm, Sweden*
[2]*Centrum für Molekulare Medizin, 13122 Berlin-Buch, Germany*
[13]*Department of Cellular Neurophysiology, Joint Institution of ASCR and Charles University, Institute of Experimental Medicine ASCR, 142 20 Prague 4, Czech Republic*
[4]*F. Hoffmann-La Roche Ltd, Department: PRPN, Building. 69/415, CH–4070 Basel, Switzerland*
[5]*Department of Human Physiology, University of Modena, 41100 Modena, Italy*

Introduction

There now exists substantial evidence that large numbers of VT signals such as transmitters and modulators travel in the microenvironment of the extracellular space (Fuxe and Agnati, 1991; Zoli et al., 1999). It then becomes important to study through which mechanisms these VT signals are cleared from the extracellular environment and if they can sometimes reach the ventricular system to act also as CSF signals. It is well-known that catabolic enzymes exist in the extracellular space and in glial and neuronal membranes facing the extracellular space for various transmitters, such as monoamines and neuropeptides (Schwartz et al., 1991; del Rio-Garcia and Smyth, 1991). These enzymes play an important role in the clearance of VT signals together with the also well-known reuptake mechanisms for various transmitters such as DA, NA and 5-HT located in the entire neuron (terminals, dendrites, axons) at the membrane level but outside the synapses. The efficient recapture of some transmitters into their source neurons when they reach the extracellular fluid is an important regulator of volume transmission (VT) (Giros et al., 1996). Glial cells also play an important role in clearance of VT signals especially ions such as K^+ (see Syková, Chapter 6, this volume). In this chapter we will, however, mainly focus on other clearance mechanisms of VT signals, namely clearance over the brain-blood barrier and over the leaky brain-CSF interface as well as through receptor mediated uptake into other discrete nerve cell systems through internalisation (Fig. 1).

*Corresponding author: Tel: 46-8-728 7028; fax: 46-8-33 79; e-mail: Kjell.Fuxe@neuro.ki.se

The brain-CSF interfaces

Recently, Lippoldt and colleagues (Lippoldt et al., 2000) have analysed the tight and adherens junctions in the ependymal cells and in the choroid plexus epithelium, since diffusion of solutes from the brain microenvironment into the ventricles would facilitate clearance and thus contribute to avoid VT signal overflow. Furthermore, this passage of chemical messages into the CSF is an important step for the endocrine-like type of VT (see Agnati and Fuxe, Chapter 1, this volume). It is known that the ependymal cells only form a leaky

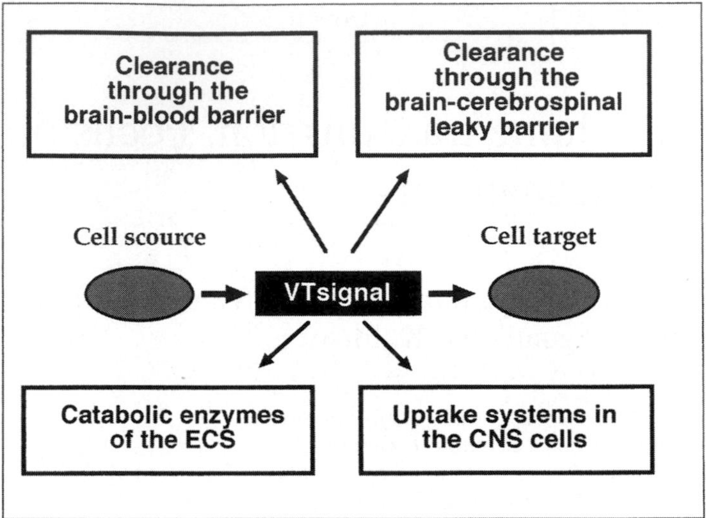

Fig. 1. Illustration of clearance mechanisms for VT signals in the CNS that may involve not only enzymatic breakdown and uptake in the cells but also clearance over the brain-blood barrier and the brain-CSF interface.

barrier at the brain-CSF interface (Peters and Swan, 1979; Rodriguez and Bouchaud, 1996). The epithelial cells of the choroid plexus instead form a true barrier between blood and CSF and are essential for the chemical homeostasis of the CSF (Brightman and Reese, 1969; Dermietzel, 1975; Dermietzel et al., 1977; van Deurs and Koehler, 1979; van Deurs, 1980; Mack et al., 1987; Dziegielewska et al., 1988; Dermietzel and Krause, 1991; Saunders et al., 1999). At barriers tight and adherens junction molecules are essential. For adherence of cells the cadherins and catenins are necessary, while for tightness molecules such as ZO-1, occludin and claudins appear responsible (Furuse et al., 1993, 1998a, b; Anderson et al., 1993; Anderson and van Itallie, 1995; Dejana et al., 1995; Lampugnani and Dejana, 1997; Mitic and Anderson, 1998; Tsukita and Furuse, 1998, 1999; Morita et al., 1999). In the study of Lippoldt et al. (Lippoldt et al., 2000) adherens junction molecules were demonstrated both between the tight epithelial cells of the plexus choroideus and between leaky ventricular ependymal cells. The tight junction molecules ZO-1, claudin and occludin were present in substantial amounts between the epithelial cells but less so between the ependymal cells in line with the existence of an ineffective barrier in the ependymal cell layer.

The most interesting results were obtained when the phorbolester phorbol myristate acetate (PMA) was given intraventricularly (1.5 μg) 24 hr earlier. PMA is known to open up cell-cell contacts (Lombardi et al., 1987; Balda et al., 1991, 1993; Wolburg et al., 1994; Citi and Denisenko, 1995; Stuart and Nigam, 1995; Kaya et al., 1996; Mullin et al., 1996, 1997) and its influence on the adherens and tight junctions were studied immunocytochemically and changes in permeability were studied by evaluating effects on dextran (mol. wt. 3000) diffusion microinjected into the striatum (Jansson et al., 1999b; Lippoldt et al., 2000).

As seen in Table 1, the spread of biotinylated dextran (mol.wt. 3000) can be measured within the rat brain parenchyma by image analysis using microdensitometrical (discrimination) and morphometrical procedures, so that total volume of spread and overall specific mean grey value (intensity) can be determined. The biotinylated dextran is a marker for the extracellular space and a streptoavidine-HRP staining procedure was used with DAB as chromogen (see Fig. 2).

The PMA was found to profoundly reduce the immunoreactivity for the adherens junction proteins cadherin and β-catenin but not for α-catenin between the ependymal cells as well as between the plexus choroideus epithelial cells (Fig. 3). Fur-

TABLE 1. Time-course and dose-response for spread of biotinylated dextran (mol wt 3000) microinjected into the neostriatum: microdensitometrical and morphometrical parameters.

		Volume V_D (mm^3)	Mean grey value (MGV)
Time (min)	2	16±2	30±3
	30	43±6[a]	30±2
	60	56±3[a]	25±1
	120	66±15[a]	21±2[b]
Dose (μg/μl)	1	20±3	24±3
	10	21±4	24±2
	30	50±3[c]	32±2[d]
	100	54±8[c]	28±1

Taken from Jansson et al. (1999b).

Volume (V_D) estimated from the individual specifically stained areas in the sections measured; mean grey value (MGV), the overall median of the mean grey values from the individual specifically stained areas in the sections measured. Data are presented as means ± S.E.M. ($n=5$ rats in each group). One way ANOVA followed by Fisher's PLSD post hoc comparison was used in the statistical analysis. $p<0.05$ was regarded as a statistically significant difference. [a]Different from 2 min; [b]Different from 2 and 30 min; [c]Different from 1 and 10 μg/μl group; [d]Different from 1 and 10 μg/μl groups. Statistically different trends ($p < 0.05$) were found for V_D in the case of the time course and the dose-response.

thermore, the occludin and claudin–5 but not ZO–1 and claudin–1 immunoreactivities (IR) were markedly reduced by the PMA treatment in the epithelial cells of the choroidal plexus (Figs. 3–5) (Lippoldt et al., 2000). These morphological results clearly indicated that protein-kinase C (PKC) activation can through reduction in both adherens and tight junction molecules reduce barrier functions at the CSF-blood and brain-CSF interfaces.

In agreement with the above morphological results PMA (i.v.t.) produced a reduction in the volume of intrastriatal biotinylated dextran spread (m.wt. 3000), which may at least in part be explained by an increased clearance over the brain-blood barrier and brain-CSF interfaces due to reduced barrier and interface functions, probably related in part to impaired junctional properties of the ependymal cells. The results certainly emphasize an important role of PKC in the brain clearance processes and thus in the regulation of the VT. In this way VT may also shift towards an endocrine-like VT involving CSF with i.a. an increased control of the subventricular zone containing e.g. the stem cells (Kuhn and Svendsen, 1999; Peretto et al., 1999).

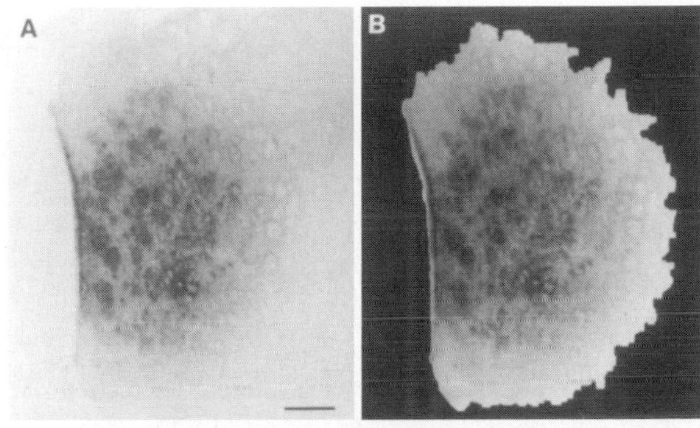

Fig. 2. The spread of biotinylated dextran (mol. wt. 3000; 100 nl, 10 μg/μl), 30 min following intrastriatal injection, detected by the streptavidin-horseradish peroxidase/DAB technique. Two pictures from the same representative section are shown with (B) and without (A) the superimposed discrimination frame used for image analysis area determination (Zeiss/ Kontron IBAS-system). Note also the preferential spread within and especially around the fibre bundles of the capsula interna. Scale bar = 500 μm.

On the role of NO in the clearance over the brain-blood barrier

NO is a wellknown lipid soluble VT signal in the nervous system, which therefore diffuses both intra- and extracellularly to reach its intracellular target, namely guanylylcyclase, leading to increased cGMP formation and altered protein phosphorylation levels. NO possesses a half-life in the order of 0.5–5 sec but in view of its fast diffusion it may diffuse for distances in the order of 100 μm, modulating a volume that may contain up to 10^6 synapses (Wood and Garthwaite, 1994). Nitric oxide synthase (NOS) is the enzyme forming

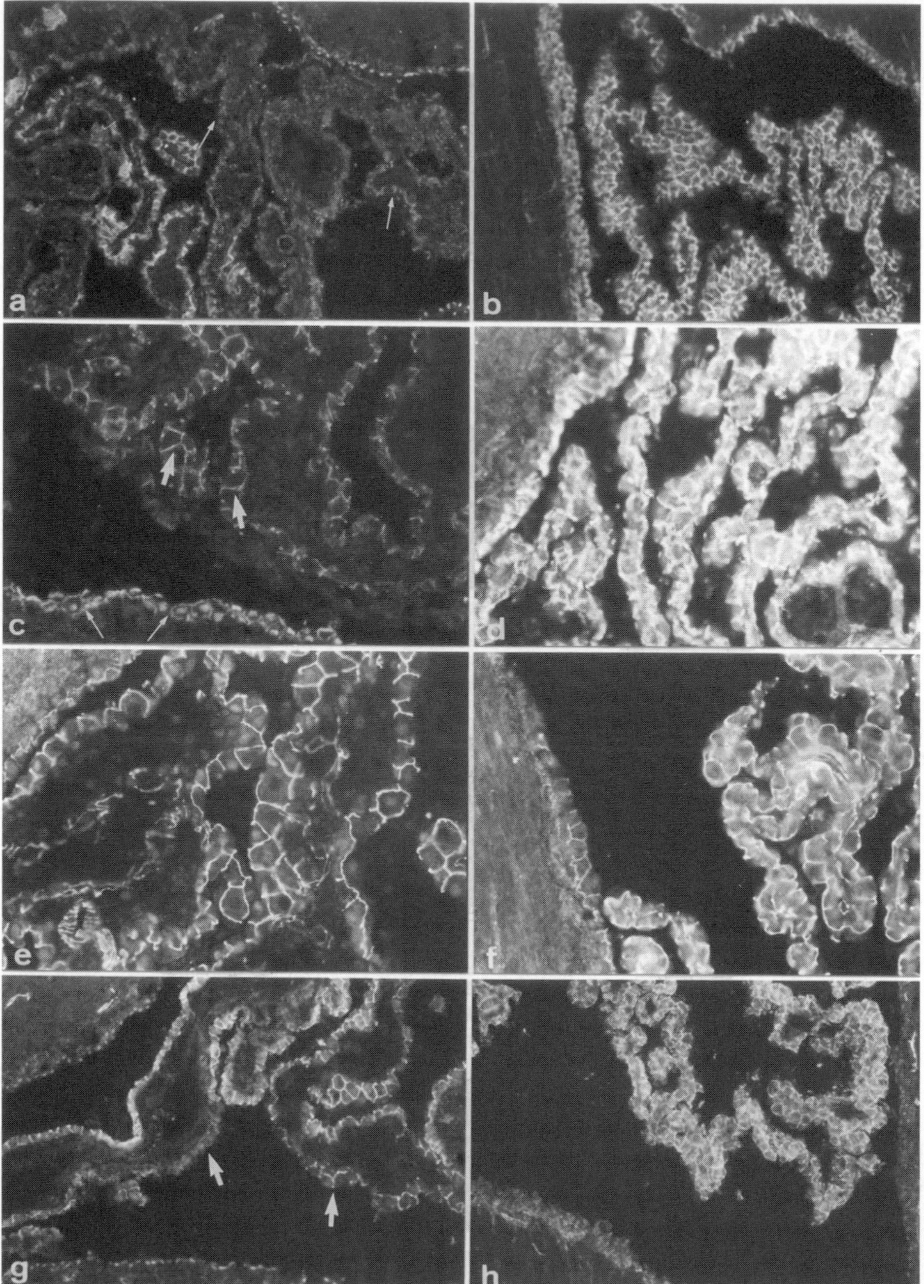

Fig. 3. Acute i.v.t. PMA-injections led to marked changes in adherens and tight junction marker immunoreactivities. The β-catenin immunoreactivity was markedly reduced between the epithelial cells of the choroid plexus (a) arrows) compared to control injected animals (b). Cadherin immunoreactivity was also markedly reduced after acute PMA injection between the ventricular wall ependymal cells (c) thin arrows) but even more pronounced between the borders of the epithelial cells of the choroid plexus (c) thick arrows) when compared to the control animal (d). No change of immunoreactivity was seen in PMA-treated (e) compared to control injected animals (f) when the sections were stained with ZO–1. In contrast, occludin immunoreactivity was markedly reduced between the epithelial cells of the choroid plexus after acute PMA-treatment (g) arrows) compared to control injected animals (h). Note the patchy pattern of occludin immunoreactivity in g. Magnification: $\times$ 200.

NO through enxymatic oxidation of L-arginine to L-citrulline. There exist three different isoforms. The endothelial NOS and neuronal NOS are constitutive and calcium/calmodulin dependent enzymes. The inducible NOS is mainly formed in astroglia and microglia upon activation by inflammatory signals. NO has previously been shown to cause a number of actions on cardiovascular events, including capillary permeability, and on astroglial cells (Jansson et al., 1999b). Therefore, NO may alter the diffusion of other VT signals through effects on clearance over the brain-blood barrier and on extracellular space properties.

In the present study we have therefore studied the effects of unspecific and neuronal specific NOS inhibitors on the diffusion parameters of the extracellular space using the tetramethylammonium (TMA^+) method (see Nicholson, Chapter 28, this volume; Syková et al., Chapter 6, this volume) and the morphological method using biotinylated dextran (M.wt. 3000) spread (see above) (Jansson et al., 1999b).

The non-specific NOS inhibitors N^G-nitro-L-arginine methylester (L-NAME) and N^G-monomethyl-L-arginine acetate (L-NMMA) produced dose-related (10–100 and 30–200 mg/kg, respectively) reductions in the total volume of striatal spread of dextran (Fig. 6).

In contrast, the neuronal specific NOS inhibitor 7-nitroindazole monosodium salt (50–100 mg/kg) did not influence the spread of dextran. The reduction in the volume of the dextran with L-NAME and L-NMMA at the 30 min time-interval was 37 and 42% (maximal effect), respectively.

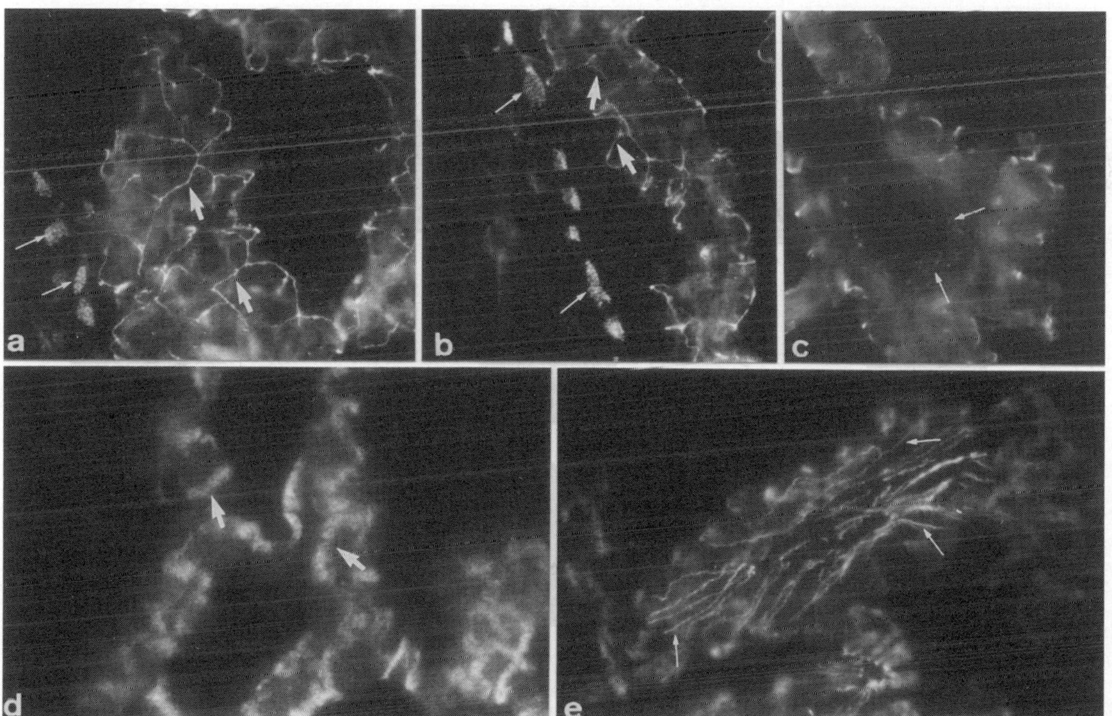

Fig. 4. (a) – (c): Claudin–1 immunoreactivity in the ventricular system of the rat brain. The pictures demonstrate the Claudin–1 punctate appearance in the ependymal cells (thin arrows) of the lateral ventricle (a) as well as staining between the choroid plexus epithelial cells (b) thick arrows). The choroidal capillaries exhibit weak Claudin–1 immunoreactivity (c) thin arrows).
(d) and (e): Claudin–5 is another transmembrane protein found in tight junctions. The ependymal cells of the lateral ventricle are not stained for Claudin–5. In contrast, the epithelial cell border areas of the choroid plexus (d) thick arrows) are strongly immunoreactive for Claudin–5, but in contrast to Claudin–1 the appearance is patchy. Strong immunoreactivity is found between the capillary endothelial cells in the choroid plexus (e) thin arrows). Magnification: × 1000.

Furthermore, with L-NAME but not with L-NMMA treatment (50 mg/kg) the specific mean grey value was reduced by 40%.

Studies with the TMA^+ method demonstrated that non-specific NOS inhibition as described above does not influence volume fraction and tortuosity of the ECS. Instead, the parameter that was altered by L-NAME (50 mg/kg) and L-NMMA (20 mg/kg) was the non-specific uptake parameter (k^1) as determined by the TMA^+ method (see Nicholson et al., this volume). 13% (L-NAME) and 16% (L-NMMA) increases were observed which correlated with the decreased volume of dextran spread. The impact of the increase in k^1 after L-NAME (50 mg/kg) on TMA^+ diffusion is illustrated in Fig. 7 by showing isoconcentration spheres of TMA^+, where TMA^+ reaches 0.1 μM 10 min after its application in the center of the sphere. A reduction of the sphere is produced by the L-NAME treatment.

As previously reported L-NAME reduced striatal blood flow and increased systemic mean arterial pressure (Jansson et al., 1999b). However, the effects on blood flow cannot explain the reduction in dextran volume spread, since it should rather increase the retention of dextran in the striatum. Also no leakage of blood proteins into the brain could be observed nor of Evans blue after NOS inhibition and after pharmacological treatment (phenylephrine) producing a similar increase of mean arterial pressure (MAP). Thus, the increases in MAP does not cause substantial disruption of the blood-brain barrier. Instead it seems likely that the changes in dextran spread and in non-specific uptake can be explained by increased brain capillary clearance from brain to blood through

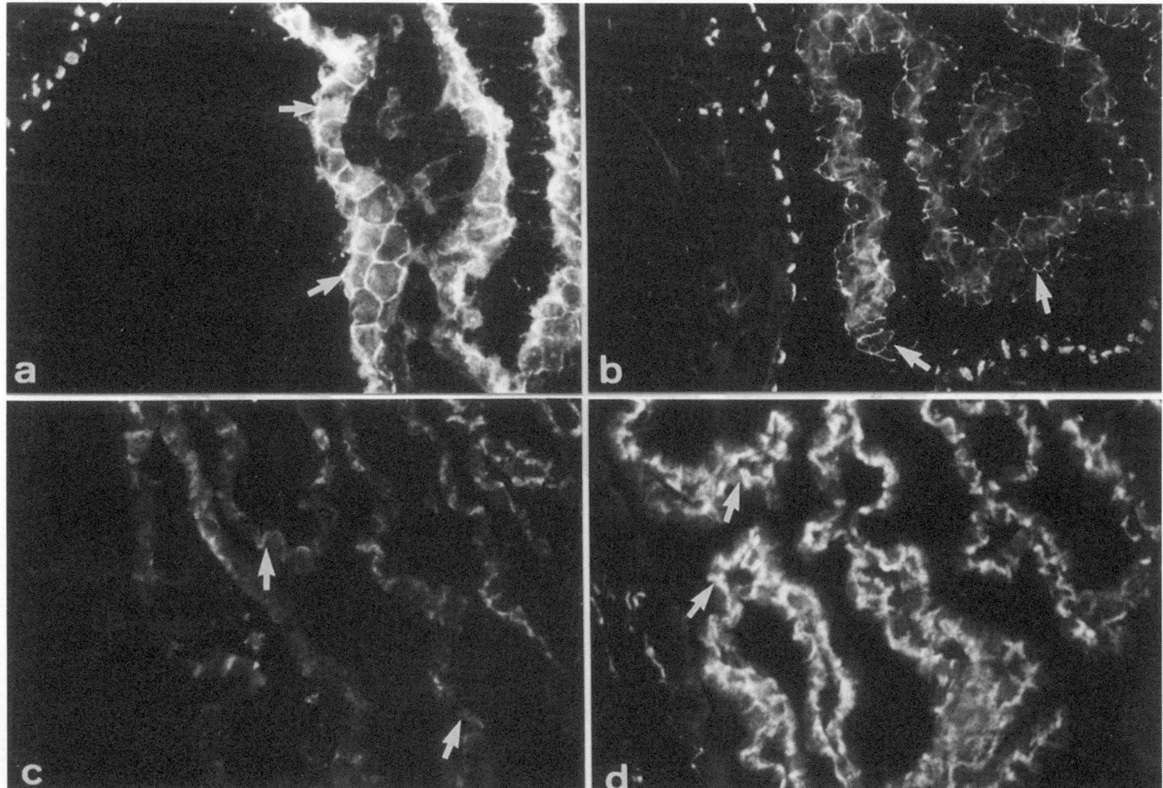

Fig. 5. Acute i.v.t. PMA-injections led to an increase in the immunoreactive area and an altered cellular localization of Claudin–1. The immunoreactivity is found to spread into cytoplasmic compartments after PMA treatment (a) arrows) compared to the sham treated animal (b). In contrast, Claudin–5 immunoreactivity decreased markedly after PMA treatment (c) arrows) compared to the control animals (d) arrows). Magnification: × 200.

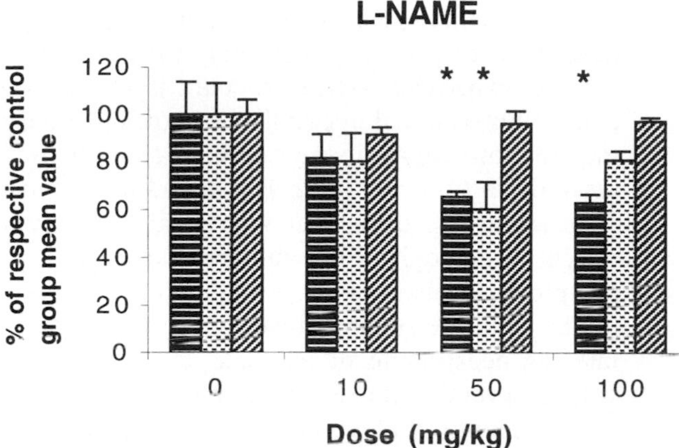

Fig. 6. Dose-related effects of L-NAME on the spread of biotinylated dextran (mol.wt. 3000), following microinjection into the neostriatum. The total volume (V_D) decreased dose-dependently and significantly following L-NAME as did the MGV ($p < 0.05$). L-NAME was given 10 min prior to the dextran microinjection and 30 min after dextran the rats were killed. The graph shows the mean relative change compared to the respective control value ± S.E.M. ($n = 3$–5 in each group). Statistical analysis was made with a one-way ANOVA followed by a PLSD test (*: $p < 0.05$). 100% for $V_D = 26$ mm^3; for MGV = 26). From Jansson et al., 1999b).

inhibition of endothelial NOS, neuronal NOS inhibition being without apparent effect. It seems possible that endothelial NO can normally restrict the transport of extracellular transmitters and modulators over the brain capillaries, while NO may not effect diffusion of tracers in the CNS (Fig. 8). Endothelial NO may exert its effects on brain capillary clearance through control of protein phosphorylation processes in junctional proteins (Rubin and Staddon, 1999).

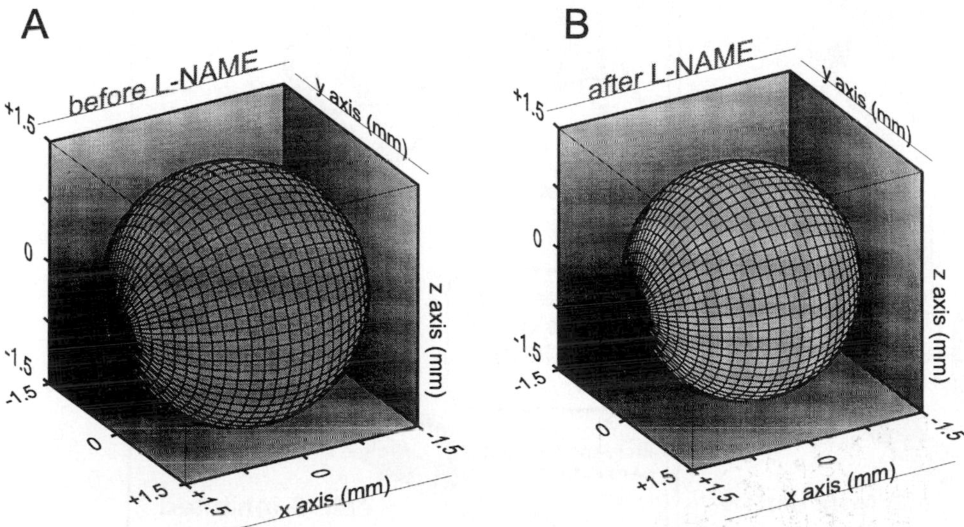

Fig. 7. Illustration of neostriatal ECS diffusion parameters, measured with the real-time iontophoretic TMA^+ method following L-NAME treatment (50 mg/kg, i.p.). Isoconcentration spheres are shown representing surfaces, where TMA^+ reaches an extracellular concentration of 0.1 μM 10 min after its application in the center of the sphere. Mean values of the ECS diffusion parameters (α, λ and κ^1) before and after L-NAME treatment as well as parameters typical for TMA^+ measurements were used. The isoconcentration sphere after L-NAME treatment is smaller than the isoconcentration sphere before the treatment (Jansson et al., 1999b).

Interleukin 1β as a long-distance endocrine-like VT signal

IL-1β is a very effective messenger in the immune system and is formed in relation to local tissue injury. In the brain its formation leads to fever, CRH release, inhibition of food intake and to the development of sickness behaviour, involving e.g. sleep promotion (Dinarello, 1994; Alheim and Bartfai, 1998). In view of its powerful actions on many neuronal-glial networks once formed and released from activated astroglia, microglia and endothelial cells it seems likely that IL-1β operates as a VT signal in brain that through diffusion and flow may reach and influence the many complex cellular networks involved in the above functions. In line with this view IL-1β extracellular levels are normally kept under rigorous control by the existence of soluble endogenous type I IL-1 receptor (R) antagonists and soluble and membrane bound type II IL–1R lacking the signalling component and acting as decoy receptors for IL-1β. In a series of experiments (Jansson et al., 1997; Jansson et al., 1999a) striatal microinjections of IL-1β have therefore been given in the halothane anaesthetised rat in doses of 5–100 ng in 100 nl and its possible diffusion analysed together with local and distant effects on microglia, FOS immunoreactivity and *c-fos* mRNA levels.

The evidence suggests that IL-1β microinjected into the neostriatum of one side can increase its volume several fold over a period of 3 hr in spite of its high m.wt (18 kDa); in this way indications for long-distance diffusion of IL-1β was obtained. Also, not only local microglia responses (Andersson et al., 1992) were observed (3 h) but also microglia activation in the paraventricular hypothalamic nucleus (PVH) bilaterally, where also hypertrophic astroglial cells were found. Furthermore, c-fos mRNA levels were only weakly

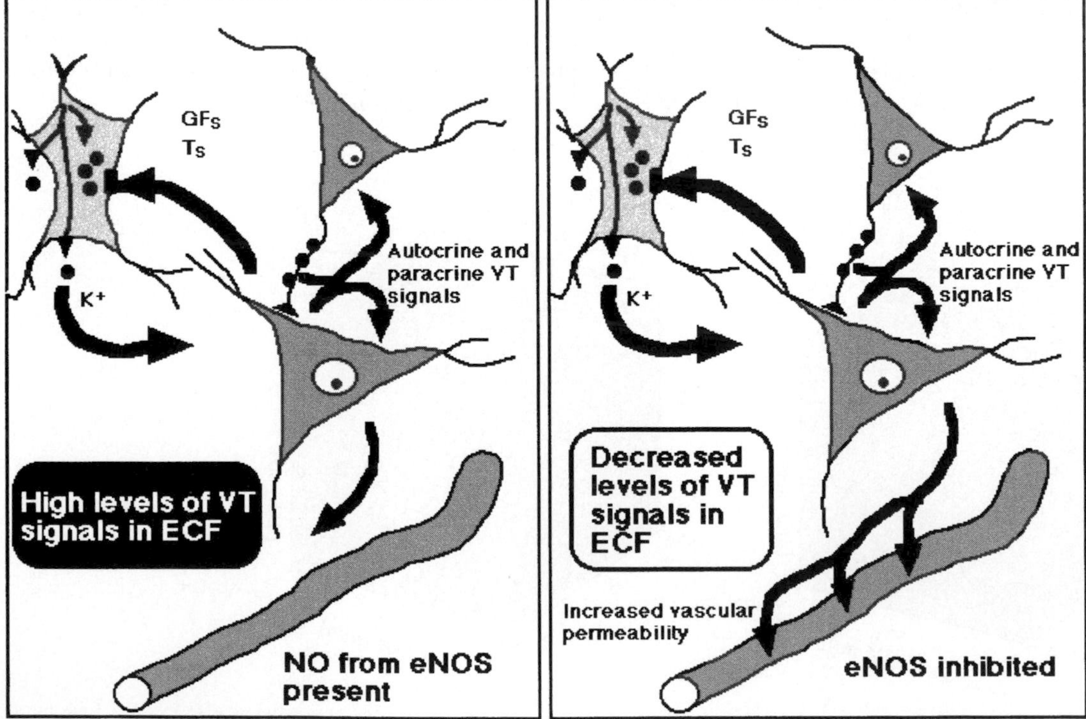

Fig. 8. The concentration of volume transmission (VT) signals present in the extracellular space (ECS) may be regulated by mechanisms controlling the brain to capillary permeability. Nitric oxide (NO) produced by endothelial nitric oxide synthase (eNOS) could be one important factor influencing VT signals in the ECS. An inhibition of eNOS may increase the capillary permeability for biotinylated dextran (M.W. 3000). GFs: growth factors, Ts: transmitters: K^+: potassium (Jansson, 1999b).

increased locally at the site of IL-1β injections but strongly increased in distant regions such as the PVH, suggesting IL-1β CSF signalling. In line with this view sleep has been found to be linked to CSF levels of IL-1β (Fang et al., 1998).

As indicated in Fig. 9, the activation of IL-1 type I receptors probably also leads to formation of prostaglandins in the meninges, in the ependymal cells, and in the plexus choroideus that contribute to the activation of FOS mechanisms in the cerebral cortex and in the PVH through diffusion via CSF into these brain areas. The results (Jansson et al., 1997; Jansson et al., 1999a) give rise to the hypothesis that IL-1β formed upon injury due, e.g. to inflammation in a local complex cellular network can via long distance diffusion and flow involving also CSF signalling (asynaptic endocrine-like VT, see Agnati and Fuxe, Chapter 1, this volume) exert its global actions on brain with production of fever, CRH release, sleep and and sickness behaviour etc. (for details, see (Jansson et al., 1999a)).

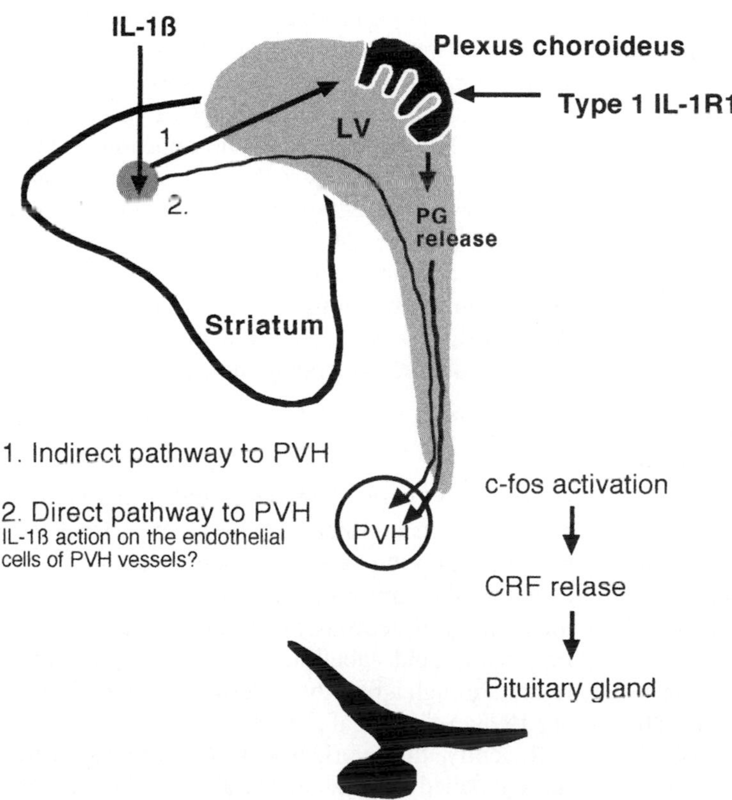

Fig. 9. Putative involvement of CSF signalling in the actions of IL–1β on the paraventricular hypothalamic nucleus, following its microinjection in the central dorsal striatum. Diffusing IL–1β may reach the ventricles over the brain-CSF interface. Thus, it may go with the CSF flow to the plexus choroideus where it may activate synthesis and release of prostaglandins that in turn via CSF can activate the paraventricular nucleus of the hypothalamus (PVH). CSF IL–1β may also directly effect the PVH through the existence of IL–1 receptors of type I in the microvessels of the PVH. The FOS activation will subsequently lead to the release of CRH (Jansson et al., 1999a).

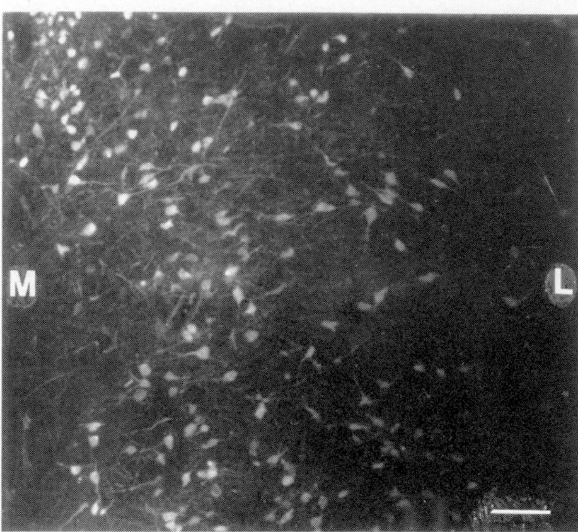

Fig. 10. Somatodendritic internalization of β-endorphin after its intraventricular injection (dose: 0.5 nmol in 5 μl; time: 10 min) in the rat into large unidentified paraventricular nerve cell body systems (Agnati et al., 1992). Scale bar: 100 μm; M = medial close to third ventricle. L = lateral.

Internalisation of intracerebrally injected β-endorphin and porcine galanin (1–29) into discrete nerve cell populations of the brain. Possible involvement in CSF signalling

β-endorphin

In 1986 Agnati et al. (Agnati et al., 1986), noticed the topographic transmitter-receptor mismatches for β-endorphin IR nerve terminals and μ and δ opioid receptors in several brain regions which led to the suggestion that β-endorphin was a VT signal. The potential involvement of β-endorphin in CSF signalling was indicated by the demonstration of high β-endorphin CSF levels after long distance running which correlated with euphoric actions (Radosevich et al., 1989).

In 1992 it was found that intraventricularly injected β-endorphin in the rat was taken up and internalised into the somadendritic regions of discrete nerve cell populations of e.g. the paraventricular hypothalamic and preoptic regions and of the hippocampal formation (Fig. 10) (Agnati et al., 1992). Subsequently, it was possible to demonstrate that intraventricular β-endorphin accumulates in dopamine and cAMP responsive phosphoprotein (DARPP-32) ir tanycytes that may represent potential transport pathways for CSF β-endorphin to reach distant opioid receptors (Bjelke and Fuxe, 1993). Bjelke (Bjelke, 1994) then demonstrated with intrastriatal β-endorphin injections that β-endorphin was internalised into the cytoplasm of DARPP-32 IR negative nerve cell bodies, giving evidence that striatal interneurons (lacking DARPP-32 IR) were selectively involved in this transient process with clearance within 30 min. It was then speculated that these events could represent a type of transcellular signalling, possibly involved in the regulation of gene transcription, but also with potential release of β-endorphin and/or its fragments into ECF from the dendritic processes and soma of these interneurons to activate striatal opioid receptors in the surround (Fuxe et al., 1994). Strong evidence for β-endorphin as a CSF signal came from the paper of Duggan's group (MacMillan et al., 1998) demonstrating that upon electrical stimulation of the arcuate nucleus (rich in β-endorphin cells) sharp rises of CSF β-endorphin level develop together with the appearance of β-endorphin IR in regions such as the cerebral cortex, where very few β-endorphin ir terminals exist (see Duggan, Chapter 24, this volume). β-endorphin therefore appears to be a CSF and ECF signal highly suited for long distance signalling with ability to globally affect brain function via the μ and δ opioid receptor populations.

Galanin

In previous work intraventricular injections of galanin or local microinjections of galanin (1.5 nmol/rat) into the ventral hippocampus have demonstrated the appearance of galanin IR in small to medium sized neurons of the hippocampal formation with rapid labelling of the somatodendritic region through a possible internalisation (Misane et al., 1998; Schött et al., 1998).

Recently, this work has been continued with a more detailed analysis of the above galanin internalisation phenomenon in hippocampal nerve cells (Jansson et al., 2000). In this study it could be demonstrated that both intraventricular and intrahippocampal microinjections of galanin (1–29) led to the selective and strong labelling of the same discrete hippocampal nerve cell population (Fig.

11). The labelling of these neurons occurs only through internalisation at the somatodendritic level with the dominant localisation in the cytoplasm of these nerve cells. The labelling was transient and disappeared between 20 to 60 min after injection (Fig. 12). A double immunolabelling analysis showed that the internalisation of galanin (1–29) was mainly restricted to GABA interneurons costoring NPY and/or somatostatin IR (Fig. 13). The galanin internalisation (1.5 nmol/rat) in these nerve cells appeared to be receptor-mediated, since it was partly blocked by the galanin receptor antagonist M35 (0.5 nmol) as evaluated following i.v.t. injections. Finally a labelling of the same hippocampal subpopulation was observed after i.v.t. fluo-galanin (1.5 nmol/rat) indicating that rapid activation of gene expression for galanin was not the explanation for the phenomenon observed. These results open up the possibility that CSF galanin signals can be internalised by this GABA/NPY/somatostatin interneuronal population where it may exert nuclear actions (gene regulation), possibly explaining the long-lasting behavioural actions of galanin (Ögren et al., 1992). Faure et al. (Faure et al., 1995) have obtained evidence that neurotensin upon somatodendritic internalisation shows targeting for perinuclear regions. Nevertheless, the exogenous galanin IR was mainly

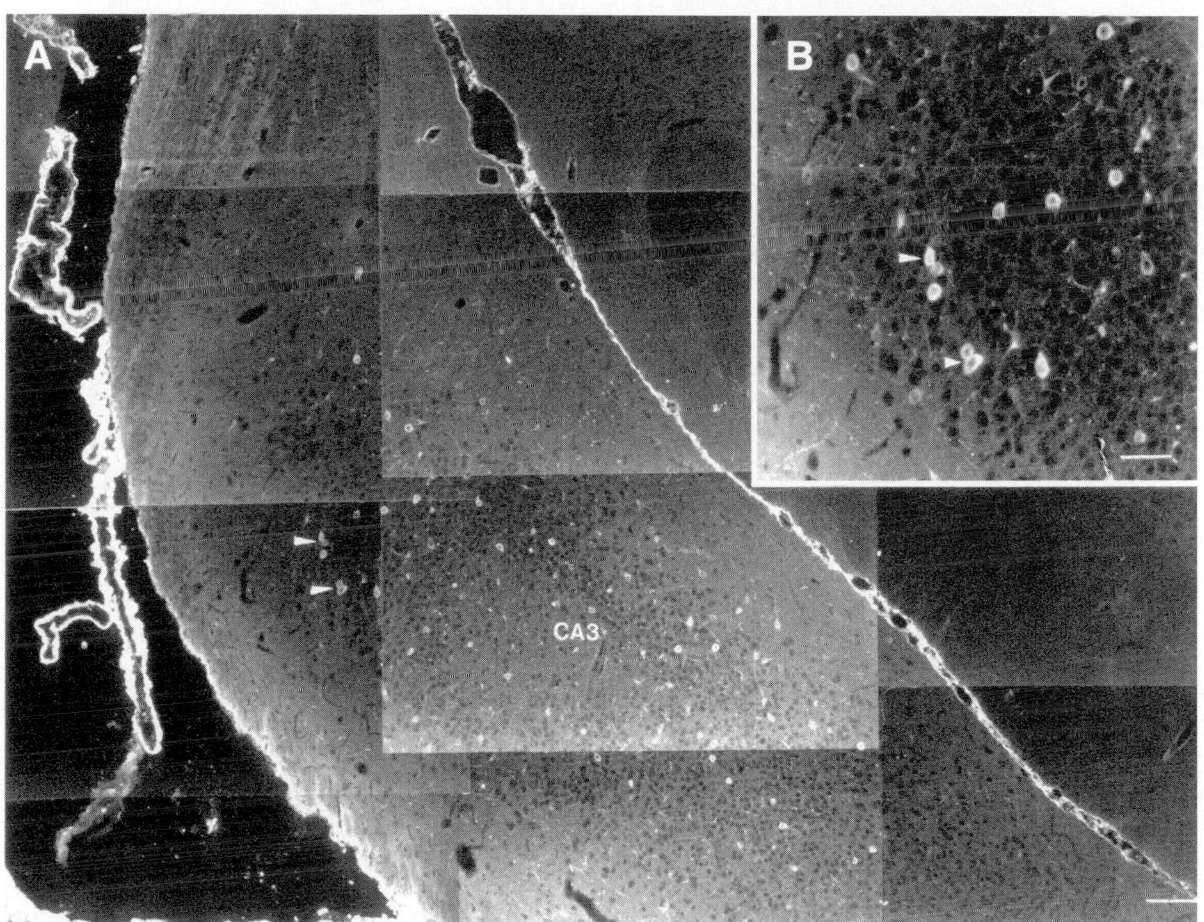

Fig. 11. Appearance of galanin-IR nerve cell bodies in ventral hippocampus 20 min following i.v.t. injection of 1.5 nmol porcine galanin. The small round galanin-IR nerve cell bodies were present mainly in the CA1 and CA3 regions of the ventral hippocampus. Arrowheads show the corresponding nerve cells in panels A and B. The epithelial cells of the plexus choroideus have become strongly galanin-IR. CA1: CA1 field of Ammon's horn. Bregma = –47 mm. Scale = 100 μm (A) and 50 μm (B) (Jansson et al., 2000).

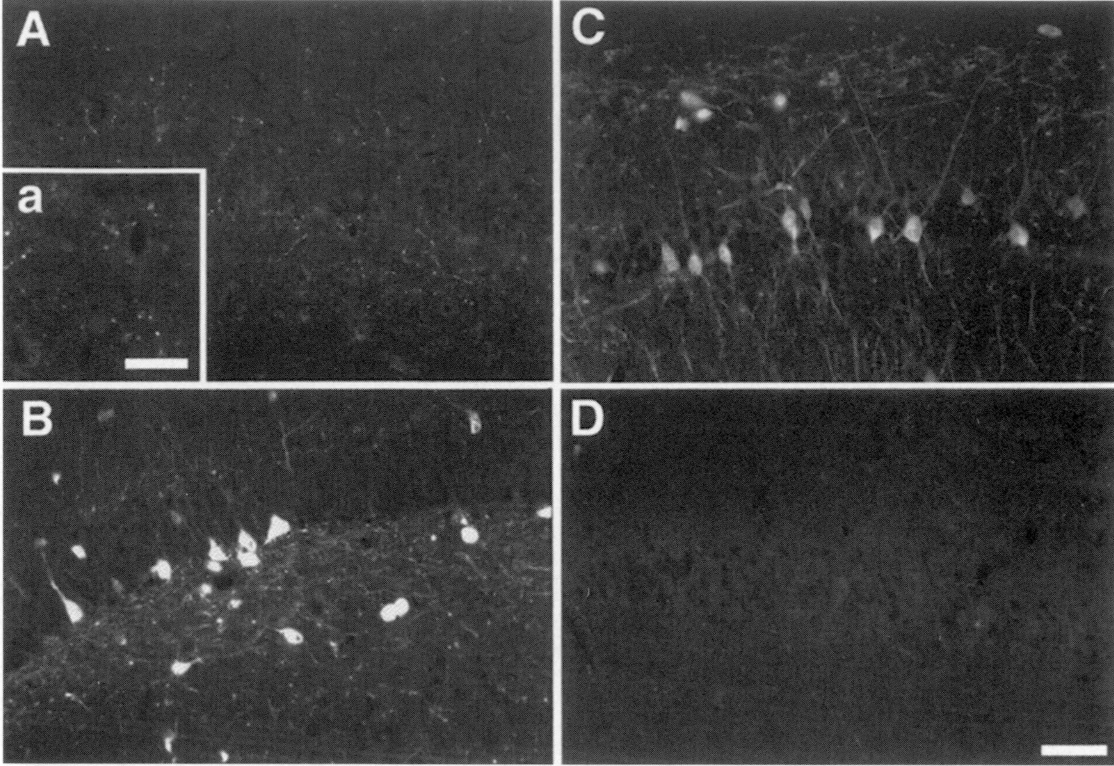

Fig. 12. Time course for the appearance of galanin-IR nerve cell bodies in the dorsal hippocampus (CA1 field) on the injected side following unilateral i.v.t. injection of porcine galanin (1.5 nmol/rat). After control injection of aCSF (20 min), only a few punctate galanin IR terminals were found(A). At 10 min (B) or 20 min (C) following galanin injections, galanin-IR developed in many small- to medium-sized rounded cell bodies and some pyramidal-like cells, mainly in the pyramidal and stratum oriens layers. At 60 min, very few if any cell bodies remained galanin-IR. Bregma = –3.6 mm. Scale bar = 50 μm (Jansson et al., 2000).

located in the cytoplasm and it must be considered that N-terminal biological active galanin fragments could be rapidly formed extra-and intracellularly from the GABA interneurons to activate the high affinity galanin N-terminal fragment binding sites in the dorsal hippocampal region (Hedlund et al., 1992). This could also contribute to the long lasting actions of galanin seen in vivo.

Conclusions

(1) Evidence has been presented that protein kinase C may be an important regulator of the tight and adherens junctions of the brain-CSF and CSF-blood interfaces as well as of endocrine-like VT.

(2) Endothelial NO may be a regulator of the clearance of VT signals over the brain-blood barrier.

(3) Brain interleukin–1β, when formed, may be a long distance endocrine-like VT signal, involving also CSF signalling, regulating in this way many neuronal-glial networks such as those for fever, food intake, sleep-wakefulness and sickness behaviours.

(4) Evidence has been obtained that exogenous β-endorphin and galanin (1–29) locally and in CSF can undergo rapid somatodendritic internalisation into discrete populations of interneuronal populations of the neostriatum and hippocampus, underlining their potential role in CSF signalling. These results support the possibility that at least in these specific interneurons β-endorphin and galanin and/or their active fragments can act as transcellular signals by effects on gene regulation and/or by rapid dendritic and somatic release into the

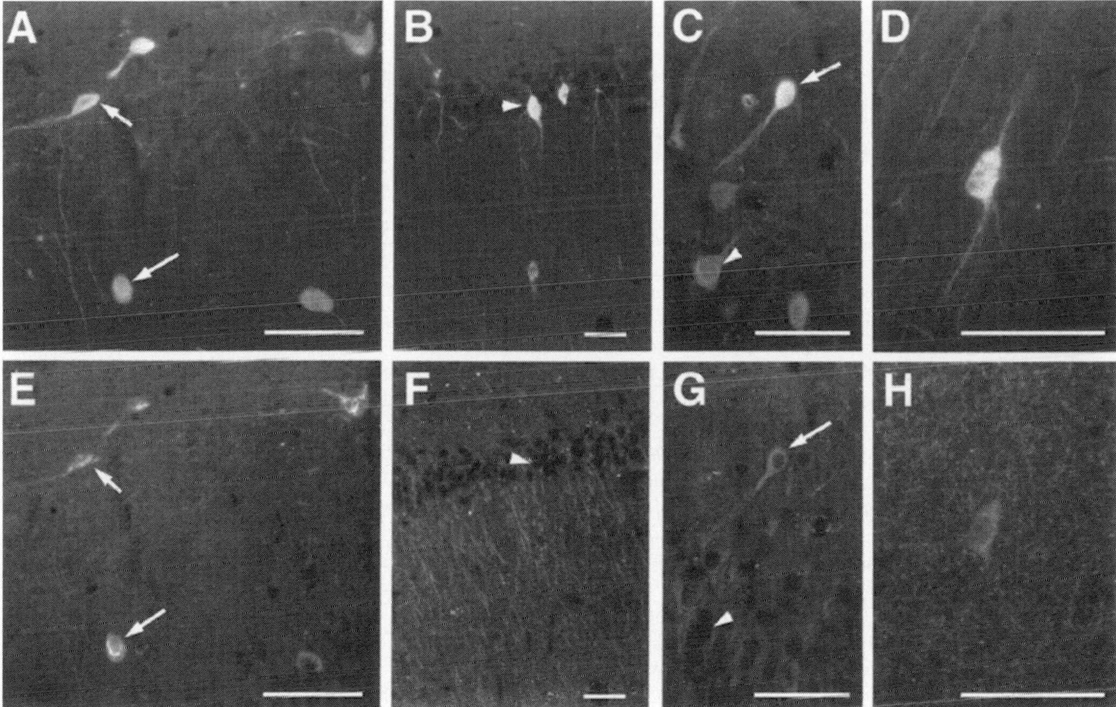

Fig. 13. Characterization of the galanin-IR nerve cell bodies appearing in the dorsal hippocampus, 10 minutes following i.v.t. injection of porcine galanin (1.5 nmol/rat) using double-labelling immunocytochemistry. In the first row galanin-IR nerve cell bodies with dendrites are shown (A–D). In panel E, some of the nerve cells are co-labelled with somatostatin-IR (arrows in A and E). The galanin-IR nerve cells were sometimes located close to 5-HT_{2A}-IR dendritic processes (arrowheads in B and F) and occasionally the galanin-IR nerve cell bodies were 5-HT_{2A}-IR (arrow in C and G). Some galanin-IR nerve cell bodies (D), were glutamic acid decarboxylase-IR. Bregma = –3.6 mm. Scale bar = 50 μm (Jansson et al., 2000).

surrounding neuropil to activate their respective high affinity peptide receptors.

Overall the present paper indicates that clearance and internalisation processes are important components of VT regulation.

Acknowledgement

This work has been supported by a grant (04X–715) from the Swedish Medical Research Council.

References

Agnati, L., Bjelke, B. and Fuxe, K. (1992) Volume transmission in the brain. Do brain cells communicate solely through synapses? A new theory proposes that information also flows in the extracellular space. *Am. Sci.*, 80: 362–374.

Agnati, L.F., Fuxe, K., Zoli, M., Ozini, I., Toffano, G. and Ferraguti, F. (1986) A correlation analysis of the regional distribution of central enkephalin and beta-endorphin immunoreactive terminals and of opiate receptors in adult and old male rats. Evidence for the existence of two main types of communication in the central nervous system: the volume transmission and the wiring transmission *Acta Physiol. Scand.*, 128: 201–207.

Alheim, K. and Bartfai, T. (1998) The interleukin-1 system: receptors, ligands, and ICE in brain and their involvement in the fever response. *Ann. NY Acad. Sci.*, 840: 51–58.

Anderson, J.M., Balda, M.S. and Fanning, A.S. (1993) The structure and regulation of tight junctions. *Curr. Opin. Cell Biol.*, 5: 772–778.

Anderson, J.M. and van Itallie, C.M. (1995) Tight junctions and the molecular basis for regulation of paracellular permeability. *Am. J. Physiol.*, 269: G467-G475.

Andersson, P.B., Perry, V.H. and Gordon, S. (1992) Intracerebral proinflammatory cytokines or leukocyte chemotaxins induces minimal myelomonocytic cell recruitment to the parenchyma of the central nervous system *J. Exp. Med.*, 176: 255–259.

Balda, M.S., Gonzales-Mariscal, L., Contreras, R.G., Macias-Silva, M., Torres-Marques, M.E., Garcia-Sainz, J.A. and Cereijido, M. (1991) Assembly and sealing of tight junctions: possible participation of G-proteins, phospholipase C, protein kinase C and calmodulin. *J. Membrane Biol.*, 122: 193–202.

Balda, M.S., Gonzales-Mariscal, L., Matter, K., Cereijido, M. and Anderson, J.M. (1993) Assembly oftight junctions: the role of diacylglycerol. *J. Cell Biol.*, 123: 293–302.

Bjelke, B. (1994) *Thesis*. Volume transmission. Experimental evidence for chemical communication via extracellular fluid pathways in the dopamine and β-endorphin neurons of the brain., Karolinska Institutet, Stockholm., ISBN 91-628-1330-7

Bjelke, B. and Fuxe, K. (1993) Intraventricular β-endorphin accumulates in DARPP-32 immunoreactive tanycytes. *NeuroReport*, 5: 265–268.

Brightman, M.W. and Reese, T.S. (1969) Junctions between intimately apposed cell membranes in the vertebrate brain. *J. Cell Biol.*, 40: 648–677.

Citi, S. and Denisenko, N. (1995) Pohsphorylation of the tight junction protein cingulin and the effects of protein kinsae inhibitors and activators in MDCK epithelial cells. *Cell Sci.*, 108: 2917–2926.

Dejana, E., Corada, M. and Lampugnani, M.G. (1995) Endothelial cell-to-cell junctions. *FASEB J.*, 9: 910–918.

del Rio-Garcia, J. and Smyth, D.G. (1991) Intracellular and extracellular processing of neuropeptides. In: K. Fuxe and L. Agnati (Eds), *Advances in Neurosciences, Volume Transmission in the Brain. Novel Mechanisms for Neural Transmission*, Vol. 1, Raven Press, New York, pp. 407–414.

Dermietzel, R. (1975) Junctions in the central nervous system of the cat. IV. Interendothelial junctions of cerebral vessels from selected areas of the brain. *Cell Tissue Res.*, 164: 45–62.

Dermietzel, R. and Krause, D. (1991) Molecular anatomy of the blood-brain barrier as defined by immunocytochemistry. *Int. Rev. Cytol.*, 127: 57–109.

Dermietzel, R., Meller, K., Tetzlaff, W. and Waelsch, M. (1977) In vivo and in vitro formation of the junctional complex in choroid epithelium. A freeze-etching study. *Cell Tissue Res.*, 181: 427–441.

Dinarello, C.A. (1994) Interleukin-1. *Adv. Pharmacol.*, 25: 21–51.

Dziegielewska, K.M., Hinds, L.A., Mollgard, K., Reynolds, M.L. and Saunders, N.R. (1988) Blood-brain, blood-cerebrospinal fluid and cerebrospinal fluid-brain barriers in a marsupial (Macropus eugenii) during development. *J. Physiol.*, 403: 367–388.

Fang, J., Wang, Y. and Krueger, J.M. (1998) Effects of interleukin-1 beta on sleep are mediated by the type I receptor *Am.J. Physiol.*, 274: R655–660.

Faure, M.P., Nouel, D. and Beaudet, A. (1995) Axonal and dendritic transport of internalized neurotensin in rat mesostriatal dopaminergic neurons *Neuroscience*, 68: 519–529.

Furuse, M., Fujita, K., Hiiragi, T., Fujimoto, K. and Tsukita, S. (1998b) Claudin-1 and-2: novel integral membrane proteins localizing at tight junctions with no sequence similarity to occludin. *J. Cell Biol.*, 141: 1539–1550.

Furuse, M., Hirase, T., Itoh, M., Nagafuchi, A., Yonemura, S., Tsukita, S. and Tsukita, S. (1993) Occludin: a novel integral membrane protein localizing at tight junctions. *J. Cell Biol.*, 123: 1777–1788.

Furuse, M., Sasaki, H., Fujimoto, K. and Tsukita, S. (1998a) A single gene product, claudin-1 or-2, reconstitutes tight junction strands and recruits occludin in fibroblasts. *J. Cell Biol.*, 143: 391–401.

Fuxe, K. and Agnati, L. (Eds) (1991) *Volume Transmission in the Brain. Novel Mechanisms for Neural Transmission*, Raven Press, New York.

Fuxe, K., Li, X.M., Bjelke, B., Hedlund, P., Biagini, G. and Agnati, L. (1994) Possible mechanisms for the powerful actions of neuropeptides. *Ann. NY Acad. Sci.*, 739: 42–59.

Giros, B., Jaber, M., Jones, S.R., Wightman, R.M. and Caron, M.G. (1996) Hyperlocomotion and indifference to cocaine and amphetamine in mice lacking the dopamine transporter *Nature*, 379: 606–612.

Hedlund, P.B., Yanaihara, N. and Fuxe, K. (1992) Evidence for specific N-terminal galanin fragment binding sites in the rat brain *Eur.J. Pharmacol.*, 224, 203–205.

Jansson, A. (1999a) *Thesis*. Experimental studies on the sources, pathways and targets for volume transmission in the rat brain., Karolinska Institute, Stockholm, ISBN 91-628-3454-1

Jansson, A., Andbjer, B., Tinner, B., Bartfai, T., Agnati, L.F. and Fuxe, K. (1997) Local injections of interleukin-1β into neostriatum and hypothalamus: implications for volume transmission. *Soc. Neurosci. Abstract*, 23: 1785.

Jansson, A., Mazel, T., Andbjer, B., Rosén, L., Guidolin, D., Zoli, M., Syková, E., Agnati, L. and Fuxe, K. (1999b) Effects of nitric oxide inhibition on the spread of biotinylated dextran and on extracellular space parameters in the neostriatum of the male rat. *Neuroscience*, 91: 69–80.

Jansson, A., Tinner, B., Andbjer, B., Razani, H., Wang, F.-H., Agnati, L.F., Ögren, S.O. and Fuxe, K. (2000) Internalization of intracerebrally administered porcine galanin (1–29) into discrete nerve cell populations of the hippocampus of the rat. *Exp. Neurol.*, 161: 153–166.

Kaya, M., Chang, L., Truong, A. and Brightman, M.W. (1996) Chemical induction of fenestrae in vessels of the blood-brain barrier. *Exp. Neurol.*, 142: 6–13.

Kuhn, H.G. and Svendsen, C.N. (1999) Origins, functions, and potential of adult neural stem cells. *Bioessays*, 21: 625–630.

Lampugnani, M.G. and Dejana, E. (1997) Interendothelial junctions: structure, signalling and functional roles. *Curr. Opin. Cell Biol.*, 9: 674–682.

Lippoldt, A., Jansson, A., Kniesel, U., Andbjer, B., Andersson, A., Wolburg, H., Haller, H. and Fuxe, K. (2000) Phorbol ester induced changes in tight and adherens junctions in the choroid plexus epithelium and in the ependyma. *Brain Res.*, 854: 197–206.

Lombardi, T., Montesano, R. and Orci, L. (1987) Phorbol ester induced diaphragmed fenestrae in large vessel endothelium in vitro. *Europ. J. Cell Biol.*, 44: 86–89.

Mack, A., Neuhaus, J. and Wolburg, H. (1987) Particular relationship between orthogonal arrays of particles and tight junctions as demonstrated in cells of the ventricular wall of the rat brain. *Cell Tissue Res.*, 248: 619–625.

MacMillan, S.A.J., Mark, M.A. and Duggan, A.W. (1998) The release of ß-endorhpin and the neuropeptide-receptor mismatch in the brain *Brain Res.*, 794: 127–136.

Misane, I., Razani, H., Wang, F.H., Jansson, A., Fuxe, K. and Ögren, S.O. (1998) Intraventricular galanin modulates a 5-HT1A receptor-mediated behavioural response in the rat *Eur.J. Neurosci.*, 10: 1230–1240.

Mitic, L.L. and Anderson, J.M. (1998) Molecular architecture of tight junctions. *Annu. Rev. Physiol.*, 60: 121–142.

Morita, K., Furuse, M., Fujimoto, K. and Tsukita, S. (1999) Claudin multigene family encoding four-transmembrane domain protein components of tight junction strands. *Proc. Natl Acad. Sci. USA*, 96: 511–516.

Mullin, J.M., Kampherstein, J.A., Laughlin, K.V., Saladik, D.T. and Soler, A.P. (1997) Transepithelial paracellular leakiness induced by chronic phorbol ester exposure correlates with polyp-like foci and redistribution of protein kinase C-alpha. *Carcinogenesis*, 18: 2339–2345.

Mullin, J.M., Soler, A.P., Laughlin, K.V., Kampherstein, J.A., Russo, L.M., Saladik, D.T., George, K., Shurina, R.D. and O'Brien, T.G. (1996) Chronic exposure of LLC-PK1 epithelia to the phorbol ester TPA produces polyp-like foci with leaky tight junctions and altered protein kinase C-alpha expression and localization. *Exp. Cell Res.*, 227: 12–22.

Ögren, S., Hökfelt, T., Kask, K., Langel, Ü. and Bartfai, T. (1992) Evidence for a role of the neuropeptide galanin in spatial learning *Neuroscience*, 51: 1–5.

Peretto, P., Merighi, A., Fasolo, A. and Bonfanti, L. (1999) The subependymal layer in rodents: a site of structural plasticity and cell migration in the adult mammalian brain. *Brain Res. Bull.*, 49: 221–243.

Peters, A. and Swan, R.C. (1979) The choroid plexus of the mature and aging rat: the choroidal epithelium. *Anat. Rec.*, 194: 325–353.

Radosevich, P.M., Nash, J.-A., Lacy, D.B., O'Donovan, C., Williams, P.E. and Abumrad, N.N. (1989) Effects of low- and high-intensity exercise on plasma and cerebrospinal fluid levels of ir-beta-endorphin, ACTH, cortisol, norepinephrine and glucose in the conscious dog. *Brain Res.*, 498: 89–98.

Rodriguez, P. and Bouchaud, C. (1996) The supra-ependymal innervation is not responsible for the repression of tight junctions in the rat cerebral ependyma. *Neurobiology*, 4: 185–201.

Rubin, L.L. and Staddon, J.M. (1999) The cell biology of the blood-brain barrier. *Ann. Rev. Neurosci.*, 22: 11–28.

Saunders, N.R., Habgood, M.D. and Dziegielewska, K.M. (1999) Barrier mechanisms in the brain.I. Adult brain. *Clin. Exp. Pharmacol. Physiol.*, 26: 11–19.

Schwartz, J.-C., Bouthenet, M.-L., Giros, B., Gros, C., Llorens-Cortes, C. and Pollard, H. (1991) Neuropeptidases and neuropeptide inactivation in the brain. In: K. Fuxe and L. Agnati (Eds), *Advances in Neurosciences, Volume Transmission in the Brain. Novel Mechanisms for Neural Transmission*, Vol. 1 Raven Press, New York, pp. 381–394.

Schött, P., Bjelke, B. and Ögren, S. (1998) Distribution and kinetics of galanin infused into the ventral hippocampus of the rat: relationship to spatial learning. *Neuroscience*, 83: 123–136.

Stuart, R.O. and Nigam, S.K. (1995) Regulated assembly of tight junctions by protein kinase C. *Proc. Natl Acad. Sci. USA*, 92: 6072–6076.

Tsukita, S. and Furuse, M. (1998) Overcoming barriers in the study of tight junction functions: from occludin to claudin. *Genes Cells*, 3: 569–573.

Tsukita, S. and Furuse, M. (1999) Occludin and claudins in tight junction strands: leading or supporting players? *Trends Cell Biol.*, 9: 268–273.

van Deurs, B. (1980) Structural aspects of brain barriers, with special reference to the permeability of the cerebral endothelium and choroidal epithelium. *Int. Rev. Cytol.*, 65: 117–191.

van Deurs, B. and Koehler, J.K. (1979) Tight junctions in the choroid plexus epithelium. A freeze-fracture study including complementary replicas. *J. Cell Biol.*, 80: 662–673.

Wolburg, H., Neuhaus, J., Kniesel, U., Krauß, B., Schmid, E.-M., Öcalan, M., Farrell, C. and Risau, W. (1994) Modulation of tight junction structure in blood-brain barrier endothelial cells. Effects of tissue culture, second messengers and cocultured astrocytes. *J. Cell Sci.*, 107: 1347–1357.

Wood, J. and Garthwaite, J. (1994) Models of the diffusional spread of nitric oxide: implications for neural niric oxide signalling and its pharmacological properties *Neuropharmacology*, 33: 1235–1244.

Zoli, M., Jansson, A., Syková, E., Agnati, L. and Fuxe, K. (1999) Volume transmission in the CNS and its relevance for neuropsychopharmacology. *Trends Pharmacol. Sci.*, 20: 142–150.

L. F. Agnati, K. Fuxe, C. Nicholson and E. Syková (Eds.)
Progress in Brain Research, Vol 125

CHAPTER 27

CSF signaling in physiology and behavior

Michael Lehman[1]* and Rae Silver[2]

[1]*Department of Cell Biology, Neurobiology and Anatomy, University of Cincinnati College of Medicine, 231 Bethesda Avenue, P.O. Box 670521, Cincinnati, OH 45267-0521, USA*
[2]*Departments of Psychology, Barnard College and Columbia University, 1190 Amsterdam Avenue, New York, NY 10027 and Department of Anatomy and Cell Biology, Columbia University College of Physicians and Surgeons, New York, NY 10032, USA*

Introduction

Cerebrospinal fluid (CSF) has traditionally been considered as a 'drainage system' for the brain, with CSF fluid components often seen as by-products of brain function, passively leaking into the ventricles. In contrast, recent evidence suggests a very different and dynamic view of CSF function, the possibility that CSF may serve as a communication pathway for substances that affect behavior and physiology. CSF signaling has sometimes been referred to as a variant of volume transmission. As discussed in other chapters, volume transmission is characterized by signal diffusion within the extracellular spaces of the brain that may occur via preferential pathways, one of these pathways being the CSF. Thus, CSF signaling has been termed 'endocrine-like' volume transmission (Agnati et al., 1995; Zoli and Agnati, 1996). Conceptually, the transmission of signals through the CSF provides a way of allowing very far reaching communication in the brain. Recent work provides compelling evidence that the CSF provides a route of communication for signals that mediate slow, global brain functions, such as sleep/wakefulness, circadian rhythms, sexual behavior and certain types of neuroendocrine responses. In this chapter, we review this evidence in light of a set of criteria for identifying a CSF signal and its physiological relevance.

*Corresponding author. Tel: 513 558 7628; Fax: 513 558 4343;
e-mail: michael.lehman@ac.edu

CSF signaling: the problem of the CSF as a 'sink'

For several reasons, the view that CSF may normally function as a signaling pathway has been considered heretical; the historical view is that CSF primarily serves as a 'sink' for brain fluids rather than a source (Davson, 1988). The CSF has also primarily been viewed as a cushion to protect the brain from traumatic injury, or a repository of nutrients for adjacent brain tissue. In addition to these assumptions, problems cited with CSF signaling include the lack of specificity of this signaling pathway due to presumed diffusion of the signal to potential targets. The ability of molecules to diffuse would be in proportion to their size, with smaller molecules travelling further into parenchyma than do larger ones; it seems unlikely that a signaling system would be based on the ability of molecules to diffuse based on their size. In addition, targets, located at a distance from the ventricles would be excluded. Finally, there is the problem of concentration: the CSF volume is large, while brain signals are released in small quantities, resulting in substantial signal dilution. Concentrations would differ with respect to the distance of the target from the source. How can it then be physiologically relevant?

Many of these objections have their root in a view of CSF dynamics based on traditional notions of bulk flow of CSF. In this view of the CSF as a 'sink', three possible states are postulated (Fig. 1). In the initial state (A), the concentration of signal in blood and choroid plexus is assumed to be maximal; Wwith infinite time and no flow of CSF (B), equal concentrations of the signal would be found in blood, brain and CSF compartments. However, with infinite time and bulk flow of CSF into venous blood (C), equilibrium could not be established between the choroid plexus blood and the CSF, so that the low concentrations in the CSF would act as a 'sink' for substances in the extracellular space of the brain. In this model, bulk flow makes it unlikely that signals could be released into the large volume of the CSF and still be capable of reaching appropriate targets.

What the traditional view does not consider is the possibility that there is regulated release of substances into CSF, as well as active uptake; in fact, in the models reviewed below there is substantial evidence that many neuropeptides and chemicals are released in the CSF in a regulated fashion. In addition, recent studies of the dynamics of CSF circulation (Greitz and Hannerz, 1996; Greitz et al., 1997) have cast doubt on the commonly accepted view of CSF bulk flow from the site of its production in the ventricles into the venous supply via the arachnoid granulations. Instead the main reabsorption of CSF appears to occur via extracellular spaces in the brain, suggesting that CSF signals released from the brain parenchymal space may readily find their way back.

Criteria for identifying a CSF signal

As in the case of synaptic transmission, it is useful to consider what might be the necessary and sufficient criteria to demonstrate the physiological role of a CSF signaling system. As summarized at a recent workshop ('The CSF as a communication pathway of the brain', 28th Annual Meeting of the Society For Neuroscience, Los Angeles, November 1998; see Nicholson, 1999), a simple but useful set of criteria might be divided into two parts: (1) documentation of the existence of the signal, and (2) confirmation that the CSF is the conduit for that specific signal. For the former, it is critical that the

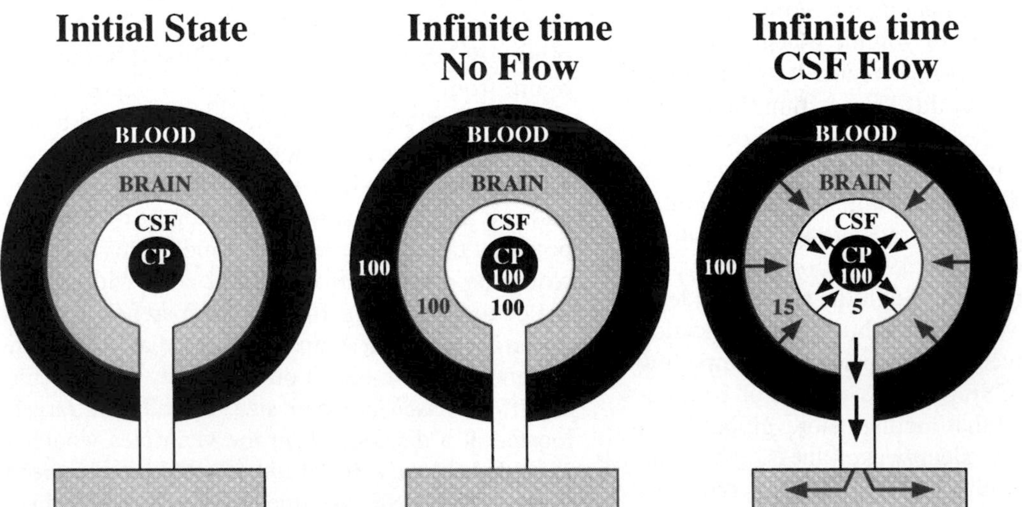

Fig. 1. Schematic of the 'sink' hypothesis of CSF flow. The black compartments represent blood in brain capillaries and the choroid plexuses (CP). In the initial state (left), the concentration of the extracellular marker is equal to 100 in blood and 0 in the other compartments. With infinite time and no flow of CSF (middle), an equilibrium is established between blood, brain, and CSF concentrations of the marker. With infinite time, the bulk flow of CSF into the venous blood is thought to produce a low concentration of tracer in the CSF, which acts as a sink for substances in the extracellular space of the brain. Modified from Davson, 1988.

signal be shown to be present in CSF and that its levels are correlated with behavior/physiology (assay evidence); in addition, one must show that removal and replacement of the signals affect physiology and/or behavior. Evidence that the CSF is the conduit for the signal include demonstration that the signal can gain access to the CSF, and that CSF fluid dynamics and turnover will allow appropriate movement of the signal. These are clearly minimal criteria, and, in addition, there are other aspects of the signaling pathway that are important to know; these include the source of the signal, its mechanism of release into the CSF, the location of targets for the signal, and the nature of its receptors.

It is important to recognize that these criteria can be applied to any signal in a large fluid volume, and might equally apply to the interstitial fluid of the brain in the context of volume transmission, and to the blood. In fact, it may be that these other compartments potentially compete with the CSF, leading to redundancy in communication channels. The possibility of redundant pathways is a general problem in evaluating the physiological role of CSF signals and will be revisited later in this chapter.

Evidence that CSF contents can alter behavior and/or physiology

In the following sections, we will consider evidence for CSF signaling in several different physiological systems: circadian rhythms, sleep, and reproduction (gonadotropin-releasing hormone and melatonin). For each we will review the strongest evidence for CSF signaling in light of the criteria outlined above. Although substantial evidence has accrued to suggest that CSF signaling is involved in these systems, we will also consider what other pieces of evidence remain to be gathered. Finally, in addition to examining these models, we will also review recent data regarding the role of mast cells in the physiological regulation of the blood brain barrier, a role of potential importance to CSF signaling as well as other types of volume transmission.

Circadian rhythms

Circadian rhythms are controlled by the suprachiasmatic nucleus. Circadian rhythms are a fundamental physiological adaptation to a changing 24 hour world. While intuitively we have the impression that daily rhythms of sleep and wakefulness are passive responses to changes in the external world, experiments in humans and other species show that these circadian rhythms are instead driven by endogenous clocks inside our bodies (Pittendrigh, 1960; Moore-Ede et al., 1982). The master clock driving behavioral and physiological rhythms in humans and mammals is located in the suprachiasmatic nucleus (SCN) of the hypothalamus. The SCN is both necessary and sufficient for a wide variety of circadian rhythms, ranging from rhythms in cognitive functions to those in autonomic and neuroendocrine events. Perhaps the most convincing evidence for the pacemaker role of the SCN comes from transplantation experiments where donor specific circadian rhythms are restored to arrhythmic SCN-lesioned hosts (Ralph et al., 1990; Lehman and Silver, 1994; Silver et al., 1996). In these studies, restored rhythms always exhibit characteristics of the donor clock, strongly suggesting that the transplanted SCN is the source of signals that communicate time of day information to the rest of the brain and body.

The suprachiasmatic nucleus produces a diffusible signal. Signals that reach the CSF have been implicated in the control of circadian rhythms in hamsters. Proof of the involvement of CSF signals derives from studies in which the SCN tissue was transplanted within a polymer capsule that permits diffusible signals to reach the host brain but does not permit outgrowth of neural efferents (Silver et al., 1996). Encapsulated SCN grafts remain viable and express neurochemical markers characteristic of the normal SCN in situ (Fig. 2); most importantly, these grafts were shown to restore donor-specific locomotor rhythms to lesioned hosts (Fig. 3). Contact with CSF appears to be essential for encapsulated graft tissue to survive and restore rhythms; capsules implanted intraparenchymally were found to contain only necrotic/gliotic remnants of the transplanted tissue. By contrast, in all cases where functional recovery was observed, capsules were located within the third ventricle. Since the SCN in the intact brain is located at the base of the hypothalamus adjacent to the third ventricle, these findings suggest that the CSF may be a normal route of communication for circadian output signals driving locomotor as well as other behavioral rhythms.

A role for diffusible signals in the control of circadian rhythms is further supported by experiments in hamsters in which knife cuts have been made to surgically isolate the SCN from the rest of the brain (Hakim et al., 1991; Nelms et al., 1999). Following creation of such hypothalamic islands, circadian rhythms in locomotor activity, drinking and heart rate persist, while neuroendocrine outputs are disrupted. Thus, the SCN may use different output pathways to control different rhythms; a diffusible signal may be sufficient for behavioral rhythms, such as activity, while neural efferents appear to be necessary for neuroendocrine rhythms.

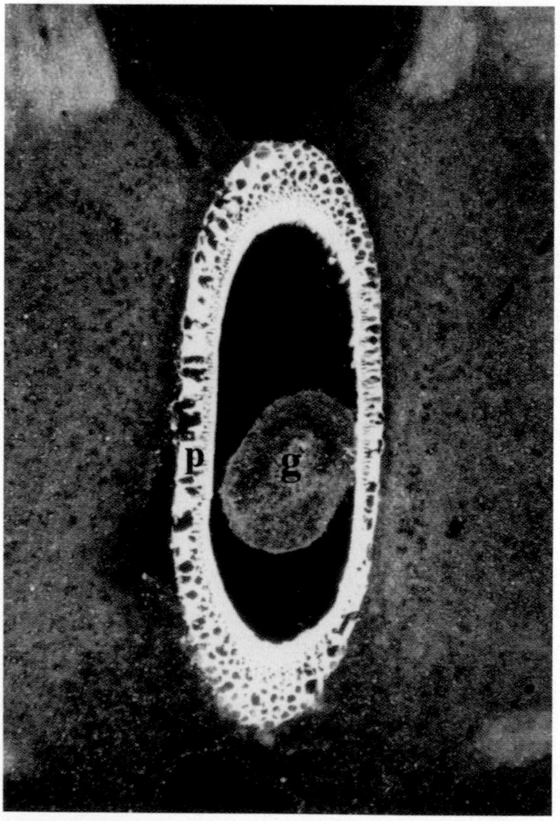

Fig. 2. Darkfield photomicrograph of a polymer-encapsulated SCN graft (g). Fetal anterior hypothalamic tissue containing the SCN was placed inside a polymer capsule, which was then heat sealed prior to implantation into the third ventricle. Encapsulated tissue survives and expresses neuropeptide markers of the SCN, such as vasopressin (yellow-gold immunolabeling in the graft). Magnification approx. 400 ×.

Summary. The strongest evidence supporting CSF signaling in the circadian systems are: (1) that the source of the signal is known; (2) the ability of polymer-encapsulated SCN grafts to restore rhythms; and (3) the persistence of rhythms after knife cuts which interrupt a majority of SCN efferents. The major missing piece of information for circadian CSF signaling is the identity of the signal itself. Although the output signal(s) remains to be identified, there is evidence that the SCN is capable of regulating secretion of signals into the CSF; for example, in rodents and primates, the SCN drives a rhythm of vasopressin secreted in the CSF (Schwartz and Reppert, 1985; Reppert et al., 1987). However, whether the precise source of CSF vasopressin is the SCN and/or other nuclei remains to be determined; furthermore, it is clear that VP is not the signal producing rhythmicity, although it may be a modulatory signal (Silver et al., 1999).

Other questions that need to be resolved include the location of the target. Results of studies varying the placement of SCN grafts in the ventricles (DeCoursey and Buggy, 1992; LeSauter et al., 1997), as well as the location of intraparenchymal implants of dissociated SCN neurons (Silver et al., 1990; Lehman et al., 1993), suggest that targets of SCN signals are located rostral to the normal location of the SCN in the anterior hypothalamus and/or medial thalamus. Consistent with being a target for CSF signals, these effective implantation sites are all located in relatively close proximity to the third ventricle.

Sleep

Sleep has two components – REM and non-REM. Modern definitions of sleep have focused on the nightly, cyclic succession of identifiable stages of sleep, characterized by recordings of EEG, eye movements and muscle tone. These stages are programmed in a predictable sequence each night, and are thought to be controlled by different but linked neural systems. The two major types of sleep seen in humans and other mammals are rapid eye movement (REM) sleep and slow-wave (non-REM) sleep; of these, the strongest evidence for CSF signals is in the regulation and control of non-REM sleep.

The idea that humoral signals regulate sleep has a very long history, and, in fact, dates back to Aristotle (see Kruger and Obal, 1997). The first experimental evidence for a sleep signal in the CSF derives from independent experiments performed in the early 1900s by Ishimori (1909) in Japan and Legendre and Pieron (1913) in France, both of whom induced sleep in dogs by injecting substances extracted from the brains of sleep-deprived

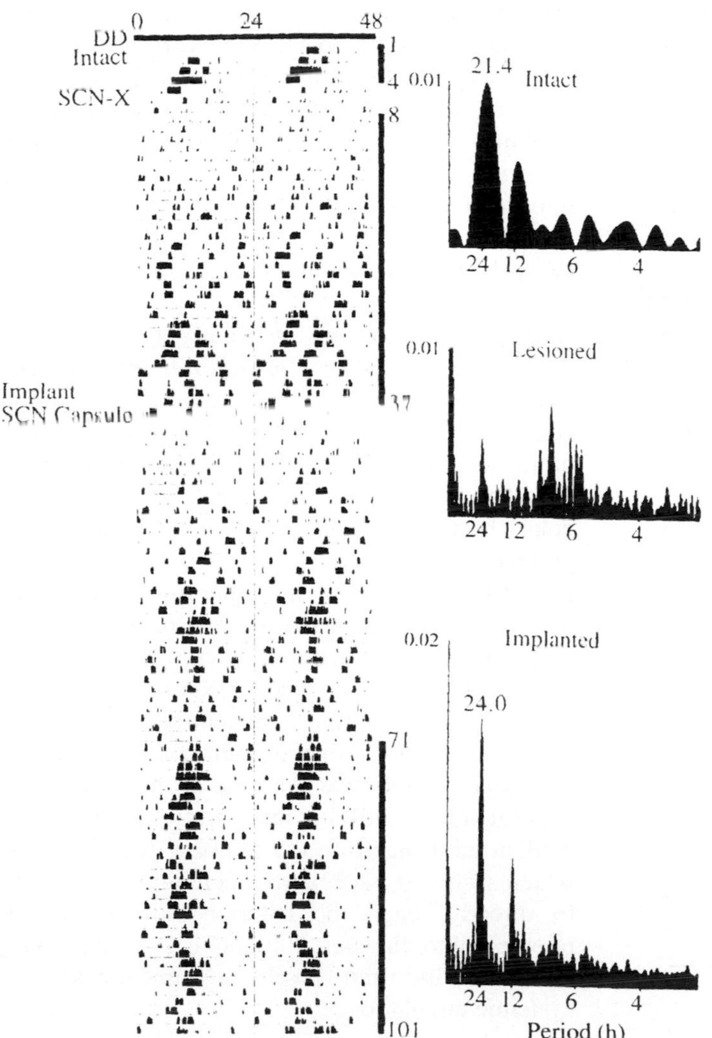

Fig. 3. Restoration of rhythmicity by a polymer-encapsulated SCN graft. The endogenous rhythm of the intact mutant host is seen at the top of the record (period 21.4 hr). After SCN lesions (SCN-x), the animal becomes arrhythmic. Implantation of the encapsulated graft occurs on day 37 of the record and the gradual emergence of circadian rhythmicity with the donor period (24 hr) is seen in the following weeks. The black vertical bars on the right of the behavioral record indicate the days (intact, lesioned and implanted) on which spectral analyses (right) were done to confirm the presence or absence of circadian rhythmicity. From Silver et al., 1996.

dogs. These early experiments were later replicated by Pappenheimer (1967), who showed that CSF taken from sleep-deprived goats was able to increase the percentage of non-REM sleep in rats (Fig. 4).

Since these earlier studies, there has been ample documentation of many factors that alter sleep propensity (Borbely, 1986; Krueger et al., 1990; Krueger and Obal, 1997). Among the best documented CSF-borne signals is the cytokine, IL-1 beta, which appears to fulfill all the criteria of a diffusible sleep signal; for example, the concentration of IL-1 beta in CSF varies with sleep and sleep deprivation (Krueger et al., 1998). Removal of IL-1 beta either by antibodies (Opp and Krueger, 1994) or receptor antagonists (Takahashi et al., 1996, 1999) inhibits non-REM sleep. Administration of IL-1 beta into CSF in a variety of species (rats, rabbits, cats, mice) induces an increase in non-REM sleep (Opp et al., 1991; Krueger et al., 1998). Finally, IL-1 receptor knockout mice do not respond to IL-1 beta, though they do respond to other substances regulating sleep, such as TNF alpha (Krueger et al., 1998). Hence the criteria for assay, removal, and replacement, are all met for IL-1 as a sleep-inducing CSF signal.

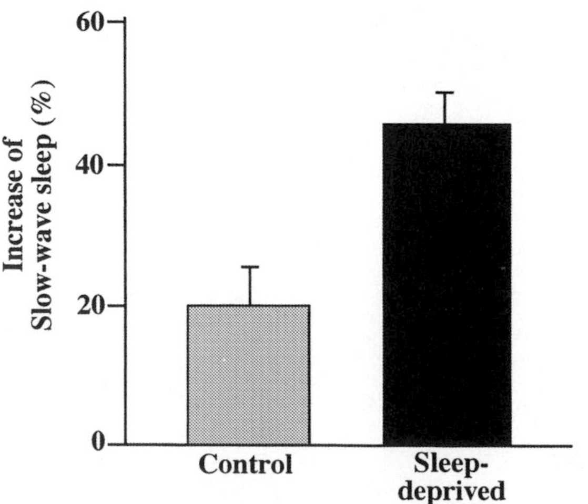

Fig. 4. Effects of CSF signals on non-rapid eye movement (REM) sleep . Graph illustrating Pappenheimer's original observation that the transfer of CSF from sleep deprived goats to rats induced excess non-REM sleep compared to the transfer of CSF from control goats. Modified from Krueger and Obal, 1997 and based on data from Fencl et al., 1971.

Summary. Despite the array of evidence supporting CSF signals in sleep, several key pieces of evidence remain to be revealed. For IL-1, as in the case of other sleep-inducing substances, the precise source of the signal, and its physiologically relevant targets, have yet to be determined. In part, this is because the potential sources and targets are so widespread, but in addition, it reflects the redundancy and complexity of the signaling pathways in the somnogenic cascade (Fig. 5). The existence of many parallel pathways in the humoral regulation of sleep presents an obstacle in clearly defining the functions of any single CSF signal; a similar problem may occur in the analysis of circadian rhythms, where both diffusible and neural output pathways serve as mechanisms for conveying information to targets (Lehman et al., 1999).

Reproduction: Gonadotropin Releasing Hormone

Gonadotropin-releasing hormone (GnRH) is a decapeptide that plays a pivotal role in the control of mammalian reproduction. Neurons that synthesize and release GnRH, and their axonal projections, represent the final common pathway by which a variety of environmental cues and endogenous hormonal signals, act to regulate the pituitary-gonadal axis (Silverman, 1988). Hence, GnRH neurons occupy a unique role in regulating the key neuroendocrine events responsible for mammalian reproduction, including puberty and the estrous/menstrual cycle. GnRH neurons control the secretion of pituitary gonadotropins by way of axonal projections that terminate in the median eminence, a neurohaemal zone at the base of the hypothalamus. GnRH is released from terminals in median eminence into the pituitary portal system, which serves as a specialized conduit for the signal to directly reach the anterior pituitary. GnRH released into the portal blood has an extremely short half-life and levels are undetectable in systemic circulation.

GnRH is secreted into both portal blood and CSF. There are two modes of secretion of GnRH in the portal blood: the pulsatile and surge modes; GnRH pulses consist of discrete episodes of elevated secretion, minutes in duration. The frequency of GnRH pulses is predominantly under negative feedback control by circulating gonadal steroids, and this so-called 'tonic' mode of secretion is present in both males and females. By contrast, the surge mode of secretion consists of a single, continuously elevated release of GnRH that lasts for hours and the amplitude of which is more than an order of magnitude greater than that of an individual GnRH pulse. The GnRH surge is the critical stimulus initiating a preovulatory gonadotropin surge, and is responsible for triggering ovulation during the estrous cycle (Goodman, 1994). Not surprisingly, the neural mechanisms controlling the occurrence of a GnRH surge are sexually differentiated and the surge mode of release occurs only in females.

In addition to its secretion into the portal blood, GnRH is also released into the CSF of the third ventricle (Van Vugt et al., 1985; Skinner et al., 1995). Measurements of CSF in sheep and cattle reveal that the pattern of release of GnRH into CSF, and its concentrations, mirrors that occurring in portal blood during both pulsatile and surge modes of secretion (Skinner et al., 1997; Gazal et al., 1998; Fig. 6. Hence, CSF GnRH is secreted in a regulated fashion by the same signals that control secretion into portal blood. Furthermore, there is a clear anatomic basis for the release of CSF GnRH, since in sheep (Lehman et al., 1986) as well as other species (Silverman, 1988), some GnRH axons penetrate the ependymal cell layer and directly contact the third ventricle (Fig. 7). CSF GnRH persists in rams in which the hypothalamus has been surgically disconnected from the pituitary (Skinner, unpublished data). Though this result suggests that ventricular GnRH may not arise from terminals of the median eminence, this experimental paradigm does not eliminate the possibility of neural regrowth of transected terminals.

The distinct functions of vascular and CSF GnRH are not known. Despite its presence and regulated release into CSF, the precise function of GnRH in this compartment remains unclear. CSF GnRH

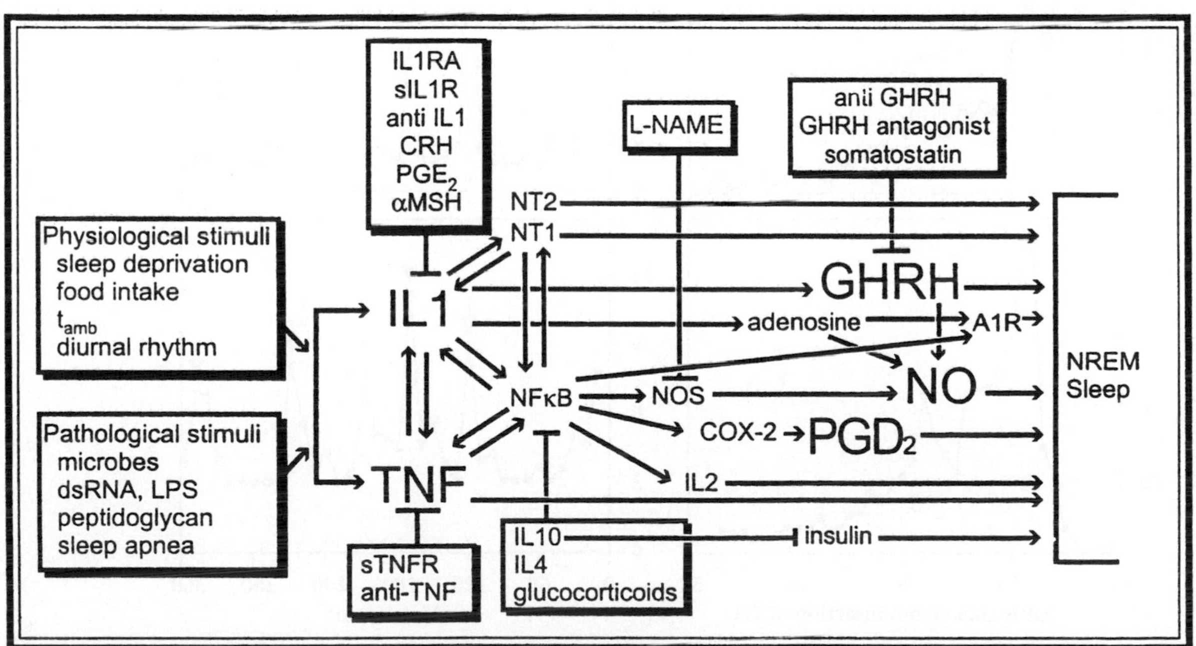

Fig. 5. Schematic diagram of the signaling pathways thought to regulate non-REM sleep. Note the presence of multiple, redundant pathways by which cytokines, such as IL–1 beta, and other sleep regulatory substances are able to modulate non-REM sleep. The presence of redundant pathways may provide stability to the sleep regulatory system in that the loss of any one step does not result in complete sleep loss. Diagram generously provided by James Krueger (Wash. State Univ.).

does not reach the pituitary, and hence is unlikely to participate in the regulation of pituitary gonadotropin secretion (Skinner et al., 1998); in addition, it has no effect on its own release, and thus does not comprise an ultrashort feedback loop regulating its own secretion (Skinner et al., 1998). One possible function is in the control of sexual behavior. In some species, GnRH may have two distinct functions: regulation of the hypothalamic-pituitary-gonadal axis and control of sexual behavior.

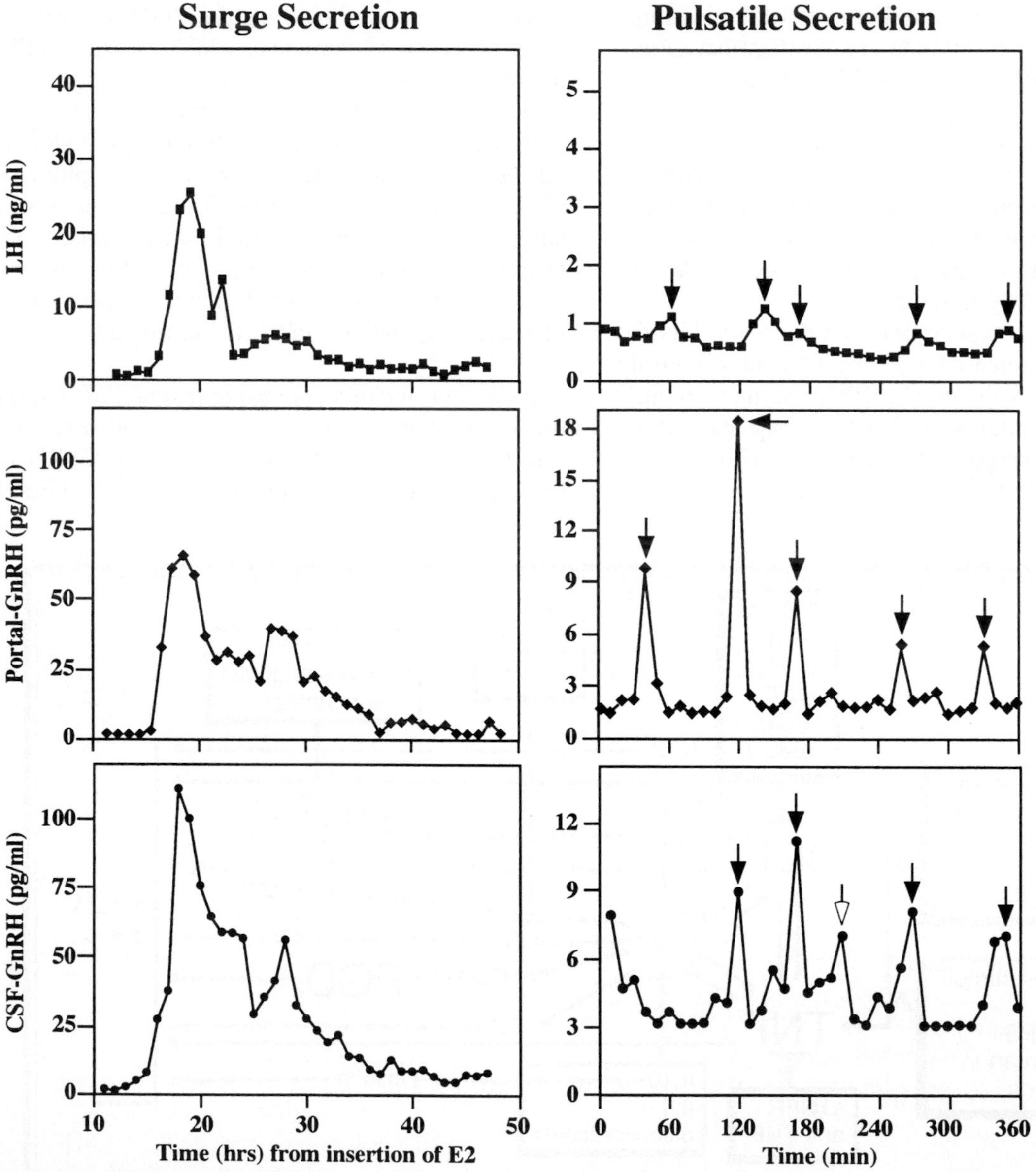

Fig. 6. Jugular LH, portal GnRH, and CSF GnRH secretory profiles in two individual ewes, one of which was sampled during an estradiol-induced surge (left panels), and the other during pulsatile secretion in the absence of steroid negative feedback (right panels). Note the coincident secretory patterns in all three compartments, but the longer duration of the portal and CSF GnRH surges compared to the LH surge. Arrows in the left panels indicate pulses; an open arrow denotes a CSF pulse not present in the other compartments. Modified from Fig. 1 and 2 in Skinner et al., 1997.

Behavioral functions of GnRH are less well documented than its neuroendocrine role, and primarily rest on experiments in which intraparenchymal or i.c.v. injections of GnRH have been shown to influence the display of sexual behavior (Sakuma and Pfaff, 1980; Fernandez-Fewell and Meredith, 1995). For example, infusion of GnRH in the midbrain central gray of female rats potentiates sexual behavior (lordosis reflex), whereas passive immunization against endogenous GnRH diminishes this behavior (Sakuma and Pfaff, 1980). In sheep, the same hormonal regimen that induces female estrous behavior, also produces a surge in CSF and portal GnRH (Skinner et al., 1999). Further, the duration of the GnRH surge observed in sheep is longer than that required for the preovulatory gonadotropin surge (Karsch et al., 1997), leading to speculation that the extended portion of the surge is instead important for priming estrous behavior. Data supporting this conclusion derives from a study in which the effect of GnRH on estrous duration was examined in hormonally treated ovariectomized ewes. Following the withdrawal of estrogen, administration of a GnRH antagonist (but not control injections) into the third ventricle resulted in a significant reduction of the duration of receptivity (Caraty et al., 1999). Therefore, one possibility is that the prolonged elevation in CSF GnRH seen during the surge is responsible for facilitating estrous behavior that normally accompanies ovulation.

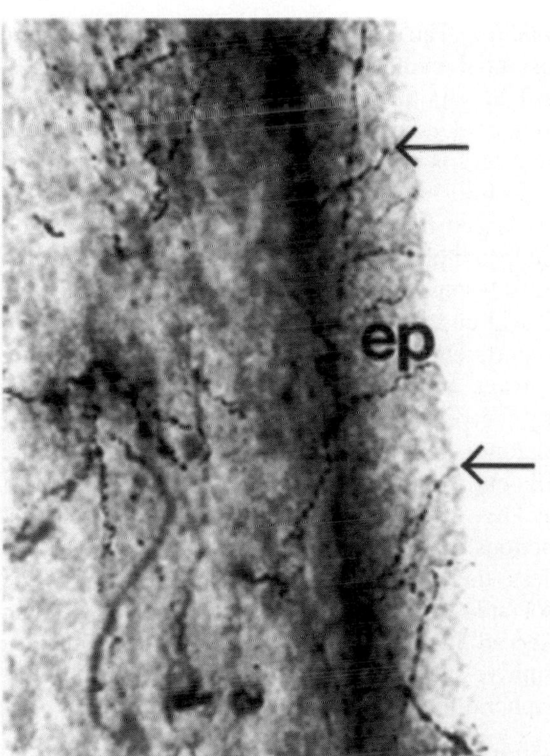

Fig. 7. Photomicrograph of gonadotropin-releasing hormone (GnRH) fibers in the sheep brain which penetrate the ependymal cell layer (ep) to directly contact the third ventricle (e.g. arrows). Modified from Lehman et al., 1986.

Summary. Strong points supporting the role of CSF GnRH as a physiological communication pathway are that both the signal and source (GnRH terminals contacting the CSF) are known; in addition, there is clear evidence of correlated changes between the signal and behavior/physiology from experiments in which both CSF and portal GnRH are sampled simultaneously. What is missing in this model is an understanding of the function of CSF GnRH; specifically, removal and replacement of CSF GnRH, and its consequences for estrous behavior, or other functions, have not yet been examined. Presumably, CSF GnRH may exert its effects upon target neurons that express GnRH receptors (Jennes et al., 1997). Interestingly, in the arcuate and ventromedial nuclei of the hypothalamus, GnRH receptor mRNA levels are highest during the early morning of proestrus and during the morning of a steroid-induced gonadotropin surge (Jennes et al., 1996), a time when estrous behavior is maximal. Hence CSF GnRH secreted at the time of the surge may diffuse to nearby targets bearing GnRH receptors in the medial hypothalamus. However, redundancy is also a potential problem in this system, since, in addition to fiber projections to the median eminence and wall of the third ventricle, GnRH neurons send axons to a number of regions, including the preoptic area and medial hypothalamus, where they form synapses (Silverman and Witkin, 1985; Langub et al., 1991; van der Beek et al., 1997). Hence, even if the relevant target is identified, determining the precise role of CSF GnRH vs. that released from local axon terminals may be problematic.

Reproduction: Melatonin

Melatonin regulates seasonal reproduction. Melatonin is a pineal hormone greatly popularized in the press for its putative role in puberty, aging and immune function. However, the best documented function of melatonin is in the regulation of reproduction in seasonally breeding mammals (see Malpaux et al., 1999). In species such as hamsters and sheep, day-length is the major environmental cue regulating reproduction, and this information is conveyed to the neuroendocrine axis by the duration of melatonin secretion. Specifically, melatonin is secreted in a diurnal rhythm in which release is highest during the night and lowest during the day. While melatoinin secretion is terminated by light, in most cases, under normal environmental conditions, melatonin secretion terminates while the animal is still in the dark; restated, the termination of melatonin secretion is under the regulation of an internal clock. The duration (not amplitude) of the melatonin signal has been shown to be the critical cue, both necessary and sufficient to control seasonal reproductive transitions (Bittman et al., 1983; Karsch et al., 1984). As noted below, the duration of melatonin secretion is identical in blood and CSF.

CSF has high melatonin levels. There is strong evidence that melatonin may exert some of its functions via a CSF route. Melatonin concentrations in the third ventricular CSF of sheep is at least 20 times higher than in jugular plasma (Skinner and Malpaux, 1999). and 100 times higher than levels in carotid artery plasma (Skinner and Malpaux, unpublished). Concentration of melatonin in the CSF of the lateral ventricle is 7-fold lower than that in third ventricle suggesting that melatonin is secreted directly into the third ventricle. The night-time melatonin rise in third ventricular CSF closely parallels that seen in peripheral circulation (Fig. 8).

This concentration gradient, with higher concentrations of CSF over blood values of melatonin, is completely unexplained by the view of the CSF as a brain drainage system. The foregoing results suggest that high melatonin levels in the CSF are a result of direct release into the third ventricle.

Sites of melanonin action in the brain are known. Brain targets of melatonin are well established in several species based on a variety of techniques including receptor distribution in the brain (Weaver et al., 1989), implants into discrete brain regions (Malpaux et al., 1998), and in situ hybridization for the receptor mRNA (Reppert et al., 1994). Many of these targets, including hypothalamic and limbic areas implicated in the control of seasonal breeding (Bartness et al., 1993; Maywood et al., 1996; Malpaux et al., 1998), are located in medial areas nearby the third ventricle; melatonin can easily diffuse out of CSF and out of vasculature, as it is highly lipophilic. Furthermore, relatively low levels of melatonin are sufficient to convey the information from day-length and to drive seasonal reproductive transitions (Bittman et al., 1983). Thus both CSF and blood are both effecive systems for delivery of the melatonin signal.

Summary. The melatonin system is one where substantial evidence has accumulated to suggest that CSF signaling may play a physiological role; strong evidence includes the fact that both the CSF signal and its source are known, that concentrations are 20 fold higher in CSF than in blood, and that there is a correlated change in the signal with a well established physiological function. What is missing in this system is evidence of the effects of selective removal and replacement of the signal in the CSF compartment; once again, redundancy of signaling pathways presents a potential problem. Specifically, given the highly lipophilic nature of melatonin, and the fact that relatively low levels are sufficient to convey the influence of daylength, it may be impossible to distinguish the distinct functions of signals of blood vs. CSF origin. Thus unless there is a specific function that can be associated with the high levels of melatonin observed in CSF, it may be difficult to make a definitive statement about the functions of CSF vs. peripheral melatonin.

Physiological role of mast cells in regulating the Blood Brain Barrier

Nature of mast cells. Mast cells (Fig. 9) are hematopoietic cells, originating in the bone marrow, first described by Paul Ehrlich in 1878; even in these early descriptions, mast cells were noteworthy for their 'stuffed' and swollen appearance. Today, mast cells are best known for their role in allergic reactions; they have the potential however, to secrete a large number of preformed and newly synthesized mediators. More recently it has become increasingly clear that mast cells have important functions in health and disease (see review by Galli et al.1999). As will be described below, among recent discoveries is their importance as a source of neuroactive and vascular mediators including NGF, histamine, and heparin; for example, in brain areas such as the thalamus, mast cells are responsible for 80% of histamine. Mast cells are the only source of endogenous heparin, and the physiological functions of mast cell heparin has now been identified (Humphries et al., 1999; Forsberg et al., 1999).

Types of mast cells, and their localization on the brain side of blood vessels. Mast cells are generally divided into serosal and mucosal sub-types. The occurrence of brain mast cells has been

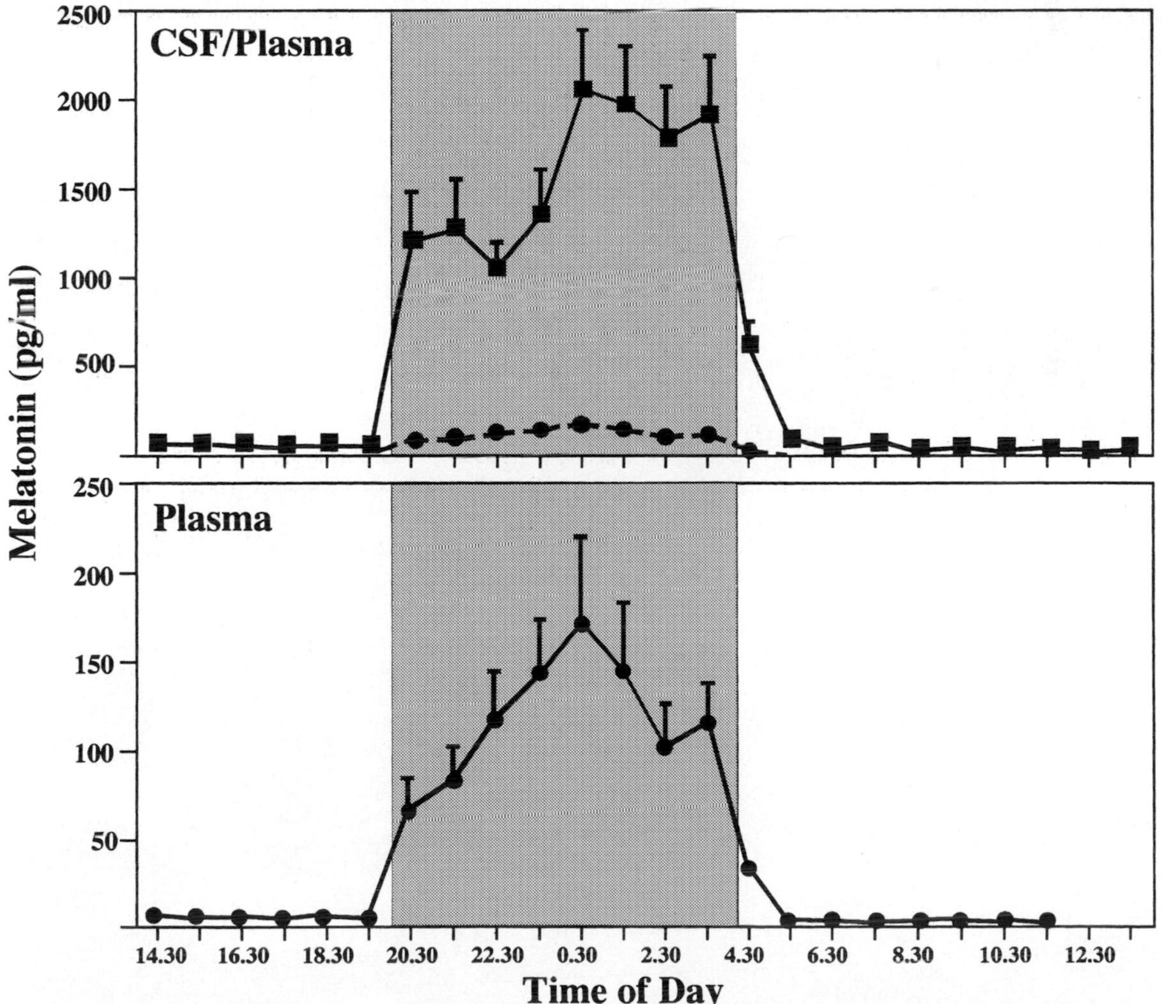

Fig. 8. Diurnal CSF and plasma melatonin levels in the sheep. Concentrations of melatonin in the third ventricle CSF (top, squares) are 20 times higher than those in the jugular plasma (bottom, circles). Plasma levels during nighttime (shaded area) are re-plotted in the top graph for direct comparison with the CSF nighttime rise. Modified from Fig. 1 in Skinner and Malpaux, 1999.

described in the older literature (Dropp, 1972) but only recently have mast cells in this region been considered as a possibly distinct sub-type. Mast cells enter the brain during the neonatal period (Lambracht Hall et al., 1990), and are frequently associated with blood vessels, generally lying on the brain side of the blood-brain barrier (BBB), in the Virchow-Robbins space, now commonly called the perivascular space. Mast cells are also found in the meninges, as well as on the ependymal walls of ventricles (Goldschmidt et al., 1984; Yang et al., 1999); they are also found within the brain parenchyma (Ronnberg et al., 1973; Edvinsson et al., 1977). CNS mast cells have been catalogued in a wide range of mammalian species, and variations in mast cell number with physiological state have been reported. For example, Kruger (1974) found that in adult Wistar rats mast cells were present in the olfactory lobes and in the diencephalon (thalamus and habenula). Alterations in mast cell number

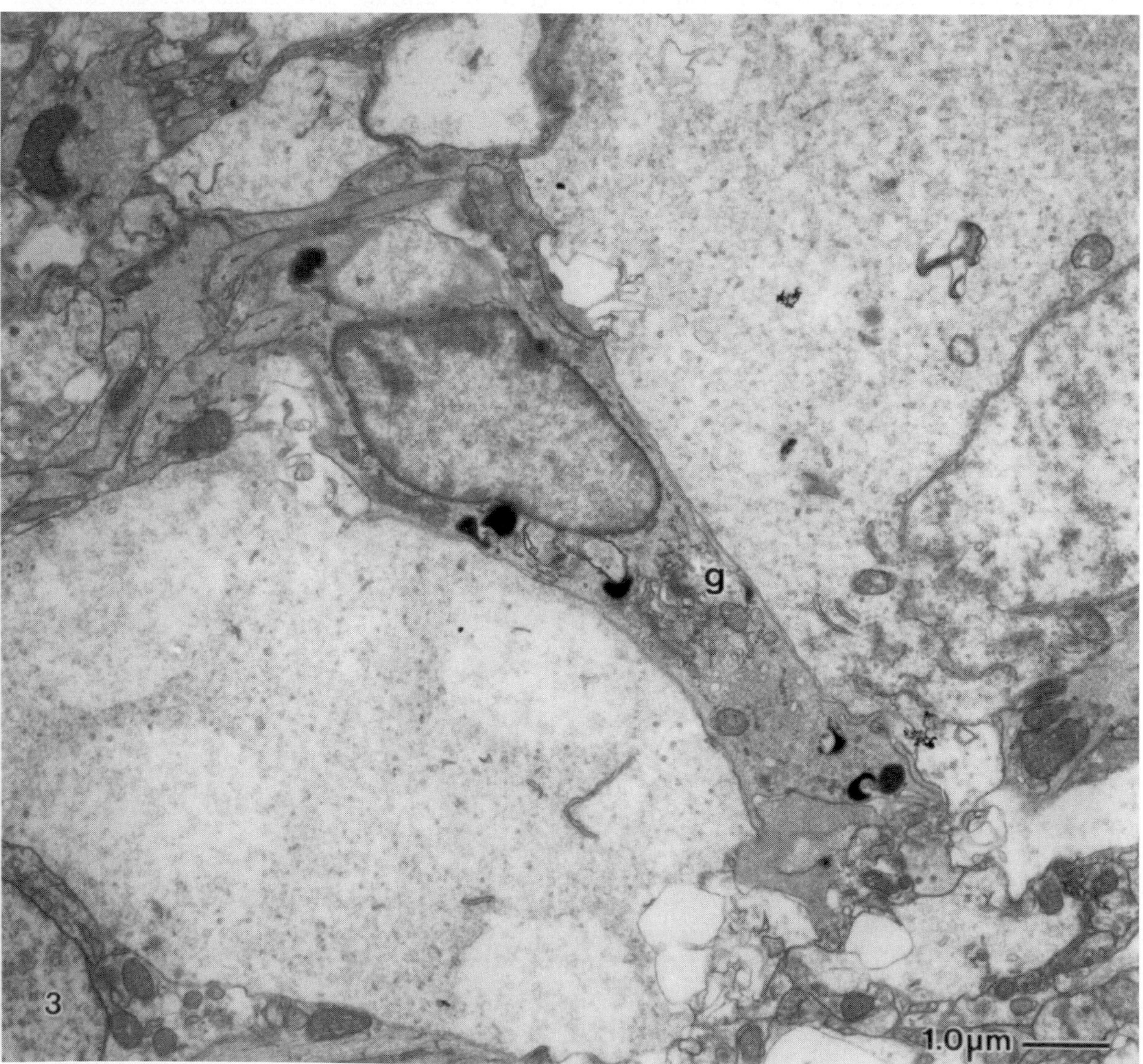

Fig. 9. Electron micrograph depicting a mast cell migrating between two cells of the ependymal wall. The presence of a large Golgi (g) apparatus in this cell suggests active biosynthesis.

with age or in response to environmental conditions have been documented in rats (Kruger 1974), musk shrew (Gill and Rissman, 1998), European hedgehog (Flood and Kruger,1970).

The possibility of a physiological role of mast cells as a mediator of volume transmission arises from the observation of a sudden increase in the appearance of brain mast cells immunostained for GnRH-like peptide, following a period of 2 hours of sexual behavior in doves (Silver et al., 1992; Zhuang et al., 1993, 1999; Silverman et al., 1994). A similar increase in the population of GnRH-positive mast cells in the thalamic region is seen in the male mouse following mating (Yang et al., 1999).

Migration of mast cells from periphery to the brain. While alterations in the population of brain mast cells have been described, it has been unclear how these changes are achieved; newly post-partum rats have significantly more mast cells in the thalamus than virgin controls (Silverman et al., 2000). Evidence from semi-thin sections from these female rats suggested that mast cells were transiting across the medium sized blood vessels. To explore whether the increases in mast cell number were due to their migration into the neural parenchyma, rat peritoneal mast cells were purified, labeled with the vital dyes PKH26 or CellTracker Green, and injected into host animals (Silverman et al., 2000). One hour after injection, dye-filled cells, containing either histamine or serotonin (stored mast cell mediators), were located close to thalamic blood vessels; injected cells represented approximately 2–20% of the total mast cell population in this brain region. Scanning confocal microscopy confirmed that the biogenic amine and the vital dye occurred in the same cell. That the donor mast cells were within the blood brain barrier was demonstrated by localization of dye-marked donor cells and either Factor VIII, a component of endothelial basal laminae, or glial fibrillary acidic protein, the intermediate filament found in astrocytes. Serial section reconstruction of confocal images demonstrated that the mast cells were deep to the basal lamina, in nests of glial processes. This is the first demonstration that mast cells can rapidly penetrate brain capillaries and may account for the rapid increases in mast cell populations following physiological manipulations.

Gonadal steroids degranulate mast cells. The increase in the dove brain mast cell population could be produced by exposure to gonadal steroid hormones (Zhuang et al., 1997; Wilhelm et al., 2000). Light microscopic immunocytochemistry, indicate an increased number of brain mast cells following exposure to either testosterone (T) or dihydrotestosterone (DHT) in the male, or 17beta estradiol (E) in the female, but not in cholesterol treated controls. Thus, an increase in the mast cell population of the habenula is produced by gonadal hormones in the absence of sexual behavior.

Electron microscopic studies further indicate that treatment with E, T, or DHT results in a significant increase in the per cent of cells in activated states compared to cholesterol treated controls; mast cell granules contain a wide range of biologically active molecules. The release of these granule contents into the neuropil of the CNS is likely to have wide ranging effects at multiple levels including vascular permeability and neuronal excitability, a subject of interest in the present context.

Mast cell degranulation alters brain leakiness. It has long been known that peripheral mast cells contain vasoactive mediators and that they can degranulate (releasing their granular contents) in a highly localized manner in response to triggering signals. In the periphery, the role of mast cell mediators in regulating vasodilation and vascular permeability has been well documented (Tharp 1989; Grega and Adamski, 1991). That the blood-brain barrier can be regulated by resident brain mast cells has been proposed, but not tested (Theoharides 1990; Manning, 1994; Purcell, 1995), although there is evidence of such a role for histamine (Domer et al., 1983; Schilling and Wahl 1994).

It has recently been demonstrated that activation of brain mast cells alters the permeability of the blood-brain barrier in the medial habenula (a brain region rich in mast cells) of doves. Since mast cells are well known for their production and secretion of vasoactive substances, animals were injected intramuscularly with the mast cell degranulator, compound 48/80, and were then injected with

Evans Blue intravenously. The distribution of the Evans blue dye in the parenchyma was examined using digital imaging. Three brain areas were analyzed: the medial habenula (which contains mast cells), the paraventricular nucleus (which abuts the third ventricle, but has no mast cells), and the lateral septal organ (a circumventricular organ with fenestrated capillaries). There was significantly more Evans blue tracer in the medial habenula of compound 48/80 treated subjects compared to saline controls (Fig. 10). While Evans blue did not enter the paraventricular nucleus in either experimental or control group, this occurred equally in the lateral septal organ in both groups. Degranulation of mast cells after compound 48/80 treatment was confirmed histochemically and ultrastructurally. The results support the hypothesis that mast cell degranulation can alter the permeability of the blood-brain barrier in the region of the medial habenula.

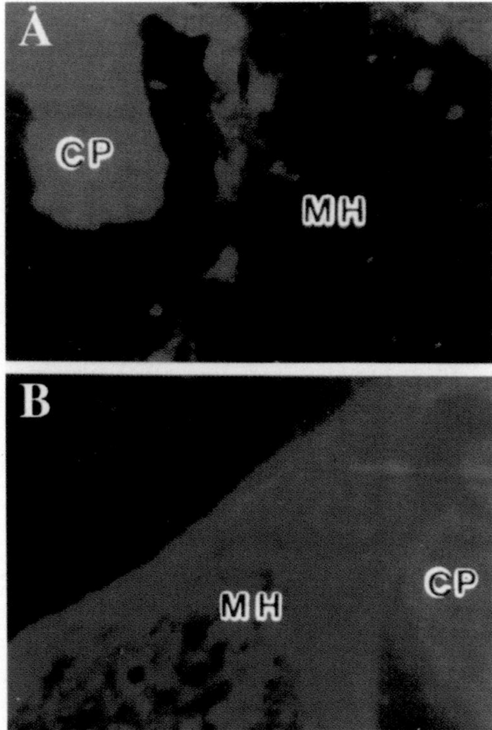

Fig. 10. Tracer leakage in the medial habenula of birds treated with a degranulating agent. A: No Evans blue tracer (red fluorescence) was seen in the medial habenula (MH) of control animals, indicating an intact blood-brain barrier in the absence of mast cell degranulation. In contrast, the tracer was seen in the adjacent choroid plexus (CP) which is known to lack the blood-brain barrier. B: In birds treated with a degranulating agent, compound 48/80 (C48/80), a significant amount of tracer can be seen in the medial habenula, as well as in the adjacent choroid plexus (CP).

Mast cells can release their contents in a highly localized fashion. Lawson et al. (1978) demonstrated that granules in a cultured mast cell could be activated to secrete histamine from those granules that contact a stimulant coupled to a large solid bead, while granules in distal parts of the mast cell remain intact. This indicates that the mast cell does not respond as a whole when it is triggered, but instead responds in a highly localized manner to localized signals; this also occurs in situ in the brain, and is the basis for highly regulated modulation of the leakiness of the brain vasculature. As shown in Fig. 11, mast cells lying within microns of each other on the same blood vessel in a mouse brain can be either activated to release their granules, or resting and filled with metachromatically stained granules.

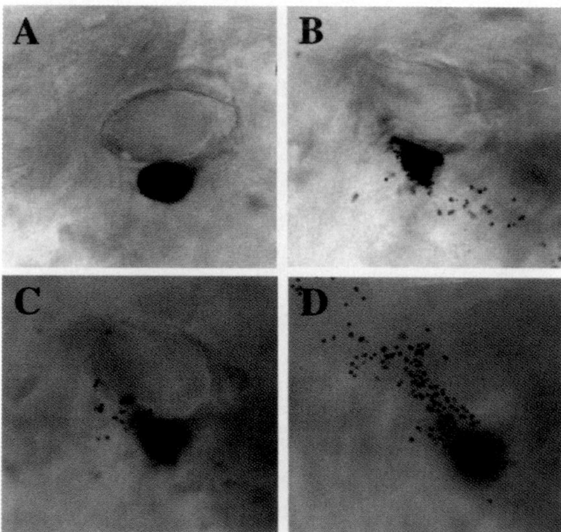

Fig. 11. Photomicrographs depicting mast cells lying near one another on the same blood vessel. The mast cell in A is fully granulated and at rest. Another lying on the same blood vessel is shown at 3 different focal planes (B-D) and is seen undergoing degranulation, with granules extending a substantial distance from cell. This pattern of granule release suggests that mast cells can be activated by local signals such as those provided by nearby neurons or glia.

Summary. Taken together, the foregoing results indicate that the BBB can be locally modulated; the physiological and/or behavioral consequences of such regulation awaits further examination. To date, the work on mast cells indicates: (1) that mast cell numbers are regulated under normal physiological conditions; (2) steroid hormones produce a two phase effect on brain mast cells; (3) initial exposure to steroids (T, DHT, E) results in an increase in the brain mast cell population; (4) ontinued exposure to steroids results in mast cell activation (degranulation); (5) mast cell degranulation opens the BBB indicating that the BBB is dynamically regulated; (6) mast cells provide a mechanism for local alterations in diffusion of the neural vasculature.

Summary of Model Systems and the Problem of Redundancy

A summary chart comparing the model systems discussed above (Table 1) reveals that while no single system meets all of the criteria for a physiological CSF signaling system, many come close. For example, in both model systems involving reproduction (GnRH and melatonin), the signal and its source are known, the signal is released into CSF in a regulated manner, and levels in the CSF are correlated with behavior/physiology. However, as discussed above, a major obstacle to determining the physiological relevance of CSF signals is the occurrence of redundant functional pathways; each signal found in CSF is necessarily also found in blood, and may have access to the same brain targets. Furthermore, even if the signal is in low concentrations in blood, it may be present in synapses (e.g. GnRH). One way of testing the role of CSF signals is by removal, but such studies by themselves are weak since they cannot account for non-specific effects of the intervention. Replacement experiments are necessary to complement removal studies, but these cannot distinguish between the physiologically relevant route by which the signal reaches it targets.

Why does such redundancy exist? One possible explanation is a phylogenetic one: during evolution, CSF signaling preceded neural signaling for specific behavioral/neuroendocrine functions. Supporting this explanation are comparative studies suggesting that CSF-contacting neurons are phylogenetically ancient and subserve similar fundamental brain functions in all vertebrates (Parent, 1981; Demski, 1987; Vigh and Vigh-Teichman, 1998). Another possible function of redundancy is that it gives stability; indeed, for critical neuroendocrine and behavioral functions, redundant signals/circuitry are common. For example, many more GnRH neurons are present in the normal mammalian brain than are necessary for pulsatile secretion (Kokoris et al., 1988). In addition, circulating levels of endogenous steroid hormones are generally higher than necessary to promote male and female sexual behavior (Damassa et al., 1977). Finally, the redundancy that exists in sleep promoting signaling pathways

TABLE 1. Summary of how model systems meet the criteria for CSF signalling

	Signal identification	Removal/ replacement	Correlation signal/ response	Signal access to CSF	Specific terminal at CSF
Circadian	?	yes	?	yes	?
Sleep	yes, many	yes	yes	yes, but source unknown	?
GnRH	yes	?	yes	yes	yes
Melatonin	yes	?	yes	yes	yes

probably functions to ensure that this critical function persists in the absence of any given signal.

Conclusions

Despite the caveats in our present knowledge, we consider the CSF route to be a viable hypothesis that should be further examined in the regulation of physiology and behavior; there are several aspects of existing evidence for CSF signaling that are particularly impressive. These include anatomical evidence of terminals contacting the CSF (GnRH system); observations of higher concentrations of signal in CSF than in plasma, which are contrary to the 'sink' hypothesis (melatonin system); and a correlation between behavior and CSF levels in cases where a correlation between behavior and blood levels is absent (CSF-GnRH and sex behavior). In addition, the ability of encapsulated tissue to restore function proves that CSF-borne signals are sufficient in at least one system, the circadian system, even though identity of the signal remains unknown.

Finally, it is interesting to speculate that communication via CSF signaling, may be particularly well suited to the coordination of global functions of the brain that occur over a relatively slower time course than that seen in synaptic communication. Hence, in addition to providing a route of communication for signals mediating sleep, circadian rhythms, and sexual behavior, the CSF may also be a conduit for signals controlling additional slow global functions of the CNS such as mood, hunger and nociceptive/antinociceptive mechanisms.

Acknowledgements

We thank Drs. James Krueger (Washington State Univ., USA), Donal Skinner (Cambridge, UK) and Benoit Malpaux (INRA, Nouzilly, FR) for sharing their unpublished data and enlightening conversations. We also thank Drs. Lique Coolen (Univ. Cincinnati, USA) Benoit Malpaux, Ann-Judith Silverman (Columbia Univ., USA) and Donal Skinner and for their excellent critical comments and Dr. Coolen for her assistance in preparing this manuscript. Dr.U. Shanas provided the light microscopic image in Fig. 11 during his post-doctoral training (RS lab). Supported by NIH NS35657 to M.N.L. and NIH MH29380 to R.S.

References

Agnati, L.F., Zoli, M., Stromberg, I. and Fuxe, K. (1995) Intercellular communication in the brain: wiring vs. volume transmission. *Neuroscience*, 69: 711–726.

Bartness, T.J., Powers, J.B., Hastings, M.H., Bittman, E.L. and Goldman, B.D. (1993) The timed infusion paradigm for melatonin delivery: what has it taught us about the melatonin signal, its reception, and the photoperiodic control of seasonal responses? *J. Pineal Res.*, 15: 161–190.

Bittman, E.L., Dempsey, R.J. and Karsch, F.J. (1983) Pineal melatonin secretion drives the reproductive response to daylength in the ewe. *Endocrinology*, 113: 2276–2283.

Borbely, A.A. (1986) Endogenous sleep-substances and sleep regulation. *J. Neural Transm. Suppl.*, 21: 243–254.

Caraty, A., Delaleu, B., Chesneau, D., Deghenghi, R. and Fabre-Nys, C. (1999) Sequential role of estradiol and the preovulatory GnRH secretion for the full expression of estrous behavior in the ewe. *Soc. Neurosc. Abstr.*, 25: 73.

Damassa, D.A., Smith, E.R., Tennent, B. and Davidson, J.M. (1977) The relationship between circulating testosterone levels and male sexual behavior in rats. *Horm. Behav.*, 8: 275–286.

Davson, H. (1988) History of the BBB Concept. In: E.A. Neuwelt (Ed.), *Implications of the Blood-Brain Barrier and Its Manipulation*, Vol. 1, Basic Science Aspects, Plenum Press, New York.

DeCoursey, P.J. and Buggy, J. (1992) Tissue and site specificity factors in restoration of circadian rhythmicity by fetal brain transplants to arrhythmic hosts: neuroendocrine aspects. In: T. Hiroshge, S. Fajimalo, and K. Honma (Eds), *Endocrine Chronobiology*. Hokkaido Univ. Press, Sapporo, Japan, pp. 46–63.

Demski, L.S. (1987) Phylogeny of luteinizing hormone-releasing hormone systems in protochordates and vertebrates. *Ann. NY Acad. Sci.*, 519: 1–14.

Domer, F.R., Boertje, S.B., Bing, E.G. and Reddix, I. (1983) Histamine- and acetylcholine-induced changes in the permeability of the blood-brain barrier of normotensive and spontaneously hypertensive rats. *Neuropharmacology*, 22: 615–619.

Edvinsson, L., Cervos-Navarro, J., Larsson, L.I., Owman, C. and Ronnberg, A.L. (1977) Regional distribution of mast cells containing histamine, dopamine, or 5-hydroxytryptamine in the mammalian brain. *Neurology*, 27: 878–883.

Ehrlich, P. (1878) *Beitrâge zur Theorie und Praxis der histologischen Fârbung* (PhD Thesis). Leipzig: University of Leipzig.

Fencl, V., Koski, G. and Pappenheimer, J.R. (1971) Factors in cerebrospinal fluid from goats that affect sleep and activity in rats. *J. Physiol.*, 216: 565–589.

Fernandez-Fewell, G.D. and Meredith, M. (1995) Facilitation of mating behavior in male hamsters by LHRH and

AcLHRH5-10: interaction with the vomeronasal system. *Physiol. Behav.*, 57: 213–221.

Flood, P.R. and Kruger, P.G. (1970) Fine structure of mast cells in the central nervous system of the hedgehog. *Acta Anat.*, 75: 443–452.

Forsberg.E., Pejler, G., Ringvall, M., Lunderius, C., Tomasini-Johansson, B., Kusche-Gullberg, M., Eriksson, I., Ledin, J., Hellman, L. and Kjellen, L. (1999) Abnormal mast cells in mice deficient in a heparin-synthesizing enzyme. *Nature*, 400: 773–776.

Galli, S.J., Maurer, M. and Lantz, C.S. (1999) Mast cells as sentinels of innate immunity. *Curr. Opin. Immunol.*, 11: 53–59.

Gazal, O.S., Leshin, L.S., Stanko, R.L., Thomas, M.G., Keisler, D.H., Anderson, L.L. and Williams, G.L. (1988) Gonadotropin-releasing hormone secretion into third-ventricle cerebrospinal fluid of cattle: correspondence with the tonic and surge release of luteinizing hormone and its tonic inhibition by suckling and neuropeptide *Y. Biol. Reprod.*, 59: 676–683.

Gill, C.J. and Rissman, E.F. (1998) Mast cells in the neonate musk shrew brain: implications for neuroendocrine immune interactions. *Dev. Brain Res.*, 111: 129–136.

Goldschmidt, R.C., Hough, L.B., Glick, S.D. and Padawer, J. (1984) Mast cells in rat thalamus: nuclear localization, sex difference and left-right asymmetry. *Brain Res.*, 323: 209–217.

Goodman, R.L. (1994) Neuroendocrine control of the ovine estrous cycle. In: E. Knobil and J.D. Neill, (Eds) *The Physiology of Reproduction*, 2nd ed., Vol. 2, New York: Raven Press Ltd., pp. 659–710.

Grega, G.J. and Adamski, S.W. (1991) Effects of local mast cell degranulation on vascular permeability to macromolecules. *Microcirc. Endo. Lymph.*, 7: 267–291.

Greitz, D. and Hannerz, J. (1996) A proposed model of cerebrospinal fluid circulation: observations with radionuclide cisternography. *Am. J. Neuroradiol.*, 17: 431–438.

Greitz, D., Greitz, T. and Hindmarsh, T. (1997) A new view on the CSF-circulation with the potential for pharmacological treatment of childhood hydrocephalus. *Acta Paediatr.*, 86: 125–132.

Hakim, H., DeBernardo, A.P. and Silver, R. (1991) Circadian locomotor rhythms, but not photoperiodic responses, survive surgical isolation of the SCN in hamsters, *J. Biol. Rhythms*, 6: 97.

Humphries, D.E., Wong, G.W., Friend, D.S., Gurish, M.F. and Stevens, R.L. (1999) Heparin-null transgenic mice are unable to store certain granule proteases in their mast cells. *J. Histochem. Cytochem.*, 47: 11645–11646.

Ishimori, K. (1909) True cause of sleep – a hypnogenic substance as evidenced in the brain of sleep-deprived animals. *Tokyo Igakkai Zasshi*, 23: 429–459.

Jennes, L., Eyigor, O., Janovick, J.A. and Conn, P.M. (1997) Brain gonadotropin releasing hormone receptors: localization and regulation. *Rec. Prog. Horm. Res.*, 52: 475–490.

Jennes, L., McShane, T., Brame, B. and Centers, A. (1996) Dynamic changes in gonadotropin releasing hormone receptor mRNA content in the mediobasal hypothalamus during the rat estrous cycle. *J. Neuroendocrinol.*, 8: 275–281.

Karsch, F.J., Bowen, J.M., Caraty, A., Evans, N.P. and Moenter, S.M. (1997) Gonadotropin-releasing hormone requirements for ovulation. *Biol. Reprod.*, 56: 303–309.

Karsch, F.J., Bittman, E.L., Foster, D.L., Goodman, R.L., Legan, S.J. and Robinson, J.E. (1984) Neuroendocrine basis of seasonal reproduction. *Rec. Prog. Horm. Res.*, 40: 185–232.

Kokoris, G.J., Lam, N.Y., Ferin, M., Silverman, A.J. and Gibson, M.J. (1988) Transplanted gonadotropin releasing hormone neurons promote pulsatile luteinizing hormone secretion in congenitally hypogonadal (hpg) male mice. *Neuroendocrinology*, 48: 45–52.

Krueger, J.M., Fang, J., Taishi, P., Chen, Z., Kushikata, T. and Gardi, J. (1998) Sleep: a physiologic role for IL–1 beta and TNF-alpha. *Ann. NY Acad. Sci.*, 856: 148–159.

Krueger, J.M., Obal, F., Opp, M., Toth, L., Johannsen, L. and Cady, A.B. (1990) Somnogenic cytokines and models concerning their effects on sleep, *Yale J. Biol. Med.*, 63: 157.

Krueger, J.M. and Obal, F. (1997) Sleep regulatory substances. In: W.J. Schwartz (Ed.), *Sleep Science: Integrating Basic Research and Clinical Practice*, Basel, Switzerland, Karger, pp. 175–194.

Kruger, P.G. (1974) Demonstration of mast cells in the albino rat brain. *Experientia*, 30, 810–811.

Lambracht Hall, M., Dimitriadou, V. and Theoharides, T.C. (1990) Migration of mast cells in the developing rat brain. *Dev. Brain Res.*, 56: 151–159.

Langub, M.C, Jr., Maley, B.E. and Watson, R.E., Jr. (1991) Ultrastructural evidence for luteinizing hormone-releasing hormone neuronal control of estrogen responsive neurons in the preoptic area. *Endocrinology*, 128: 27–36.

Lawson, D., Fewtrell, C. and Raff, M.C. (1978) Localized mast cell degranulation induced by concanavalin A-sepharose beads. Implications for the Ca^{2+} hypothesis of stimulus-secretion coupling. *J. Cell Biol.*, 79: 394–400.

Legendre, R. and Pieron, H. (1913) Recherches sur le besoin de sommeil consecutif a une vielle prolongee. *Z. Allg. Physiol.*, 14: 235–262.

Lehman, M.N., Nelms, J.L., LeSauter, J. and Silver, R. (1999) *Neural transplants and surgical isolation of the suprachiasmatic nucleus (SCN) as tools to study the 'hands' of the circadian clock.* 7th International Symposium on Neural Transplantation. Odense, Denmark.

Lehman, M.N. and Silver, R. (1994) Restoration of circadian rhythms by neural transplants. In: J. Marwah (Ed.), *Neuronal Transplantation*, CNS Neuronal Injury and Regeneration. CRC Press Inc. pp. 141–160.

Lehman, M.N., Tresco, P., LeSauter, J., and Silver, R. 1997. *Fetal brain grafts differentiate and restore behavioral function in the absence of neural connections or vascularization by the host brain.* Abstracts of the Sixth International Neural Tranplantation Meeting., San Diego, California.

Lehman, M.N., Zimmer, K.A., and Strother, W.N. (1993) Influence of the site of implantation on the restoration of circadian rhythmicity by dissociated cell grafts of the suprachiasmatic nucleus. *Soc. Neurosci. Abstr.*, 19: 1055.

LeSauter, J., Romero, P., Cascio, M. and Silver, R. (1997) Attachment site of grafted SCN influences precision of restored circadian rhythm. *J. Biol. Rhythms*, 12: 327–338.

Malpaux, B., Daveau, A., Maurice-Mandon, F., Duarte, G. and Chemineau, P. (1998) Evidence that melatonin acts in the premammillary hypothalamic area to control reproduction in the ewe: presence of binding sites and stimulation of luteinizing hormone secretion by in situ microimplant delivery. *Endocrinology*, 139: 1508–1516.

Malpaux, B., Daveau, A., Maurice, F., Gayrard, V. and Thiery, JC. (1993) Short-day effects of melatonin on luteinizing hormone secretion in the ewe: evidence for central sites of action in the mediobasal hypothalamus. *Biol. Reprod.*, 48: 752–760.

Malpaux, B., Thiery, J.C. and Chemineau, P. (1999) Melatonin and the seasonal control of reproduction. *Reprod. Nutr. Dev.*, 39: 355–366.

Manning, K.A. (1994) Histaminergic and non-histamine-immunoreactive mast cells within the cat lateral geniculate complex examined with light and electron microscopy. *Neuroscience*, 63: 191–206.

Maywood, E.S., Bittman, E.L. and Hastings, M.H. (1996) Lesions of the melatonin- and androgen-responsive tissue of the dorsomedial nucleus of the hypothalamus block the gonadal response of male Syrian hamsters to programmed infusions of melatonin. *Biol. Reprod.*, 54: 470–477.

Meyer-Bernstein, E.S., Jetton, A.E., Matsumoto, S., Markuns, J., Arendt, J., Lehman, M.N. and Bittman, E.L. (1999) Effects of suprachiasmatic transplants upon the circadian rhythms of neuroendocrine function in golden hamsters. *Endocrinology*, 140: 207–218.

Moore-Ede, M.C., Sulzman, F.M. and Fuller, C.A. (1982) *The Clocks That Time Us*, Harvard University Press, Cambridge.

Nelms, J.L., Silver, R. and Lehman, M.N. (1999) Do circadian rhythms of heart rate, body temperature, drinking and activity persist after isolation of the hamster SCN? *Soc. Neurosci. Abstr.*, 25:

Nicholson, C. (1999) Signals that go with the flow. *Trends Neurosci.*, 22: 143–145.

Opp, M.R. and Krueger, J.M. (1994) Anti-interleukin–1 beta reduces sleep and sleep rebound after sleep deprivation in rats. *Am. J. Physiol.*, 266: R688–695.

Opp, M.R., Obal, F. Jr. and Krueger, J.M. (1991) Interleukin 1 alters rat sleep: temporal and dose-related effects. *Am. J. Physiol.*, 260: R52–58.

Pappenheimer, J.R., Miller, T.B. and Goodrich, C.A. (1967) Sleep-promoting effects of cerebrospinal fluid from sleep-deprived goats. *Proc. Natl Acad. Sci.*, USA 1967 Aug; 58: 513–517.

Parent, A. (1981) Comparative anatomy of the serotoninergic systems. *J. Physiol. (Paris)*, 77: 147–156.

Pittendrigh, C.S. (1960) Circadian rhythms and circadian organization of living systems, *Cold Spring Harbor Symp. Quant. Biol.*, 25: 159.

Purcell, W.M. and Atterwill, C.K. (1995) Mast cells in neuroimmune function: neurotoxicological and neuropharmacological perspectives. *Neurochem. Res.*, 20: 521–532.

Ralph, M.R., Foster, R.G., Davis, F.C. and Menaker, M. (1990) Transplanted suprachiasmatic nucleus determines circadian period, *Science*, 247: 975–978.

Reppert, S.M., Schwartz, W.J. and Uhl, G.R. (1987) Arginine vasopressin: a novel peptide rhythm in cerebrospinal fluid. *Trends Neurosci.*, 10: 76–80.

Reppert, S.M., Weaver, D.R. and Ebisawa, T. (1994) Cloning and characterization of a mammalian melatonin receptor that mediates reproductive and circadian responses. *Neuron*, 13: 1177–1185.

Ronnberg, A.L., Edvinsson, L., Larsson, L.I., Nielsen, K.C. and Owman, C. (1973) Regional variation in the presence of mast cells in the mammalian brain. *Agents Act.*, 3: 191.

Sakuma, Y. and Pfaff, D.W. (1980) LH–RH in the mesencephalic central grey can potentiate lordosis reflex of female rats. *Nature*, 283: 566–567.

Schilling, L. and Wahl, M. (1994) Opening of the blood-brain barrier during cortical superfusion with histamine. *Brain Res.*, 653: 289–296.

Schwartz, W.J. and Reppert, S.M. (1985) Neural regulation of the circadian vasopressin rhythm in cerebrospinal fluid: a pre-eminent role for the suprachiasmatic nuclei. *J. Neurosci.*, 5: 2771–2778.

Silver, R., Lehman, M.N., Gibson, M., Gladstone, W.R. and Bittman, E.L. (1990) Dispersed cell suspension of fetal SCN restore circadian rhythmicity in SCN-lesioned adult hamsters. *Brain Res.*, 525: 45–58.

Silver, R., LeSauter, J., Tresco, P. and M. Lehman. (1996) A diffusible coupling signal from the transplanted suprachiasmatic nucleus controlling circadian locomotor rhythms. *Nature*, 382: 810–813.

Silver, R., Ramos, C.L. and Silverman, A-J. (1992) Sex behavior triggers appearance of non-neural cells containing gonadotropin-releasing hormone in doves. *J. Neuroendocrinol.*, 4: 207–210.

Silver, R.S., Sookhoo, A.I., LeSauter, J., Stevens, P., Jansen, H.T. and M.N. Lehman. (1999) Multiple regulatory elements result in regional specificity in circadian rhythms of neuropeptide expression in the mouse SCN. *NeuroReport*, 10: 3165–3174.

Silverman, A.J. (1988) The Gonadotropin-Releasing Hormone (GnRH) Neuronal Systems: Immunocytochemistry. In: E. Knobil and J. Neill (Eds), *The Physiology of Reproduction*. New York: Raven Press, Ltd., pp. 1283–1304.

Silverman, A.J., Millar, R.P. King, J.A., Zhuang, X. and Silver, R. (1994) Mast cells with gonadotropin-releasing hormone-like immunoreactivity in the brain of doves. *Proc. Natl Acad. Sci.*, USA, 91: 3695–3699.

Silverman, A.J., Sutherland, A.K., Wilhelm, M. and Silver, R. (2000) Mast cells migrate from blood to brain. *J. Neurosci.*, 20: 401–408.

Silverman, A.J. and Witkin, J.W. (1985) Synaptic interactions of luteinizing hormone-releasing hormone (LHRH) neurons in the guinea pig preoptic area. *J. Histochem. Cytochem.*, 33: 69–72.

Skinner, D.C., Caraty, A. and Evans, N.P. (1998) Does gonadotropin-releasing hormone in the cerebrospinal fluid modulate luteinizing hormone release? *Neuroendocrinology*, 67: 37–44.

Skinner, D.C., Caraty, A., Malpaux, B. and Evans, N.P. (1997) Simultaneous measurement of gonadotropin-releasing hormone in the third ventricular cerebrospinal fluid and hypophyseal portal blood of the ewe. *Endocrinology*, 138: 4699–4704.

Skinner, D.C. and Malpaux, B. (1999) High melatonin concentrations in third ventricular cerebrospinal fluid are not due to Galen vein blood recirculating through the choroid plexus. *Endocrinology*, 140: 4399–4405.

Takahashi, S., Kapas, L., Fang, J., Seyer, J.M., Wang, Y. and Krueger, J.M. (1996) An interleukin–1 receptor fragment inhibits spontaneous sleep and muramyl dipeptide-induced sleep in rabbits. *Am. J. Physiol.*, 271: R101–108.

Takahashi, S., Kapas, L., Fang, J. and Krueger J.M, (1999) Somnogenic relationships between tumor necrosis factor and interleukin–1. *Am. J. Physiol.*, 276: R1132–1140.

Tharp, M.D. (1989) The interaction between mast cells and endothelial cells. *J. Invest. Dermatol.*, 93 (2 Suppl.): 107S–112S.

Theoharides, T.C, (1990) Mast cells: the immune gate to the brain. *Life Sci.*, 46, 607–617.

van der Beek, E.M., Wiegant, V.M., van Oudheusden, H.J., van der Donk, H.A., van den Hurk, R. and Buijs, R.M. (1997) Synaptic contacts between gonadotropin-releasing hormone-containing fibers and neurons in the suprachiasmatic nucleus and perichiasmatic area: an anatomical substrate for feedback regulation? *Brain Res.*, 755: 101–111.

Van Vugt, D.A., Diefenbach, W.D., Alston, E. and Ferin, M. (1985) Gonadotropin-releasing hormone pulses in third ventricular cerebrospinal fluid of ovariectomized rhesus monkeys: correlation with luteinizing hormone pulses. *Endocrinology*, 117: 1550–1558.

Vigh, B. and Vigh-Teichman, I. (1998) Actual problems of the cerebrospinal fluid-contacting neurons. *Microsc. Res. Tech.*, 41: 57–83.

Weaver, D.R., Rivkees, S.A. and Reppert, S.M. (1989) Localization and characterization of melatonin receptors in rodent brain by in vitro autoradiography. *J. Neurosci.*, 9: 2581–2590.

Wilhelm, M., King, B., Silverman, A-J. and Silver, R. (2000) Gonadal steroids regulate the number and activational state of mast cells in the medial habenula. *Endocrinology*, in press.

Yang, M., Chien, C.L. and Lu, K-S. (1999) Morphological, immunohistochemical and quantitative studies of murine brain mast cells after mating. *Brain Res.*, 846: 30–39.

Zhuang, X., Silverman, A.J. and Silver, R. (1996) Brain mast cell degranulation regulates blood-brain barrier. *J. Neurobiol.*, 31: 393–403.

Zhuang, X., Silverman, A.J. and Silver, R. (1997) Mast cell number and maturation in the central nervous system: influence of tissue type, location and exposure to steroid hormones. *Neuroscience*, 80: 1237–1245.

Zhuang, X., Silverman, A.J. and Silver, R. (1999) Distribution and local differentiation of mast cells in the parenchyma of the forebrain. *J. Comp. Neurol.*, 408: 477–488.

Zhuang, X., Silverman, A.J. and Silver, R. (1993) Reproductive behavior, endocrine state, and the distribution of GnRH-like immunoreactive mast cells in dove brain. *Horm. Behav.*, 27: 283–295.

Zoli, M. and Agnati, L.F. (1996) Wiring and volume transmission in the central nervous system: the concept of closed and open synapses. *Prog. Neurobiol.*, 49: 363–380.

SECTION VI

Summary

L. F. Agnati, K. Fuxe, C. Nicholson and E. Syková (Eds.)
Progress in Brain Research, Vol 125

CHAPTER 28

Volume transmission in the year 2000

Charles Nicholson

Department of Physiology and Neuroscience, New York University School of Medicine, 550 First Avenue, New York, NY 10016, USA

Introduction

Since the publication of 'Volume Transmission in the Brain', edited by Fuxe and Agnati in 1991, the concept has undergone a metamorphosis. A decade ago, volume transmission (VT) was a stimulating but controversial concept in search of experimental evidence; today VT has become a largely accepted idea with many compelling examples. This final chapter will comment on some of the issues arising from recent studies, many of them described in the previous pages, that constitute the thriving topic of VT.

The development of the concepts of VT have been outlined elsewhere (Agnati et al., 1995; Bloom, 2000) and it may even be too early in the evolution of this idea to attempt a definitive history. It is generally agreed, though, that VT is primarily associated with signaling via the extracellular space (ECS) of the brain.

What exactly is the ECS?

The ECS consists simply of the spaces between the cells; as noted by Kuffler and Potter (1964), "the system of clefts ... in three dimensions they resemble the water phase of a foam." There is no doubt that we owe a debt to Kuffler and other pioneering investigators, such as Van Harreveld (1972) and Schmitt and Sampson (1969) who, more than 30 years ago, sought to study this rather unfashionable domain. The ECS is often incorporated into the broader concept of the brain cell microenvironment (Schmitt and Samson, 1969; Nicholson, 1980a; Cserr 1986).

The ECS varies in size but is thought to average about 20 nm in width; despite the narrow spaces, the whole ECS still manages to occupy some 20% of the total volume of the brain (Nicholson et al., 2000; Syková et al., 2000). The interstitial spaces are filled with a solution closely resembling CSF in composition (Bretag, 1969; Syková et al., 2000) and therefore mainly composed of salt solution.

In addition to the dilute saline, as noted in several chapters (Nicholson et al. 2000; Syková et al., 2000; Viaggiano et al., 2000) and elsewhere (Bignami et al., 1993; Margolis and Margolis, 1993) the ECS contains a matrix composed of macromolecules, including glycosaminoglycans and proteoglycans. Recent studies have revealed that the matrix is not uniform, for example some cells have dense pericellular nets around them (Viaggiano et al., 2000). Potentially this matrix has the capacity to modify the way substances migrate through the ECS and even to favor the movement of some molecules over others. Studies in non-nervous tissues, like cartilage and connective tissue (Laurent, 1998) suggest additional possibilities for the roles of the matrix in the brain.

Corresponding author: Tel: +1 (212) 263 5421;
Fax: +1 (212) 689 9060;
e-mail: cn7@is.nyu.edu

The morphology of the ECS may also differ in its properties from one place to another (heterogeneity) or may vary along different geometrical axes (anisotropy) (see Nicholson et al., 2000; Syková et al., 2000).

Where do glia fit in?

We are accustomed to regarding neurons as the main actors in the brain with glia playing a supporting but ill-defined role and this mindset carries over to VT. As the chapter by Chvátal and Syková (2000) illustrates, however, the many roles of glia are coming into focus; glia may indeed be very important and certainly in the context of K^+ and H^+-mediated VT they are an essential element (Ransom, 2000). Glia are also intimately involved with the extracellular matrix (Viaggiano et al., 2000).

This volume also provides convincing evidence that the exchange of signals between glia and neurons is essential for maintaining the metabolic status of brain cells (Coles et al., 2000; Magistretti and Pellerin, 2000) and the regulation of cell volume (Leibfritz et al., 2000)

Glia are also the cells that are predominantly formed into syncitia by gap junctions; these junctions may form a basis for a coupled intercellular communication that itself has many of the characteristics of extracellular VT (Dermietzel, 1998). Here perhaps we shall soon encounter another paradigm shift as the ideas emerging about VT inform the discussion of glial networks. Beyond that, glia are ideal elements in a reaction-diffusion mechanism, but evidence for such a role is not yet available.

What's in a name?

In order for a concept to achieve a distinct identity, it needs a suitable name. This has been a somewhat difficult issue for VT; different authors have used several competing terms. Schmitt (1984) wrote of 'informational substances', a term suggested to him by Fred Samson, being "delivered in a non-conventional parasynaptic mode." The most popular terms have been 'extrasynaptic' and 'non-synaptic'. Extrasynaptic has generally implied a short range effect such as when transmitters leave the synaptic cleft and encounter extrasynaptic receptors a few micrometers away (Pickel, 2000). Non-synaptic is a negative definition of potentially enormous scope but which still may exclude the action of transmitter released from a synapse but acting outside it. Such action is designated by the terms 'spillover' and 'crosstalk' and features in several recent studies (Barbour and Häuser, 1997; Kullmann, 2000). Bach-y-Rita (1994) suggested 'non-synaptic diffusion neurotransmission' while Bunin and Wightman (1999) recently returned to the designation 'paracrine'.

VT places emphasis on the channel by which information is transmitted. The term 'volume transmission' has a rational origin (see Fuxe and Agnati et al., 1991; Agnati et al., 1995) through its analogy with 'volume conduction', a term that applies to the way current distributes in tissue that has been in use since the early days of electrophysiology. This goes back at least to the definitive paper entitled 'Analysis of the distribution of the action currents of nerve in volume conductors' by Lorente di Nó (1947), part of a 1000 page treatise on nerve physiology by that investigator. It is worth noting that the connection between volume conduction and volume transmission is quite profound. The potential distribution in a conductor with distributed current sources and sinks is described by Poisson's partial differential equation (Lorente di Nó 1947; Nicholson 1973), which in turn describes many electrostatic situations; this is also precisely the form of the diffusion equation in steady-state conditions. Consequently, many steady-state diffusion problems can be solved by considering their electrical analogues (e.g. Berg and Purcell, 1977).

Despite the logical origin of the designation VT, the term has not been to everyone's taste; there is merit, however, in using a commonly agreed term to identify and categorize diverse research, which might otherwise be lost in the vast sea of modern publications. However, the reader should be warned that searching a database for 'volume transmission' today might still fail to locate important papers.

It is beginning to appear reasonable to divide the category of volume transmission into sub-classes, although there is no agreement at this time on how these sub-classes should be defined. This goes to the heart of the issue, and is also a testament to the progress in the field, that we are now able to identify different types of VT. All classification schemes ultimately fail, as they should, in the face of continued research that reveals new details and complexity; but classification, 'naming', is crucial to thinking about problems and it will help us refine our understanding of VT.

How far is 'far'?

One of the features of VT, and indeed all other types of communication, is that the signal is sent over some distance. In the case of classical wiring transmission (WT) that distance can be a few micrometers to a meter or more, depending on the length of the intervening nerve fiber and nobody asks the question "how far can the signal go and still be considered WT"? But for VT this is a question that seems to lurk in the back of everyone's mind; the reason for this may be threefold. First there is no visible one-to-one connection between the source of the signal and the recipient; second, there is a significant time delay between the emission of the signal and the arrival at the receptor; this can amount to tens of seconds or more; third, there is usually a substantial decrement in amplitude with distance from the signal source. Taken together these factors introduce uncertainty into the observation process, as to whether signaling has actually taken place. Essentially, there is not the same evidence of causality associated with VT as is present in WT.

The chapters of this book, and many other pertinent studies, show that VT takes place over different distance scales. One can begin with spillover of glutamate and other transmitters from the synapse (Barbour and Häuser, 1997; Vizi and Kiss, 1998; Kullmann 2000) with the potential of acting on extrasynaptic receptors (Pickel, 2000) taking place over a distance of the order of a micrometer. Following that, one moves to dopaminergic (Agnati et al., 2000; Descarries and Mechawar, 2000; Gonon et al., 2000; Kehr et al., 2000; Parker and Grillner, 2000; Rice, 2000; Strömberg et al., 2000; Zigmond, 2000), and seroninergic (Bunin and Wightman, 1999) signaling over distances ranging from tens to hundreds of micrometers. A similar propagation distance is expected for neuropeptides (Duggan, 2000; Jansson et al., 2000), opioids (Chavkin, 2000) and adenosine (Ferré and Fuxe, 2000); isolated varicosities (Descarries and Mechawar, 2000) also come into play at this distance scale. These distances can all be confined within a brain region, for example, retina, cortex, and striatum; when the distances exceed a millimeter or more, then communication is taking place between structures. Here we arrive at the realm of long distance VT mediated perhaps by periaxonal (Bjelke et al., 1995), perivascular systems (Ichimura et al., 1991) or CSF and the ventriculo-cisternal cavities (Nicholson, 1999; Lehman and Silver, 2000).

It is evident that there must be both quantitative and qualitative differences in the signaling that takes place just around a synapse and signaling that takes place over distances of millimeters or centimeters.

What are the signals?

Starting first with the simplest signaling compounds, one can begin with the inorganic ions. Nothing could be simpler than H^+, otherwise known as pH, which has the ability to modulate many facets of neuronal activity (Chesler, 1990; Agnati et al., 1994; Ransom, 2000). K^+ has long been a candidate for VT (Nicholson 1971; Syková 1983; Jefferys, 1995; Chvátal and Syková, 2000; Syková et al., 2000; Voipio and Kaila, 2000). It is clearly established that glia have a major role in the control of extracellular K^+, either through spatial buffering or active transport (Chvátal and Syková, 2000, Nicholson et al., 2000). Finally one notes that extracellular Ca^{2+} has considerable ability to influence neuronal activity (Nicholson, 1980, Egelman and Montague, 1998; Rusakov et al., 1999). One unique feature of all these ions is that there are no specific receptors in the usual sense of the word; instead virtually all cells are depolarized to some extent by a rise in extracellular K^+, or affected in more complex ways by H^+ or Ca^{2+}.

Moving on to slightly more complex substances, we encounter the neuroactive gases, particularly, NO (Bredt and Snyder, 1994; Schulman, 1997), CO

(Verma et al., 1993), and CO_2 (Agnati et al., 1994). Due to their lipid solubility, these compounds have the special ability to move through cells and so are not constrained by the geometry of the ECS. Being small they also move quickly (see below) but it is not clear how far the compound can travel before being destroyed (Philippides et al., 2000). By far the most studied compound is NO, which is involved in vast numbers of physiological process in neural and non-neural tissues.

Next in the hierarchy are the classical transmitter substances, glutamate (Bergles et al., 1999), acetylcholine and GABA (Isaacson et al., 1993; Chronwall et al., 1998). All of these are normally released from conventional synapses but may leave the synapse and reach extrasynaptic receptors (Kullmann, 2000) or conventional postsynaptic receptors but located at a synapse other than the releasing one i.e. crosstalk (Barbour and Häusser, 1997; Kullmann, 2000). It is the prevailing opinion at this time that the VT actions of these transmitters are limited to a short range from the site of release (Vizi and Kiss, 1998); active transport or dilution (concentration falling below the level necessary to activate a receptor) terminates the action. In pathological situations there is good evidence that a reversal of the glutamate transporter may flood brain tissue with this excitatory amino acid (Szatkowski et al., 1990) but there is less evidence for this type of mechanism under normal conditions.

The next group of compounds is the biogenic amines, most notably dopamine and serotonin. These have also attained the status of classical transmitters but their action is usually mediated via a G-protein cascade (Hille, 1992), leading to a multitude of effects and a certain time delay, compared to the transmitters mentioned above which usually act directly on receptors that are directly coupled to ionophores. As is abundantly evident in this book, dopamine is a compelling candidate for VT (Agnati et al., 2000; Descarries and Meschawar, 2000; Gonon et al., 2000; Kehr et al., 2000; Parker and Grillner, 2000; Pickel, et al., 2000; Rice, 2000; Strömberg, 2000; Zigmond, 2000). Potentially its action can extend from juxtasynaptic sites to receptors that are tens or even hundreds of micrometers from the synapse. Furthermore, dopamine can be released from non-synaptic sites in dendrites (Geffen et al., 1976; Rice et al., 1994; Rice, 2000). Serotonin has many properties in common with dopamine and is coming under increasing scrutiny as a major mediator of VT (Ridet et al., 1993; Bunin and Wightman, 1998; Zhou et al. 1998).

Beyond the amines comes a vast group of potential signaling compounds including the neuropeptides (Beck et al., 1995; Duggan, 2000; Jansson et al., 2000; Parker and Grillner, 2000) opioids (Cheng et al., 1996; Wang et al., 1996, Chavkin, 2000) and adenosine (Ferré and Fuxe, 2000). The distribution of many of these substances has been described by Nieuwenhuys (1985) in his book on the chemoarchitecture of the nervous system; since that publication, the candidates have considerably increased. Very recently it has been suggested that extracellular ATP mediates the well known, but little understood, calcium waves of glial syncitia (Guthrie et al., 1999).

What are the functional roles of VT?

VT can convey almost any type of information, except that requiring the fastest response time; fast motor commands are handled by myelinated, rapidly conducting nerves and are a hallmark of WT; otherwise both cognitive and non-cognitive information are likely to be handled by VT. For example, the well-known involvement of dopamine and serotonin in setting mood and the extensive evidence for VT involvement of these two substances, strongly suggests a link. On the non-cognitive side, several papers have emphasized the role of VT in metabolic control (Coles et al., 2000; Leibfritz, 2000; Magistretti and Pellerin, 2000). There is also reason to think that VT is involved in brain plasticity (Parker and Grillner, 2000) including the best known example of LTP (Schulman, 1997).

Another area where the concepts of diffusion and VT are applicable is during development or repair of the CNS; this vast subject is largely ignored in the present book. Suffice to say here that diffusing compounds clearly play a major role in many phases of development (e.g. Crick, 1970; Goodhill, 1998; Zheng et al., 1996) and may involve many of the substances mentioned above as well as compounds like nerve growth factor (Krewson et al., 1995).

How do signals get to their destinations?

Thinking again about WT, the mechanism of signal transmission is clear; over very short distances, decremental electrotonic conduction can suffice, but in the vast majority of instances, action potentials convey the signals. At the conclusion of the journey, a synapse converts the electrical signal into a very short-range chemical mediator.

For VT, over distances of, say, less than 200 μm, the transmission mechanism is generally thought to be diffusion. This automatically introduces both a decremental element into the process and a time-dependence, so that, as is very well known (Nicholson et al., 2000), the process works well over short distances but is ineffective when the source and receptor are far apart. Furthermore, processes such as uptake into cells, removal across the BBB and degradation can seriously attenuate the strength of the signal and sometimes, as in the case of DA and the striatum, are the dominant factor (Wightman and Zimmerman, 1990; Nicholson, 1995; Rice, 2000; Kehr et al., 2000). The narrow interconnecting structure of the ECS hinders diffusion compared to that in a free medium (a beaker of solution for example) and this reduces the diffusion coefficient typically by a factor of about 2.5 (Nicholson et al., 2000). This slows the rate of rise of a signal but can result in a higher amplitude, if the release is sustained.

The decremental element means that the amplitude of the signal, that is the concentration in the ECS, diminishes with distance so that the question of receptor sensitivity becomes an important factor in the viability of the communication process. The time delay for a molecule to diffuse a certain distance means that rapid signaling is not possible.

So where is the benefit in a diffusion-mediated signaling system? A major gain is that no wiring is needed; this makes the system flexible, allowing new signaling systems, even involving different molecules, to come into play quite easily. Diffusion-mediated systems also allow for multiplicity of signals in a small volume; so long as the receptors are specific many chemical signals can share the same spaces.

For VT to overcome the limitation of distance signaling, a physical process other than diffusion is required. This most probably involves directed flow in the juxta-axonal spaces (Bjelke et al., 1995) or Robbins-Virchow spaces around blood vessels (Ichimura et al., 1991) aided perhaps by peristaltic movement; another flow-based transport system is via the motion of the CSF (Lehman and Silver, 2000) or the blood itself. Bulk flow is a powerful method of overcoming the limitations of diffusion (Bobo et al., 1994) but it is more complex than diffusion, because the flow has to be established and maintained before it can be used for reliable communication.

There are other possible modes of VT communication which have yet to be explored in detail; reaction-diffusion is one example. This means that a substances diffuses over a short range, it then stimulates an appropriate receptor and effector to either release more of the same substance or another species; this can lead to a propagating wave and has been the topic of much theoretical and experimental research. Familiar examples are the Belosov–Zhabotinsky reaction in chemistry and other disciplines (Swinney and Krinsky, 1992) and the action potential itself in a single axon (Hastings, 1975). In the brain, the best-known example of reaction-diffusion is probably spreading depression (Nicholson, 1993; Martins-Ferreira et al., 1999); some forms of calcium waves may also have the requisite features (Martins-Ferreira et al., 1999). As has been pointed out on several occasions (e.g. Scriven, 1974; Nicholson, 1993), a reaction diffusion mechanism can allow signals to propagate faster than does diffusion alone, while at the same time maintaining the amplitude of the signal.

How have technical advances contributed to understanding VT?

There is a close and inevitable link between the emergence and refinement of techniques and the understanding of new phenomena; this has been true for VT. The last decade has seen substantial progress in our ability to label molecules, receptors and transporters (Pickel, 2000; Duggan, 2000). This led initially to ideas of receptor mismatch (Herkenham, 1987) and then the realization that many receptors and transporters are not intimately associated with classical synapses (Pickel, 2000). Steady advances in microscopy and stereology have provided undeniable proof that isolated varicositites, bereft of postsynaptic structures, do exist (Descarries and Mechawar, 2000).

Functionally, we have arrived at better methods of measuring local diffusion properties based on ion-sensitive microelectrodes (Nicholson et al., 2000; Syková et al., 2000) or microdialysis (Kehr et al., 2000). Other applications of these methods, and of in vivo and in vitro voltammetry (Boulton et al., 1995), have led to detailed studies of DA behavior in tissue as well as other amines and transmitters (see several chapters in this book).

Radiotracer techniques remain versatile and powerful tools (Krewson et al., 1995; Patlak et al., 1998; Kehr et al., 2000); but electrophysiology, (Gonon, 1997; Gonon et al., 2000; Parker and Grillner, 2000; Voipio and Kaila, 2000) is still the final arbiter of a VT signaling process in many instances.

Despite this progress in techniques, it has to be admitted that there is no single method that can reveal VT, but rather that it usually requires a careful application of several approaches in combination to arrive at unassailable evidence. Furthermore, in contrast to classical WT where nerve fiber and synapse provide unmistakable anatomical evidence, there is a general lack of any characteristic structural basis for VT.

Another technical approach that has only recently begun to be used is modeling of putative VT processes; even simple diffusion concepts can lead to quite complex spatial patterns. But, as always with modeling, there has been little incentive until a critical mass of solid experimental data became available. This book shows that this point has been reached and models of short-range VT signaling by diffusion are beginning to appear (Barbour and Häusser, 1997; Goodhill, 1998; Egelman and Montague, 1998; Rusakov and Kulliman, 1998; Philippides et al., 2000) as well as new conceptual advances (Agnati and Fuxe, 2000).

What does a comparative approach tell us?

It is fitting to end this summary by returning to the chapter where Nieuwenhuys (2000) draws attention to the power and value of the comparative approach. In his survey, Nieuwenhuys provides an extraordinarily broad perspective ranging over the vertebrate and invertebrate kingdoms; this reminds us, as has been true so often in the past (Bullock and Horridge, 1965), that the invertebrates provide clues to the origin and meaning of VT. Another chapter, that of Parker and Grillner (2000), covers a single species but shows how a detailed knowledge of the neurophysiology of an ancient vertebrate, the lamprey, can set the stage for understanding where and when VT enters the picture. Doubtless, many other vertebrates (Nieuwenhuys et al., 1998) have unique mechanisms that could greatly extend our understanding of the ramifications of VT; but, as Nieuwenhuys shows us, we may have to return to the invertebrates to truly understand the origin of VT.

Where will we be 2010?

The forays, just mentioned, into species far removed from the rat and mouse remind us that the majority of organisms are engaged in ceaseless chemical sensing and signaling; this is as true for the amoebae, slime mold and the moth as it is for hagfish, the rat and even humans. As noted also by Nieuwenhuys in the conclusion to his chapter, one can view VT as an internalization of this chemical communication and, seen in this light, it is hardly surprising that it is so ubiquitous.

Reading through this book one is certainly struck by the solid evidence for VT, yet one also sees that it is still tied to classical synaptic transmission in many instances; much of the evidence is based on seeking transmitter-filled vesicles in unlikely places or presynaptic structures without post synaptic. As some of the studies involving metabolites suggest, there is no reason that some forms of VT should not be entirely liberated from the classical synapse structures, perhaps involving the reversal of transporters or other release mechanisms.

All those working in this area of research are privileged to be involved in bringing about a paradigm shift in our understanding of communication in the CNS, possibly equal in importance to the revolution that occurred when the classical synapse was defined. Perhaps in another decade, we will all view VT as the normal mode of intercellular communication and see WT as just a specialization of this modality.

Acknowledgment

This work was supported by NIH Grants NS28642 and NS34115.

References

Agnati, L.F., Cortelli, P., Biagini, G., Bjelke, B. and Fuxe, K. (1994) Different classes of volume transmission signals exist in the central nervous system and are affected by metabolic signals, temperature gradients and pressure waves. *NeuroReport*, 6: 9–12.

Agnati, L.F. and Fuxe, K. (2000) Volume transmission as a key feature of information handling in the central nervous system. Possible new interpretive value of the Turing's B-type machine. In: L.F. Agnati, K. Fuxe, C. Nicholson and E. Syková, (Eds.), *Volume Transmission Revisited*, Elsevier, Amsterdam, pp. 3–19.

Agnati, L.F., Zoli, M., Ferrari, R., Di Paola.L., Torri, C., Fuxe K. and Zini, I. (2000) Evidence for the existence of pulses of dopamine in the extracellular space of the rat striatum. In: *Volume Transmission Revisited*, L.F. Agnati, K. Fuxe, C. Nicholson and E. Syková, (Eds.), Elsevier, Amsterdam, pp. 303–308.

Agnati, L.F., Zoli, M., Strömberg, I. and Fuxe, K. (1995) Intercellular communication in the brain: Wiring versus volume transmission. *Neuroscience* 69: 711–726.

Bach-y-Rita, P. (1994) Non-synaptic diffusion neurotransmission: a novel concept for future migraine research. *Cephalalgia*, 14: 396–400.

Barbour, B. and Häusser, M. (1997) Intersynaptic diffusion of neurotransmitter. *Trends Neurosci.*, 20: 377–384, 509.

Beck, H., Schröck, H. and Sandkühler, J. (1995) Controlled superfusion of the rat spinal cord for studying non-synaptic transmission: an autoradiographic analysis. *J. Neurosci. Meth.*, 58: 193–202.

Berg, H.C. and Purcell, E.M. (1977) Physics of chemoreception. *Biophys. J.*, 20: 193–219.

Bergles, D.E., Diamond, J.S. and Jahr, C.E. (1999) Clearance of glutamate inside the synapse and beyond. *Curr. Opin. Neurobiol.*, 9: 293–298.

Bignami, A., Hosley, M. and Dahl, D. (1993) Hyaluronic acid and hyaluronic acid-binding proteins in brain extracellular matrix. *Anat. Embryol.*, 188: 419–433.

Bjelke, B., England, R., Nicholson, C., Rice, M.E., Lindberg, J., Zoli, M., Agnati, L.F. and Fuxe, K. (1995) Long distance pathways of diffusion for dextran along fibre bundles in brain. Relevance for volume transmission. *NeuroReport*, 6: 1005–1009.

Bloom, F.E. (2000) Integration of wiring transmission and volume transmission. In: L.F. Agnati, K. Fuxe, C. Nicholson and E. Syková, (Eds.), *Volume Transmission Revisited*, Elsevier, Amsterdam, pp. 21–26.

Bobo, R.H., Laske, D.W., Akbasak, A., Morrison, P.F., Dedrick, R.L. and Oldfield, E.H. (1994) Convection-enhanced delivery of macromolecules in the brain. *Proc. Natl Acad. Sci. USA*, 91: 2076–2080.

Boulton, A.A., Baker, G.B. and Adams, R.N. (1995) *Voltammetric Methods in Brain System*. Neuromethods 27. Humana Press, Totowa, New Jersey.

Bredt, D.S. and Snyder, S.H. (1994) Nitric oxide: a physiologic messenger molecule. *Ann. Rev. Biochem.*, 63: 175–195.

Bretag, A.H. (1969) Synthetic interstitial fluid for isolated mammalian tissue. *Life Sci.*, 8: 319–329.

Bullock, T.H. and Horridge, G.A. (1965) *Structure and Function in the Nervous Systems of Invertebrates*. W. H. Freeman, San Francisco.

Bunin, M.A. and Wightman, R.M. (1998) Quantitative evaluation of 5-hydroxytryptamine (serotonin) neuronal release and uptake: an investigation of extrasynaptic transmission. *J. Neurosci.*, 18: 4854–4860.

Bunin, M.A. and Wightman, R.M. (1999) Paracrine neurotransmission in the CNS: involvement of 5-HT. *Trends Neurosci.*, 22: 377–382.

Chavkin, C. (2000) Dynorphins are endogenous opioid peptides released from granule cells to act neurohumorly and inhibit excitatory neurotransmission in the hippocampus. In: L.F. Agnati, K. Fuxe, C. Nicholson and E. Syková, (Eds.), *Volume Transmission Revisited*, Elsevier, Amsterdam, pp. 363–367.

Cheng, P.Y., Moriwaki, A., Wang, J.B., Uhl, G.R. and Pickel, V.M. (1996) Ultrastructural localization of μ-opioid receptors in the superficial layers of the rat cervical spinal cord: extrasynaptic localization and proximity to Leu^5-enkephalin. *Brain Res.*, 731: 141–154.

Chesler, M. (1990) The regulation and modulation of pH in the nervous system. *Prog. Neurobiol.*, 34: 401–427.

Chronwall, B.M., Sands, S.A., Cummings, K.C., Hagler, K.E., Norberg, M., Morris, S.J. and Gary, K.A. (1998) Differential innervation of individual melanotropes suggests a role for non-synaptic inhibitory regulation of the developing and adult rat pituitary intermediate lobe. *Synapse* 28: 227–243.

Chvátal, A., and Syková, E. (2000) Glial influence on neuronal signaling. In: L. F. Agnati, K. Fuxe, C. Nicholson and E. Syková, (Eds.), *Volume Transmission Revisited*, Elsevier, Amsterdam, pp. 199–216.

Coles, J.A. Véga C. and Marcaggi P. (2000) Metabolic trafficking between cells in nervous tissue. In: L.F. Agnati, K. Fuxe, C. Nicholson and E. Syková, (Eds.), *Volume Transmission Revisited*, Elsevier, Amsterdam, pp. 241–254.

Crick, F. (1970) Diffusion in embryogenesis. *Nature*, 225: 420–422.

Cserr, H.F., Ed. (1986) The Neuronal Microenvironment. *Ann. NY Acad. Sci.*, 481: 1–391.

Dermietzel, R. (1998) Gap junction wiring: a 'new' principle in cell-to-cell communication in the nervous system? *Brain Res. Rev.*, 26: 176–183.

Descarries, L., Mechawar, N. (2000) Ultrastructural evidence for diffuse transmission by monoamine and acetlycholine neurons of the central nervous system. In: L.F. Agnati, K. Fuxe, C. Nicholson and E. Syková, (Eds.), *Volume Transmission Revisited*, Elsevier, Amsterdam, pp. 27–47.

Duggan, A.W. (2000) Neuropeptide spread in the brain and spinal cord. In: L. F. Agnati, K. Fuxe, C. Nicholson and E. Syková, (Eds.), *Volume Transmission Revisited*, Elsevier, Amsterdam, pp. 369–380.

Egelman, D.M. and Montague, P.R. (1998) Computational properties of peri-dendritic calcium fluctuations. *J. Neurosci.*, 18: 8580–8589.

Ferré, S. and Fuxe K. (2000) Adenosine as a volume transmission signal: a feedback detector of neuronal activation. In: L.F. Agnati, K. Fuxe, C. Nicholson and E. Syková, (Eds.), *Volume Transmission Revisited*, Elsevier, Amsterdam, pp. 353–361.

Fuxe, K. and Agnati, L.F., Eds. (1991) *Volume Transmission in the Brain. Advances in Neuroscience*, 1. Raven Press, New York.

Geffen, L.B., Jessell, T.M., Cuello, A.C. and Iversen, L.L. (1976) Release of dopamine from dendrites in rat substantia nigra. *Nature*, 260: 258–260.

Goodhill, G.J. (1998) Mathematical guidance for axons. *Trends Neurosci.*, 21: 226–231.

Gonon, F. (1997) Prolonged and extrasynaptic excitatory action of dopamine mediated by D1 receptors in the rat striatum in vivo. *J. Neurosci.*, 17: 5972–5978.

Gonon, F., Burie, J.B., Jaber, M., Benoit-Marand, M., Dumartin, B. and Bloch, B. (2000) Geometry and kinetics of dopaminergic transmission in the rat striatum and in mice lacking the dopamine transporter. In: L.F. Agnati, K. Fuxe, C. Nicholson and E. Syková, (Eds.), *Volume Transmission Revisited*, Elsevier, Amsterdam, pp. 291–302.

Guthrie, P.B., Knappenberger, J., Segal, M., Bennett, M.V. L., Charles, A.C. and Kater, S.B. (1999) ATP released from astrocytes mediates glial calcium waves. *J. Neurosci.*, 19: 520–528.

Hastings, S.P. (1975) Some mathematical problems from neurobiology. *Am. Math. Monthly*, 82: 881–895.

Herkenham, M. (1987) Mismatches between neurotransmitter and receptor localizations in brain: observations and implications. *Neuroscience*, 23: 1–38.

Hille, B. (1992) G protein-coupled mechanisms and nervous signaling. *Neuron*, 9: 187–195.

Ichimura, T., Fraser, P.A. and Cserr, H.F. (1991) Distribution of extracellular tracers in perivascular spaces of the rat brain. *Brain Res.*, 545: 103–113.

Isaacson, J.S., Solis, J.M. and Nicoll, R.A. (1993) Local and diffuse synaptic actions of GABA in the hippocampus. *Neuron*, 10: 165–175.

Jansson, A., Lippoldt, A., Mazel, T., Bartfai, T., Ogren, S-V., Syková, E., Agnati, L.F. and Fuxe, K. (2000) Long distance signalling in volume transmission. Focus on clearance mechanisms. In: L.F. Agnati, K. Fuxe, C. Nicholson and E. Syková, (Eds.), *Volume Transmission Revisited*, Elsevier, Amsterdam, pp. 399–413.

Jefferys, J.G.R. (1995) Non-synaptic modulation of neuronal activity in the brain: electric currents and extracellular ions. *Physiol. Rev.*, 75: 689–723.

Kehr, J, Höistad, M., Fuxe, K. (2000) Diffusion of radiolabeled dopamine, its metabolites and mannitol in the rat striatum studied by dual-probe microdialysis. In: L.F. Agnati, K. Fuxe, C. Nicholson and E. Syková, (Eds.), *Volume Transmission Revisited*, Elsevier, Amsterdam, pp. 179–190.

Krewson, C.E., Klarman M.L. and Saltzman, W.M. (1995) Distribution of nerve growth factor following direct delivery to brain interstitium. *Brain Res.*, 680: 196–206.

Kuffler, S.W. and Potter, D.D. (1964) Glia in the leech central nervous system: physiological properties and neuron-glia relationship. *J. Neurophysiol.*, 27: 290–320.

Kullmann, D. (2000) Spillover and synaptic crosstalk mediated by glutamate and GABA in the mammalian brain. In: L.F. Agnati, K. Fuxe, C. Nicholson and E. Syková, (Eds.), *Volume Transmission Revisited*, Elsevier, Amsterdam, pp. 339–351.

Laurent, T.C. (1998) The common origin of matrix biology and capillary physiology. In: K. Reed, K. Rubin (Eds.), *Connective Tissue Biology. Integration and Reductionism, Wenner-Gren International Series*, 71.R. Portland Press, London, pp. 1–7.

Lehman, M. and Silver, R. (2000) CSF signaling in physiology and behavior. In: L.F. Agnati, K. Fuxe, C. Nicholson and E. Syková, (Eds.), *Volume Transmission Revisited*, Elsevier, Amsterdam, pp. 415–433.

Leibfritz, D., Pfeuffer, J., Flögel, U., Meier, C. and Bröer, S. (2000) Cell volume and water exchange in neural cells monitored by diffusion weighted ^{1}H NMR spectroscopy. In: L.F. Agnati, K. Fuxe, C. Nicholson and E. Syková, (Eds.), *Volume Transmission Revisited*, Elsevier, Amsterdam, pp. 255–263.

Lorente de Nó, R. (1947) Analysis of the distribution of the action currents of nerve in volume conductors. *Stud. Rockefeller Inst.*, 132: 384–477.

Magistretti, P.J. and Pellerin, L. (2000) The astrocyte-mediated coupling between synaptic activity and energy metabolism operates through volume transmission. In: L.F. Agnati, K. Fuxe, C. Nicholson and E. Syková, (Eds.), *Volume Transmission Revisited*, Elsevier, Amsterdam, pp. 229–240.

Margolis, R.K. and Margolis, R.U. (1993) Nervous tissue proteoglycans. *Experientia*, 49: 429–446.

Martins-Ferreira, H., Nedergaard, M., Nicholson, C. (2000) Perspectives on spreading depression. *Brain Res. Rev.*, 32: 215–234.

Nicholson, C. (1999) Signals that go with the flow. *Trends Neurosci.*, 22: 143–145.

Nicholson, C. (1973) Theoretical analysis of field potentials in anisotropic ensembles of neuronal elements. *IEEE Trans. Biomed. Eng.*, 20: 278–288.

Nicholson, C. (1979) Brain cell microenvironment as a communication channel. In: F.O. Schmitt and F.G. Worden (Eds). *The Neurosciences Fourth Study Program*, MIT Press, Cambridge, MA, pp. 457–476.

Nicholson, C. (1980a) Dynamics of the brain cell microenvironment. *Neurosci. Res. Prog. Bull.*, 18: 177–322.

Nicholson, C. (1980b) Modulation of extracellular calcium and its functional implications. *Fed. Proc.*, 39: 1519–1523.

Nicholson, C. (1993) Volume transmission and the propagation of spreading depression. In: A. Lehmenkühler, K.H. Grotemeyer, F. Tegtmeier (Eds.), *Migraine: Basic Mechanisms and Treatment*, Urban & Schwarzenberg, Munich, pp. 293–308.

Nicholson, C. (1995) Interaction between diffusion and Michaelis–Menten uptake of dopamine after iontophoresis in striatum. *Biophys. J.*, 68: 1699–1715.

Nicholson, C., Chen, K.C., Hrabetová, S. and Tao, L. (2000) Diffusion of molecules in brain extracellular space: theory and experiment. In: L.F. Agnati, K. Fuxe, C. Nicholson and E. Syková, (Eds.), *Volume Transmission Revisited*, Elsevier, Amsterdam, pp. 129–154.

Nieuwenhuys, R. (1985) *Chemoarchitecture of the Brain*. Springer-Verlag, Berlin.

Nieuwenhuys, R. (2000) Comparative aspects of volume transmission; with sidelight on other forms of intercellular communication. In: L.F. Agnati, K. Fuxe, C. Nicholson and E. Syková, (Eds.), *Volume Transmission Revisited*, Elsevier, Amsterdam, pp. 49–126.

Nieuwenhuys, R., ten Donkelaar, H.J. and Nicholson, C. (1998) *The Central Nervous System of Vertebrates*. Springer, Berlin.

Parker, D. and Grillner, S. (2000) Neuronal mechanisms of synaptic and network plasticity in the lamprey spinal cord. In: L.F. Agnati, K. Fuxe, C. Nicholson and E. Syková, (Eds.), *Volume Transmission Revisited*, Elsevier, Amsterdam, pp. 381–398.

Patlak, C.S., Hospod, F.E., Trowbridge, S.D. and Newman, G.C. (1998) Diffusion of radiotracers in normal and ischemic brain slices. *J. Cereb. Blood Flow Metab.*, 18: 776–802.

Philippides, A., Husbands, P. and O'Shea, M. (2000) Four-dimensional neuronal signaling by nitric oxide: a computational analysis. *J. Neurosci.*, 20: 1199–1207.

Pickel, V.M. (2000) Extrasynaptic distribution of monoamine transporters and receptors. In: L.F. Agnati, K. Fuxe, C. Nicholson and E. Syková, (Eds.), *Volume Transmission Revisited*, Elsevier, Amsterdam, pp. 267–276.

Ransom, B. (2000) Glial modulation of neural excitability mediated by extracellular pH: a hypothesis revisited. In: L.F. Agnati, K. Fuxe, C. Nicholson and E. Syková, (Eds.), *Volume Transmission Revisited*, Elsevier, Amsterdam, pp. 217–228.

Rice, M.E. (2000) Distinct regional differences in dopamine-mediated volume transmission. In: L.F. Agnati, K. Fuxe, C. Nicholson and E. Syková, (Eds.), *Volume Transmission Revisited*, Elsevier, Amsterdam, pp. 277–290.

Rice, M.E., Richards, C.D., Nedergaard, S., Hounsgaard, J., Nicholson, C. and Greenfield, S.A. (1994) Direct monitoring of dopamine and 5-HT release in substantia nigra and ventral tegmental area in vitro. *Exp. Brain Res.*, 100: 395–406.

Ridet, J-L., Rajaofetra, N., Teilhac, J-R., Geffard, M. and Privat, A. (1993) Evidence for non-synaptic serotonergic and noradrenergic innervation of the rat dorsal horn and possible involvement of neuron-glia interactions. *Neuroscience*, 52: 143–157.

Rusakov D.A. and Kullmann D.M. (1998) Extrasynaptic glutamate diffusion in the hippocampus: ultrastructural constraints, uptake, and receptor activation. *J. Neurosci.*, 18: 3158–3170.

Rusakov, D.A., Kullmann, D.M. and Stewart, M.G. (1999) Hippocampal synapses: do they talk to their neighbours? *Trends Neurosci.*, 22: 382–388.

Schmitt, F.O. (1984) Molecular regulators of brain functioning: a new view. *Neuroscience*, 13: 991–1001.

Schmitt, F.O. and Samson, F.E. (1969) The brain cell microenvironment. *Neurosci. Res. Prog. Bull.*, 7: 277–417.

Schulman, H. (1997) Nitric oxide: a spatial second messenger. *Molec. Psych.*, 2: 296–299.

Scriven, L.E. (1974) Analysis of pattern and rhythm. *Neurosci. Res. Prog. Bull.*, 12: 36–39.

Strömberg, I., Kehr, J. and Fuxe, K. (2000) Restoration of dopamine transmission in graft reinnervated striatum. Evidence for regulation of dopamine D_2 receptor function in regions lacking dopamine. In: L.F. Agnati, K. Fuxe, C. Nicholson and E. Syková, (Eds.), *Volume Transmission Revisited*, Elsevier, Amsterdam, pp. 309–315.

Swinney, H.L. and Krinsky, V.I. (1992) *Waves and Patterns in Chemical and Biological Media*. MIT Press, Cambridge, Massachusetts.

Syková, E. (1983) Extracellular K^+ accumulation in the central nervous system. *Prog. Biophys. Molec. Biol.*, 42: 135–189.

Syková, M.E., Mazel, T., Vargová, L., Voisek, I. and Prokopová S. (2000) Extracellular space diffusion and pathological states. In: L.F. Agnati, K. Fuxe, C. Nicholson and E. Syková, (Eds.), *Volume Transmission Revisited*, Elsevier, Amsterdam, pp. 155–178.

Szatkowski, M., Barbour, B. and Attwell, D. (1990) Non-vesicular release of glutamate from glial cells by reversed electrogenic glutamate uptake. *Nature*, 348: 443–446.

Van Harreveld, A. (1972) The extracellular space in the vertebrate central nervous system. In: *The Structure and Function of Nervous Tissue*, G.H. Bourne (Ed.), Academic Press, New York, pp. 447–511.

Verma, A., Hirsch, D.J., Glatt, C.E., Ronnett, G.V. and Snyder, S.H. (1993) Carbon monoxide – a putative neural messenger. *Science*, 259: 381–384.

Viaggiano, D., Ibrahim, M. and Celio, M.R. (2000) Relationship between glia and the perineuronal nets of extracellular matrix in the rat cerebral cortex. In: L.F. Agnati, K. Fuxe, C. Nicholson and E. Syková, (Eds.), *Volume Transmission Revisited*, Elsevier, Amsterdam, pp. 193–198.

Vizi, E.S. and Kiss, J.P. (1998) Neurochemistry and pharmacology of the major hippocampal transmitter systems: Synaptic and nonsynaptic interactions. *Hippocampus*, 8: 566–607.

Voipio, J. and Kaila, K. (2000) GABAergic excitation and K^+-mediated volume transmission in the hippocampus. In: L.F. Agnati, K. Fuxe, C. Nicholson and E. Syková, (Eds.), *Volume Transmission Revisited*, Elsevier, Amsterdam, pp. 329–338.

Wang, H., Moriwaki, A., Wang, J.B., Uhl, G.R. and Pickel, V.M. (1996) Ultrastructural immunocytochemical localization of μ opioid receptors and Leu^5-enkephalin in the patch compartment of the rat caudate-putamen nucleus. *J. Comp. Neurol.*, 375: 659–674.

Wightman, R.M. and Zimmerman, J.B. (1990) Control of dopamine extracellular concentration in rat striatum by impulse flow and uptake. *Brain Res. Rev.*, 15: 135–144.

Zhou, F.C., Tao-Cheng, J.H., Segu, L., Patel, T. and Wang, Y. (1998) Serotonin transporters are located on the axons beyond the synaptic junctions: anatomical and functional evidence. *Brain Res.*, 805: 241–254.

Zheng, J.Q., Wan, J.J. and Poo M.M. (1996) Essential role of filopodia in chemotropic turning of nerve growth cone induced by a glutamate gradient. *J. Neurosci.*, 16: 1140–1149.

Zigmond, M.J. (2000) When it comes to communication between neurons, synapses are over-rated: Insights from animal model of Parkinsonism. In: L.F. Agnati, K. Fuxe, C. Nicholson and E. Syková, (Eds.), *Volume Transmission Revisited*, Elsevier, Amsterdam, pp. 317–327.

Subject Index

Abbreviations used in this index

5-HT	5-hydroxytryptamine = serotonin
6-OHDA	6-hydroxydopamine
BBB	blood-brain barrier
CGRP	calcitonin gene-related peptide
CSF	cerebrospinal fluid
DAT	dopamine transporter
DOPAC	3,4-dihydroxyphenylacetic acid
ECS	extracellular space
fMRI	functional magnetic resonance imaging
GABA	gamma-aminobutyric acid
GFAP	glial fibrillary acidic protein
GnRH	gonadotrophin releasing hormone
HVA	homovanillic acid
MR	magnetic resonance
MRI	magnetic resonance imaging
NMDA	N-methyl-D-asparate
NMR	nuclear magnetic resonance
NO	nitric oxide
PET	positron emission tomography
TMA	tetramethylammonium